TEXTS AND READINGS IN MATHEMATICS 19

Linear Algebra

Second Edition

Texts and Readings in Mathematics

Already Published Volumes
R. B. Bapat: Linear Algebra and Linear Models (Second Edition)
R. Bhatia: Fourier Series
C. Musili: Representations of Finite Groups
H. Helson: Linear Algebra (Second Edition)
D. Sarason: Notes on Complex Function Theory
M. G. Nadkarni: Basic Ergodic Theory (Second Edition)
H. Helson: Harmonic Analysis (Second Edition)
K. Chandrasekharan: A Course on Integration Theory
K. Chandrasekharan: A Course on Topological Groups
R. Bhatia (ed.): Analysis, Geometry and Probability
K. R. Davidson: C* - Algebras by Example
M. Bhattacharjee *et al.*: Notes on Infinite Permutation Groups
V. S. Sunder: Functional Analysis - Spectral Theory
V. S. Varadarajan: Algebra in Ancient and Modern Times
M. G. Nadkarni: Spectral Theory of Dynamical Systems
A. Borel: Semisimple Groups and Riemannian Symmetric Spaces
M. Marcolli: Seiberg-Witten Gauge Theory
A. Bottcher and S. M. Grudsky: Toeplitz Matrices, Asymptotic Linear Algebra and Functional Analysis

Linear Algebra

Second Edition

A. Ramachandra Rao
Indian Statistical Institute
Calcutta

and

P. Bhimasankaram
Indian Statistical Institute
Hyderabad

HINDUSTAN
BOOK AGENCY

Published by

Hindustan Book Agency (India)
P 19 Green Park Extension
New Delhi 110 016
India

email: hba@vsnl.com
http://www.hindbook.com

Produced from camera ready copy supplied by the Author.

ISBN 81-85931-61-5

To

my wife

Vani

and children

Rajeev

Padmaja

— A.R.R.

To

my wife

Vijaya

and children

Chandu

Chandana

— P.B.

Preface

There have been broadly three approaches to the treatment of Linear Algebra, viz., those emphasizing vector spaces, determinants and elementary operations. The books following the first approach (for example, those by Halmos and Hoffman and Kunze) mainly use the setting of linear transformations with a view to possible extensions of the results to Banach and Hilbert spaces. The books following the second approach (for example that by Mirsky) study Matrix Algebra making heavy use of determinants. The last approach is generally used in *service courses* where the emphasis is on the basic operations and computations involving matrices with not much importance given for proofs and concepts.

For several applications in Science and Engineering, matrix setting is of importance. However, the vector space approach is more suitable for geometric intuition leading to transparent proofs besides being amenable for generalization to infinite-dimensional spaces. The Indian School led by Professors C. R. Rao and S. K. Mitra has successfully used this approach to great advantage. We follow them in this book and systematically develop the elemenatry parts of matrix theory exploiting the properties of row and column spaces of matrices. The book is meant to be a text book at the honours level for students of Mathematics and/or Statistics though it can be used with advantage by students of other subjects like Physics, Computer Science, Engineering, Operational Research, etc. It can also serve as a reference book for scientists in various other disciplines.

The developments in Linear Algebra during the past few decades have brought into focus several techniques like rank-factorization, generalized inverses and singular value decomposition, which are hitherto not included in elementary text books. These techniques are actually simple enough to be taught at undergraduate (honours) level and, when properly used, provide a better understanding of the topic besides giving simpler proofs of results, thus enabling the student to learn the subject without tears. One of our aims in writing this book is to provide a treatment incorporating these.

The book is organized in two parts. The results in *Chapters* 1 through

6 are valid over any field while in *Chapters* 7, 8 and 9 the field is assumed to be the real field $\mathbb{R}$ or the complex field $\mathbb{C}$. We think this is very useful because, without this separation, one may have to spend a huge amount of time to decide whether a (frequently used) result is true over a general field. Fields other than $\mathbb{R}$ and $\mathbb{C}$ are very much used in subjects like Statistics, Computer Science, Coding Theory, Combinatorics, etc. However, a student not interested in general fields would not lose much by taking the field to be $\mathbb{R}$ or $\mathbb{C}$ throughout. We assure the reader that the proofs in *Chapters* 1–6 are no more complicated than they would be if the field were assumed to be $\mathbb{R}$ or $\mathbb{C}$.

In an earlier version of the book, there were two chapters, one on Linear Programming and the other on Statistical Applications, which we have dropped for the sake of uniformity and brevity. We have, moreover, simplified some of the proofs and the presentation of the material in several places and included a discussion on the locus of a general equation of the second degree in the plane.

We mention a few general features of the book. Many of the procedures are given in the form of algorithms which can easily be converted to computer programmes in any high level language. However we have not tried to give the best computationally stable algorithms as they usually need a lot more theory and are available in computer packages. We have included a large number of examples illustrating almost every definition, result, procedure, etc. However, the real highlight of the book is the large collection of exercises given at the ends of the sections. Hints are provided along with the exercises for a few tricky ones; more hints and solutions are given at the end of the book for the rest. We urge each student to attempt as many exercises as possible without looking at the solutions. However, one should not feel discouraged if he or she needs frequent help of the solutions as there are many exercises which are either tough or lengthy. We have marked with an asterisk exercises and topics which, we think, are somewhat difficult and may be omitted in a first reading.

We had to omit some important topics like non-negative matrices and matrix analysis mainly because of the constraint on the length of the book. Though we have corrected the errors in the earlier version of the book which were brought to our notice, some errors (typographical or even conceptual) are bound to remain in spite of our best efforts. We shall be grateful if the readers inform us of any errors found or suggestions for improvement.

The list of references given at the end of the book is by no means exhaustive and includes mainly those which have influenced us considerably in one form or another. More references can be obtained from some of these.

Regarding the numbering of results: we have numbered theorems, lemmas, definitions, examples, algorithms etc. (but not corollaries) in the same sequence for ease in tracing them. *Section* 4.5 refers to the fifth section in *Chapter* 4. *Exercise* 4.5.2 is the second exercise in *Section* 4.5. (4.5.2) refers to the second numbered equation (or displayed item) in *Section* 4.5 and is different from *Theorem* 4.5.2. An index of symbols and notations and a fairly detailed subject index are given at the end of the book.

Though this book contains material which can be covered in just about one year, it can also be used to teach a one-semester course as outlined on the following page.

It is a pleasure to acknowledge the help received from various persons. P. Bhimasankaram is particularly grateful to Professor C. R. Rao and Professor S. K. Mitra from whom he learnt Linear Algebra. Their thoughts on the subject greatly influenced his own understanding of the subject and are reflected to some extent in the present work. However, the present authors are solely responsible for any shortcomings in this book.

We are thankful to the authorities of the Indian Statistical Institute for providing us the necessary facilities while writing this book. We also thank the colleagues with whom we had discussions and the students of several batches of B. Stat. (Hons.) and M. Stat. at the Institute for their intelligent querries and for bearing with our experimentation using preliminary versions of this book.

Finally we thank Hindustan Book Agency for their efforts in bringing out this book in a neat form.

A. Ramachandra Rao
P. Bhimasankaram

Suggestions for a One-Semester Course

A one-semester course of about 50 lectures can be taught from the book by covering the following portions. For shorter courses, one has to carefully choose the topics. Here, $i.j.k$ refers to the item $i.j.k$, which may be a theorem, example or definition, etc., and $i.j$ refers to the whole of *Section $i.j$ excluding difficult examples or results indicated by an asterisk, if any.*

1.1–1.5, 1.6.1–1.6.4, 1.6.6, 1.7.1–1.7.4, 1.8.1, 1.8.2.

2.1–2.2, 2.3.1, 2.3.4, 2.4–2.6, Idea of partitioned matrices using (2.7.2).

3.1–3.2, 3.3.1–3.3.7, 3.4.1, 3.5.1–3.5.12, 3.6.1–3.6.3, (3.7.1), 3.8.1, 3.8.2, (3.8.8), 3.9.2, 3.9.3, 3.9.7.

4.1–4.2, 4.3.1, 4.4.1–4.4.3, 4.4.7–4.4.10, 4.5.1–4.5.3.

5.1–5.3, 5.5.1, 5.5.2.

6.3 (assume properties of permutations), 6.4.1–6.4.4 (omit proofs), 6.5.1, 6.6.1–6.6.4, 6.7.1.

7.1, 7.2.1–7.2.3, 7.2.7, 7.3.1, 7.3.8–7.3.10, 7.4.1–7.4.4, 7.4.6–7.4.9, 7.5.1–7.5.6, 7.6.1–7.6.6 (excluding 7.6.3).

8.1, 8.2.1–8.2.7, 8.3.1, 8.3.2, 8.3.7, 8.3.10, 8.4.1, 8.4.2, 8.7.1, 8.7.2.

9.1, 9.2 , 9.3.1–9.3.5, 9.4.1, 9.4.2, 9.4.5, 9.4.8, 9.4.9, 9.6.1.

The algorithms in the above topics may be omitted except 4.3.1, 4.4.10 and 7.4.8.

Contents

Chapter 0
Preliminaries

In this chapter we explain a few preliminary concepts. Some of the results are stated without proofs. Those who are familiar with the topics may skip this chapter and go to the next.

0.1 Relations

We assume that the reader is familiar with the basic operations on sets and their properties. We will use the notation $\mathbb{R}$ for the set of (all) real numbers and $\mathbb{C}$ for the set of complex numbers.

If A and B are two (not necessarily distinct) sets, we define their *cartesian product* to be

$$A \times B = \{(x, y) : x \in A,\ y \in B\}$$

Thus $A \times B$ is the set of all ordered pairs with the first component coming from A and the second component from B. For example,

$$\{1,2\} \times \{2,3,4\} = \{(1,2),(1,3),(1,4),(2,2),(2,3),(2,4)\}$$

$\mathbb{R} \times \mathbb{R}$ is the set of all ordered pairs of real numbers and is usually denoted by $\mathbb{R}^2$. Since each point in the plane can be represented uniquely by an ordered pair of real numbers (once the axes are fixed) and each ordered pair corresponds to a unique point, the plane is usually denoted by $\mathbb{R}^2$. Similarly $\mathbb{R}^3 = \mathbb{R} \times \mathbb{R} \times \mathbb{R}$ denotes the 3-dimensional Euclidean space.

A (binary) *relation* on a set X is a subset R of $X \times X$. If $(x, y) \in R$ we say that x stands in the relation R to y. We will then write xRy for convenience. A relation is usually specified by a statement about x and y. For example, $x \leq y$ is a relation on $\mathbb{R}$, viz., $\{(x,y) : x \in \mathbb{R},\ y \in \mathbb{R},\ x \leq y\}$. '$x$ and y have the same age' is a relation on the set of all people. A relation R on X is said to be *reflexive* if xRx for all $x \in X$. R is said to be *symmetric* if yRx holds whenever xRy holds. R is said to be *transitive* if

$$xRy, yRz \Longrightarrow xRz$$

The relation $\leq$ on $\mathbb{R}$ is reflexive and transitive but not symmetric. The relation $|x - y| \leq 1$ on $\mathbb{R}$ is reflexive and symmetric but not transitive. Notice that $\{(x, y) : x + y > 0\}$ is the empty relation on the set X of all negative numbers. This relation is symmetric and transitive (we say the definitions are satisfied vacuously here) but is not reflexive. The relation 'x and y have the same age' is reflexive, symmetric and transitive. Such a relation is called an *equivalence relation*.

Suppose $x \sim y$ is an equivalence relation on X. Then the *equivalence class of* x (with respect to $\sim$) is $E_x = \{y : x \sim y\}$. By an *equivalence class* we mean the equivalence class of some element of X. We now show that the distinct equivalence classes form a partition of X. We first observe that each equivalence class is non-empty since $x \in E_x$. We next show that any two distinct equivalence classes are disjoint. Suppose $z \in E_x \cap E_y$. Then $x \sim z$, $y \sim z$. So $x \sim z$ and $z \sim y$ and hence $x \sim y$. If now $u \in E_x$ then $x \sim u$ and $y \sim x$, so $y \sim u$ and $u \in E_y$. Thus $E_x \subseteq E_y$. By symmetry the other inclusion follows, so $E_x = E_y$. Thus any two distinct equivalence classes are disjoint. Finally every element x of X belongs to an equivalence class, viz., E_x. Thus the distinct equivalence classes form a partition of X.

Conversely, given any partition of X into non-empty subsets (called blocks) we may define a relation $\sim$ on X by stipulating that $x \sim y$ iff x, y belong to the same block of the partition. It is easy to check that this is an equivalence relation and that its equivalence classes are precisely the blocks of the given partition.

Thus the concept of equivalence relation is the same as that of a partition.

0.2 Functions

Given two non-empty sets X and Y, a *function* f *from* X *to* Y is a rule (which may or may not be simple) associating with each element x of X, a single element $f(x)$ of Y. We also use the term *mapping* (or *map* for short) for a function. X is called the *domain* and Y the *codomain*. We use the notation $f : X \to Y$ to denote that f is a function from X to Y. The element $f(x)$ of Y associated with the element x of X by f is called the *image of* x *under* f. We also say that f maps x to $f(x)$ and write $f : x \mapsto f(x)$. The *range of* f is the subset $\{f(x) : x \in X\}$ of Y, i.e., the set $\{y \in Y : y = f(x)$ for at least one $x \in X\}$. For any

$A \subseteq X$, $f(A)$ is defined as $\{y \in Y : y = f(x) \text{ for at least one } x \in A\}$. If $f : X \to Y$ and if $A \subseteq X$, the *restriction* $f|_A$ *of* f *to* A is the function $g : A \to Y$ defined by $g(x) = f(x)$ for all $x \in A$.

To give examples, $f(x) = x^2$ is a function from $\mathbb{R}$ to $\mathbb{R}$. $f|_{\{0,1\}}$ is the map with domain $\{0, 1\}$ taking 0 to 0 and 1 to 1. Note that $f([-1, 2]) = [0, 4]$. $g : z \mapsto |z|$ is a function from $\mathbb{C}$ to $\mathbb{R}$. The following is another function from $\mathbb{R}$ to $\mathbb{R}$:

$$h(x) = \begin{cases} x^2 & \text{if } x \geq 0 \\ 2x + 5 & \text{if } x < 0 \end{cases}$$

Some other functions from $\mathbb{R}$ to $\mathbb{R}$ are k and u defined by

$$k(x) = \begin{cases} x + 1 & \text{if } x \geq 0 \\ x & \text{if } x < 0 \end{cases}$$

and $u(x) = 3x$ for all x.

Note that two functions f_1 and f_2 from X to Y are equal iff $f_1(x) = f_2(x)$ for *all* $x \in X$.

If f is a function from X to Y and g is a function from Y to Z then it is easy to verify that h defined by

$$h(x) = g(f(x)) \quad \text{for all } x \in X$$

is a function from X to Z. We denote this function by $g \circ f$ and call it the *resultant* or *composite* of g and f. Note that $g \circ f$ can, in fact, be defined whenever the range of f is a subset of the domain of g. One should be careful about the order of f and g in the resultant. It is possible that $g \circ f$ is defined while $f \circ g$ is not. Even if both are defined, they may not be equal. For example, for the functions f and u given above,

$$(f \circ u)(x) = f(3x) = 9x^2 \quad \text{for all } x \in \mathbb{R}$$

while

$$(u \circ f)(x) = u(x^2) = 3x^2 \quad \text{for all } x \in \mathbb{R}$$

So $f \circ u \neq u \circ f$.

A map f from X to Y is said to be 1-1 (read *one-to-one*) or an *injection* if

$$f(x_1) = f(x_2) \Rightarrow x_1 = x_2$$

or, equivalently, if different elements of X have different images under f. The maps k and u above are 1-1 but the others are not. A map f from X to Y is said to be *onto* or a *surjection* if the range of f is Y

or, equivalently, if every element of the codomain is the image of some element under f. The maps h and u above are onto maps but the others are not. A map f from X to Y is said to be a 1-1 *correspondence* or a *bijection* if it is both 1-1 and onto. The map u above is a bijection but none of the others is. A bijection from X to itself is also called a *permutation* of X when X is finite. It is easy to check that if $f : X \to Y$ and $g : Y \to Z$ are bijections then $g \circ f$ is a bijection from X to Z.

The *identity map* on X, denoted by I_X (or 1_X), is the map from X to X defined by

$$I_X(x) = x \quad \text{for all } x \in X$$

If f is a map from X to Y, we say that a map g from Y to X is a *left inverse of* f if $g \circ f = I_X$. Similarly a map h from Y to X is said to be a *right inverse of* f if $f \circ h = I_Y$.

Theorem 0.2.1 A map f has a left inverse iff it is 1-1.

Proof Let f be a map from X to Y. Suppose first that f has a left inverse g. Then for $x_1, x_2 \in X$,

$$\begin{aligned} f(x_1) = f(x_2) &\Rightarrow g(f(x_1)) = g(f(x_2)) \\ &\Rightarrow I_X(x_1) = I_X(x_2) \\ &\Rightarrow x_1 = x_2 \end{aligned}$$

So f is 1-1. Conversely suppose f is 1-1. Let Y_0 be the range of f. Choose and fix an element x_0 of X. Then construct a map g from Y to X thus: for any $y \in Y$,

$$g(y) = \begin{cases} x & \text{if } y \in Y_0 \text{ and } f(x) = y \\ x_0 & \text{if } y \notin Y_0 \end{cases}$$

Note that if $y \in Y_0$, then there exists an $x \in X$ such that $f(x) = y$ since y is in the range. Further the x is unique since f is 1-1. Thus g is a map from Y to X. Now for all $x \in X$,

$$(g \circ f)(x) = g(f(x)) = x$$

(note that $f(x) \in Y_0$), hence $g \circ f = I_X$. ∎

Theorem 0.2.2 A map f has a right inverse iff it is onto.

Proof Let f be a map from X to Y. Suppose first that f has a right inverse h. Then for any $y \in Y$, we have

$$y = I_Y(y) = (f \circ h)(y) = f(h(y))$$

Since $h(y) \in X$, it follows that y belongs to the range of f. Thus $Y \subseteq$ range of f. Since range of $f \subseteq Y$, we have equality and f is onto.

Conversely let f be onto. Then for each $y \in Y$, choose and fix one $x \in X$ such that $f(x) = y$, and write $h(y) = x$. (For some y there may be more than one x such that $f(x) = y$; then choose one such x.) This defines a map h from Y to X. Also for each $y \in Y$,

$$(f \circ h)(y) = f(x) = y$$

where $h(y) = x$. Thus $f \circ h = I_y$. ∎

From the two preceding theorems and their proofs, we see that given a map f from X to X, there exists a map g from X to X such that

$$f \circ g = g \circ f = I_X$$

iff f is a bijection. Such a map g is called an *inverse* of f. Note that inverse means a map which is both a left inverse and a right inverse.

The map h given on page 3 is onto, so it has a right inverse. But h is not 1-1 and so has no left inverse. The map $v : \mathbb{R} \to \mathbb{R}$ defined by

$$v(y) = \begin{cases} \sqrt{y} & \text{if } y \geq 5 \\ (y-5)/2 & \text{if } y < 5 \end{cases}$$

is a right inverse of h. (Find out a different right inverse of h.)

The map $w : \mathbb{R} \to \mathbb{R}$ defined by

$$w(y) = \begin{cases} y - 1 & \text{if } y \geq 1 \\ y & \text{if } y < 0 \\ 12 & \text{if } 0 \leq y < 1 \end{cases}$$

is a left inverse of k (find another) but k has no right inverse.

The map $s : y \mapsto y/3$ from $\mathbb{R}$ to $\mathbb{R}$ is the inverse of the map u.

Theorem 0.2.3 If $f : X \to Y$, $g : Y \to Z$ and $h : Z \to W$ are maps then

$$(h \circ g) \circ f = h \circ (g \circ f)) \tag{0.2.1}$$

Proof It is easy to see that both sides of (0.2.1) are defined and are maps from X to W. To show equality, we have for all $x \in X$,

$$((h \circ g) \circ f)(x) = (h \circ g)(f(x)) = h(g(f(x)))$$

and

$$(h \circ (g \circ f))(x) = h((g \circ f)(x)) = h(g(f(x)))$$

Thus (0.2.1) follows. ∎

0.3 Groups

A *binary operation on* a set S is a map from $S \times S$ to S. It associates an element of S with every ordered pair of elements of S. We will usually denote the image of the ordered pair (x, y) by $x \circ y$. Sometimes other notations like $x + y$, $x \cdot y$, etc. are used.

For example, usual addition, usual multiplication and usual subtraction are binary operations on $\mathbb{R}$. Usual division is not a binary operation on $\mathbb{R}$ since we cannot associate any real number with the pair $(3, 0)$ (note that division by 0 is not defined). On any set S,

$$x \circ y = x$$

is a binary operation.

A binary operation $\circ$ on S is said to be *associative* if

$$(x \circ y) \circ z = x \circ (y \circ z) \quad \text{for all } x, y, z \in S$$

For example, addition and multiplication in $\mathbb{R}$ are associative but subtraction in not. For any set X, composition is an associative binary operation on the set of all maps from X to X. If the binary operation $\circ$ is associative, then we can talk of the element $x \circ y \circ z$ without any ambiguity (note, however, that this element may not be equal to $y \circ z \circ x$ or even $x \circ z \circ y$). For example, if f, h and u are the maps from $\mathbb{R}$ to $\mathbb{R}$ considered in the preceding section and $\circ$ denotes composition then $f \circ h \circ u$ is the map taking x to

$$\begin{cases} 81x^4 & \text{if } x \geq 0 \\ (6x+5)^2 & \text{if } x < 0 \end{cases}$$

But $f \circ u \circ h$ takes x to

$$\begin{cases} 81x^4 & \text{if } x \geq 0 \\ (6x+15)^2 & \text{if } x < 0 \end{cases}$$

and $u \circ h \circ f$ takes x to $3x^4$ for all $x \in \mathbb{R}$.

Note that the operation defined by $x \circ y = x$ is associative.

An element e of S is said to be an *identity element* for a binary operation $\circ$ on S if

$$x \circ e = e \circ x = x \quad \text{for all } x \in S$$

Another common notation for an identity element is 1. If we use the notation $+$ for the binary operation, then an identity element is called a *zero element* and is denoted 0. For (usual) addition on $\mathbb{R}$, the number 0 is the zero element. For multiplication on $\mathbb{R}$, 1 is the identity element. For composition on the set of all maps from X to X, the identity element is I_X. There is no identity element for the operation $x \circ y = x$ on any set with at least 2 elements, or for the operation of multiplication on the set of all even integers.

Theorem 0.3.1 If an identity element exists for a binary operation, it is unique.

Proof Suppose e_1 and e_2 are two identity elements. Then $e_1 \circ e_2 = e_2$ since e_1 is an identity element. Also $e_1 \circ e_2 = e_1$ since e_2 is an identity element. Hence $e_1 = e_2$. ∎

Suppose $\circ$ is an associative binary operation on S with identity element e. Then $y \in S$ is said to be an *inverse* of $x \in S$ if

$$x \circ y = y \circ x = e$$

We will then denote y by x^{-1}. If we use the notation $+$ for the operation and consequently denote the identity element by 0, we will call an inverse of x a *negative* of x and denote it by $-x$. With respect to the operation of usual addition on $\mathbb{R}$, the negative of x is the usual $-x$. For multiplication on $\mathbb{R}$, the inverse of any non-zero element x is $\frac{1}{x}$ and 0 has no inverse. For composition on the set of all maps from X to X, f has an inverse iff f is a bijection and then the inverse of f is the inverse map as defined earlier.

Theorem 0.3.2 Let $\circ$ be an associative binary operation on S with identity element e. If $x \in S$ has an inverse y then y is unique.

Proof Let y_1 and y_2 be inverses of x. Then

$$y_1 = y_1 \circ e = y_1 \circ (x \circ y_2) = (y_1 \circ x) \circ y_2 = e \circ y_2 = y_2.$$ ∎

Definition 0.3.3 A *group* is a set S together with an associative binary operation on S such that there is an identity element and every element has an inverse.

Thus a group is $(S, \circ)$ where the following axioms are satisfied:

(i) $x, y \in S \Rightarrow x \circ y \in S$,

(ii) $x \circ (y \circ z) = (x \circ y) \circ z$ for all $x, y, z \in S$,

(iii) there exists an element e of S such that $e \circ x = x \circ e = x$ for all $x \in S$

(iv) for each element x of S there exists an element y of S such that $x \circ y = y \circ x = e$.

Note that e is unique and x^{-1} is unique for a given x.

We now give a few examples of groups. $(\mathbb{R}, +)$ is a group with 0 as the zero element and $-x$ as the negative of x. $(\mathbb{R}-\{0\}, \cdot)$ is a group with 1 as the identity element and $1/x$ as the inverse of x. Note that $(\mathbb{R}, \cdot)$ is not a group. Similarly $(\mathbb{Z}, \cdot)$ is not a group, where $\mathbb{Z}$ is the set of all integers. $(\{1, -1\}, \cdot)$ is a group. It can be checked that $(\{0, 1, \ldots, n-1\}, + \bmod n)$ is a group, where the operation is addition mod n and is defined thus: the sum of x and y in the group is the remainder obtained when the ordinary sum $x + y$ is divided by n. For example, if $n = 6$ then in the group $3 + 5$ is 2 while $1 + 2$ is 3. To prove associativity here, one shows that both $x + (y + z)$ and $(x + y) + z$ in the group are equal to the remainder obtained when the ordinary sum $x + y + z$ is divided by n. 0 is the zero element, the negative of 0 is 0 and the negative of any other x is $n - x$. Finally it can be checked that the set of all bijections from X to X forms a group under composition.

A binary operation $\circ$ on S is said to be *commutative* if $x \circ y = y \circ x$ for all $x, y \in S$. A group $(S, \circ)$ is *commutative* if $\circ$ is commutative. All the groups given in the preceding paragraph are commutative except the last one.

From now on, we will usually drop the symbol $\circ$ in $x \circ y$ and write xy for convenience. In a group we have the cancellation laws:

$$xy = xz \Rightarrow y = z$$
$$yx = zx \Rightarrow y = z$$

To prove the first one we 'premultiply' both sides of $xy = xz$ by x^{-1}. To Prove the second we 'postmultiply' both sides of $yx = zx$ by x^{-1}.

In a group, we have the following results.

(i) If $xx = x$ then $xx = xe$, so $x = e$.

(ii) If $xy = e$ then y is x^{-1}. This is because $xy = e = xx^{-1}$ and x can be cancelled on the left. Similarly if $zx = e$ then $z = x^{-1}$.

(iii) We have $(x^{-1})^{-1} = x$. To prove this we note that $xx^{-1} = e$ by the definition of x^{-1}. So by what was proved in the preceding paragraph, x is the inverse of x^{-1}.

(iv) $(xy)^{-1} = y^{-1}x^{-1}$ for any two elements x and y (note the reversal of the order). For this, it is enough to see that $(xy)(y^{-1}x^{-1}) = e$.

0.4 Rings and fields

In the preceding section we studied an algebraic structure with one binary operation. In this section we consider some structures with two binary operations.

Definition 0.4.1 A *ring* is a set R together with two binary operations $+$ and $\cdot$ on R satisfying the following axioms:

(i) $(R, +)$ is a commutative group

(ii) $\cdot$ is associative

(iii) $x \cdot (y + z) = (x \cdot y) + (x \cdot z)$ for all $x, y, z \in R$,
$(y + z) \cdot x = (y \cdot x) + (z \cdot x)$ for all $x, y, z \in R$

Let us consider a few examples. Each of the sets $\mathbb{R}$, $\mathbb{C}$, $\mathbb{Z}$, $\mathbb{Q}$ forms a ring under the operations of usual addition and usual multiplication. Here $\mathbb{Q}$ denotes the set of all rational numbers. The set $\{0, 1, \ldots, n-1\}$ forms a ring under the operations of addition mod n and multiplication mod n (where the latter is defined thus: the product of x and y mod n is the remainder obtained when the ordinary product xy is divided by n). This ring is denoted by $\mathbb{Z}_n$. The polynomials in a variable t (see *Section* 0.5) with real coefficients form a ring under the operations of usual addition and usual multiplication. The set of all real-valued functions of a real variable forms a ring under usual addition and usual multiplication defined by

$$(f+g)(x) = f(x) + g(x) \quad \text{for all } x \in \mathbb{R},$$
$$(fg)(x) = f(x) \cdot g(x) \quad \text{for all } x \in \mathbb{R}$$

Note that this multiplication has nothing to do with composition of maps.

From now on we drop the dot and write xy for the product $x \cdot y$ in a ring. We denote the zero element (with respect to $+$) by 0 and the negative of x (with respect to $+$) by $-x$.

In a ring, $0x = x0 = 0$ for all x. To prove this, note that

$$0x + 0x = (0+0)x = 0x$$

Hence $0x = 0$. Similarly '$x0 = 0$' can be proved. It can also be shown using the distributive laws that $x(-y) = (-x)y = -(xy)$ and

$(-x)(-y) = xy$. For example, $xy + x(-y) = x(y + (-y)) = x0 = 0$, so $x(-y) = -(xy)$. We will write $x - y$ for $x + (-y)$. Properties like $x - (y - z) = (x - y) + z$ and $x(y - z) = xy - xz$ can be proved easily.

A ring is said to be *commutative* if $xy = yx$ for all x and y. A *unity* or *identity* of a ring is an identity for the multiplication and is denoted by 1 if it exists. A ring is said to be an *integral domain* if it is commutative, has a unity and if the product of any two non-zero elements is non-zero. A *field* is a commutative ring such that the non-zero elements form a group under multiplication.

$\mathbb{R}$, $\mathbb{C}$, $\mathbb{Q}$ form fields. $\mathbb{Z}$ is an integral domain but not a field. It can be shown that the ring $\mathbb{Z}_n$ of residues mod n is a field iff n is prime. If n is not prime, this ring is not even an integral domain. For example, the product of 2 and 3 is 0 in $\mathbb{Z}_6$. Polynomials form an integral domain which is not a field. Under usual addition and multiplication, the set of all even integers forms a ring without unity. The ring of real valued functions of a real variable is not even an integral domain since the product of two non-zero functions can be the zero function.

In a field we can divide any element x by any non-zero element y and form the ratio $\frac{x}{y}$ as $xy^{-1} = y^{-1}x$. Note that x/y is an element of the field. For example, in the field $\mathbb{Z}_5$, $\frac{4}{3} = 4 \cdot 2 = 3$. We thus have all the four *arithmetic operations* (addition, subtraction, multiplication and division by a non-zero element) in a field and these have all the usual properties. One should however note that there is no concept of positivity or order in a field (as is clear from the examples of $\mathbb{C}$ and the field of residues mod p, where p is a prime). It is also not true that $x_1^2 + \cdots + x_n^2 = 0$ implies $x_1 = \cdots = x_n = 0$. For example, $1^2 + i^2 = 0$ in $\mathbb{C}$ and $1^2 + 1^2 + \cdots + 1^2$ (p times) $= 0$ in $\mathbb{Z}_p$.

We finally mention that there exists a (finite) field with exactly n elements iff n is a power of a prime. Also for any n which is a power of a prime, there is essentially only one field (which is $\mathbb{Z}_n$ if n is prime) with n elements and it is denoted by $\mathrm{GF}(n)$.

0.5 Polynomials

Fix any field F (e.g., $\mathbb{R}$ or $\mathbb{C}$). A *polynomial over* F (or with coefficients from F) is an expression of the form

$$p(t) = a_0 + a_1t + a_2t^2 + \cdots + a_kt^k \tag{0.5.1}$$

where $k \geq 0$ and $a_0, a_1, \ldots, a_k \in F$. For example $2 - 3t - t^3$ is a polynomial over $\mathbb{R}$. The polynomial $p(t)$ in (0.5.1) is 0 if $a_0 = a_1 = \cdots = a_k = 0$. If at least one $a_i \neq 0$ then $p(t) \neq 0$. If $p(t) \neq 0$, the *degree* of $p(t)$ is the largest non-negative integer i such that the coefficient of t^i in $p(t)$ is non-zero. We also define the *degree of the zero polynomial* to be $-\infty$ (minus infinity). Note that the polynomials of degree 0 are the non-zero constant polynomials. If $p(t)$ is as in (0.5.1) with $a_k \neq 0$ we say that a_k is the *leading coefficient*. A non-zero polynomial is said to be *monic* if its leading coefficient is 1.

Two polynomials are *equal* if for each $i \geq 0$, t^i has the same coefficient in both of them. We define the *sum* $p + q$ of two polynomials p and q thus: the coefficient of t^i in $p + q$ is the sum of the coefficients of t^i in p and in q. For example

$$(2 - 2t - t^3) + (-2 + t + 3t^2) = -t + 3t^2 - t^3$$

We define the *product* pq also in a natural way: expand the product in the usual way and collect the coefficients of each power of t. For example

$$\begin{aligned} &(2 - 2t - t^2)(1 + 2t + t^2 + 3t^4) \\ &= 2 + 2t - 3t^2 - 4t^3 + 5t^4 - 6t^5 - 3t^6 \end{aligned}$$

Finally, if $b \in F$ and $p(t)$ is given by (0.5.1) then $b \cdot p(t)$ is the polynomial $(ba_0) + (ba_1)t + \cdots + (ba_k)t^k$, obtained by treating b as a (constant) polynomial.

It is easy to verify the following:

$$\begin{aligned} \text{degree of } p + q &\leq \max(\text{degree of } p,\ \text{degree of } q), \\ \text{degree of } pq &= \text{degree of } p + \text{degree of } q \end{aligned}$$

To prove the second relation note that if $p(t)$ is as in (0.5.1), $q(t) = b_0 + b_1 t + \cdots + b_\ell t^\ell$ and if $a_k \neq 0$ and $b_\ell \neq 0$ then the coefficient of $t^{k+\ell}$ in pq is $a_k b_\ell \neq 0$ and is the leading coefficient of pq. It can be easily verified that the polynomials in t with coefficients from F form an integral domain under addition and multiplication as defined above. This integral domain is denoted by $F[t]$.

If $p(t)$ is the polynomial (0.5.1), then the *value of* $p(t)$ *at* $t = b$ (where $b \in F$) is

$$p(b) = a_0 + a_1 b + \cdots + a_k b^k$$

It is easy to see that if $p(t) = q(t)$ as polynomials then $p(b) = q(b)$ for all $b \in F$. We note that the converse of this statement is not true. For example, if $F = \mathbb{Z}_2$ and if $p(t) = t^2 + t$ then $p(b) = 0$ for all $b \in F$ but $p \neq 0$. However if $p(t)$ is of degree k and $p(b) = 0$ for more than k distinct b's in F then $p(t) = 0$. It can be easily seen that if $f(t) = p(t) + q(t)$ then $f(b) = p(b) + q(b)$ for all $b \in F$. Similarly if $g(t) = p(t) \cdot q(t)$ then $g(b) = p(b) \cdot q(b)$.

Given any non-zero polynomial $p(t)$ with degree k and any polynomial $f(t)$, we can divide $f(t)$ by $p(t)$ and get a *quotient* $q(t)$ and a *remainder* $r(t)$ such that

$$f(t) = p(t) \cdot q(t) + r(t), \text{ degree of } r(t) < k$$

To prove this, let ℓ be the degree of f. If $\ell < k$, take $q = 0$ and $r = f$. Suppose next $\ell \geq k$, the coefficient of t^ℓ in f is b^ℓ and the coefficient of t^k in p is a_k. Then clearly

$$f(t) = p(t) \cdot b_\ell a_k^{-1} t^{\ell - k} + s(t)$$

for some polynomial $s(t)$ with degree $< \ell$. If the degree of $s(t)$ is less than k we are done. Otherwise, repeat the above procedure with s in place of f. This procedure is the same as that taught in school algebra and is known as the *division algorithm*.

The quotient q and the remainder r are unique for given p and f. To see this, suppose

$$f = pq_1 + r_1 = pq_2 + r_2$$

where degree of $r_1 < k$ and degree of $r_2 < k$. Then $r_1 - r_2 = p(q_2 - q_1)$. The degree of LHS $\leq k - 1$. If $q_2 \neq q_1$ then the degree of RHS $\geq k$, a contradiction. So $q_2 = q_1$ and $r_2 = r_1$.

We say that $p(t)$ *divides* $f(t)$ if there exists a polynomial $q(t)$ such that $f = pq$. If p divides f then $f = 0$ or degree of $f \geq$ degree of p. A polynomial p is *irreducible* if degree of p is at least 1 and if p cannot be written as p_1p_2 with degrees of p_1 and p_2 less than the degree of p. Clearly every polynomial of degree 1 is irreducible. The irreducible polynomials in $F[t]$ play a role similar to that of prime numbers in $\mathbb{Z}$. It can be proved that any monic polynomial can be expressed uniquely in the form $p_1^{n_1} p_2^{n_2} \cdots p_k^{n_k}$ where $p_1, \ldots, p_k$ are distinct irreducible monic polynomials and $n_1, \ldots, n_k$ are positive integers. Given any two non-zero polynomials f and g there exists a unique monic polynomial h such that h divides f, h divides g and every common divisor of f and g divides h. This h is called the *greatest common divisor* of f and g. Similarly any two non-zero polynomials have an *l.c.m.* Two non-zero polynomials f

and g are said to be *relatively prime* if their g.c.d. is 1. If a polynomial p divides the product of two polynomials f and g and p is relatively prime to f then p divides g. In particular, if p divides fg and p is irreducible then p divides f or p divides g.

Taking the divisor as $t-a$ in the division algorithm we see that any polynomial $f(t)$ is $(t-a)q(t)+r$ for some polynomial $q(t)$ and some $r \in F$ (note that the degree of $t-a$ is 1). Substituting a for t we get $f(a)=r$. This is the well-known remainder theorem.

We say that $a \in F$ is a *root* of $f(t)$ if $f(a)=0$. A polynomial over F need not always have a root in F. For example, x^2+1 has no root in $\mathbb{R}$. But every polynomial a_0+a_1x with degree 1 has a root, viz., $-a_0a_1^{-1}$. By the remainder theorem, a is a root of $f(t)$ iff $t-a$ divides $f(t)$. If $a_1,\ldots,a_k$ are distinct elements of F then $t-a_1,\ldots,t-a_k$ are distinct irreducible polynomials. So if $a_1,\ldots,a_k$ are distinct roots of $f(t)$ then $(t-a_1)\cdots(t-a_k)$ divides $f(t)$. It follows that a non-zero polynomial of degree n can have at most n distinct roots.

The *fundamental theorem of algebra* says that every polynomial over $\mathbb{C}$ with degree ≥ 2 is a product of polynomials of degree 1 (and so has a root in $\mathbb{C}$). In other words, the irreducible polynomials in $\mathbb{C}[t]$ are those of degree 1.

Just as we form rational numbers from integers we can form rational functions from polynomials. A *rational function* is an expression of the form $f(t)/g(t)$ where $g \neq 0$. We can define equality, sum and product for these just as for rational numbers. The rational functions form a field.

Chapter 1
Vector spaces

1.1 Introduction

In Physics we learn that a force applied at a point O has both magnitude and direction. It is represented by an arrow OP as in *Figure* 1.1.1, where the length OP represents the magnitude and O to P the direction of the force. If we now apply another force OQ at the point O, the resultant (also called the sum) of the two forces is obtained by the *parallelogram law*: it is OR where $OPRQ$ is a parallelogram. Also, if the strength of the force OP is doubled without changing the direction, the new force is OS where S is the point on the line OP such that $OS = 2\,OP$. If the direction of the force OP is reversed without altering the magnitude, the new force is OT where T is the point on OP such that $OT = -OP$ with the usual convention. In general, α times the force OP is OW where W is a point on OP (extended either way, if necessary) such that $OW/OP = \alpha$, where α may be positive, negative or zero.

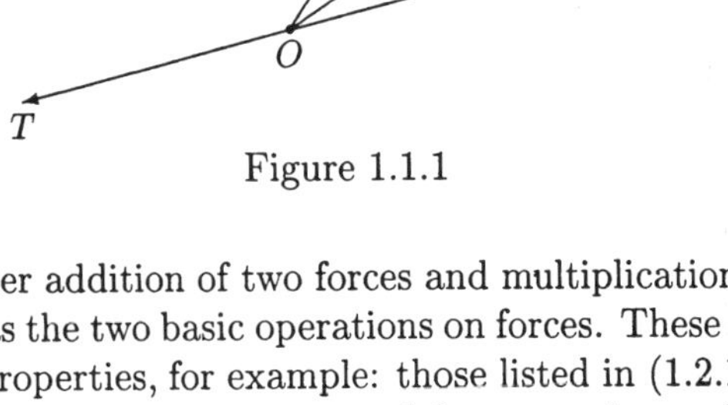

Figure 1.1.1

We consider addition of two forces and multiplication of a force by a real number as the two basic operations on forces. These operations have several nice properties, for example: those listed in (1.2.1). Arrows such as OP and OQ are called *vectors*, and forces can be studied by studying them. The set of all vectors together with the two basic operations is called a *vector space*. This concept will be defined formally in the next section.

Now, each arrow OP can be represented by the point P once the point O, which is taken to be the origin, is fixed. If we are studying vectors in a plane, we may introduce a coordinate system (x-axis, y-axis and the unit of distance) in the plane and represent P by the ordered pair (x, y) where x and y are the x-coordinate and the y-coordinate of P. This is a 1-1 correspondence between points of the plane and the set $\mathbb{R}^2 = \mathbb{R} \times \mathbb{R}$ of all ordered pairs of real numbers. It can be proved that if $P = (x_1, y_1)$ and $Q = (x_2, y_2)$, then $R = (x_1 + x_2, y_1 + y_2)$ where $OR = OP + OQ$ and $W = (\alpha x_1, \alpha y_1)$, where $OW = \alpha\, OP$. Similar statements hold for the 3-dimensional space and $\mathbb{R}^3$. Note that once we represent vectors in the plane and in space as ordered pairs and triples, addition of two vectors and multiplication of a vector by a number (also called a scalar) are easy and we can use the powerful tools of algebra. In this book we do not consider operations on vectors like vector product which are studied in Vector Calculus.

Another natural way in which vectors arise is in the simultaneous study of several characteristics of an individual. Suppose we want to study x_1 = height, x_2 = weight, x_3 = age and x_4 = blood pressure, of an individual. The values of all these can together be represented by an ordered 4-tuple $\mathbf{x} = (x_1, x_2, x_3, x_4)$ which can be viewed as an element of $\mathbb{R}^4$. Such 4-tuples may be called vectors in analogy with $\mathbb{R}^2$ and $\mathbb{R}^3$. If we now have the vectors $\mathbf{x} = (x_1, x_2, x_3, x_4)$ and $\mathbf{w} = (w_1, w_2, w_3, w_4)$ for two individuals then the average values of the characteristics for the two individuals are given by the vector $\frac{1}{2}\mathbf{x} + \frac{1}{2}\mathbf{w}$ where addition of vectors and multiplication of a vector by a scalar are defined in a natural way.

We give one other motivation. Suppose we want to solve the system of linear equations:

$$\begin{aligned} x_1 - 2x_2 - x_3 &= 3 \\ 2x_1 + x_2 + 3x_3 &= 2 \\ 3x_1 - x_2 + 4x_3 &= 1 \end{aligned}$$

For this, we subtract twice the first equation from the second to eliminate x_1, etc. Note that this amounts to adding -2 times the vector $(1, -2, -1, 3)$ to the vector $(2, 1, 3, 2)$ in $\mathbb{R}^4$, where the four components are the coefficients of x_1, x_2, x_3 and the RHS. Thus in solving a system of linear equations, we actually use vectors.

Exercise

1. If $P = (x_1, y_1)$ and $Q = (x_2, y_2)$ in *Figure* 1.1.1, verify that $R = (x_1+x_2, y_1+y_2)$ and $W = (\alpha x_1, \alpha y_1)$. State and prove the corresponding results in the 3-dimensional space.

1.2 Vector space: axiomatic definition

As we have already seen in the *Introduction*, each point of the plane can be thought of as a vector, namely, the ordered pair $\mathbf{x} = (x_1, x_2)$ where x_1 and x_2 are the 'x- and y-coordinates' of the point. If $\mathbf{x} = (x_1, x_2)$ and $\mathbf{y} = (y_1, y_2)$ then the sum $\mathbf{x}+\mathbf{y}$ is (x_1+y_1, x_2+y_2) and $\alpha\mathbf{x}$ is $(\alpha x_1, \alpha x_2)$. It can easily be checked that these operations have the following properties:

$$\begin{aligned}
\mathbf{x}+\mathbf{y} &= \mathbf{y}+\mathbf{x} \\
(\mathbf{x}+\mathbf{y})+\mathbf{z} &= \mathbf{x}+(\mathbf{y}+\mathbf{z}) \\
\mathbf{x}+\mathbf{0} &= \mathbf{x} \\
\mathbf{x}+(-\mathbf{x}) &= \mathbf{0} \\
\alpha(\beta\mathbf{x}) &= (\alpha\beta)\mathbf{x} \\
(\alpha+\beta)\mathbf{x} &= \alpha\mathbf{x}+\beta\mathbf{x} \\
\alpha(\mathbf{x}+\mathbf{y}) &= \alpha\mathbf{x}+\alpha\mathbf{y} \\
1\mathbf{x} &= \mathbf{x}
\end{aligned} \tag{1.2.1}$$

Here, $\mathbf{0} = (0,0)$ stands for the vector corresponding to the origin and $-\mathbf{x}$ is $(-x_1, -x_2)$.

Let us look at another example. Consider the set V of all polynomials with real coefficients. If we call the elements of V as vectors, we can define addition of two vectors and multiplication of a vector by a scalar (i.e., a real number) in the usual way. It is then easy to check that the properties listed above are satisfied.

We mention one more example† which looks different from all the above examples. Let Ω be a fixed non-empty set and let V be the set of all subsets of Ω, usually known as the *power set* of Ω. Vectors are the elements of V, i.e., subsets of Ω. We define the sum of two vectors $\mathbf{A}$ and $\mathbf{B}$ to be their symmetric difference

$$\mathbf{A} \triangle \mathbf{B} = (\mathbf{A}-\mathbf{B}) \cup (\mathbf{B}-\mathbf{A})$$

†This example may be omitted in the first reading

We next define multiplication of a vector **A** by a scalar α. Here we consider only the scalars 0 and 1 and assume that they form a field of order 2, so the sum and product of scalars are defined by

$$0+0=0,\ 0+1=1+0=1,\ 1+1=0,$$
$$0\cdot 0=0\cdot 1=1\cdot 0=0,\ 1\cdot 1=1$$

Now we define the scalar multiple $\alpha\mathbf{A}$ to be **A** if $\alpha = 1$ and $\emptyset$ (the null set) if $\alpha = 0$. With addition and scalar multiplication thus defined, it can be checked that all the properties listed in (1.2.1) are satisfied with $\emptyset$ as the zero vector and **A** itself serving as $-\mathbf{A}$.

It is easy to see that the three structures considered above have some common features which we propose to study in this chapter. Of course each of them has its own special characteristics. For example we have the concepts of distance and angle in $\mathbb{R}^2$ which are lacking in the other structures. In the vector space of polynomials, we may define in a natural way the product of two vectors, but there is no natural analogue in $\mathbb{R}^2$. The last structure we considered above has some peculiar properties which are not shared by the others, for example: the sum of any vector with itself is the zero vector. Also note that in the field $\{0, 1\}$ as well as in any finite field and the complex field $\mathbb{C}$, there is no notion of positivity or order.

In this chapter we study the properties of such structures as above with respect to the common features. It turns out that the properties listed in (1.2.1) enable us to develop a sound theory which also has wide applicability. We thus take these properties as the axioms for the definition of a vector space. Essentially, a vector space is a linear structure and is also called a *linear space*.

We give below the definition of a vector space over a *general* field. The main reason for this is that vector spaces over finite fields are of great interest in Computer Science, Coding Theory, Combinatorics, Design of Experiments and Abstract Algebra; vector spaces over the field of rational numbers are useful in Number Theory and Design of Experiments and vector spaces over the field of complex numbers are needed for the study of eigenvalues. Thus vector spaces over fields other than $\mathbb{R}$ are useful in many contexts. Besides, the theory of vector spaces over a general field is no more complicated than that over $\mathbb{R}$. So throughout *Part* I (*Chapters* 1 through 6), with the exception of *Section* 3.8, we consider vector spaces over a general field. *However, many important examples of vector spaces take the field to be* $\mathbb{R}$ *or* $\mathbb{C}$ *and the reader would* *not lose much by making this assumption. In fact, for easy visualization, one can take the field to be $\mathbb{R}$ in most cases.*

In all numerical examples and exercises the numbers will be assumed to be real unless specified otherwise.

Definition 1.2.1 *A vector space over a field F* is a quadruple $(V, +, \cdot, F)$[†] satisfying the following axioms for all $\alpha, \beta \in F$ and $\mathbf{x}, \mathbf{y}, \mathbf{z} \in V$:

1. $(V, +)$ is a commutative group, that is,
   (a) $+$ is a map from $V \times V$ to V (we write the image of $(\mathbf{x}, \mathbf{y})$ as $\mathbf{x} + \mathbf{y}$ for convenience) (Closure with respect to $+$)
   (b) $(\mathbf{x} + \mathbf{y}) + \mathbf{z} = \mathbf{x} + (\mathbf{y} + \mathbf{z})$ (Associativity of $+$)
   (c) there exists an element $\mathbf{0}$ of V such that $\mathbf{x} + \mathbf{0} = \mathbf{0} + \mathbf{x} = \mathbf{x}$ for all $\mathbf{x} \in V$ (Existence of $\mathbf{0}$)
   (d) for each $\mathbf{x}$ in V there exists an element $-\mathbf{x}$ in V such that $\mathbf{x} + (-\mathbf{x}) = (-\mathbf{x}) + \mathbf{x} = \mathbf{0}$ (Existence of negative)
   (e) $\mathbf{x} + \mathbf{y} = \mathbf{y} + \mathbf{x}$ (Commutativity of $+$)
2. $\cdot$ is a map from $F \times V$ to V (we write the image of $(\alpha, \mathbf{x})$ as $\alpha \cdot \mathbf{x}$ for convenience) (Closure with respect to $\cdot$)
3. $\alpha \cdot (\beta \cdot \mathbf{x}) = (\alpha\beta) \cdot \mathbf{x}$
4. $1 \cdot \mathbf{x} = \mathbf{x}$
5. $(\alpha + \beta) \cdot \mathbf{x} = (\alpha \cdot \mathbf{x}) + (\beta \cdot \mathbf{x})$ (Distributivity)
6. $\alpha \cdot (\mathbf{x} + \mathbf{y}) = (\alpha \cdot \mathbf{x}) + (\alpha \cdot \mathbf{y})$ (Distributivity)

Here $+$ denotes *vector addition* except in the LHS of (5) where it denotes addition in F. The elements of V are called *vectors* and the elements of F are called *scalars*. F itself is called the *base field* or *ground field* of the vector space. Note that $\cdot$ denotes *scalar multiplication* of a vector. Also $\mathbf{0}$ of axiom 1(c) is called the *null vector* or *zero vector* and $-\mathbf{x}$ of axiom 1(d) the *negative of* $\mathbf{x}$.

We will normally use bold face lower case Roman letters (like $\mathbf{x}$, $\mathbf{u}$ *and* $\mathbf{x}_1$*) to denote vectors and lower case Greek letters (like* α, β *and* ξ_1*) to denote scalars.*

[†]We may also refer to this vector space as $(V, +, \cdot)$ when F is clear from the context.

A *real* (resp. *complex*) *vector space* is a vector space over $\mathbb{R}$ (resp. $\mathbb{C}$). Note that $\mathbb{C}$ with the usual addition and multiplication of a real number with a complex number, can be considered to be a real vector space.

Before studying the properties of general vector spaces we give some more examples.

Example 1.2.2 Let n be a fixed integer ≥ 1. Then $(\mathbb{R}^n, +, \cdot)$ is a vector space over $\mathbb{R}$, where

$$\begin{aligned}(x_1, x_2, \ldots, x_n) + (y_1, y_2, \ldots, y_n) &= (x_1 + y_1, x_2 + y_2, \ldots, x_n + y_n) \\ \alpha \cdot (x_1, x_2, \ldots, x_n) &= (\alpha x_1, \alpha x_2, \ldots, \alpha x_n)\end{aligned}$$

We remind the reader that the elements of $\mathbb{R}^n$ are the ordered n-tuples from $\mathbb{R}$. Note that if $\mathbf{x} = (x_1, x_2, \ldots, x_n) \in \mathbb{R}^n$ and $\mathbf{y} = (y_1, y_2, \ldots, y_n) \in \mathbb{R}^n$ then $\mathbf{x} = \mathbf{y}$ iff $x_i = y_i$ for $i = 1, 2, \ldots, n$. We refer to x_i as the i-th component of $\mathbf{x}$. The zero vector here is the n-tuple $(0, 0, \ldots, 0)$. The negative of $\mathbf{x} = (x_1, x_2, \ldots, x_n)$ is the vector $(-x_1, -x_2, \ldots, -x_n)$.

More generally, for any field F, the vector space F^n over F is defined in an exactly analogous fashion. Notice that the addition of vectors and scalar multiplication in the vector space F^1 coincide respectively with the addition and multiplication in the field F. So we sometimes use F to denote F^1.

By 'the vector space F^n', we mean F^n over F.

We mention that the vector space $\mathbb{R}^2$ as defined here is the same as that introduced geometrically in *Section* 1.1.

Example 1.2.3 The set $\mathbb{C}^n$ becomes a vector space over $\mathbb{R}$ if vector addition and scalar multiplication are defined as in the preceding example. Note that this is different from the vector space $\mathbb{C}^n$.

Example 1.2.4 The set of all ordered triplets (x_1, x_2, x_3) of real numbers such that

$$\frac{x_1}{3} = \frac{x_2}{4} = \frac{x_3}{2} \tag{1.2.2}$$

forms a real vector space, where the operations $+$ and $\cdot$ are as in $\mathbb{R}^3$. Notice that (1.2.2) is the equation of the line passing through the origin and the point $(3, 4, 2)$. In fact the set of all points on any line passing through the origin in $\mathbb{R}^3$ forms a vector space.

Example 1.2.5 $\mathbb{R}$ with usual addition and multiplication, is a vector space over the field of rational numbers $\mathbb{Q}$. The zero vector here is the real number $\mathbf{0}$ and the negative of $\mathbf{x}$ is the real number $-\mathbf{x}$.

Example 1.2.6 For any non-empty set X and any field F, F^X is a vector space over F. Here F^X is the set of all functions from X to F and the operations $+$ and $\cdot$ are defined as follows:

$$\begin{aligned}(f+g)(x) &= f(x)+g(x) \quad &&\text{for all } x \in X\\ (\alpha \cdot f)(x) &= \alpha f(x) &&\text{for all } x \in X \text{ and } \alpha \in F\end{aligned}$$

The zero vector is the function f_0 where

$$f_0(x) = 0 \quad \text{for all } x \in X$$

and the negative of f is the function g where

$$g(x) = -f(x) \quad \text{for all } x \in X$$

One can see that F^n is the special case of F^X corresponding to $X = \{1, 2, \ldots, n\}$, $f \in F^X$ being identified with $(f(1), f(2), \ldots, f(n)) \in F^n$.

Example 1.2.7 The set $\mathcal{P}_n$ of all polynomials with real coefficients and of degree $\leq n-1$ is a vector space over $\mathbb{R}$ with the operations

$$\left(\sum_{i=0}^{n-1} \alpha_i t^i\right) + \left(\sum_{i=0}^{n-1} \beta_i t^i\right) = \sum_{i=0}^{n-1} (\alpha_i + \beta_i) t^i$$

and

$$\gamma \cdot \left(\sum_{i=0}^{n-1} \alpha_i t^i\right) = \sum_{i=0}^{n-1} (\gamma \alpha_i) t^i$$

Here the zero vector is the zero polynomial which is taken to be of degree $\leq n-1$ (the degree of the zero polynomial may be defined as $-\infty$). Also the negative of $\sum_{i=0}^{n-1} \alpha_i t^i$ is $\sum_{i=0}^{n-1} (-\alpha_i) t^i$.

***Example 1.2.8** The set of all real-valued random variables on a fixed sample space forms a vector space over $\mathbb{R}$ under the usual operations.

***Example 1.2.9** The power set of a set Ω forms a vector space over $F = \{0, 1\}$ with the operations as defined on page 16.

We shall now prove a few properties of the operations of a vector space $(V, +, \cdot, F)$. The four statements given below follow from the fact that $(V, +)$ is a group, see *Section* 0.3.

(i) $\mathbf{0}$ is unique,

(ii) $-\mathbf{x}$ is unique for a given $\mathbf{x}$ in V,

(iii) $\mathbf{u}+\mathbf{x}=\mathbf{u}+\mathbf{y} \Rightarrow \mathbf{x}=\mathbf{y}$,

(iv) $\mathbf{x}+\mathbf{x}=\mathbf{x} \Rightarrow \mathbf{x}=\mathbf{0}$.

Theorem 1.2.10 The following hold for all $\mathbf{x} \in V$ and $\alpha \in F$:

(i) $0 \cdot \mathbf{x} = \alpha \cdot \mathbf{0} = \mathbf{0}$,

(ii) $\alpha \cdot \mathbf{x} = \mathbf{0} \Rightarrow \alpha = 0$ or $\mathbf{x} = \mathbf{0}$,

(iii) $(-1) \cdot \mathbf{x} = -\mathbf{x}$.

Proof (i) $0 \cdot \mathbf{x} = (0+0) \cdot \mathbf{x} = (0 \cdot \mathbf{x}) + (0 \cdot \mathbf{x})$. Hence $0 \cdot \mathbf{x} = \mathbf{0}$ by observation (iv) preceding this theorem. Also, $\alpha \cdot \mathbf{0} = \alpha \cdot (\mathbf{0}+\mathbf{0}) = (\alpha \cdot \mathbf{0}) + (\alpha \cdot \mathbf{0})$, hence $\alpha \cdot \mathbf{0} = \mathbf{0}$.

(ii) Let $\alpha \cdot \mathbf{x} = \mathbf{0}$. If $\alpha = 0$ we are done. So let $\alpha \neq 0$. Then α^{-1} exists and

$$\mathbf{x} = 1 \cdot \mathbf{x} = (\alpha^{-1} \cdot \alpha) \cdot \mathbf{x} = \alpha^{-1} \cdot (\alpha \cdot \mathbf{x}) = \alpha^{-1} \cdot \mathbf{0} = \mathbf{0}$$

(iii) $\mathbf{x} + [(-1) \cdot \mathbf{x}] = (1 \cdot \mathbf{x}) + [(-1) \cdot \mathbf{x}] = [1 + (-1)] \cdot \mathbf{x} = 0 \cdot \mathbf{x} = \mathbf{0}$. ∎

Before proceeding further we shall simplify some of our notations. *For brevity we refer to the vector space* $(V, +, \cdot, F)$ *as the vector space* V*. Thus* V *will denote both the vector space and the underlying set. For the same reason, we will denote the product* $\alpha \cdot \mathbf{x}$ *simply by* $\alpha\mathbf{x}$ *where* α *is a scalar and* $\mathbf{x}$ *is a vector.*

Exercises

1. Verify that the vector spaces in *Examples* 1.2.2 through 1.2.8 satisfy the axioms of *Definition* 1.2.1.
2. In each of the following, find precisely which axioms in the definition of a vector space are violated. Take $V = \mathbb{R}^2$ and $F = \mathbb{R}$ throughout.
   (a) $(x_1, x_2) + (y_1, y_2) = (x_1 + y_1, 0)$, $\alpha(x_1, x_2) = (\alpha x_1, 0)$
   (b) $(x_1, x_2) + (y_1, y_2) = (x_1 + y_1, x_2 + y_2)$, $\alpha(x_1, x_2) = (\alpha x_1, 0)$
   (c) $(x_1, x_2) + (y_1, y_2) = (x_1 + y_1, x_2 + y_2)$, $\alpha(x_1, x_2) = (\alpha x_1, 2\alpha x_2)$
   (d) $(x_1, x_2) + (y_1, y_2) = (x_1 + y_1, x_2 + y_2)$, $\alpha(x_1, x_2) = (\alpha + x_1, \alpha + x_2)$.

*3. Verify the axioms of a vector space for *Example* 1.2.9. Also find which of the axioms will be violated if addition of vectors is changed to $\mathbf{A} + \mathbf{B} = \mathbf{A} \cup \mathbf{B}$.

4. Show that the set of all positive real numbers forms a vector space over $\mathbb{R}$ if the sum of $\mathbf{x}$ and $\mathbf{y}$ is defined to be the usual product $\mathbf{xy}$ and α times $\mathbf{x}$ is defined to be $\mathbf{x}^\alpha$.
5. Show that the set $\mathcal{P}$ of all polynomials (with no upper bound on the degree) in a variable t, with coefficients from a field F, forms a vector space over F under the usual operations of addition and scalar multiplication.
6. Show that $(-\alpha)\mathbf{x} = \alpha(-\mathbf{x}) = -(\alpha\mathbf{x})$.
7. The vector $\mathbf{u}+(-\mathbf{v})$ is denoted by $\mathbf{u}-\mathbf{v}$. Prove the following: (i) $\mathbf{u}-\mathbf{v}$ is the unique solution of $\mathbf{v}+\mathbf{x}=\mathbf{u}$, (ii) $(\mathbf{u}-\mathbf{v})+\mathbf{w}=(\mathbf{u}+\mathbf{w})-\mathbf{v}$ and (iii) $\alpha(\mathbf{u}-\mathbf{v}) = \alpha\mathbf{u}-\alpha\mathbf{v}$.
8. Let O be the origin in the following parallelopiped in $\mathbb{R}^3$.

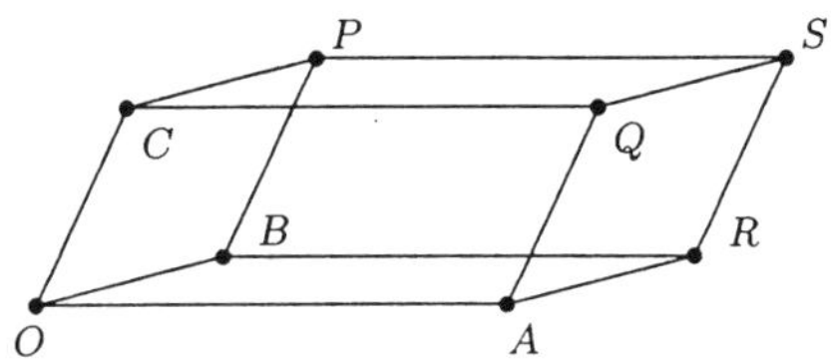

Figure 1.2.1

Considering the vertices as vectors, express P, Q, R and S in terms of A, B and C. Also express S in terms of the other three vertices of the parallelogram $CQSP$ and show that $C = B - P$.

9. In the vector space F^3 where $F = \mathbb{Z}_3$, compute: $(1,1,2)+(0,2,2)$, the negative of $(0,1,2)$ and $2(1,1,2)$.

*10. Let $\Omega = \{1,2,\ldots,n\}$ and let $F = \mathbb{Z}_2$. Establish a one-to-one correspondence between the power set of Ω and F^n under which addition and scalar multiplication in the former vector space (*Example* 1.2.9) correspond to those in the latter.

11. If G is a field and $F \subseteq G$ forms a subfield, show that G is a vector space over F. (What are the operations of this vector space?)

1.3 Subspaces

One way of getting new vector spaces from a given vector space V is to look at subsets S of V which form vector spaces by themselves. For example, the points of $\mathbb{R}^2$ lying on the x-axis themselves form a vector space and we call this a subspace of $\mathbb{R}^2$. Notice that the vector space in *Example* 1.2.4 is a subspace of $\mathbb{R}^3$.

Definition 1.3.1 Let $(V, +, \cdot, F)$ be a vector space and let $S \subseteq V$. If S forms a vector space over the same field F under the operations: the restriction of $+$ to $S \times S$ and the restriction of $\cdot$ to $F \times S$, then we say that S *forms a subspace* of V. Also then we denote the restricted operations by the same symbols $+$ and $\cdot$ and say that $(S, +, \cdot, F)$ is a *subspace* of $(V, +, \cdot, F)$.

Theorem 1.3.2 Let $(V, +, \cdot, F)$ be a vector space and $S \subseteq V$. Then S forms a subspace iff

(i) $\mathbf{0} \in S$,

(ii) $\mathbf{x}, \mathbf{y} \in S \Rightarrow \mathbf{x} + \mathbf{y} \in S$,

(iii) $\alpha \in F,\ \mathbf{x} \in S \Rightarrow \alpha\mathbf{x} \in S$.

Proof *Only if part* Let S form a subspace of V. Let $\overline{\mathbf{0}}$ be the zero vector of the subspace S. Then $\overline{\mathbf{0}} = \overline{\mathbf{0}} + \overline{\mathbf{0}}$ holds in the subspace. Since $\overline{\mathbf{0}} \in V$, this holds in V also. So $\overline{\mathbf{0}} = \mathbf{0}$ and $\mathbf{0} \in S$. Statements (ii) and (iii) hold trivially since S is a subspace.

If part Let S satisfy (i), (ii) and (iii). By (ii), S has the closure property with respect to $+$. Associativity and commutativity of $+$ hold in V and since $S \subseteq V$, they hold in S also. By (i), $\mathbf{0}$ belongs to S and serves as the zero element of S.

Let $\mathbf{x} \in S$. Then by (iii), $-\mathbf{x} = (-1)\mathbf{x} \in S$, where $-\mathbf{x}$ is the negative of $\mathbf{x}$ in V. Hence $-\mathbf{x}$ serves as the negative of $\mathbf{x}$ in S. This shows that $(S, +)$ is a commutative group.

By (iii), we have closure of S with respect to scalar multiplication.

Axioms (3) to (6) of *Definition* 1.2.1 hold in S because they hold in V. Thus S forms a subspace of V. ∎

We leave it to the reader to prove that a *non-empty* subset S of a vector space over F forms a subspace iff

$$\mathbf{x}, \mathbf{y} \in S \text{ and } \alpha \in F \Longrightarrow \alpha\mathbf{x} + \mathbf{y} \in S \qquad (1.3.1)$$

(See *Exercise* 1.3.2.)

The set $\{\mathbf{0}\}$ and the set V are obviously subspaces of the vector space V. These are called *trivial subspaces* of V. We now give several examples of nontrivial subspaces of vector spaces.

Example 1.3.3 The following subsets of the vector space $\mathbb{R}^n$ form subspaces:

(i) $\{(x_1, \ldots, x_n) : x_1 = \cdots = x_m = 0\}$ for any fixed m, $1 \leq m < n$,
(ii) $\{(x_1, \ldots, x_n) : x_1 + \cdots + x_n = 0\}$,
(iii) $\{(x_1, x_2, x_3) : 2x_1 - 3x_2 + \sqrt{2}x_3 = 0,\ x_1 - 5x_3 = 0\}$ when $n = 3$.

Example 1.3.4 The following subsets of $\mathbb{R}$ form subspaces of $\mathbb{R}$ over $\mathbb{Q}$ (see *Example* 1.2.5): (i) $\mathbb{Q}$, and (ii) $\{\alpha + \beta\sqrt{2} + \gamma\sqrt{3} : \alpha, \beta, \gamma \in \mathbb{Q}\}$.

Example 1.3.5 If $Y \subseteq X$, $\{f : f \in F^X$ and $f(x) = 0$ for all $x \in Y\}$ is a subspace of F^X. Also, the set of all continuous functions and the set of all differentiable functions form subspaces of $\mathbb{R}^{\mathbb{R}}$.

Example 1.3.6 If $0 \leq m \leq n$, $\mathcal{P}_m$ forms a subspace of $\mathcal{P}_n$. Moreover, the subset of even polynomials as well as the subset of odd polynomials form subspaces of $\mathcal{P}_n$. Recall that $\sum_{i=0}^{n-1} \alpha_i t^i$ is even or odd according as $\alpha_i = 0$ whenever i is odd or even.

Example 1.3.7 Let $\mathbf{x}, \mathbf{y}$ be two fixed vectors in a vector space V over F. Then

$$S = \{\alpha\mathbf{x} + \beta\mathbf{y} : \alpha, \beta \in F\}$$

is a subspace of V. Note that $\mathbf{x}$ and $\mathbf{y}$ belong to S since $\mathbf{x} = 1 \cdot \mathbf{x} + 0 \cdot \mathbf{y}$ and $\mathbf{y} = 0 \cdot \mathbf{x} + 1 \cdot \mathbf{y}$.

***Example 1.3.8** Consider the vector space in *Example* 1.2.9. For any non-empty subset $\mathbf{A}$ of Ω, $\{\emptyset, \mathbf{A}\}$ is a subspace. For any distinct non-empty subsets $\mathbf{A}$ and $\mathbf{B}$ of Ω, $\{\emptyset, \mathbf{A}, \mathbf{B}, \mathbf{A} \triangle \mathbf{B}\}$ is another subspace.

That all the above are subspaces follows easily by verifying (1.3.1) after observing that they contain the null vector.

We have seen in *Example* 1.3.7 that given two vectors $\mathbf{x}$ and $\mathbf{y}$, the set S of all vectors of the form $\alpha\mathbf{x} + \beta\mathbf{y}$ is a subspace containing $\mathbf{x}$ and $\mathbf{y}$. If T is any subspace containing $\mathbf{x}$ and $\mathbf{y}$ then clearly $T \supseteq S$, thus S is the smallest subspace containing $\mathbf{x}$ and $\mathbf{y}$. Generalizing this argument, we will now find the smallest subspace containing any given set A. For this we need the following

Definition 1.3.9 A *linear combination* of the vectors $\mathbf{x}_1, \ldots, \mathbf{x}_k$ in a vector space V, where $k \geq 1$, is a vector or an expression of the form $\alpha_1\mathbf{x}_1 + \cdots + \alpha_k\mathbf{x}_k$, where $\alpha_1, \ldots, \alpha_k$ are scalars. A *linear combination from a non-empty set* A of vectors is a linear combination of finitely many vectors belonging to A. As a matter of convention, *we define* $\mathbf{0}$ *to be the linear combination from the empty set.*

We note that, whereas an expression $\sum \alpha_i \mathbf{x}_i$ determines a unique vector, a vector may have different representations in the form $\sum \alpha_i \mathbf{x}_i$. For example, if $\mathbf{x}_1 = (1,0,2)$, $\mathbf{x}_2 = (-1,1,0)$ and $\mathbf{x}_3 = (3,-1,4)$, then the vector $\mathbf{x} = (0,1,2)$ can be expressed in different ways as a linear combination from $A = \{\mathbf{x}_1, \mathbf{x}_2, \mathbf{x}_3\}$ thus:

$$\mathbf{x} = \mathbf{x}_1 + \mathbf{x}_2 = 3\mathbf{x}_1 - \mathbf{x}_3 = 5\mathbf{x}_1 - \mathbf{x}_2 - 2\mathbf{x}_3$$

On the other hand, the vector $\mathbf{y} = (1,1,1)$ is not a linear combination from A since, if $\mathbf{y} = \alpha_1\mathbf{x}_1 + \alpha_2\mathbf{x}_2 + \alpha_3\mathbf{x}_3$ then equating corresponding components we get

$$\begin{aligned} \alpha_1 - \alpha_2 + 3\alpha_3 &= 1 \\ \alpha_2 - \alpha_3 &= 1 \\ 2\alpha_1 + 4\alpha_3 &= 1 \end{aligned}$$

from which a contradiction can be obtained easily.

Definition 1.3.10 For any subset A of V, the *linear span* (or simply *span*) of A is the set of all linear combinations from A and is denoted $\mathrm{Sp}(A)$ or $\overline{A}$. Thus, when $A \neq \emptyset$,

$$\mathrm{Sp}(A) = \{\textstyle\sum_{i=1}^k \alpha_i \mathbf{x}_i : \mathbf{x}_1, \mathbf{x}_2, \ldots, \mathbf{x}_k \in A;\ \alpha_1, \alpha_2, \ldots, \alpha_k \in F \text{ and } k \text{ is a positive integer}\}$$

We clarify that the RHS here is, really, the set of all vectors $\mathbf{x} \in V$ such that $\mathbf{x}$ can be written as $\sum_{i=1}^k \alpha_i \mathbf{x}_i$ for some $\mathbf{x}_1, \ldots, \mathbf{x}_k \in A$ and some $\alpha_1, \ldots, \alpha_k \in F$ where k is some positive integer. Note that $\mathrm{Sp}(\emptyset) = \{\mathbf{0}\}$, the singleton-set containing the zero vector.

Theorem 1.3.11 For any subset A of V, $\mathrm{Sp}(A)$ is a subspace of V and $\mathrm{Sp}(A) \supseteq A$. Moreover, $\mathrm{Sp}(A)$ is the smallest subspace of V containing A in the sense that every subspace of V containing A also contains $\mathrm{Sp}(A)$.

Proof The theorem follows easily when $A = \emptyset$ since $\mathrm{Sp}(\emptyset) = \{\mathbf{0}\}$. So let $A \neq \emptyset$. Then $\mathbf{0} \in \mathrm{Sp}(A)$ since $\mathbf{0} = 0\mathbf{x}$ for any $\mathbf{x} \in A$. Moreover, the sum of two linear combinations from A is again a linear combination from A and a scalar times a linear combination from A is also a linear combination from A. Hence $\mathrm{Sp}(A)$ is a subspace of V. Since $\mathbf{x} = 1 \cdot \mathbf{x}$, it follows that $A \subseteq \mathrm{Sp}(A)$. If T is any subspace of V and $T \supseteq A$, then $T \supseteq \mathrm{Sp}(A)$ since T is closed under linear combinations. ∎

Definition 1.3.12 In view of the preceding theorem, $\mathrm{Sp}(A)$ is called the *subspace generated by* A and we say that A is a *generating set* of the subspace $S = \mathrm{Sp}(A)$.

A subspace may have several generating sets as illustrated in the following example.

Example 1.3.13 Let $A = \{(1,0),(0,1)\}$ and $B = \{(1,2),(2,4),(4,5)\}$. It is easy to verify that A generates $\mathbb{R}^2$. We will show that B also generates $\mathbb{R}^2$. Let us denote $\mathbf{x} = (1,2)$, $\mathbf{y} = (2,4)$ and $\mathbf{z} = (4,5)$. Let $\mathbf{u} = (u_1, u_2)$ be an arbitrary vector in $\mathbb{R}^2$. Then we have to show that there exist scalars α, β and γ such that $\mathbf{u} = \alpha\mathbf{x} + \beta\mathbf{y} + \gamma\mathbf{z}$, that is, we want a solution of the following two linear equations

$$\begin{aligned} \alpha + 2\beta + 4\gamma &= u_1 \\ 2\alpha + 4\beta + 5\gamma &= u_2 \end{aligned} \tag{1.3.2}$$

Since we have two equations in three unknowns we try to get a solution by fixing one of the variables at 0. Fixing $\alpha = 0$ and solving (1.3.2) for β and γ, it is easy to see that $\alpha = 0$, $\beta = (4u_2 - 5u_1)/6$ and $\gamma = (2u_1 - u_2)/3$ form a solution of (1.3.2) for arbitrary u_1, u_2, hence B generates $\mathbb{R}^2$. ∎

For further examples of generating sets, see *Exercise* 1.3.4. In order to decide whether a vector belongs to the span of a given set we essentially have to solve linear equations which will be discussed in detail in *Chapter* 5. We give some more basic properties of span in the following

Theorem 1.3.14 For any subsets A and B of a vector space V,

(i) A is a subspace of V iff $A = \mathrm{Sp}(A)$.

(ii) If $A \supseteq B$ then $\mathrm{Sp}(A) \supseteq \mathrm{Sp}(B)$

(iii) $\mathrm{Sp}(\mathrm{Sp}(A)) = \mathrm{Sp}(A)$.

Proof The *if part* of (i) follows since $\mathrm{Sp}(A)$ is a subspace. To prove the *only if part* of (i), let A be a subspace. Then A is closed under linear combinations, so $A \supseteq \mathrm{Sp}(A)$. Since $\mathrm{Sp}(A)$ contains A, equality follows. This proves (i). To prove (ii), let $A \supseteq B$. Then every linear combination from B is also a linear combination from A, hence $\mathrm{Sp}(A) \supseteq \mathrm{Sp}(B)$. Statement (iii) follows from (i) since $\mathrm{Sp}(A)$ is a subspace. ∎

We illustrate the use of the preceding theorem by showing that if $A \subseteq B$ and $\mathrm{Sp}(A) \supseteq B$ then $\mathrm{Sp}(A) = \mathrm{Sp}(B)$. By (ii) we get $\mathrm{Sp}(\mathrm{Sp}(A)) \supseteq \mathrm{Sp}(B)$. By (iii), $\mathrm{Sp}(\mathrm{Sp}(A)) = \mathrm{Sp}(A)$, so $\mathrm{Sp}(A) \supseteq \mathrm{Sp}(B)$. Since $A \subseteq B$, we also have $\mathrm{Sp}(A) \subseteq \mathrm{Sp}(B)$, so equality follows.

It is easy to see that the union of two subspaces of a vector space V need not be a subspace. Consider for example the x-axis and the y-axis in $\mathbb{R}^2$. We now prove that the intersection of any two subspaces is a subspace. This can be extended to any family of subspaces, see *Exercise* 1.3.9.

Theorem 1.3.15 The intersection of any two subspaces S and T of a vector space V is a subspace of V.

Proof Let $W = S \cap T$. Since $\mathbf{0} \in S$ and $\mathbf{0} \in T$, it follows that $\mathbf{0} \in W$. Next let $\mathbf{x}, \mathbf{y} \in W$ and $\alpha \in F$. Then $\mathbf{x}, \mathbf{y} \in S$, so $\alpha\mathbf{x} + \mathbf{y} \in S$. Similarly $\alpha\mathbf{x} + \mathbf{y} \in T$, hence $\alpha\mathbf{x} + \mathbf{y} \in W$. Thus W is a subspace. ∎

We now give the geometric meaning of linear combinations, subspaces and span in $\mathbb{R}^2$ and $\mathbb{R}^3$.

If $\mathbf{x}$ is a non-null vector (in $\mathbb{R}^2$ or $\mathbb{R}^3$) then clearly $\mathrm{Sp}(\{\mathbf{x}\}) = \{\alpha\mathbf{x} : \alpha \in \mathbb{R}\}$ is the set of all points on the line passing through the origin O and the point $\mathbf{x}$. If $\mathbf{x}_1, \mathbf{x}_2$ are non-zero vectors such that neither is a scalar multiple of the other (that is, the points O, $\mathbf{x}_1$ and $\mathbf{x}_2$ are not collinear) then it is not difficult to see that $\mathrm{Sp}(\{\mathbf{x}_1, \mathbf{x}_2\}) = \{\alpha_1\mathbf{x}_1 + \alpha_2\mathbf{x}_2 : \alpha_1, \alpha_2 \in \mathbb{R}\}$ is the set of all points in the plane containing $O, \mathbf{x}_1$ and $\mathbf{x}_2$, see *Figure* 1.3.1. If $\mathbf{x}_1, \mathbf{x}_2$ and $\mathbf{x}_3$ are non-zero vectors (in $\mathbb{R}^3$) such that $O, \mathbf{x}_1, \mathbf{x}_2$ and $\mathbf{x}_3$ are not coplanar, then it can be shown that $\mathrm{Sp}(\{\mathbf{x}_1, \mathbf{x}_2, \mathbf{x}_3\}) = \{\alpha_1\mathbf{x}_1 + \alpha_2\mathbf{x}_2 + \alpha_3\mathbf{x}_3 : \alpha_1, \alpha_2, \alpha_3 \in \mathbb{R}\}$ is $\mathbb{R}^3$.

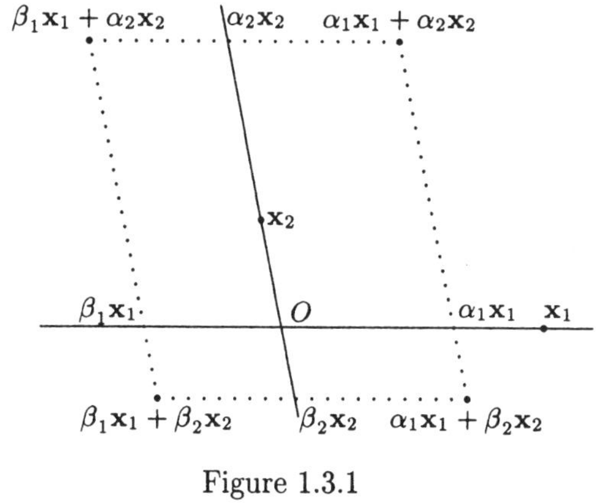

Figure 1.3.1

Theorem 1.3.16 The subspaces of $\mathbb{R}^2$ are $\{\mathbf{0}\}$, the lines through the origin and $\mathbb{R}^2$ itself. The subspaces of $\mathbb{R}^3$ are $\{\mathbf{0}\}$, the lines through the origin, the planes through the origin and $\mathbb{R}^3$ itself.

Proof It is easy to see that $\{\mathbf{0}\}$, any line through the origin, any plane through the origin and the vector space itself are subspaces. To prove the converse, let S be a subspace. Clearly $\mathbf{0} \in S$. If $S = \{\mathbf{0}\}$, we are done. So let $S \neq \{\mathbf{0}\}$. Then there exists an $\mathbf{x}_1 \in S$ such that $\mathbf{x}_1 \neq \mathbf{0}$. Now S contains the line $O\mathbf{x}_1 = \text{Sp}(\{\mathbf{x}_1\})$. If $S = O\mathbf{x}_1$, we are done. So let $S \neq O\mathbf{x}_1$. Then there exists an $\mathbf{x}_2 \in S$ outside the line $O\mathbf{x}_1$. Now S contains the plane $O\mathbf{x}_1\mathbf{x}_2 = \text{Sp}(\{\mathbf{x}_1, \mathbf{x}_2\})$. If $S = O\mathbf{x}_1\mathbf{x}_2$, we are done (this happens automatically if the vector space is $\mathbb{R}^2$). If $S \neq O\mathbf{x}_1\mathbf{x}_2$, then there exists an $\mathbf{x}_3 \in S$ not belonging to the plane $O\mathbf{x}_1\mathbf{x}_2$. Now S contains $\text{Sp}(\{\mathbf{x}_1, \mathbf{x}_2, \mathbf{x}_3\}) = \mathbb{R}^3$, so $S = \mathbb{R}^3$. ∎

The span of a set $A \subseteq \mathbb{R}^3$ can now be described as follows. It is the smallest among the following sets which contains A: $\{\mathbf{0}\}$, lines through the origin, planes through the origin and $\mathbb{R}^3$.

Exercises

1. In each of the following, find out whether the subsets given form subspaces of the vector space V.
   (a) $V = \mathbb{R}^3$, $S = \{(x_1, x_2, x_3) : 2x_1 + x_2 + x_3 = 1\}$.
   (b) $V = \mathbb{R}^2$, $S =$ the set of all (x_1, x_2) such that $x_1 \geq 0$ and $x_2 \geq 0$ and $T =$ the set of all (x_1, x_2) such that $x_1x_2 \geq 0$.
   (c) $V = \mathbb{R}^{\mathbb{R}}$, $S = \{f : f \text{ is monotone}\}$, $T = \{f : f(2) = (f(5))^2\}$ and $W = \{f : f(2) = f(5)\}$. Note that monotone means either non-decreasing or non-increasing.
   (d) $V = \mathbb{R}^{\mathbb{R}}$, $S =$ the set of all those functions whose range is finite (i.e., the function takes finitely many values).
   (e) $V = \mathbb{R}^X$ where X is the set of all positive integers and $S =$ the set of all f such that the sequence $(f(1), f(2), \ldots)$ converges.
   (f) $V = \mathcal{P}_5$, $S = \{p \in V : p = 0 \text{ or degree } p \geq 2\}$.
   (g) $V = \mathcal{P}$ over $\mathbb{R}$ and $S = \{p(t) \in \mathcal{P} : p(5) = 0\}$.
   (h) $V = \mathcal{P}$ over $\mathbb{R}$ and $S = \{p(t) \in \mathcal{P} : p(5) \neq 2\}$.
   *(i) $V =$ power set of $\mathbb{R}$ (see *Example* 1.2.9), $S =$ the set of all finite subsets of $\mathbb{R}$.
   *(j) V is the vector space of *Example* 1.2.8, S is the set of all discrete random variables belonging to V and T is the set of all continuous random variables belonging to V.

(k) $V = \mathbb{C}^n$ over $\mathbb{R}$, $S = \{(x_1, x_2, \ldots, x_n) \in \mathbb{C}^n : x_1$ is real$\}$.

2. Let V be a vector space over F and S a *non-empty* subset of V. Prove that the following are equivalent: (a) S is a subspace, (b) $\mathbf{x}, \mathbf{y} \in S \Rightarrow \mathbf{x}+\mathbf{y} \in S$; $\mathbf{x} \in S, \alpha \in F \Rightarrow \alpha\mathbf{x} \in S$, (c) $\mathbf{x}, \mathbf{y} \in S, \alpha \in F \Rightarrow \alpha\mathbf{x}+\mathbf{y} \in S$, (d) $\mathbf{x}_1, \mathbf{x}_2 \in S$, $\alpha_1, \alpha_2 \in F \Rightarrow \alpha_1\mathbf{x}_1 + \alpha_2\mathbf{x}_2 \in S$, (e) $\mathbf{x}_1, \mathbf{x}_2, \ldots, \mathbf{x}_k \in S$ and $\alpha_1, \alpha_2, \ldots, \alpha_k \in F \Rightarrow \alpha_1\mathbf{x}_1 + \cdots + \alpha_k\mathbf{x}_k \in S$. Where is $S \neq \emptyset$ used?

3. (a) Consider the vectors $\mathbf{x}_1 = (1, 3, 2)$ and $\mathbf{x}_2 = (-2, 4, 3)$ in $\mathbb{R}^3$. Show that the span of $\{\mathbf{x}_1, \mathbf{x}_2\}$ is

$$\{(\xi_1, \xi_2, \xi_3) : \xi_1 - 7\xi_2 + 10\xi_3 = 0\} = \{(\alpha, \beta, (-\alpha + 7\beta)/10) : \alpha, \beta \in \mathbb{R}\}$$

(b) Consider the vectors $\mathbf{x}_1 = (1, 2, 1, -1)$, $\mathbf{x}_2 = (2, 4, 1, 1)$, $\mathbf{x}_3 = (-1, -2, -2, -4)$ and $\mathbf{x}_4 = (3, 6, 2, 0)$ in $\mathbb{R}^4$. Show that the span of $\{\mathbf{x}_1, \ldots, \mathbf{x}_4\}$ is

$$\{(\xi_1, \xi_2, \xi_3, \xi_4) : 2\xi_1 - \xi_2 = 0, 2\xi_1 - 3\xi_3 - \xi_4 = 0\}$$

Show that this subspace can also be written as

$$\{(\alpha, 2\alpha, \beta, 2\alpha - 3\beta) : \alpha, \beta \in \mathbb{R}\}$$

4. (a) For the subspaces S in *Examples* 1.3.3(i) and (iii), show that the set of all vectors in S with non-negative components is a generating set.

(b) For the subspace (i) in *Example* 1.3.4, show that any subset containing a non-zero vector is a generating set. For the subspace (ii), show that $\{1, \sqrt{2}, \sqrt{3}\}$ is a generating set.

(c) For the subspace in *Example* 1.3.7, show that each of $\{\mathbf{x}, \mathbf{y}\}$ and $\{\mathbf{x}+\mathbf{y}, \mathbf{x}+2\mathbf{y}\}$ is a generating set.

*(d) For the vector space F^X show that the set $\{f : 0 \in$ range of $f\}$ is a generating set provided X has at least two elements and $\{f : 0 \notin$ range of $f\}$ is a generating set provided F has at least 3 elements.

*5. In the vector space of *Example* 1.2.9 with $\Omega = \{1, 2, \ldots, 5\}$, find the subspace generated by $\{\{1, 2, 3\}, \{2, 3, 4\}, \{1, 5\}\}$.

6. (a) For any two subsets A and B of a vector space V, show that

i. $\mathrm{Sp}(A) \cup \mathrm{Sp}(B) \subseteq \mathrm{Sp}(A \cup B)$,

ii. $\mathrm{Sp}(A \cap B) \subseteq \mathrm{Sp}(A) \cap \mathrm{Sp}(B)$,

and that proper inclusion is possible in each.

(b) Prove or disprove: $\mathrm{Sp}(A) \cap \mathrm{Sp}(B) \neq \{\mathbf{0}\} \Longrightarrow A \cap B \neq \emptyset$.

7. Let S be a subspace of a vector space V. If $\mathbf{x}$ and $\mathbf{y}$ are vectors such that $\mathbf{x}+\mathbf{y} \in S$ then show that either both $\mathbf{x}$ and $\mathbf{y}$ belong to S or none of $\mathbf{x}$ and $\mathbf{y}$ belongs to S. If $\mathbf{x}$ is a vector and α is a non-zero scalar such that $\alpha\mathbf{x} \in S$ then show that $\mathbf{x} \in S$. What can you say if $\mathbf{x}+\mathbf{y}+\mathbf{z} \in S$?

*8. Show that the intersection of any family of subspaces is a subspace (the intersection of the empty family of subsets of V is defined to be V).

9. Show that for any set $A \subseteq V$, $\mathrm{Sp}(A)$ is the intersection of all subspaces of V containing A.

10. Let S and T be subspaces of V. Then prove that $S \cup T$ is a subspace iff either $S \subseteq T$ or $T \subseteq S$. (Hint for *only if part*: If $S \not\subseteq T$ and $T \not\subseteq S$, consider $\mathbf{x} + \mathbf{y}$ where $\mathbf{x} \in S - T$ and $\mathbf{y} \in T - S$.)

*11. Write down the 6 different subspaces of F^2 where F is the field of residues mod 3. Draw a figure but note that geometric intuition may not be entirely correct here.

*12. Let X be the set of all positive integers. In the vector space $\mathbb{R}^X$, what is the span of the set $A = \{f_i : i \geq 1\}$, where f_i is the function in $\mathbb{R}^X$ taking value 1 at $x = i$ and 0 elsewhere? Show that if $f \in \mathrm{Sp}(A)$ then the range of f is finite but the converse is not true.

1.4 Linear independence

Suppose we are given a generating set A for a subspace S. One question that arises naturally is: is there any redundancy in A in the sense that some proper subset of A generates S? To answer this, we start with

Definition 1.4.1 Vectors $\mathbf{x}_1, \mathbf{x}_2, \ldots, \mathbf{x}_k$ are *linearly independent* if

$$\alpha_1\mathbf{x}_1 + \alpha_2\mathbf{x}_2 + \cdots + \alpha_k\mathbf{x}_k = \mathbf{0} \Rightarrow \alpha_1 = \alpha_2 = \cdots = \alpha_k = 0. \qquad (1.4.1)$$

Otherwise, i.e., if there exist scalars $\alpha_1, \alpha_2, \ldots, \alpha_k$, not all zero, with

$$\alpha_1\mathbf{x}_1 + \alpha_2\mathbf{x}_2 + \cdots + \alpha_k\mathbf{x}_k = \mathbf{0}$$

then $\mathbf{x}_1, \mathbf{x}_2, \ldots, \mathbf{x}_k$ are said to be *linearly dependent.* A finite set of vectors $A = \{\mathbf{x}_1, \mathbf{x}_2, \ldots, \mathbf{x}_k\}$ is *linearly independent* if $\mathbf{x}_1, \mathbf{x}_2, \ldots, \mathbf{x}_k$ are linearly independent. An infinite set $A \subseteq V$ is *linearly independent* if every finite subset of A is linearly independent. A set $A \subseteq V$ is *linearly dependent* if it is not linearly independent.

Several comments are in order. If $\mathbf{x}_i = \mathbf{x}_j$ for some i and j with $i \neq j$, then $\mathbf{x}_1, \mathbf{x}_2, \ldots, \mathbf{x}_k$ are linearly dependent since we can take $\alpha_i = 1$, $\alpha_j = -1$ and every other α equal to 0. The empty set of vectors is linearly independent since the implication (1.4.1) holds vacuously. If $\mathbf{x}_1, \mathbf{x}_2, \ldots, \mathbf{x}_k$ are linearly independent, then $\mathbf{x}_{i_1}, \mathbf{x}_{i_2}, \ldots, \mathbf{x}_{i_k}$ are also linearly independent for any permutation $(i_1, i_2, \ldots, i_k)$ of $(1, 2, \ldots, k)$, so the linear independence of a set of vectors is well-defined. Finally, 'not all zero' cannot be replaced by 'all non-zero' in the definition of linear dependence. For example, if $\mathbf{x}_1 = (0,0)$ and $\mathbf{x}_2 = (1,2)$, then $3\mathbf{x}_1 + 0\mathbf{x}_2 = \mathbf{0}$ but there do not exist non-zero scalars α_1 and α_2 such that $\alpha_1\mathbf{x}_1 + \alpha_2\mathbf{x}_2 = \mathbf{0}$.

It is immediate from the definition that any set containing $\mathbf{0}$ is linearly dependent. Since $\alpha\mathbf{x} = \mathbf{0}$ and $\mathbf{x} \neq \mathbf{0}$ imply $\alpha = 0$, it follows that the singleton set $\{\mathbf{x}\}$ is linearly independent iff $\mathbf{x} \neq \mathbf{0}$. Since $\alpha_1\mathbf{x}_1 + \cdots + \alpha_k\mathbf{x}_k = \mathbf{0}$ can also be written as $\alpha_1\mathbf{x}_1 + \cdots + \alpha_k\mathbf{x}_k + 0\mathbf{x}_{k+1} + \cdots + 0\mathbf{x}_m = \mathbf{0}$ when $k < m$, it follows that a superset of a linearly dependent set is linearly dependent and a subset of a linearly independent set is linearly independent.

Example 1.4.2 Consider the vectors

$$\begin{aligned}
\mathbf{x}_1 &= (1,-1,0,0,\ldots,0),\\
\mathbf{x}_2 &= (1,0,-1,0,\ldots,0),\\
\mathbf{x}_3 &= (1,0,0,-1,\ldots,0),\\
\cdots & \quad \cdots\\
\mathbf{x}_{n-1} &= (1,0,0,\ldots,0,-1),\\
\mathbf{x}_n &= (n-1,-1,-1,\ldots,-1)
\end{aligned}$$

in $\mathbb{R}^n$ where $n \geq 3$. The set $\{\mathbf{x}_1, \ldots, \mathbf{x}_n\}$ is linearly dependent because $\mathbf{x}_n = \mathbf{x}_1 + \cdots + \mathbf{x}_{n-1}$. We next show that $\{\mathbf{x}_1, \mathbf{x}_2, \ldots, \mathbf{x}_{n-1}\}$ is linearly independent. For this let $\alpha_1\mathbf{x}_1 + \cdots + \alpha_{n-1}\mathbf{x}_{n-1} = \mathbf{0}$. Equating the corresponding components on the two sides we get $\alpha_1 + \alpha_2 + \cdots + \alpha_{n-1} = 0$, $-\alpha_1 = 0$, $-\alpha_2 = 0, \ldots, -\alpha_{n-1} = 0$. Hence $\alpha_1 = \cdots = \alpha_{n-1} = 0$. ∎

Example 1.4.3 In the vector space $\mathbb{R}$ over the field $\mathbb{Q}$, the sets $\{1, \sqrt{2}\}$ and $\{\sqrt{2}, \sqrt{3}\}$ are linearly independent and the set $\{\sqrt{2}, \sqrt{3}, \sqrt{12}\}$ is linearly dependent.

***Example 1.4.4** Let $X = \{a_1, a_2, \ldots\}$ where $a_1, a_2, \ldots$ form a sequence of distinct real numbers. Consider the function f_i belonging to the vector space $\mathbb{R}^X$ taking value 1 at $x = a_i$ and 0 elsewhere. Then the infinite set $\{f_1, f_2, \ldots\}$ is linearly independent. To see this, let $\sum_{i=1}^k \alpha_i f_{n_i} = 0$. Equating the values of both the sides at a_{n_j}, we get $\alpha_j = 0$ for $j = 1, \ldots, k$. If f is the function taking value 1 at $x = a_1$ and a_2 and 0 elsewhere, then $\{f, f_1, f_2\}$ is linearly dependent.

Example 1.4.5 The vectors $1, t, t^2, \ldots, t^{n-1}$ in the vector space $\mathcal{P}_n$

form a linearly independent set. The vectors $1+t+t^2, 2-3t+4t^2$ and $1-9t+5t^2$ form a linearly dependent set in $\mathcal{P}_n$ when $n \geq 3$. The vectors $1, t, t^2, \ldots$ in the vector space $\mathcal{P}$ form an infinite linearly independent set.

To find out whether a given set of vectors in F^n is linearly independent we have to essentially find out whether a homogeneous system of linear equations (that is, the RHS of each of these equations is zero) has a non-zero solution. Another way is to use the concept of *rank* which can be evaluated using elementary operations, see, for example, *Algorithm* 4.4.4.

Theorem 1.4.6 The vectors $\mathbf{x}_1, \mathbf{x}_2, \ldots, \mathbf{x}_k$ are linearly dependent iff $\mathbf{x}_j$ belongs to the span of $\mathbf{x}_1, \ldots, \mathbf{x}_{j-1}$ for some j such that $1 \leq j \leq k$.

Proof We note that when $j = 1$, the condition stated means $\mathbf{x}_1 = \mathbf{0}$. To prove the *if part*, suppose $\mathbf{x}_j = \alpha_1\mathbf{x}_1 + \cdots + \alpha_{j-1}\mathbf{x}_{j-1}$. Then $\alpha_1\mathbf{x}_1 + \cdots + \alpha_{j-1}\mathbf{x}_{j-1} + 1\mathbf{x}_j + 0\mathbf{x}_{j+1} + \cdots + 0\mathbf{x}_k = \mathbf{0}$ (this is true even if $j = 1$), so $\mathbf{x}_1, \ldots, \mathbf{x}_k$ are linearly dependent. To prove the *only if part*, let $\mathbf{x}_1, \mathbf{x}_2, \ldots, \mathbf{x}_k$ be linearly dependent. Then there exist scalars $\alpha_1, \alpha_2, \ldots, \alpha_k$, not all zero, such that $\alpha_1\mathbf{x}_1 + \cdots + \alpha_k\mathbf{x}_k = \mathbf{0}$. Let j be the largest suffix such that $\alpha_j \neq 0$. If $j = 1$ then $\alpha_1\mathbf{x}_1 = \mathbf{0}$, so $\mathbf{x}_1 = \mathbf{0}$ and the stated condition holds for $j = 1$. If $j \geq 2$, we have

$$\mathbf{x}_j = \sum_{i=1}^{j-1}(-\alpha_j^{-1}\alpha_i)\mathbf{x}_i$$

so $\mathbf{x}_j$ belongs to the span of $\{\mathbf{x}_1, \ldots, \mathbf{x}_{j-1}\}$. ∎

Corollary A set $A \subseteq V$ is linearly dependent iff there exists an $\mathbf{x} \in A$ such that $\mathbf{x} \in \mathrm{Sp}(A - \{\mathbf{x}\})$.

Proof To prove the *only if part*, let A be linearly dependent. Then some finite subset $\{\mathbf{x}_1, \ldots, \mathbf{x}_k\}$ of A is linearly dependent. Now take $\mathbf{x}$ to be an $\mathbf{x}_j$ given by the theorem. To prove the *if part*, suppose $\mathbf{x} \in \mathrm{Sp}(A - \{\mathbf{x}\})$. Then $\mathbf{x} = \alpha_1\mathbf{y}_1 + \cdots + \alpha_p\mathbf{y}_p$ for some $\mathbf{y}_1, \ldots, \mathbf{y}_p \in A - \{\mathbf{x}\}$ and some $\alpha_1, \ldots, \alpha_p \in F$. Now $1\mathbf{x} - \alpha_1\mathbf{y}_1 - \cdots - \alpha_p\mathbf{y}_p = \mathbf{0}$, hence $\{\mathbf{x}, \mathbf{y}_1, \ldots, \mathbf{y}_p\}$ and so A are linearly dependent. ∎

This corollary shows why linear dependence of infinite sets is defined the way it was done. We sometimes say that $\mathbf{x}$ *depends linearly* on B if $\mathbf{x} \in \mathrm{Sp}(B)$. Note that A is linearly dependent iff some vector in it depends linearly on others.

It is immediate from the preceding theorem that a doubleton set $\{\mathbf{x}, \mathbf{y}\}$ is linearly independent iff none of $\mathbf{x}$ and $\mathbf{y}$ is a scalar multiple of the other. The theorem is also very useful in situations like the following: let $\mathbf{x}_1, \ldots, \mathbf{x}_4$ be the polynomials $2 - 3t^2 + t^3$, $3 + t - t^2$, 2 and $5 - t$. Then none of $\mathbf{x}_3$, $\mathbf{x}_4$, $\mathbf{x}_2$, $\mathbf{x}_1$ can be a linear combination of the preceding vectors, so the four vectors are linearly independent. We next prove an important result which will be used repeatedly later on.

Theorem 1.4.7 Let A be linearly independent and $\mathbf{y} \notin A$. Then $A \cup \{\mathbf{y}\}$ is linearly dependent iff $\mathbf{y} \in \mathrm{Sp}(A)$.

Proof The *if part* is trivial. To prove the *only if part*, let $A \cup \{\mathbf{y}\}$ be linearly dependent. Then some finite subset $\{\mathbf{x}_1, \ldots, \mathbf{x}_k\}$ of $A \cup \{\mathbf{y}\}$ is linearly dependent, where $\mathbf{x}_k = \mathbf{y}$. So by the preceding theorem, some $\mathbf{x}_j$ is a linear combination of the preceding vectors. Since $\mathbf{x}_1, \ldots, \mathbf{x}_{k-1}$ are linearly independent, it follows that $j = k$ and $\mathbf{y} \in \mathrm{Sp}(A)$. ∎

We now prove an important property of linearly independent sets.

Theorem 1.4.8 (*Steinitz Exchange Theorem*) If $A = \{\mathbf{x}_1, \mathbf{x}_2, \ldots, \mathbf{x}_k\}$ and $B = \{\mathbf{y}_1, \mathbf{y}_2, \ldots, \mathbf{y}_{k+1}\}$ are linearly independent subsets of a vector space V, then there exists a $\mathbf{y}_j \in B - A$ such that $A \cup \{\mathbf{y}_j\}$ is linearly independent.

Proof We prove the theorem by the method of contradiction. Suppose, if possible, that $A \cup \{\mathbf{y}_j\}$ is linearly dependent for all $\mathbf{y}_j \in B - A$. Then we claim that $\mathrm{Sp}(A) \supseteq B$. To prove this, let $\mathbf{y}_j \in B$. If $\mathbf{y}_j \in A$ then trivially $\mathbf{y}_j \in \mathrm{Sp}(A)$. If $\mathbf{y}_j \notin A$ then $\mathbf{y}_j \in B - A$ and $A \cup \{\mathbf{y}_j\}$ is linearly dependent. So, by the preceding theorem, $\mathbf{y}_j \in \mathrm{Sp}(A)$. Thus $\mathrm{Sp}(A) \supseteq B$.

Since A and B are linearly independent sets, it follows that $\mathbf{x}_1, \ldots, \mathbf{x}_k$ and $\mathbf{y}_1, \ldots, \mathbf{y}_{k+1}$ are all non-null vectors. Now $\mathrm{Sp}(A) \supseteq B$, so $\mathbf{y}_1, \mathbf{x}_1, \ldots, \mathbf{x}_k$ are linearly dependent and it follows by *Theorem* 1.4.6 that

$$\mathbf{x}_{j_1} \in \mathrm{Sp}(\{\mathbf{y}_1, \mathbf{x}_1, \ldots, \mathbf{x}_{j_1-1}\})$$

for some j_1 such that $1 \le j_1 \le k$. Let

$$A_1 = \{\mathbf{y}_1, \mathbf{x}_1, \ldots, \mathbf{x}_{j_1-1}, \mathbf{x}_{j_1+1}, \ldots, \mathbf{x}_k\} = \{\mathbf{y}_1\} \cup (A - \{\mathbf{x}_{j_1}\})$$

Then $\mathrm{Sp}(A_1) \supseteq A$, hence $\mathrm{Sp}(A_1) \supseteq \mathrm{Sp}(A) \supseteq B$.

Since $\mathrm{Sp}(A_1) \supseteq B$, the vectors $\mathbf{y}_2, \mathbf{y}_1, \mathbf{x}_1, \ldots, \mathbf{x}_{j_1-1}, \mathbf{x}_{j_1+1}, \ldots, \mathbf{x}_k$ are linearly dependent. By *Theorem* 1.4.6, one of these vectors is a linear combination of the preceding vectors. Notice that this vector cannot be $\mathbf{y}_2$ or $\mathbf{y}_1$ because $\{\mathbf{y}_2, \mathbf{y}_1\}$ is linearly independent. Thus

$$\mathbf{x}_{j_2} \in \mathrm{Sp}(\{\mathbf{y}_2, \mathbf{y}_1, \mathbf{x}_1, \ldots, \mathbf{x}_{j_2-1}\} - \{\mathbf{x}_{j_1}\})$$

for some $j_2 \neq j_1$, $1 \leq j_2 \leq k$. Now let

$$A_2 = \{\mathbf{y}_2\} \cup (A_1 - \{\mathbf{x}_{j_2}\})$$

Then $\mathrm{Sp}(A_2) \supseteq A_1$, so $\mathrm{Sp}(A_2) \supseteq \mathrm{Sp}(A_1) \supseteq B$.

Proceeding like this (introducing a $\mathbf{y}$ and deleting an $\mathbf{x}$ each time), we finally get $\mathrm{Sp}(A_k) \supseteq B$, where $A_k = \{\mathbf{y}_k, \mathbf{y}_{k-1}, \ldots, \mathbf{y}_1\}$. So $\mathbf{y}_{k+1} \in \mathrm{Sp}(A_k)$, a contradiction since B is linearly independent. ∎

Theorem 1.4.9 Let X be a set of vectors in a vector space and A a maximal† linearly independent subset of X. If A is finite and has size k, then no linearly independent subset of X has size more than k and every maximal linearly independent subset of X has size exactly k.

Proof Suppose there is a linearly independent subset C of X with size more than k. Let B be a subset of C with size $k+1$. Then B is also linearly independent, so by the preceding theorem, there exists $\mathbf{y} \in B - A$ such that $A \cup \{\mathbf{y}\}$ is linearly independent. Clearly $A \cup \{\mathbf{y}\}$ is a subset of X and contains A properly, a contradiction to the maximality of A. This proves that no linearly independent subset of A has size more than k. If, now, D is any maximal linearly independent subset of X then the size of D is not more than the size of A. Interchanging the roles of A and D we also see that the size of A is not more than the size of D. Hence D has size exactly k. ∎

We note here that the existence of a maximal linearly independent subset of X is obvious if X is finite. It exists even if X is infinite but the proof of this is beyond the scope of this book. A proof can be found in Halmos (1957).

Exercises

1. Find all subsets of B in *Example* 1.3.13 which generate $\mathbb{R}^2$.
2. Find whether each of the following sets is linearly dependent or independent (over $\mathbb{R}$):

†This means simply that A is a linearly independent subset of X and there is no proper superset of A which is a linearly independent subset of X. It does *not* mean that A has maximum size among linearly independent subsets of X.

(a) $\{(3,5,-4),\ (2,7,-8),\ (5,1,-4)\}$,
(b) $\{(1,0,2),\ (1,1,-1),\ (3,1,3),\ (1,3,4)\}$,
(c) $\{(1,2,-1,0),\ (0,1,2,1),\ (2,3,-4,-1)\}$,
(d) $\{(1,1,1,1),\ (1,2,3,4),\ (1,3,1,2),\ (2,1,1,2)\}$,

3. Show that $\mathbf{x}_1, \mathbf{x}_2, \ldots, \mathbf{x}_k$ are linearly independent iff

$$\alpha_1\mathbf{x}_1 + \cdots + \alpha_k\mathbf{x}_k = \beta_1\mathbf{x}_1 + \cdots + \beta_k\mathbf{x}_k \Longrightarrow \alpha_1 = \beta_1, \ldots, \alpha_k = \beta_k$$

4. (a) Show that in *Example* 1.4.2 every proper subset of $\{\mathbf{x}_1, \mathbf{x}_2, \ldots, \mathbf{x}_n\}$ is linearly independent.
   (b) Show that $\{1, \sqrt{2}\}$ and $\{\sqrt{2}, \sqrt{3}, \sqrt{6}\}$ are linearly independent over $\mathbb{Q}$ and that $\{\sqrt{2}, \sqrt{3}, \sqrt{12}\}$ is linearly dependent over $\mathbb{Q}$.
   (c) Verify that $1+t+t^2, 2-3t+4t^2$ and $1-9t+5t^2$ form a linearly dependent set in $\mathcal{P}_3$.
5. (a) If $\mathbf{x}, \mathbf{y}$ are linearly independent show that $\mathbf{x}+\alpha\mathbf{y}$ and $\mathbf{x}+\beta\mathbf{y}$ are linearly independent whenever $\alpha \neq \beta$.
   (b) If $\mathbf{x}, \mathbf{y}, \mathbf{z}$ are linearly independent show that $\mathbf{x}+\mathbf{y}$, $\mathbf{y}+\mathbf{z}$ and $\mathbf{z}+\mathbf{x}$ are linearly independent, provided $1+1 \neq 0$. Show by an example that the condition $1+1 \neq 0$ cannot be dropped.
6. Let $\mathbf{x}_1, \mathbf{x}_2, \ldots, \mathbf{x}_k$ be vectors in F^n and let $\mathbf{y}_i$ be the vector in F^{n-1} formed by the first $n-1$ components of $\mathbf{x}_i$ for $i = 1, \ldots, k$. Show that if $\mathbf{y}_1, \mathbf{y}_2, \ldots, \mathbf{y}_k$ are linearly independent in F^{n-1} then $\mathbf{x}_1, \mathbf{x}_2, \ldots, \mathbf{x}_k$ are linearly independent in F^n. Is the converse true? Why?
7. Show by an example that linear independence of A cannot be dropped in *Theorem* 1.4.7.
8. Find all the maximal linearly independent subsets of $\{\mathbf{x}_1, \mathbf{x}_2, \ldots, \mathbf{x}_5\}$ where $\mathbf{x}_1 = (1,1,0,1)$, $\mathbf{x}_2 = (1,2,-1,0)$, $\mathbf{x}_3 = (1,0,1,2)$, $\mathbf{x}_4 = (0,1,1,1)$ and $\mathbf{x}_5 = (2,0,2,4)$ in $\mathbb{R}^4$.
9. Let A be a linearly independent subset of a subspace S. If $\mathbf{x} \notin S$, show that $A \cup \{\mathbf{x}\}$ is linearly independent. If $B \subseteq V - S$ and B is linearly independent, does it follow that $A \cup B$ is linearly independent?
10. Let $\mathrm{Sp}(A) = S$. Then show that no proper subset of A generates S iff A is linearly independent.
11. Give geometric characterizations of $\{\mathbf{x}_1, \mathbf{x}_2\}$ and $\{\mathbf{x}_1, \mathbf{x}_2, \mathbf{x}_3\}$ being linearly independent in $\mathbb{R}^3$.
12. (a) For what values of α are the vectors $(0,1,\alpha), (\alpha,1,0)$ and $(1,\alpha,1)$ in $\mathbb{R}^3$ linearly independent?
   (b) Determine all the values of α and β for which the vectors $(\alpha,\beta,\beta,\beta)$, $(\beta,\alpha,\beta,\beta)$, $(\beta,\beta,\alpha,\beta)$ and $(\beta,\beta,\beta,\alpha)$ of $\mathbb{R}^4$ are linearly dependent.

1.5 Basis and dimension

A vector space may have many generating sets. For instance we have seen in *Example* 1.3.13 that $A = \{(1,0),(0,1)\}$ and $B = \{(1,2),(2,4),(4,5)\}$ are generating sets of $\mathbb{R}^2$. In fact $\mathbb{R}^2$ itself is a generating set of $\mathbb{R}^2$. However there is some redundancy in B and $\mathbb{R}^2$ in the sense that some proper subsets of these also generate $\mathbb{R}^2$ whereas A and $C = \{(2,4),(4,5)\}$ do not have this redundancy. We study the latter type of generating sets in this section.

Definition 1.5.1 A subset A of a subspace S is said to be a *basis of* S if A is a minimal† generating set of S. A *basis of the vector space* V is a basis of V considered as a subspace of itself.

We will prove in *Theorem* 1.5.8 that if a vector space V has a finite basis then every subspace of V has a (finite) basis. It can actually be proved that every vector space has a basis but the proof of this is beyond the scope of this book.

Example 1.5.2 Consider the vector space F^n. The set $\{\mathbf{e}_1,\ldots,\mathbf{e}_n\}$ is a basis of F^n, where

$$\begin{aligned} \mathbf{e}_1 &= (1,0,\ldots,0) \\ \mathbf{e}_2 &= (0,1,0,\ldots,0) \\ &\cdots\cdots\cdots\cdots \\ \mathbf{e}_n &= (0,\ldots,0,1) \end{aligned}$$

We call this basis the *canonical basis of* F^n.

Example 1.5.3 Consider the subspaces of $\mathbb{R}^n$ given in *Example* 1.3.3. The set

$$\{(0,\ldots,0,\underset{\substack{\uparrow\\(m+1)\text{-th}}}{1},0,\ldots,0),(0,\ldots,0,0,\underset{\substack{\uparrow\\(m+2)\text{-th}}}{1},0,\ldots,0),\ldots,(0,\ldots,0,\underset{\substack{\uparrow\\n\text{-th}}}{1})\}$$

is a basis of the subspace (i). The set

$$\{(1,-1,0,\ldots,0),\ (1,0,-1,0,\ldots,0),\ldots;(1,0,\ldots,0,-1)\}$$

is a basis of the subspace (ii). The singleton set $\{(15, 10+\sqrt{2}, 3)\}$ is a basis of the subspace (iii).

†See the footnote on page 34.

Example 1.5.4 Consider the subspaces in *Example* 1.3.4. The singleton set consisting of any non-zero rational number is a basis of the subspace (i) and the set $\{1, \sqrt{2}, \sqrt{3}\}$ is a basis of the subspace (ii).

Example 1.5.5 Consider the subspaces in *Example* 1.3.6. The set $\{1, t, \ldots, t^m\}$ is a basis of $\mathcal{P}_m$. The set $\{1, t^2, t^4, \ldots, t^{2k}\}$ is a basis of the subspace of even polynomials, where $2k$ is the largest even integer $\leq n-1$. The set $\{t, t^3, \ldots, t^{2\ell+1}\}$ is a basis of the subspace of odd polynomials, where $2\ell + 1$ is the largest odd integer $\leq n - 1$.

Example 1.5.6 Consider the vector space $\mathbb{C}^n$ over the field $\mathbb{R}$ (see *Example* 1.2.3). Notice that $\mathbf{e}_1, \mathbf{e}_2, \ldots, \mathbf{e}_n$ do not form a basis of this vector space (why?). In fact, $\mathbf{e}_1, \mathbf{e}_2, \ldots, \mathbf{e}_n$ together with $(i, 0, \ldots, 0)$, $(0, i, 0, \ldots, 0)$, $\ldots$, $(0, \ldots, 0, i)$ form a basis, where $i = \sqrt{-1}$. However, $\mathbf{e}_1, \mathbf{e}_2, \ldots, \mathbf{e}_n$ form a basis of the vector space $\mathbb{C}^n$ over $\mathbb{C}$.

In the next theorem, we use the following convention: two linear combinations (treated as expressions) from a set A are considered to be the same if each vector of A has equal coefficients in the two linear combinations, where the coefficient of a vector is taken to be zero if the vector is not used. For example, $2\mathbf{x}_1 + 0\mathbf{x}_2 + \mathbf{x}_3 - 5\mathbf{x}_5$ is considered to be the same linear combination as $\mathbf{x}_3 + 2\mathbf{x}_1 + 0\mathbf{x}_4 - 5\mathbf{x}_5 + 0\mathbf{x}_6$. We now give several characterizations of a basis.

Theorem 1.5.7 Let $A \subseteq S$ where S is a subspace of a vector space V. Then the following statements are equivalent:

(i) A is a minimal generating set of S,

(ii) every element of S can be expressed uniquely as a linear combination from A,

(iii) A generates S and A is linearly independent,

(iv) A is a maximal linearly independent subset of S.

Proof (i) $\Rightarrow$ (ii) Suppose not. Then some vector $\mathbf{x}$ in S can be expressed as

$$\mathbf{x} = \alpha_1 \mathbf{x}_1 + \cdots + \alpha_k \mathbf{x}_k = \beta_1 \mathbf{x}_1 + \cdots + \beta_k \mathbf{x}_k$$

where $\mathbf{x}_1, \mathbf{x}_2, \ldots, \mathbf{x}_k$ are distinct vectors in A and $\alpha_i \neq \beta_i$ for some i. We then have

$$\mathbf{x}_i = (\alpha_i - \beta_i)^{-1} \sum_{j \neq i} (\beta_j - \alpha_j) \mathbf{x}_j$$

Clearly now every linear combination from A is also a linear combination from $A-\{\mathbf{x}_i\}$, so $A-\{\mathbf{x}_i\}$ generates S, a contradiction to the minimality of A.

(ii) $\Rightarrow$ (iii) That A generates S is clear. To prove that A is linearly independent, let $\mathbf{x}_1, \mathbf{x}_2, \ldots, \mathbf{x}_k \in A$ and $\alpha_1\mathbf{x}_1 + \cdots + \alpha_k\mathbf{x}_k = \mathbf{0}$. Since $\mathbf{0} \in S$ and $\mathbf{0} = 0 \cdot \mathbf{x}_1 + \cdots + 0 \cdot \mathbf{x}_k$, it follows by hypothesis that $\alpha_i = 0$ for all i.

(iii) $\Rightarrow$ (iv) That A is linearly independent is given. Suppose A is not maximal. Then there exists a vector $\mathbf{y}$ in $S - A$ such that $A \cup \{\mathbf{y}\}$ is linearly independent. Clearly then $\mathbf{y} \notin \mathrm{Sp}(A)$, a contradiction since A generates S.

(iv) $\Rightarrow$ (i) First we show that A generates S. Let $\mathbf{y} \in S$. If $\mathbf{y} \in A$ then $\mathbf{y} \in \mathrm{Sp}(A)$. If $\mathbf{y} \notin A$, then by hypothesis, $A \cup \{\mathbf{y}\}$ is linearly dependent, so by *Theorem* 1.4.7, $\mathbf{y} \in \mathrm{Sp}(A)$. Thus A generates S.

Next we show that A is a minimal generating set for S. Suppose not. Then $A - \{\mathbf{x}\}$ generates S for some $\mathbf{x} \in A$. Hence $\mathbf{x} \in \mathrm{Sp}(A - \{\mathbf{x}\})$, a contradiction since A is linearly independent. ∎

We call a vector space *finite-dimensional* if it has a finite basis. Clearly any vector space with a finite generating set A is finite-dimensional since A contains a minimal generating set. *From now on we will consider only finite-dimensional vector spaces unless otherwise specified.*

Theorem 1.5.8 Let S be a subspace of a finite-dimensional vector space V. Then S has a finite basis and all bases of S have the same size.

Proof By hypothesis, V has a finite basis A. Let k be the size of A. By the preceding theorem, A is a maximal linearly independent subset of V. So by *Theorem* 1.4.9, no linearly independent subset of V and so of S has size more than k. Since there exists at least one linearly independent subset of S, viz., $\emptyset$, there exists a linearly independent subset B of S with the maximum size. Clearly B is a maximal linearly independent subset of S and so is a basis of S. That all bases of S have the same size also follows from *Theorem* 1.4.9. ∎

Definition 1.5.9 The *dimension* of a subspace S is the size of any basis of S. We denote it by $d(S)$.

In view of *Examples* 1.5.2 through 1.5.6 we have the following. The vector space F^n has dimension n. The subspaces in *Example* 1.3.3 have dimensions $n-m, n-1$ and 1 respectively. The subspaces in *Example*

1.3.4 have dimensions 1, 3 and 2 respectively. The subspaces in *Example* 1.3.6 have dimensions m, $k+1$ and $\ell+1$ respectively, where $2k$ is the largest even integer $\leq n-1$ and $2\ell+1$ is the largest odd integer $\leq n-1$. The dimension of $\mathbb{C}^n$ over $\mathbb{R}$ is $2n$.

Notice that if X is an infinite set, the vector space F^X is not finite-dimensional since it has an infinite linearly independent subset as we saw in *Example* 1.4.4. The vector space of all polynomials is also not finite-dimensional.

The subspace $\{\mathbf{0}\}$ containing only the null vector has dimension 0 since $\emptyset$ is a basis for it. In fact, the only subspace of dimension 0 is $\{\mathbf{0}\}$ (why?)

Theorem 1.5.10 Let S be a subspace of a finite-dimensional vector space. Then

(a) every linearly independent subset A of S can be extended to a basis of S,

(b) every generating set C of S contains a basis of S.

Proof Since no linearly independent subset of S has size greater than $d(S)$, there exists a maximum size linearly independent subset B of S containing A. It follows as in the proof of *Theorem* 1.5.8 that B is a basis of S. This proves (a).

To prove (b), let B be a maximal linearly independent subset of C (note that it exists). We now show $\mathrm{Sp}(B) \supseteq C$. Let $\mathbf{x} \in C$. If $\mathbf{x} \in B$ then $\mathbf{x} \in \mathrm{Sp}(B)$. If $\mathbf{x} \notin B$ then $B \cup \{\mathbf{x}\}$ is linearly dependent, so by *Theorem* 1.4.7, $\mathbf{x} \in \mathrm{Sp}(B)$. Thus $\mathrm{Sp}(B) \supseteq C$. Taking span of both sides, we get $\mathrm{Sp}(B) \supseteq S$. But $B \subseteq S$, so $\mathrm{Sp}(B) = S$. Since B is linearly independent, it follows that B is a basis of S. ∎

Theorem 1.5.11 Let S be a subspace of dimension k. Then

(a) any k linearly independent vectors in S form a basis of S,

(b) any generating set of S with size k is a basis of S.

This follows immediately from the last two theorems.

The proof given above for the result that any linearly independent subset of S can be extended to a basis of S does not give a constructive procedure. We will now give a procedure (see also *Exercise* 4.4.14a) for such an extension when we know a basis of S. (It is usually easy to obtain some basis of S.) Let $A = \{\mathbf{x}_1, \mathbf{x}_2, \ldots, \mathbf{x}_r\}$ be a linearly independent subset of S and $B = \{\mathbf{y}_1, \mathbf{y}_2, \ldots, \mathbf{y}_s\}$ be a basis of S.

Algorithm 1.5.12 (*Extension of a linearly independent set to a basis*)

Step 1 Set $i = 1$ and $A_1 = A$.

Step 2 Is the size of A_i equal to s? If yes, stop and note that A_i is the required basis. If no, go to *Step* 3.

Step 3 Is $\mathbf{y}_i$ a linear combination from A_i? If yes, set $A_{i+1} = A_i$. If no, set $A_{i+1} = A_i \cup \{\mathbf{y}_i\}$.

Step 4 Increase i by 1 and go to *Step* 2.

To prove that this algorithm gives a basis of S which is an extension of A, we first note that each A_i is a subset of S (in fact, a subset of $A \cup B$), and contains A. We next show that each A_i is linearly independent. This is obvious for $i = 1$; if it is assumed for i, it follows for $i + 1$ by *Theorem* 1.4.7. Finally, we note that $\mathbf{y}_1, \ldots, \mathbf{y}_{i-1} \in \mathrm{Sp}(A_i)$ for all i. Hence the size of A_i becomes s at least when $i = s + 1$. Thus the algorithm stops and the final A_i is a linearly independent subset of S with size $d(S)$ and hence a basis of S. Further, this basis contains A as required.

We illustrate the above procedure with

Example 1.5.13 Let

$$S = \{(u_1, u_2, u_3, u_4) \in \mathbb{R}^4 : 2u_1 + u_3 + u_4 = 0\},$$

$A = \{\mathbf{x}_1\}$ where $\mathbf{x}_1 = (1, 0, -1, -1)$ and $B = \{\mathbf{y}_1, \mathbf{y}_2, \mathbf{y}_3\}$ where $\mathbf{y}_1 = (0, 1, 0, 0)$, $\mathbf{y}_2 = (1, 2, -1, -1)$ and $\mathbf{y}_3 = (1, 0, 0, -2)$.

Note that A is a linearly independent subset of S and B is a basis of S. We set $i = 1$ and $A_1 = \{\mathbf{x}_1\}$. Clearly $\mathbf{y}_1$ is not a linear combination from A_1, so we set $A_2 = A_1 \cup \{\mathbf{y}_1\} = \{\mathbf{x}_1, \mathbf{y}_1\}$. Now $\mathbf{y}_2 = \mathbf{x}_1 + 2\mathbf{y}_1$, obtained by solving for the coefficients. Thus $\mathbf{y}_2 \in \mathrm{Sp}(A_2)$, so we set $A_3 = A_2$. Now, if $\mathbf{y}_3 = \alpha\mathbf{x}_1 + \beta\mathbf{y}_1$ then, comparing the components, we get $\alpha = 1$, $\beta = 0$, $-\alpha = 0$ and $-\alpha = -2$, a contradiction. So $\mathbf{y}_3$ is not a linear combination from A_3 and we set $A_4 = A_3 \cup \{\mathbf{y}_3\} = \{\mathbf{x}_1, \mathbf{y}_1, \mathbf{y}_3\}$. Since the size of A_4 is 3, which is the dimension of S, we conclude that A_4 is the required basis. ∎

We next give a procedure (see also page 169) to construct a basis of a subspace S when we know a finite generating set for S. Let $\mathbf{x}_1, \mathbf{x}_2, \ldots, \mathbf{x}_k$ form a generating set of S.

Algorithm 1.5.14 (*Finding a basis contained in a finite generating set* $\{\mathbf{x}_1, \mathbf{x}_2, \ldots, \mathbf{x}_k\}$)

Step 1 Set $i = 1, j = 1$ and $A_1 = \emptyset$.

Step 2 Does $\mathbf{x}_i \in \mathrm{Sp}(A_j)$? If yes, go to *Step* 3. If no, go to *Step* 4.

Step 3 If $i = k$, stop and declare that A_j is the required basis. If $i < k$, increase i by 1 and go to *Step* 2.

Step 4 Set $A_{j+1} = A_j \cup \{\mathbf{x}_i\}$, increase j by 1 and go to *Step* 3.

To show that the algorithm works, we first note that $\mathrm{Sp}(A_j) = \mathrm{Sp}(\{\mathbf{x}_1, \ldots, \mathbf{x}_{i-1}\})$ just before each execution of *Step* 2, so the final A_j generates S. Also, for any i, $\mathbf{x}_i$ is included in the chosen set if $\mathbf{x}_i \notin \mathrm{Sp}(\{\mathbf{x}_1, \ldots, \mathbf{x}_{i-1}\})$, so the final A_j is linearly independent by *Theorem* 1.4.6 and is a basis of S.

Note If the dimension d of S is known, then the above algorithm can be made faster by modifying *Step* 3 thus: "If $i = j - 1 + k - d$, stop and declare that $A_j \cup \{\mathbf{x}_{i+1}, \mathbf{x}_{i+2}, \ldots, \mathbf{x}_k\}$ is the required basis. Otherwise increase i by 1 and go to *Step* 2."

We illustrate the use of the above procedure with

Example 1.5.15 Let S be as in *Example* 1.5.13. Suppose we are given the generating set $\{\mathbf{x}_1, \ldots, \mathbf{x}_4\}$ of S where $\mathbf{x}_1 = (1, 0, 1, -3)$, $\mathbf{x}_2 = (0, 0, 1, -1)$, $\mathbf{x}_3 = (4, 0, 1, -9)$ and $\mathbf{x}_4 = (0, 1, 0, 0)$.

We start the preceding algorithm with $i = 1$, $j = 1$ and $A_1 = \emptyset$. Since $\mathbf{x}_1 \notin \mathrm{Sp}(\emptyset) = \{\mathbf{0}\}$, we set $A_2 = \{\mathbf{x}_1\}$. We next increase j to 2 and i to 2. Now $\mathbf{x}_2 \notin \mathrm{Sp}(A_2)$ since $\mathbf{x}_2$ is not a scalar multiple of $\mathbf{x}_1$. Hence we set $A_3 = A_2 \cup \{\mathbf{x}_2\} = \{\mathbf{x}_1, \mathbf{x}_2\}$, increase j to 3 and i to 3. Now $\mathbf{x}_3 = 4\mathbf{x}_1 - 3\mathbf{x}_2 \in \mathrm{Sp}(A_3)$, so we increase i to 4. If $\mathbf{x}_4 \in \mathrm{Sp}(A_3)$ then there exist scalars α and β such that $\alpha\mathbf{x}_1 + \beta\mathbf{x}_2 = \mathbf{x}_4$. Comparing the second components we get $\alpha \cdot 0 + \beta \cdot 0 = 1$, a contradiction. Thus $\mathbf{x}_4 \notin \mathrm{Sp}(A_3)$. So we set $A_4 = A_3 \cup \{\mathbf{x}_4\} = \{\mathbf{x}_1, \mathbf{x}_2, \mathbf{x}_4\}$ and increase j to 4. Since $i = k$ now, the algorithm stops and A_4 is the required basis of S. Note that if we knew beforehand that $d(S) = 3$, then we could have stopped the algorithm after finding that $\mathbf{x}_3 \in \mathrm{Sp}(A_3)$ since the condition $i = j - 1 + k - d$ is satisfied at this stage. ∎

Clearly $d(S) = d(T)$ does not imply $S = T$ for subspaces S and T. For example, consider the x-axis and the y-axis in $\mathbb{R}^2$. However, we have the following very useful result.

Theorem 1.5.16 Let S and T be subspaces such that $S \subseteq T$. Then

(i) $d(S) \leq d(T)$ and

(ii) $d(S) = d(T)$ iff $S = T$.

Proof Let B be a basis of S. Then B is a linearly independent subset of T. Hence B can be extended to a basis of T and (i) follows. The *if part* of (ii) is trivial. To prove the *only if part*, let $d(S) = d(T)$. Then by *Theorem* 1.5.11(a), B forms a basis of T, hence $\mathrm{Sp}(B) = T$. Since B is a basis of S we have $\mathrm{Sp}(B) = S$ and so $S = T$. ∎

It follows from the preceding theorem that if V is of dimension n, the only subspace of V with dimension n is V itself.

Coordinate System

We will now show that a basis gives a coordinate system and vice versa.

We recall that the plane becomes a vector space as soon as we fix the origin at any point in the plane since we can then define vector addition and scalar multiplication as explained at the beginning of *Section* 1.1.

Now, a coordinate system in the plane consists of the axes OX and OY meeting in the origin O and the points P and Q at unit distance from the origin along OX and OY, see *Figure* 1.5.1. Then the coordinates α and β of any point A are obtained geometrically as follows. Draw a line through A parallel to OY to meet OX at B. Then $\alpha = OB$. Similarly draw a line through A parallel to OX to meet OY at C. Then $\beta = OC$. Note that O, P and Q are not collinear. So the vectors $\mathbf{s}$ and $\mathbf{t}$ representing the points P and Q are linearly independent and form a basis of $\mathbb{R}^2$. Now the vector representing B is $\alpha\mathbf{s}$ since $OB/OP = \alpha$. Similarly the vector representing C is $\beta\mathbf{t}$ and, since $ACOB$ is a parallelogram, it follows that the vector $\mathbf{x}$ representing A is $\alpha\mathbf{s} + \beta\mathbf{t}$. Thus the coordinates of A are the unique scalars α and β such that $\mathbf{x} = \alpha\mathbf{s} + \beta\mathbf{t}$.

Conversely, given any basis $\{\mathbf{s}, \mathbf{t}\}$ of the plane (where the origin is already fixed), we can construct a coordinate system with P corresponding to $\mathbf{s}$ and Q corresponding to $\mathbf{t}$ as the reference points on the axes OP and OQ. Note that length OP need not be equal to length OQ. Now any vector A in the plane can be expressed uniquely as $\alpha\mathbf{s} + \beta\mathbf{t}$ and in this coordinate system α and β are the first and second coordinates of the point A.

Notice that the parallelogram $ACOB$ will be a rectangle if the coordinate system is rectangular, i.e., if OX and OY are perpendicular.

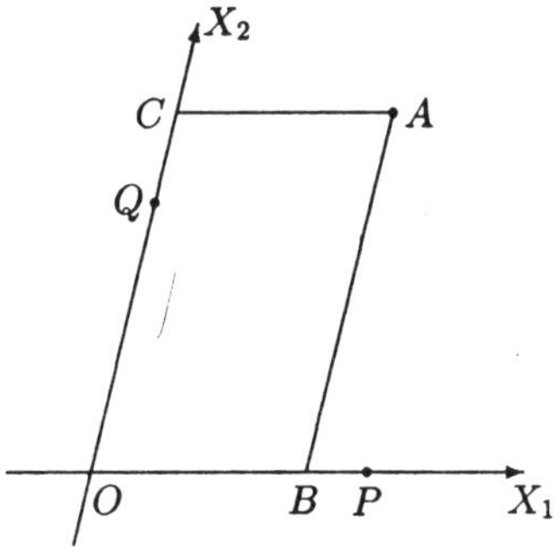

Figure 1.5.1

The above discussion can be extended to any finite-dimensional vector space. Let $\{\mathbf{x}_1, \mathbf{x}_2, \ldots, \mathbf{x}_n\}$ be a basis of V and $P_1, P_2, \ldots, P_n$ the points corresponding to $\mathbf{x}_1, \mathbf{x}_2, \ldots, \mathbf{x}_n$ respectively. We may then take the lines $OP_1, OP_2, \ldots, OP_n$ (each extended indefinitely on both sides) as the coordinate axes and $P_1, P_2, \ldots, P_n$ as the reference points along these axes. Now every $\mathbf{x} \in V$ can be expressed uniquely as $\alpha_1\mathbf{x}_1 + \cdots + \alpha_n\mathbf{x}_n$. The ordered n-tuple $(\alpha_1, \alpha_2, \ldots, \alpha_n)$ thus obtained is called the *coordinate vector of* $\mathbf{x}$ and α_i the i-th coordinate of $\mathbf{x}$. Notice that the i-th coordinate of a point can be determined only from the entire coordinate system and not from $\mathbf{x}_i$ alone.

Example 1.5.17 Let $V = \mathbb{R}^2$ and $\mathbf{x}_1 = (1,1)$, $\mathbf{x}_2 = (-1,2)$. Then for any point (u,v) in V it can be checked that $(u,v) = \alpha_1\mathbf{x}_1 + \alpha_2\mathbf{x}_2$ iff $\alpha_1 = (2u+v)/3$ and $\alpha_2 = (v-u)/3$. Thus every vector in V is a unique linear combination of $\mathbf{x}_1$ and $\mathbf{x}_2$, so $\mathbf{x}_1, \mathbf{x}_2$ form a basis of V.

Let us now consider the new coordinate system with the same origin O and with $P = \mathbf{x}_1$ and $Q = \mathbf{x}_2$ as the reference points along the new axes OP and OQ. Then $\alpha_1 = (2u+v)/3$ and $\alpha_2 = (v-u)/3$ are the coordinates of the point (u,v) in the new coordinate system. For example, the coordinates of $(0,1)$ in the new system are $\frac{1}{3}$, $\frac{1}{3}$, see *Figure* 1.5.2.

Exercises

1. Show that $f_1(t) = 1$, $f_2(t) = t - 2$ and $f_3(t) = (t-2)^2$ form a basis of $\mathcal{P}_3$. Express $3t^2 - 5t + 4$ as a linear combination of f_1, f_2, f_3.
2. If $\{\mathbf{x}_1, \mathbf{x}_2, \ldots, \mathbf{x}_k\}$ is a basis of a subspace S, show that

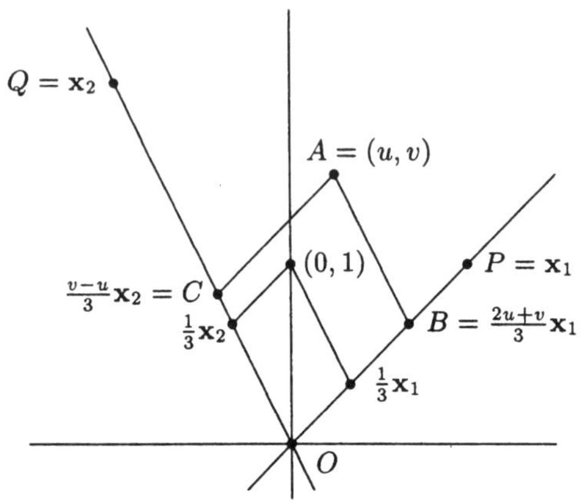

Figure 1.5.2

(a) $\{\alpha\mathbf{x}_1, \mathbf{x}_2, \ldots, \mathbf{x}_k\}$ is a basis of S iff $\alpha \neq 0$,

(b) $\{\mathbf{x}_1 + \beta\mathbf{x}_2, \mathbf{x}_2, \ldots, \mathbf{x}_k\}$ is a basis of S for any scalar β,

(c) $\{\mathbf{x}_1 + \beta\mathbf{x}_2, \alpha\mathbf{x}_1 + \mathbf{x}_2, \mathbf{x}_3, \ldots, \mathbf{x}_k\}$ is a basis of S iff $\alpha\beta \neq 1$.

3. If a subspace S of $\mathbb{R}^n$ has a basis $\{\mathbf{x}_1, \mathbf{x}_2, \ldots, \mathbf{x}_k\}$ such that all components of $\mathbf{x}_1$ are strictly positive, show that S has a basis B such that all components of each vector in B are strictly positive.

4. Find a basis and hence the dimension of the vector space in *Exercise* 1.2.4.

5. Let $A \subseteq V$. If one vector in $\mathrm{Sp}(A)$ can be expressed uniquely as a linear combination from A then show that A is linearly independent and, so, is a basis of $\mathrm{Sp}(A)$.

6. Show that a vector space V over F has a unique basis iff either '$d(V) = 0$' or '$d(V) = 1$ and $|F| = 2$'.

7. Prove or disprove: if A, B and C are pair-wise disjoint subsets of V such that $A \cup B$ and $A \cup C$ are bases of V, then $\mathrm{Sp}(B) = \mathrm{Sp}(C)$.

8. Prove or disprove: if B is a basis of V and S is a subspace of V then B contains a basis of S.

9. Find a basis of the span of the vectors $(1, 0, 1, 2, -1), (2, 1, -1, 3, 0)$, $(0, -1, 3, 1, -2)$, $(3, 1, 0, 1, 1)$, $(3, 1, 0, 3, 0)$ in $\mathbb{R}^5$.

*10. (a) Find a basis of the vector space of *Example* 1.2.9 when Ω is an arbitrary non-empty finite set.

(b) Find all the bases of the subspaces in *Example* 1.3.8.

11. Consider the basis

$$B = \{(1, -1, 0, 0, 0), (1, 0, -1, 0, 0), (1, 0, 0, -1, 0), (1, 0, 0, 0, -1)\}$$

of the subspace

$$S = \{(u_1, u_2, u_3, u_4, u_5) : u_1 + u_2 + \cdots + u_5 = 0\}$$

of $\mathbb{R}^5$. Using B, extend the linearly independent subset $\{\mathbf{x}_1, \mathbf{x}_2\}$ of S to a basis of S, where $\mathbf{x}_1 = (1, 0, 0, 2, -3)$ and $\mathbf{x}_2 = (1, 1, 0, 4, -6)$.

12. Extend $A = \{(1, 1, \ldots, 1)\}$ to a basis of $\mathbb{R}^n$.

13. Show that $\mathbf{x}_1 = 2+3t$, $\mathbf{x}_2 = 3+5t$, $\mathbf{x}_3 = 5-8t^2+t^3$ and $\mathbf{x}_4 = 4t-t^2$ form a basis of $\mathcal{P}_4$. Find the coordinates of the vector $a_0 + a_1t + a_2t^2 + a_3t^3$ with respect to this coordinate system.

14. Let $x_1, x_2, \ldots, x_n$ be fixed distinct real numbers.

    (a) Show that $\ell_1(t), \ell_2(t) \ldots, \ell_n(t)$ form a basis of $\mathcal{P}_n$, where $\ell_i(t) = \prod_{j \neq i}(t - x_j)$. This basis leads to what is known as Lagrange's interpolation formula. If $f(t) \in \mathcal{P}_n$ is written as $\sum_{i=1}^{n} \alpha_i \ell_i(t)$, show that $\alpha_i = f(x_i)/\ell_i(x_i)$.

    (b) Show that $\psi_1(t), \psi_2(t), \ldots, \psi_n(t)$ form a basis of $\mathcal{P}_n$, where $\psi_1(t) = 1$ and $\psi_i(t) = \prod_{j=1}^{i-1}(t - x_j)$ for $i = 2, \ldots, n$. This basis leads to what is known as Newton's divided difference formula.

15. Prove *Theorem* 1.5.7 after making the following changes. Take S to be any subset (instead of a subspace) of V. Replace (i) by: A is a minimal set with the property $\mathrm{Sp}(A) \supseteq S$. In (iii), replace '$A$ generates S' by '$\mathrm{Sp}(A) \supseteq S$'.

16. Find a basis of each of the following subspaces of $\mathbb{R}^4$. Also express S_3 in the form

$$\left\{(x_1, x_2, x_3, x_4) : \frac{x_1}{a_1} = \frac{x_2}{a_2} = \frac{x_3}{a_3} = \frac{x_4}{a_4}\right\}$$

    (a) $S_1 = \{(x_1, x_2, x_3, x_4) : x_1 - 2x_3 + x_4 = 0\}$,

    (b) $S_2 = \{(x_1, x_2, x_3, x_4) : x_1 + x_2 - x_3 = 0,\ x_2 + 2x_3 - x_4 = 0,\ 2x_1 + 3x_2 - x_4 = 0\}$,

    (c) $S_3 = \{(x_1, x_2, x_3, x_4) : x_1 + x_2 - x_3 = 0,\ x_1 + x_2 + 2x_3 + x_4 = 0,\ x_1 - 3x_2 - x_3 + 2x_4 = 0\}$.

17. Let $B = \{\mathbf{x}_1, \mathbf{x}_2, \ldots, \mathbf{x}_k\}$ be a basis of S and $\mathbf{x} = \alpha_1\mathbf{x}_1 + \cdots + \alpha_k\mathbf{x}_k \notin B$. Obtain a necessary and sufficient condition for $(B \cup \{\mathbf{x}\}) - \{\mathbf{x}_i\}$ to be a basis of S.

18. Show that the subspaces of continuous functions and differentiable functions in *Example* 1.3.5 are not finite-dimensional.

*19. Let F be a finite field with q elements and V an n-dimensional vector space over F.

(a) Show that $|V| = q^n$.

(b) Show (using *Theorem* 1.4.6) that the number of ordered k-tuples $(\mathbf{x}_1, \mathbf{x}_2, \ldots, \mathbf{x}_k)$ such that $\mathbf{x}_1, \mathbf{x}_2, \ldots, \mathbf{x}_k$ are linearly independent vectors in V, is

$$(q^n - 1)(q^n - q)(q^n - q^2) \cdots (q^n - q^{k-1})$$

(c) Show that the number of distinct (unordered) bases of V is

$$(q^n - 1)(q^n - q) \cdots (q^n - q^{n-1})/n!$$

(d) Show that the number of k-dimensional subspaces of V is

$$\frac{(q^n - 1)(q^n - q) \cdots (q^n - q^{k-1})}{(q^k - 1)(q^k - q) \cdots (q^k - q^{k-1})} = \frac{(q^n - 1)(q^{n-1} - 1) \cdots (q^{n-k+1} - 1}{(q^k - 1)(q^{k-1} - 1) \cdots (q - 1)}$$

This number is usually denoted by $\begin{bmatrix} n \\ k \end{bmatrix}_q$.

(e) Show that the number of ℓ-dimensional subspaces of V containing a given k-dimensional subspace is $\begin{bmatrix} n-k \\ \ell-k \end{bmatrix}_q$.

20. If F is a subfield of a finite field G, prove that the number of elements in G is a power of the number of elements in F.

21. Let F be a subfield of a field G and let $\mathbf{x}_1, \mathbf{x}_2, \ldots, \mathbf{x}_k \in F^n$. Show that $\mathbf{x}_1, \mathbf{x}_2, \ldots, \mathbf{x}_k$ are linearly independent in F^n over F iff they are linearly independent in G^n over G. (Hint: first consider the case $k = n$.)

1.6 Calculus of subspaces

In *Section* 1.3 we have seen that the intersection $S \cap T$ of subspaces S and T is always a subspace. Clearly $S \cap T$ is the largest subspace contained in both S and T. We have also noted that $S \cup T$ need not be a subspace. However, it is easy to see that the span of $S \cup T$ is the smallest subspace containing both S and T. We give a simpler characterization of this in

Theorem 1.6.1 Let S and T be subspaces. Then

$$\mathrm{Sp}(S \cup T) = \{\mathbf{x} + \mathbf{y} : \mathbf{x} \in S \text{ and } \mathbf{y} \in T\} \tag{1.6.1}$$

Proof Denote the LHS and the RHS of (1.6.1) by X and W respectively. To prove that $X \supseteq W$, let $\mathbf{u} \in W$. Then $\mathbf{u} = \mathbf{x} + \mathbf{y}$ for some $\mathbf{x} \in S$ and $\mathbf{y} \in T$. Now $\mathbf{x} \in S \cup T$, so $\mathbf{x} \in X$. Similarly $\mathbf{y} \in X$. Since X is a subspace it follows that $\mathbf{u} \in X$. Thus $X \supseteq W$.

To prove the other inclusion, we first show that W is a subspace. Let $\mathbf{u}_1 = \mathbf{x}_1 + \mathbf{y}_1$ and $\mathbf{u}_2 = \mathbf{x}_2 + \mathbf{y}_2$ be arbitrary elements of W where $\mathbf{x}_1, \mathbf{x}_2 \in S$ and $\mathbf{y}_1, \mathbf{y}_2 \in T$. Then for any scalar α,

$$\alpha \mathbf{u}_1 + \mathbf{u}_2 = (\alpha \mathbf{x}_1 + \mathbf{x}_2) + (\alpha \mathbf{y}_1 + \mathbf{y}_2)$$

belongs to W. Clearly $\mathbf{0} \in W$, so W is a subspace. Now $S \subseteq W$ since $\mathbf{x} = \mathbf{x} + \mathbf{0}$ for any $\mathbf{x}$ in S. Similarly $T \subseteq W$, so $S \cup T \subseteq W$. Since W is a subspace, it follows that $X \subseteq W$. ∎

Definition 1.6.2 The *sum* of any two subsets A and B of a vector space is

$$A + B = \{\mathbf{x} + \mathbf{y} : \mathbf{x} \in A,\ \mathbf{y} \in B\}$$

If A is a singleton $\{\mathbf{x}\}$, we write $\mathbf{x} + B$ for $A + B$ and say that $\mathbf{x} + B$ is a *translate of* B.

Theorem 1.6.1 shows that if S and T are subspaces, then the sum $S+T$ is the smallest subspace containing both S and T. Clearly, $S+T = T+S$.

Example 1.6.3 The sum of any two distinct lines through the origin in $\mathbb{R}^2$ is $\mathbb{R}^2$. The sum of the 'x-y plane' and the 'y-z plane' in $\mathbb{R}^3$ is $\mathbb{R}^3$.

Example 1.6.4 In $\mathbb{R}^4$, let

$$S = \{(\xi_1, \xi_2, \xi_3, \xi_4) : \xi_1 + \xi_2 = 0,\ \xi_3 + \xi_4 = 0\}$$

and

$$T = \{(\xi_1, \xi_2, \xi_3, \xi_4) : \xi_1 + \xi_3 = 0,\ \xi_2 + \xi_4 = 0\}$$

Then we will show that

$$S + T = \{(\xi_1, \xi_2, \xi_3, \xi_4) : \xi_1 + \xi_2 + \xi_3 + \xi_4 = 0\} \tag{1.6.2}$$

Note that W, the RHS of (1.6.2), is a subspace. Since it contains S and T it contains $S+T$. To prove that $W \subseteq S+T$, let $\mathbf{u} = (u_1, u_2, u_3, u_4) \in W$. Then $u_4 = -(u_1 + u_2 + u_3)$. We now try to express $\mathbf{u}$ as $\mathbf{x} + \mathbf{y}$ where $\mathbf{x} \in S$ and $\mathbf{y} \in T$. Any vector $\mathbf{x}$ in S is of the form $(\alpha, -\alpha, \beta, -\beta)$ and any vector $\mathbf{y}$ in T is of the form $(\gamma, \delta, -\gamma, -\delta)$. So we have to solve the equations $\alpha + \gamma = u_1$, $-\alpha + \delta = u_2$, $\beta - \gamma = u_3$ and $-\beta - \delta = u_4$ for α, β, γ and δ. Notice that the last equation is redundant since it is obtained by adding the first three equations. It is now easy to see that $\alpha = 0$, $\beta = u_1 + u_3$, $\gamma = u_1$, $\delta = u_2$ is a solution. Thus $\mathbf{u} = (0, 0, u_1 + u_3, -u_1 - u_3) + (u_1, u_2, -u_1, -u_2)$. Hence $W \subseteq S + T$ and (1.6.2) follows. We incidentally note that, here, a vector in $S+T$ is expressible as $\mathbf{x}+\mathbf{y}$ with $\mathbf{x}\in S$ and $\mathbf{y}\in T$ in more than one way. For example,

$$\mathbf{u}=(1,-1,u_1+u_3-1,1-u_1-u_3)+(u_1-1,u_2+1,1-u_1,-u_2-1). \blacksquare$$

***Example 1.6.5** In the vector space of *Example* 1.2.9, consider the subspaces $S=\{\emptyset, A\}$ and $T=\{\emptyset, B\}$ where A and B are two subsets of Ω. Then $S+T=\{\emptyset, A, B, A\mathbin{\triangle} B\}$.

Theorem 1.6.6 (*Modular Law*) For any two subspaces S and T,

$$d(S+T)+d(S\cap T)=d(S)+d(T)$$

Proof Let $d(S)=s$, $d(T)=t$ and $d(S\cap T)=r$. Then we have to show that $d(S+T)=s+t-r$.

Let $C=\{\mathbf{x}_1,\ldots,\mathbf{x}_r\}$ be a basis of $S\cap T$. Then C is a linearly independent subset of S and can be extended to a basis $\{\mathbf{x}_1,\ldots,\mathbf{x}_r,\mathbf{y}_1,\ldots,\mathbf{y}_{s-r}\}$ of S. Similarly C can be extended to a basis $\{\mathbf{x}_1,\ldots,\mathbf{x}_r,\mathbf{z}_1,\ldots,\mathbf{z}_{t-r}\}$ of T. We show that the vectors $\mathbf{x}_1,\ldots,\mathbf{x}_r,\mathbf{y}_1,\ldots,\mathbf{y}_{s-r},\mathbf{z}_1,\ldots,\mathbf{z}_{t-r}$ form a basis of $S+T$. That they generate $S+T$ is easy to verify. We shall show that they are linearly independent. So let

$$\sum_{i=1}^{r}\alpha_i\mathbf{x}_i+\sum_{j=1}^{s-r}\beta_j\mathbf{y}_j+\sum_{k=1}^{t-r}\gamma_k\mathbf{z}_k=\mathbf{0}$$

Then

$$\sum\alpha_i\mathbf{x}_i+\sum\beta_j\mathbf{y}_j=-\sum\gamma_k\mathbf{z}_k \tag{1.6.3}$$

Clearly the LHS of (1.6.3) belongs to S and the RHS to T. Thus $-\sum\gamma_k\mathbf{z}_k$ belongs to $S\cap T$ and so can be expressed as $\sum\delta_i\mathbf{x}_i$. Then $\sum\delta_i\mathbf{x}_i+\sum\gamma_k\mathbf{z}_k=\mathbf{0}$. Since $\mathbf{x}_1,\ldots,\mathbf{x}_r,\mathbf{z}_1,\ldots,\mathbf{z}_{t-r}$ form a basis of T, it follows that $\gamma_1=\cdots=\gamma_{t-r}=0$ (and $\delta_1=\cdots=\delta_r=0$ but we will not need this fact). Substituting the values of γ's in (1.6.3) and using the fact that $\mathbf{x}_1,\ldots,\mathbf{x}_r,\mathbf{y}_1,\ldots,\mathbf{y}_{s-r}$ form a basis of S, we get $\alpha_1=\cdots=\alpha_r=0$ and $\beta_1=\cdots=\beta_{s-r}=0$. Thus $\mathbf{x}_1,\ldots,\mathbf{x}_r,\mathbf{y}_1,\ldots,\mathbf{y}_{s-r},\mathbf{z}_1,\ldots,\mathbf{z}_{t-r}$ form a linearly independent set and so a basis of $S+T$. Hence $d(S+T)=s+t-r$. ∎

Corollary For any two subspaces S and T of a vector space,

$$d(S+T)\leq d(S)+d(T)$$

Further, equality holds iff $S\cap T=\{\mathbf{0}\}$.

The concept of the sum of two subspaces can be extended to the sum of any finite number of subspaces as follows:

$$S_1 + \cdots + S_k = \{\mathbf{x}_1 + \cdots + \mathbf{x}_k : \mathbf{x}_i \in S_i,\ 1 \le i \le k\}$$

Notice that if $(i_1, i_2, \ldots, i_k)$ is a permutation of $(1, 2, \ldots, k)$ then $S_{i_1} + S_{i_2} + \cdots + S_{i_k} = S_1 + S_2 + \cdots + S_k$. It is also easy to see that $(S_1 + S_2) + S_3 = S_1 + S_2 + S_3$. Thus the sum of several subspaces can be found by repeated addition.

*Flats

We shall now briefly discuss a concept closely related to that of a subspace. Recall that the subspaces of $\mathbb{R}^3$ are $\{\mathbf{0}\}$, the lines through the origin, the planes through the origin and $\mathbb{R}^3$ itself. In geometry, we are also interested in the lines and planes not necessarily passing through the origin. These are called flats.

Definition 1.6.7 A set $A \subseteq V$ is said to be a *flat* or an *affine set* if either $A = \emptyset$ or A is a translate of a subspace (i.e., $A = \mathbf{x} + S$ for some subspace S of V and some $\mathbf{x} \in V$).

To characterize flats through linear combinations we give

Definition 1.6.8 An *affine combination* of $\mathbf{x}_1, \ldots, \mathbf{x}_k$ is a linear combination $\alpha_1\mathbf{x}_1 + \cdots + \alpha_k\mathbf{x}_k$ where $\alpha_1 + \cdots + \alpha_k = 1$.

Theorem 1.6.9 $A \subseteq V$ is a flat iff A is closed under affine combinations, that is,

$$\mathbf{x}_1, \ldots, \mathbf{x}_k \in A,\ \alpha_1, \ldots, \alpha_k \in F,\ \alpha_1 + \cdots + \alpha_k = 1 \Rightarrow \alpha_1\mathbf{x}_1 + \cdots + \alpha_k\mathbf{x}_k \in A \qquad (1.6.4)$$

Proof *Only if part* Let $A = \mathbf{x}_0 + S$ where S is a subspace. To prove (1.6.4), let the antecedent (i.e. LHS) be given. Then for $i = 1, 2, \ldots, k$, $\mathbf{x}_i = \mathbf{x}_0 + \mathbf{s}_i$ for some $\mathbf{s}_i \in S$. So $\alpha_1\mathbf{x}_1 + \cdots + \alpha_k\mathbf{x}_k = \mathbf{x}_0 + \alpha_1\mathbf{s}_1 + \cdots + \alpha_k\mathbf{s}_k \in \mathbf{x}_0 + S = A$.

If part If $A = \emptyset$, we are done. So let $A \neq \emptyset$. Choose and fix any $\mathbf{x}_0 \in A$. Let $S = -\mathbf{x}_0 + A = \{\mathbf{x} - \mathbf{x}_0 : \mathbf{x} \in A\}$. Clearly then $A = \mathbf{x}_0 + S$. Since $\mathbf{x}_0 \in A$, we have $\mathbf{0} \in S$. Let $\mathbf{u}, \mathbf{v} \in S$ and α, $\beta \in F$. Then $\mathbf{u} = \mathbf{x} - \mathbf{x}_0$ and $\mathbf{v} = \mathbf{y} - \mathbf{x}_0$ for some $\mathbf{x}, \mathbf{y} \in A$. So by (1.6.4), $\alpha\mathbf{u} + \beta\mathbf{v} + \mathbf{x}_0 = \alpha\mathbf{x} + \beta\mathbf{y} + (1 - \alpha - \beta)\mathbf{x}_0 \in A$ and $\alpha\mathbf{u} + \beta\mathbf{v} \in S$. Thus S is a subspace and $A = \mathbf{x}_0 + S$ is a flat. ∎

It can be proved (see *Exercise* 1.6.9) that (1.6.4) holds for all $k \geq 1$ iff it holds for $k = 2$, provided $1 + 1 \neq 0$ in F. *Geometrically*, a vector $\mathbf{z}$ is an affine combination of the (distinct) vectors $\mathbf{x}$ and $\mathbf{y}$ iff $\mathbf{z}$ lies on the line passing through $\mathbf{x}$ and $\mathbf{y}$. Thus A is a flat iff it contains, along with any two points P and Q, the entire line PQ. Similarly if $\mathbf{x}$, $\mathbf{y}$ and $\mathbf{z}$ are non-collinear points, then $\mathbf{w}$ is an affine combination of $\mathbf{x}$, $\mathbf{y}$ and $\mathbf{z}$ iff $\mathbf{w}$ lies on the plane containing $\mathbf{x}$, $\mathbf{y}$ and $\mathbf{z}$.

Theorem 1.6.10 Let S and T be subspaces of V and $\mathbf{x}, \mathbf{v} \in V$. Then $\mathbf{x} + S \subseteq \mathbf{y} + T$ iff $S \subseteq T$ and $\mathbf{x} - \mathbf{y} \in T$.

Proof *If part* Let $\mathbf{z} \in \mathbf{x} + S$. Then $\mathbf{z} = \mathbf{x} + \mathbf{u}$ for some $\mathbf{u} \in S$. Now $\mathbf{x} - \mathbf{y} \in T$ and $\mathbf{u} \in S \subseteq T$. So $\mathbf{x} - \mathbf{y} + \mathbf{u} \in T$. Hence $\mathbf{z} = \mathbf{y} + (\mathbf{x} - \mathbf{y} + \mathbf{u}) \in \mathbf{y} + T$.

Only if part $\mathbf{x} = \mathbf{x} + \mathbf{0} \in \mathbf{x} + S \subseteq \mathbf{y} + T$. So $\mathbf{x} - \mathbf{y} \in T$. Now let $\mathbf{u} \in S$. Then $\mathbf{x} + \mathbf{u} \in \mathbf{x} + S \subseteq \mathbf{y} + T$. So $\mathbf{x} + \mathbf{u} = \mathbf{y} + \mathbf{z}$ for some $\mathbf{z} \in T$. Since $\mathbf{y} - \mathbf{x} \in T$ and $\mathbf{z} \in T$, it follows that $\mathbf{u} = \mathbf{y} - \mathbf{x} + \mathbf{z} \in T$. ∎

Corollary For subspaces S and T, $\mathbf{x} + S = \mathbf{y} + T$ implies $S = T$.

Thus, the subspace S associated with a flat $\mathbf{x} + S$ is unique though $\mathbf{x}$ is not. For further discussion on flats, see *Section* 5.3.

Exercises

1. Prove the statements made in *Examples* 1.6.3 and 1.6.5.
2. Let S, T and W be subspaces. If $W \subseteq T$, prove that $S + W \subseteq S + T$. Is the converse true?
3. When is $S + T$ equal to $S \cup T$?
4. Prove that the intersection of any two planes through the origin in $\mathbb{R}^3$ contains a line through the origin.
5. Let S and T be subspaces of a vector space V with $d(S) = 2$, $d(T) = 3$ and $d(V) = 5$. Find the minimum and maximum possible values of $d(S + T)$ and show that every (integer) value between these can be attained.
6. Show that the distributive law

$$S \cap (T + W) = (S \cap T) + (S \cap W)$$

is false for subspaces. However prove that it holds whenever $S \supseteq T$ or $S \supseteq W$. This latter result is also known as the modular law.

7. Let S and T be as given in *Example* 1.6.4 and let $W = \{(\xi_1, \xi_2, \xi_3, \xi_4) : \xi_1 + \xi_3 = 0,\ \xi_2 + \xi_4 = 0,\ \xi_1 + \xi_4 = 0\}$. Show that any vector in $S + W$ can be expressed in only one way as $\mathbf{s}+\mathbf{w}$ with $\mathbf{s} \in S$ and $\mathbf{w} \in W$. Show also that a similar statement is false for $S+T$ though $S+T = S+W$.
8. Let S and T be subspaces of $\mathbb{R}^4$ given by

$$S = \{(x_1, x_2, x_3, x_4) : 3x_1 + x_2 + x_3 + x_4 = 0,\ x_1 - x_3 + 2x_4 = 0\}$$

and

$$T = \{(x_1, x_2, x_3, x_4) : 5x_1 + 2x_2 + 3x_3 = 0,\ x_2 + x_3 + x_4 = 0\}$$

   (a) Obtain a basis for each of $S \cap T$, S, T and $S + T$.
   (b) Verify the modular law for S and T.
   (c) Extend the basis of $S + T$ you obtained in (a) to a basis of $\mathbb{R}^4$.
   (d) Express $S + T$ and $S \cap T$ in the same form as S and T.

*9. Define the *affine set generated by* $A \subseteq V$ to be the set of all affine combinations from A. Also call a set A *affine independent* if $\mathbf{x}$ does not belong to the affine set generated by $A - \{\mathbf{x}\}$ for all $\mathbf{x} \in A$.
   (a) Let $1 + 1 \neq 0$ in F. Show that (1.6.4) for $k = 2$ implies (1.6.4) for all $k \geq 1$.
   (b) Prove the analogues of *Theorems* 1.3.11, 1.3.14 and 1.3.15 and *Exercises* 1.3.9, 1.3.10 and 1.3.11(a).
   (c) Show that the sum of two flats is a flat.
   (d) Show that the union of two flats is a flat iff one of them is contained in the other or '$|F| = 2$ and one is a translate of the other'.
   (e) Prove that A is affine independent iff

$$\alpha_1\mathbf{x}_1 + \cdots + \alpha_k\mathbf{x}_k = \mathbf{0},\ \alpha_1 + \cdots + \alpha_k = 0 \Rightarrow \alpha_1 = \cdots = \alpha_k = 0$$

   If $\mathbf{x}_0 \in A$, prove that A is affine independent iff $\{\mathbf{x} - \mathbf{x}_0 : \mathbf{x} \in A,\ \mathbf{x} \neq \mathbf{x}_0\}$ is linearly independent. Also prove the analogues of *Theorems* 1.4.6, 1.4.8 and 1.4.9.
   (f) Prove the analogue of *Theorem* 1.5.7 and define affine basis and affine dimension of an affine set. Show that the affine dimension of $\mathbf{x} + S$ is $d(S) + 1$ if S is a subspace.

1.7 Direct sum and complement

We have seen in *Example* 1.6.3 that if S is the 'x-y plane' and T is the 'y-z plane', then $S + T = \mathbb{R}^3$. But a vector $\mathbf{u} = (u_1, u_2, u_3)$ in $S + T$ can be expressed in more than one way as $\mathbf{x} + \mathbf{y}$ with $\mathbf{x} \in S$ and $\mathbf{y} \in T$.

For example, $\mathbf{u} = (u_1, u_2, 0) + (0, 0, u_3) = (u_1, u_2 + 1, 0) + (0, -1, u_3)$. Consider now $W = \{(\gamma, \gamma, \gamma) : \gamma \in \mathbb{R}\}$. Then $S + W$ is also equal to $\mathbb{R}^3$. Suppose $\mathbf{u} = (\alpha, \beta, 0) + (\gamma, \gamma, \gamma)$ is a representation of $\mathbf{u} = (u_1, u_2, u_3)$ as $\mathbf{v} + \mathbf{w}$ with $\mathbf{v} \in S$ and $\mathbf{w} \in W$. Then $\gamma = u_3$, $\alpha = u_1 - u_3$ and $\beta = u_2 - u_3$ and $\mathbf{v}$ and $\mathbf{w}$ are unique. Motivated by this, we give

Definition 1.7.1 The sum of two subspaces S and T is said to be *direct* (or S and T *independent*) if any vector in $S + T$ can be expressed in a unique way as $\mathbf{x} + \mathbf{y}$ with $\mathbf{x} \in S$ and $\mathbf{y} \in T$. We use $S \oplus T$ to denote $S + T$ when the sum is direct.

Theorem 1.7.2 Let S and T be subspaces of a vector space. Then the following statements are equivalent.

(i) $S + T$ is direct,

(ii) $S \cap T = \{\mathbf{0}\}$,

(iii) If $\mathbf{x} \in S - \{\mathbf{0}\}$ and $\mathbf{y} \in T - \{\mathbf{0}\}$ then $\mathbf{x}, \mathbf{y}$ are linearly independent,

(iv) $\mathbf{0} = \mathbf{x} + \mathbf{y}$, $\mathbf{x} \in S$, $\mathbf{y} \in T \Rightarrow \mathbf{x} = \mathbf{0}$ and $\mathbf{y} = \mathbf{0}$,

(v) $d(S + T) = d(S) + d(T)$.

Proof From the corollary to *Theorem* 1.6.6 it is clear that (ii) and (v) are equivalent. We now prove that (i) through (iv) are equivalent.

(i) $\Rightarrow$ (ii) Let $S + T$ be direct and $\mathbf{z} \in S \cap T$. Then $\mathbf{z} = \mathbf{z} + \mathbf{0}$ and $\mathbf{z} \in S$ and $\mathbf{0} \in T$. Also, $\mathbf{z} = \mathbf{0} + \mathbf{z}$ and $\mathbf{0} \in S$ and $\mathbf{z} \in T$. Since $S + T$ is direct, it follows that $\mathbf{z} = \mathbf{0}$. Thus $S \cap T \subseteq \{\mathbf{0}\}$. The opposite inclusion is trivial and equality follows.

(ii) $\Rightarrow$ (iii) Let $\mathbf{x} \in S$ and $\mathbf{y} \in T$ be non-null and let $\alpha\mathbf{x} + \beta\mathbf{y} = \mathbf{0}$. Then $\alpha\mathbf{x} = -\beta\mathbf{y} \in S \cap T$, so $\alpha\mathbf{x} = \mathbf{0}$ and $\beta\mathbf{y} = \mathbf{0}$. Since $\mathbf{x}$ and $\mathbf{y}$ are non-null, it follows that $\alpha = \beta = 0$.

(iii) $\Rightarrow$ (iv) Let $\mathbf{x} + \mathbf{y} = \mathbf{0}$ where $\mathbf{x} \in S$ and $\mathbf{y} \in T$. If $\mathbf{x} \neq \mathbf{0}$ then $\mathbf{y} \neq \mathbf{0}$ and $\mathbf{x}, \mathbf{y}$ are linearly dependent, a contradiction. So $\mathbf{x} = \mathbf{y} = \mathbf{0}$.

(iv) $\Rightarrow$ (i) Suppose $\mathbf{u} \in S + T$ can be written as $\mathbf{u} = \mathbf{x}_1 + \mathbf{y}_1 = \mathbf{x}_2 + \mathbf{y}_2$ where $\mathbf{x}_1, \mathbf{x}_2 \in S$ and $\mathbf{y}_1, \mathbf{y}_2 \in T$. Then $\mathbf{0} = (\mathbf{x}_1 - \mathbf{x}_2) + (\mathbf{y}_1 - \mathbf{y}_2)$, so by (iv), $\mathbf{x}_1 - \mathbf{x}_2 = \mathbf{0}$ and $\mathbf{y}_1 - \mathbf{y}_2 = \mathbf{0}$, hence (i) follows. ∎

Motivated by the preceding theorem, we introduce the following terminology: we say that two subspaces S and T are *virtually disjoint* if $S \cap T = \{\mathbf{0}\}$. Notice that two subspaces cannot be disjoint in the strict sense.

Definition 1.7.3 Let S be a subspace of a vector space V. Then a subspace T of V is said to be a *complement* of S if $S \oplus T = V$.

For example, if S is a line through the origin in $\mathbb{R}^2$ then any other line through the origin is a complement of S. If S is a line through the origin in $\mathbb{R}^3$ then any plane containing the origin and not containing S is a complement of S. Thus in general a complement of a subspace is not unique. However, see *Exercise* 1.7.13.

If T is a complement of S then clearly S is a complement of T and we say that S and T are *complementary subspaces*.

Theorem 1.7.4 Every subspace of a vector space has a complement.

Proof Let S be a subspace of a vector space V and let A be a basis of S. Extend A to a basis D of V and let $T = \mathrm{Sp}(D - A)$. Suppose now $\mathbf{0} = \mathbf{x} + \mathbf{y}$, $\mathbf{x} \in S$ and $\mathbf{y} \in T$. Then $\mathbf{x}$ is a linear combination from A and $\mathbf{y}$ is a linear combination from $D - A$ and $\mathbf{x} + \mathbf{y} = \mathbf{0}$. Since D is linearly independent, it follows that $\mathbf{x} = \mathbf{y} = \mathbf{0}$. So, by the preceding theorem, $S + T$ is direct. Since $A \cup (D - A)$ generates V, it follows that $S \oplus T = V$ and T is a complement of S. ∎

Definition 1.7.5 Let T be a complement of S and $\mathbf{u} \in V$. Let $\mathbf{u} = \mathbf{x} + \mathbf{y}$, $\mathbf{x} \in S$ and $\mathbf{y} \in T$. Then $\mathbf{x}$ is called *the projection of* $\mathbf{u}$ *into* S *along* T.

Since the representation of $\mathbf{u}$ as $\mathbf{x} + \mathbf{y}$ with $\mathbf{x} \in S$ and $\mathbf{y} \in T$ exists and is unique, the projection of $\mathbf{u}$ into S along T is well-defined. Clearly, if $\mathbf{x}$ is the projection of $\mathbf{u}$ into S along T, then $\mathbf{u} - \mathbf{x}$ is the projection of $\mathbf{u}$ into T along S. By definition, the projection of any vector into S along T belongs to S.

To give an example, let S be the ξ_1-axis and let T be the line $\{(\xi_1, \xi_2) : \xi_1 = \xi_2\}$ in $\mathbb{R}^2$. Then T is a complement of S and the projection of $\mathbf{u} = (-1, 2)$ into S along T is $(-3, 0)$ since $(-1, 2) = (-3, 0) + (2, 2)$, $(-3, 0) \in S$ and $(2, 2) \in T$. The projection of $\mathbf{u}$ into T along S is $(2, 2)$.

From *Figure* 1.7.1, it is clear that, geometrically, projection into S along T means projection into S parallel to T.

We next give a more non-trivial example of complement and projection.

Example 1.7.6 Consider the vector space $\mathbb{R}^5$. Let

$$S = \{(\xi_1, \xi_2, \xi_3, \xi_4, \xi_5) : \xi_1 + \xi_4 = 0,\ 2\xi_1 + \xi_3 + \xi_5 = 0\}$$

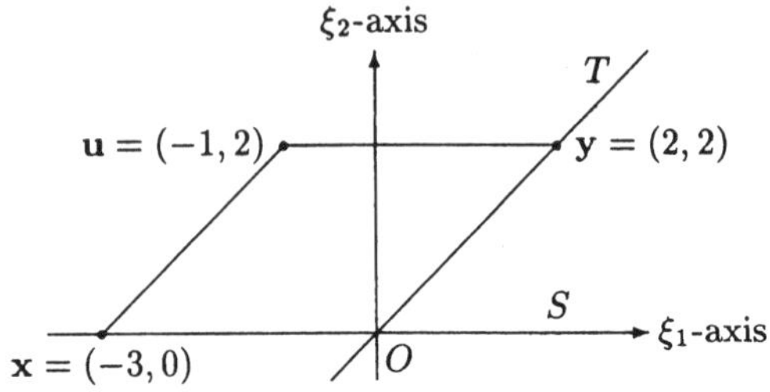

Figure 1.7.1

We shall use the method of proof of *Theorem* 1.7.4 to get a complement of S. It is easy to check that $\mathbf{x}_1 = (1, 0, 0, -1, -2), \mathbf{x}_2 = (0, 0, 1, 0, -1)$ and $\mathbf{x}_3 = (0, 1, 0, 0, 0)$ form a basis of S. We shall extend $\{\mathbf{x}_1, \mathbf{x}_2, \mathbf{x}_3\}$ to a basis of $\mathbb{R}^5$. For this we use *Algorithm* 1.5.12. Consider the canonical basis of $\mathbb{R}^5$, viz., $\{\mathbf{e}_1, \mathbf{e}_2, \mathbf{e}_3, \mathbf{e}_4, \mathbf{e}_5\}$.

We take $A_1 = \{\mathbf{x}_1, \mathbf{x}_2, \mathbf{x}_3\}$. If $\mathbf{e}_1$ is a linear combination of $\mathbf{x}_1, \mathbf{x}_2, \mathbf{x}_3$, then there exist $\alpha_1, \alpha_2, \alpha_3$ such that $\alpha_1\mathbf{x}_1 + \alpha_2\mathbf{x}_2 + \alpha_3\mathbf{x}_3 = \mathbf{e}_1$. Comparing the first components and the fourth components we easily get a contradiction. Hence $\mathbf{e}_1$ is not a linear combination from A_1 and we set $A_2 = \{\mathbf{x}_1, \mathbf{x}_2, \mathbf{x}_3, \mathbf{e}_1\}$. Next we check if $\mathbf{e}_2$ is a linear combination from A_2. This is indeed so since $\mathbf{e}_2 = \mathbf{x}_3$. So we set $A_3 = A_2$. It is easy to verify that $\mathbf{e}_3$ is not a linear combination from A_3. So we set $A_4 = \{\mathbf{x}_1, \mathbf{x}_2, \mathbf{x}_3, \mathbf{e}_1, \mathbf{e}_3\}$. Since the size of A_4 is $d(\mathbb{R}^5)$, A_4 is the required basis and

$$T_1 = \mathrm{Sp}(\{\mathbf{e}_1, \mathbf{e}_3\}) = \{(\eta_1, 0, \eta_3, 0, 0) : \eta_1, \eta_3 \in \mathbb{R}\}$$

is a complement of S.

We incidentally note that by considering $\mathbf{e}_1, \mathbf{e}_2, \mathbf{e}_3, \mathbf{e}_4, \mathbf{e}_5$ in different orders and applying the same algorithm, one can get $T_2 = \mathrm{Sp}(\{\mathbf{e}_4, \mathbf{e}_5\})$ as well as $T_3 = \mathrm{Sp}(\{\mathbf{e}_1, \mathbf{e}_4\})$ and $T_4 = \mathrm{Sp}(\{\mathbf{e}_3, \mathbf{e}_4\})$ as complements of S.

Consider $\mathbf{u} = (1, 1, 1, 1, 1) \in \mathbb{R}^5$. We shall now obtain the projection of $\mathbf{u}$ into S along T_1. For this we first find $\alpha_1, \alpha_2, \ldots, \alpha_5$ such that

$$\alpha_1\mathbf{x}_1 + \alpha_2\mathbf{x}_2 + \alpha_3\mathbf{x}_3 + \alpha_4\mathbf{e}_1 + \alpha_5\mathbf{e}_3 = \mathbf{u},$$

that is such that $\alpha_1 + \alpha_4 = 1,\ \alpha_3 = 1,\ \alpha_2 + \alpha_5 = 1,\ -\alpha_1 = 1$ and $-2\alpha_1 - \alpha_2 = 1$. It is easy to see that these equations have the unique solution

$$\alpha_1 = -1,\ \alpha_2 = 1,\ \alpha_3 = 1,\ \alpha_4 = 2 \text{ and } \alpha_5 = 0$$

Now $\alpha_1\mathbf{x}_1+\alpha_2\mathbf{x}_2+\alpha_3\mathbf{x}_3 \in S$ and $\alpha_4\mathbf{e}_1+\alpha_5\mathbf{e}_3 \in T_1$, so the projection of $\mathbf{u}$ into S along T_1 is $\alpha_1\mathbf{x}_1+\alpha_2\mathbf{x}_2+\alpha_3\mathbf{x}_3 = -\mathbf{x}_1+\mathbf{x}_2+\mathbf{x}_3 = (-1,1,1,1,1)$. Similarly the projection of $\mathbf{u}$ into T_1 along S is $\alpha_4\mathbf{e}_1+\alpha_5\mathbf{e}_3 = 2\mathbf{e}_1 = (2,0,0,0,0)$.

We leave it to the reader to show that the projection of $\mathbf{u}$ into S along T_2 is $(1,1,1,-1,-3)$ and the projection of $\mathbf{u}$ into T_2 along S is $(0,0,0,2,4)$. ∎

Notice that *the projections of the same vector* $\mathbf{u}$ *into the subspace* S *are different along different complements.* Thus, *when we talk about the projection of a vector into a subspace* S*, it is important to mention the complement* T *along which the projection is taken.*

Theorem 1.7.7 Let T be a complement of S and let $P(\mathbf{u})$ denote the projection of $\mathbf{u}$ into S along T. Then

(i) the projection of $\mathbf{u}$ into T along S is $\mathbf{u}-P(\mathbf{u})$,

(ii) $\mathbf{u} \in S$ iff $P(\mathbf{u}) = \mathbf{u}$,

(iii) $\mathbf{u} \in T$ iff $P(\mathbf{u}) = \mathbf{0}$,

(iv) $P(P(\mathbf{u})) = P(\mathbf{u})$,

(v) $P(\mathbf{u}+\mathbf{v}) = P(\mathbf{u})+P(\mathbf{v})$,

(vi) $P(\alpha\mathbf{u}) = \alpha P(\mathbf{u})$.

Proof (i) Let $\mathbf{u} = \mathbf{x}+\mathbf{y}$ where $\mathbf{x} \in S$ and $\mathbf{y} \in T$. Then $P(\mathbf{u}) = \mathbf{x}$. The projection of $\mathbf{u}$ into T along S is, by definition, $\mathbf{y} = \mathbf{u}-\mathbf{x} = \mathbf{u}-P(\mathbf{u})$.

(ii) If $\mathbf{u} \in S$ then $\mathbf{u} = \mathbf{u}+\mathbf{0}$, $\mathbf{u} \in S$ and $\mathbf{0} \in T$. Hence $P(\mathbf{u}) = \mathbf{u}$. Conversely, if $P(\mathbf{u}) = \mathbf{u}$ then $\mathbf{u} \in S$ since $P(\mathbf{u})$ always belongs to S.

(iii) is easy to prove. Since $P(\mathbf{u}) \in S$, (iv) follows from (ii).

(v) Let $\mathbf{u} = \mathbf{x}_1+\mathbf{y}_1$ and $\mathbf{v} = \mathbf{x}_2+\mathbf{y}_2$ where $\mathbf{x}_1, \mathbf{x}_2 \in S$ and $\mathbf{y}_1, \mathbf{y}_2 \in T$. Then $\mathbf{u}+\mathbf{v} = (\mathbf{x}_1+\mathbf{x}_2)+(\mathbf{y}_1+\mathbf{y}_2)$, $\mathbf{x}_1+\mathbf{x}_2 \in S$ and $\mathbf{y}_1+\mathbf{y}_2 \in T$. So $P(\mathbf{u}+\mathbf{v}) = \mathbf{x}_1+\mathbf{x}_2 = P(\mathbf{u})+P(\mathbf{v})$.

(vi) Let $\mathbf{u} = \mathbf{x}+\mathbf{y}$ where $\mathbf{x} \in S$ and $\mathbf{y} \in T$. Then $\alpha\mathbf{u} = (\alpha\mathbf{x}) + (\alpha\mathbf{y})$, $\alpha\mathbf{x} \in S$ and $\alpha\mathbf{y} \in T$. So $P(\alpha\mathbf{u}) = \alpha\mathbf{x} = \alpha P(\mathbf{u})$. ∎

We now extend the concept of direct sum to several subspaces.

We say that the sum $S_1+\cdots+S_k$ of the subspaces $S_1,\ldots,S_k$ is *direct* (or $S_1,\ldots,S_k$ are *independent*) if any vector $\mathbf{x}$ in $S_1+\cdots+S_k$ can be expressed uniquely as $\mathbf{x}_1+\cdots+\mathbf{x}_k$ with $\mathbf{x}_i \in S_i$, $1 \le i \le k$. Also then we denote $S_1+\cdots+S_k$ by $S_1 \oplus \cdots \oplus S_k$.

Theorem 1.7.8 Let $S_1, S_2, \ldots, S_k$ be subspaces of a vector space.

Then the following statements are equivalent.

(i) $S_1 + \cdots + S_k$ is direct,

(ii) $(S_1 + \cdots + S_i) \cap S_{i+1} = \{\mathbf{0}\}$ for $i = 1, \ldots, k-1$,

(iii) $\mathbf{0} = \mathbf{x}_1 + \cdots + \mathbf{x}_k$, $\mathbf{x}_i \in S_i$ $(i = 1, \ldots, k) \Rightarrow \mathbf{x}_i = \mathbf{0}$ for $i = 1, \ldots, k$,

(iv) $d(S_1 + \cdots + S_k) = d(S_1) + \cdots + d(S_k)$.

Proof (i) $\Rightarrow$ (ii) Fix i, $1 \le i \le k-1$. Let $\mathbf{x} \in (S_1 + \cdots + S_i) \cap S_{i+1}$. Then $\mathbf{x} = \mathbf{x}_1 + \cdots + \mathbf{x}_i$ for some $\mathbf{x}_1 \in S_1, \ldots, \mathbf{x}_i \in S_i$. Hence

$$\begin{aligned} \mathbf{x} &= \mathbf{x}_1 + \cdots + \mathbf{x}_i + \mathbf{0} + \mathbf{0} + \cdots + \mathbf{0} \\ &= \mathbf{0} + \cdots + \mathbf{0} + \mathbf{x} + \mathbf{0} + \cdots + \mathbf{0} \end{aligned}$$

Since $S_1 + \cdots + S_k$ is direct we get $\mathbf{x} = \mathbf{0}$ by comparing the $(i+1)$-th terms. Now (ii) follows.

(ii) $\Rightarrow$ (iii) Let $\mathbf{0} = \mathbf{x}_1 + \cdots + \mathbf{x}_k$ where $\mathbf{x}_i \in S_i$ for $i = 1, \ldots, k$. Then $\mathbf{x}_k = -(\mathbf{x}_1 + \cdots + \mathbf{x}_{k-1}) \in (S_1 + \cdots + S_{k-1}) \cap S_k$, so $\mathbf{x}_k = \mathbf{0}$. Next it can be shown similarly that $\mathbf{x}_{k-1} = \mathbf{0}$. Proceeding thus, we get (iii).

(iii) $\Rightarrow$ (i) The proof of this is similar to that of the corresponding statement in *Theorem* 1.7.2.

(ii) $\Rightarrow$ (iv) Using (ii) for $i = k-1$, $i = k-2$, etc., we get

$$\begin{aligned} d(S_1 + \cdots + S_k) &= d(S_1 + \cdots + S_{k-1}) + d(S_k) \\ &= d(S_1 + \cdots + S_{k-2}) + d(S_{k-1}) + d(S_k) \\ &\cdots \;\; \cdots \\ &= d(S_1) + \cdots + d(S_k) \end{aligned}$$

(iv) $\Rightarrow$ (ii) By the corollary to *Theorem* 1.6.6, we have

$$\begin{aligned} d(S_1 + \cdots + S_k) &\le d(S_1 + \cdots + S_{k-1}) + d(S_k) \\ &\le d(S_1 + \cdots + S_{k-2}) + d(S_{k-1}) + d(S_k) \\ &\cdots \;\; \cdots \\ &\le d(S_1) + \cdots + d(S_k) \end{aligned}$$

By hypothesis, the first and the last terms are equal, hence equality holds throughout. Now by *Theorem* 1.7.2, (ii) follows. ∎

Example 1.7.9 Consider the subspace $S = \{(\xi_1, \xi_2, \xi_3, \xi_4, \xi_5) : 2\xi_1 + \xi_3 + \xi_5 = 0\}$ of $\mathbb{R}^5$. It is easy to check that $\mathbf{x}_1 = (1, 0, -2, 0, 0)$, $\mathbf{x}_2 = (1, 0, 0, 0, -2)$, $\mathbf{x}_3 = (0, 1, 0, 0, 0)$ and $\mathbf{x}_4 = (0, 0, 0, 1, 0)$ form a basis of S. Write $S_1 = \mathrm{Sp}(\{\mathbf{x}_1\})$, $S_2 = \mathrm{Sp}(\{\mathbf{x}_2\})$ and $S_3 = \mathrm{Sp}(\{\mathbf{x}_3, \mathbf{x}_4\})$. Then $S = S_1 + S_2 + S_3$. Further since $d(S) = 4 = d(S_1) + d(S_2) + d(S_3)$ it follows that $S_1 + S_2 + S_3$ is direct.

Suppose next $T_1 = S_1$, $T_2 = \mathrm{Sp}(\{\mathbf{x}_2, \mathbf{x}_1 + \mathbf{x}_4\})$ and $T_3 = S_3$. Clearly then $S = T_1 + T_2 + T_3$. It is also easy to check that T_1, T_2 and T_3 are pair-wise virtually disjoint. However, $T_1 + T_2 + T_3$ is not direct since $d(T_1 + T_2 + T_3) = 5 \neq d(S)$. That the sum is not direct can also be seen from: $\mathbf{0} = \mathbf{x}_1 - (\mathbf{x}_1 + \mathbf{x}_4) + \mathbf{x}_4$, $\mathbf{x}_1 \in T_1$, $\mathbf{x}_1 + \mathbf{x}_4 \in T_2$, $\mathbf{x}_4 \in T_3$ and $\mathbf{x}_1 \neq \mathbf{0}$. ∎

The preceding example shows that $S_i \cap S_j = \{\mathbf{0}\}$ for all $i, j,\ i \neq j$, does not imply that $S_1 + \cdots + S_k$ is direct. However, if $S_1 + \cdots + S_k$ is direct then $(S_{i_1} + \cdots + S_{i_r}) \cap (S_{i_{r+1}} + \cdots + S_{i_s}) = \{\mathbf{0}\}$ for any two disjoint subsets $\{i_1, \ldots, i_r\}$ and $\{i_{r+1}, \ldots, i_s\}$ of $\{1, 2, \ldots, k\}$.

Exercises

1. Let S and T be subspaces of a vector space. Show that $S + T$ is direct iff there exist bases A of S and B of T such that A, B are disjoint and $A \cup B$ is linearly independent (and so a basis of $S + T$). Show also that the words "A, B are disjoint and" cannot be omitted.
2. Show that $S + T$ is direct iff A and B are disjoint for all bases A of S and B of T.
3. Find a complement of the subspace (i) of *Example* 1.3.3 in $\mathbb{R}^n$.
4. Find a complement of the subspace S in V with S and V as in *Exercise* 1.3.1(k).
5. If S and T are two subspaces of a vector space having a common complement W, does it follow that $S = T$? Justify your answer.
6. If $S + T$ is direct, show that there exists a complement of S which contains T.
7. For any two subspaces S and T of a vector space, show that there exists a subspace $W \subseteq T$ such that $S \oplus W = S + T$. Deduce that if $S \subseteq T$, there exists a subspace W such that $S \oplus W = T$.
8. In the vector space $\mathbb{R}^4$, find two different complements of the subspace $S = \{(x_1, x_2, x_3, x_4) : x_3 - x_4 = 0\}$.
*9. Show that a non-trivial subspace S of V has two virtually disjoint complements iff $d(S) \geq d(V)/2$.
10. Show that the subspace S of even polynomials is a complement of the subspace T of odd polynomials in the vector space $\mathcal{P}_n$. When $n = 5$, what is the projection of $3 - t + 2t^2 + t^3 - 5t^4$ into S along T?
11. Let $\mathbf{x}_1 = (2, 0, 1, 3)$, $\mathbf{x}_2 = (0, 3, 1, 1)$ and $\mathbf{x}_3 = (2, -6, -1, 1)$. Find a basis of $S = \mathrm{Sp}(\{\mathbf{x}_1, \mathbf{x}_2, \mathbf{x}_3\})$. Find a complement T of S which contains the vector $\mathbf{u} = (4, 0, 8, 0)$. By modifying T get a complement W of S which does not contain $\mathbf{u}$ and find the projection of $\mathbf{u}$ into S along W.

12. Prove that every complement of S can be obtained as in the proof of *Theorem* 1.7.4.
13. Prove that a subspace S has a unique complement iff $S = \{\mathbf{0}\}$ or $S = V$.
14. If S is a subspace of V and $\mathbf{x}$ and $\mathbf{y}$ are fixed vectors such that $\mathbf{x} \in S$ and $\mathbf{y} \notin S$, show that there exists a complement T of S such that $\mathbf{x}$ is the projection of $\mathbf{y}$ into S along T.
15. If $S_i \neq \{0\}$ for all i, show that $S_1 + \cdots + S_k$ is direct iff $\mathbf{x}_1, \ldots, \mathbf{x}_k$ are linearly independent whenever $\mathbf{x}_i$ is a non-zero vector from S_i, $i = 1, \ldots, k$.
16. Prove that if one vector in $S_1 + \cdots + S_k$ can be expressed uniquely as $\mathbf{x}_1 + \cdots + \mathbf{x}_k$ with $\mathbf{x}_i \in S_i$ for $i = 1, \ldots, k$, then $S_1 + \cdots + S_k$ is direct.
17. Consider the subspaces

$$\begin{aligned} S_1 &= \{(\alpha, \beta, \alpha, \beta, -2\alpha - 2\beta) : \alpha, \beta \in \mathbb{R}\} \\ S_2 &= \{(\alpha, \alpha, \beta, \beta, -2\alpha - 2\beta) : \alpha, \beta \in \mathbb{R}\} \\ S_3 &= \{(\alpha, \beta, \beta, 2\beta - \alpha, -4\beta) : \alpha, \beta \in \mathbb{R}\} \\ S_4 &= \{(0, \alpha, 0, \beta, -\alpha - \beta) : \alpha, \beta \in \mathbb{R}\} \end{aligned}$$

of $\mathbb{R}^5$. Find an ordered basis of $S_1 + \cdots + S_4$ such that the first r_i vectors form a basis of $S_1 + \cdots + S_i$ (for some r_i) for each i.
18. If S and Y are subspaces such that $S \subseteq Y$ and Z is a complement of S, show using *Exercise* 1.6.6 that $Z \cap Y$ is a complement of S relative to Y (i.e., $S \oplus (Z \cap Y) = Y$).

1.8 Isomorphism

Consider the vector spaces $\mathbb{R}^n$ and $\mathcal{P}_n$. Though these are different, they have the same structure as explained below. We can associate the element $(\alpha_0, \alpha_1, \ldots, \alpha_{n-1})$ of $\mathbb{R}^n$ with the element $\sum_{i=0}^{n-1} \alpha_i t^i$ of $\mathcal{P}_n$. Note that this is a one-to-one correspondence. Moreover, this mapping preserves vector addition and scalar multiplication. Thus $\mathcal{P}_n$ and $\mathbb{R}^n$ have the same structure as vector spaces and we say they are isomorphic. We formalize this concept in the following

Definition 1.8.1 Let V_1 and V_2 be two vector spaces over the same field F. Then a map φ from V_1 to V_2 is said to be an *isomorphism* if

(i) φ is a bijection (i.e., φ is one-to-one and onto),

(ii) $\varphi(\mathbf{x} + \mathbf{y}) = \varphi(\mathbf{x}) + \varphi(\mathbf{y})$ for all $\mathbf{x}, \mathbf{y} \in V_1$, and

(iii) $\varphi(\alpha\mathbf{x}) = \alpha\,\varphi(\mathbf{x})$ for all $\alpha \in F$ and $\mathbf{x} \in V_1$.

V_1 is said to be *isomorphic* to V_2 if there is an isomorphism from V_1 to V_2. Also then we write $V_1 \simeq V_2$.

Suppose V_1 is isomorphic to V_2 and φ is an isomorphism from V_1 to V_2. Then it is easy to see that the inverse map φ^{-1} (it exists because φ is a bijection) is an isomorphism from V_2 to V_1, so V_2 is isomorphic to V_1 and we say that V_1 and V_2 are isomorphic.

It is easy to prove that if φ is an isomorphism, $\varphi(\mathbf{0}) = \mathbf{0}$ and $\varphi(-\mathbf{x}) = -\varphi(\mathbf{x})$. Moreover, if $\varphi(\mathbf{x}) = \mathbf{0}$ then $\mathbf{x} = \mathbf{0}$ since φ^{-1} is an isomorphism.

The conditions (ii) and (iii) of the preceding definition together can be written in several equivalent forms, e.g.:

$$\varphi(\alpha\mathbf{x} + \mathbf{y}) = \alpha\varphi(\mathbf{x}) + \varphi(\mathbf{y}) \quad \text{for all } \mathbf{x}, \mathbf{y} \in V_1 \text{ and all } \alpha \in F,$$
$$\varphi(\alpha\mathbf{x} + \beta\mathbf{y}) = \alpha\varphi(\mathbf{x}) + \beta\varphi(\mathbf{y}) \quad \text{for all } \mathbf{x}, \mathbf{y} \in V_1 \text{ and all } \alpha \in F$$

We leave the proof of this to the reader. The last condition can also be extended to finitely many vectors in V_1. A map φ satisfying these conditions is said to be *linear*. Such maps are of interest by themselves and will be studied in *Chapter* 2.

Theorem 1.8.2 Two vector spaces over a field F are isomorphic iff they have the same dimension.

Proof *If part* Let V_1 and V_2 be two n-dimensional vector spaces over F. Let $\{\mathbf{x}_1, \ldots, \mathbf{x}_n\}$ and $\{\mathbf{y}_1, \ldots, \mathbf{y}_n\}$ be bases of V_1 and V_2 respectively. Define a map φ from V_1 to V_2 thus: $\varphi(\mathbf{x}) = \alpha_1\mathbf{y}_1 + \cdots + \alpha_n\mathbf{y}_n$, where $\mathbf{x} = \alpha_1\mathbf{x}_1 + \cdots + \alpha_n\mathbf{x}_n$. This map is well-defined, 1-1 and onto since a vector can be expressed uniquely as a linear combination from a basis. Next let $\mathbf{u} = \alpha_1\mathbf{x}_1 + \cdots + \alpha_n\mathbf{x}_n$ and $\mathbf{v} = \beta_1\mathbf{x}_1 + \cdots + \beta_n\mathbf{x}_n$ be arbitrary vectors in V_1. Then

$$\varphi(\gamma\mathbf{u} + \mathbf{v}) = (\gamma\alpha_1 + \beta_1)\mathbf{y}_1 + \cdots + (\gamma\alpha_n + \beta_n)\mathbf{y}_n = \gamma\varphi(\mathbf{u}) + \varphi(\mathbf{v})$$

Thus φ is an isomorphism between V_1 and V_2.

Only if part Let V_1 and V_2 be isomorphic vector spaces over F. Let $\{\mathbf{x}_1, \ldots, \mathbf{x}_m\}$ be a basis of V_1 and let φ be an isomorphism from V_1 to V_2. Then

$$\begin{aligned}\alpha_1\varphi(\mathbf{x}_1) + \cdots + \alpha_m\varphi(\mathbf{x}_m) = \mathbf{0} &\Rightarrow \varphi(\alpha_1\mathbf{x}_1 + \cdots + \alpha_m\mathbf{x}_m) = \mathbf{0}\\ &\Rightarrow \alpha_1\mathbf{x}_1 + \cdots + \alpha_m\mathbf{x}_m = \mathbf{0}\\ &\Rightarrow \alpha_1 = \cdots = \alpha_m = 0\end{aligned}$$

Hence $\varphi(\mathbf{x}_1), \ldots, \varphi(\mathbf{x}_m)$ are linearly independent and $d(V_2) \geq d(V_1)$. The opposite inequality follows similarly. ∎

Remark Our proof of the *if part* actually shows that for any bases $\{\mathbf{x}_1, \ldots, \mathbf{x}_n\}$ and $\{\mathbf{y}_1, \ldots, \mathbf{y}_n\}$ of V_1 and V_2 respectively, there is an isomorphism from V_1 to V_2 which takes $\mathbf{x}_i$ to $\mathbf{y}_i$ for $i = 1, \ldots, n$.

It is immediate from the preceding theorem that any n-dimensional vector space over F is isomorphic to F^n.

We now mention some more examples of isomorphic vector spaces. $\mathbb{C}$ over $\mathbb{R}$ and $\mathbb{R}^2$ are isomorphic (as vector spaces). Let X be a finite set of size n. Then F^X is isomorphic to F^n. Consider the vector space V in *Example* 1.2.9 when Ω is a finite set of size n. Then V is isomorphic to F^n where $F = \mathbb{Z}_2$. To see this, let $\Omega = \{\omega_1, \omega_2, \ldots, \omega_n\}$. Define the map φ from V to F^n as: $\varphi(A) = (\xi_1, \xi_2, \ldots, \xi_n)$ where ξ_i is 1 if $\omega_i \in A$ and 0 otherwise. Then it is easy to check that φ is an isomorphism.

We can also consider an isomorphism from a vector space onto itself. For example, rotation about the origin (see *Figure* 1.8.1) by an angle θ is an isomorphism from $\mathbb{R}^2$ to itself. To prove this we first note that this map f takes the point $P = (\xi_1, \xi_2)$ to the point $Q = (\xi_1 \cos\theta - \xi_2 \sin\theta,\ \xi_1 \sin\theta + \xi_2 \cos\theta)$. To prove that f is one-to-one and onto, we obtain its inverse. From geometry, it is clear that rotation about the origin by the angle $-\theta$ is the inverse of f. In fact, the inverse image of (η_1, η_2) under f is $(\eta_1 \cos\theta + \eta_2 \sin\theta, -\eta_1 \sin\theta + \eta_2 \cos\theta)$. Thus f is a bijection. Now,

$$\begin{aligned} f(\alpha(\xi_1, \xi_2) + (\zeta_1, \zeta_2)) &= f((\alpha\xi_1 + \zeta_1, \alpha\xi_2 + \zeta_2)) \\ &= ((\alpha\xi_1 + \zeta_1)\cos\theta - (\alpha\xi_2 + \zeta_2)\sin\theta, \\ &\qquad (\alpha\xi_1 + \zeta_1)\sin\theta + (\alpha\xi_2 + \zeta_2)\cos\theta) \\ &= \alpha(\xi_1 \cos\theta - \xi_2 \sin\theta,\ \xi_1 \sin\theta + \xi_2 \cos\theta) \\ &\quad + (\zeta_1 \cos\theta - \zeta_2 \sin\theta,\ \zeta_1 \sin\theta + \zeta_2 \cos\theta) \\ &= \alpha f((\xi_1, \xi_2)) + f((\zeta_1, \zeta_2)) \end{aligned}$$

Hence f is linear and is an isomorphism from $\mathbb{R}^2$ to itself.

Another example of an isomorphism from F^n to itself is obtained as follows. Let $(i_1, i_2, \ldots, i_n)$ be a fixed permutation of $(1, 2, \ldots, n)$. Then the map φ defined by

$$\varphi((x_1, x_2, \ldots, x_n)) = (x_{i_1}, x_{i_2}, \ldots, x_{i_n})$$

is an isomorphism from F^n onto itself. We call such a φ a *permutation transformation*.

The importance of isomorphism lies in the fact that an isomorphic image of a vector space may be easier to study or to visualize than

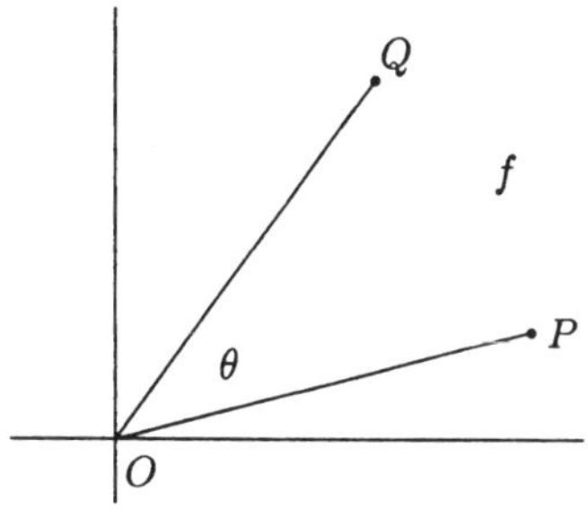

Figure 1.8.1

the original vector space. For example, $\mathbb{R}^n$ is easier to visualize than a general n-dimensional real vector space. However it is important to note that not all properties of a vector space are reflected in its isomorphic images. For one thing, to study a vector space through its isomorphic image, we need to fix a basis of the vector space. Secondly there may be operations and properties which are specific to a given vector space but are not relevant in an isomorphic image. For example, the product of two elements of $\mathcal{P}_n$ is defined in a natural way but has no counterpart in $\mathbb{R}^n$. Similarly angles and distances may not be preserved by an isomorphism from $\mathbb{R}^n$ to itself.

Of course, all the concepts and properties defined only through vector addition and scalar multiplication are valid in any isomorphic image of a vector space (see *Exercise* 1.8.2).

Exercises

1. Let φ be an isomorphism. Then prove that $\varphi(\mathbf{x}) = \mathbf{0}$ iff $\mathbf{x} = \mathbf{0}$ and that $\varphi(-\mathbf{x}) = -\varphi(\mathbf{x})$.
2. Let φ be an isomorphism from V_1 onto V_2. Prove the following:
   (a) If $\mathbf{x}_1, \mathbf{x}_2, \ldots, \mathbf{x}_k$ are linearly independent vectors in V_1 then $\varphi(\mathbf{x}_1)$, $\varphi(\mathbf{x}_2), \ldots, \varphi(\mathbf{x}_k)$ are linearly independent in V_2.
   (b) If $\mathbf{x}_1, \mathbf{x}_2, \ldots, \mathbf{x}_k$ generate S in V_1 then $\varphi(\mathbf{x}_1), \varphi(\mathbf{x}_2), \ldots, \varphi(\mathbf{x}_k)$ generate $\varphi(S)$ in V_2.
   (c) If S is a subspace of V_1 then $\varphi(S)$ is a subspace of V_2 with the same dimension.

(d) If T is a complement of S in V_1 and $\mathbf{x}$ is the projection of $\mathbf{u}$ into S along T, then $\varphi(T)$ is a complement of $\varphi(S)$ in V_2 and $\varphi(\mathbf{x})$ is the projection of $\varphi(\mathbf{u})$ into $\varphi(S)$ along $\varphi(T)$.

3. Show that each of the following maps is an isomorphism from F^n to itself.

   (a) $\varphi((x_1, x_2, \ldots, x_n)) = (\alpha x_1, x_2, \ldots, x_n)$ where α is a fixed non-zero scalar.

   (b) $\varphi((x_1, x_2, \ldots, x_n)) = (x_1 + \beta x_2, x_2, \ldots, x_n)$ where β is a fixed scalar.

4. Prove that if V_1 and V_2 are isomorphic and V_2 and V_3 are isomorphic then V_1 and V_3 are isomorphic. Deduce that being isomorphic is an equivalence relation.

5. Show that the map f:
$$(x_1, x_2) \mapsto (3x_1 + x_2,\ x_1 - 2x_2)$$
is an isomorphism from $\mathbb{R}^2$ to itself. What is the distance between $f(0,0)$ and $f(1,0)$? Does f take points on every straight line to those on another straight line? (Consider lines passing through the origin as well as those not passing through the origin.) Show that f takes a parallelogram to a parallelogram.

*6. Prove that the map φ from the power set of Ω to F^n given on page 60 is indeed an isomorphism.

*7. Is the vector space of *Example* 1.2.9 isomorphic to a subspace of $\mathbb{R}^\Omega$?

8. Find an isomorphism from the vector space of *Example* 1.2.3 to $\mathbb{R}^{2n}$.

9. Find an isomorphism between the vector space of *Example* 1.2.4 and $\mathbb{R}^1$.

10. Show that the vector space of *Exercise* 1.2.4 is isomorphic to $\mathbb{R}^1$.

11. Show that $f : (x_1, x_2, \ldots, x_n) \mapsto (\overline{x}_1, \overline{x}_2, \ldots, \overline{x}_n)$ is not an isomorphism from $\mathbb{C}^n$ to $\mathbb{C}^n$ but is an isomorphism from the vector space $\mathbb{C}^n$ over $\mathbb{R}$ to itself. Here $\overline{x}$ denotes the complex conjugate of x.

12. Call a sequence $(a_1, a_2, a_3, \ldots)$ of real numbers a *Fibonacci sequence* if $a_n = a_{n-1} + a_{n-2}$ for all $n \geq 3$. Show that the set of all Fibonacci sequences forms a vector space under component-wise addition and scalar multiplication defined in a natural way and that it is isomorphic to $\mathbb{R}^2$.

*13. Let V be a finite-dimensional vector space over F. A *linear functional on* V is a map $\mathbf{y} : V \to F$ such that
$$\mathbf{y}(\mathbf{x}_1 + \mathbf{x}_2) = \mathbf{y}(\mathbf{x}_1) + \mathbf{y}(\mathbf{x}_2) \text{ and } \mathbf{y}(\alpha\mathbf{x}_1) = \alpha\mathbf{y}(\mathbf{x}_1)$$

for all $\mathbf{x}_1, \mathbf{x}_2 \in V$ and $\alpha \in F$. (See also *Example* 2.2.4.) We define the sum of two linear functionals and a scalar multiple of a linear functional by

$$\begin{aligned}(\mathbf{y}_1 + \mathbf{y}_2)(\mathbf{x}) &= \mathbf{y}_1(\mathbf{x}) + \mathbf{y}_2(\mathbf{x}) \quad &\text{for all } \mathbf{x} \in V \\ (\alpha\mathbf{y})(\mathbf{x}) &= \alpha \cdot \mathbf{y}(\mathbf{x}) \quad &\text{for all } \mathbf{x} \in V.\end{aligned}$$

(a) Prove that the set V' of all linear functionals on V forms a vector space over F. V' is called the *dual* of V.

(b) Let $\mathcal{X} = \{\mathbf{x}_1, \mathbf{x}_2, \ldots, \mathbf{x}_n\}$ be a basis of V and $\beta_1, \beta_2, \ldots, \beta_n \in F$. Then show that there is a unique $\mathbf{y} \in V'$ such that $\mathbf{y}(\mathbf{x}_i) = \beta_i$ for $i = 1, \ldots, n$. Deduce that there exists a unique basis $\mathcal{X}' = \{\mathbf{y}_1, \mathbf{y}_2, \ldots, \mathbf{y}_n\}$ of V' such that $\mathbf{y}_i(\mathbf{x}_j) = 1$ or 0 according as $i = j$ or $i \neq j$.

(c) If $\mathbf{x} \in V$ and $\mathbf{x} \neq \mathbf{0}$ then show that there exists a $\mathbf{y} \in V'$ such that $\mathbf{y}(\mathbf{x}) \neq 0$.

(d) With $\mathbf{x} \in V$, associate the element $\widehat{\mathbf{x}}$ of $(V')'$ defined as $\widehat{\mathbf{x}}(\mathbf{y}) = \mathbf{y}(\mathbf{x})$. Show that $\mathbf{x} \mapsto \widehat{\mathbf{x}}$ is an isomorphism between V and V''. (Hint: prove that the map is 1-1 and linear. Use dimensions to get 'onto'.) This is called the *natural isomorphism*. We identify V'' with V and say that V is reflexive.

(e) If S is a subset of V, the *annihilator of* S is defined as

$$S^0 = \{\mathbf{y} \in V' : \mathbf{y}(\mathbf{x}) = 0 \text{ for all } \mathbf{x} \in S\}$$

Prove that S^0 is a subspace of V'.

(f) If S and T are subspaces of V then show that $(S+T)^0 = S^0 \cap T^0$. Deduce that $(S \cap T)^0 = S^0 + T^0$ by using *Theorem* 1.5.16.

(g) If $d(V) = n$ and S is an m-dimensional subspace of V then prove that $d(S^0) = n - m$. (Hint: Take a basis of S, extend it to a basis $\mathcal{X}$ of V and consider the basis $\mathcal{X}'$ of V' as in (b).) Deduce that $S \simeq (S^0)^0$ under the natural isomorphism.

1.9 Quotient space*

In *Section* 1.7, we saw that a subspace S of a vector space V can have many complements. In this section, we study a vector space defined in a unique and natural manner for given V and S and which, in a sense, behaves like a complement of S. We finally take a brief look at 'external direct sum'.

We start with a simple example. Let $V = \mathbb{R}^2$ and $S =$ the x_1-axis. Consider the set W consisting of the x_1-axis and all lines parallel to it.

Notice that every such line can be written as $\mathbf{x}+S$ for some $\mathbf{x}$, e.g., the line $x_2 = c$ is $(0,c)+S$. Of course such a representation is not unique since $(a,c)+S = (b,c)+S$ for any a and b. In fact, $\mathbf{x}+S = \mathbf{y}+S$ for any $\mathbf{y} \in \mathbf{x}+S$. Now, using the operations in V, we try to define, in a natural way, operations in W so that it becomes a vector space over $\mathbb{R}$. Consider the lines $\mathbf{x}+S$ and $\mathbf{y}+S$ (containing $\mathbf{x}$ and $\mathbf{y}$ respectively). We define their sum to be the line containing $\mathbf{x}+\mathbf{y}$, i.e., $(\mathbf{x}+\mathbf{y})+S$. We similarly define α times the line $\mathbf{x}+S$ to be the line containing $\alpha\mathbf{x}$, i.e., $(\alpha\mathbf{x})+S$. It is easy to check that these operations on W are well-defined and that W is a vector space. We now show that every complement of S is isomorphic to W under a natural isomorphism. Fix any complement T of S. Then T is a line through the origin different from the x_1-axis. Let φ be the map taking the point $\mathbf{x}$ of T to the line parallel to the x_1-axis and containing $\mathbf{x}$. It is easy to see that φ is an isomorphism from T onto W.

We shall now generalize the concepts of the above example to a general vector space.

Definition 1.9.1 Let V be a vector space over F and let S be a subspace of V. A subset of V of the form $\mathbf{x}+S$ is called a *coset of* S. Let W be the set of all distinct cosets of S. Define addition and scalar multiplication on W by:

$$(\mathbf{x}+S)+(\mathbf{y}+S) = (\mathbf{x}+\mathbf{y})+S \tag{1.9.1}$$

and

$$\alpha(\mathbf{x}+S) = (\alpha\mathbf{x})+S \tag{1.9.2}$$

Then W becomes a vector space over F, known as the *quotient of* V *and* S or the *quotient space of* V *modulo* S and is denoted V/S.

We have to prove that the operations in V/S are well-defined and that V/S is a vector space. For this we need the following result which is an immediate consequence of *Theorem* 1.6.10.

Lemma Let $\mathbf{x},\mathbf{y} \in V$. Then the following statements are equivalent: (a) $\mathbf{x}+S = \mathbf{y}+S$, (b) $\mathbf{x} \in \mathbf{y}+S$ (or $\mathbf{y} \in \mathbf{x}+S$) and (c) $\mathbf{x}-\mathbf{y} \in S$.

Theorem 1.9.2 The operations in V/S (viz., (1.9.1) and (1.9.2)) are well-defined and V/S is a vector space over F.

Proof To prove that the sum is well-defined, suppose $\mathbf{x}+S = \mathbf{x}'+S$ and $\mathbf{y}+S = \mathbf{y}'+S$. Then $\mathbf{x}-\mathbf{x}' \in S$ and $\mathbf{y}-\mathbf{y}' \in S$, so $(\mathbf{x}+\mathbf{y})-(\mathbf{x}'+\mathbf{y}') \in$

S and $(\mathbf{x}+\mathbf{y})+S = (\mathbf{x}'+\mathbf{y}')+S$. Similarly it can be proved that scalar multiplication is well-defined.

That V/S is a vector space is easy to verify. We only mention that $S = \mathbf{0}+S$ is the zero vector and the negative of $\mathbf{x}+S$ is $(-\mathbf{x})+S$. ∎

Theorem 1.9.3 Let S be a subspace of V. Then $\mathbf{x}_1+S, \ldots, \mathbf{x}_k+S$ form a basis of V/S iff $\mathbf{x}_1, \ldots, \mathbf{x}_k$ form a basis of some complement of S.

Proof *If part* Let $\mathbf{x}_{k+1}, \ldots, \mathbf{x}_n$ form a basis of S. Then $\mathbf{x}_1, \ldots, \mathbf{x}_n$ form a basis of V. So every $\mathbf{x}$ in V is $\sum_{i=1}^n \alpha_i \mathbf{x}_i$ for some scalars $\alpha_1, \ldots, \alpha_n$. Now

$$\mathbf{x}+S = \sum_{i=1}^k \alpha_i \mathbf{x}_i + S = \sum_{i=1}^k \alpha_i(\mathbf{x}_i + S)$$

Thus $\mathbf{x}_1+S, \ldots, \mathbf{x}_k+S$ generate V/S. Suppose next $\sum_{i=1}^k \alpha_i(\mathbf{x}_i+S) = S$ (note that the null vector in V/S is S). Then $\sum_{i=1}^k \alpha_i \mathbf{x}_i \in S$. Since $\mathbf{x}_1, \ldots, \mathbf{x}_k$ belong to a complement of S, it follows that $\sum_{i=1}^k \alpha_i \mathbf{x}_i = \mathbf{0}$ and $\alpha_1 = \cdots = \alpha_k = 0$. Thus $\mathbf{x}_1+S, \ldots, \mathbf{x}_k+S$ are linearly independent and form a basis of V/S.

Only if part Let $T = \mathrm{Sp}(\{\mathbf{x}_1, \ldots, \mathbf{x}_k\})$. Suppose $\sum_{i=1}^k \alpha_i \mathbf{x}_i \in S$. Then $\sum_{i=1}^k \alpha_i(\mathbf{x}_i + S) = S$. Since S is the null vector it follows that $\alpha_1 = \cdots = \alpha_k = 0$. Thus $S \cap T = \{\mathbf{0}\}$ and $\mathbf{x}_1, \ldots, \mathbf{x}_k$ are linearly independent. Next let $\mathbf{x} \in V$. Then $\mathbf{x}+S = \sum_{i=1}^k \alpha_i(\mathbf{x}_i+S)$ for some $\alpha_1, \ldots, \alpha_k$. So $\mathbf{x} = \sum_{i=1}^k \alpha_i \mathbf{x}_i + \mathbf{s}$ for some $\mathbf{s} \in S$. Hence $\mathbf{x} \in S+T$. Thus $S \oplus T = V$ and $\mathbf{x}_1, \ldots, \mathbf{x}_k$ form a basis of T. ∎

Corollary $d(V/S) = d(V) - d(S)$.

Theorem 1.9.4 Every complement T of S is isomorphic to V/S under the natural isomorphism $\mathbf{x} \mapsto \mathbf{x}+S$.

Proof If $\mathbf{x}, \mathbf{y} \in T$ and $\mathbf{x}+S = \mathbf{y}+S$ then $\mathbf{x}-\mathbf{y} \in S \cap T$, so $\mathbf{x} = \mathbf{y}$. Thus the map is 1-1. To prove that the map is onto, consider $\mathbf{x}+S$ where $\mathbf{x} \in V$. Then $\mathbf{x} = \mathbf{s}+\mathbf{t}$ for some $\mathbf{s} \in S$ and $\mathbf{t} \in T$. Now $\mathbf{x}+S = \mathbf{t}+S$ and the map is onto. The rest is easy to prove. ∎

From the preceding theorem it follows that $\{\mathbf{x}+S : \mathbf{x} \in T\}$ is the same for all complements T of S and is V/S. In this sense V/S may be treated as a canonical representation of all the complements of S.

We now discuss briefly the concept of 'external direct sum'. Suppose

V_1 and V_2 are two vector spaces over the same field F. Then we can form a new vector space on $V_1 \times V_2$ as follows: define $(\mathbf{u}_1, \mathbf{u}_2) + (\mathbf{v}_1, \mathbf{v}_2) = (\mathbf{u}_1 + \mathbf{v}_1, \mathbf{u}_2 + \mathbf{v}_2)$ and $\alpha(\mathbf{u}_1, \mathbf{u}_2) = (\alpha\mathbf{u}_1, \alpha\mathbf{u}_2)$. Then it is easy to verify that $V_1 \times V_2$ becomes a vector space over F, called the *external direct sum of* V_1 *and* V_2 and denoted $V_1 \oplus V_2$. To distinguish this from the direct sum considered in *Section* 1.7, we may call the latter 'internal direct sum'.

It is easy to see that $\mathbf{u}_1 \mapsto (\mathbf{u}_1, \mathbf{0})$ is an isomorphism from V_1 to the subspace $V_1' = \{(\mathbf{u}_1, \mathbf{u}_2) : \mathbf{u}_2 = \mathbf{0}\}$ of $V_1 \oplus V_2$ and, similarly, $\mathbf{u}_2 \mapsto (\mathbf{0}, \mathbf{u}_2)$ is an isomorphism from V_2 to the subspace $V_2' = \{(\mathbf{u}_1, \mathbf{u}_2) : \mathbf{u}_1 = \mathbf{0}\}$ of $V_1 \oplus V_2$. It is also clear that $V_1 \oplus V_2$ is the internal direct sum of V_1' and V_2'. Hence we identify $\mathbf{u}_1$ with $(\mathbf{u}_1, \mathbf{0})$ and $\mathbf{u}_2$ with $(\mathbf{0}, \mathbf{u}_2)$ and then the external direct sum can also be viewed as an internal direct sum and there is no need to distinguish between the two concepts.

We also note that each coset of V_1' in $V_1 \oplus V_2$ is of the form $\{(\mathbf{u}_1, \mathbf{u}_2) : \mathbf{u}_1 \in V_1\}$ for some $\mathbf{u}_2 \in V_2$ and corresponds naturally to $\mathbf{u}_2$. Thus $(V_1 \oplus V_2)/V_1'$ corresponds in a natural way to V_2. Similarly $(V_1 \oplus V_2)/V_2'$ corresponds in a natural way to V_1.

Exercises

1. Prove *Theorem* 1.9.2 completely.
2. If S is a subspace of V, show that $\sim$ is an equivalence relation on V, where $\mathbf{x} \sim \mathbf{y}$ iff $\mathbf{x} - \mathbf{y} \in S$. What are the equivalence classes?
3. Consider the vector space $V = \mathcal{P}_6$ over $\mathbb{R}$. Let S be the subspace of all even polynomials. Show that $V/S = \{\mathcal{P}_{\alpha_1,\alpha_3,\alpha_5} : \alpha_1, \alpha_3, \alpha_5 \in \mathbb{R}\}$ where, for any fixed $\alpha_1, \alpha_3, \alpha_5$,
$$\mathcal{P}_{\alpha_1,\alpha_3,\alpha_5} = \left\{ \sum_{i=0}^{5} \alpha_i t^i : \alpha_0, \alpha_2, \alpha_4 \in \mathbb{R} \right\}$$
4. (a) Show that every subspace of V/S is of the form $\{\mathbf{w} + S : \mathbf{w} \in W\}$ for some subspace W of V and conversely.

   (b) In (a), show that (i) W can be chosen to contain S, and (ii) W can be chosen to be virtually disjoint from S.
5. Show that $\mathbf{x}_1 + S, \ldots, \mathbf{x}_k + S$ are linearly independent in V/S iff $\mathbf{x}_1, \ldots, \mathbf{x}_k$ are linearly independent and $\mathrm{Sp}(\{\mathbf{x}_1, \ldots, \mathbf{x}_k\}) \cap S = \{\mathbf{0}\}$ or, equivalently, $d(\mathrm{Sp}(S \cup \{\mathbf{x}_1, \ldots, \mathbf{x}_k\})) = d(S) + k$.
6. Prove the modular law (*Theorem* 1.6.6) by showing that $T/(S \cap T)$ is isomorphic to $(S + T)/S$.

Chapter 2
Algebra of matrices

2.1 Introduction

Linear Algebra is mainly the study of linear transformations on vector spaces and matrices which can be used to represent them. These occur in almost every branch of knowledge. One can cite any number of examples: rotations, reflections, projections, differentiation and integration operators, rigid body motion, shear transformation, linear models occurring in Statistics, Economics, etc. They can also be used as approximations to certain non-linear transformations.

In this chapter we introduce matrices as representations of linear transformations and define and study the basic operations on them.

2.2 Linear transformations and matrices

In *Section* 1.8, we studied maps φ from a vector space V_1 to a vector space V_2 having the following two properties:

(i) $\varphi(\mathbf{x}+\mathbf{y}) = \varphi(\mathbf{x}) + \varphi(\mathbf{y})$ and $\varphi(\alpha\mathbf{x}) = \alpha\varphi(\mathbf{x})$ for all $\mathbf{x}, \mathbf{y} \in V_1$ and all scalars α,

(ii) φ is a bijection.

We now study the wider class of maps obtained by omitting condition (ii).

Definition 2.2.1 Let V_1 and V_2 be vector spaces over a field F. A map φ from V_1 to V_2 is said to be a *linear transformation* if it satisfies condition (i) above. Then we also say that φ is linear. A linear transformation from V to itself is called a *linear operator on* V.

It is easy to see that $\varphi : V_1 \to V_2$ is linear iff

$$\varphi(\alpha_1\mathbf{x}_1 + \cdots + \alpha_k\mathbf{x}_k) = \alpha_1\varphi(\mathbf{x}_1) + \cdots + \alpha_k\varphi(\mathbf{x}_k)$$

for all $\mathbf{x}_1, \ldots, \mathbf{x}_k \in V_1$ and $\alpha_1, \ldots, \alpha_k \in F$, where k is any positive integer. A linear transformation preserves the two operations in a vector space and is the analogue of a homomorphism for vector spaces.

We now give several examples of linear transformations. It is trivial to verify that the map 0 which takes every vector in V_1 to $\mathbf{0}$ of V_2 is a linear transformation. This is called the *null transformation*. Also the identity map on any vector space V is a linear operator on V. We give below some non-trivial examples.

Example 2.2.2 The rotation R_θ of the plane by an angle θ about the origin is a linear operator on $\mathbb{R}^2$, see *Section* 1.8. It takes the point (ξ_1, ξ_2) to $(\xi_1 \cos\theta - \xi_2 \sin\theta,\ \xi_1 \sin\theta + \xi_2 \cos\theta)$.

The reflection M_m of the plane in the line $y = mx$ (imagine a two-sided mirror placed along this line) is a linear operator on $\mathbb{R}^2$. The image of (ξ_1, ξ_2) under this map is

$$\left(\frac{2m\xi_2 - (m^2-1)\xi_1}{1+m^2}, \frac{2m\xi_1 + (m^2-1)\xi_2}{1+m^2}\right)$$

(What is M_∞, the reflection in the line $x = 0$?)

Both R_θ and M_m are special cases of the map φ from $\mathbb{R}^2$ to itself defined as

$$\varphi(\xi_1, \xi_2) = (a\xi_1 + b\xi_2,\ c\xi_1 + d\xi_2)$$

where a, b, c, d are any fixed real numbers. We now prove that φ is a linear transformation. Let (ξ_1, ξ_2) and (η_1, η_2) be any two points in the plane and let α be a real number. Then

$$\begin{aligned}
&\varphi((\xi_1, \xi_2) + \alpha(\eta_1, \eta_2)) \\
&\quad = \varphi(\xi_1 + \alpha\eta_1,\ \xi_2 + \alpha\eta_2) \\
&\quad = (a(\xi_1 + \alpha\eta_1) + b(\xi_2 + \alpha\eta_2), c(\xi_1 + \alpha\eta_1) + d(\xi_2 + \alpha\eta_2)) \\
&\quad = (a\xi_1 + b\xi_2, c\xi_1 + d\xi_2) + \alpha(a\eta_1 + b\eta_2, c\eta_1 + d\eta_2) \\
&\quad = \varphi((\xi_1, \xi_2)) + \alpha\varphi((\eta_1, \eta_2))
\end{aligned}$$

Thus, φ is a linear transformation. We note that R_θ and M_m are isomorphisms but rotation of the plane by a point different from the origin and reflection of the plane in any line not passing through the origin are *not even* linear transformations.

Example 2.2.3 The projector (or projection operator) into the ξ_1-axis along the ξ_2-axis is a linear operator on $\mathbb{R}^2$. It takes the vector (ξ_1, ξ_2) to $(\xi_1, 0)$.

More generally, the projector P into a subspace S of V along a complement T is a linear operator on V (see *Theorem* 1.7.7 (v) and (vi)).

Projectors form an important class of linear transformations as we will see later.

Example 2.2.4 The map φ from F^n to $F(= F^1)$ defined by

$$\varphi((x_1, x_2, \ldots, x_n)) = a_1x_1 + a_2x_2 + \cdots + a_nx_n$$

is a linear transformation for any fixed scalars $a_1, a_2, \ldots, a_n$. Such a linear transformation is called a *linear functional on* F^n or a *linear form* in $x_1, x_2, \ldots, x_n$. (See also *Exercise* 1.8.13.) We note that linear functionals occur in many real life situations. For example, if a_i denotes the price per unit of the i-th commodity and x_i is the amount of the i-th commodity purchased then $a_1x_1 + a_2x_2 + \cdots + a_nx_n$ is the total cost.

Example 2.2.5 Let $m < n$. Then the map $\varphi : F^n \to F^m$ defined by

$$\varphi((x_1, x_2, \ldots, x_n)) = (x_1, x_2, \ldots, x_m)$$

and the map $\psi : F^m \to F^n$ defined by

$$\psi((x_1, x_2, \ldots, x_m)) = (x_1, x_2, \ldots, x_m, 0, \ldots, 0)$$

are linear transformations.

Example 2.2.6 Differentiation is a linear transformation from $\mathcal{P}_n$ to $\mathcal{P}_{n-1}$. Also, integration defined by

$$\int (p_0 + p_1t + \cdots + p_{n-2}t^{n-2})\, dt = p_0t + \frac{p_1}{2}t^2 + \cdots + \frac{p_{n-2}}{n-1}t^{n-1}$$

is a linear transformation from $\mathcal{P}_{n-1}$ to $\mathcal{P}_n$. Note the absence of the constant term on the right hand side.

Example 2.2.7 A permutation transformation is a linear transformation as noted in *Section* 1.8.

We now study some basic operations on linear transformations.

Theorem 2.2.8 Let f, g be linear transformations from V_1 to V_2 and let h be a linear transformation from V_2 to V_3, where V_1, V_2, V_3 are vector spaces over the same field F. Then

(i) αf and $f + g$ are linear transformations from V_1 to V_2 and

(ii) $h \circ f$ is a linear transformation from V_1 to V_3.

Proof (i) is easy. To prove (ii), let $\mathbf{x}, \mathbf{y}$ be arbitrary vectors in V_1 and let $\alpha \in F$. Then $f(\mathbf{x}), f(\mathbf{y})$ belong to V_2 and

$$\begin{aligned}(h \circ f)(\mathbf{x} + \alpha\mathbf{y}) &= h(f(\mathbf{x} + \alpha\mathbf{y})) \\ &= h(f(\mathbf{x}) + \alpha f(\mathbf{y})) \quad \text{since } f \text{ is linear} \\ &= h(f(\mathbf{x})) + \alpha\, h(f(\mathbf{y})) \quad \text{since } h \text{ is linear} \\ &= (h \circ f)(\mathbf{x}) + \alpha\,(h \circ f)(\mathbf{y}).\end{aligned}$$ ∎

Consider a linear transformation f from a vector space V_1 to a vector space V_2. Choose and fix an *ordered basis* $\mathcal{X} = \{\mathbf{x}_1, \mathbf{x}_2, \ldots, \mathbf{x}_n\}$ of V_1. (Actually, $\mathcal{X}$ is a list of vectors which form a basis but we use set notation.) Then any vector $\mathbf{x}$ in V_1 can be expressed uniquely as $\mathbf{x} = \alpha_1\mathbf{x}_1 + \cdots + \alpha_n\mathbf{x}_n$. Now $f(\mathbf{x}) = \alpha_1 f(\mathbf{x}_1) + \cdots + \alpha_n f(\mathbf{x}_n)$ since f is linear. Thus, to specify f it is enough to know the vectors $f(\mathbf{x}_1), \ldots, f(\mathbf{x}_n)$. Now choose and fix an ordered basis $\mathcal{Y} = \{\mathbf{y}_1, \mathbf{y}_2, \ldots, \mathbf{y}_m\}$ of V_2. Then each $f(\mathbf{x}_j)$ can be expressed uniquely as a linear combination of $\mathbf{y}_1, \ldots, \mathbf{y}_m$, say:

$$f(\mathbf{x}_j) = \sum_{i=1}^{m} a_{ij}\mathbf{y}_i, \quad j = 1, \ldots, n \tag{2.2.1}$$

Then

$$f(\mathbf{x}) = \sum_{j=1}^{n} \alpha_j f(\mathbf{x}_j) = \sum_{j=1}^{n}\sum_{i=1}^{m} a_{ij}\alpha_j\mathbf{y}_i \tag{2.2.2}$$

where $\mathbf{x} = \alpha_1\mathbf{x}_1 + \cdots + \alpha_n\mathbf{x}_n$. Thus the linear transformation f is completely specified by the mn scalars a_{ij} ($i = 1, \ldots, m$; $j = 1, \ldots, n$) once $\mathcal{X}$ and $\mathcal{Y}$ are fixed. These mn scalars can be written conveniently in the form of a rectangular array as follows:

$$\begin{bmatrix} a_{11} & a_{12} & \cdots & a_{1j} & \cdots & a_{1n} \\ a_{21} & a_{22} & \cdots & a_{2j} & \cdots & a_{2n} \\ \cdots & \cdots & \cdots & \cdots & \cdots & \cdots \\ a_{i1} & a_{i2} & \cdots & a_{ij} & \cdots & a_{in} \\ \cdots & \cdots & \cdots & \cdots & \cdots & \cdots \\ a_{m1} & a_{m2} & \cdots & a_{mj} & \cdots & a_{mn} \end{bmatrix} \tag{2.2.3}$$

where a_{ij} is written at the junction of the i-th row and the j-th column.

Definition 2.2.9 *A matrix of order* $m \times n$ *over* F is an array of mn (not necessarily distinct) scalars arranged in m rows and n columns as in (2.2.3) above. Such a matrix is also called *an* $m \times n$ *matrix*.

We usually denote matrices by bold face capital letters like $\mathbf{A}, \mathbf{B}, \mathbf{X}, \mathbf{\Lambda}, \mathbf{\Sigma}$. We use either square brackets or parentheses to enclose the elements of a matrix. We will write $\mathbf{A} = ((a_{ij}))$ if, for all i and j, a_{ij} is the element in the i-th row and j-th column (which we call the (i, j)-th *element* of $\mathbf{A}$). We generally use the lower case letter with two suffixes to denote the elements of a matrix denoted by the corresponding capital letter, e.g., the (i, j)-th element of $\mathbf{B}$ is usually denoted b_{ij}. Sometimes we denote the (i, j)-th element of $\mathbf{A}$ by $(\mathbf{A})_{ij}$ (*but not by* $\mathbf{A}_{ij}$ or A_{ij}).

We draw attention to the fact that each row of an $m \times n$ *matrix has* n *components and each column has* m *components.*

A matrix with just one column is called a *column matrix* or a *column vector* and a matrix with just one row is called a *row matrix* or a *row vector*.

Note that two matrices $\mathbf{A} = ((a_{ij}))$ and $\mathbf{B} = ((b_{ij}))$ are *equal* iff they are of the same order and $a_{ij} = b_{ij}$ for all i and j.

We usually study matrices over a fixed field F. So from now on, we shall drop the reference to F and talk simply of a matrix. In *Chapters* 1 to 6 of this book, F may be any field while we take it to be $\mathbb{R}$ or $\mathbb{C}$ in *Chapters* 7 to 9. However, the reader may initially think in terms of the real field since it is the most familiar and the most used. A matrix over the field of real numbers is called a *real matrix* while that over $\mathbb{C}$ is called a *complex matrix*. In our illustrations we will be using real matrices unless otherwise specified.

To illustrate the above notations, consider the following real matrix:

$$\begin{bmatrix} 2 & -3 & 0 & 1 & 2 \\ 5 & 0 & 1 & 2 & -3 \\ -4 & -1 & 0 & 0 & \frac{1}{2} \\ \sqrt{2} & -2 & 0 & 4 & 2 \end{bmatrix}$$

It has 4 rows and 5 columns and is thus a 4×5 matrix. Its third row is

$$\begin{bmatrix} -4 & -1 & 0 & 0 & \frac{1}{2} \end{bmatrix}$$

and its second column is

$$\begin{bmatrix} -3 \\ 0 \\ -1 \\ -2 \end{bmatrix}$$

The (3,2)-th element is -1.

Definition 2.2.10 The *matrix of the linear transformation* $f : V_1 \to V_2$ *with respect to the ordered bases* $\mathcal{X} = \{\mathbf{x}_1, \mathbf{x}_2, \ldots, \mathbf{x}_n\}$ *of* V_1 *and* $\mathcal{Y} = \{\mathbf{y}_1, \mathbf{y}_2, \ldots, \mathbf{y}_n\}$ *of* V_2 is (2.2.3) if (2.2.1) holds. If $V_1 = V_2$ and $\mathcal{X} = \mathcal{Y}$ then we refer to the matrix of f with respect to $\mathcal{X}$ and $\mathcal{Y}$ as *the matrix of* f *with respect to* $\mathcal{X}$.

Even if $V_1 = V_2$ we may consider the matrix of f with respect to different ordered bases $\mathcal{X}$ and $\mathcal{Y}$. The matrix of a linear transformation is always with respect to ordered bases. However, we usually drop the word 'ordered' as the order is either the natural order or that in which the vectors in the basis are listed.

Note that the matrix $\mathbf{A}$ of f with respect to $\mathcal{X} = \{\mathbf{x}_1, \ldots, \mathbf{x}_n\}$ and $\mathcal{Y} = \{\mathbf{y}_1, \ldots, \mathbf{y}_m\}$ is obtained thus: $f(\mathbf{x}_j)$ is expressed as a linear combination of $\mathbf{y}_1, \ldots, \mathbf{y}_m$ and the coefficients are written as the j-th column of $\mathbf{A}$. We illustrate this procedure with a few examples.

Example 2.2.11 Consider the rotation R_θ of *Example* 2.2.2. Let $\mathcal{X} = \mathcal{Y} = \{(1,0),(0,1)\}$ and let $\mathbf{A}$ be the matrix of R_θ with respect to $\mathcal{X}$ and $\mathcal{Y}$. Then $R_\theta(\mathbf{x}_1) = R_\theta((1,0)) = (\cos\theta, \sin\theta) = \cos\theta \cdot \mathbf{y}_1 + \sin\theta \cdot \mathbf{y}_2$ and $R_\theta(\mathbf{x}_2) = R_\theta((0,1)) = (-\sin\theta, \cos\theta) = -\sin\theta \cdot \mathbf{y}_1 + \cos\theta \cdot \mathbf{y}_2$. Writing the coefficients as the first and second columns, we get

$$\mathbf{A} = \begin{bmatrix} \cos\theta & -\sin\theta \\ \sin\theta & \cos\theta \end{bmatrix}$$

We will next find the matrix of R_θ with respect to a different pair of bases. Let $\mathcal{X} = \{(1,0),(0,1)\}$ and $\mathcal{Y} = \{(1,1),(1,-2)\}$. Then

$$R_\theta(\mathbf{x}_1) = (\cos\theta, \sin\theta) = \frac{2\cos\theta + \sin\theta}{3}(1,1) + \frac{\cos\theta - \sin\theta}{3}(1,-2)$$

and

$$R_\theta(\mathbf{x}_2) = (-\sin\theta, \cos\theta) = \frac{-2\sin\theta + \cos\theta}{3}(1,1) - \frac{\sin\theta + \cos\theta}{3}(1,-2)$$

Hence the matrix of R_θ with respect to the above bases is

$$\begin{bmatrix} \dfrac{2\cos\theta + \sin\theta}{3} & \dfrac{-2\sin\theta + \cos\theta}{3} \\ \dfrac{\cos\theta - \sin\theta}{3} & \dfrac{-\sin\theta - \cos\theta}{3} \end{bmatrix}$$

Example 2.2.12 The matrix of M_m of *Example* 2.2.2 with respect to the canonical basis (i.e., $\mathcal{X} = \mathcal{Y} = \{(1,0),(0,1)\}$) is

$$\begin{bmatrix} \dfrac{1-m^2}{1+m^2} & \dfrac{2m}{1+m^2} \\ \dfrac{2m}{1+m^2} & \dfrac{m^2-1}{1+m^2} \end{bmatrix}$$

The matrix of the linear transformation φ of *Example* 2.2.2 with respect to the canonical basis is

$$\begin{bmatrix} a & b \\ c & d \end{bmatrix}$$

Example 2.2.13 The matrix of the identity transformation on F^n with respect to any basis of F^n is the $n \times n$ matrix

$$\begin{bmatrix} 1 & 0 & \cdots & 0 \\ 0 & 1 & \cdots & 0 \\ \cdots & \cdots & \cdots & \cdots \\ 0 & 0 & \cdots & 1 \end{bmatrix},$$

called an *identity matrix* and denoted by $\mathbf{I}_n$ or simply by $\mathbf{I}$ when the order is clear.

The matrix of the null transformation ψ from F^n to F^m (defined by $\psi(\mathbf{x}) = \mathbf{0}$ for all $\mathbf{x}$) with respect to any basis of F^n and any basis of F^m is the $m \times n$ matrix

$$\begin{bmatrix} 0 & 0 & \cdots & 0 \\ 0 & 0 & \cdots & 0 \\ \cdots & \cdots & \cdots & \cdots \\ 0 & 0 & \cdots & 0 \end{bmatrix},$$

called a *zero matrix* or *null matrix* and denoted by $\mathbf{0}$.

Example 2.2.14 Let φ be the projector projecting vectors in $\mathbb{R}^2$ into the ξ_1-axis along the ξ_2-axis. Then the matrix of φ with respect to the canonical basis is

$$\begin{bmatrix} 1 & 0 \\ 0 & 0 \end{bmatrix}$$

Example 2.2.15 The matrix of the linear functional φ of *Example* 2.2.4 with respect to the canonical bases in F^n and F^1 is

$$\begin{bmatrix} a_1 & a_2 & \cdots & a_n \end{bmatrix}$$

Example 2.2.16 Let φ be the permutation transformation from F^n to itself taking $(x_1, x_2, \ldots, x_n)$ to $(x_{i_1}, x_{i_2}, \ldots, x_{i_n})$. The matrix of φ with respect to the canonical basis is the $n \times n$ matrix $\mathbf{A} = ((a_{pq}))$

where a_{pq} is 1 if $i_p = q$ and 0 otherwise. For instance, if $n = 4$ and $(i_1, i_2, i_3, i_4) = (2, 4, 3, 1)$, then the matrix is

$$\begin{bmatrix} 0 & 1 & 0 & 0 \\ 0 & 0 & 0 & 1 \\ 0 & 0 & 1 & 0 \\ 1 & 0 & 0 & 0 \end{bmatrix}$$

To see this, we note that $\mathbf{e}_1 = (1, 0, 0, 0)$ goes to the vector

$$(0, 0, 0, 1) = 0 \cdot \mathbf{e}_1 + 0 \cdot \mathbf{e}_2 + 0 \cdot \mathbf{e}_3 + 1 \cdot \mathbf{e}_4$$

So the first column of **A** has 0's in the first three places and 1 in the last place. The other columns are found similarly. ∎

In *Theorem* 2.2.8, we have seen how combinations of linear transformations lead to new linear transformations. We will now derive their matrices.

Theorem 2.2.17 Let f, g, h be linear transformations as in *Theorem* 2.2.8. Let $\mathcal{X} = \{\mathbf{x}_1, \ldots, \mathbf{x}_n\}$, $\mathcal{Y} = \{\mathbf{y}_1, \ldots, \mathbf{y}_m\}$ and $\mathcal{Z} = \{\mathbf{z}_1, \ldots, \mathbf{z}_p\}$ be ordered bases of V_1, V_2, V_3 respectively. Let $\mathbf{A} = ((a_{ij}))$ and $\mathbf{B} = ((b_{ij}))$ be the matrices of f and g respectively with respect to the bases $\mathcal{X}$ and $\mathcal{Y}$. Let $\mathbf{C} = ((c_{\ell k}))$ be the matrix of h with respect to the bases $\mathcal{Y}$ and $\mathcal{Z}$. Then

(i) the matrix of αf with respect to the bases $\mathcal{X}$ and $\mathcal{Y}$ is $((\alpha a_{ij}))$

(ii) the matrix of $f + g$ with respect to $\mathcal{X}$ and $\mathcal{Y}$ is $((a_{ij} + b_{ij}))$

(iii) the matrix of $h \circ f$ with respect to $\mathcal{X}$ and $\mathcal{Z}$ is $((d_{\ell j}))$ where $d_{\ell j} = \sum_{i=1}^{m} c_{\ell i} a_{ij}$.

Proof For $j = 1, \ldots, n$, we have

$$(\alpha f)(\mathbf{x}_j) = \alpha \cdot f(\mathbf{x}_j) = \alpha \cdot \sum_{i=1}^{m} a_{ij} \mathbf{y}_i = \sum_{i=1}^{m} (\alpha a_{ij}) \mathbf{y}_i$$

This proves (i). To prove (ii), notice that for $j = 1, \ldots, n$, we have

$$\begin{aligned} (f+g)(\mathbf{x}_j) &= f(\mathbf{x}_j) + g(\mathbf{x}_j) \\ &= \sum_{i=1}^{m} a_{ij} \mathbf{y}_i + \sum_{i=1}^{m} b_{ij} \mathbf{y}_i \\ &= \sum_{i=1}^{m} (a_{ij} + b_{ij}) \mathbf{y}_i \end{aligned}$$

This proves (ii). To prove (iii), we have for $j = 1, \ldots, n$,

$$\begin{aligned}(h \circ f)(\mathbf{x}_j) &= h(f(\mathbf{x}_j)) \\ &= h(\textstyle\sum_{i=1}^{m} a_{ij}\mathbf{y}_i) \quad \text{using (2.2.2)} \\ &= \textstyle\sum_{i=1}^{m} a_{ij}h(\mathbf{y}_i) \quad \text{since } h \text{ is linear} \\ &= \textstyle\sum_{i=1}^{m} a_{ij} \sum_{\ell=1}^{p} c_{\ell i}\mathbf{z}_\ell \\ &= \textstyle\sum_{\ell=1}^{p}(\sum_{i=1}^{m} c_{\ell i}a_{ij})\mathbf{z}_\ell.\end{aligned}$$

■

Example 2.2.18 By *Examples* 2.2.11 and 2.2.12, the matrices of the rotation R_θ and the reflection M_0 with respect to the canonical basis are

$$\mathbf{A} = \begin{bmatrix} \cos\theta & -\sin\theta \\ \sin\theta & \cos\theta \end{bmatrix} \text{ and } \mathbf{C} = \begin{bmatrix} 1 & 0 \\ 0 & -1 \end{bmatrix}$$

We now find the matrix $\mathbf{D}$ of $M_0 \circ R_\theta$ with respect to the canonical basis, using (iii) of the preceding theorem. Clearly

$$\begin{aligned}d_{11} &= c_{11}a_{11} + c_{12}a_{21} = 1 \cdot \cos\theta + 0 \cdot \sin\theta = \cos\theta \\ d_{12} &= c_{11}a_{12} + c_{12}a_{22} = -\sin\theta \\ d_{21} &= c_{21}a_{11} + c_{22}a_{21} = -\sin\theta \\ d_{22} &= c_{21}a_{12} + c_{22}a_{22} = -\cos\theta\end{aligned}$$

Thus

$$\mathbf{D} = \begin{bmatrix} \cos\theta & -\sin\theta \\ -\sin\theta & -\cos\theta \end{bmatrix}$$

Hence

$$\begin{aligned}&(M_m \circ R_\theta)((\xi_1, \xi_2)) \\ &\quad = (M_m \circ R_\theta)(\xi_1\mathbf{e}_1 + \xi_2\mathbf{e}_2) \\ &\quad = \textstyle\sum_{j=1}^{2}\sum_{i=1}^{2} d_{ij}\xi_j\mathbf{e}_i \qquad \text{by (2.2.2)} \\ &\quad = \xi_1\cos\theta \cdot \mathbf{e}_1 - \xi_2\sin\theta \cdot \mathbf{e}_1 - \xi_1\sin\theta \cdot \mathbf{e}_2 - \xi_2\cos\theta \cdot \mathbf{e}_2 \\ &\quad = (\xi_1\cos\theta - \xi_2\sin\theta, -\xi_1\sin\theta - \xi_2\cos\theta) \qquad (2.2.4)\end{aligned}$$

(2.2.4) can also be verified directly by using *Example* 2.2.2.

Exercises

1. In each of the following, determine whether the maps given are linear transformations from V_1 to V_2.

   (a) $V_1 = \mathbb{R}^3$, $V_2 = \mathbb{R}^2$, $f(x_1, x_2, x_3) = (x_1 - x_2, 2x_1 + x_2 + 4x_3)$

(b) $V_1 = V_2 = \mathbb{R}^2$ and $f(x_1, x_2)$ is (x_1, x_2) or $(x_1^2/x_2, x_2)$ according as $x_2 = 0$ or not

(c) $V_1 = V_2 = \mathbb{R}^3$, $f(x_1, x_2, x_3) = (x_3, x_1, 0)$

(d) $V_1 = \mathbb{R}^2$, $V_2 = \mathbb{R}$, $f(x_1, x_2) = c$ (a constant)

(e) $V_1 = \mathcal{P}_n$, $V_2 = \mathbb{R}$, $f(p) = \int_2^3 p(t)\,dt$

(f) $V_1 = \mathcal{P}_2$, $V_2 = \mathcal{P}_3$, $f(a_0 + a_1 t) = 2 + a_0 t + \frac{1}{2} a_1 t^2$

(g) $V_1 = \mathcal{P}_n$, $V_2 = \mathcal{P}_{n+k}$, $f(p(t)) = h(t)p(t)$ where $h(t)$ is a fixed polynomial of degree k.

2. Prove the statements made without proof in *Example* 2.2.2.

3. Consider the parallelogram $ABCD$ in the plane, where $A = (0, 0)$, $B = (1, 0)$, $C = (2, 1)$ and $D = (1, 1)$. Exhibit a linear operator f on $\mathbb{R}^2$ which transforms $ABCD$ to the rectangle $ABPQ$ where $P = (1, \frac{1}{2})$ and $Q = (0, \frac{1}{2})$. Show that the points inside $ABCD$ are taken by f to points inside $PQRS$. Obtain the matrix of f with respect to the canonical basis.

4. Show that each of the following mappings from the plane $\mathbb{R}^2$ to itself is a linear transformation and obtain its matrix with respect to the canonical basis: (a) Reflection in the origin defined by $\varphi(\mathbf{x}) = -\mathbf{x}$ and (b) *Shear transformation* representing shear parallel to x-axis, viz., $f(x, y) = (x + ky, y)$ where k is a fixed real number.

5. In each of the following, find the matrix of the linear transformation f with respect to the bases $\mathcal{X}$ and $\mathcal{Y}$.

   (a) $V_1 = \mathbb{R}^2$, $V_2 = \mathbb{R}^3$, $f(x_1, x_2) = (2x_1 - 3x_2, x_1, x_2 + 5x_1)$, $\mathcal{X}$ and $\mathcal{Y}$ are the canonical bases of $\mathbb{R}^2$ and $\mathbb{R}^3$ respectively.

   (b) V_1, V_2 and f as in (a) above, $\mathcal{X} = \{(1, -1), (4, 0)\}$ and $\mathcal{Y} = \{(1, 1, 0), (1, 0, 1), (0, 1, 1)\}$.

6. Find the matrix of φ of *Example* 2.2.5 with respect to the canonical bases of F^n and F^m. Also find the matrix of ψ of the same example with respect to the canonical bases of F^m and F^n.

7. Consider the differentiation transformation f from $\mathcal{P}_4$ to $\mathcal{P}_3$ and the integration transformation g from $\mathcal{P}_3$ to $\mathcal{P}_4$ as given in *Example* 2.2.6. Let $\mathcal{X}$ be the basis $\{1, t, t^2, t^3\}$ of $\mathcal{P}_4$ and $\mathcal{Y}$ be the basis $\{1, t, t^2\}$ of $\mathcal{P}_3$.

   (a) Find the matrix of f with respect to $\mathcal{X}$ and $\mathcal{Y}$.

   (b) Find the matrix of g with respect to $\mathcal{Y}$ and $\mathcal{X}$.

   (c) Find the matrix of $f \circ g$ with respect to $\mathcal{Y}$ and the matrix of $g \circ f$ with respect to $\mathcal{X}$.

8. Consider the following linear operators on $\mathbb{R}^2$:

$$f(x_1, x_2) = (2x_1 + 3x_2, x_1 - x_2)$$
$$g(x_1, x_2) = (x_1, 2x_1 - 5x_2)$$

   (a) Find the functional forms of $f + g$, $f \circ g$, $g \circ f$ and $3f$.

   (b) Find the matrices of f and g with respect to the canonical basis.

   (c) Find the matrices of $f + g$, $f \circ g$, $g \circ f$ and $3f$ with respect to the canonical basis using the functional forms obtained in (a).

   (d) Find the matrices of $f + g$, $f \circ g$, $g \circ f$ and $3f$ using (b) and *Theorem* 2.2.17.

9. Show that the set S of all linear transformations from F^n to F^m forms a vector space over F and determine its dimension.

10. Let $V_1 = V_2 = \mathbb{R}^2$ and $f(x_1, x_2)$ be the projection of (x_1, x_2) into the line $x_1 + 2x_2 = 0$ along $2x_1 + x_2 = 0$. Find the matrix $\mathbf{A}$ of f with respect to the canonical basis. What is the matrix of $f^2 := f \circ f$ with respect to the canonical basis?

11. Find the functional form of the linear operator f on $\mathbb{R}^2$ whose matrix with respect to $\mathcal{X} = \{(1,1), (1,-1)\}$ is $\begin{bmatrix} 2 & -1 \\ 3 & 5 \end{bmatrix}$.

*12. Let S be a subspace of a vector space V.

   (a) Show that $\mathbf{x} \mapsto \mathbf{x} + S$ is a linear transformation from V to V/S (see *Section* 1.9).

   (b) If f is a linear operator on V such that $f(S) \subseteq S$, show that $\mathbf{x} + S \mapsto f(\mathbf{x}) + S$ is a well defined linear operator on V/S.

13. Let f be a linear transformation from V_1 to V_2. Prove the following:

   (a) If S is a subspace of V_1 then $f(S)$ is a subspace of V_2. Moreover, if $\mathbf{x}_1, \ldots, \mathbf{x}_k$ generate S then $f(\mathbf{x}_1), \ldots, f(\mathbf{x}_k)$ generate $f(S)$.

   (b) If $f(\mathbf{x}_1), \ldots, f(\mathbf{x}_r)$ are linearly independent then so are $\mathbf{x}_1, \ldots, \mathbf{x}_r$.

   (c) $d(f(S)) \leq d(S)$ for any subspace S of V_1.

14. Let V_1 and V_2 be vector spaces over F and let $\{\mathbf{x}_1, \ldots, \mathbf{x}_n\}$ be a basis of V_1. Then for *any* given vectors $\mathbf{y}_1, \ldots, \mathbf{y}_n$ in V_2, show that there is a unique linear transformation f from V_1 to V_2 such that $f(\mathbf{x}_i) = \mathbf{y}_i$, $i = 1, \ldots, n$.

*15. If V_1 and V_2 are vector spaces over the field of rational numbers and f is a map from V_1 to V_2 such that $f(\mathbf{x}+\mathbf{y}) = f(\mathbf{x}) + f(\mathbf{y})$ for all $\mathbf{x}, \mathbf{y} \in V_1$, show that f is a linear transformation.

2.3 Operations on matrices

In the preceding section we saw how operations on linear transformations lead to corresponding operations on their matrices. In this section we formally define operations on matrices and study them.

Definition 2.3.1 If $\mathbf{A} = ((a_{ij}))$ and $\mathbf{B} = ((b_{ij}))$ are $m \times n$ matrices and $\mathbf{C} = ((c_{\ell i}))$ is a $p \times m$ matrix then

(i) $\alpha\mathbf{A}$ is the $m \times n$ matrix $((\alpha a_{ij}))$

(ii) $\mathbf{A} + \mathbf{B}$ is the $m \times n$ matrix $((a_{ij} + b_{ij}))$ and

(iii) $\mathbf{CA}$ is the $p \times n$ matrix $((d_{\ell j}))$ where $d_{\ell j} = \sum_{i=1}^{m} c_{\ell i} a_{ij}$

We note that in order to obtain the element in the ℓ-th row and the j-th column of $\mathbf{CA}$, we multiply the elements $c_{\ell 1}, c_{\ell 2}, \ldots, c_{\ell m}$ of the ℓ-th row of the first matrix $\mathbf{C}$ with the corresponding elements $a_{1j}, a_{2j}, \ldots, a_{mj}$ of the j-th column of the second matrix $\mathbf{A}$ and take their sum. For example, if

$$\mathbf{A} = \begin{bmatrix} 1 & 0 \\ 2 & -3 \\ -4 & 1 \end{bmatrix}, \ \mathbf{B} = \begin{bmatrix} 2 & 1 \\ -2 & 0 \\ 5 & -1 \end{bmatrix} \text{ and } \mathbf{C} = \begin{bmatrix} 1 & 0 & -1 \\ -2 & -5 & -1 \\ -3 & 1 & 4 \\ 1 & -1 & 0 \end{bmatrix}$$

then

$$(-2)\mathbf{A} = \begin{bmatrix} -2 & 0 \\ -4 & 6 \\ 8 & -2 \end{bmatrix}, \ \mathbf{A}+\mathbf{B} = \begin{bmatrix} 3 & 1 \\ 0 & -3 \\ 1 & 0 \end{bmatrix} \text{ and } \mathbf{CA} = \begin{bmatrix} 5 & -1 \\ -8 & 14 \\ -17 & 1 \\ -1 & 3 \end{bmatrix}$$

Note that the sum $\mathbf{A}+\mathbf{B}$ is defined iff $\mathbf{A}$ and $\mathbf{B}$ are of the same order. The product $\mathbf{CA}$ is defined iff the number of columns of $\mathbf{C}$ equals the number of rows of $\mathbf{A}$. Also then, $\mathbf{CA}$ has the same number of rows as $\mathbf{C}$ and the same number of columns as $\mathbf{A}$. While the addition of matrices is *component-wise*, the multiplication of matrices is in terms of rows of the first matrix and columns of the second matrix.

The conclusions of *Theorem* 2.2.17 can now be stated thus:

(i) the matrix of αf is $\alpha \cdot$ (matrix of f)

(ii) the matrix of $f + g$ is (matrix of f) + (matrix of g)

(iii) the matrix of $h \circ f$ is (matrix of h) $\cdot$ (matrix of f)

We next use the concept of the product of matrices to obtain an explicit expression for the image of a vector under a linear transformation. Let f be a linear transformation from V_1 to V_2 and $\mathbf{A} = ((a_{ij}))$ the matrix of f with respect to the bases $\mathcal{X} = \{\mathbf{x}_1, \ldots, \mathbf{x}_n\}$ and $\mathcal{Y} = \{\mathbf{y}_1, \ldots, \mathbf{y}_m\}$ of V_1 and V_2. Let $\mathbf{x} = \sum_{j=1}^{n} \alpha_j \mathbf{x}_j \in V_1$. Then by (2.2.2)

$$f(\mathbf{x}) = \sum_{i=1}^{m} \left(\sum_{j=1}^{n} a_{ij}\alpha_j \right) \mathbf{y}_i = \sum_{i=1}^{m} \beta_i \mathbf{y}_i$$

where $\beta_i = \sum_{j=1}^{n} a_{ij}\alpha_j$. Writing

$$\boldsymbol{\alpha} = \begin{bmatrix} \alpha_1 \\ \alpha_2 \\ \vdots \\ \alpha_n \end{bmatrix} \quad \text{and} \quad \boldsymbol{\beta} = \begin{bmatrix} \beta_1 \\ \beta_2 \\ \vdots \\ \beta_m \end{bmatrix}$$

we have $\boldsymbol{\beta} = \mathbf{A}\boldsymbol{\alpha}$. Recall that $\boldsymbol{\alpha}$ is called the coordinate vector of $\mathbf{x}$ with respect to $\mathcal{X}$ and $\boldsymbol{\beta}$ the coordinate vector of $f(\mathbf{x})$ with respect to $\mathcal{Y}$. We have thus proved

Theorem 2.3.2 Let $\mathbf{A}$ be the matrix of f with respect to $\mathcal{X}$ and $\mathcal{Y}$. If $\boldsymbol{\alpha}$ is the coordinate vector (written as a column) of $\mathbf{x}$ with respect to $\mathcal{X}$, then $\mathbf{A}\boldsymbol{\alpha}$ is the coordinate vector of $f(\mathbf{x})$ with respect to $\mathcal{Y}$.

Corollary If f is a linear transformation from F^n to F^m and $\mathbf{A}$ is the matrix of f with respect to the canonical bases of F^n and F^m, then $f(\mathbf{x}) = \mathbf{A}\mathbf{x}$ for all $\mathbf{x} \in F^n$. It is assumed here that $\mathbf{x}$ is written as a column vector. In particular, every linear functional f on F^n is of the form $f(\mathbf{x}) = \alpha_1 x_1 + \cdots + \alpha_n x_n$ for some scalars $\alpha_1, \ldots, \alpha_n$.

This corollary follows on observing that the coordinate vector of $\mathbf{x}$ with respect to the canonical basis is $\mathbf{x}$ itself. Our next result is a sort of converse of the preceding corollary and shows that every matrix is the matrix of some linear transformation.

Theorem 2.3.3 Let $\mathbf{A}$ be an $m \times n$ matrix. Then $f : F^n \to F^m$ defined by $f(\mathbf{x}) = \mathbf{A}\mathbf{x}$ (where $\mathbf{x}$ is written as a column vector) is a linear transformation and $\mathbf{A}$ is the matrix of f with respect to the canonical bases of F^n and F^m.

The proof of this theorem is left as an exercise to the reader. (It also follows from (10) and (11) of *Theorem* 2.4.2 and *Corollary* 2 to *Theorem* 2.6.3.)

We may have to consider a vector in F^n sometimes as a column matrix and sometimes as a row matrix. The need for this arises partly from the fact that we can multiply an $n \times n$ matrix $\mathbf{A}$ on the left by a row matrix but not by a column matrix and we can multiply $\mathbf{A}$ on the right by a column matrix but not by a row matrix. Motivated by this, we introduce another operation on matrices in the following

Definition 2.3.4 If $\mathbf{A} = ((a_{ij}))$ is an $m \times n$ matrix, the *transpose* $\mathbf{A}^{\mathrm{T}}$ of $\mathbf{A}$ is the $n \times m$ matrix with (k,ℓ)-th element equal to $a_{\ell k}$ for $k = 1, 2, \ldots, n$ and $\ell = 1, 2, \ldots, m$. Another common notation for $\mathbf{A}^{\mathrm{T}}$ is $\mathbf{A}'$ but we will not use this.

For example, if

$$\mathbf{A} = \begin{bmatrix} 1 & 0 \\ 2 & -3 \\ -4 & 1 \end{bmatrix} \quad \text{and} \quad \mathbf{x} = \begin{bmatrix} 1 \\ 2 \\ 3 \end{bmatrix}$$

then

$$\mathbf{A}^{\mathrm{T}} = \begin{bmatrix} 1 & 2 & -4 \\ 0 & -3 & 1 \end{bmatrix} \quad \text{and} \quad \mathbf{x}^{\mathrm{T}} = [\, 1 \;\; 2 \;\; 3 \,]$$

We observe that the i-th row of $\mathbf{A}^{\mathrm{T}}$ is the i-th column of $\mathbf{A}$, written as a row. Similarly the j-th column of $\mathbf{A}^{\mathrm{T}}$ is the j-th row of $\mathbf{A}$, written as a column. Clearly the transpose of a row (resp. column) matrix is a column (resp. row) matrix.

For some practical uses of matrix operations, see *Exercises* 2.3.11 through 2.3.13.

Exercises

1. Consider the matrices

$$\mathbf{A}_1 = \begin{bmatrix} 1 & 0 \\ 2 & 5 \\ 3 & -1 \\ 0 & 2 \end{bmatrix}, \quad \mathbf{A}_2 = \begin{bmatrix} 2 & 1 \\ 1 & 1 \\ 0 & 4 \\ 1 & -1 \end{bmatrix}, \quad \mathbf{A}_3 = \begin{bmatrix} -2 \\ 0 \\ 3 \\ 4 \end{bmatrix}$$

and $\mathbf{A}_4 = \begin{bmatrix} 2 & 0 & -3 & 0 \\ 1 & 3 & 1 & 0 \end{bmatrix}$. Find $\mathbf{A}_i + \mathbf{A}_j$ $(i \neq j)$ and $\mathbf{A}_k\mathbf{A}_\ell$ whenever they are defined. Also find $\mathbf{A}_3^{\mathrm{T}}\mathbf{A}_1$ and $\mathbf{A}_3^{\mathrm{T}}\mathbf{A}_4^{\mathrm{T}}$.

2. For each of the following pairs, find **AB** and **BA**.

(a) $\mathbf{A} = \begin{bmatrix} 2 & 0 & 0 \\ 0 & 3 & 0 \\ 0 & 0 & 4 \end{bmatrix}$ and $\mathbf{B} = \begin{bmatrix} 1 & -1 & 2 \\ 2 & 3 & -2 \\ 2 & 5 & 6 \end{bmatrix}$

(b) $\mathbf{A} = \begin{bmatrix} 0 & 0 & 0 & \alpha_1 \\ 0 & 0 & \alpha_2 & 0 \\ 0 & \alpha_3 & 0 & 0 \\ \alpha_4 & 0 & 0 & 0 \end{bmatrix}$ and $\mathbf{B} = \begin{bmatrix} 0 & 0 & 0 & \beta_1 \\ 0 & 0 & \beta_2 & 0 \\ 0 & \beta_3 & 0 & 0 \\ \beta_4 & 0 & 0 & 0 \end{bmatrix}$

(c) $\mathbf{A} = \begin{bmatrix} 3 & 1 & 0 \\ 0 & 4 & 2 \\ 0 & 0 & 3 \end{bmatrix}$ and $\mathbf{B} = \begin{bmatrix} 5 & 1 & 12 \\ 0 & 7 & -1 \\ 0 & 0 & 2 \end{bmatrix}$

(d) $\mathbf{A} = \begin{bmatrix} 0 & 1 & 0 \\ 0 & 0 & 1 \\ 1 & 0 & 0 \end{bmatrix}$ and $\mathbf{B} = \begin{bmatrix} 1 & 2 & 3 \\ 4 & 5 & 6 \\ 7 & 8 & 9 \end{bmatrix}$

(e) $\mathbf{A} = \begin{bmatrix} 2 \\ -1 \\ 0 \end{bmatrix}$ and $\mathbf{B} = [3 \ \ 6 \ \ 5]$

(f) $\mathbf{A} = \begin{bmatrix} 0 & 0 \\ 0 & 1 \end{bmatrix}$ and $\mathbf{B} = \begin{bmatrix} 1 & 0 \\ 1 & 0 \end{bmatrix}$.

3. If $\mathbf{A} = \begin{bmatrix} 1 & 0 \\ -1 & 0 \\ 2 & 1 \end{bmatrix}$ and $\mathbf{B} = \begin{bmatrix} 0 & 1 \\ 0 & 0 \end{bmatrix}$, find $\mathbf{AA}^{\mathrm{T}}$, $\mathbf{A}^{\mathrm{T}}\mathbf{A}$ and $\mathbf{B}^2$.

4. How many multiplications (of scalars) are needed to compute the product **AB**, where **A** is an $m \times n$ matrix and **B** is an $n \times p$ matrix?

5. If $\mathbf{A} = \begin{bmatrix} 2 \\ -1 \\ 0 \end{bmatrix}$, $\mathbf{B} = (3 \ \ 0 \ \ 1 \ \ -5)$ and $\mathbf{C} = \begin{bmatrix} 2 \\ 5 \\ 8 \\ 1 \end{bmatrix}$, compute **ABC**.

(Which of the two possible ways of doing this is easier?)

6. For each of the following matrices, find $\mathbf{A}^k$ for all $k \geq 2$, where $\mathbf{A}^k = \mathbf{A}^{k-1}\mathbf{A}$.

(a) $\mathbf{A} = \begin{bmatrix} 0 & 2 & 5 \\ 0 & 0 & 3 \\ 0 & 0 & 0 \end{bmatrix}$, (b) $\mathbf{A} = \begin{bmatrix} 0 & 0 & 0 \\ 2 & 0 & 0 \\ 0 & 3 & 0 \end{bmatrix}$, (c) $\mathbf{A} = \begin{bmatrix} 0 & 2 & 0 \\ 3 & 0 & 0 \\ 0 & 0 & 0 \end{bmatrix}$,

(d) $\mathbf{A} = \begin{bmatrix} 0 & 1 & 0 & 0 \\ 0 & 0 & 0 & 1 \\ 1 & 0 & 0 & 0 \\ 0 & 0 & 1 & 0 \end{bmatrix}$, (e) $\mathbf{A} = \begin{bmatrix} 0 & 0 & 1 & 0 \\ 0 & 0 & 0 & 1 \\ 1 & 0 & 0 & 0 \\ 0 & 1 & 0 & 0 \end{bmatrix}$,

(f) $\mathbf{A} = \begin{bmatrix} \alpha & 1 & 0 & \cdots & 0 & 0 \\ 0 & \alpha & 1 & \cdots & 0 & 0 \\ 0 & 0 & \alpha & \cdots & 0 & 0 \\ \cdots & \cdots & \cdots & \cdots & \cdots & \cdots \\ 0 & 0 & 0 & \cdots & \alpha & 1 \\ 0 & 0 & 0 & \cdots & 0 & \alpha \end{bmatrix}$ of order n and

(g) $\mathbf{A} = \begin{bmatrix} 1 & 1 & 1 \\ 1 & \omega & \omega^2 \\ 1 & \omega^2 & \omega \end{bmatrix}$ when $\omega = e^{2\pi i/3}$.

7. (a) Determine all 2×2 real matrices $\mathbf{A}$ such that $\mathbf{A}^2 = \mathbf{0}$.
   (b) Determine all 2×2 real matrices $\mathbf{A}$ such that $\mathbf{A}^2 = \mathbf{I}$.
8. Give an example of a 2×2 *non-null* matrix $\mathbf{A}$ over GF(2) such that $\mathbf{A}\mathbf{A}^{\mathrm{T}} = \mathbf{0}$.
9. Show that if $\mathbf{A}$ is a non-null real matrix, then $\mathbf{A}\mathbf{A}^{\mathrm{T}} \neq \mathbf{0}$.
10. Consider the linear operator $f(x_1, x_2, x_3) = (x_2 + x_3, x_3, 0)$ on $\mathbb{R}^3$. Obtain the functional forms of $f \circ f$ and $f \circ f \circ f$ and find the matrices of f, $f \circ f$ and $f \circ f \circ f$ with respect to the canonical basis.

*11. Consider an economic system using and producing n commodities $G_1, G_2, \ldots, G_n$. Suppose a_{ij} units of G_i $(i = 1, 2, \ldots, n)$ are required at time 0 to produce one unit of G_j at time 1.

   (a) To achieve the output vector $\mathbf{x} = [x_1 \ x_2 \ \cdots \ x_n]^{\mathrm{T}}$ at time 1 show that the vector of material requirements at time 0 is $\mathbf{Ax}$.
   (b) Show that the j-th column of $\mathbf{A}^2$ gives the material requirements at time 0 to produce a unit of G_j at time 2.
   (c) If $\mathbf{p} = (p_1 \ p_2 \ \cdots \ p_n)$ is the vector of (per unit) prices, show that the j-th component of $\mathbf{pA}$ is the material cost to produce one unit of G_j and that $(\mathbf{pA})\mathbf{x}$ is the material cost for producing the output vector $\mathbf{x}$.

*12. Consider road and rail links between four cities C_1, C_2, C_3 and C_4. Let

$$\mathbf{A} = \begin{bmatrix} 0 & 3 & 1 & 2 \\ 3 & 0 & 1 & 0 \\ 1 & 1 & 0 & 4 \\ 2 & 0 & 4 & 0 \end{bmatrix} \quad \text{and} \quad \mathbf{B} = \begin{bmatrix} 0 & 1 & 1 & 0 \\ 1 & 0 & 2 & 0 \\ 1 & 2 & 0 & 1 \\ 0 & 0 & 1 & 0 \end{bmatrix}$$

where a_{ij} and b_{ij} are the number of road links and the number of rail links between C_i and C_j. $\mathbf{A}$ and $\mathbf{B}$ are called the *adjacency matrices* for the road and rail networks.

   (a) If the number of road links between each pair of cities is doubled show that the new adjacency matrix for the road network would be $2\mathbf{A}$.

(b) Show that $\mathbf{A}+\mathbf{B}$ is the adjacency matrix for the total network (i.e., the union of the road and rail networks).

(c) Show that $a_{ik}a_{kj}$ is the number of 2-step paths from C_i to C_j via C_k and that the (i,j)-element of $\mathbf{A}^2$ is the number of distinct 2-step paths from C_i to C_j. Generalize to $\mathbf{A}^p$.

(d) Show that the total number of road links at any C_i, which is the same as the i-th row sum of $\mathbf{A}$, can be obtained as the i-th element of $\mathbf{Ax}$ where $\mathbf{x}$ is the 4×1 column matrix with all entries 1.

(e) Till now, we assumed that a link between C_i and C_j is two-way. Generalize to one-way links. Note that paths have to be properly directed now. How do you get the total number of road links *from* C_i *to* the other cities and the total number of road links *to* C_i *from* the other cities? What is the interpretation of $\mathbf{A}^{\mathrm{T}}$?

*13. Consider an organism which passes through 4 stages S_1, S_2, S_3, S_4 and then dies. At time 0 let there be f_{i0} organisms just entering stage S_i $(i = 1,\ldots,4)$. Assume that a proportion p_i of the organisms entering stage S_i at time t die and the rest enter stage S_{i+1} at time $t+1$ $(i=1,2,3)$. Also, a proportion r_i of the organisms entering stage S_i at time t give birth to one organism each belonging to stage S_1 at time $t+1$ $(i=1,\ldots,4)$. Let f_{in} be the number of organisms entering stage S_i at time n. Let

$$\mathbf{f}_n = (f_{1n}, f_{2n}, f_{3n}, f_{4n}) \quad \text{for } n = 0,1,2,\ldots$$

and

$$\mathbf{A} = \begin{bmatrix} r_1 & 1-p_1 & 0 & 0 \\ r_2 & 0 & 1-p_2 & 0 \\ r_3 & 0 & 0 & 1-p_3 \\ r_4 & 0 & 0 & 0 \end{bmatrix}$$

Then show that $\mathbf{f}_n = \mathbf{f}_0\mathbf{A}^n$ where $\mathbf{A}^n$.

2.4 Properties of matrix operations

In this section we will prove several elementary but important properties of the four operations on matrices that we have introduced in the preceding section.

We recall that an $m\times n$ matrix $\mathbf{A}$ is called a *zero-matrix* or *null matrix* if all its elements are 0. It is denoted by $\mathbf{0}_{m\times n}$ or simply $\mathbf{0}$ when the order is clear from the context.

An $n\times n$ matrix $\mathbf{A} = ((a_{ij}))$ is called an *identity matrix* if a_{ij} is 1 or 0 according as $i=j$ or $i\neq j$. It is then denoted by $\mathbf{I}_n$ or simply $\mathbf{I}$ when the order is clear from the context. The (i,j)-th element of an identity matrix is δ_{ij} where the *Kronecker symbol* δ_{ij} is defined by

$$\delta_{ij} = \begin{cases} 1 & \text{if } i = j \\ 0 & \text{if } i \neq j \end{cases}$$

From now on, we will denote the i-th *column of* $\mathbf{I}_n$ *by* $\mathbf{e}_i$. Thus $\mathbf{e}_i$ has 1 in the i-th position and 0's elsewhere. Clearly $\mathbf{e}_1, \mathbf{e}_2, \ldots, \mathbf{e}_n$ form the canonical basis of F^n. It is easy to see that $\mathbf{I}_n{}^{\mathrm{T}} = \mathbf{I}_n$, thus the rows of $\mathbf{I}_n$ are $\mathbf{e}_1^{\mathrm{T}}, \mathbf{e}_2^{\mathrm{T}}, \ldots, \mathbf{e}_n^{\mathrm{T}}$.

Definition 2.4.1 If $\mathbf{A} = ((a_{ij}))$ is an $m \times n$ matrix, $-\mathbf{A}$ is the $m \times n$ matrix $((-a_{ij}))$. If $\mathbf{A}$ and $\mathbf{B}$ are matrices of the same order, $\mathbf{A} - \mathbf{B} = \mathbf{A} + (-\mathbf{B})$.

We note that $-\mathbf{A} = (-1)\mathbf{A}$. We next give several elementary properties of the matrix operations. In the next theorem, $\mathbf{A}, \mathbf{B}, \mathbf{C}$ are matrices of arbitrary orders subject to the condition that the expressions considered are defined and α, β are arbitrary scalars.

Theorem 2.4.2 We have

(1) $\mathbf{A} + \mathbf{B} = \mathbf{B} + \mathbf{A}$ (commutativity of addition)
(2) $\mathbf{A} + (\mathbf{B} + \mathbf{C}) = (\mathbf{A} + \mathbf{B}) + \mathbf{C}$ (associativity of addition)
(3) $\mathbf{0} + \mathbf{A} = \mathbf{A} + \mathbf{0} = \mathbf{A}$
(4) $\mathbf{A} + (-\mathbf{A}) = (-\mathbf{A}) + \mathbf{A} = \mathbf{0}$
(5) $\alpha(\mathbf{A} + \mathbf{B}) = \alpha\mathbf{A} + \alpha\mathbf{B}$
(6) $(\alpha + \beta)\mathbf{A} = \alpha\mathbf{A} + \beta\mathbf{A}$
(7) $\alpha(\beta\mathbf{A}) = (\alpha\beta)\mathbf{A}$
(8) $1 \cdot \mathbf{A} = \mathbf{A}$
(9) $\mathbf{A}(\mathbf{B}\mathbf{C}) = (\mathbf{A}\mathbf{B})\mathbf{C}$ (associativity of matrix multiplication)
(10) $\mathbf{A}(\alpha\mathbf{B}) = \alpha(\mathbf{A}\mathbf{B})$
(11) $\mathbf{A}(\mathbf{B} + \mathbf{C}) = \mathbf{A}\mathbf{B} + \mathbf{A}\mathbf{C}$ (distributivity)
(12) $(\mathbf{B} + \mathbf{C})\mathbf{A} = \mathbf{B}\mathbf{A} + \mathbf{C}\mathbf{A}$ (distributivity)
(13) $\mathbf{A} \cdot \mathbf{I} = \mathbf{A}$ where $\mathbf{A}$ is of order $m \times n$ and $\mathbf{I}$ is of order $n \times n$
(14) $\mathbf{I} \cdot \mathbf{A} = \mathbf{A}$ where $\mathbf{I}$ is of order $m \times m$ and $\mathbf{A}$ is of order $m \times n$
(15) $0 \cdot \mathbf{A} = \mathbf{0}$ (0 in $0 \cdot \mathbf{A}$ is a scalar)
(16) $\mathbf{0} \cdot \mathbf{A} = \mathbf{0}$
(17) $\mathbf{A} \cdot \mathbf{0} = \mathbf{0}$

(18) $(\mathbf{A}+\mathbf{B})^{\mathrm{T}} = \mathbf{A}^{\mathrm{T}} + \mathbf{B}^{\mathrm{T}}$

(19) $(\alpha\mathbf{A})^{\mathrm{T}} = \alpha\,\mathbf{A}^{\mathrm{T}}$

(20) $(\mathbf{AB})^{\mathrm{T}} = \mathbf{B}^{\mathrm{T}}\mathbf{A}^{\mathrm{T}}$

(21) $(\mathbf{A}^{\mathrm{T}})^{\mathrm{T}} = \mathbf{A}$

Before we prove the above properties we note that in the equalities (1), (2), (5), (9), (10), (11), (12), (18) and (20), whenever either side of the equality sign is defined, both sides are defined and are matrices of the same order. (Verification of this is left to the reader.) So, to complete the proof it suffices to show the equality of the corresponding elements of the two sides. We give the proofs for some; the rest are left as an exercise to the reader.

Proof (1) We have, for all i and j,

$$(\mathbf{A}+\mathbf{B})_{ij} = (\mathbf{A})_{ij} + (\mathbf{B})_{ij} = (\mathbf{B})_{ij} + (\mathbf{A})_{ij} = (\mathbf{B}+\mathbf{A})_{ij}$$

(2) We have, for all i and j,

$$\begin{aligned}(\mathbf{A}+(\mathbf{B}+\mathbf{C}))_{ij} &= (\mathbf{A})_{ij} + (\mathbf{B}+\mathbf{C})_{ij} = (\mathbf{A})_{ij} + ((\mathbf{B})_{ij} + (\mathbf{C})_{ij}) \\ &= ((\mathbf{A})_{ij} + (\mathbf{B})_{ij}) + (\mathbf{C})_{ij} = (\mathbf{A}+\mathbf{B})_{ij} + (\mathbf{C})_{ij} = ((\mathbf{A}+\mathbf{B})+\mathbf{C})_{ij}\end{aligned}$$

(3) By (1), $\mathbf{0}+\mathbf{A} = \mathbf{A}+\mathbf{0}$. Now, $(\mathbf{A}+\mathbf{0})_{ij} = (\mathbf{A})_{ij} + 0 = (\mathbf{A})_{ij}$. Hence $\mathbf{A}+\mathbf{0} = \mathbf{A}$.

(4) By (1), $\mathbf{A}+(-\mathbf{A}) = (-\mathbf{A})+\mathbf{A}$. Now, $(\mathbf{A}+(-\mathbf{A}))_{ij} = (\mathbf{A})_{ij} + (-\mathbf{A})_{ij} = (\mathbf{A})_{ij} - (\mathbf{A})_{ij} = 0$.

(5) We have, for all i and j,

$$\begin{aligned}(\alpha(\mathbf{A}+\mathbf{B}))_{ij} &= \alpha(\mathbf{A}+\mathbf{B})_{ij} = \alpha((\mathbf{A})_{ij} + (\mathbf{B})_{ij}) \\ &= \alpha(\mathbf{A})_{ij} + \alpha(\mathbf{B})_{ij} = (\alpha\mathbf{A})_{ij} + (\alpha\mathbf{B})_{ij} = (\alpha\mathbf{A}+\alpha\mathbf{B})_{ij}\end{aligned}$$

(9) Since either side is defined, the number of columns in $\mathbf{A}$ equals the number of rows in $\mathbf{B}$ and the number of columns in $\mathbf{B}$ equals the number of rows in $\mathbf{C}$. Without loss of generality let the orders of $\mathbf{A}, \mathbf{B}, \mathbf{C}$ to be $m \times n$, $n \times p$ and $p \times q$ respectively. Then

$$\begin{aligned}(\mathbf{A}(\mathbf{BC}))_{i\ell} &= \textstyle\sum_{j=1}^{n} (\mathbf{A})_{ij} (\mathbf{BC})_{j\ell} \\ &= \textstyle\sum_{j=1}^{n} (\mathbf{A})_{ij} \sum_{k=1}^{p} (\mathbf{B})_{jk} (\mathbf{C})_{k\ell} \\ &= \textstyle\sum_{j=1}^{n} \sum_{k=1}^{p} (\mathbf{A})_{ij} (\mathbf{B})_{jk} (\mathbf{C})_{k\ell} \\ &= \textstyle\sum_{k=1}^{p} \left(\sum_{j=1}^{n} (\mathbf{A}_{ij}) (\mathbf{B})_{jk} \right) (\mathbf{C})_{k\ell} \\ &= \textstyle\sum_{k=1}^{p} (\mathbf{AB})_{ik} (\mathbf{C})_{k\ell} \\ &= ((\mathbf{AB})\mathbf{C})_{i\ell}\end{aligned}$$

for $i = 1, 2, \ldots, m$ and $\ell = 1, 2, \ldots, q$.

(11) Suppose $\mathbf{A}$ is of order $m \times n$ and $\mathbf{B}$ and $\mathbf{C}$ are of order $n \times p$. Then

$(\mathbf{A}(\mathbf{B}+\mathbf{C}))_{ij} = \sum_{k=1}^{n} (\mathbf{A})_{ik}((\mathbf{B})_{kj} + (\mathbf{C})_{kj}) =$

$\sum_{k=1}^{n} (\mathbf{A})_{ik}(\mathbf{B})_{kj} + \sum_{k=1}^{n} (\mathbf{A})_{ik}(\mathbf{C})_{kj} = (\mathbf{AB})_{ij} + (\mathbf{AC})_{ij} = (\mathbf{AB}+\mathbf{AC})_{ij}$

(13) Recall that $(\mathbf{I})_{ij}$ is 1 or 0 according as $i = j$ or not. So

$(\mathbf{AI})_{ij} = (\mathbf{A})_{i1}(\mathbf{I})_{1j} + \cdots + (\mathbf{A})_{ij}(\mathbf{I})_{jj} + \cdots + (\mathbf{A})_{in}(\mathbf{I})_{nj} = (\mathbf{A})_{ij}(\mathbf{I})_{jj} = (\mathbf{A})_{ij}$

(18) We have, for all i and j,

$((\mathbf{A}+\mathbf{B})^{\mathrm{T}})_{ij} = (\mathbf{A}+\mathbf{B})_{ji} = (\mathbf{A})_{ji} + (\mathbf{B})_{ji} = (\mathbf{A}^{\mathrm{T}})_{ij} + (\mathbf{B}^{\mathrm{T}})_{ij} = (\mathbf{A}^{\mathrm{T}}+\mathbf{B}^{\mathrm{T}})_{ij}$

(20) Let $\mathbf{A}$ and $\mathbf{B}$ be of orders $m \times n$ and $n \times p$ respectively. Then

$$\begin{aligned} ((\mathbf{AB})^{\mathrm{T}})_{ij} &= (\mathbf{AB})_{ji} \\ &= \sum_{k=1}^{n} (\mathbf{A})_{jk}(\mathbf{B})_{ki} \\ &= \sum_{k=1}^{n} (\mathbf{B})_{ki}(\mathbf{A})_{jk} \\ &= \sum_{k=1}^{n} (\mathbf{B}^{\mathrm{T}})_{ik}(\mathbf{A}^{\mathrm{T}})_{kj} \\ &= (\mathbf{B}^{\mathrm{T}}\mathbf{A}^{\mathrm{T}})_{ij} \end{aligned}$$

for $i = 1, 2, \ldots, p$ and $j = 1, 2, \ldots, m$. ∎

As shown in (1) above, matrix addition is commutative. *However, matrix multiplication is not commutative.* If $\mathbf{A}$ is a 2×3 matrix and $\mathbf{B}$ is a 3×4 matrix, $\mathbf{AB}$ is defined but $\mathbf{BA}$ is not. If $\mathbf{A}$ is a 2×3 matrix and $\mathbf{B}$ is a 3×2 matrix, both $\mathbf{AB}$ and $\mathbf{BA}$ are defined but $\mathbf{AB}$ is of order 2 whereas $\mathbf{BA}$ is order 3. Even if both $\mathbf{AB}$ and $\mathbf{BA}$ are defined and are of the same order, they may not be equal. For example, if

$$\mathbf{A} = \begin{bmatrix} 1 & -2 \\ -3 & 6 \end{bmatrix} \text{ and } \mathbf{B} = \begin{bmatrix} 2 & 10 \\ 1 & 5 \end{bmatrix}$$

then

$$\mathbf{AB} = \begin{bmatrix} 0 & 0 \\ 0 & 0 \end{bmatrix} \text{ and } \mathbf{BA} = \begin{bmatrix} -28 & 56 \\ -14 & 28 \end{bmatrix} \tag{2.4.1}$$

This fact is not surprising in view of *Theorem* 2.2.17 as the composition of linear transformations is not commutative. In *Example* 2.2.18, $h \circ f$ takes the point $(1,0)$ of $\mathbb{R}^2$ to $(\cos\theta, -\sin\theta)$ whereas $f \circ h$ takes $(1,0)$ to $(\cos\theta, \sin\theta)$.

Note that if α, β are scalars then $\beta\alpha = 0$ whenever $\alpha\beta = 0$. Further, if $\alpha\beta = 0$ then at least one of α and β is 0. However, (2.4.1) shows that both these statements are false for matrices. In fact

$$\mathbf{A} \neq \mathbf{0},\ \mathbf{AB} = \mathbf{AC} \not\Rightarrow \mathbf{B} = \mathbf{C}$$

To see this, take $\mathbf{A}, \mathbf{B}$ as above and $\mathbf{C} = \mathbf{0}$.

In the product $\mathbf{AB}$ we say that $\mathbf{B}$ *is premultiplied by* $\mathbf{A}$ and $\mathbf{A}$ *is postmultiplied by* $\mathbf{B}$. We also say that $\mathbf{A}$ and $\mathbf{B}$ *commute* if $\mathbf{AB} = \mathbf{BA}$.

We now extend some of the properties (1) to (21) of the preceding theorem to the case of several matrices. Since matrix addition is associative we can define $\mathbf{A} + \mathbf{B} + \mathbf{C}$ as the matrix $\mathbf{A} + (\mathbf{B} + \mathbf{C})$ which is the same as $(\mathbf{A} + \mathbf{B}) + \mathbf{C}$. The parentheses can be dropped. Similarly we can define $\sum_{i=1}^{k} \mathbf{A}_i = \mathbf{A}_1 + \mathbf{A}_2 + \cdots + \mathbf{A}_k$ and its (i, j)-th element is $(\mathbf{A}_1)_{ij} + (\mathbf{A}_2)_{ij} + \cdots + (\mathbf{A}_k)_{ij}$. Further, since matrix addition is commutative, the matrices in the sum can be written in any order.

Since matrix multiplication is associative, we can define the product $\mathbf{A}_1\mathbf{A}_2 \cdots \mathbf{A}_k$ unambiguously (provided the number of columns of $\mathbf{A}_s$ equals the number of rows of $\mathbf{A}_{s+1}$ for all s) and its (i, j)-th element is (see the proof of (9) in *Theorem* 2.4.2)

$$\sum_{\ell_1} \sum_{\ell_2} \cdots \sum_{\ell_{k-1}} (\mathbf{A}_1)_{i\ell_1} (\mathbf{A}_2)_{\ell_1\ell_2} \cdots (\mathbf{A}_k)_{\ell_{k-1}j}$$

where ℓ_s runs from 1 to the number of columns in $\mathbf{A}_s$ $(s = 1, 2, \ldots, k-1)$. We also note the useful fact that *if* $\mathbf{A}_1\mathbf{A}_2$ *and* $\mathbf{A}_2\mathbf{A}_3$ *are defined then* $\mathbf{A}_1\mathbf{A}_2\mathbf{A}_3$ *is automatically defined*. However, since matrix multiplication is not commutative, the matrices cannot be rearranged in the product $\mathbf{A}_1\mathbf{A}_2 \cdots \mathbf{A}_k$ (the rearranged product may not be defined; even if it is, it may not yield the same matrix).

For any matrix $\mathbf{A}$, we can form the product $\mathbf{AA}$ iff the number of rows of $\mathbf{A}$ equals the number of columns of $\mathbf{A}$. Such a matrix is called a *square matrix*. For a square matrix $\mathbf{A}$, we define $\mathbf{A}^2$ to be $\mathbf{AA}$ and $\mathbf{A}^p$ to be $\mathbf{A}^{p-1}\mathbf{A}$ for any integer $p \geq 2$. When $\mathbf{A}$ is an $n \times n$ non-null matrix, we also define $\mathbf{A}^0 = \mathbf{I}_n$ in analogy with real numbers. It is easy to see that for a square matrix $\mathbf{A}$,

$$\begin{aligned} \mathbf{A}^p \cdot \mathbf{A}^q &= \mathbf{A}^{p+q} \\ (\mathbf{A}^p)^q &= \mathbf{A}^{pq} \end{aligned} \tag{2.4.2}$$

where p and q are non-negative integers. From (2.4.2) it also follows that $\mathbf{A}^p \cdot \mathbf{A}^q = \mathbf{A}^q \cdot \mathbf{A}^p$. Thus powers of $\mathbf{A}$ commute.

Using the distributive laws (11) and (12), we can prove that

$$\left(\sum_{i=1}^{k}\mathbf{A}_i\right)\left(\sum_{j=1}^{\ell}\mathbf{B}_j\right)=\sum_{i=1}^{k}\sum_{j=1}^{\ell}\mathbf{A}_i\mathbf{B}_j$$

where the summations on the RHS can be taken in any order. However, in the product $\mathbf{A}_i\mathbf{B}_j$, $\mathbf{A}_i$ and $\mathbf{B}_j$ cannot be interchanged.

We draw attention to the fact that (20) of *Theorem* 2.4.2 says that the transpose of the product of two matrices is the product of their transposes *taken in the reverse order*. This statement can be extended to several matrices as

$$(\mathbf{A}_1\mathbf{A}_2\cdots\mathbf{A}_k)^{\mathrm{T}}=\mathbf{A}_k^{\mathrm{T}}\mathbf{A}_{k-1}^{\mathrm{T}}\cdots\mathbf{A}_1^{\mathrm{T}} \tag{2.4.3}$$

This can be proved by induction on k as follows. For $k=1$ there is nothing to prove. So, assuming the result for $k=s$, we shall prove it for $k=s+1$, where $s\geq 1$. Now

$$\begin{aligned}(\mathbf{A}_1\cdots\mathbf{A}_s\mathbf{A}_{s+1})^{\mathrm{T}} &= ((\mathbf{A}_1\cdots\mathbf{A}_s)\mathbf{A}_{s+1})^{\mathrm{T}}\\ &= \mathbf{A}_{s+1}^{\mathrm{T}}(\mathbf{A}_1\cdots\mathbf{A}_s)^{\mathrm{T}} \quad \text{by (20) of } \textit{Theorem } 2.4.2\\ &= \mathbf{A}_{s+1}^{\mathrm{T}}(\mathbf{A}_s^{\mathrm{T}}\cdots\mathbf{A}_1^{\mathrm{T}}) \quad \text{by induction hypothesis}\\ &= \mathbf{A}_{s+1}^{\mathrm{T}}\mathbf{A}_s^{\mathrm{T}}\cdots\mathbf{A}_1^{\mathrm{T}}\end{aligned}$$

This proves (2.4.3). From this result it follows that if $\mathbf{A}$ is a square matrix then $(\mathbf{A}^k)^{\mathrm{T}}=(\mathbf{A}^{\mathrm{T}})^k$ for any positive integer k.

Before we conclude this section we define and give some simple properties of polynomials in a square matrix. Let $\mathbf{A}$ be an $n\times n$ matrix over F and let

$$f(t)=\alpha_0+\alpha_1 t+\cdots+\alpha_k t^k \tag{2.4.4}$$

be a polynomial over F (i.e., the coefficients belong to F). We then define $f(\mathbf{A})$ to be the $n\times n$ matrix

$$f(\mathbf{A})=\alpha_0\mathbf{I}+\alpha_1\mathbf{A}+\cdots+\alpha_k\mathbf{A}^k \tag{2.4.5}$$

where $\mathbf{I}$ is the identity matrix of order n. We emphasize that α_0 should be replaced by $\alpha_0\mathbf{I}$ in forming $f(\mathbf{A})$. A matrix of the form (2.4.5) is said to be a *polynomial in* $\mathbf{A}$. From (2.4.2) and the properties of matrix operations proved above, it is not difficult to deduce the following

Theorem 2.4.3 Any two polynomials in a square matrix commute.

Theorem 2.4.4 Let $\mathbf{A}$ be a square matrix, let $f(t)$ and $g(t)$ be two polynomials and let α be a scalar. Then we have the following:

(i) If $h(t) = \alpha f(t)$ then $h(\mathbf{A}) = \alpha f(\mathbf{A})$

(ii) If $h(t) = f(t) + g(t)$ then $h(\mathbf{A}) = f(\mathbf{A}) + g(\mathbf{A})$

(iii) If $h(t) = f(t) \cdot g(t)$ then $h(\mathbf{A}) = f(\mathbf{A}) \cdot g(\mathbf{A})$

(iv) $f(\mathbf{A}^{\mathrm{T}}) = [f(\mathbf{A})]^{\mathrm{T}}$

Proof We give the proof of (iii); the proofs of (i), (ii) and (iv) are relatively simpler and are left to the reader. Let

$$f(t) = \alpha_0 + \alpha_1 t + \cdots + \alpha_k t^k$$

and

$$g(t) = \beta_0 + \beta_1 t + \cdots + \beta_\ell t^\ell$$

Then $h(t) := f(t)g(t) = \gamma_0 + \gamma_1 t + \cdots + \gamma_{k+\ell} t^{k+\ell}$ where

$$\gamma_j = \sum_{i=0}^{j} \alpha_i \beta_{j-i}, \quad j = 0, 1, \ldots, k + \ell$$

and it is understood that $\alpha_i = 0$ if $i > k$ and $\beta_i = 0$ if $i > \ell$. Now

$$\begin{aligned} f(\mathbf{A})g(\mathbf{A}) &= \left(\textstyle\sum_{i=0}^{k} \alpha_i \mathbf{A}^i\right)\left(\textstyle\sum_{p=0}^{\ell} \beta_p \mathbf{A}^p\right) \\ &= \textstyle\sum_{i=0}^{k} \sum_{p=0}^{\ell} \alpha_i \beta_p \mathbf{A}^{i+p} \\ &= \textstyle\sum_{j=0}^{k+\ell} \left(\sum_{i=0}^{j} \alpha_i \beta_{j-i}\right) \mathbf{A}^j \\ &= h(\mathbf{A}). \end{aligned}$$

∎

Exercises

1. Prove that in (1), (2), (5), (9), (10), (11), (12), (18) and (20) of *Theorem* 2.4.2, whenever either side of the equality sign is defined, both sides are defined and are matrices of the same order.
2. Deduce property 12 of *Theorem* 2.4.2 from property 11, property 14 from property 13 and property 17 from property 16.
3. What is the simplest way of computing the product $\mathbf{A}\mathbf{x}\mathbf{y}^{\mathrm{T}}\mathbf{B}$ where $\mathbf{x}$ and $\mathbf{y}$ are column vectors and $\mathbf{A}$ and $\mathbf{B}$ are matrices?
4. Let $\mathbf{x} = (2, 1, -1, 0)^{\mathrm{T}}$, $\mathbf{y} = (-1, 1, 1, 3)^{\mathrm{T}}$, $\mathbf{A} = \begin{bmatrix} 2 & 1 & 3 \\ -1 & 0 & 1 \end{bmatrix}$, $\mathbf{u} = (-1, 4, 2)^{\mathrm{T}}$ and $\mathbf{v} = (1, 4, 3)^{\mathrm{T}}$. Show that the matrix product $\mathbf{A}\mathbf{u}\mathbf{x}^{\mathrm{T}}\mathbf{y}\mathbf{v}^{\mathrm{T}}$ is defined and evaluate it. What is its order?

5. Prove or disprove:

   (a) $\mathbf{A}(\mathbf{B}+\mathbf{C})\mathbf{D} = \mathbf{ABD}+\mathbf{ADC}$.

   (b) $\mathbf{A}(\mathbf{B}+\mathbf{C})\mathbf{D} = \mathbf{ACD}+\mathbf{ABD}$.

   (c) $\mathbf{Axy}^{\mathrm{T}}\mathbf{uv}^{\mathrm{T}}\mathbf{B} = \mathbf{y}^{\mathrm{T}}\mathbf{u}\cdot\mathbf{Axv}^{\mathrm{T}}\mathbf{B}$ where $\mathbf{x}, \mathbf{y}, \mathbf{u}$ and $\mathbf{v}$ are column vectors.

   (d) Given any non-null column vector $\mathbf{x}$, there exists a column vector $\mathbf{y}$ such that $\mathbf{y}^{\mathrm{T}}\mathbf{x} = 1$.

   (e) If $\mathbf{x}$ and $\mathbf{y}$ are column vectors then all columns of $\mathbf{xy}^{\mathrm{T}}$ are scalar multiples of $\mathbf{x}$.

   (f) Let $\mathbf{x}$ and $\mathbf{y}$ be non-null vectors in F^n such that $\mathbf{y}^{\mathrm{T}}\mathbf{x} = 0$. Then $\mathbf{z}^{\mathrm{T}}\mathbf{x} = 0$ implies $\mathbf{z} = \alpha\mathbf{y}$ for some scalar α.

6. Though '$\mathbf{AB} = \mathbf{0} \Rightarrow \mathbf{A} = \mathbf{0}$ or $\mathbf{B} = \mathbf{0}$' is false in general, prove that it is true when $\mathbf{A}$ is a column matrix and $\mathbf{B}$ is a row matrix.

7. (a) Show that $(\mathbf{A}+\mathbf{B})^2 = \mathbf{A}^2 + 2\mathbf{AB} + \mathbf{B}^2$ iff $\mathbf{AB} = \mathbf{BA}$.

   (b) Show that $(\mathbf{A}+\mathbf{B})(\mathbf{A}-\mathbf{B}) = \mathbf{A}^2 - \mathbf{B}^2$ iff $\mathbf{AB} = \mathbf{BA}$.

8. If $\mathbf{AB} = \mathbf{BA}$, prove that

   (a) $\mathbf{A}^k$ and $\mathbf{B}^\ell$ commute for any non-negative integers k and ℓ,

   (b) $\mathbf{A}$ commutes with $f(\mathbf{B})$ for every polynomial f,

   (c) for any given positive integer k there exists a matrix $\mathbf{C}$ such that $\mathbf{A}^k - \mathbf{B}^k = (\mathbf{A}-\mathbf{B})\mathbf{C}$.

9. Deduce *Theorem* 2.4.3 from *Theorem* 2.4.4.

10. Show that the set $F^{m\times n}$ of all $m \times n$ matrices over F forms a vector space over F under the natural operations. Find a basis of this vector space and hence its dimension.

11. Consider the set $B = \{\mathbf{I}, \mathbf{A}, \mathbf{A}^2, \ldots\}$ in the vector space $\mathbb{R}^{2\times 2}$ of all 2×2 real matrices, where $\mathbf{A} = \begin{bmatrix} 0 & 1 \\ 0 & 0 \end{bmatrix}$.

   (a) Is the span of B equal to the set of all upper triangular matrices of order 2×2?

   (b) Obtain a basis of $\mathrm{Sp}(B)$. Extend it to a basis of $\mathbb{R}^{2\times 2}$.

*12. Show that given any square matrix $\mathbf{A}$, there exists a non-zero polynomial $f(t)$ such that $f(\mathbf{A}) = \mathbf{0}$.

2.5 Matrices with special structures

In this section we introduce some matrices with special (simple) structures and study them.

As already stated in the preceding section, a matrix of order $m \times n$ is said to be a *square matrix* if $m = n$. We also refer to such a matrix as a (square) *matrix of order n*. Some authors use the term *rectangular matrix* to denote a non-square matrix.

A $1 \times n$ matrix is called a *row vector* and an $m \times 1$ matrix is called a *column vector*.

> *From now on, whenever we want to denote a column vector by a symbol, we will use* $\mathbf{x}, \mathbf{y}, \mathbf{z}$, *etc. For row vectors we will use* $\mathbf{x}^{\mathrm{T}}$, $\mathbf{y}^{\mathrm{T}}$, $\mathbf{z}^{\mathrm{T}}$, *etc., where* $\mathbf{x}, \mathbf{y}, \mathbf{z}$ *are the corresponding column vectors.*

We note that if a row vector $\mathbf{x}^{\mathrm{T}}$ can be postmultiplied by a column vector $\mathbf{y}$ then $\mathbf{x}$ and $\mathbf{y}$ must have the same number of components and the product $\mathbf{x}^{\mathrm{T}}\mathbf{y}$ is a 1×1 matrix. However, any column vector $\mathbf{y}$ can be postmultiplied by any row vector $\mathbf{z}^{\mathrm{T}}$; the product $\mathbf{y}\mathbf{z}^{\mathrm{T}}$ is a matrix. For example, if

$$\mathbf{x} = \begin{bmatrix} 1 \\ 0 \\ -2 \end{bmatrix}, \ \mathbf{y} = \begin{bmatrix} 2 \\ 3 \\ 5 \end{bmatrix} \text{ and } \mathbf{z} = \begin{bmatrix} 1 \\ 0 \\ -3 \\ 4 \end{bmatrix}$$

then

$$\mathbf{x}^{\mathrm{T}}\mathbf{y} = (1 \ \ 0 \ \ -2)\begin{bmatrix} 2 \\ 3 \\ 5 \end{bmatrix} = [-8]$$

and

$$\mathbf{y}\mathbf{z}^{\mathrm{T}} = \begin{bmatrix} 2 \\ 3 \\ 5 \end{bmatrix}(1, 0, -3, 4) = \begin{bmatrix} 2 & 0 & -6 & 8 \\ 3 & 0 & -9 & 12 \\ 5 & 0 & -15 & 20 \end{bmatrix}$$

However, $\mathbf{x}\mathbf{y}$, $\mathbf{x}^{\mathrm{T}}\mathbf{y}^{\mathrm{T}}$, $\mathbf{y}^{\mathrm{T}}\mathbf{z}$, $\mathbf{z}^{\mathrm{T}}\mathbf{y}$ are not defined.

Since the operations on 1×1 matrices (viz., addition and multiplication) coincide with the corresponding operations in the base field, we usually do not distinguish between 1×1 matrices and scalars.

For a square matrix $\mathbf{A} = ((a_{ij}))$ of order n, the elements $a_{11}, a_{22}, \ldots,$ a_{nn} form the *principal diagonal* or, simply, the *diagonal* of $\mathbf{A}$. We call

a_{kk} the k-th *diagonal element of* **A**. The elements not on the diagonal are called the *off-diagonal elements.* For example, in the square matrix shown below, the elements in the box are the diagonal elements and the remaining elements are the off-diagonal elements.

$$\begin{bmatrix} \boxed{2} & 1 & 3 \\ -3 & \boxed{0} & 4 \\ 0 & 2 & \boxed{-1} \end{bmatrix}$$

Definition 2.5.1 The *trace* of a square matrix **A** is the sum of all the diagonal elements of **A** and is denoted $\text{tr}(\mathbf{A})$.

Clearly $\text{tr}(\mathbf{A}+\mathbf{B}) = \text{tr}(\mathbf{A}) + \text{tr}(\mathbf{B})$ if **A** and **B** are square matrices of the same order and $\text{tr}(\alpha\mathbf{A}) = \alpha \cdot \text{tr}(\mathbf{A})$ if α is a scalar.

Theorem 2.5.2 If both the matrices **AB** and **BA** are defined, $\text{tr}(\mathbf{AB}) = \text{tr}(\mathbf{BA})$.

Proof Let **A** be of order $m \times n$. Then by hypothesis, **B** is of order $n \times m$, so both **AB** and **BA** are square matrices. Now

$$\begin{aligned} \text{tr}(\mathbf{AB}) &= \textstyle\sum_{i=1}^{m} (\mathbf{AB})_{ii} \\ &= \textstyle\sum_{i=1}^{m} \sum_{j=1}^{n} a_{ij} b_{ji} \\ &= \textstyle\sum_{j=1}^{n} \sum_{i=1}^{m} b_{ji} a_{ij} \\ &= \textstyle\sum_{j=1}^{n} (\mathbf{BA})_{jj} \\ &= \text{tr}(\mathbf{BA}) \end{aligned}$$ ∎

A square matrix $\mathbf{A} = ((a_{ij}))$ is said to be *symmetric* if $a_{ij} = a_{ji}$ for all i and j. For example,

$$\begin{bmatrix} 1 & 0 & 3 \\ 0 & -1 & -4 \\ 3 & -4 & 0 \end{bmatrix}$$

is a symmetric matrix. Note that symmetry of a matrix refers to symmetry *about the principal diagonal*. Clearly **A** is symmetric iff $\mathbf{A} = \mathbf{A}^{\mathrm{T}}$.

A square matrix in which all the off-diagonal elements are 0 is called a *diagonal matrix*. Thus $\mathbf{A} = ((a_{ij}))$ is a diagonal matrix iff $a_{ij} = 0$

whenever $i \neq j$. The $n \times n$ diagonal matrix with k-th diagonal entry d_k for $k = 1, \ldots, n$, is denoted by

$$\mathrm{diag}(d_1, d_2, \ldots, d_n)$$

For example,

$$\mathrm{diag}(1,0,1,-3) = \begin{bmatrix} 1 & 0 & 0 & 0 \\ 0 & 0 & 0 & 0 \\ 0 & 0 & 1 & 0 \\ 0 & 0 & 0 & -3 \end{bmatrix}$$

A diagonal matrix in which all diagonal elements are equal is called a *scalar matrix*. Such a matrix can be written as $\alpha\mathbf{I}$ for some scalar α and behaves like a scalar (e.g., $\alpha\mathbf{I} + \beta\mathbf{I} = (\alpha+\beta)\mathbf{I}$ and $\alpha\mathbf{I} \cdot \mathbf{A} = \alpha\mathbf{A}$.)

A square matrix $\mathbf{A} = ((a_{ij}))$ is called an *upper triangular matrix* if all elements below the diagonal are 0, that is,

$$a_{ij} = 0 \quad \text{whenever } i > j$$

A square matrix $\mathbf{B} = ((b_{ij}))$ is called a *lower triangular matrix* if all elements above the diagonal are 0, that is, $b_{ij} = 0$ whenever $i < j$. A square matrix is said to be *triangular* if it is either upper triangular or lower triangular. In the following two matrices, $\mathbf{A}$ is upper triangular and $\mathbf{B}$ is lower triangular.

$$\mathbf{A} = \begin{bmatrix} 1 & 2 & 0 \\ 0 & 0 & -3 \\ 0 & 0 & -1 \end{bmatrix}, \ \mathbf{B} = \begin{bmatrix} -2 & 0 & 0 \\ 2 & 1 & 0 \\ 1 & -7 & 0.3 \end{bmatrix}$$

The statements given below follow directly from the above definitions.

1. Each of $\mathbf{0}_{n\times n}$ and $\mathbf{I}_{n\times n}$ is a symmetric matrix, a diagonal matrix, a scalar matrix, an upper triangular matrix and a lower triangular matrix.

2. If $\mathbf{A}, \mathbf{B}$ are symmetric matrices of the same order and α is any scalar then $\alpha\mathbf{A}$ and $\mathbf{A}+\mathbf{B}$ are symmetric matrices. Note, however, that $\mathbf{AB}$ may not be symmetric. For example:

$$\text{If } \mathbf{A} = \begin{bmatrix} 2 & 1 \\ 1 & 4 \end{bmatrix} \text{ and } \mathbf{B} = \begin{bmatrix} 3 & 4 \\ 4 & 1 \end{bmatrix} \text{ then } \mathbf{AB} = \begin{bmatrix} 10 & 9 \\ 19 & 8 \end{bmatrix}.$$

In fact, when $\mathbf{A}$ and $\mathbf{B}$ are symmetric, $\mathbf{AB}$ is symmetric iff $\mathbf{A}$ and $\mathbf{B}$ commute.

3. $\mathbf{A}$ is upper triangular iff $\mathbf{A}^{\mathrm{T}}$ is lower triangular.

4. $\mathbf{A}$ is diagonal iff $\mathbf{A}$ is both upper triangular and lower triangular. Also $\mathbf{A}$ is diagonal iff it is symmetric and upper triangular.

5. The sum of two upper (resp. lower) triangular matrices is upper (resp. lower) triangular. A scalar multiple of an upper (resp. lower) triangular matrix is upper (resp. lower) triangular.

Theorem 2.5.3 If $\mathbf{A} = ((a_{ij}))$ and $\mathbf{B} = ((b_{ij}))$ are upper triangular matrices of order n, then $\mathbf{AB}$ is upper triangular with $a_{kk}b_{kk}$ as the k-th diagonal entry.

Proof Let $\mathbf{AB} = \mathbf{C} = ((c_{ij}))$. Fix $i \geq k$. Then

$$\begin{aligned} c_{ik} &= \textstyle\sum_{j=1}^{m} a_{ij}b_{jk} \\ &= \textstyle\sum_{j=1}^{k-1} a_{ij}b_{jk} + a_{ik}b_{kk} + \sum_{j=k+1}^{n} a_{ij}b_{jk} \end{aligned} \tag{2.5.1}$$

Since $\mathbf{A}$ is upper triangular, $a_{ij} = 0$ for $j = 1, \ldots, k-1$, thus the first sum on the RHS of (2.5.1) is 0. Since $\mathbf{B}$ is upper triangular, $b_{jk} = 0$ for $j = k+1, \ldots, n$, so the last sum on the RHS of (2.5.1) is 0. Thus $c_{ik} = a_{ik}b_{kk}$. Taking $i = k$, we get $c_{kk} = a_{kk}b_{kk}$. When $i > k$, we get $c_{ik} = 0$ since $a_{ik} = 0$. Thus $\mathbf{C}$ is upper triangular. ∎

Theorem 2.5.4 The product of two lower triangular matrices is lower triangular.

This theorem can be proved easily by imitating the proof of the preceding theorem. We now give an interesting alternative proof using the concept of transpose. Let $\mathbf{A}$ and $\mathbf{B}$ be two $n \times n$ lower triangular matrices. Then $\mathbf{A}^{\mathrm{T}}$ and $\mathbf{B}^{\mathrm{T}}$ are upper triangular, hence $\mathbf{B}^{\mathrm{T}}\mathbf{A}^{\mathrm{T}}$ is upper triangular by the preceding theorem. So

$$\mathbf{AB} = ((\mathbf{AB})^{\mathrm{T}})^{\mathrm{T}} = (\mathbf{B}^{\mathrm{T}}\mathbf{A}^{\mathrm{T}})^{\mathrm{T}}$$

is lower triangular, which proves the present theorem. This technique (of using transposes) is quite powerful and can be used for proving several results. For example, results 12, 14 and 17 of *Theorem* 2.4.2 can be deduced from 11, 13 and 16 respectively by this technique. We will come across several other uses in the sequel.

Exercises

1. Prove that an $n \times n$ matrix $\mathbf{A}$ commutes with every $n \times n$ matrix iff $\mathbf{A}$ is a scalar matrix.

2. Find all matrices which commute with $\begin{bmatrix} 0 & 1 \\ 0 & 0 \end{bmatrix}$.

3. (a) If $\mathbf{x}$ is a column vector with n components and $\mathbf{A}$ is an $n \times n$ matrix, show that $\mathbf{x}^{\mathrm{T}}\mathbf{A}\mathbf{x} = \mathrm{tr}(\mathbf{A}\mathbf{x}\mathbf{x}^{\mathrm{T}})$.
   (b) Show that $\mathrm{tr}(\mathbf{ABC})$ equals $\mathrm{tr}(\mathbf{BCA})$ but not $\mathrm{tr}(\mathbf{BAC})$.
   (c) Show that trace is a linear functional on the vector space of all $n \times n$ matrices.
   (d) Do there exist square matrices $\mathbf{A}$ and $\mathbf{B}$ of the same order such that $\mathbf{AB} - \mathbf{BA} = \mathbf{I}$? Why?

4. If $\mathbf{A} = \begin{bmatrix} a & b \\ 0 & 1 \end{bmatrix}$, find $\mathbf{A}^k$ for all $k \geq 2$.

5. If $\mathbf{A}$ is a square matrix of order n such that $a_{ij} = 0$ whenever $i \geq j$, show that $\mathbf{A}^n = \mathbf{0}$.

*6. A square matrix $\mathbf{A}$ is said to be an *upper k-diagonal matrix* if $a_{ij} = 0$ whenever $i > j$ or $j > i + k - 1$. Show that if $\mathbf{A}$ and $\mathbf{B}$ are upper k-diagonal matrices then $\mathbf{AB}$ is an upper $(2k-1)$-diagonal matrix.

7. If $\mathbf{A} = \begin{bmatrix} 3 & 1 \\ 2 & 0 \end{bmatrix}$, find $\mathbf{B} = \mathbf{A}^3 - 3\mathbf{A}^2 + 2\mathbf{A}$, $\mathbf{C} = -2\mathbf{A}^2 + 3\mathbf{A} + \mathbf{I}$, $\mathbf{BC}$ and $\mathbf{CB}$.

8. Show that the upper triangular matrices form a subspace S of the vector space of all $n \times n$ matrices and that the set T of all strictly lower triangular (i.e., $a_{ij} = 0$ whenever $i \leq j$) matrices is a complement of S. When $n = 3$, find the projection of $\begin{bmatrix} 2 & 1 & 0 \\ 3 & 0 & 2 \\ 0 & 0 & 4 \end{bmatrix}$ into S along T.

9. Show that the set S of all 3×3 symmetric matrices forms a subspace of the vector space of all 3×3 matrices. What is its dimension? Obtain a complement of S.

10. Let $\mathbf{A}$ be a real $m \times n$ matrix. Show that each diagonal element of $\mathbf{AA}^{\mathrm{T}}$ is non-negative. If the i-th diagonal element of $\mathbf{AA}^{\mathrm{T}}$ is 0, show that the i-th rows of $\mathbf{A}$ and $\mathbf{AA}^{\mathrm{T}}$ are null and the i-th column of $\mathbf{AA}^{\mathrm{T}}$ is null.

11. Let $\mathbf{B}$ be a real symmetric matrix with positive diagonal elements. Does it follow that $\mathbf{B} = \mathbf{AA}^{\mathrm{T}}$ for some matrix $\mathbf{A}$? Justify your answer.

12. Let $\mathbf{C}$ be an $m \times n$ matrix. If $\mathbf{A}$ is a symmetric matrix of order n, show that $\mathbf{CAC}^{\mathrm{T}}$ is symmetric. Deduce that $\mathbf{CC}^{\mathrm{T}}$ and $\mathbf{C}^{\mathrm{T}}\mathbf{C}$ are symmetric. Is $\mathbf{CC}^{\mathrm{T}} = \mathbf{C}^{\mathrm{T}}\mathbf{C}$? Why?

13. Let $\mathbf{A}$ be an $n \times n$ matrix such that $\mathbf{CAC}^{\mathrm{T}}$ is symmetric for all matrices $\mathbf{C}$ of order $m \times n$ where m is a fixed positive integer ≥ 2. Show that $\mathbf{A}$ is symmetric.

14. Let $\mathbf{A}$ be an arbitrary (not necessarily symmetric) $n \times n$ matrix and $\mathbf{x}$ an $n \times 1$ vector. Show that $\mathbf{x}^{\mathrm{T}}\mathbf{A}\mathbf{x}$ is symmetric.

*15. Define an $n \times n$ matrix $\mathbf{A}$ to be pseudo-symmetric if $a_{ij} = a_{n+1-j,n+1-i}$ for all i and j. Show that if $\mathbf{A}$ is pseudo-symmetric then so is $\mathbf{A}^2$.

2.6 Rows and columns of a matrix product

We start with a notation for the rows and columns of a matrix which will be used in the sequel.

Notation $\mathbf{A}_{i*}$ *denotes the* i-th *row of* $\mathbf{A}$ *and* $\mathbf{A}_{*j}$ *denotes the* j-th *column of* $\mathbf{A}$.

Theorem 2.6.1 If $\mathbf{A}$ is an $m \times n$ matrix and $\mathbf{x} = (x_1, x_2, \ldots, x_n)^{\mathrm{T}}$ is a column vector with n components then

$$\mathbf{Ax} = x_1\mathbf{A}_{*1} + x_2\mathbf{A}_{*2} + \cdots + x_n\mathbf{A}_{*n}$$

Proof Clearly both sides of the conclusion are column vectors with m components. Thus to prove equality, it is enough to show that the i-th components of the two are equal for $i = 1, \ldots, m$. Now it is easy to see that the i-th components of both sides are $a_{i1}x_1 + a_{i2}x_2 + \cdots + a_{in}x_n$.∎

We draw attention to the fact that even though $\mathbf{A}$ is postmultiplied by $\mathbf{x}$ in forming the product $\mathbf{Ax}$, $\mathbf{Ax}$ *is a linear combination of the columns of* $\mathbf{A}$ *and not the rows of* $\mathbf{A}$. In fact $\mathbf{Ax}$ is a (column) vector with m components whereas each row of $\mathbf{A}$ is a (row) vector with n components.

Theorem 2.6.2 If $\mathbf{A}$ is an $m \times n$ matrix and $\mathbf{y}^{\mathrm{T}} = (y_1, y_2, \ldots, y_m)$ is a row vector with m components then

$$\mathbf{y}^{\mathrm{T}}\mathbf{A} = y_1\mathbf{A}_{1*} + y_2\mathbf{A}_{2*} + \cdots + y_m\mathbf{A}_{m*}$$

This theorem follows from the preceding theorem. To illustrate the above theorems, let

$$\mathbf{A} = \begin{bmatrix} 1 & 0 & -1 & 2 \\ -3 & 1 & 0 & 1 \\ 2 & 5 & -1 & 6 \end{bmatrix}, \ \mathbf{x} = \begin{bmatrix} 2 \\ 0 \\ -3 \\ 1 \end{bmatrix} \text{ and } \mathbf{y} = \begin{bmatrix} 1 \\ -1 \\ 3 \end{bmatrix}$$

Then

$$\mathbf{Ax} = 2\begin{bmatrix} 1 \\ -3 \\ 2 \end{bmatrix} + 0\begin{bmatrix} 1 \\ 0 \\ 5 \end{bmatrix} + (-3)\begin{bmatrix} -1 \\ 0 \\ -1 \end{bmatrix} + 1\begin{bmatrix} 2 \\ 1 \\ 6 \end{bmatrix} = \begin{bmatrix} 7 \\ -5 \\ 13 \end{bmatrix} \text{ and}$$

$\mathbf{y}^{\mathrm{T}}\mathbf{A} = 1\cdot(1,0,-1,2)+(-1)(-3,1,0,1)+3(2,5,-1,6) = (10,14,-4,19)$

Theorem 2.6.3 Let $\mathbf{A}, \mathbf{B}, \mathbf{C}$ be matrices of orders $m \times n$, $n \times p$, and $p \times q$ respectively. Then

(i) $(\mathbf{AB})_{ij} = \mathbf{A}_{i*}\mathbf{B}_{*j}$,

(ii) $(\mathbf{AB})_{i*} = \mathbf{A}_{i*}\mathbf{B}$,

(iii) $(\mathbf{AB})_{*j} = \mathbf{AB}_{*j}$,

(iv) $(\mathbf{ABC})_{ij} = \mathbf{A}_{i*}\mathbf{BC}_{*j}$.

Proof Statement (i) follows from definition. Statements (ii) and (iii) can be verified easily. Statement (iv) follows from (i) and (ii) thus: $(\mathbf{ABC})_{ij} = (\mathbf{AB})_{i*}\mathbf{C}_{*j} = \mathbf{A}_{i*}\mathbf{BC}_{*j}$. ∎

Corollary 1 The columns of $\mathbf{AB}$ are linear combinations of the columns of $\mathbf{A}$ and the rows of $\mathbf{AB}$ are linear combinations of the rows of $\mathbf{B}$.

Corollary 2 For any $m \times n$ matrix $\mathbf{A}$, we have

$$\mathbf{A}_{i*} = \mathbf{e}_i^{\mathrm{T}}\mathbf{A} = (0,\ldots,0,1,0,\ldots,0)\mathbf{A},$$

$$\mathbf{A}_{*j} = \mathbf{Ae}_j = \mathbf{A}\begin{bmatrix}0\\ \vdots\\ 0\\ 1\\ 0\\ \vdots\\ 0\end{bmatrix} \quad\text{and}\quad (\mathbf{A})_{ij} = \mathbf{e}_i^{\mathrm{T}}\mathbf{Ae}_j = (0,\ldots,0,1,0,\ldots,0)\,\mathbf{A}\begin{bmatrix}0\\ \vdots\\ 0\\ 1\\ 0\\ \vdots\\ 0\end{bmatrix}$$

Here the $\mathbf{e}_i^{\mathrm{T}}$ which premultiplies $\mathbf{A}$ has m components and the $\mathbf{e}_j$ which postmultiplies $\mathbf{A}$ has n components. The 1's in the row vectors occur in the i-th position and the 1's in the column vectors occur in the j-th position.

Corollary 2 follows easily from the theorem on noting that $\mathbf{e}_i^{\mathrm{T}}$ is the i-th row of $\mathbf{I}_m$ and $\mathbf{e}_j$ is the j-th column of $\mathbf{I}_n$. As an illustration of the use of this corollary we can immediately deduce the following

Remark Let $\mathbf{A}$ be a matrix over F with n columns.

(i) If $\mathbf{Ax} = \mathbf{0}$ for all $\mathbf{x} \in F^n$ then $\mathbf{A} = \mathbf{0}$

(ii) If $\mathbf{Ax} = \mathbf{x}$ for all $\mathbf{x} \in F^n$ then $\mathbf{A} = \mathbf{I}$

Theorem 2.6.4 If $\mathbf{A}$ is an $m \times n$ matrix then

(i) $\text{diag}(\alpha_1, \ldots, \alpha_m) \cdot \mathbf{A} = \begin{bmatrix} \alpha_1 \mathbf{A}_{1*} \\ \alpha_2 \mathbf{A}_{2*} \\ \vdots \\ \alpha_m \mathbf{A}_{m*} \end{bmatrix}$

(ii) $\mathbf{A} \cdot \text{diag}(\beta_1, \ldots, \beta_n) = [\beta_1 \mathbf{A}_{*1} : \beta_2 \mathbf{A}_{*2} : \cdots : \beta_n \mathbf{A}_{*n}]$.

Proof By (ii) of the preceding theorem, the i-th row of the LHS of (i) is $(0, \ldots, 0, \alpha_i, 0, \ldots, 0)\mathbf{A}$. By *Theorem* 2.6.2, this is $0\mathbf{A}_{1*} + \cdots + 0\mathbf{A}_{i-1,*} + \alpha_i \mathbf{A}_{i*} + 0\mathbf{A}_{i+1,*} + \cdots + 0\mathbf{A}_{m*} = \alpha_i \mathbf{A}_{i*}$, which is the i-th row of the RHS of (i). Since i is arbitrary, (i) follows. The proof of (ii) is similar. ∎

The preceding theorem can be restated in words thus: *premultiplying (resp. postmultiplying) a matrix* **A** *by a diagonal matrix is equivalent to multiplying the i-th row (resp. column) of* **A** *by the i-th diagonal entry of the diagonal matrix*. For example, if

$$\mathbf{A} = \begin{bmatrix} 3 & 1 & -2 & 0 \\ 0 & 1 & 5 & 8 \\ 2 & 0 & 0 & 3 \end{bmatrix}, \ \mathbf{D}_1 = \text{diag}(2, 4, -3) \text{ and } \mathbf{D}_2 = \text{diag}(1, -2, 3, 2)$$

then

$$\mathbf{D}_1 \mathbf{A} = \begin{bmatrix} 6 & 2 & -4 & 0 \\ 0 & 4 & 20 & 32 \\ -6 & 0 & 0 & -9 \end{bmatrix} \text{ and } \mathbf{A}\mathbf{D}_2 = \begin{bmatrix} 3 & -2 & -6 & 0 \\ 0 & -2 & -15 & 16 \\ 2 & 0 & 0 & 6 \end{bmatrix}$$

Definition 2.6.5 A *submatrix* of a matrix **A** is the matrix obtained from **A** by deleting a (possibly empty) set of rows and a (possibly empty) set of columns.

Let **A** be an $m \times n$ matrix and let $1 \le i_1 < i_2 < \cdots < i_k \le m$ and $1 \le j_1 < j_2 < \cdots < j_\ell \le n$. Then the submatrix of **A** obtained by deleting all the rows other than the i_1-th,i_2-th,...,i_k-th rows and deleting all the columns other than the j_1-th,j_2-th,...,j_ℓ-th columns is denoted by $\mathbf{A}(i_1, i_2, \ldots, i_k \mid j_1, j_2, \ldots, j_\ell)$. Clearly this is

$$\begin{bmatrix} a_{i_1 j_1} & a_{i_1 j_2} & \cdots & a_{i_1 j_\ell} \\ a_{i_2 j_1} & a_{i_2 j_2} & \cdots & a_{i_2 j_\ell} \\ \cdots & \cdots & \cdots & \cdots \\ a_{i_k j_1} & a_{i_k j_2} & \cdots & a_{i_k j_\ell} \end{bmatrix}$$

Notice that the rows and columns of a submatrix of $\mathbf{A}$ *appear in the same order as in* $\mathbf{A}$. Clearly any row or column of $\mathbf{A}$ can be viewed as a submatrix of $\mathbf{A}$.

Definition 2.6.6 A submatrix $\mathbf{A}(i_1, i_2, \ldots, i_k \,|\, j_1, j_2, \ldots, j_\ell)$ of a square matrix $\mathbf{A}$ is said to be a *principal submatrix* if $k = \ell$ and $i_1 = j_1$, $i_2 = j_2, \ldots, i_k = j_k$. A *leading principal submatrix of* $\mathbf{A}$ is a submatrix of the form $\mathbf{A}(1, 2, \ldots, k \,|\, 1, 2, \ldots, k)$.

A principal submatrix of a square matrix is obtained by deleting some rows and the *corresponding* columns. A leading principal submatrix is obtained by deleting the last few rows and the corresponding columns.

Example 2.6.7 Let

$$\mathbf{A} = \begin{bmatrix} 2 & 0 & -1 & 4 & 1 \\ 1 & -1 & 2 & 4 & 0 \\ 0 & 1 & -2 & 1 & 1 \\ 2 & 5 & 6 & 0 & 1 \end{bmatrix}$$

Then

$$\mathbf{B} = \mathbf{A}(1, 2, 4 \,|\, 1, 3, 5) = \begin{bmatrix} 2 & -1 & 1 \\ 1 & 2 & 0 \\ 2 & 6 & 1 \end{bmatrix}$$

is a submatrix of $\mathbf{A}$. The principal submatrix $\mathbf{B}(1, 3 \,|\, 1, 3)$ of $\mathbf{B}$ and the leading 2×2 principal submatrix of $\mathbf{B}$ are

$$\begin{bmatrix} 2 & 1 \\ 2 & 1 \end{bmatrix} \text{ and } \begin{bmatrix} 2 & -1 \\ 1 & 2 \end{bmatrix}$$

Clearly the submatrix $\mathbf{A}(1 \,|\, 1, 2, 3, 4, 5)$ is the first row of $\mathbf{A}$ and $\mathbf{A}(1, 2, 3, 4 \,|\, 3)$ is the third column of $\mathbf{A}$.

Exercises

1. If $\mathbf{A} = \begin{bmatrix} 5 & 4 & 0 \\ 1 & 3 & 8 \\ 2 & 6 & 12 \end{bmatrix}$, find column vectors $\mathbf{u}$ and $\mathbf{v}$ such that $\mathbf{u}^{\mathrm{T}}\mathbf{A}\mathbf{v} = 8$. Are $\mathbf{u}$ and $\mathbf{v}$ unique?

2. If $\mathbf{u}$ and $\mathbf{v}$ are column vectors with n and m components respectively and if $\mathbf{u} \neq \mathbf{0}$, show that there exists an $m \times n$ matrix $\mathbf{A}$ such that $\mathbf{A}\mathbf{u} = \mathbf{v}$.

3. A *permutation matrix* is a square matrix with entries 0's and 1's such that each row and each column contains exactly one 1. Show that pre- (resp. post-) multiplying a matrix $\mathbf{A}$ by a permutation matrix permutes the rows (resp. columns) of $\mathbf{A}$.
4. Let $\mathbf{A}$ be an $n \times n$ matrix. If $\operatorname{tr}(\mathbf{AB}) = 0$ for all $n \times n$ matrices $\mathbf{B}$, show that $\mathbf{A} = \mathbf{0}$.
5. Let f be a linear functional on the vector space $F^{n\times n}$ of all $n\times n$ matrices such that (i) $f(\mathbf{AB}) = f(\mathbf{BA})$ for all $\mathbf{A}, \mathbf{B} \in F^{n\times n}$ and (ii) $f(\mathbf{I}) = n$. Then prove that $f(\mathbf{A}) = \operatorname{tr}(\mathbf{A})$ for all $\mathbf{A}$. If condition (ii) is dropped, what can you say about $f(\mathbf{A})$?

*6. Let f be a linear functional on $F^{n\times n}$ such that $f(\mathbf{A}) = 0$ whenever $\mathbf{A}^2 = \mathbf{0}$. Show that $f(\mathbf{A}) = c\operatorname{tr}(\mathbf{A})$ for some scalar c.

7. If $\mathbf{A}$ is an $n \times n$ matrix and $\mathbf{x}, \mathbf{y}$ are $n \times 1$ vectors, show that $\mathbf{x}^{\mathrm{T}}\mathbf{A}\mathbf{y} = \sum_{i=1}^{n}\sum_{j=1}^{n} a_{ij}x_iy_j$. *Without* using this, prove that $\mathbf{x}^{\mathrm{T}}\mathbf{A}\mathbf{y} = \mathbf{y}^{\mathrm{T}}\mathbf{A}\mathbf{x}$ if $\mathbf{A}$ is symmetric.
8. Let $\mathbf{A}$ be a square matrix and $\mathbf{B} = \frac{1}{2}(\mathbf{A} + \mathbf{A}^{\mathrm{T}})$. Prove the following.
   (a) $\mathbf{B}$ is symmetric.
   (b) $\mathbf{x}^{\mathrm{T}}\mathbf{B}\mathbf{x} = \mathbf{x}^{\mathrm{T}}\mathbf{A}\mathbf{x}$ for all $n \times 1$ vectors $\mathbf{x}$.
   (c) If $\mathbf{C}$ is a symmetric matrix such that $\mathbf{x}^{\mathrm{T}}\mathbf{C}\mathbf{x} = \mathbf{x}^{\mathrm{T}}\mathbf{A}\mathbf{x}$ for all $\mathbf{x}$, then $\mathbf{C} = \mathbf{B}$.
9. Find the effect of premultiplying a $2 \times n$ matrix by each of the matrices $\begin{bmatrix} 1 & 0 \\ \alpha & 1 \end{bmatrix}$, $\begin{bmatrix} 1 & \alpha \\ 0 & 1 \end{bmatrix}$ and $\begin{bmatrix} \alpha & 0 \\ 0 & 1 \end{bmatrix}$. Obtain corresponding results for postmultiplication.
10. Let $\mathbf{A}$ and $\mathbf{B}$ be $m \times n$ matrices. Show that $\mathbf{A} = \mathbf{BC}$ for some upper triangular matrix $\mathbf{C}$ iff the j-th column of $\mathbf{A}$ is a linear combination of the first j columns of $\mathbf{B}$ for $j = 1, \ldots, n$.
11. (a) If $\mathbf{A}$ is an $m \times n$ matrix and if $\mathbf{A}\mathbf{x}_1 = \mathbf{0}, \ldots, \mathbf{A}\mathbf{x}_n = \mathbf{0}$ for some basis $\{\mathbf{x}_1, \mathbf{x}_2, \ldots, \mathbf{x}_n\}$ of F^n, show that $\mathbf{A} = \mathbf{0}$.
    (b) If $\mathbf{A}$ is an $n \times n$ matrix and if $\mathbf{A}\mathbf{x}_1 = \mathbf{x}_1, \ldots, \mathbf{A}\mathbf{x}_n = \mathbf{x}_n$ for some basis $\{\mathbf{x}_1, \mathbf{x}_2, \ldots, \mathbf{x}_n\}$ of F^n, show that $\mathbf{A} = \mathbf{I}$.
12. (a) If $\mathbf{y}^{\mathrm{T}}\mathbf{A}\mathbf{x} = 0$ for all $\mathbf{A}$ and if $\mathbf{x} \neq \mathbf{0}$, prove that $\mathbf{y} = \mathbf{0}$.
    (b) If $\mathbf{y}^{\mathrm{T}}\mathbf{A}\mathbf{x} = 0$ for all $\mathbf{A}$ and if $\mathbf{y} \neq \mathbf{0}$, prove that $\mathbf{x} = \mathbf{0}$.
13. Prove that if $\mathbf{A}$ is an $m \times n$ matrix, $1 \le i_1 < i_2 < \cdots < i_k \le m$, and $1 \le j_1 < j_2 < \cdots < j_\ell \le n$, then
$$\mathbf{A}(i_1, \ldots, i_k \,|\, j_1, \ldots, j_\ell) = \mathbf{I}_m(i_1, \ldots, i_k \,|\, 1, \ldots, m)\mathbf{A}\mathbf{I}_n(1, \ldots, n \,|\, j_1, \ldots, j_\ell)$$
14. Let $\mathbf{P}$ be the $n \times n$ matrix

$$\mathbf{P} = \begin{bmatrix} 0 & 0 & \cdots & 0 & 1 \\ 0 & 0 & \cdots & 1 & 0 \\ \cdots & \cdots & \cdots & \cdots & \cdots \\ 0 & 1 & \cdots & 0 & 0 \\ 1 & 0 & \cdots & 0 & 0 \end{bmatrix}$$

(a) If $\mathbf{x} = (x_1, x_2, \ldots, x_n)^{\mathrm{T}}$, show that $\mathbf{Px} = (x_n, x_{n-1}, \ldots, x_1)^{\mathrm{T}}$.

(b) Show that $\mathbf{P}^{\mathrm{T}} = \mathbf{P}$ and $\mathbf{P}^2 = \mathbf{I}$.

(c) If $\mathbf{A} = ((a_{ij}))$ is an $n \times n$ matrix, what is the (i, j)-th element of $\mathbf{PAP}$? Write down $\mathbf{PAP}$ when $n = 3$ and when $n = 4$.

(d) If $\mathbf{A}$ is upper triangular, what can you say about $\mathbf{PAP}$?

2.7 Partitioned matrices

Suppose we want to find the product $\mathbf{AB}$ where

$$\mathbf{A} = \begin{bmatrix} 1 & 0 & -3 & 0 \\ 0 & 1 & 4 & 0 \end{bmatrix} \text{ and } \mathbf{B} = \begin{bmatrix} 4 & -1 & 6 \\ 3 & 8 & 3 \\ 0 & 0 & 0 \\ 4 & 3 & -5 \end{bmatrix}$$

It can be checked by direct multiplication that

$$\mathbf{AB} = \begin{bmatrix} 4 & -1 & 6 \\ 3 & 8 & 3 \end{bmatrix}$$

However, there is an easy way to get the product here. We observe that the first two columns of $\mathbf{A}$ form $\mathbf{I}$ and the fourth column is $\mathbf{0}$ whereas the third row of $\mathbf{B}$ is $\mathbf{0}$. Thus if we write

$$\mathbf{A} = \left[\mathbf{I} \;\middle|\; \begin{matrix} -3 \\ 4 \end{matrix} \;\middle|\; \mathbf{0} \right] \text{ and } \mathbf{B} = \left[\begin{array}{ccc} 4 & -1 & 6 \\ 3 & 8 & 3 \\ \hline & \mathbf{0} & \\ \hline 4 & 3 & -5 \end{array} \right] \tag{2.7.1}$$

and then multiply $\mathbf{A}$ and $\mathbf{B}$ as if $\mathbf{A}$ were a row and $\mathbf{B}$ a column we get

$$\mathbf{AB} = \mathbf{I}\begin{bmatrix} 4 & -1 & 6 \\ 3 & 8 & 0 \end{bmatrix} + \begin{bmatrix} -3 \\ 4 \end{bmatrix}\mathbf{0} + \mathbf{0}\begin{bmatrix} 4 & 3 & -5 \end{bmatrix} \tag{2.7.2}$$

and the RHS is clearly the same as that obtained earlier. This type of multiplication is valid provided the columns of $\mathbf{A}$ and the rows of $\mathbf{B}$ are partitioned in the same way. Thus by considering suitable partitions of rows or columns (or both) we can sometimes achieve economy in forming sums and products.

Let $\mathbf{A}$ be an $m \times n$ matrix. Then a *row partition* P of $\mathbf{A}$ is a partition of the rows obtained by drawing horizontal lines after certain rows. Note that there has to be at least one row before the first line, after the last line and between any two successive lines. A *column partition* Q of $\mathbf{A}$ is defined similarly.

Definition 2.7.1 A *partitioned matrix* is a matrix $\mathbf{A}$ together with a row partition P and a column partition Q of $\mathbf{A}$. We denote it by $\mathbf{A}_{PQ}$ and call $\mathbf{A}$ the *underlying matrix* of $\mathbf{A}_{PQ}$.

We usually represent $\mathbf{A}_{PQ}$ by drawing horizontal lines and vertical lines as already indicated. Given such a partitioned matrix, it is easy to get the underlying matrix: just erase the horizontal and vertical lines. Sometimes we use a colon instead of a vertical line, as in $[\mathbf{C} : \mathbf{D}]$.

Example 2.7.2 Let

$$\mathbf{A} = \begin{bmatrix} 2 & 1 & -1 & 0 & 0 & 0 \\ 4 & 0 & 3 & -1 & 2 & 5 \\ 3 & 1 & 1 & 0 & 1 & -2 \end{bmatrix}, \tag{2.7.3}$$

$$P = (\{1\}, \{2,3\}) \text{ and } Q = (\{1,2\}, \{3\}, \{4,5,6\})$$

Then

$$\mathbf{A}_{PQ} = \left[\begin{array}{cc|c|ccc} 2 & 1 & -1 & 0 & 0 & 0 \\ \hline 4 & 0 & 3 & -1 & 2 & 5 \\ 3 & 1 & 1 & 0 & 1 & -2 \end{array}\right] \tag{2.7.4}$$

Conversely, given (2.7.4), $\mathbf{A}$, P and Q have to be as given in (2.7.3).

The usefulness of partitioned matrices stems from the fact that a partitioned matrix $\mathbf{A}_{PQ}$ can be viewed in a natural way as a 'matrix' $\mathbf{C} = ((\mathbf{C}_{pq}))$ of a smaller order. For example, the partitioned matrix $\mathbf{A}_{PQ}$ given in (2.7.4) can be viewed as the 2×3 'matrix'

$$\begin{bmatrix} \mathbf{C}_{11} & \mathbf{C}_{12} & \mathbf{C}_{13} \\ \mathbf{C}_{21} & \mathbf{C}_{22} & \mathbf{C}_{23} \end{bmatrix} \tag{2.7.5}$$

where

$$\mathbf{C}_{11} = \begin{bmatrix} 2 & 1 \end{bmatrix}, \quad \mathbf{C}_{12} = \begin{bmatrix} -1 \end{bmatrix} \quad \mathbf{C}_{13} = \begin{bmatrix} 0 & 0 & 0 \end{bmatrix}$$

$$\mathbf{C}_{21} = \begin{bmatrix} 4 & 0 \\ 3 & 1 \end{bmatrix}, \quad \mathbf{C}_{22} = \begin{bmatrix} 3 \\ 1 \end{bmatrix}, \quad \mathbf{C}_{23} = \begin{bmatrix} -1 & 2 & 5 \\ 0 & 1 & -2 \end{bmatrix} \tag{2.7.6}$$

Note that $\mathbf{C}_{pq}$ and $\mathbf{C}_{p'q'}$ have the same number of rows whenever $p = p'$ and the same number of columns whenever $q = q'$.

We often indicate the fact that the partitioned matrix $\mathbf{A}_{PQ}$ can be viewed as (2.7.5) by dropping the reference to P and Q and writing

$$\mathbf{A} = \begin{bmatrix} \mathbf{C}_{11} & \mathbf{C}_{12} & \mathbf{C}_{13} \\ \mathbf{C}_{21} & \mathbf{C}_{22} & \mathbf{C}_{23} \end{bmatrix}$$

The proof of the following theorem is by routine verification and is omitted.

Theorem 2.7.3 We have

(i) If $\mathbf{A}_{PQ}$ corresponds to $((\mathbf{C}_{pq}))$ and $\mathbf{B} = \alpha\mathbf{A}$ then $\mathbf{B}_{PQ}$ corresponds to $((\alpha\mathbf{C}_{pq}))$.

(ii) If $\mathbf{A}_{PQ}$ corresponds to $((\mathbf{C}_{pq}))$, $\mathbf{B}_{PQ}$ corresponds to $((\mathbf{D}_{pq}))$ and $\mathbf{E} = \mathbf{A} + \mathbf{B}$ then $\mathbf{E}_{PQ}$ corresponds to $((\mathbf{C}_{pq} + \mathbf{D}_{pq}))$.

(iii) If $\mathbf{A}_{PQ}$ corresponds to $((\mathbf{C}_{pq}))$, $\mathbf{B}_{QR}$ corresponds to $((\mathbf{D}_{qr}))$ and $\mathbf{G} = \mathbf{AB}$ then $\mathbf{G}_{PR}$ corresponds to $((\sum_q \mathbf{C}_{pq}\mathbf{D}_{qr}))$.

(iv) If $\mathbf{A}_{PQ}$ corresponds to $((\mathbf{C}_{pq}))$ and $\mathbf{H} = \mathbf{A}^{\mathrm{T}}$ then the (q,p)-element of the 'matrix' corresponding to $\mathbf{H}_{QP}$ is ${\mathbf{C}_{pq}}^{\mathrm{T}}$.

We emphasize that in (ii), the row and column partitions used for $\mathbf{B}$ are the same as those used for $\mathbf{A}$. In (iii), the row partition used for $\mathbf{B}$ is the same as the column partition used for $\mathbf{A}$. In (iv), the row and column partitions are interchanged while going from $\mathbf{A}$ to $\mathbf{A}^{\mathrm{T}}$ and the (q,p)-th element of the 'matrix' corresponding to $\mathbf{H}_{QP}$ is ${\mathbf{C}_{pq}}^{\mathrm{T}}$ and not $\mathbf{C}_{pq}$ (why?)

To illustrate (iii) above, let $\mathbf{A}_{PQ}$ be as given in (2.7.4) so that the corresponding $\mathbf{C}_{pq}$'s are given by (2.7.5) and (2.7.6). Let

$$\mathbf{B}_{QR} = \left[\begin{array}{rr|rr} 1 & 0 & 0 & 0 \\ 0 & 1 & 0 & 0 \\ \hline 1 & -2 & 2 & 5 \\ \hline 2 & -1 & 2 & 0 \\ -1 & 3 & 0 & -1 \\ 0 & 1 & 0 & 0 \end{array}\right]$$

which can be viewed as

$$\begin{bmatrix} \mathbf{D}_{11} & \mathbf{D}_{12} \\ \mathbf{D}_{21} & \mathbf{D}_{22} \\ \mathbf{D}_{31} & \mathbf{D}_{32} \end{bmatrix}$$

where

$$\mathbf{D}_{11} = \begin{bmatrix} 1 & 0 \\ 0 & 1 \end{bmatrix}, \qquad \mathbf{D}_{12} = \begin{bmatrix} 0 & 0 \\ 0 & 0 \end{bmatrix}$$

$$\mathbf{D}_{21} = \begin{bmatrix} 1 & -2 \end{bmatrix}, \qquad \mathbf{D}_{22} = \begin{bmatrix} 2 & 5 \end{bmatrix}$$

$$\mathbf{D}_{31} = \begin{bmatrix} 2 & -1 \\ -1 & 3 \\ 0 & 1 \end{bmatrix}, \quad \mathbf{D}_{32} = \begin{bmatrix} 2 & 0 \\ 0 & -1 \\ 0 & 0 \end{bmatrix}$$

Then we get

$$(\mathbf{AB})_{PR} = \left[\begin{array}{cc|cc} 1 & 3 & -2 & -5 \\ \hline 3 & 6 & 4 & 13 \\ 3 & 0 & 2 & 4 \end{array}\right]$$

by partitioning the product $\mathbf{AB}$ using P and R. This can also be obtained using (iii) above. For example, the (2,1)-entry of $(\mathbf{AB})_{PR}$ is

$$\begin{aligned} &\mathbf{C}_{21}\mathbf{D}_{11} + \mathbf{C}_{22}\mathbf{D}_{21} + \mathbf{C}_{23}\mathbf{D}_{31} \\ &= \begin{bmatrix} 4 & 0 \\ 3 & 1 \end{bmatrix}\begin{bmatrix} 1 & 0 \\ 0 & 1 \end{bmatrix} + \begin{bmatrix} 3 \\ 1 \end{bmatrix}[1 \;\; -2] + \begin{bmatrix} -1 & 2 & 5 \\ 0 & 1 & -2 \end{bmatrix}\begin{bmatrix} 2 & -1 \\ -1 & 3 \\ 0 & 1 \end{bmatrix} \\ &= \begin{bmatrix} 3 & 6 \\ 3 & 0 \end{bmatrix} \end{aligned}$$

To illustrate (iv), let $\mathbf{A}_{PQ}$ be as given in (2.7.4). Then by partitioning $\mathbf{A}^{\mathrm{T}}$ using Q and P, we get

$$(\mathbf{A}^{\mathrm{T}})_{QP} = \left[\begin{array}{c|cc} 2 & 4 & 3 \\ 1 & 0 & 1 \\ \hline -1 & 3 & 1 \\ \hline 0 & -1 & 0 \\ 0 & 2 & 1 \\ 0 & 5 & -2 \end{array}\right]$$

This can also be verified using (iv). For example, the $(2,2)$-entry of $(\mathbf{A}^{\mathrm{T}})_{QP}$ is $\mathbf{C}_{22}{}^{\mathrm{T}} = [3 \;\; 1]$ (and not $\mathbf{C}_{22}$).

We note that if $\mathbf{A}$ is an $m \times n$ matrix and P is the row partition with m blocks and Q is the column partition with n blocks then the 'matrix' associated with $\mathbf{A}_{PQ}$ coincides with $\mathbf{A}$ itself since we identify 1×1 matrices with scalars.

Sometimes we use only a row partition or only a column partition for a matrix, as in (2.7.1). If we use only a row partition in a partitioned matrix, it is to be understood that the column partition is the trivial one with all columns put into one block. Similarly when a row partition is not explicitly mentioned in a partitioned matrix it is to be taken as the partition with just one block.

To illustrate the use of partitioned matrices, we prove some theorems given in *Section* 2.6 using them. To prove *Theorem* 2.6.1, we take the column partition of $\mathbf{A}$ as well as the row partition of $\mathbf{x}$ to be Q with n blocks. Also let P be the row partition of $\mathbf{A}$ with just one block and let R be the column partition of $\mathbf{x}$ with just one block. Then

$$\mathbf{A}_{PQ} = [\mathbf{A}_{*1} : \mathbf{A}_{*2} : \cdots : \mathbf{A}_{*n}] \text{ and } \mathbf{x}_{QR} = \begin{bmatrix} x_1 \\ x_2 \\ \vdots \\ x_n \end{bmatrix}$$

So by *Theorem* 2.7.3 (iii),

$$(\mathbf{Ax})_{PR} = \mathbf{A}_{*1}x_1 + \mathbf{A}_{*2}x_2 + \cdots + \mathbf{A}_{*n}x_n$$

Since P and R are the trivial partitions with one block each, it follows that $(\mathbf{Ax})_{PR} = \mathbf{Ax}$. Note that here $\mathbf{A}_{*j}x_j$ is a matrix product (of the $m \times 1$ matrix $\mathbf{A}_{*j}$ and the 1×1 matrix x_j). It can easily be checked that for any column vector $\mathbf{y}$ and scalar α, $\mathbf{y}\alpha = \alpha\mathbf{y}$ where the product on the left is the matrix product and the product on the right is scalar multiple of a vector. Thus we have, finally, $\mathbf{Ax} = x_1\mathbf{A}_{*1} + \cdots + x_n\mathbf{A}_{*n}$.

Theorem 2.6.2 can be proved similarly. We incidentally note that $y_i\mathbf{A}_{i*}$ may be looked upon either as a matrix product or as a scalar multiple of a vector.

Exercises

1. If $\mathbf{M} = \begin{bmatrix} \mathbf{A} & \mathbf{B} \\ \mathbf{C} & \mathbf{D} \end{bmatrix}$ is a partitioned matrix of order $m \times n$, where $\mathbf{A}$ is of order $k \times \ell$, find the orders of $\mathbf{B}, \mathbf{C}$ and $\mathbf{D}$.

2. (a) Let $\mathbf{A}$ and $\mathbf{B}$ be matrices of the same order and let $\mathbf{A} = \mathbf{PQ}$ and $\mathbf{B} = \mathbf{RS}$. Write $\mathbf{A} + \mathbf{B}$ as the product of two partitioned matrices involving $\mathbf{P}, \mathbf{Q}, \mathbf{R}$ and $\mathbf{S}$.

   (b) If $\mathbf{A}$ and $\mathbf{B}$ are matrices of the same order, find a matrix $\mathbf{D}$ such that $\mathbf{A} + \mathbf{B} = (\mathbf{A} : \mathbf{B})\mathbf{D}$.

*3. Let $\mathcal{A} = \{\mathbf{A}_{\alpha,\beta} : \alpha, \beta \in F\}$ and $\mathcal{B} = \{\mathbf{B}_{\alpha,\beta,\gamma,\delta} : \alpha, \beta, \gamma, \delta \in F\}$, where

$$\mathbf{A}_{\alpha,\beta} = \begin{bmatrix} \alpha & \beta \\ -\beta & \alpha \end{bmatrix} \text{ and } \mathbf{B}_{\alpha,\beta,\gamma,\delta} = \begin{bmatrix} \alpha & \beta & \gamma & \delta \\ -\beta & \alpha & -\delta & \gamma \\ -\gamma & \delta & \alpha & -\beta \\ -\delta & -\gamma & \beta & \alpha \end{bmatrix}$$

Show that if $\mathbf{A}_1, \mathbf{A}_2 \in \mathcal{A}$, then $\mathbf{A}_1^{\mathrm{T}}$, $\mathbf{A}_1 + \mathbf{A}_2$ and $\mathbf{A}_1\mathbf{A}_2 \in \mathcal{A}$. Prove a similar result for $\mathcal{B}$ also. If $\mathbf{B} \in \mathcal{B}$, what is $\mathbf{BB}^{\mathrm{T}}$?

4. Using the best way of partitioning **A** and **C**, evaluate **ABC** when **B** is given to be $\begin{bmatrix} \mathbf{B}_1 & \mathbf{0} \\ \mathbf{0} & \mathbf{B}_2 \end{bmatrix}$.

5. Find the following products:

(a) $\begin{bmatrix} \mathbf{A} & \mathbf{B} \\ \mathbf{C} & \mathbf{D} \end{bmatrix} \begin{bmatrix} \mathbf{G} & \mathbf{0} \\ \mathbf{0} & \mathbf{0} \end{bmatrix} \begin{bmatrix} \mathbf{P} & \mathbf{Q} \\ \mathbf{R} & \mathbf{S} \end{bmatrix}$, (b) $\begin{bmatrix} \mathbf{A} \\ \mathbf{C} \end{bmatrix} \mathbf{G} \begin{bmatrix} \mathbf{P} & \mathbf{Q} \end{bmatrix}$,

(c) $\begin{bmatrix} \mathbf{I} & \mathbf{0} \\ \mathbf{D} & \mathbf{I} \end{bmatrix} \begin{bmatrix} \mathbf{B} & \mathbf{0} \\ \mathbf{0} & \mathbf{0} \end{bmatrix} \begin{bmatrix} \mathbf{I} & \mathbf{C} \\ \mathbf{0} & \mathbf{0} \end{bmatrix}$, (d) $\begin{bmatrix} \mathbf{A} & \mathbf{0} \\ \mathbf{B} & \mathbf{C} \end{bmatrix} \begin{bmatrix} \mathbf{D} & \mathbf{0} \\ \mathbf{E} & \mathbf{G} \end{bmatrix}$,

(e) $\begin{bmatrix} \mathbf{P}_1 & \mathbf{0} \\ \mathbf{0} & \mathbf{Q}_1 \end{bmatrix} \begin{bmatrix} \mathbf{P}_2 & \mathbf{0} \\ \mathbf{0} & \mathbf{Q}_2 \end{bmatrix} \begin{bmatrix} \mathbf{P}_3 & \mathbf{0} \\ \mathbf{0} & \mathbf{Q}_3 \end{bmatrix}$.

6. Prove the results corresponding to (14), (18) and (20) of *Theorem* 2.4.2 for 'matrices' whose elements are matrices.

*7. Prove *Theorem* 2.7.3.

8. Prove statement (ii) of *Theorem* 2.6.4 by taking the row partition of **A** with one block, row and column partitions of the diagonal matrix with n blocks and using Theorem 2.7.3.

9. A partitioned matrix $\mathbf{A} = ((\mathbf{A}_{ij}))$ is said to be a *block upper triangular matrix* if the row and column partition have the same number of blocks, $\mathbf{A}_{ij} = \mathbf{0}$ whenever $i > j$ and $\mathbf{A}_{ii}$ is a square matrix for all i. Let **A** and **B** be block upper triangular matrices of the same order partitioned in the same way.

   (a) Show that $\mathbf{A} + \mathbf{B}$ and $\mathbf{AB}$ are also block upper triangular.

   (b) Show that if α is a scalar then $\alpha \cdot \mathbf{A}$ is also block upper triangular.

   (c) If $\mathbf{A}_{ii} = \mathbf{0}$ for all i (in addition to **A** being block upper triangular), show that $\mathbf{A}^k = \mathbf{0}$ where k is the number of blocks in the row and column partitions.

10. Let **A** be a square matrix. Show that the columns of $\mathbf{A}^3$ are linear combinations of the columns of **A**.

11. Let $\mathbf{A}$ and $\mathbf{B}$ be the partitioned matrices

$$\mathbf{A} = \begin{bmatrix} \mathbf{I} & \mathbf{0} \\ \mathbf{C} & \mathbf{I} \end{bmatrix} \quad \text{and} \quad \mathbf{B} = \begin{bmatrix} \mathbf{I} & \mathbf{0} \\ -\mathbf{C} & \mathbf{I} \end{bmatrix}$$

where $\mathbf{C}$ is arbitrary. Show that $\mathbf{AB} = \mathbf{BA} = \mathbf{I}$.

*12. The *Kronecker product* of an $m \times n$ matrix $\mathbf{A}$ and a $p \times q$ matrix $\mathbf{B}$ is defined to be the $mp \times nq$ matrix

$$\mathbf{A} \otimes \mathbf{B} = ((a_{ij}\mathbf{B}))$$

Prove the following:

(a) $\mathbf{A} \otimes \mathbf{B}$ may not equal $\mathbf{B} \otimes \mathbf{A}$

(b) $\mathbf{A} \otimes (\alpha\mathbf{B}) = (\alpha\mathbf{A}) \otimes \mathbf{B} = \alpha(\mathbf{A} \otimes \mathbf{B})$

(c) $\mathbf{A} \otimes (\mathbf{B} \otimes \mathbf{C}) = (\mathbf{A} \otimes \mathbf{B}) \otimes \mathbf{C}$

(d) $\mathbf{A} \otimes (\mathbf{B} + \mathbf{C}) = (\mathbf{A} \otimes \mathbf{B}) + (\mathbf{A} \otimes \mathbf{C})$

(e) $(\mathbf{B} + \mathbf{C}) \otimes \mathbf{A} = (\mathbf{B} \otimes \mathbf{A}) + (\mathbf{C} \otimes \mathbf{A})$

(f) $\mathbf{I} \otimes \mathbf{A} = \text{diag}(\mathbf{A}, \mathbf{A}, \ldots, \mathbf{A})$ and $\mathbf{I}_n \otimes \mathbf{I}_p = \mathbf{I}_{np}$

(g) $\mathbf{0} \otimes \mathbf{A} = \mathbf{0}, \ \mathbf{A} \otimes \mathbf{0} = \mathbf{0}$

(h) $(\mathbf{AB}) \otimes (\mathbf{CD}) = (\mathbf{A} \otimes \mathbf{C})(\mathbf{B} \otimes \mathbf{D})$

(i) $(\mathbf{A} \otimes \mathbf{B})^{\mathrm{T}} = \mathbf{A}^{\mathrm{T}} \otimes \mathbf{B}^{\mathrm{T}}$

Chapter 3
Rank and inverse

3.1 Introduction

Some important problems in linear algebra are those of solving a system of linear equations, finding the dimension of a subspace or a flat, finding the inverse of a matrix, determining whether a set of vectors is linearly independent, finding a basis contained in a generating set of a subspace, etc. It turns out that rank of a matrix, which is the maximum number of linearly independent rows (or columns), plays a crucial role in the solution of most of these problems. Some other closely related concepts are those of nullity and inverse of a matrix, both useful in solving linear equations. In this chapter we define rank and study its basic properties. We also study nullity, existence and properties of inverse and a few other topics like projectors and change of bases. Computational procedures will be taken up in the next chapter.

3.2 Row space and column space

Let $\mathbf{A}$ be an $m \times n$ matrix over a field F. Then the range of the linear transformation $f : F^n \to F^m$ defined by $f(\mathbf{x}) = \mathbf{Ax}$ is, by *Theorem* 2.6.1, the set of all linear combinations of the columns of $\mathbf{A}$ or the span of the columns of $\mathbf{A}$. Motivated by this, we give

Definition 3.2.1 Let $\mathbf{A}$ be an $m \times n$ matrix. Then the *column space* of $\mathbf{A}$ is $\mathcal{C}(\mathbf{A}) := \{\mathbf{Ax} : \mathbf{x} \in F^n\}$ and the *row space* of $\mathbf{A}$ is $\mathcal{R}(\mathbf{A}) := \{\mathbf{y}^{\mathrm{T}}\mathbf{A} : \mathbf{y} \in F^m\}$.

Note that $\mathcal{C}(\mathbf{A})$ is a subspace of F^m (and *not* F^n) and $\mathcal{R}(\mathbf{A})$ is a subspace of F^n. Some authors use $\mathcal{R}(\mathbf{A})$ for column space of $\mathbf{A}$, $\mathcal{R}$ standing for the range of the corresponding linear transformation. However, we use $\mathcal{R}$ for row space. Another notation used for column space of $\mathbf{A}$ is $\mathcal{M}(\mathbf{A})$, $\mathcal{M}$ standing for manifold.

Example 3.2.2 Let

$$\mathbf{A} = \begin{bmatrix} 2 & 4 & 1 & -1 \\ 3 & 6 & 0 & 1 \\ -1 & -2 & -2 & 3 \end{bmatrix}$$

We will show that $\mathcal{C}(\mathbf{A})$ is

$$\{(\alpha,\beta,\gamma)^{\mathrm{T}} : 2\alpha - \beta + \gamma = 0\} = \{(\alpha,\beta,\beta-2\alpha)^{\mathrm{T}} : \alpha,\beta \in F\} \quad (3.2.1)$$

Let S and T denote the LHS and the RHS of (3.2.1). Then it is easy to verify that every column of $\mathbf{A}$ belongs to S and that S is a subspace. So it follows that $\mathcal{C}(\mathbf{A}) \subseteq S$. It is trivial to see that $S = T$. Finally notice that a general vector in T, viz.

$$\begin{bmatrix} \alpha \\ \beta \\ \beta - 2\alpha \end{bmatrix} = (\alpha+\beta)\begin{bmatrix} 1 \\ 0 \\ -2 \end{bmatrix} + \beta\begin{bmatrix} -1 \\ 1 \\ 3 \end{bmatrix}$$

belongs to $\mathcal{C}(\mathbf{A})$. This proves (3.2.1). It can be shown similarly that

$$\begin{aligned} \mathcal{R}(\mathbf{A}) &= \{(\alpha,\beta,\gamma,\delta) : 2\alpha - \beta = 0 \text{ and } \alpha - 5\gamma - 3\delta = 0\} \\ &= \{(\alpha, 2\alpha, \gamma, \tfrac{\alpha}{3} - \tfrac{5\gamma}{3}) : \alpha,\gamma \in F\}. \end{aligned}$$ ∎

Clearly $\mathcal{C}(\mathbf{A}) = \mathcal{R}(\mathbf{A}^{\mathrm{T}})$ (a more precise statement is: $\mathbf{y} \in \mathcal{C}(\mathbf{A})$ iff $\mathbf{y}^{\mathrm{T}} \in \mathcal{R}(\mathbf{A}^{\mathrm{T}})$). Similarly, $\mathcal{R}(\mathbf{A}) = \mathcal{C}(\mathbf{A}^{\mathrm{T}})$.

We call $d(\mathcal{R}(\mathbf{A}))$ the *row rank* of $\mathbf{A}$ and $d(\mathcal{C}(\mathbf{A}))$ the *column rank* of $\mathbf{A}$, though we will presently prove that they are equal. Let us recall a few facts about bases of subspaces in this context. If $\mathbf{A}$ has column rank r then

(i) any r linearly independent columns of $\mathbf{A}$ form a basis for $\mathcal{C}(\mathbf{A})$ (see *Theorem* 1.5.11),

(ii) every maximal linearly independent set of columns of $\mathbf{A}$ contains exactly r vectors (see *Theorem* 1.5.8),

(iii) any r columns of $\mathbf{A}$ which generate $\mathcal{C}(\mathbf{A})$ (i.e., every other column of $\mathbf{A}$ can be expressed as a linear combination of these) form a basis of $\mathcal{C}(\mathbf{A})$ (see *Theorem* 1.5.11).

We refer to a basis of $\mathcal{C}(\mathbf{A})$ consisting of columns of $\mathbf{A}$ as a *column basis*. A *row basis* is defined similarly.

Theorem 3.2.3 For any matrix $\mathbf{A}$, the row rank of $\mathbf{A}$ equals the column rank of $\mathbf{A}$.

Proof Let $\mathbf{A}$ be an $m \times n$ matrix with row rank r and column rank s. If $\mathbf{A} = \mathbf{0}$, then $\mathcal{R}(\mathbf{A}) = \{\mathbf{0}\}$ and $\mathcal{C}(\mathbf{A}) = \{\mathbf{0}\}$, so $r = s = 0$ and we are done. So let $\mathbf{A} \neq \mathbf{0}$ and let $\mathbf{B} = [\,\mathbf{x}_1 : \mathbf{x}_2 : \cdots : \mathbf{x}_s\,]$ be an $m \times s$ matrix whose columns form a basis of $\mathcal{C}(\mathbf{A})$. Then $\mathbf{A}_{*j}$ is a linear combination of the columns of $\mathbf{B}$, so there exists an $s \times 1$ vector $\mathbf{y}_j$ such that $\mathbf{A}_{*j} = \mathbf{B}\mathbf{y}_j$ for $j = 1, \ldots, n$. Now

$$\begin{aligned}\mathbf{A} = [\,\mathbf{A}_{*1} : \cdots : \mathbf{A}_{*n}\,] &= [\,\mathbf{B}\mathbf{y}_1 : \cdots : \mathbf{B}\mathbf{y}_n\,] \\ &= \mathbf{B}[\,\mathbf{y}_1 : \cdots : \mathbf{y}_n\,] \\ &= \mathbf{B}\mathbf{C}\end{aligned}$$

where $\mathbf{C} = [\,\mathbf{y}_1 : \cdots : \mathbf{y}_n\,]$. So $\mathbf{A}_{i*} = \mathbf{B}_{i*} \cdot \mathbf{C}$ and, by *Theorem* 2.6.2, this is a linear combination of the rows of $\mathbf{C}$. Thus every row of $\mathbf{A}$ belongs to $\mathcal{R}(\mathbf{C})$, hence $\mathcal{R}(\mathbf{A}) \subseteq \mathcal{R}(\mathbf{C})$. Taking dimensions, we get $r \leq \text{row rank}(\mathbf{C}) \leq s$ where the last inequality follows from the fact that $\mathbf{C}$ has only s rows. Interchanging the roles of row rank and column rank, we get $r \geq s$ and equality follows. ∎

Definition 3.2.4 The *rank* of a matrix $\mathbf{A}$ is the common value of the row rank of $\mathbf{A}$ and the column rank of $\mathbf{A}$ and is denoted $\rho(\mathbf{A})$.

Clearly, rank of $\mathbf{A}$ is the dimension of the range of the linear transformation $f : \mathbf{x} \mapsto \mathbf{A}\mathbf{x}$. The rank of the null matrix of any order is 0. Conversely if $\rho(\mathbf{A}) = 0$ then $\mathcal{R}(\mathbf{A}) = \{\mathbf{0}\}$ and $\mathbf{A} = \mathbf{0}$. Since row rank equals column rank it follows that for any matrix $\mathbf{A}$, $\rho(\mathbf{A}) = \rho(\mathbf{A}^{\mathrm{T}})$. It is easy to see that the rank of the matrix $\mathbf{A}$ in *Example* 3.2.2 is 2.

The rank of an $m \times n$ matrix obviously lies between 0 and $\min(m, n)$. Conversely, given any non-negative integer $r \leq \min(m, n)$, there exists an $m \times n$ matrix $\mathbf{A}$ with rank r. For example, take

$$\mathbf{A} = \begin{bmatrix} \mathbf{I}_r & \mathbf{0} \\ \mathbf{0} & \mathbf{0} \end{bmatrix}$$

Theorem 3.2.5 Let $\mathbf{A}$ be an $m \times n$ matrix of rank r and $\mathbf{B}$ a submatrix of $\mathbf{A}$. Then $\rho(\mathbf{B}) \leq \rho(\mathbf{A})$.

Proof The result follows on considering row rank if $\mathbf{B}$ is obtained from $\mathbf{A}$ by omitting only some rows and by considering column rank if $\mathbf{B}$ is obtained from $\mathbf{A}$ by omitting only some columns. Now, any submatrix can be obtained by omitting some rows and then some columns. The theorem follows. ∎

Exercises

1. Consider the real matrix

$$\mathbf{A} = \begin{bmatrix} 4 & 1 & 6 & 0 \\ 0 & 1 & 2 & -4 \\ 1 & 0 & 1 & 1 \end{bmatrix}$$

Show that $\mathcal{R}(\mathbf{A}) = \{(\alpha, \beta, \gamma, \delta) : 3\alpha - 2\gamma - \delta = 0, 6\alpha + 6\beta - 5\gamma - \delta = 0\}$ and $\mathcal{C}(\mathbf{A}) = \{(\alpha, \beta, \gamma)^{\mathrm{T}} : \alpha - \beta - 4\gamma = 0\}$. Obtain a basis of $\mathcal{R}(\mathbf{A})$ and a basis of $\mathcal{C}(\mathbf{A})$.

2. If $\mathbf{A}$ is a matrix of rank r, show that $\rho(\mathbf{B}) = r$ where $\mathbf{B} = \begin{bmatrix} \mathbf{A} & \mathbf{0} \\ \mathbf{0} & \mathbf{0} \end{bmatrix}$.

*3. (a) Show using *Exercise* 1.8.3(b) that adding a scalar multiple of one column to another leaves both the column rank and the row rank of a matrix unaltered.

   (b) Deduce from (a) that row rank equals column rank. (Hint: reduce the columns not belonging to a column basis to null vectors.)

4. Prove without using *Theorem* 3.2.3 that the row rank of a submatrix of $\mathbf{A}$ cannot exceed the row rank of $\mathbf{A}$. (Hint: use *Exercise* 1.4.6).

5. Prove that $\rho\begin{bmatrix} \mathbf{A} & \mathbf{B} \\ \mathbf{0} & \mathbf{C} \end{bmatrix} \geq \rho(\mathbf{A}) + \rho(\mathbf{C})$. Show that strict inequality can occur. Deduce that the rank of an upper triangular matrix is not less than the number of non-zero diagonal elements.

6. Let $\mathbf{A}$ be an $m \times n$ matrix of rank r. Determine the possible values for the rank of the matrix obtained by (i) changing exactly one element and (ii) changing two elements.

7. If $\mathbf{A}$ is an $m \times n$ matrix of rank r, determine the possible values for the rank of the submatrix obtained by deleting a row and a column.

8. If $\mathbf{A}$ is an $m \times n$ matrix with rank r, show that for every k such that $1 \leq k \leq r$, $\mathbf{A}$ has a $k \times k$ submatrix with rank k. (Hint: first show that $\mathbf{A}$ has a $k \times n$ submatrix with rank k.)

9. Show that an $m \times n$ matrix $\mathbf{A}$ has rank at most 1 iff $\mathbf{A} = \mathbf{x}\mathbf{y}^{\mathrm{T}}$ for some column vectors $\mathbf{x}$ and $\mathbf{y}$. Show further that $\rho(\mathbf{A}) = 1$ iff both $\mathbf{x}$ and $\mathbf{y}$ are non-null.

*10. Let p and q be fixed positive integers. Consider the pq row vectors $\mathbf{x}_{ij}^{\mathrm{T}} \in \mathbb{R}^{p+q+1}$, $i = 1, \ldots, p; j = 1, \ldots, q$, where $\mathbf{x}_{ij}^{\mathrm{T}}$ has 1's in the first, the $(i+1)^{\text{th}}$ and the $(p+j+1)^{\text{th}}$ positions and 0's elsewhere. Show that the dimension of the span of these vectors is $p + q - 1$.

3.3 Inverse of a matrix

We know that every non-zero real number a has an inverse, $a^{-1} = \frac{1}{a}$ with the property $a \cdot a^{-1} = a^{-1} \cdot a = 1$. Any non-zero element of a general field also has inverse in the same sense (see *Chapter* 0). The inverse of a is useful in solving the equation $ax = b$ since $x = a^{-1}b$ is the solution. In *Chapter* 0, we also noticed that a one-to-one map has a left inverse, an onto map has a right inverse and a bijection has a (two-sided) inverse. We shall now investigate the inverses of matrices. Inverses of linear transformations will be studied in *Section* 3.11.

Definition 3.3.1 An $m \times n$ matrix $\mathbf{A}$ is said to be of *full row rank* if its rank is m, that is, if its rows are linearly independent. Similarly $\mathbf{A}$ is said to be of *full column rank* if its columns are linearly independent.

Definition 3.3.2 A *left inverse* of a matrix $\mathbf{A}$ is any matrix $\mathbf{B}$ such that $\mathbf{BA} = \mathbf{I}$. A *right inverse* of $\mathbf{A}$ is any matrix $\mathbf{C}$ such that $\mathbf{AC} = \mathbf{I}$. A matrix $\mathbf{B}$ is said to be an *inverse* of $\mathbf{A}$ if it is both a left inverse and a right inverse of $\mathbf{A}$.

We shall use $\mathbf{A}_L^{-1}$, $\mathbf{A}_R^{-1}$ and $\mathbf{A}^{-1}$ to denote a left inverse of $\mathbf{A}$, a right inverse of $\mathbf{A}$ and an inverse of $\mathbf{A}$ when they exist.

We observe that if $\mathbf{A}$ is an $m \times n$ matrix and $\mathbf{B}$ is a left inverse of $\mathbf{A}$ then $\mathbf{B}$ must be an $n \times m$ matrix and $\mathbf{BA} = \mathbf{I}_n$. Similarly if $\mathbf{C}$ is a right inverse of $\mathbf{A}$ then $\mathbf{C}$ is an $n \times m$ matrix and $\mathbf{AC} = \mathbf{I}_m$.

Theorem 3.3.3 Let $\mathbf{A}$ be an $m \times n$ matrix over F. Then the following statements are equivalent.

(i) $\mathbf{A}$ has a right inverse

(ii) $\mathbf{XA} = \mathbf{YA} \Rightarrow \mathbf{X} = \mathbf{Y}$

(iii) $\mathbf{XA} = \mathbf{0} \Rightarrow \mathbf{X} = \mathbf{0}$

(iv) $\rho(\mathbf{A}) = m$ (that is, $\mathbf{A}$ is of full row rank)

(v) $\mathcal{C}(\mathbf{A}) = F^m$

Proof (i) $\Rightarrow$ (ii) Let $\mathbf{C}$ be a right inverse of $\mathbf{A}$. If $\mathbf{XA} = \mathbf{YA}$, then by postmultiplying by $\mathbf{C}$ we get $\mathbf{XAC} = \mathbf{YAC}$, that is, $\mathbf{X} = \mathbf{Y}$.

(ii) $\Rightarrow$ (iii) Take $\mathbf{Y} = \mathbf{0}$.

(iii) $\Rightarrow$ (iv) We shall show that the rows of $\mathbf{A}$ are linearly independent. So let $\alpha_1\mathbf{A}_{1*} + \alpha_2\mathbf{A}_{2*} + \cdots + \alpha_m\mathbf{A}_{m*} = \mathbf{0}$. Then $\mathbf{XA} = \mathbf{0}$ where $\mathbf{X} = (\alpha_1, \alpha_2, \ldots, \alpha_m)$. So $\mathbf{X} = \mathbf{0}$, i.e., $\alpha_i = 0$ for all i.

(iv) $\Rightarrow$ (v) Clearly $\mathcal{C}(\mathbf{A}) \subseteq F^m$. By hypothesis, both sides have the same dimension, so equality follows.

(v) $\Rightarrow$ (i) Consider the vectors $\mathbf{e}_j$ in the canonical basis of F^m. Since $F^m = \mathcal{C}(\mathbf{A})$, it follows that $\mathbf{e}_j = \mathbf{A}\mathbf{c}_j$ for some $n \times 1$ vector $\mathbf{c}_j$. So

$$\mathbf{A}[\,\mathbf{c}_1 : \mathbf{c}_2 : \cdots : \mathbf{c}_m\,] = [\,\mathbf{e}_1 : \mathbf{e}_2 : \cdots : \mathbf{e}_m\,] = \mathbf{I}$$

and $\mathbf{C} = [\,\mathbf{c}_1 : \mathbf{c}_2 : \cdots : \mathbf{c}_m\,]$ is a right inverse of $\mathbf{A}$. ∎

We note that the significance of (ii) in the preceding theorem is that we can cancel $\mathbf{A}$ on the right provided $\mathbf{A}$ satisfies any one of the other conditions. We also note that in statements (ii) and (iii), the matrices $\mathbf{X}$ and $\mathbf{Y}$ with m columns and an arbitrary number of rows can be replaced by row vectors. Statement (v) says that the linear transformation $f : \mathbf{x} \mapsto \mathbf{A}\mathbf{x}$ is *onto* F^m.

Example 3.3.4 Let

$$\mathbf{A} = \begin{bmatrix} 1 & 2 & 3 \\ 0 & 1 & 2 \end{bmatrix}$$

Clearly $\mathbf{A}$ is of full row rank. So it has a right inverse. The proof of the preceding theorem in fact gives a way of computing a right inverse by solving simultaneous linear equations. Observe that the first column $(x, y, z)^{\mathrm{T}}$ of a right inverse has to satisfy $\mathbf{A}(x, y, z)^{\mathrm{T}} = (1, 0)^{\mathrm{T}}$, that is,

$$\begin{aligned} x + 2y + 3z &= 1 \\ y + 2z &= 0 \end{aligned}$$

To solve these equations we fix x arbitrarily and try to solve for y and z. Fixing $x = 1$ we get the solution $(1, 0, 0)^{\mathrm{T}}$ and fixing $x = 0$ we get the solution $(0, 2, -1)^{\mathrm{T}}$. Similarly the second column of a right inverse is any solution of the two equations: $x + 2y + 3z = 0$ and $y + 2z = 1$. It can be checked that $(-2, 1, 0)^{\mathrm{T}}$ and $(1, -5, 3)^{\mathrm{T}}$ are two solutions. Thus

$$\begin{bmatrix} 1 & -2 \\ 0 & 1 \\ 0 & 0 \end{bmatrix} \quad \text{and} \quad \begin{bmatrix} 0 & 1 \\ 2 & -5 \\ -1 & 3 \end{bmatrix}$$

are two right inverses of $\mathbf{A}$. (How many right inverses does $\mathbf{A}$ have?) ∎

It is easy to see that a matrix $\mathbf{B}$ is a left inverse of a matrix $\mathbf{A}$ iff $\mathbf{B}^{\mathrm{T}}$ is a right inverse of $\mathbf{A}^{\mathrm{T}}$. Hence analogues of many of the results for right inverses can be obtained for left inverses and vice versa. In particular, the next theorem follows from *Theorem* 3.3.3.

Theorem 3.3.5 Let $\mathbf{A}$ be an $m \times n$ matrix over F. Then the following statements are equivalent.

(i) $\mathbf{A}$ has a left inverse

(ii) $\mathbf{AX} = \mathbf{AY} \Rightarrow \mathbf{X} = \mathbf{Y}$

(iii) $\mathbf{AX} = \mathbf{0} \Rightarrow \mathbf{X} = \mathbf{0}$

(iv) $\rho(\mathbf{A}) = n$ (that is, $\mathbf{A}$ is of full column rank)

(v) $\mathcal{R}(\mathbf{A}) = F^n$

We note that statement (ii) of the above theorem says that the linear transformation $x \mapsto \mathbf{Ax}$ is 1-1 (or $\mathbf{A}$ can be cancelled on the left).

Theorem 3.3.6 If a matrix $\mathbf{A}$ has a left inverse $\mathbf{B}$ and a right inverse $\mathbf{C}$ then $\mathbf{A}$ is square, $\mathbf{B} = \mathbf{C}$ and $\mathbf{A}$ has a unique left inverse, a unique right inverse and a unique inverse.

Proof That $\mathbf{A}$ is square follows since rank of $\mathbf{A}$ is both m and n by the two preceding theorems. Hence $\mathbf{B} = \mathbf{BI} = \mathbf{BAC} = \mathbf{IC} = \mathbf{C}$. If, now, $\mathbf{D}$ is any left inverse of $\mathbf{A}$ then $\mathbf{D} = \mathbf{C} = \mathbf{B}$, so $\mathbf{A}$ has a unique left inverse. Similarly $\mathbf{A}$ has a unique right inverse. Clearly $\mathbf{B} = \mathbf{C}$ is an inverse of $\mathbf{A}$. ∎

By the preceding theorem, *if a matrix* $\mathbf{A}$ *has an inverse, then* $\mathbf{A}^{-1}$ *is unique,* $\mathbf{A}$ *is square and* $\mathbf{AA}^{-1} = \mathbf{A}^{-1}\mathbf{A} = \mathbf{I}$. The next theorem follows easily from the three preceding theorems.

Theorem 3.3.7 Let $\mathbf{A}$ be a square matrix of order n. Then the following statements are equivalent:

(i) $\mathbf{A}$ has a right inverse

(ii) $\rho(\mathbf{A}) = n$

(iii) $\mathbf{A}$ has a left inverse

(iv) $\mathbf{A}$ has an inverse.

Corollary For a square matrix, any left (resp. right) inverse is also a right (resp. left) inverse and is the inverse.

It is easy to see that a diagonal matrix is invertible (i.e., has an inverse) iff all its diagonal entries are non-zero. Generalizing this, we now give an important class of matrices having inverse over the fields $\mathbb{R}$ and $\mathbb{C}$. We start with

Definition 3.3.8 An $n \times n$ complex (or real) matrix $\mathbf{A}$ is said to be *strictly diagonally dominated* if

$$|a_{ii}| > \sum_{\substack{j=1 \\ j \neq i}}^{n} |a_{ij}| \quad \text{for } i = 1, 2, \ldots, n$$

***Theorem 3.3.9** Every strictly diagonally dominated matrix has an inverse.

Proof Let $\mathbf{A}$ be an $n \times n$ strictly diagonally dominated matrix. In view of *Theorems* 3.3.5 and 3.3.7 it is enough to show that $\mathbf{Ax} = \mathbf{0} \Rightarrow \mathbf{x} = \mathbf{0}$. So let $\mathbf{Ax} = \mathbf{0}$. Let i be such that $|x_i| = \max_j |x_j|$. Suppose $|x_i| > 0$. Now $\mathbf{Ax} = \mathbf{0}$ gives

$$a_{ii}x_i = -\sum_{j \neq i} a_{ij}x_j$$

Taking moduli on both sides, we get

$$|a_{ii}||x_i| \leq \sum_{j \neq i} |a_{ij}||x_j| \leq \Big(\sum_{j \neq i} |a_{ij}|\Big)|x_i| < |a_{ii}||x_i|$$

This contradiction proves that $|x_i| = 0$ and so $\mathbf{x} = \mathbf{0}$. ∎

Exercises

1. (a) Show that every non-null matrix of order $m \times 1$ has a left inverse and every non-null matrix of order $1 \times n$ has a right inverse.
   (b) Find four right inverses of $[\,2 \;\; 1 \;\; 3 \;\; 4\,]$.
2. Let $\mathbf{A}$ be an $n \times n$ matrix with an inverse and let $\mathbf{B}$ be an $n \times p$ matrix. Then show that $[\,\mathbf{A} : \mathbf{B}\,]$ has a right inverse by exhibiting one.
3. Let

$$\mathbf{A} = \begin{bmatrix} 3 & 1 \\ 1 & 0 \\ 0 & 1 \end{bmatrix} \quad \text{and} \quad \mathbf{B} = \begin{bmatrix} 0 & 1 & 2 \\ 1 & 0 & -1 \end{bmatrix}$$

   (a) Compute two left inverses of $\mathbf{A}$ and two right inverses of $\mathbf{B}$.
   (b) Compute $\mathbf{C} = \mathbf{AB}$ and $\mathbf{D} = \mathbf{GH}$, where $\mathbf{G}$ is some left inverse of $\mathbf{A}$ and $\mathbf{H}$ is some right inverse of $\mathbf{B}$ and check that $\mathbf{CDC} = \mathbf{C}$.

*4. If an $m \times n$ matrix $\mathbf{A}$ has a unique left (or right) inverse, prove that $m = n$ and that $\mathbf{A}$ has an inverse.

5. If $\mathbf{B}$ and $\mathbf{C}$ are left inverses of $\mathbf{A}$, show that $\alpha\mathbf{B} + (1 - \alpha)\mathbf{C}$ is also a left inverse of $\mathbf{A}$.

6. Let $\mathbf{A}$ and $\mathbf{B}$ be $n \times n$ matrices such that $\mathbf{AB}$ is diagonal with non-zero diagonal entries. Show that in general, $\mathbf{A}$ and $\mathbf{B}$ may not commute but if the diagonal entries of $\mathbf{AB}$ are all equal then $\mathbf{A}$ and $\mathbf{B}$ commute.
7. Prove or disprove: if $\mathbf{A}$ is an $m \times n$ matrix with rank m and $\mathbf{B}$ is an $n \times m$ matrix with rank m then $\mathbf{AB}$ has an inverse.
8. Show that for any square matrix $\mathbf{A}$ over $\mathbb{R}$ or $\mathbb{C}$, there exists a scalar α such that $\alpha\mathbf{I} + \mathbf{A}$ has an inverse and that this is false over GF(2).
9. If $\mathbf{A}$ is a square matrix such that $3\mathbf{A}^4 - 4\mathbf{A}^3 + 2\mathbf{A} + 5\mathbf{I} = \mathbf{0}$, prove that $\mathbf{A}$ has an inverse. Extend this to a polynomial of degree k.
10. Let $\mathbf{A}$ be a non-null matrix.
    (a) Show that a submatrix of $\mathbf{A}$ formed by the rows in a row basis of $\mathbf{A}$ and the columns in a column basis of $\mathbf{A}$, is invertible. Deduce that the rank of $\mathbf{A}$ is the maximum k for which $\mathbf{A}$ has a $k \times k$ invertible submatrix.
    (b) Show that a submatrix of $\mathbf{A}$ formed by k linearly independent rows and k linearly independent columns need not be invertible. However, see *Exercise* 3.2.8.
11. (a) Show that the rank of a symmetric matrix is the maximum order of a principal submatrix which is invertible.
    (b) Prove or disprove: if $\mathbf{A}$ is symmetric and $k \leq \rho(\mathbf{A})$, $\mathbf{A}$ has a $k \times k$ invertible principal submatrix. (Compare *Exercise* 3.2.8.)
12. Let $m > n$. Prove that there cannot exist (i) a linear transformation from F^n onto F^m, (ii) a 1–1 linear transformation from F^m to F^n. (Note that such maps exist if F is infinite and we do not insist on linearity.)

*13. If $m \leq n$, find the number of $m \times n$ matrices with full row rank over a field of order q. Also find the number of $m \times n$ matrices with full column rank when $m \geq n$.

3.4 Properties of inverse

In this section we give some important properties of inverses of matrices when they exist.

Definition 3.4.1 A square matrix $\mathbf{A}$ is said to be *non-singular* if it has an inverse. A square matrix which does not possess an inverse is said to be *singular*.

By *Theorem* 3.3.7, an $n \times n$ matrix $\mathbf{A}$ is non-singular iff its columns (or rows) form a basis of F^n.

It follows from the definition that if $\mathbf{A}$ is a non-singular matrix of order n then for any $n \times p$ matrix $\mathbf{B}$, there is a unique matrix $\mathbf{X}$ such that $\mathbf{AX} = \mathbf{B}$, viz., $\mathbf{A}^{-1}\mathbf{B}$. Similarly, for any $m \times n$ matrix $\mathbf{C}$, there is a unique matrix $\mathbf{Y}$ such that $\mathbf{YA} = \mathbf{C}$, viz., $\mathbf{CA}^{-1}$.

Theorem 3.4.2 If $\mathbf{A}$ is non-singular, then $\mathbf{A}^{-1}$ and $\mathbf{A}^{\mathrm{T}}$ are also non-singular, $(\mathbf{A}^{-1})^{-1} = \mathbf{A}$ and $(\mathbf{A}^{\mathrm{T}})^{-1} = (\mathbf{A}^{-1})^{\mathrm{T}}$.

Proof Since $(\mathbf{A}^{-1})\mathbf{A} = \mathbf{I}$, it follows that $\mathbf{A}$ is a right inverse and so the inverse of $\mathbf{A}^{-1}$. Similarly, $\mathbf{A}^{\mathrm{T}}(\mathbf{A}^{-1})^{\mathrm{T}} = (\mathbf{A}^{-1}\mathbf{A})^{\mathrm{T}} = \mathbf{I}^{\mathrm{T}} = \mathbf{I}$, so $(\mathbf{A}^{-1})^{\mathrm{T}}$ is the inverse of $\mathbf{A}^{\mathrm{T}}$. ∎

It is easy to see that if $\mathbf{A}$ is non-singular and α is a non-zero scalar then $\alpha\mathbf{A}$ is non-singular and $(\alpha\mathbf{A})^{-1} = \frac{1}{\alpha}\mathbf{A}^{-1}$.

Theorem 3.4.3 Let $\mathbf{A}$ and $\mathbf{B}$ be square matrices of the same order. Then $\mathbf{AB}$ is non-singular iff both $\mathbf{A}$ and $\mathbf{B}$ are non-singular. Also then $(\mathbf{AB})^{-1} = \mathbf{B}^{-1}\mathbf{A}^{-1}$.

Proof If $\mathbf{AB}$ is non-singular and $\mathbf{C} = (\mathbf{AB})^{-1}$ then $\mathbf{ABC} = \mathbf{I}$, so $\mathbf{BC}$ is an inverse of $\mathbf{A}$ and $\mathbf{A}$ is non-singular. Also, $\mathbf{CAB} = \mathbf{I}$, so $\mathbf{B}$ is non-singular. Conversely, let $\mathbf{A}$ and $\mathbf{B}$ be non-singular. Then

$$(\mathbf{AB}) \cdot (\mathbf{B}^{-1}\mathbf{A}^{-1}) = \mathbf{A} \cdot (\mathbf{BB}^{-1}) \cdot \mathbf{A}^{-1} = \mathbf{AIA}^{-1} = \mathbf{I}$$

So $\mathbf{B}^{-1}\mathbf{A}^{-1}$ is the inverse of $\mathbf{AB}$ and $\mathbf{AB}$ is non-singular. ∎

By repeated application of the preceding theorem, it can be shown that if $\mathbf{A}_1, \mathbf{A}_2, \ldots, \mathbf{A}_k$ are non-singular matrices of the same order then

$$(\mathbf{A}_1\mathbf{A}_2 \cdots \mathbf{A}_k)^{-1} = \mathbf{A}_k^{-1}\mathbf{A}_{k-1}^{-1} \cdots \mathbf{A}_1^{-1}$$

Thus, to get the inverse of the product of $\mathbf{A}_1, \mathbf{A}_2, \ldots, \mathbf{A}_k$, we take the inverses $\mathbf{A}_1^{-1}, \mathbf{A}_2^{-1}, \ldots, \mathbf{A}_k^{-1}$ and multiply them in the *reverse order*. In particular, if $\mathbf{A}$ is non-singular and k is a positive integer then $\mathbf{A}^k$ is non-singular and its inverse is $(\mathbf{A}^{-1})^k$. This is denoted by $\mathbf{A}^{-k}$. It is easy to see that the *laws of indices* (2.4.2) hold for any integers p and q (positive, negative or zero) and any non-singular matrix $\mathbf{A}$.

The sum of two non-singular matrices need not be non-singular. Find an example for this!

We will now obtain expressions for the inverses of some simple types of matrices. Clearly a 1×1 matrix $\mathbf{A} = (a_{11})$ is non-singular iff $a_{11} \neq 0$ and then $\mathbf{A}^{-1} = (1/a_{11})$. For a 2×2 matrix we have the following

Theorem 3.4.4 Let $\mathbf{A} = ((a_{ij}))$ be a 2×2 matrix. Then $\mathbf{A}$ is non-singular iff $\Delta := a_{11}a_{22} - a_{12}a_{21} \neq 0$. Also then

$$\mathbf{A}^{-1} = \frac{1}{\Delta}\begin{bmatrix} a_{22} & -a_{12} \\ -a_{21} & a_{11} \end{bmatrix} \tag{3.4.1}$$

Proof *If part* Suppose $\Delta \neq 0$ and $\mathbf{B}$ is the RHS of (3.4.1). Then it can be verified that $\mathbf{AB} = \mathbf{I}$, so (3.4.1) holds and $\mathbf{A}$ is non-singular.[†]

Only if part Let $\mathbf{A}$ be non-singular. Suppose $\Delta = 0$. Since the first column of $\mathbf{A}$ is non-zero, we may take without loss of generality that $a_{11} \neq 0$ (the proof is similar if $a_{21} \neq 0$). Then $\Delta = 0$ gives $a_{22} = (a_{12}/a_{11})a_{21}$ and $a_{12} = (a_{12}/a_{11})a_{11}$. Hence the second column of $\mathbf{A}$ is a_{12}/a_{11} times the first column, a contradiction since $\mathbf{A}$ is non-singular. ∎

For general matrices of higher orders, explicit expressions like (3.4.1) for the inverse are too cumbersome to be useful, though a formula can be given using determinants, see *Theorem* 6.6.3. The following result is trivial and its proof is left to the reader.

Theorem 3.4.5 A diagonal matrix $\mathrm{diag}(d_1, d_2, \ldots, d_n)$ is non-singular iff each d_i is non-zero. Also then its inverse is $\mathrm{diag}(\frac{1}{d_1}, \frac{1}{d_2}, \ldots, \frac{1}{d_n})$.

Clearly $\mathbf{I}_n$ is non-singular and its inverse is itself. We will prove later (see *Theorem* 3.8.2) that a lower (or upper) triangular matrix $\mathbf{A}$ is non-singular iff all the diagonal elements of $\mathbf{A}$ are non-zero and then $\mathbf{A}^{-1}$ is lower (resp. upper) triangular. We now obtain the inverse of a permutation matrix. Recall that a permutation matrix is a matrix with entries 0's and 1's such that each row and each column contains exactly one 1.

Theorem 3.4.6 Let $\mathbf{P}$ be a permutation matrix. Then $\mathbf{P}$ is non-singular and $\mathbf{P}^{-1} = \mathbf{P}^{\mathrm{T}}$.

Proof Let the order of $\mathbf{P}$ be n. Since $\mathbf{P}$ is a permutation matrix, there exists a permutation $(i_1, i_2, \ldots, i_n)$ of $(1, 2, \ldots, n)$ such that $\mathbf{P} =$

[†]The reader can actually derive the formula (3.4.1) by solving $\mathbf{AB} = \mathbf{I}$. Initially it may be assumed that all entries of $\mathbf{A}$ are non-zero while solving for $\mathbf{B}$ and once the formula (3.4.1) is obtained, it can be verified that it holds even if some entries in $\mathbf{A}$ are zero, provided $\Delta \neq 0$. We incidentally note that Δ is the determinant of $\mathbf{A}$ (see *Chapter* 6).

$[\,\mathbf{e}_{i_1} : \mathbf{e}_{i_2} : \cdots : \mathbf{e}_{i_n}\,]$. So

$$\mathbf{P}^{\mathrm{T}}\mathbf{P} = \begin{bmatrix} \mathbf{e}_{i_1}^{\mathrm{T}} \\ \vdots \\ \mathbf{e}_{i_n}^{\mathrm{T}} \end{bmatrix} [\,\mathbf{e}_{i_1} : \cdots : \mathbf{e}_{i_n}\,] = ((\delta_{k\ell})) = \mathbf{I} \qquad \blacksquare$$

Corollary If $\mathbf{P}$ is a permutation matrix obtained from $\mathbf{I}$ by interchanging two rows, $\mathbf{P}^{-1} = \mathbf{P}$.

Let $\mathbf{A}$ be a non-singular matrix whose inverse is of interest. Sometimes it happens that it is easier to compute the inverse of a matrix $\mathbf{B}$ obtained from $\mathbf{A}$ by permuting the columns (or rows). How do we get the inverse of $\mathbf{A}$ from that of $\mathbf{B}$? The answer is given in the following

***Theorem 3.4.7** Let $\mathbf{B}$ be obtained from a non-singular matrix $\mathbf{A}$ by permuting the columns so that the j-th column of $\mathbf{B}$ is the i_j-th column of $\mathbf{A}$ for $j = 1, 2, \ldots, n$, where $(i_1, i_2, \ldots, i_n)$ is a permutation of $(1, 2, \ldots, n)$. Then $\mathbf{A}^{-1}$ can be obtained from $\mathbf{B}^{-1}$ by permuting the rows thus: the i_j-th row of $\mathbf{A}^{-1}$ is the j-th row of $\mathbf{B}^{-1}$.

Proof Let $\mathbf{P}$ be the permutation matrix $[\,\mathbf{e}_{i_1} : \cdots : \mathbf{e}_{i_n}\,]$. Then

$$\mathbf{AP} = [\,\mathbf{Ae}_{i_1} : \mathbf{Ae}_{i_2} : \cdots : \mathbf{Ae}_{i_n}\,] = [\,\mathbf{A}_{*i_1} : \mathbf{A}_{*i_2} : \cdots : \mathbf{A}_{*i_n}\,] = \mathbf{B}$$

Hence $\mathbf{A}^{-1} = \mathbf{PB}^{-1}$. Now the i_j-th row of $\mathbf{P}$ is $\mathbf{e}_j^{\mathrm{T}}$ since it has 1 in the j-th place. So the i_j-th row of $\mathbf{A}^{-1}$ is $\mathbf{e}_j^{\mathrm{T}}\mathbf{B}^{-1} = (\mathbf{B}^{-1})_{j*}$. $\blacksquare$

It can similarly be shown that if $\mathbf{C}$ is obtained from $\mathbf{A}$ by permuting the *rows* such that the j-th row of $\mathbf{C}$ is the i_j-th row of $\mathbf{A}$ for $j = 1, 2, \ldots, n$ where $(i_1, i_2, \ldots, i_n)$ is a permutation of $(1, 2, \ldots, n)$, then $\mathbf{A}^{-1}$ can be obtained from $\mathbf{C}^{-1}$ by permuting the *columns* thus: the i_j-th column of $\mathbf{A}^{-1}$ is the j-th column of $\mathbf{C}^{-1}$ for $j = 1, 2, \ldots n$.

Exercises

1. For each of the following matrices $\mathbf{A}$, find (i) whether $\mathbf{A}$ is non-singular and (ii) $\mathbf{A}^{-1}$ if $\mathbf{A}$ is non-singular.

$$(a) \begin{bmatrix} 2 & 1 \\ 1 & 2 \end{bmatrix},\ (b) \begin{bmatrix} 0 & 1 \\ 0 & 2 \end{bmatrix},\ (c) \begin{bmatrix} 0 & 1 \\ 1 & 0 \end{bmatrix},\ (d) \begin{bmatrix} \alpha & \beta \\ \gamma & 0 \end{bmatrix},\ (e) \begin{bmatrix} 1 & 2 \\ 2 & 4 \end{bmatrix}$$

$$(f) \begin{bmatrix} 3 & 0 & 0 \\ 0 & 2 & 0 \\ 0 & 0 & -1 \end{bmatrix},\quad (g) \begin{bmatrix} 2 & 1 & 0 \\ 0 & 2 & 3 \\ 0 & 0 & 2 \end{bmatrix},\quad (h) \begin{bmatrix} 3 & 4 & 1 \\ 0 & 0 & 2 \\ 0 & 0 & 7 \end{bmatrix}$$

2. Determine all 2×2 (real) matrices $\mathbf{A}$ such that $\mathbf{A}^{-1} = \mathbf{A}$. Is this class the same as the class of all 2×2 matrices $\mathbf{A}$ such that $\mathbf{A}^2 = \mathbf{I}$?
3. A square matrix $\mathbf{A}$ is said to be *idempotent* if $\mathbf{A}^2 = \mathbf{A}$. Show that if $\mathbf{A}$ is a non-singular idempotent matrix then $\mathbf{A} = \mathbf{I}$.
4. Show that the sum of two non-singular matrices need not be non-singular.
5. (a) If $\mathbf{A}$ and $\mathbf{B}$ commute and $\mathbf{B}$ is non-singular, show that $\mathbf{A}$ and $\mathbf{B}^{-1}$ commute.

   (b) If $f(\lambda)$ and $g(\lambda)$ be polynomials such that $g(\mathbf{A})$ is non-singular, show that $f(\mathbf{A})[g(\mathbf{A})]^{-1} = [g(\mathbf{A})]^{-1}f(\mathbf{A})$. This common matrix is called the value $h(\mathbf{A})$ of the rational function $h(\lambda) = f(\lambda)/g(\lambda)$ at $\mathbf{A}$.
6. Show that the inverse of a non-singular symmetric (resp. skew-symmetric) matrix is symmetric (resp. skew-symmetric). A matrix $\mathbf{A}$ is said to be *skew-symmetric* if $\mathbf{A}^{\mathrm{T}} = -\mathbf{A}$.
7. Find 2×2 singular matrices $\mathbf{A}, \mathbf{B}, \mathbf{C}$ and $\mathbf{D}$ such that $\begin{bmatrix} \mathbf{A} & \mathbf{B} \\ \mathbf{C} & \mathbf{D} \end{bmatrix}$ is non-singular.
8. Find the inverse of the $n \times n$ matrix $\alpha\mathbf{I} + \beta\mathbf{J}$ assuming $\alpha \neq 0$ and $\alpha + n\beta \neq 0$. Here $\mathbf{J}$ denotes the matrix with all elements equal to 1. (Hint: make a guess by assuming that the inverse is of the same type.)

*9. If $\mathbf{A}$ and $\mathbf{B}$ are non-singular matrices, show that $\mathbf{A} \otimes \mathbf{B}$ is also non-singular and that $(\mathbf{A} \otimes \mathbf{B})^{-1} = \mathbf{A}^{-1} \otimes \mathbf{B}^{-1}$. (See *Exercise* 2.7.12.)

10. Prove that the rank of a $k \times \ell$ submatrix of a non-singular matrix of order n is at least $k + \ell - n$. (Hint: first consider the case $\ell = n$.)
11. Show that the set of all non-singular matrices of order n forms a non-abelian group under matrix multiplication. (This group is called *General Linear group* and is denoted $GL(n)$.) Show also that the permutation matrices of order n form a subgroup of $GL(n)$.

*12. (a) Show that the set of all matrices of the form $\begin{bmatrix} \alpha & 0 \\ 0 & 0 \end{bmatrix}$ with $\alpha \neq 0$ forms a group under matrix multiplication. What is the 'identity element' of this group? What is the 'inverse' of an element?

   (b) Do matrices of the form $\begin{bmatrix} 0 & \beta \\ 0 & 0 \end{bmatrix}$ form a group under multiplication? Why?

3.5 Rank of a product of matrices

In this section we obtain bounds for the rank of the product $\mathbf{AB}$ in terms of the ranks of $\mathbf{A}$ and $\mathbf{B}$. We start with some preliminary results which are also of independent interest. In fact, these will be used repeatedly throughout this book.

Upper bound for $\rho(\mathbf{AB})$

Theorem 3.5.1 Let $\mathbf{A}$ and $\mathbf{B}$ be matrices with the same number of rows. Then $\mathcal{C}(\mathbf{B}) \subseteq \mathcal{C}(\mathbf{A})$ iff $\mathbf{B} = \mathbf{AC}$ for some matrix $\mathbf{C}$.

Proof *If part* Let $\mathbf{x} \in \mathcal{C}(\mathbf{B})$. Then $\mathbf{x} = \mathbf{By}$ for some $\mathbf{y}$. So $\mathbf{x} = \mathbf{A}(\mathbf{Cy}) \in \mathcal{C}(\mathbf{A})$.

Only if part Let $\mathbf{B}$ have p columns. Then for $j = 1, \ldots, p$, we have $\mathbf{B}_{*j} \in \mathcal{C}(\mathbf{A})$, so $\mathbf{B}_{*j} = \mathbf{Ac}_j$ for some $\mathbf{c}_j$. Now $\mathbf{B} = \mathbf{A}\,[\,\mathbf{c_1} : \cdots : \mathbf{c_p}\,]$. ∎

Theorem 3.5.2 Let $\mathbf{A}$ and $\mathbf{B}$ be matrices with the same number of columns. Then $\mathcal{R}(\mathbf{B}) \subseteq \mathcal{R}(\mathbf{A})$ iff $\mathbf{B} = \mathbf{DA}$ for some matrix $\mathbf{D}$.

This theorem can be deduced from the preceding theorem by using transposes. We now obtain an upper bond for the rank of $\mathbf{AB}$.

Theorem 3.5.3 $\rho(\mathbf{AB}) \leq \min(\rho(\mathbf{A}), \rho(\mathbf{B}))$.

Proof By *Theorem* 3.5.1, $\mathcal{C}(\mathbf{AB}) \subseteq \mathcal{C}(\mathbf{A})$. By taking dimensions we get $\rho(\mathbf{AB}) \leq \rho(\mathbf{A})$. That $\rho(\mathbf{AB}) \leq \rho(\mathbf{B})$ follows similarly from *Theorem* 3.5.2. ∎

Corollary (i) $\rho(\mathbf{AB}) = \rho(\mathbf{B})$ if $\mathbf{A}$ is of full column rank and (ii) $\rho(\mathbf{AB}) = \rho(\mathbf{A})$ if $\mathbf{B}$ is of full row rank.

Proof We prove only the first statement since the second is proved similarly. By the theorem, $\rho(\mathbf{AB}) \leq \rho(\mathbf{B})$. Since $\mathbf{A}$ is of full column rank, it has a left inverse $\mathbf{X}$. Now $\rho(\mathbf{B}) = \rho(\mathbf{XAB}) \leq \rho(\mathbf{AB})$ by the theorem. Hence equality follows. ∎

From the above corollary we immediately have the following

Theorem 3.5.4 The rank of a matrix is not altered by postmultiplication or premultiplication by a non-singular matrix.

Theorem 3.5.5 The following are equivalent:

(i) $\rho(\mathbf{AB}) = \rho(\mathbf{A})$,

(ii) $\mathcal{C}(\mathbf{AB}) = \mathcal{C}(\mathbf{A})$ and

(iii) $\mathbf{A} = \mathbf{ABC}$ for some matrix $\mathbf{C}$.

Proof (i) ⇒ (ii) We have $\mathcal{C}(\mathbf{AB}) \subseteq \mathcal{C}(\mathbf{A})$ by *Theorem* 3.5.1. By hypothesis, the two subspaces have the same dimension, so equality follows.

(ii) $\Rightarrow$ (iii) This follows from *Theorem* 3.5.1.

(iii) $\Rightarrow$ (i) $\rho(\mathbf{AB}) \leq \rho(\mathbf{A}) = \rho(\mathbf{ABC}) \leq \rho(\mathbf{AB})$. Hence equality holds throughout. ∎

The next theorem can be proved similarly. It can also be deduced from the preceding theorem using transposes.

Theorem 3.5.6 The following are equivalent:

(i) $\rho(\mathbf{AB}) = \rho(\mathbf{B})$,

(ii) $\mathcal{R}(\mathbf{AB}) = \mathcal{R}(\mathbf{B})$ and

(iii) $\mathbf{B} = \mathbf{DAB}$ for some matrix $\mathbf{D}$.

In *Theorems* 3.3.3 and 3.3.5 we have seen that if $\mathbf{A}$ is of full row rank then $\mathbf{A}$ can be cancelled on the right and if $\mathbf{A}$ is of full column rank, it can be cancelled on the left. We now prove the cancellation laws under weaker hypotheses.

Theorem 3.5.7 (*Rank cancellation laws*)

(i) If $\mathbf{ABC} = \mathbf{ABD}$ and $\rho(\mathbf{AB}) = \rho(\mathbf{B})$, then $\mathbf{BC} = \mathbf{BD}$.

(ii) If $\mathbf{CAB} = \mathbf{DAB}$ and $\rho(\mathbf{AB}) = \rho(\mathbf{A})$, then $\mathbf{CA} = \mathbf{DA}$.

Proof We shall prove only (i) since the proof of (ii) is similar. By the preceding theorem, there exists a matrix $\mathbf{E}$ such that $\mathbf{B} = \mathbf{EAB}$. On premultiplying $\mathbf{ABC} = \mathbf{ABD}$ by $\mathbf{E}$, the result follows. ∎

Nullity of a matrix

We have obtained an upper bound for the rank of $\mathbf{AB}$ in *Theorem* 3.5.3. To obtain a lower bound for the same, we need the concept of nullity of a matrix.

Let $\mathbf{A}$ be an $m \times n$ matrix. Consider the linear transformation $f : \mathbf{x} \mapsto \mathbf{Ax}$. The set $\{\mathbf{x} : f(\mathbf{x}) = \mathbf{0}\} = \{\mathbf{x} : \mathbf{Ax} = \mathbf{0}\}$ is called the *kernel* of f and is denoted $\mathcal{K}(f)$. It is easy to check that this is a subspace of F^n. By *Theorem* 3.3.5, f is 1-1 iff its kernel is $\{\mathbf{0}\}$. The subspace $\{\mathbf{x} : \mathbf{Ax} = \mathbf{0}\}$ also arises in the study of the set of all solutions of a system of linear equations $\mathbf{Ax} = \mathbf{b}$ as we will see in *Chapter* 5. Motivated by these, we give the following

Definition 3.5.8 The *null space* of a matrix $\mathbf{A}$ is $\{\mathbf{x} : \mathbf{Ax} = \mathbf{0}\}$ and is denoted by $\mathcal{N}(\mathbf{A})$. The dimension of $\mathcal{N}(\mathbf{A})$ is called the *nullity* of $\mathbf{A}$

and is denoted by $\nu(\mathbf{A})$.

There is a close connection between rank and nullity which is given in the following

Theorem 3.5.9 For any matrix $\mathbf{A}$ with n columns, $\nu(\mathbf{A}) = n - \rho(\mathbf{A})$.

Proof Let $B = \{\mathbf{x}_1, \mathbf{x}_2, \ldots, \mathbf{x}_\nu\}$ be a basis of $\mathcal{N}(\mathbf{A})$. Since $\mathcal{N}(\mathbf{A}) \subseteq F^n$, B can be extended to a basis $\{\mathbf{x}_1, \ldots, \mathbf{x}_\nu, \mathbf{x}_{\nu+1}, \ldots, \mathbf{x}_n\}$ of F^n. We will prove that $\mathbf{A}\mathbf{x}_{\nu+1}, \ldots, \mathbf{A}\mathbf{x}_n$ form a basis of $\mathcal{C}(\mathbf{A})$. Clearly they belong to $\mathcal{C}(\mathbf{A})$. Take any vector $\mathbf{z} \in \mathcal{C}(\mathbf{A})$. Then $\mathbf{z} = \mathbf{A}\mathbf{x}$ for some $\mathbf{x}$ in F^n. Now $\mathbf{x}$ can be written as $\mathbf{x} = \sum_{i=1}^n \alpha_i \mathbf{x}_i$ for some scalars $\alpha_1, \ldots, \alpha_n$. Then

$$\mathbf{z} = \mathbf{A}\mathbf{x} = \sum_{i=\nu+1}^{n} \alpha_i\, \mathbf{A}\mathbf{x}_i$$

since $\mathbf{A}\mathbf{x}_1 = \cdots = \mathbf{A}\mathbf{x}_\nu = \mathbf{0}$. Thus $\mathbf{A}\mathbf{x}_{\nu+1}, \ldots, \mathbf{A}\mathbf{x}_n$ generate $\mathcal{C}(\mathbf{A})$. To prove linear independence, let $\beta_1 \mathbf{A}\mathbf{x}_{\nu+1} + \cdots + \beta_{n-\nu}\mathbf{A}\mathbf{x}_n = \mathbf{0}$. Then $\beta_1 \mathbf{x}_{\nu+1} + \cdots + \beta_{n-\nu}\mathbf{x}_n \in \mathcal{N}(\mathbf{A})$. Hence

$$\beta_1 \mathbf{x}_{\nu+1} + \cdots + \beta_{n-\nu}\mathbf{x}_n = \theta_1 \mathbf{x}_1 + \cdots + \theta_\nu \mathbf{x}_\nu$$

for some scalars $\theta_1, \ldots, \theta_\nu$. Since $\{\mathbf{x}_1, \ldots, \mathbf{x}_n\}$ is linearly independent it follows that $\beta_1 = \cdots = \beta_{n-\nu} = 0$. Thus $\mathbf{A}\mathbf{x}_{\nu+1}, \ldots, \mathbf{A}\mathbf{x}_n$ are linearly independent and so form a basis of $\mathcal{C}(\mathbf{A})$. Hence $\rho(\mathbf{A}) = d(\mathcal{C}(\mathbf{A})) = n - \nu = n - \nu(\mathbf{A})$. ∎

We mention that the number of rows of $\mathbf{A}$ has no role to play in the statement of the preceding theorem.

The preceding theorem can also be proved by exhibiting a basis of $\mathcal{N}(\mathbf{A})$. Let $\rho(\mathbf{A}) = r$. Without any loss of generality, we assume that the first r columns of $\mathbf{A}$ form a basis of $\mathcal{C}(\mathbf{A})$. Then for $j = r+1, \ldots, n$,

$$\mathbf{A}_{*j} = \alpha_{j1}\mathbf{A}_{*1} + \cdots + \alpha_{jr}\mathbf{A}_{*r} \tag{3.5.1}$$

for some scalars $\alpha_{j1}, \ldots, \alpha_{jr}$. For $j = r+1, \ldots, n$, consider the vectors

$$\mathbf{y}_j = (\alpha_{j1}, \ldots, \alpha_{jr}, 0, \ldots, 0, -1, 0, \ldots, 0)^{\mathrm{T}}$$

where the -1 occurs in the j-th place. It is easy to see that $\mathbf{y}_{r+1}, \ldots, \mathbf{y}_n$ belong to $\mathcal{N}(\mathbf{A})$ and are linearly independent. We now show that they generate $\mathcal{N}(\mathbf{A})$. For this, let $\mathbf{x} = (x_1, x_2, \ldots, x_n)^{\mathrm{T}} \in \mathcal{N}(\mathbf{A})$. Then $\sum_{j=1}^n x_j \mathbf{A}_{*j} = \mathbf{0}$. Substituting for $\mathbf{A}_{*,r+1}, \ldots, \mathbf{A}_{*n}$ from (3.5.1), we get

$$\sum_{i=1}^{r}\left(x_i + \sum_{j=r+1}^{n} x_j\alpha_{ji}\right)\mathbf{A}_{*i} = \mathbf{0}$$

Since $\mathbf{A}_{*1}, \ldots, \mathbf{A}_{*r}$ are linearly independent, it follows that the coefficient of $\mathbf{A}_{*i}$ is 0 for $i = 1, \ldots, r$. It is now easy to verify that $\mathbf{x} = -x_{r+1}\mathbf{y}_{r+1} - \cdots - x_n\mathbf{y}_n$. This proves that $\mathbf{y}_{r+1}, \ldots, \mathbf{y}_n$ generate $\mathcal{N}(\mathbf{A})$ and so form a basis for $\mathcal{N}(\mathbf{A})$, hence $\nu(\mathbf{A}) = n - \rho(\mathbf{A})$.

Lower bound for $\rho(\mathbf{AB})$

We shall now obtain a lower bound for $\rho(\mathbf{AB})$. For this we need the following theorem which is also of independent interest. Note that the preceding theorem is a special case of this obtained by taking $T = F^p$.

Theorem 3.5.10 Let $\mathbf{B}$ be an $n \times p$ matrix and T a subspace of F^p. Let $S = \{\mathbf{Bx} : \mathbf{x} \in T\}$ and $W = \mathcal{N}(\mathbf{B})$. Then S is a subspace of F^n and $d(S) = d(T) - d(T \cap W)$.

Proof It is easy to see that S is a subspace of F^n. We obtain its dimension by imitating the proof of the preceding theorem.

Extend a basis $\{\mathbf{x}_1, \ldots, \mathbf{x}_k\}$ of $T \cap W$ to a basis $\{\mathbf{x}_1, \ldots, \mathbf{x}_k, \mathbf{x}_{k+1}, \ldots, \mathbf{x}_t\}$ of T. We will show that $\mathbf{Bx}_{k+1}, \ldots, \mathbf{Bx}_t$ form a basis for S. Clearly $\mathbf{Bx}_i \in S$ for $i = k+1, \ldots, t$. To show that $\mathbf{Bx}_{k+1}, \ldots, \mathbf{Bx}_t$ span S, let $\mathbf{z} \in S$. Then $\mathbf{z} = \mathbf{Bx}$ for some $\mathbf{x} \in T$. Now $\mathbf{x} = \sum_{i=1}^{t} \alpha_i\mathbf{x}_i$ for some scalars $\alpha_1, \ldots, \alpha_t$. Then

$$\mathbf{Bx} = \alpha_{k+1}\mathbf{Bx}_{k+1} + \cdots + \alpha_t\mathbf{Bx}_t$$

since $\mathbf{Bx}_i = \mathbf{0}$ for $i = 1, \ldots, k$. Thus $\mathbf{Bx}_{k+1}, \ldots, \mathbf{Bx}_t$ span S. To prove linear independence. let

$$\beta_1\mathbf{Bx}_{k+1} + \cdots + \beta_{t-k}\mathbf{Bx}_t = \mathbf{0}$$

Then $\beta_1\mathbf{x}_{k+1} + \cdots + \beta_{t-k}\mathbf{x}_t \in \mathcal{N}(\mathbf{B})$. Now $\mathbf{x}_{k+1}, \ldots, \mathbf{x}_t \in T$ and T is a subspace. Hence $\beta_1\mathbf{x}_{k+1} + \cdots + \beta_{t-k}\mathbf{x}_t$ belongs to T and so to $T \cap W$. Therefore

$$\beta_1\mathbf{x}_{k+1} + \cdots + \beta_{t-k}\mathbf{x}_t = \gamma_1\mathbf{x}_1 + \cdots + \gamma_k\mathbf{x}_k$$

for some scalars $\gamma_1, \ldots, \gamma_k$. Since $\{\mathbf{x}_1, \ldots, \mathbf{x}_t\}$ is linearly independent, it follows that $\beta_1 = \cdots = \beta_{t-k} = 0$. Thus $\{\mathbf{Bx}_{k+1}, \ldots, \mathbf{Bx}_t\}$ is linearly independent and so a basis of S. The theorem follows. ∎

Theorem 3.5.11 If **AB** is defined,

$$\rho(\mathbf{AB}) = \rho(\mathbf{B}) - d(\mathcal{C}(\mathbf{B}) \cap \mathcal{N}(\mathbf{A}))$$

Proof We have

$$\begin{aligned} d(\mathcal{C}(\mathbf{B}) \cap \mathcal{N}(\mathbf{A})) \\ &= d(\{\mathbf{Bx} : \mathbf{x} \in \mathcal{N}(\mathbf{AB})\}) \\ &= \nu(\mathbf{AB}) - d(\mathcal{N}(\mathbf{AB}) \cap \mathcal{N}(\mathbf{B})) \quad \text{by the preceding theorem} \\ &= \nu(\mathbf{AB}) - \nu(\mathbf{B}) \quad \text{since } \mathcal{N}(\mathbf{AB}) \supseteq \mathcal{N}(\mathbf{B}) \\ &= \rho(\mathbf{B}) - \rho(\mathbf{AB}) \quad \text{by } \textit{Theorem } 3.5.9. \end{aligned}$$

∎

Theorem 3.5.12 (*Sylvester's inequality*) Let **A** and **B** be matrices of orders $m \times n$ and $n \times p$ respectively. Then

$$\rho(\mathbf{AB}) \geq \rho(\mathbf{A}) + \rho(\mathbf{B}) - n$$

Further, equality holds iff $\mathcal{N}(\mathbf{A}) \subseteq \mathcal{C}(\mathbf{B})$.

Proof Since $d(\mathcal{C}(\mathbf{B}) \cap \mathcal{N}(\mathbf{A})) \leq d(\mathcal{N}(\mathbf{A})) = n - \rho(\mathbf{A})$, the present theorem follows from the preceding theorem. ∎

Sylvester's inequality can also be stated as: $\nu(\mathbf{AB}) \leq \nu(\mathbf{A}) + \nu(\mathbf{B})$ since $\nu(\mathbf{AB}) = p - \rho(\mathbf{AB})$ and $\nu(\mathbf{A}) + \nu(\mathbf{B}) = n - \rho(\mathbf{A}) + p - \rho(\mathbf{B})$. This inequality can be extended to the product of several matrices as

$$\nu(\mathbf{A}_1\mathbf{A}_2 \cdots \mathbf{A}_k) \leq \nu(\mathbf{A}_1) + \nu(\mathbf{A}_2) + \cdots + \nu(\mathbf{A}_k)$$

Now, $\nu(\mathbf{AB}) \geq \nu(\mathbf{B})$ since $\rho(\mathbf{AB}) \leq \rho(\mathbf{B})$. If **A** and **B** are square matrices of the same order then we also have $\nu(\mathbf{AB}) \geq \nu(\mathbf{A})$ and

$$\max(\nu(\mathbf{A}), \nu(\mathbf{B})) \leq \nu(\mathbf{AB}) \leq \nu(\mathbf{A}) + \nu(\mathbf{B})$$

These inequalities are known as *Sylvester's law of nullity.*

*__Theorem 3.5.13__ (*Frobenius inequality*) For any three matrices **A**, **B** and **C** such that **ABC** is defined,

$$\rho(\mathbf{ABC}) \geq \rho(\mathbf{AB}) + \rho(\mathbf{BC}) - \rho(\mathbf{B})$$

Proof Clearly, $\mathcal{C}(\mathbf{BC}) \cap \mathcal{N}(\mathbf{A}) \subseteq \mathcal{C}(\mathbf{B}) \cap \mathcal{N}(\mathbf{A})$. On taking dimensions, the present theorem follows from *Theorem* 3.5.11. ∎

We now give another form of *Theorem* 3.5.10 which plays an important role in the theory of linear statistical models (see Rao (1973)).

***Theorem 3.5.14** If $\mathbf{A}$ and $\mathbf{B}$ are matrices with n columns and $S = \{\mathbf{Bx} : \mathbf{Ax} = \mathbf{0}\}$, then

$$d(S) = \rho\begin{bmatrix}\mathbf{A}\\ \mathbf{B}\end{bmatrix} - \rho(\mathbf{A})$$

Proof Let $T = \mathcal{N}(\mathbf{A})$. Then

$$d(T \cap \mathcal{N}(\mathbf{B})) = d(\mathcal{N}(\mathbf{A}) \cap \mathcal{N}(\mathbf{B})) = d\left(\mathcal{N}\begin{bmatrix}\mathbf{A}\\ \mathbf{B}\end{bmatrix}\right) = n - \rho\begin{bmatrix}\mathbf{A}\\ \mathbf{B}\end{bmatrix}$$

Hence the present theorem follows from *Theorem* 3.5.10. ∎

Exercises

1. If $\mathcal{R}(\mathbf{A}) \subseteq \mathcal{R}(\mathbf{B})$, show that $\mathcal{R}(\mathbf{AE}) \subseteq \mathcal{R}(\mathbf{BE})$. Deduce that if $\mathcal{R}(\mathbf{A}) = \mathcal{R}(\mathbf{B})$, then $\rho(\mathbf{AE}) = \rho(\mathbf{BE})$. State and prove corresponding results for column space.
2. Show that if $\mathbf{A}^2 = \mathbf{A}^3$ and $\rho(\mathbf{A}) = \rho(\mathbf{A}^2)$ then $\mathbf{A} = \mathbf{A}^2$. Show also that the condition $\rho(\mathbf{A}) = \rho(\mathbf{A}^2)$ cannot be dropped even for a 2×2 matrix.
3. Prove or disprove: $\rho(\mathbf{ABC}) \leq \rho(\mathbf{AC})$.
4. Show by examples that in *Theorem* 3.5.7, $\mathbf{C} = \mathbf{D}$ may not follow.
5. Prove that the kernel of any linear transformation is a subspace of the domain.
6. For a square matrix $\mathbf{A}$ of order n, prove that (i) $\mathcal{N}(\mathbf{A}) \subseteq \mathcal{C}(\mathbf{I} - \mathbf{A})$ and (ii) $\mathcal{C}(\mathbf{A}) + \mathcal{C}(\mathbf{I} - \mathbf{A}) = F^n$.
7. (a) If $\mathbf{A}$ and $\mathbf{B}$ are matrices with the same number of rows then show that $\mathcal{C}[\,\mathbf{A} : \mathbf{B}\,] = \mathcal{C}(\mathbf{A}) + \mathcal{C}(\mathbf{B})$.
   (b) If $\mathbf{A}$ and $\mathbf{C}$ have the same number of columns then show that $\mathcal{R}\left[\begin{smallmatrix}\mathbf{A}\\ \mathbf{C}\end{smallmatrix}\right] = \mathcal{R}(\mathbf{A}) + \mathcal{R}(\mathbf{C})$ and $\mathcal{N}\left[\begin{smallmatrix}\mathbf{A}\\ \mathbf{C}\end{smallmatrix}\right] = \mathcal{N}(\mathbf{A}) \cap \mathcal{N}(\mathbf{C})$.
8. (a) Show that $\rho[\,\mathbf{A} : \mathbf{B}\,] = \rho(\mathbf{A})$ iff $\mathbf{B} = \mathbf{AC}$ for some matrix $\mathbf{C}$.
   (b) Show that $\rho\left[\begin{smallmatrix}\mathbf{A}\\ \mathbf{D}\end{smallmatrix}\right] = \rho(\mathbf{A})$ iff $\mathbf{D} = \mathbf{EA}$ for some matrix $\mathbf{E}$.
9. For $n \times n$ matrices $\mathbf{A}$ and $\mathbf{B}$, show that the rank of $\begin{bmatrix}\mathbf{A} & \mathbf{I}\\ \mathbf{I} & \mathbf{B}\end{bmatrix}$ is n iff $\mathbf{B} = \mathbf{A}^{-1}$.
10. Prove that $\mathcal{N}(\mathbf{A}) \subseteq \mathcal{N}(\mathbf{B})$ iff $\mathcal{R}(\mathbf{A}) \supseteq \mathcal{R}(\mathbf{B})$. Deduce that null space is not altered by premultiplication by a non-singular matrix.
11. Consider an $n \times n$ non-singular matrix $\mathbf{M} = \left[\begin{smallmatrix}\mathbf{P}\\ \mathbf{Q}\end{smallmatrix}\right]$ where $\mathbf{P}$ is of order $k \times n$. Let $\mathbf{M}^{-1} = [\,\mathbf{C} : \mathbf{D}\,]$ where $\mathbf{C}$ is of order $n \times k$.
    (a) Show that $\mathbf{C}$ is a right inverse of $\mathbf{P}$.

(b) Show that $\mathcal{N}(\mathbf{P}) = \mathcal{C}(\mathbf{D})$. Hence prove that $\nu(\mathbf{P}) = n - \rho(\mathbf{P})$ and deduce *Theorem* 3.5.9.

12. Show that $\mathcal{N}(\mathbf{A}) = \mathcal{C}(\mathbf{B})$ iff $\mathcal{C}(\mathbf{A}^{\mathrm{T}}) = \mathcal{N}(\mathbf{B}^{\mathrm{T}})$. Deduce that every subspace S of F^n is the null space of a matrix as follows: take a matrix $\mathbf{B}$ whose columns form a basis of S. Let $\mathbf{A}$ be such that the columns of $\mathbf{A}^{\mathrm{T}}$ form a basis of $\mathcal{N}(\mathbf{B}^{\mathrm{T}})$. Then $S = \mathcal{N}(\mathbf{A})$.
13. From *Theorem* 3.5.9 deduce that if $\mathbf{AB} = \mathbf{0}$ then $\rho(\mathbf{A}) + \rho(\mathbf{B}) \leq n$, where n is the number of columns of $\mathbf{A}$.
14. For each positive integer k, find a matrix $\mathbf{A}$ of order $2k$ and rank k such that $\mathbf{A}^2 = \mathbf{0}$.
15. Let $\mathbf{A}$ and $\mathbf{B}$ be 5×5 matrices. For each k, $0 \leq k \leq 5$, find all possible values for $\rho(\mathbf{BA})$ given that $\rho(\mathbf{AB}) = k$. Prove your statement.
16. Prove that

$$\rho(\mathbf{A}_1\mathbf{A}_2\cdots\mathbf{A}_k) \geq \sum_{i=1}^{k} \rho(\mathbf{A}_i) - \sum_{i=1}^{k-1} n_i$$

where n_i is the number of columns of $\mathbf{A}_i$.

17. Prove that if $\rho(\mathbf{A}^k) = \rho(\mathbf{A}^{k+1})$ then $\rho(\mathbf{A}^{k+1}) = \rho(\mathbf{A}^{k+2})$. Deduce that for any square matrix $\mathbf{A}$ of order n there exists an integer p (called the *index* of $\mathbf{A}$) such that $1 \leq p \leq n-1$ and

$$\rho(\mathbf{A}) > \cdots > \rho(\mathbf{A}^p) = \rho(\mathbf{A}^{p+1}) = \cdots$$

Also then show that

$$\mathcal{C}(\mathbf{A}) \supset \cdots \supset \mathcal{C}(\mathbf{A}^p) = \mathcal{C}(\mathbf{A}^{p+1}) = \cdots$$

and

$$\mathcal{N}(\mathbf{A}) \subset \cdots \subset \mathcal{N}(\mathbf{A}^p) = \mathcal{N}(\mathbf{A}^{p+1}) = \cdots$$

18. Prove that for any square matrix $\mathbf{A}$ and any positive integer k, $\rho(\mathbf{A}^k) \leq \frac{1}{2}[\rho(\mathbf{A}^{k-1}) + \rho(\mathbf{A}^{k+1})]$. Deduce the preceding exercise from this.

*19. Prove that $\rho(\mathbf{PAQ}) = \rho(\mathbf{A})$ iff $\rho(\mathbf{A}) = \rho(\mathbf{PA}) = \rho(\mathbf{AQ})$.

20. Let g be a linear transformation from V_1 to V_2.

    (a) Using *Theorem* 3.5.9 prove that $d(\mathcal{K}(g)) = d(V_1) - d(g(V_1))$.

    (b) Show that $d(\{g(\mathbf{x}) : \mathbf{x} \in W\}) = d(W) - d(W \cap \mathcal{K}(g))$ for any subspace W of V_1 by considering the restriction h of g to W. Hence deduce *Theorem* 3.5.10 from *Theorem* 3.5.9.

*21. Let $\mathcal{G}$ be a set of $n \times n$ matrices which forms a group under matrix multiplication. If $\mathbf{A}, \mathbf{B} \in \mathcal{G}$, show that $\mathcal{R}(\mathbf{A}) = \mathcal{R}(\mathbf{B})$, $\mathcal{C}(\mathbf{A}) = \mathcal{C}(\mathbf{B})$ and $\rho(\mathbf{A}) = \rho(\mathbf{A}^2)$.

*22. If $\mathbf{R}$, $\mathbf{Q}$ and $\mathbf{S}$ are non-singular matrices, show that $\mathcal{C}(\mathbf{B}) \subseteq \mathcal{C}(\mathbf{A})$ iff $\mathcal{C}(\mathbf{RBQ}) \subseteq \mathcal{C}(\mathbf{RAS})$.

*23. Let $\mathcal{N}(\mathbf{A} : \mathbf{B}) = \mathcal{C}\begin{bmatrix}\mathbf{C}\\ \mathbf{D}\end{bmatrix}$ where $\mathbf{AC}$ is defined. Then show that $\mathcal{C}(\mathbf{A}) \cap \mathcal{C}(\mathbf{B}) = \mathcal{C}(\mathbf{AC}) = \mathcal{C}(\mathbf{BD})$.

3.6 Rank-factorization of a matrix

We have seen in the last three sections that matrices which are of full rank (either full row rank or full column rank) have several nice properties. In this section we show that every non-null matrix can be written as a product of two full rank matrices. The earlier results can then be used for the study of general matrices. This technique is very powerful and will be used repeatedly throughout this book. We have, in fact, used it in the proof of *Theorem* 3.2.3.

Definition 3.6.1 Let $\mathbf{A}$ be an $m \times n$ matrix with rank $r \geq 1$. Then $(\mathbf{P}, \mathbf{Q})$ is said to be a *rank-factorization* of $\mathbf{A}$ if $\mathbf{P}$ is of order $m \times r$, $\mathbf{Q}$ is of order $r \times n$ and $\mathbf{A} = \mathbf{PQ}$.

From the proof of *Theorem* 3.2.3 it follows that every non-null matrix has a rank-factorization. Note that a null matrix cannot have a rank-factorization since there cannot be a matrix with 0 rows. A simple computational procedure for finding a rank-factorization of any non-null matrix will be given in *Section* 4.4.

It is easy to see that if $(\mathbf{P}, \mathbf{Q})$ is a rank-factorization of $\mathbf{A}$, then $(\mathbf{Q}^{\mathrm{T}}, \mathbf{P}^{\mathrm{T}})$ is a rank-factorization of $\mathbf{A}^{\mathrm{T}}$. In the next result we study when a factorization is a rank-factorization.

Theorem 3.6.2 Let $\mathbf{A} = \mathbf{PQ}$ where $\mathbf{P}$ is an $m \times k$ matrix and $\mathbf{Q}$ a $k \times n$ matrix. Then $k \geq \rho(\mathbf{A})$. Moreover, the following are equivalent:

(i) $k = \rho(\mathbf{A})$, i.e., $(\mathbf{P}, \mathbf{Q})$ is a rank-factorization of $\mathbf{A}$,

(ii) $\rho(\mathbf{P}) = \rho(\mathbf{Q}) = k$, i.e., $\mathbf{P}$ is of full column rank and $\mathbf{Q}$ is of full row rank,

(iii) the columns of $\mathbf{P}$ form a basis of $\mathcal{C}(\mathbf{A})$

(iv) the rows of $\mathbf{Q}$ form a basis of $\mathcal{R}(\mathbf{A})$

Proof The first conclusion follows from

$$k \geq \rho(\mathbf{P}) \geq \rho(\mathbf{PQ}) = \rho(\mathbf{A}) \tag{3.6.1}$$

We next prove the equivalence of (i)–(iv).

(i) ⇒ (ii) (3.6.1) gives $\rho(\mathbf{P}) = k$; $\rho(\mathbf{Q}) = k$ follows similarly.

(ii) ⇒ (iii) By *Theorem* 3.5.5 and the corollary to *Theorem* 3.5.3 it follows that $\mathcal{C}(\mathbf{A}) = \mathcal{C}(\mathbf{P})$. Since the columns of $\mathbf{P}$ are linearly independent it follows that they form a basis of $\mathcal{C}(\mathbf{A})$.

That (iii) implies (i) is trivial. The proofs of (ii) ⇒ (iv) and (iv) ⇒ (i) are similar to those of (ii) ⇒ (iii) and (iii) ⇒ (i) respectively. ∎

Corollary If $(\mathbf{P}, \mathbf{Q})$ is a rank-factorization of $\mathbf{A}$ then $\mathcal{C}(\mathbf{P}) = \mathcal{C}(\mathbf{A})$, $\mathcal{R}(\mathbf{Q}) = \mathcal{R}(\mathbf{A})$ and $\mathcal{N}(\mathbf{Q}) = \mathcal{N}(\mathbf{A})$.

It follows from the above theorem that $\rho(\mathbf{A})$ is the minimum number of columns of $\mathbf{P}$ (or rows of $\mathbf{Q}$) in a factorization $\mathbf{A} = \mathbf{PQ}$ of $\mathbf{A}$.

As an application of rank-factorization, we now show that the rank of a square matrix $\mathbf{A}$ can be obtained easily if $\mathbf{A} = \mathbf{A}^2$. Such matrices will be studied in the next section.

Theorem 3.6.3 If $\mathbf{A} = \mathbf{A}^2$, rank of $\mathbf{A}$ equals trace of $\mathbf{A}$.

Proof The result is trivial if the rank r of $\mathbf{A}$ is 0, so let $r \geq 1$ and let $(\mathbf{P}, \mathbf{Q})$ be a rank-factorization of $\mathbf{A}$. Then $\mathbf{PQPQ} = \mathbf{PQ} = \mathbf{PI}_r\mathbf{Q}$. Since $\mathbf{P}$ can be cancelled on the left and $\mathbf{Q}$ can be cancelled on the right, we get $\mathbf{QP} = \mathbf{I}_r$. Now $\operatorname{tr}(\mathbf{I}_r) = r$ and, by *Theorem* 2.5.4, trace of $\mathbf{QP}$ equals trace of $\mathbf{PQ}$. Hence the present theorem follows. ∎

Example 3.6.4 Consider the matrix $\mathbf{A}$ of *Example* 3.2.2. It is easy to see that $(2, 3, -1)^{\mathrm{T}}$ and $(1, 0, -2)^{\mathrm{T}}$ form a basis of $\mathcal{C}(\mathbf{A})$. So let

$$\mathbf{P} = \begin{bmatrix} 2 & 1 \\ 3 & 0 \\ -1 & -2 \end{bmatrix}$$

Then there exists a unique $\mathbf{Q}$ such that $\mathbf{A} = \mathbf{PQ}$. Solving for each column of $\mathbf{Q}$, we get

$$\mathbf{Q} = \begin{bmatrix} 1 & 2 & 0 & \frac{1}{3} \\ 0 & 0 & 1 & -\frac{5}{3} \end{bmatrix}$$

Clearly now $(\mathbf{P}, \mathbf{Q})$ is a rank-factorization of $\mathbf{A}$. ∎

In the following theorem we show that a matrix of rank r can be represented in a nice form. It will be easy to find a rank-factorization of a matrix given in this form.

Theorem 3.6.5 Let $\mathbf{A}$ be an $m \times n$ matrix of rank $r \geq 1$. Then there exist permutation matrices $\mathbf{P}$ and $\mathbf{Q}$ such that

$$\mathbf{A} = \mathbf{P}\begin{bmatrix} \mathbf{B} & \mathbf{BC} \\ \mathbf{DB} & \mathbf{DBC} \end{bmatrix}\mathbf{Q} \tag{3.6.2}$$

where $\mathbf{B}$ is a non-singular matrix of order r and $\mathbf{C}$ and $\mathbf{D}$ are some matrices of orders $r \times (n-r)$ and $(m-r) \times r$ respectively.

Proof By permuting the columns of $\mathbf{A}$ we can bring the columns in a column basis to the first r positions. Thus there exists a permutation matrix $\mathbf{R}$ such that $\mathbf{AR} = [\,\mathbf{G} : \mathbf{H}\,]$ where $\mathbf{G}$ is an $m \times r$ matrix of rank r. Now, the columns of $\mathbf{H}$ are linear combinations of the columns of $\mathbf{G}$, so $\mathbf{H} = \mathbf{GC}$ for some matrix $\mathbf{C}$ of order $r \times (n-r)$. Thus we have $\mathbf{AR} = [\,\mathbf{G} : \mathbf{GC}\,]$. Now, by permuting the rows of $\mathbf{G}$ we can bring the rows in a row basis of $\mathbf{G}$ to the first r positions. Thus there exists a permutation matrix $\mathbf{S}$ such that

$$\mathbf{SG} = \begin{bmatrix} \mathbf{B} \\ \mathbf{K} \end{bmatrix}$$

where $\mathbf{B}$ is an $r \times r$ matrix of rank r. Since the rows of $\mathbf{K}$ are linear combinations of the rows of $\mathbf{B}$, we have $\mathbf{K} = \mathbf{DB}$ for some $\mathbf{D}$. Now $\mathbf{B}$ is non-singular and

$$\mathbf{SAR} = \mathbf{S}[\,\mathbf{G} : \mathbf{GC}\,] = [\,\mathbf{SG} : \mathbf{SGC}\,] = \begin{bmatrix} \mathbf{B} & \mathbf{BC} \\ \mathbf{DB} & \mathbf{DBC} \end{bmatrix}$$

If $\mathbf{P} = \mathbf{S}^{-1}$ and $\mathbf{Q} = \mathbf{R}^{-1}$, then $\mathbf{P}$ and $\mathbf{Q}$ are also permutation matrices and (3.6.2) holds. ∎

Remark Notice that the matrix $\mathbf{A}$ in (3.6.2) can be factorized as $\mathbf{A} = \mathbf{P}_1\mathbf{Q}_1$ where

$$\mathbf{P}_1 = \mathbf{P}\begin{bmatrix} \mathbf{B} \\ \mathbf{DB} \end{bmatrix} \quad \text{and} \quad \mathbf{Q}_1 = [\,\mathbf{I}_r : \mathbf{C}\,]\mathbf{Q}$$

Since $\mathbf{P}_1$ is of order $m \times r$, it follows that $(\mathbf{P}_1, \mathbf{Q}_1)$ is a rank-factorization of $\mathbf{A}$.

Exercises

1. If $(\mathbf{P}, \mathbf{Q})$ is a rank-factorization of $\mathbf{A}$, show that $(\mathbf{PT}, \mathbf{T}^{-1}\mathbf{Q})$ is a rank-factorization of $\mathbf{A}$ for all non-singular $\mathbf{T}$ and that every rank-factorization of $\mathbf{A}$ is of this form.

2. Show that $\mathbf{Q}_1 = \mathbf{Q}_2$ if $(\mathbf{P}, \mathbf{Q}_1)$ and $(\mathbf{P}, \mathbf{Q}_2)$ are rank-factorizations of $\mathbf{A}$ and $\mathbf{P}_1 = \mathbf{P}_2$ if $(\mathbf{P}_1, \mathbf{Q})$ and $(\mathbf{P}_2, \mathbf{Q})$ are rank-factorizations of $\mathbf{A}$.
3. Obtain a rank-factorization for each of the following matrices and describe all the rank-factorizations of $\mathbf{A}$.

$$\text{(a) } \mathbf{A} = \begin{bmatrix} 3 & 1 & 0 \\ 2 & 0 & 1 \end{bmatrix}, \quad \text{(b) } \mathbf{B} = \begin{bmatrix} 5 & 1 & 3 \\ 0 & 0 & 2 \\ 10 & 2 & 4 \end{bmatrix}$$

4. Show that a matrix $\mathbf{A}$ is of rank 1 iff $\mathbf{A} = \mathbf{x}\mathbf{y}^{\mathrm{T}}$ for some non-null column vectors $\mathbf{x}$ and $\mathbf{y}$.
5. Prove that if $\mathbf{A}$ is a symmetric matrix of rank 1, then $\mathbf{A} = \alpha\, \mathbf{u}\mathbf{u}^{\mathrm{T}}$ for some non-zero scalar α and some non-null vector $\mathbf{u}$. Can α be dropped?

*6. (a) If $\mathcal{C}(\mathbf{A}) \subseteq \mathcal{C}(\mathbf{B})$ and $\mathcal{R}(\mathbf{A}) \subseteq \mathcal{R}(\mathbf{D})$, prove that $\mathbf{A} = \mathbf{BCD}$ for some matrix $\mathbf{C}$.

(b) If $\mathbf{A}$ is symmetric and $\mathcal{C}(\mathbf{A}) \subseteq \mathcal{C}(\mathbf{B})$, prove that $\mathbf{A} = \mathbf{BCB}^{\mathrm{T}}$ for some symmetric matrix $\mathbf{C}$, assuming $1 + 1 \neq 0$. (Hint: first get a matrix $\mathbf{C}$ such that $\mathbf{A} = \mathbf{BCB}^{\mathrm{T}}$ and then get a symmetric $\mathbf{C}$).

*7. (a) If $\mathbf{A}$ is an $m \times n$ matrix with rank $< \min(m, n)$, show that the rank can be increased by changing one element.

(b) Show that every non-singular matrix $\mathbf{A}$ can be made singular by changing one element but the rank of a non-null matrix cannot always be reduced by changing one element.

8. Deduce Sylvester's inequality from *Exercise* 3.4.10 using rank-factorization
9. Deduce Sylvester's inequality from *Theorem* 3.5.9 as follows: Let $(\mathbf{P}, \mathbf{Q})$ be a rank-factorization of $\mathbf{AB}$. Then $[\mathbf{A} : \mathbf{P}]\left[\begin{smallmatrix} \mathbf{B} \\ -\mathbf{Q} \end{smallmatrix}\right] = \mathbf{0}$. Now use *Exercise* 3.5.13.

*10. Using a rank-factorization of the middle matrix, deduce Frobenius inequality from Sylvester's inequality.

11. Let $(\mathbf{P}, \mathbf{Q})$ be a rank-factorization of a non-null square matrix $\mathbf{A}$. Show that $\mathbf{A} = \mathbf{A}^2$ iff $\mathbf{QP} = \mathbf{I}$ and that $\rho(\mathbf{A}) = \rho(\mathbf{A}^2)$ iff $\mathbf{QP}$ is non-singular.
12. Prove the following converse of *Theorem* 3.6.5: if (3.6.2) holds, where $\mathbf{P}$ and $\mathbf{Q}$ are permutation matrices and $\mathbf{B}$ is an $r \times r$ non-singular matrix, then $\rho(\mathbf{A}) = r$.
13. If $\mathbf{A}$ is an $m \times n$ matrix with rank $r > 0$, show that there exist non-singular matrices $\mathbf{P}$ and $\mathbf{Q}$ of orders m and n respectively such that

$$\mathbf{A} = \mathbf{P}\begin{bmatrix} \mathbf{I}_r & \mathbf{0} \\ \mathbf{0} & \mathbf{0} \end{bmatrix}\mathbf{Q}$$

3.7 Rank of a sum and projectors

In *Section* 3.5 we studied the relation between the rank of a product of matrices and the rank of each of them. In this section we study the rank of the sum of matrices. We also characterize projectors and study some of their properties.

Theorem 3.7.1 For any two matrices $\mathbf{A}$ and $\mathbf{B}$ of the same order,

$$\rho(\mathbf{A}+\mathbf{B}) \leq \rho(\mathbf{A}) + \rho(\mathbf{B}) \tag{3.7.1}$$

Further, equality holds iff $\mathcal{C}(\mathbf{A}) \cap \mathcal{C}(\mathbf{B}) = \{\mathbf{0}\}$ and $\mathcal{R}(\mathbf{A}) \cap \mathcal{R}(\mathbf{B}) = \{\mathbf{0}\}$.

Proof Since $(\mathbf{A}+\mathbf{B})\mathbf{x} = \mathbf{A}\mathbf{x} + \mathbf{B}\mathbf{x}$ for all $\mathbf{x}$, we have $\mathcal{C}(\mathbf{A}+\mathbf{B}) \subseteq \mathcal{C}(\mathbf{A}) + \mathcal{C}(\mathbf{B})$. Taking dimensions we get

$$\begin{aligned}
\rho(\mathbf{A}+\mathbf{B}) &\leq d(\mathcal{C}(\mathbf{A}) + \mathcal{C}(\mathbf{B})) \\
&= d(\mathcal{C}(\mathbf{A})) + d(\mathcal{C}(\mathbf{B})) - d(\mathcal{C}(\mathbf{A}) \cap \mathcal{C}(\mathbf{B})) \\
&\leq d(\mathcal{C}(\mathbf{A})) + d(\mathcal{C}(\mathbf{B})) \\
&= \rho(\mathbf{A}) + \rho(\mathbf{B})
\end{aligned} \tag{3.7.2}$$

This proves (3.7.1).

Suppose now that equality holds in (3.7.1). Then equality holds throughout the string of inequalities in (3.7.2), hence $d(\mathcal{C}(\mathbf{A}) \cap \mathcal{C}(\mathbf{B})) = 0$ and so $\mathcal{C}(\mathbf{A}) \cap \mathcal{C}(\mathbf{B}) = \{\mathbf{0}\}$. Similarly, $\mathcal{R}(\mathbf{A}) \cap \mathcal{R}(\mathbf{B}) = \{\mathbf{0}\}$. This proves the *only if part* of the second statement in the theorem. To prove the *if part*, let $\rho(\mathbf{A}) = r$ and $\rho(\mathbf{B}) = s$. If $r = 0$ or $s = 0$, the result is trivial, so let $r \geq 1$ and $s \geq 1$. Let $(\mathbf{P}_1, \mathbf{Q}_1)$ and $(\mathbf{P}_2, \mathbf{Q}_2)$ be rank-factorizations of $\mathbf{A}$ and $\mathbf{B}$ respectively. Then $\mathbf{A}+\mathbf{B} = \mathbf{P}_1\mathbf{Q}_1 + \mathbf{P}_2\mathbf{Q}_2 = \mathbf{P}\mathbf{Q}$ where

$$\mathbf{P} = [\,\mathbf{P}_1 : \mathbf{P}_2\,] \quad \text{and} \quad \mathbf{Q} = \begin{bmatrix} \mathbf{Q}_1 \\ \mathbf{Q}_2 \end{bmatrix}$$

(Note that the partitioned matrices are well-defined.) Now $\mathcal{C}(\mathbf{P}_1) \cap \mathcal{C}(\mathbf{P}_2) = \mathcal{C}(\mathbf{A}) \cap \mathcal{C}(\mathbf{B}) = \{\mathbf{0}\}$, the columns of $\mathbf{P}_1$ are linearly independent and the columns of $\mathbf{P}_2$ are linearly independent. Hence it is easy to see that the $r+s$ columns of $\mathbf{P}$ are linearly independent. Similarly the $r+s$ rows of $\mathbf{Q}$ are linearly independent. So $(\mathbf{P}, \mathbf{Q})$ is a rank-factorization of $\mathbf{A}+\mathbf{B}$ and $\rho(\mathbf{A}+\mathbf{B}) = r+s = \rho(\mathbf{A}) + \rho(\mathbf{B})$. ∎

The preceding theorem can be extended to several matrices as follows.

Theorem 3.7.2 If $\mathbf{A}_1, \mathbf{A}_2, \ldots, \mathbf{A}_k$ are $m \times n$ matrices then

$$\rho(\mathbf{A}_1 + \mathbf{A}_2 + \cdots + \mathbf{A}_k) \leq \rho(\mathbf{A}_1) + \rho(\mathbf{A}_2) + \cdots + \rho(\mathbf{A}_k) \qquad (3.7.3)$$

Further, equality holds in (3.7.3) iff the sums $\mathcal{C}(\mathbf{A}_1) + \mathcal{C}(\mathbf{A}_2) + \cdots + \mathcal{C}(\mathbf{A}_k)$ and $\mathcal{R}(\mathbf{A}_1) + \mathcal{R}(\mathbf{A}_2) + \cdots + \mathcal{R}(\mathbf{A}_k)$ are direct.

Proof The inequality (3.7.3) follows easily from

$$\mathcal{C}(\mathbf{A}_1 + \cdots + \mathbf{A}_k) \subseteq \mathcal{C}(\mathbf{A}_1) + \cdots + \mathcal{C}(\mathbf{A}_k)$$

and the modular law. Suppose now equality holds in (3.7.3). Then it is easy to see that

$$d(\mathcal{C}(\mathbf{A}_1) + \cdots + \mathcal{C}(\mathbf{A}_k)) = d(\mathcal{C}(\mathbf{A}_1)) + \cdots + d(\mathcal{C}(\mathbf{A}_k))$$

So by *Theorem* 1.7.9, $\mathcal{C}(\mathbf{A}_1) + \cdots + \mathcal{C}(\mathbf{A}_k)$ is direct. By considering transposes, we get $\mathcal{R}(\mathbf{A}_1) + \cdots + \mathcal{R}(\mathbf{A}_k)$ is direct. To prove the converse, let the sum of the column spaces and the sum of the row spaces be direct. We assume that $\mathbf{A}_1, \mathbf{A}_2, \ldots, \mathbf{A}_k$ are non-null since null $\mathbf{A}_i$'s can be ignored without affecting (3.7.3). Let $(\mathbf{P}_i, \mathbf{Q}_i)$ be a rank-factorization of $\mathbf{A}_i$, $i = 1, \ldots, k$. Then $\mathbf{A}_1 + \cdots + \mathbf{A}_k = \mathbf{P}_1\mathbf{Q}_1 + \cdots + \mathbf{P}_k\mathbf{Q}_k = \mathbf{PQ}$ where

$$\mathbf{P} = [\mathbf{P}_1 : \cdots : \mathbf{P}_k] \text{ and } \mathbf{Q} = \begin{bmatrix} \mathbf{Q}_1 \\ \vdots \\ \mathbf{Q}_k \end{bmatrix}$$

Now the columns of $\mathbf{P}_i$ form a basis of $\mathcal{C}(\mathbf{A}_i)$ for $i = 1, \ldots, k$. Since $\mathcal{C}(\mathbf{A}_1) + \cdots + \mathcal{C}(\mathbf{A}_k)$ is direct, it follows that the columns of $\mathbf{P}$ are linearly independent. Similarly the rows of $\mathbf{Q}$ are linearly independent, so

$$\rho(\mathbf{A}_1 + \cdots + \mathbf{A}_k) = \rho[\,\mathbf{P}_1 : \cdots : \mathbf{P}_k\,] = \rho(\mathbf{A}_1) + \cdots + \rho(\mathbf{A}_k). \qquad \blacksquare$$

We now study projectors.

Definition 3.7.3 Let S be a subspace of F^n and T a complement of S. Then an $n \times n$ matrix $\mathbf{A}$ is said to be the *projector into S along T* if, for all $\mathbf{x} \in F^n$, $\mathbf{Ax}$ is the projection of $\mathbf{x}$ into S along T. We say that an $n \times n$ matrix $\mathbf{A}$ is *a projector* if $\mathbf{A}$ is the projector into S along T for some S and T.

We now show that a projector into S along T exists and is unique. Uniqueness follows from the fact that if $\mathbf{A}$ is a projector into S along T, then $\mathbf{A}_{*j} = \mathbf{A}\mathbf{e}_j$ is the projection of $\mathbf{e}_j$ into S along T for all j. Existence can be proved thus: let f be the linear operator on F^n taking $\mathbf{x}$ to the projection of $\mathbf{x}$ into S along T, see *Theorem* 1.7.7. Then the matrix of f with respect to the canonical basis is a projector into S along T. For a constructive proof of existence, see *Exercise* 3.7.16.

Suppose $\mathbf{A}$ is the projector into S along T. Then T, being the set of all vectors whose projection into S along T is $\mathbf{0}$, equals $\mathcal{N}(\mathbf{A})$. Since any subspace of F^n can be taken to be T, this proves

Theorem 3.7.4 Every subspace of F^n is the set of all solutions of finitely many homogeneous linear equations.

Geometrically, this means that every subspace is an intersection of finitely many 'hyperplanes' through the origin (the set of all points satisfying an equation of the form $\mathbf{a}^{\mathrm{T}}\mathbf{x} = b$ with $\mathbf{a} \neq \mathbf{0}$ is called a hyperplane in F^n).

Theorem 3.7.5 The following statements about an $n \times n$ matrix $\mathbf{A}$ are equivalent.

(i) $\mathbf{A}$ is a projector,

(ii) $\mathbf{A}$ is idempotent, i.e., $\mathbf{A}^2 = \mathbf{A}$,

(iii) $\mathcal{N}(\mathbf{A}) = \mathcal{C}(\mathbf{I} - \mathbf{A})$,

(iv) $\rho(\mathbf{A}) + \rho(\mathbf{I} - \mathbf{A}) = n$,

(v) $\mathcal{C}(\mathbf{A}) + \mathcal{C}(\mathbf{I} - \mathbf{A})$ is direct.

Proof (i) $\Rightarrow$ (ii) Let $\mathbf{A}$ be the projector into S along T. Then for every $\mathbf{x} \in F^n$, $\mathbf{A}\mathbf{x} \in S$, so $\mathbf{A}\mathbf{A}\mathbf{x} = \mathbf{A}\mathbf{x}$. Hence $\mathbf{A}^2 = \mathbf{A}$.

(ii) $\Rightarrow$ (iii) By (ii), $\mathbf{A}(\mathbf{I} - \mathbf{A}) = \mathbf{0}$, so $\mathcal{C}(\mathbf{I} - \mathbf{A}) \subseteq \mathcal{N}(\mathbf{A})$. The reverse inclusion is trivial since $\mathbf{x} \in \mathcal{N}(\mathbf{A}) \Rightarrow \mathbf{x} = (\mathbf{I} - \mathbf{A})\mathbf{x}$. Hence equality follows.

(iii) $\Rightarrow$ (iv) Equate the dimensions of the two sides of (iii).

(iv) $\Rightarrow$ (v) By hypothesis, $\rho(\mathbf{A}) + \rho(\mathbf{I} - \mathbf{A}) = \rho(\mathbf{A} + \mathbf{I} - \mathbf{A})$, so (v) follows from *Theorem* 3.7.1.

(v) $\Rightarrow$ (i) We have $\mathcal{C}(\mathbf{A}) + \mathcal{C}(\mathbf{I} - \mathbf{A}) = F^n$ since any vector $\mathbf{x}$ in F^n can be written as $\mathbf{A}\mathbf{x} + (\mathbf{I} - \mathbf{A})\mathbf{x}$, So, it follows that $\mathcal{C}(\mathbf{I} - \mathbf{A})$ is a complement of $\mathcal{C}(\mathbf{A})$ and $\mathbf{A}\mathbf{x}$ is the projection of $\mathbf{x}$ into $\mathcal{C}(\mathbf{A})$ along $\mathcal{C}(\mathbf{I} - \mathbf{A})$. ∎

Corollary Let $(\mathbf{P}, \mathbf{Q})$ be a rank-factorization of $\mathbf{A}$. Then $\mathbf{A}$ is a projector iff $\mathbf{Q}\mathbf{P} = \mathbf{I}$.

The *if part* is trivial and the *only if part* follows from the proof of *Theorem* 3.6.3. The next remark follows from the proof of the preceding theorem.

Remark Let $\mathbf{A}$ be a projector. Then $\mathbf{A}$ is the projector into $\mathcal{C}(\mathbf{A})$ along $\mathcal{N}(\mathbf{A}) = \mathcal{C}(\mathbf{I}-\mathbf{A})$. Moreover, $\mathbf{I}-\mathbf{A}$ is the projector into $\mathcal{N}(\mathbf{A})$ along $\mathcal{C}(\mathbf{A})$ and $\mathbf{A}^{\mathrm{T}}$ is the projector into $\mathcal{R}(\mathbf{A})$ along $\mathcal{R}(\mathbf{I}-\mathbf{A}) = \mathcal{N}(\mathbf{A}^{\mathrm{T}})$.

We next see when the sum of projectors is a projector. We start with a comprehensive result.

Theorem 3.7.6 Consider the following four statements, where $\mathbf{A} = \mathbf{A}_1 + \mathbf{A}_2 + \cdots + \mathbf{A}_k$ and $\mathbf{A}_1, \mathbf{A}_2, \ldots, \mathbf{A}_k$ are $n \times n$ matrices:
(a) $\mathbf{A}^2 = \mathbf{A}$,
(b) $\mathbf{A}_i^2 = \mathbf{A}_i$ for $i = 1, \ldots, k$,
(c) $\mathbf{A}_i\mathbf{A}_j = \mathbf{0}$ whenever $i \neq j$,
(d) $\rho(\mathbf{A}) = \rho(\mathbf{A}_1) + \cdots + \rho(\mathbf{A}_k)$.
Each of the pairs (a) and (b), (a) and (d), and (b) and (c) implies all of (a)–(d). Moreover, (c*) implies (d), and (a) and (c*) imply (b) and (d) where
(c*): $\mathbf{A}_i\mathbf{A}_j = \mathbf{0}$ whenever $i \neq j$ and $\rho(\mathbf{A}_i) = \rho(\mathbf{A}_i^2)$ for all i.

Proof Since null $\mathbf{A}_i$'s can be ignored, we assume that $\mathbf{A}_1, \mathbf{A}_2, \ldots, \mathbf{A}_k$ are all non-null. Let $(\mathbf{P}_i, \mathbf{Q}_i)$ be a rank-factorization of $\mathbf{A}_i$ for $i = 1, \ldots, k$. Then $\mathbf{A} = \mathbf{PQ}$ where

$$\mathbf{P} = [\mathbf{P}_1 : \cdots : \mathbf{P}_k] \text{ and } \mathbf{Q} = \begin{bmatrix} \mathbf{Q}_1 \\ \vdots \\ \mathbf{Q}_k \end{bmatrix} \tag{3.7.4}$$

Given (a) and (d), it follows from the proof of *Theorem* 3.7.2 that $(\mathbf{P}, \mathbf{Q})$ is a rank-factorization of $\mathbf{A}$. Since $\mathbf{A}^2 = \mathbf{A}$ it follows from the preceding corollary that $\mathbf{QP} = \mathbf{I}$. Thus $\mathbf{Q}_i\mathbf{P}_j = \delta_{ij}\mathbf{I}$ and so (b) and (c) follow.

Given (a) and (b), (d) follows since rank equals trace for $\mathbf{A}$ as well as for each $\mathbf{A}_i$. Now (c) follows from what was proved in the preceding paragraph.

Given (b) and (c), (a) follows trivially and then (d) follows.

Given (c*), it follows that $\mathbf{Q}_i\mathbf{P}_j = \mathbf{0}$ whenever $i \neq j$. Also

$$\rho(\mathbf{A}_i) = \rho(\mathbf{A}_i^2) = \rho(\mathbf{P}_i\mathbf{Q}_i\mathbf{P}_i\mathbf{Q}_i) = \rho(\mathbf{Q}_i\mathbf{P}_i),$$

so $\mathbf{Q}_i\mathbf{P}_i$ is non-singular. Now for any i, we have

$$\sum_{j=1}^{k} \mathbf{P}_j\mathbf{x}_j = \mathbf{0} \Rightarrow \sum_{j=1}^{k} \mathbf{Q}_i\mathbf{P}_j\mathbf{x}_j = \mathbf{0} \Rightarrow \mathbf{Q}_i\mathbf{P}_i\mathbf{x}_i = \mathbf{0} \Rightarrow \mathbf{x}_i = \mathbf{0}$$

Hence $\mathcal{C}(\mathbf{P}_1) + \mathcal{C}(\mathbf{P}_2) + \cdots + \mathcal{C}(\mathbf{P}_k)$ is direct. It can similarly be proved that $\mathcal{R}(\mathbf{Q}_1) + \mathcal{R}(\mathbf{Q}_2) + \cdots + \mathcal{R}(\mathbf{Q}_k)$ is direct. Hence (d) follows by *Theorem* 3.7.2. This completes the proof of the theorem. ∎

Corollary Let $\mathbf{A}_1, \mathbf{A}_2, \ldots, \mathbf{A}_k$ be idempotent matrices of order n. Then $\mathbf{A}_1 + \mathbf{A}_2 + \cdots + \mathbf{A}_k$ is idempotent iff $\mathbf{A}_i\mathbf{A}_j = \mathbf{0}$ whenever $i \neq j$.

Theorem 3.7.7 (*Fisher-Cochran Theorem*): Let $\mathbf{A}_1, \mathbf{A}_2, \ldots, \mathbf{A}_k$ be $n \times n$ matrices such that $\mathbf{A}_1 + \mathbf{A}_2 + \cdots + \mathbf{A}_k = \mathbf{I}$. Then the following three statements are equivalent: (i) $\mathbf{A}_i$ is idempotent for $i = 1, \ldots, k$, (ii) $\mathbf{A}_i\mathbf{A}_j = \mathbf{0}$ whenever $i \neq j$, and (iii) $\rho(\mathbf{A}_1) + \cdots + \rho(\mathbf{A}_k) = n$.

Proof Given (ii), multiplying both sides of $\mathbf{A}_1 + \mathbf{A}_2 + \cdots + \mathbf{A}_k = \mathbf{I}$ by $\mathbf{A}_i$ we get (i). The rest follows from the preceding theorem. ∎

We note in passing that if $\mathbf{A}_i$ is real symmetric, then $\rho(\mathbf{A}_i) = \rho(\mathbf{A}_i^2)$.

Exercises

1. Show that the rank of $\mathbf{A}$ is the minimum k for which $\mathbf{A}$ can be written as the sum of k matrices of rank 1.
2. Let $\rho(\mathbf{A}_1 + \mathbf{A}_2 + \cdots + \mathbf{A}_k) = \rho(\mathbf{A}_1) + \rho(\mathbf{A}_2) + \cdots + \rho(\mathbf{A}_k)$. If $\mathbf{B} = \mathbf{A}_{i_1} + \cdots + \mathbf{A}_{i_s}$ and $\mathbf{C} = \mathbf{A}_{j_1} + \cdots + \mathbf{A}_{j_t}$, where $\{i_1, \ldots, i_s\}$ and $\{j_1, \ldots, j_t\}$ are disjoint, show that $\rho(\mathbf{B} + \mathbf{C}) = \rho(\mathbf{B}) + \rho(\mathbf{C})$.
3. If $\mathbf{A}$ and $\mathbf{B}$ are square matrices of the same order, show that $\rho(\mathbf{AB} - \mathbf{I}) \leq \rho(\mathbf{A} - \mathbf{I}) + \rho(\mathbf{B} - \mathbf{I})$. (Hint: Add and subtract $\mathbf{A}$ to $\mathbf{AB} - \mathbf{I}$).
4. If $\mathbf{A}$ is an $n \times n$ matrix with rank $r < n$, show that there exists an $n \times n$ matrix $\mathbf{B}$ with rank $n - r$ such that $\mathbf{A} + \mathbf{B}$ is non-singular.
5. Let $\mathbf{A} = \begin{bmatrix} 2 & 1 & 4 \\ 1 & \frac{1}{2} & 2 \\ 0 & 0 & 0 \end{bmatrix}$ and $\mathbf{B} = \begin{bmatrix} 1 & 2 & -1 \\ 0 & 0 & 0 \\ 1 & 2 & -1 \end{bmatrix}$.
   (a) Obtain rank-factorizations of $\mathbf{A}$ and $\mathbf{B}$. Hence or otherwise show that $\rho(\mathbf{A}) = \rho(\mathbf{B}) = 1$ and $\rho(\mathbf{A} + \mathbf{B}) = \rho(\mathbf{A}) + \rho(\mathbf{B})$.
   (b) Find a matrix $\mathbf{C}$ with rank 1 such that $\mathbf{A} + \mathbf{B} + \mathbf{C}$ is non-singular.
   (c) Find $(\mathbf{A} + \mathbf{B} + \mathbf{C})^{-1}$ and verify that $\mathbf{A}(\mathbf{A} + \mathbf{B} + \mathbf{C})^{-1}\mathbf{A} = \mathbf{A}$ and $\mathbf{A}(\mathbf{A} + \mathbf{B} + \mathbf{C})^{-1}\mathbf{B} = \mathbf{0}$.

6. Show that if $\mathbf{A} := \mathbf{A}_1 + \mathbf{A}_2 + \cdots + \mathbf{A}_k$ is non-singular and $\rho(\mathbf{A}) = \rho(\mathbf{A}_1) + \rho(\mathbf{A}_2) + \cdots + \rho(\mathbf{A}_k)$, then $\mathbf{A}_i(\mathbf{A}_1 + \cdots + \mathbf{A}_k)^{-1}\mathbf{A}_j$ is $\mathbf{A}_i$ or $\mathbf{0}$ according as $i = j$ or not.
7. (a) If $\rho(\mathbf{A}) = 5$ and $\rho(\mathbf{B}) = 2$, obtain a lower bound for $\rho(\mathbf{A}+\mathbf{B})$.
   (b) Obtain upper and lower bounds for $\rho(\mathbf{A}-\mathbf{B})$ in terms of $\rho(\mathbf{A})$ and $\rho(\mathbf{B})$.
8. Show that $\rho(\mathbf{A}+\mathbf{B}) = \rho(\mathbf{A})+\rho(\mathbf{B})$ iff $\mathcal{C}(\mathbf{A}) \cap \mathcal{C}(\mathbf{B}) = \{\mathbf{0}\}$ and $\mathcal{C}(\mathbf{A}) \subseteq \mathcal{C}(\mathbf{A}+\mathbf{B})$. What is the corresponding result in terms of row spaces?
9. Let $\mathbf{A}$ be an idempotent matrix. Then show that (a) $\mathcal{C}(\mathbf{B}) \subseteq \mathcal{C}(\mathbf{A})$ iff $\mathbf{AB} = \mathbf{B}$ and (b) $\mathcal{R}(\mathbf{B}) \subseteq \mathcal{R}(\mathbf{A})$ iff $\mathbf{BA} = \mathbf{B}$.
10. Show that for any matrix $\mathbf{B}$, $\begin{bmatrix} \mathbf{I} & \mathbf{B} \\ \mathbf{0} & \mathbf{0} \end{bmatrix}$ is idempotent.
11. Let $\mathbf{A}$ and $\mathbf{C}$ be square matrices. Show that $\mathbf{H} = \begin{bmatrix} \mathbf{A} & \mathbf{B} \\ \mathbf{0} & \mathbf{C} \end{bmatrix}$ is idempotent iff $\mathbf{A}$ and $\mathbf{C}$ are idempotent, $\mathbf{ABC} = \mathbf{0}$ and $(\mathbf{I}-\mathbf{A})\mathbf{B}(\mathbf{I}-\mathbf{C}) = \mathbf{0}$.
12. Show that if $\begin{bmatrix} \mathbf{I} & \mathbf{B} \\ \mathbf{C} & \mathbf{D} \end{bmatrix}$ is idempotent, then $\mathbf{D} - \mathbf{CB}$ is also idempotent.
13. Determine all projectors of order 2×2 over $\mathbb{R}$.
14. If $\mathbf{A}$ is idempotent and upper (or lower) triangular, show that each diagonal entry of $\mathbf{A}$ is 0 or 1.

*15. Show that if $1 + 1 \neq 0$, then $\mathbf{P} \mapsto 2\mathbf{P} - \mathbf{I}$ is a bijection from the set of all projectors of order n to the set of all $n \times n$ matrices $\mathbf{A}$ such that $\mathbf{A}^2 = \mathbf{I}$.

16. Let T be a complement of S in F^n. Let $\mathbf{A} = [\,\mathbf{B} : \mathbf{C}\,]$ where the columns of $\mathbf{B}$ form a basis of S and the columns of $\mathbf{C}$ form a basis of T. Partition $\mathbf{A}^{-1}$ as $\begin{bmatrix} \mathbf{G} \\ \mathbf{H} \end{bmatrix}$ such that $\mathbf{BG}$ is defined. Then show that $\mathbf{BG}$ is the projector into S along T.

*17. In *Theorem* 3.7.6, show that each of the pairs: (a) and (c), (b) and (d), and (c*) and (d) do not imply the other statements.

*18. If $\rho(\mathbf{A}) = \rho(\mathbf{A}^2)$ and $\mathbf{AB} = \mathbf{BA} = \mathbf{0}$, prove that $\rho(\mathbf{A}+\mathbf{B}) = \rho(\mathbf{A}) + \rho(\mathbf{B})$. Show that none of '$\mathbf{AB} = \mathbf{0}$' and '$\mathbf{BA} = \mathbf{0}$' can be dropped here.

*19. Let $\mathbf{A}$ and $\mathbf{B}$ be projectors of the same order. Then show that $\mathbf{A}+\mathbf{B}$ is a projector iff $\mathcal{C}(\mathbf{A}) \subseteq \mathcal{N}(\mathbf{B})$ and $\mathcal{C}(\mathbf{B}) \subseteq \mathcal{N}(\mathbf{A})$.

*20. Let $\mathbf{A}$ and $\mathbf{B}$ be projectors of the same order. Show that the following statements are equivalent:

   (a) $\mathbf{A} - \mathbf{B}$ is a projector
   (b) $\mathbf{AB} = \mathbf{BA} = \mathbf{B}$
   (c) $\rho(\mathbf{A}-\mathbf{B}) = \rho(\mathbf{A}) - \rho(\mathbf{B})$

(d) $\mathcal{C}(\mathbf{B}) \subseteq \mathcal{C}(\mathbf{A})$ and $\mathcal{R}(\mathbf{B}) \subseteq \mathcal{R}(\mathbf{A})$.

*21. Let $\mathbf{A}$ and $\mathbf{B}$ be projectors of the same order. Assume that $1+1 \neq 0$ in the base field.

(a) If $\mathbf{A}+\mathbf{B}$ is a projector, show that it is the projector into $\mathcal{C}(\mathbf{A}) \oplus \mathcal{C}(\mathbf{B})$ along $\mathcal{N}(\mathbf{A}) \cap \mathcal{N}(\mathbf{B})$.

(b) If $\mathbf{A}-\mathbf{B}$ is a projector, show that it is the projector into $\mathcal{C}(\mathbf{A}) \cap \mathcal{N}(\mathbf{B})$ along $\mathcal{N}(\mathbf{A}) \oplus \mathcal{C}(\mathbf{B})$.

(c) Let $\mathcal{C}(\mathbf{AB}) \subseteq \mathcal{C}(\mathbf{B})$. Then show that $\mathbf{AB}$ is a projector and that $\mathbf{BA}$ need not be a projector.

(d) If $\mathbf{AB}=\mathbf{BA}$, show that $\mathbf{AB}$ is the projector into $\mathcal{C}(\mathbf{A}) \cap \mathcal{C}(\mathbf{B})$ along $\mathcal{N}(\mathbf{A}) + \mathcal{N}(\mathbf{B})$. Show also that $\mathbf{A}+\mathbf{B}-\mathbf{AB}$ is a projector.

(e) Show by means of an example that '$\mathbf{AB}$ and $\mathbf{BA}$ are projectors' does not imply that $\mathcal{C}(\mathbf{AB}) \subseteq \mathcal{C}(\mathbf{B})$ (and hence does not imply $\mathbf{AB}=\mathbf{BA}$).

3.8 Inverse of a partitioned matrix

Consider a partitioned matrix

$$\mathbf{M} = \begin{bmatrix} \mathbf{A} & \mathbf{B} \\ \mathbf{C} & \mathbf{D} \end{bmatrix} \tag{3.8.1}$$

where $\mathbf{A}, \mathbf{B}, \mathbf{C}, \mathbf{D}$ are of orders $p \times p$, $p \times q$, $q \times p$ and $q \times q$ respectively. In this section we investigate the conditions under which (i) $\mathbf{M}$ is non-singular given that $\mathbf{A}$ is non-singular and (ii) $\mathbf{A}$ is non-singular given that $\mathbf{M}$ is non-singular. We shall also obtain explicit expressions for $\mathbf{M}^{-1}$ and $\mathbf{A}^{-1}$ in terms of each other.

Theorem 3.8.1 Let $\mathbf{M}$ be as in (3.8.1) and let $\mathbf{A}$ be non-singular. Then $\mathbf{M}$ is non-singular iff $\mathbf{F} := \mathbf{D} - \mathbf{CA}^{-1}\mathbf{B}$ is non-singular. Also then

$$\begin{aligned} \mathbf{M}^{-1} &= \begin{bmatrix} \mathbf{A}^{-1} + \mathbf{A}^{-1}\mathbf{BF}^{-1}\mathbf{CA}^{-1} & -\mathbf{A}^{-1}\mathbf{BF}^{-1} \\ -\mathbf{F}^{-1}\mathbf{CA}^{-1} & \mathbf{F}^{-1} \end{bmatrix} \\ &= \begin{bmatrix} \mathbf{A}^{-1} & \mathbf{0} \\ \mathbf{0} & \mathbf{0} \end{bmatrix} + \begin{bmatrix} \mathbf{A}^{-1}\mathbf{B} \\ -\mathbf{I} \end{bmatrix} \mathbf{F}^{-1} [\mathbf{CA}^{-1} : -\mathbf{I}] \end{aligned} \tag{3.8.2}$$

Proof Write

$$\mathbf{G} = \begin{bmatrix} \mathbf{P} & \mathbf{Q} \\ \mathbf{R} & \mathbf{S} \end{bmatrix}$$

where $\mathbf{P}, \mathbf{Q}, \mathbf{R}, \mathbf{S}$ are matrices of orders $p \times p$, $p \times q$, $q \times p$ and $q \times q$ respectively. Then $\mathbf{G}$ is the inverse of $\mathbf{M}$ iff

$$\begin{bmatrix} \mathbf{A} & \mathbf{B} \\ \mathbf{C} & \mathbf{D} \end{bmatrix} \cdot \begin{bmatrix} \mathbf{P} & \mathbf{Q} \\ \mathbf{R} & \mathbf{S} \end{bmatrix} = \begin{bmatrix} \mathbf{I} & \mathbf{0} \\ \mathbf{0} & \mathbf{I} \end{bmatrix}$$

which holds iff the following four conditions hold:

$$\mathbf{AP} + \mathbf{BR} = \mathbf{I} \tag{3.8.3}$$

$$\mathbf{AQ} + \mathbf{BS} = \mathbf{0} \tag{3.8.4}$$

$$\mathbf{CP} + \mathbf{DR} = \mathbf{0} \tag{3.8.5}$$

$$\mathbf{CQ} + \mathbf{DS} = \mathbf{I} \tag{3.8.6}$$

To prove the *only if part* of the theorem, let $\mathbf{M}$ be non-singular and let $\mathbf{G}$ as above be the inverse of $\mathbf{M}$. Then the relations (3.8.3)–(3.8.6) hold for $\mathbf{P}, \mathbf{Q}, \mathbf{R}, \mathbf{S}$. From (3.8.4) we have $\mathbf{Q} = -\mathbf{A}^{-1}\mathbf{BS}$ since $\mathbf{A}$ is non-singular. Substituting this in (3.8.6) we get $(\mathbf{D} - \mathbf{CA}^{-1}\mathbf{B})\mathbf{S} = \mathbf{I}$. Since $\mathbf{F} = (\mathbf{D} - \mathbf{CA}^{-1}\mathbf{B})$ is a square matrix, it follows that $\mathbf{F}$ is non-singular.

To prove the *if part*, let $\mathbf{F}$ be non-singular. Then it can be verified that the RHS of (3.8.2) is $\mathbf{M}^{-1}$. However, we shall solve (3.8.3)–(3.8.6) for $\mathbf{P}, \mathbf{Q}, \mathbf{R}, \mathbf{S}$ to derive (3.8.2). Now, (3.8.4) gives $\mathbf{Q} = -\mathbf{A}^{-1}\mathbf{BS}$. Substituting this in (3.8.6) we get $\mathbf{FS} = \mathbf{I}$, so $\mathbf{S} = \mathbf{F}^{-1}$. Hence $\mathbf{Q} = -\mathbf{A}^{-1}\mathbf{BF}^{-1}$. Also, (3.8.3) gives $\mathbf{P} = \mathbf{A}^{-1}(\mathbf{I} - \mathbf{BR})$. Substituting this in (3.8.5) we get $\mathbf{FR} = -\mathbf{CA}^{-1}$. Since $\mathbf{F}$ is non-singular, we get $\mathbf{R} = -\mathbf{F}^{-1}\mathbf{CA}^{-1}$. So $\mathbf{P} = \mathbf{A}^{-1}(\mathbf{I} + \mathbf{BF}^{-1}\mathbf{CA}^{-1}) = \mathbf{A}^{-1} + \mathbf{A}^{-1}\mathbf{BF}^{-1}\mathbf{CA}^{-1}$. It is easy to see that $\mathbf{P}, \mathbf{Q}, \mathbf{R}, \mathbf{S}$ thus obtained satisfy (3.8.3)–(3.8.6) and so (3.8.2) follows. ∎

Corollary Consider $\mathbf{M}$ as in (3.8.1) with $\mathbf{C} = \mathbf{0}$. Then $\mathbf{M}$ is non-singular iff $\mathbf{A}$ and $\mathbf{D}$ are non-singular. Also then

$$\begin{bmatrix} \mathbf{A} & \mathbf{B} \\ \mathbf{0} & \mathbf{D} \end{bmatrix}^{-1} = \begin{bmatrix} \mathbf{A}^{-1} & -\mathbf{A}^{-1}\mathbf{BD}^{-1} \\ \mathbf{0} & \mathbf{D}^{-1} \end{bmatrix} \tag{3.8.7}$$

Proof We show that if $\mathbf{M}$ is non-singular, then $\mathbf{A}$ and $\mathbf{D}$ are non-singular. The remaining part follows from the theorem since $\mathbf{C} = \mathbf{0}$ and $\mathbf{F} = \mathbf{D}$. Equating the $(2,2)$-entry of the partitioned matrix $\mathbf{MM}^{-1}$ to $\mathbf{I}$, we see that $\mathbf{D}$ is non-singular and equating the $(1,1)$-entry of $\mathbf{M}^{-1}\mathbf{M}$ to $\mathbf{I}$, we get $\mathbf{A}$ is non-singular. ∎

The next result follows immediately from the preceding corollary by induction on the order of the matrix:

Theorem 3.8.2 Let $\mathbf{A}$ be an upper (or lower) triangular matrix of order n. Then $\mathbf{A}$ is non-singular iff all diagonal elements of $\mathbf{A}$ are non-zero. Also then $\mathbf{A}^{-1}$ is upper (resp. lower) triangular with $1/a_{ii}$ as the i-th diagonal entry for $i = 1, 2, \ldots, n$.

Theorem 3.8.1 tells us that if $\mathbf{A}^{-1}$ is known, $\mathbf{M}^{-1}$ can be found easily. We only need to compute the inverse of the $q \times q$ matrix $\mathbf{F} = \mathbf{D} - \mathbf{C}\mathbf{A}^{-1}\mathbf{B}$ and make a few multiplications as given in (3.8.2) which is much easier than inverting the $(p+q) \times (p+q)$ matrix $\mathbf{M}$ directly.

The case $q = 1$ is of special interest. We are then appending an extra row and an extra column to $\mathbf{A}$ to get $\mathbf{M}$. Such a partitioned matrix

$$\mathbf{M} = \begin{bmatrix} \mathbf{A} & \mathbf{b} \\ \mathbf{c}^{\mathrm{T}} & d \end{bmatrix}$$

is called a *bordered matrix*. Its inverse can be found by

Algorithm 3.8.3 (*Inverse of a bordered matrix*)

Step 1 Compute $\mathbf{u} = \mathbf{A}^{-1}\mathbf{b}$ and $\mathbf{v}^{\mathrm{T}} = \mathbf{c}^{\mathrm{T}}\mathbf{A}^{-1}$.

Step 2 Compute $f = d - \mathbf{c}^{\mathrm{T}}\mathbf{A}^{-1}\mathbf{b} = d - \mathbf{c}^{\mathrm{T}}\mathbf{u}$ and $s = \frac{1}{f}$.

Step 3 Compute $\mathbf{q} = -s\,\mathbf{u}$ and $\mathbf{r}^{\mathrm{T}} = -s\,\mathbf{v}^{\mathrm{T}}$.

Step 4 Compute $\mathbf{P} = \mathbf{A}^{-1} + s\,\mathbf{u}\mathbf{v}^{\mathrm{T}} = \mathbf{A}^{-1} - \mathbf{q}\mathbf{v}^{\mathrm{T}}$.

Step 5 Stop, for, $\mathbf{M}^{-1} = \begin{bmatrix} \mathbf{P} & \mathbf{q} \\ \mathbf{r}^{\mathrm{T}} & s \end{bmatrix}$.

In *Theorem* 3.8.1 we assumed that $\mathbf{A}$ is non-singular. We now state the corresponding result when it is assumed that $\mathbf{D}$ is non-singular. It can be shown by imitating the proof of *Theorem* 3.8.1 that $\mathbf{M}$ is non-singular iff $\mathbf{H} := \mathbf{A} - \mathbf{B}\mathbf{D}^{-1}\mathbf{C}$ is non-singular and then the inverse of $\mathbf{M}$ is given by

$$\begin{bmatrix} \mathbf{H}^{-1} & -\mathbf{H}^{-1}\mathbf{B}\mathbf{D}^{-1} \\ -\mathbf{D}^{-1}\mathbf{C}\mathbf{H}^{-1} & \mathbf{D}^{-1} + \mathbf{D}^{-1}\mathbf{C}\mathbf{H}^{-1}\mathbf{B}\mathbf{D}^{-1} \end{bmatrix} \tag{3.8.8}$$

Note that if $\mathbf{A}$, $\mathbf{D}$ and $\mathbf{M}$ are all non-singular, it follows that $\mathbf{H}^{-1} = \mathbf{A}^{-1} + \mathbf{A}^{-1}\mathbf{B}\mathbf{F}^{-1}\mathbf{C}\mathbf{A}^{-1}$.

So far we have considered the computation of $\mathbf{M}^{-1}$ when $\mathbf{A}^{-1}$ or $\mathbf{D}^{-1}$ is known. Sometimes it is of importance to find the inverse of $\mathbf{A}$ when $\mathbf{M}^{-1}$ is known. The following theorem is particularly useful if q is small when compared to p.

Theorem 3.8.4 Let $\mathbf{M}$ be as in (3.8.1) and let $\mathbf{M}^{-1} = \begin{bmatrix} \mathbf{P} & \mathbf{Q} \\ \mathbf{R} & \mathbf{S} \end{bmatrix}$ where $\mathbf{P}, \mathbf{Q}, \mathbf{R}, \mathbf{S}$ are of orders $p \times p$, $p \times q$, $q \times p$ and $q \times q$ respectively. Then $\mathbf{A}$ is non-singular iff $\mathbf{S}$ is non-singular. If $\mathbf{S}^{-1}$ is known, $\mathbf{A}^{-1}$ can be obtained as $\mathbf{P} - \mathbf{Q}\mathbf{S}^{-1}\mathbf{R}$.

Proof To prove the *if part*, let $\mathbf{S}$ be non-singular. Then using (3.8.8) for $\mathbf{M}^{-1}$ instead of $\mathbf{M}$, we get $\mathbf{A} = (\mathbf{P} - \mathbf{Q}\mathbf{S}^{-1}\mathbf{R})^{-1}$. Hence $\mathbf{A}$ is non-singular and $\mathbf{A}^{-1} = \mathbf{P} - \mathbf{Q}\mathbf{S}^{-1}\mathbf{R}$.

To prove the *only if part*, let $\mathbf{A}$ be non-singular. By *Theorem* 3.8.1, $\mathbf{S}$ is $(\mathbf{D} - \mathbf{C}\mathbf{A}^{-1}\mathbf{B})^{-1}$ and so is non-singular. ∎

Exercises

1. Show that $\begin{bmatrix} \mathbf{I} & \mathbf{B} \\ \mathbf{0} & \mathbf{I} \end{bmatrix}^{-1} = \begin{bmatrix} \mathbf{I} & -\mathbf{B} \\ \mathbf{0} & \mathbf{I} \end{bmatrix}$

2. (a) Find the inverse of $\begin{bmatrix} \mathbf{A} & \mathbf{0} \\ \mathbf{C} & \mathbf{D} \end{bmatrix}$ where $\mathbf{A}$ and $\mathbf{D}$ are non-singular.

   (b) Find the inverse of $\begin{bmatrix} \mathbf{0} & \mathbf{B} \\ \mathbf{C} & \mathbf{D} \end{bmatrix}$ where $\mathbf{B}$ and $\mathbf{C}$ are non-singular.

3. Let $\mathbf{M} = \begin{bmatrix} 2 & 1 & 1 \\ 2 & 4 & 0 \\ 1 & 1 & 4 \end{bmatrix}$. Find the inverse of $\begin{bmatrix} 2 & 1 \\ 2 & 4 \end{bmatrix}$. Check that $\mathbf{M}$ is non-singular using *Theorem* 3.8.1 and find $\mathbf{M}^{-1}$ using *Algorithm* 3.8.3.

4. Find the inverse of

$$\begin{bmatrix} 2 & 0 & 0 & 2 & -1 \\ 0 & 1 & 0 & 0 & 2 \\ 0 & 0 & 3 & 1 & 3 \\ 1 & 1 & 1 & 2 & -1 \\ 0 & -1 & 2 & 1 & 3 \end{bmatrix}$$

   by partitioning it suitably.

*5. Show that if $\mathbf{M}$ is non-singular then by a permutation of rows, $\mathbf{M}$ can be converted to a matrix with all leading principal submatrices non-singular.

6. Let $\mathbf{M}$ be as in (3.8.1) and let $\mathbf{M}^{-1} = \begin{bmatrix} \mathbf{P} & \mathbf{Q} \\ \mathbf{R} & \mathbf{S} \end{bmatrix}$ where $\mathbf{P}, \mathbf{Q}, \mathbf{R}$ and $\mathbf{S}$ have the same orders as $\mathbf{A}, \mathbf{B}, \mathbf{C}$ and $\mathbf{D}$ respectively. Then show that $\mathbf{D}$ is non-singular iff $\mathbf{P}$ is non-singular. Also then show that $\mathbf{D}^{-1} = \mathbf{S} - \mathbf{R}\mathbf{P}^{-1}\mathbf{Q}$.

7. Let $\mathbf{M} = \begin{bmatrix} \mathbf{A} & \mathbf{B} \\ \mathbf{C} & \mathbf{D} \end{bmatrix}$ where $\mathbf{B}$ is non-singular. Then show that $\mathbf{M}$ is non-singular iff $\mathbf{K} := \mathbf{C} - \mathbf{D}\mathbf{B}^{-1}\mathbf{A}$ is non-singular. Also then find $\mathbf{M}^{-1}$ in terms of $\mathbf{A}, \mathbf{B}, \mathbf{C}, \mathbf{D}$, $\mathbf{B}^{-1}$ and $\mathbf{K}^{-1}$.

8. Given that the inverse of $\begin{bmatrix} 2 & 1 & 3 \\ 1 & 4 & 2 \\ 3 & 2 & 6 \end{bmatrix}$ is $\begin{bmatrix} 2 & 0.0 & -1.0 \\ 0 & 0.3 & -0.1 \\ -1 & -0.1 & 0.7 \end{bmatrix}$, find $\begin{bmatrix} 2 & 1 \\ 1 & 4 \end{bmatrix}^{-1}$ using *Theorem* 3.8.4.

9. Let $\mathbf{A}$ be non-singular. Then show that $\mathbf{A} + \mathbf{u}\mathbf{v}^{\mathrm{T}}$ is non-singular iff $\mathbf{v}^{\mathrm{T}}\mathbf{A}^{-1}\mathbf{u} \neq -1$. Also then show that

$$(\mathbf{A} + \mathbf{u}\mathbf{v}^{\mathrm{T}})^{-1} = \mathbf{A}^{-1} - \frac{\mathbf{A}^{-1}\mathbf{u}\mathbf{v}^{\mathrm{T}}\mathbf{A}^{-1}}{1 + \mathbf{v}^{\mathrm{T}}\mathbf{A}^{-1}\mathbf{u}}$$

(Hint: consider $\mathbf{M} = \begin{bmatrix} \mathbf{A} & -\mathbf{u} \\ \mathbf{v}^{\mathrm{T}} & 1 \end{bmatrix}$.)

10. Using the result of the preceding exercise, obtain a formula for $\mathbf{B}^{-1}$ where $\mathbf{B}$ is obtained from $\mathbf{A}$ by changing one row or one column.

11. Let $\mathbf{A}_{\alpha,\beta}$ denote the $n \times n$ matrix

$$\begin{bmatrix} \alpha & \beta & \cdots & \beta \\ \beta & \alpha & \cdots & \beta \\ \cdots & \cdots & \cdots & \cdots \\ \beta & \beta & \cdots & \alpha \end{bmatrix}$$

   (a) Show that $\mathbf{A}_{\alpha,\beta}$ can be written in the form $\mathbf{A} + \mathbf{u}\mathbf{v}^{\mathrm{T}}$ where $\mathbf{A}$ is a scalar matrix.

   (b) Deduce the following from (a) and *Exercise* 3.8.9:

      i. $\mathbf{A}_{\alpha,\beta}$ is non-singular iff $\Delta := (\alpha - \beta)(\alpha + (n-1)\beta) \neq 0$.

      ii. $\mathbf{A}_{\alpha,\beta}^{-1} = \mathbf{A}_{\gamma,\delta}$ where $\gamma = (\alpha + (n-2)\beta)/\Delta$ and $\delta = -\beta/\Delta$.

   (When $\alpha = 1$ and $|\beta| \leq 1$, $\mathbf{A}_{\alpha,\beta}$ is called the *intraclass correlation matrix* and plays an important role in Statistics.)

*12. Let $\mathbf{A}$ and $\mathbf{D}$ be non-singular matrices of possibly different orders and let the matrix $\mathbf{A}+\mathbf{B}\mathbf{D}\mathbf{C}$ be defined. Show that $\mathbf{A}+\mathbf{B}\mathbf{D}\mathbf{C}$ is non-singular iff $\mathbf{D}^{-1} + \mathbf{C}\mathbf{A}^{-1}\mathbf{B}$ is non-singular. Also then, show that $(\mathbf{A} + \mathbf{B}\mathbf{D}\mathbf{C})^{-1} = \mathbf{A}^{-1} - \mathbf{A}^{-1}\mathbf{B}(\mathbf{D}^{-1} + \mathbf{C}\mathbf{A}^{-1}\mathbf{B})^{-1}\mathbf{C}\mathbf{A}^{-1}$.

*13. Consider the $n \times n$ matrix $\mathbf{A}_\theta = \begin{bmatrix} \mathbf{I} & \theta\mathbf{1} \\ \theta\mathbf{1}^{\mathrm{T}} & 1 \end{bmatrix}$ where $\mathbf{1}$ is a vector with all entries 1.

   (a) For what values of θ is $\mathbf{A}_\theta$ non-singular?

   (b) Find $\mathbf{A}_\theta^{-1}$ whenever it exists. If $\mathbf{A}_\theta^{-1}$ is written as $\mathbf{I}+\mathbf{B}$, show that $\rho(\mathbf{B}) = 2$ and find a rank-factorization of $\mathbf{B}$ if $\theta \neq 0$.

*14. Let $\mathbf{A}$ be an $m \times n$ matrix of rank n and let $\mathbf{G}$ be a left inverse of $\mathbf{A}$. Let $\mathbf{x}$ be an $m \times 1$ vector.

(a) Show that $[\,\mathbf{A} : \mathbf{x}\,]$ has a left inverse iff $(\mathbf{I} - \mathbf{AG})\mathbf{x} \neq \mathbf{0}$.

(b) If $\mathbf{y}^{\mathrm{T}}$ is a row of $\mathbf{I}-\mathbf{AG}$ such that $\mathbf{y}^{\mathrm{T}}\mathbf{x} \neq 0$, show that $\begin{bmatrix} \mathbf{G}(\mathbf{I} - \mathbf{x}\mathbf{z}^{\mathrm{T}}) \\ \mathbf{z}^{\mathrm{T}} \end{bmatrix}$ is a left inverse of $[\,\mathbf{A} : \mathbf{x}\,]$, where $\mathbf{z} = (1/\mathbf{y}^{\mathrm{T}}\mathbf{x})\mathbf{y}$.

3.9 Rank of real and complex matrices

The results proved till now hold for matrices over any field. In this section we prove some results on rank and inverse of real and complex matrices which use special properties of $\mathbb{R}$ and $\mathbb{C}$ and are very useful.

Notice that in $\mathbb{R}$ we have

$$\alpha_1^2 + \alpha_2^2 + \cdots + \alpha_k^2 = 0 \Rightarrow \alpha_1 = \cdots = \alpha_k = 0$$

Thus for any $\mathbf{x} \in \mathbb{R}^k$, $\mathbf{x}^{\mathrm{T}}\mathbf{x} = 0 \Rightarrow \mathbf{x} = \mathbf{0}$. Similarly in $\mathbb{C}$ we have

$$\overline{\alpha}_1\alpha_1 + \cdots + \overline{\alpha}_k\alpha_k = 0 \Rightarrow \alpha_1 = \cdots = \alpha_k = 0$$

because $\overline{\alpha}\alpha = |\alpha|^2$ for any complex number α. In terms of vectors this means $\overline{\mathbf{x}}^{\mathrm{T}}\mathbf{x} = 0 \Rightarrow \mathbf{x} = \mathbf{0}$ for any $\mathbf{x} \in \mathbb{C}^k$.

We start with a preliminary result which says that the null space of a real matrix is a complement of its row space (they are actually orthogonal complements, see *Section* 7.5).

Theorem 3.9.1 For any real matrix $\mathbf{A}$ with n columns, $\mathcal{N}(\mathbf{A})$ is a complement of $\mathcal{C}(\mathbf{A}^{\mathrm{T}})$ in $\mathbb{R}^n$.

Proof Let $\mathbf{x} \in \mathcal{N}(\mathbf{A}) \cap \mathcal{C}(\mathbf{A}^{\mathrm{T}})$. Then $\mathbf{x} = \mathbf{A}^{\mathrm{T}}\mathbf{y}$ for some $\mathbf{y}$ and $\mathbf{Ax} = \mathbf{AA}^{\mathrm{T}}\mathbf{y} = \mathbf{0}$. So $\mathbf{x}^{\mathrm{T}}\mathbf{x} = \mathbf{y}^{\mathrm{T}}\mathbf{AA}^{\mathrm{T}}\mathbf{y} = 0$ and $\mathbf{x} = \mathbf{0}$. Thus $\mathcal{N}(\mathbf{A})$ and $\mathcal{C}(\mathbf{A}^{\mathrm{T}})$ are virtually disjoint. Since their dimensions add up to n, the theorem follows. ∎

Theorem 3.9.2 For any real matrix $\mathbf{A}$, we have the following: $\rho(\mathbf{A}) = \rho(\mathbf{A}^{\mathrm{T}}) = \rho(\mathbf{AA}^{\mathrm{T}}) = \rho(\mathbf{A}^{\mathrm{T}}\mathbf{A})$, $\mathcal{R}(\mathbf{A}) = \mathcal{R}(\mathbf{A}^{\mathrm{T}}\mathbf{A})$, $\mathcal{C}(\mathbf{A}) = \mathcal{C}(\mathbf{AA}^{\mathrm{T}})$ and $\mathcal{N}(\mathbf{A}) = \mathcal{N}(\mathbf{A}^{\mathrm{T}}\mathbf{A})$.

Proof We first prove the last statement. Let $\mathbf{x} \in \mathcal{N}(\mathbf{A}^{\mathrm{T}}\mathbf{A})$. Then $\mathbf{A}^{\mathrm{T}}\mathbf{Ax} = \mathbf{0}$, so $\mathbf{x}^{\mathrm{T}}\mathbf{A}^{\mathrm{T}}\mathbf{Ax} = \mathbf{0}$. Hence $\mathbf{Ax} = \mathbf{0}$ and $\mathbf{x} \in \mathcal{N}(\mathbf{A})$. Thus $\mathcal{N}(\mathbf{A}^{\mathrm{T}}\mathbf{A}) \subseteq \mathcal{N}(\mathbf{A})$. The reverse inclusion is trivial, so equality follows. Taking dimensions, we get $\rho(\mathbf{A}^{\mathrm{T}}\mathbf{A}) = \rho(\mathbf{A})$ since the two matrices have the same number of columns. On replacing $\mathbf{A}$ by $\mathbf{A}^{\mathrm{T}}$ and using *Theorems* 3.5.5 and 3.5.6, the remaining parts of the theorem follow easily. ∎

Corollary For any real matrix $\mathbf{A}$, there exist real matrices $\mathbf{C}$ and $\mathbf{D}$ such that $\mathbf{A} = \mathbf{A}\mathbf{A}^{\mathrm{T}}\mathbf{C}$ and $\mathbf{A} = \mathbf{D}\mathbf{A}^{\mathrm{T}}\mathbf{A}$.

Theorem 3.9.3 For real matrices $\mathbf{A}$, $\mathbf{B}$ and $\mathbf{C}$ we have the (rank) cancellation laws:

(i) If $\mathbf{A}\mathbf{A}^{\mathrm{T}}\mathbf{B} = \mathbf{A}\mathbf{A}^{\mathrm{T}}\mathbf{C}$ then $\mathbf{A}^{\mathrm{T}}\mathbf{B} = \mathbf{A}^{\mathrm{T}}\mathbf{C}$.

(ii) If $\mathbf{B}\mathbf{A}^{\mathrm{T}}\mathbf{A} = \mathbf{C}\mathbf{A}^{\mathrm{T}}\mathbf{A}$ then $\mathbf{B}\mathbf{A}^{\mathrm{T}} = \mathbf{C}\mathbf{A}^{\mathrm{T}}$.

Using *Theorem* 3.9.2, we can give a left inverse of a real matrix with full column rank and a right inverse of a real matrix with full row rank.

Theorem 3.9.4 Let $\mathbf{A}$ be an $m \times n$ real matrix. Then

(i) $\mathbf{A}$ is of full column rank iff $\mathbf{A}^{\mathrm{T}}\mathbf{A}$ is non-singular. Also then $(\mathbf{A}^{\mathrm{T}}\mathbf{A})^{-1}\mathbf{A}^{\mathrm{T}}$ is a left inverse of $\mathbf{A}$ and

(ii) $\mathbf{A}$ is of full row rank iff $\mathbf{A}\mathbf{A}^{\mathrm{T}}$ is non-singular. Also then $\mathbf{A}^{\mathrm{T}}(\mathbf{A}\mathbf{A}^{\mathrm{T}})^{-1}$ is a right inverse of $\mathbf{A}$.

Proof We give the proof of (i) and leave the proof of (ii) which is similar. The first statement in (i) follows from *Theorem* 3.9.2 and the second statement is trivial. ∎

Example 3.9.5 Consider the real matrix

$$\mathbf{A} = \begin{bmatrix} 1 & -1 \\ 0 & 2 \\ 4 & 1 \\ -1 & 2 \end{bmatrix}$$

It can be checked that

$$\mathbf{A}^{\mathrm{T}}\mathbf{A} = \begin{bmatrix} 18 & 1 \\ 1 & 10 \end{bmatrix}$$

Clearly $\mathbf{A}^{\mathrm{T}}\mathbf{A}$ is non-singular, so $\mathbf{A}$ is of full column rank. It can also be checked that

$$\begin{aligned} (\mathbf{A}^{\mathrm{T}}\mathbf{A})^{-1}\mathbf{A}^{\mathrm{T}} &= \tfrac{1}{179}\begin{bmatrix} 10 & -1 \\ -1 & 18 \end{bmatrix} \cdot \begin{bmatrix} 1 & 0 & 4 & -1 \\ -1 & 2 & 1 & 2 \end{bmatrix} \\ &= \tfrac{1}{179}\begin{bmatrix} 11 & -2 & 39 & -12 \\ -19 & 36 & 14 & 37 \end{bmatrix} \end{aligned}$$

is a left inverse of $\mathbf{A}$. The matrix $\mathbf{A}$ has, of course, many left inverses. For example,

$$\begin{bmatrix} 1 & \frac{1}{2} & 0 & 0 \\ 0 & \frac{1}{2} & 0 & 0 \end{bmatrix}$$

is also a left inverse of $\mathbf{A}$ (note that the submatrix of this formed by the first two columns is the inverse of the corresponding submatrix of $\mathbf{A}$). We incidentally note that $\mathbf{A}\mathbf{A}^{\mathrm{T}}$ is of rank 2 and $(\mathbf{A}, \mathbf{A}^{\mathrm{T}})$ is a rank-factorization of $\mathbf{A}\mathbf{A}^{\mathrm{T}}$. ∎

We next give the results for complex matrices. We start with some preliminary results. For an $m \times n$ complex matrix $\mathbf{A} = ((a_{ij}))$, we define $\overline{\mathbf{A}}$ to be the $m \times n$ matrix the (i, j)-th element of which is $\overline{a}_{ij}$. It is easy to check that $\overline{\alpha \mathbf{A}} = \overline{\alpha}\, \overline{\mathbf{A}}$, $\overline{\mathbf{A} + \mathbf{B}} = \overline{\mathbf{A}} + \overline{\mathbf{B}}$ and $\overline{\mathbf{AB}} = \overline{\mathbf{A}}\, \overline{\mathbf{B}}$.

Theorem 3.9.6 For any complex matrix $\mathbf{A}$, $\rho(\overline{\mathbf{A}}) = \rho(\mathbf{A})$.

Proof We claim that vectors $\mathbf{x}_1, \mathbf{x}_2, \ldots, \mathbf{x}_k$ in $\mathbb{C}^n$ are linearly independent iff $\overline{\mathbf{x}}_1, \overline{\mathbf{x}}_2, \ldots, \overline{\mathbf{x}}_k$ are linearly independent. This follows since

$$\alpha_1 \mathbf{x}_1 + \cdots + \alpha_k \mathbf{x}_k = \mathbf{0} \Longleftrightarrow \overline{\alpha}_1 \overline{\mathbf{x}}_1 + \cdots + \overline{\alpha}_k \overline{\mathbf{x}}_k = \mathbf{0}$$

and

$$\alpha_1 = \cdots = \alpha_k = 0 \Longleftrightarrow \overline{\alpha}_1 = \cdots = \overline{\alpha}_k = 0$$

So, a set of rows of $\mathbf{A}$ is linearly independent iff the corresponding rows of $\overline{\mathbf{A}}$ are linearly independent. The theorem follows. ∎

We have seen at the beginning of this section that the complex analogue of the result '$\mathbf{x}^{\mathrm{T}}\mathbf{x} = 0 \Rightarrow \mathbf{x} = \mathbf{0}$ for $\mathbf{x} \in \mathbb{R}^n$' replaces $\mathbf{x}^{\mathrm{T}}$ by $\overline{\mathbf{x}}^{\mathrm{T}}$. Motivated by this, we denote $\overline{\mathbf{A}}^{\mathrm{T}}$ by $\mathbf{A}^*$ for any complex matrix $\mathbf{A}$ and call it the *adjoint* of $\mathbf{A}$. It is easy to verify that

$$(\alpha \mathbf{A})^* = \overline{\alpha} \mathbf{A}^*, \quad (\mathbf{A} + \mathbf{B})^* = \mathbf{A}^* + \mathbf{B}^*, \quad (\mathbf{AC})^* = \mathbf{C}^* \mathbf{A}^* \qquad (3.9.1)$$

Theorem 3.9.7 Theorems 3.9.2, 3.9.3 and 3.9.4 hold for complex matrices if $\mathbf{A}^{\mathrm{T}}$ is replaced by $\mathbf{A}^*$.

We omit the proofs as the earlier proofs go through *verbatim*.

Exercises

1. Find a 2×2 complex matrix $\mathbf{A}$ such that $\rho(\mathbf{A}^{\mathrm{T}}\mathbf{A}) < \rho(\mathbf{A})$.

2. (a) Let $\mathbf{A}$ be a real matrix and let $\mathbf{b} \in \mathcal{C}(\mathbf{A})$. Then show that $\mathbf{u}$ is a solution of the system $\mathbf{Ax} = \mathbf{b}$ iff $\mathbf{u}$ is a solution of $\mathbf{A}^{\mathrm{T}}\mathbf{Ax} = \mathbf{A}^{\mathrm{T}}\mathbf{b}$.
   (b) For any real $m \times n$ matrix $\mathbf{A}$ and any real $m \times 1$ vector $\mathbf{b}$, show that the system $\mathbf{A}^{\mathrm{T}}\mathbf{Ax} = \mathbf{A}^{\mathrm{T}}\mathbf{b}$ has a solution.
3. Show that $\rho(\mathbf{A}^{\mathrm{T}}\mathbf{AB}) = \rho(\mathbf{AB})$ and $\rho(\mathbf{ABB}^{\mathrm{T}}) = \rho(\mathbf{AB})$ if $\mathbf{A}$ and $\mathbf{B}$ are real. Deduce that $\rho(\mathbf{A}^{\mathrm{T}}\mathbf{AA}^{\mathrm{T}}) = \rho(\mathbf{A})$ for any real matrix $\mathbf{A}$.
4. Let $\mathbf{A}$ and $\mathbf{D} = \mathrm{diag}(d_1, d_2, \ldots, d_m)$ be real matrices such that $\mathbf{A}^{\mathrm{T}}\mathbf{DA}$ is defined. Prove that $\rho(\mathbf{A}^{\mathrm{T}}\mathbf{DA}) = \rho(\mathbf{A})$ if all d_i's are positive. Find out whether this result remains true if the condition '$d_i > 0$ for all i' is replaced by the following (one at a time): (i) $d_i < 0$ for all i, (ii) d_i's are arbitrary and (iii) $d_i \geq 0$ for all i.
5. If $\mathbf{A}$ and $\mathbf{B}$ are real matrices, show that the condition $\mathcal{C}(\mathbf{B}) \subseteq \mathcal{C}(\mathbf{A}^{\mathrm{T}})$ is sufficient but not necessary for $\rho(\mathbf{AB}) = \rho(\mathbf{B})$.
6. Find all real $m \times n$ matrices $\mathbf{A}$ such that $\mathrm{tr}(\mathbf{A}^{\mathrm{T}}\mathbf{A}) = 0$.
7. Let $\mathbf{A}$ be a matrix over $\mathbb{C}$. Then show that $\mathbf{y} \in \mathcal{C}(\overline{\mathbf{A}})$ iff $\overline{\mathbf{y}} \in \mathcal{C}(\mathbf{A})$.
8. Prove (3.9.1) and *Theorem* 3.9.7.
9. Let $\mathbf{A} = \begin{bmatrix} 1 & 3 & 0 & 2 & 5 \\ 2 & 6 & 3 & 0 & 2 \end{bmatrix}$. Find a right inverse of $\mathbf{A}$ using *Theorem* 3.9.4. Hence obtain a left inverse of $\mathbf{A}^{\mathrm{T}}$.

*10. If $\mathbf{A}$ and $\mathbf{B}$ are real or complex matrices with the same number of rows such that $\mathcal{C}(\mathbf{B})$ is a complement of $\mathcal{C}(\mathbf{A})$, show that $\mathbf{AA}^*\mathbf{C}^{-1}$ is defined and is the projector into $\mathcal{C}(\mathbf{A})$ along $\mathcal{C}(\mathbf{B})$, where $\mathbf{C} = \mathbf{AA}^* + \mathbf{BB}^*$.

11. A complex matrix $\mathbf{A}$ is said to be *hermitian* (resp. *skew-hermitian*) if $\mathbf{A}^* = \mathbf{A}$ (resp. $\mathbf{A}^* = -\mathbf{A}$). Show that any complex matrix can be written uniquely as $\mathbf{B}+\mathbf{C}$ where $\mathbf{B}$ is hermitian and $\mathbf{C}$ is skew-hermitian. Obtain the corresponding result for real matrices.

3.10 Change of basis

In geometry, we often change the origin and the coordinate axes to simplify the equation of a curve. The latter amounts to a change of basis as we have seen in *Section* 1.5. In this section we study change of bases and their effect on the coordinate vector of a point and the matrix of a linear transformation.

We note that change of origin is easy to deal with. If the origin is shifted to the point $\mathbf{w}$ and the coordinate axes are shifted parallel to themselves so as to pass through the new origin, it is easy to see that the coordinate vector of a point P with respect to the new coordinate system is $\mathbf{x} - \mathbf{w}$ where $\mathbf{x}$ is the coordinate vector of P with respect to the old coordinate system. We now turn to change of basis without changing the origin.

Let

$$\mathcal{X} = \{\mathbf{x}_1, \mathbf{x}_2, \ldots, \mathbf{x}_n\} \tag{3.10.1}$$

and

$$\mathcal{X}' = \{\mathbf{x}'_1, \mathbf{x}'_2, \ldots, \mathbf{x}'_n\} \tag{3.10.2}$$

be two (ordered) bases of a vector space V. Then there exist unique scalars p_{ij} $(i, j = 1, \ldots, n)$ such that

$$\mathbf{x}'_j = \sum_{i=1}^{n} p_{ij} \mathbf{x}_i, \quad j = 1, \ldots, n \tag{3.10.3}$$

We call the $n \times n$ matrix $\mathbf{P} = ((p_{ij}))$ the *matrix of transition from* $\mathcal{X}$ *to* $\mathcal{X}'$. Clearly, $\mathbf{P}$ is the matrix of the identity transformation 1_V on V with respect to $\mathcal{X}'$ and $\mathcal{X}$. Also, if $\boldsymbol{\alpha}$ and $\boldsymbol{\beta}$ are the coordinate vectors of any $\mathbf{x} \in V$ with respect to $\mathcal{X}$ and $\mathcal{X}'$, then by *Theorem* 2.3.2,

$$\boldsymbol{\alpha} = \mathbf{P}\boldsymbol{\beta} \tag{3.10.4}$$

This gives the relation between the coordinate vectors of a point with respect to the two coordinate systems $\mathcal{X}$ and $\mathcal{X}'$.

Now let $\mathbf{S}$ be the matrix of transition from $\mathcal{X}'$ to $\mathcal{X}$. Then $\mathbf{S}$ is the matrix of 1_V with respect to $\mathcal{X}$ and $\mathcal{X}'$ and, by *Theorem* 2.2.17, the matrix of 1_V $(= 1_V \circ 1_V)$ with respect to $\mathcal{X}$ is $\mathbf{PS}$. Thus $\mathbf{PS} = \mathbf{I}$ and $\mathbf{S} = \mathbf{P}^{-1}$ since $\mathbf{S}$ is a square matrix. We have thus proved

Theorem 3.10.1 Let $\mathcal{X}$ and $\mathcal{X}'$ be bases of V and let $\mathbf{P}$ be the matrix of transition from $\mathcal{X}$ to $\mathcal{X}'$. Then $\mathbf{P}$ is non-singular and $\mathbf{P}^{-1}$ is the matrix of transition from $\mathcal{X}'$ to $\mathcal{X}$.

In the converse direction, we have

Theorem 3.10.2 Let $\mathcal{X} = \{\mathbf{x}_1, \ldots, \mathbf{x}_n\}$ be a basis of V and let $\mathcal{X}' = \{\mathbf{x}'_1, \ldots, \mathbf{x}'_n\}$ where $\mathbf{x}'_j$ is defined by (3.10.3) and $\mathbf{P} = ((p_{ij}))$ is a non-singular matrix. Then $\mathcal{X}'$ is also a basis of V.

Proof It is enough to prove that $\mathcal{X}'$ is linearly independent. So let $\alpha_1 \mathbf{x}'_1 + \cdots + \alpha_n \mathbf{x}'_n = \mathbf{0}$. Then

$$\mathbf{0} = \sum_{j=1}^{n} \alpha_j \mathbf{x}'_j = \sum_{j=1}^{n} \alpha_j \sum_{i=1}^{n} p_{ij} \mathbf{x}_i = \sum_{i=1}^{n} \Big(\sum_{j=1}^{n} p_{ij} \alpha_j \Big) \mathbf{x}_i$$

Since $\mathcal{X}$ is a basis, we get $\sum_{j=1}^{n} p_{ij}\alpha_j = 0$ for all i, which can be written as $\mathbf{P}\boldsymbol{\alpha} = \mathbf{0}$, where $\boldsymbol{\alpha} = (\alpha_1, \ldots, \alpha_n)^{\mathrm{T}}$. Since $\mathbf{P}$ is non-singular, we get $\boldsymbol{\alpha} = \mathbf{0}$ which proves that $\mathcal{X}'$ is linearly independent and so a basis. ∎

For example, let $\mathcal{X}$ be the basis $\{(1,1),(-1,2)\}$ of $\mathbb{R}^2$ and let $\mathcal{X}'$ be the canonical basis of $\mathbb{R}^2$. Then from *Example* 1.5.17, we get $(1,0) = \frac{2}{3}\mathbf{x}_1 - \frac{1}{3}\mathbf{x}_2$ and $(0,1) = \frac{1}{3}\mathbf{x}_1 + \frac{1}{3}\mathbf{x}_2$. Thus the matrix of transition from $\mathcal{X}$ to $\mathcal{X}'$ is

$$\mathbf{P} = \frac{1}{3}\begin{bmatrix} 2 & 1 \\ -1 & 1 \end{bmatrix}$$

So by (3.10.4) it follows that if β_1 and β_2 are the coordinates of a point P with respect to $\mathcal{X}'$ then the coordinates of P with respect to $\mathcal{X}$ are $\alpha_1 = \frac{1}{3}(2\beta_1 + \beta_2)$ and $\alpha_2 = \frac{1}{3}(-\beta_1 + \beta_2)$. Now

$$\mathbf{P}^{-1} = \begin{bmatrix} 1 & -1 \\ 1 & 2 \end{bmatrix}$$

so $\beta_1 = \alpha_1 - \alpha_2$ and $\beta_2 = \alpha_1 + 2\alpha_2$.

Theorem 3.10.3 Let f be a linear transformation from V_1 to V_2. Let $\mathcal{X}$ and $\mathcal{X}'$ be bases of V_1 and let $\mathcal{Y}$ and $\mathcal{Y}'$ be bases of V_2. Let $\mathbf{A}$ be the matrix of f with respect to $\mathcal{X}, \mathcal{Y}$ and $\mathbf{B}$ the matrix of f with respect to $\mathcal{X}', \mathcal{Y}'$. Let $\mathbf{P}$ be the matrix of transition from $\mathcal{X}$ to $\mathcal{X}'$ and $\mathbf{Q}$ the matrix of transition from $\mathcal{Y}$ to $\mathcal{Y}'$. Then

$$\mathbf{B} = \mathbf{Q}^{-1}\mathbf{A}\mathbf{P}$$

Proof We have $f = 1_{V_2} \circ f \circ 1_{V_1}$ (see *Figure* 3.10.1). So the present theorem follows from *Theorem* 2.2.17 since the matrix of 1_{V_2} with respect to $\mathcal{Y}$ and $\mathcal{Y}'$ is $\mathbf{Q}^{-1}$; the matrix of f with respect to $\mathcal{X}$ and $\mathcal{Y}$ is $\mathbf{A}$ and the matrix of 1_{V_1} with respect to $\mathcal{X}'$ and $\mathcal{X}$ is $\mathbf{P}$. ∎

$$\begin{array}{ccccccc} V_1 & \xrightarrow{\;1_{V_1}\;} & V_1 & \xrightarrow{\;f\;} & V_2 & \xrightarrow{\;1_{V_2}\;} & V_2 \\ \mathcal{X}' & \mathbf{P} & \mathcal{X} & \mathbf{A} & \mathcal{Y} & \mathbf{Q}^{-1} & \mathcal{Y}' \end{array}$$

Figure 3.10.1

In the set-up of the preceding theorem, if $\mathcal{X}' = \mathcal{X}$ then $\mathbf{B} = \mathbf{Q}^{-1}\mathbf{A}$ because $\mathbf{P} = \mathbf{I}$. Similarly if $\mathcal{Y}' = \mathcal{Y}$ then $\mathbf{B} = \mathbf{A}\mathbf{P}$ because $\mathbf{Q} = \mathbf{I}$. The next result follows on taking $V_1 = V_2$, $\mathcal{Y} = \mathcal{X}$ and $\mathcal{Y}' = \mathcal{X}'$.

Theorem 3.10.4 Let f be a linear operator on V and $\mathbf{A}$ the matrix of f with respect to a basis $\mathcal{X}$ of V. If $\mathcal{X}'$ is another basis of V and $\mathbf{P}$ is the matrix of transition from $\mathcal{X}$ to $\mathcal{X}'$ then the matrix of f with respect to $\mathcal{X}'$ is $\mathbf{P}^{-1}\mathbf{AP}$.

In *Theorem* 3.10.3 we have seen that the matrix of f with respect to $\mathcal{X}'$ and $\mathcal{Y}'$ is $\mathbf{Q}^{-1}\mathbf{AP}$, where $\mathbf{A}$ is the matrix of f with respect to $\mathcal{X}$ and $\mathcal{Y}$ and $\mathbf{P}$ and $\mathbf{Q}$ are the matrices of transition from $\mathcal{X}$ to $\mathcal{X}'$ and $\mathcal{Y}$ to $\mathcal{Y}'$ respectively. In the converse direction, suppose $\mathcal{X} = \{\mathbf{x}_1, \ldots, \mathbf{x}_n\}$ and $\mathcal{Y} = \{\mathbf{y}_1, \ldots, \mathbf{y}_m\}$ are bases of V_1 and V_2 respectively and $\mathbf{A}$ is the matrix of f with respect to $\mathcal{X}$ and $\mathcal{Y}$. Let $\mathbf{G} = ((g_{ij}))$ be a non-singular matrix of order m, $\mathbf{H} = ((h_{ij}))$ a non-singular matrix of order n and let $\mathbf{B} = \mathbf{GAH}$. Then $\mathbf{B}$ is the matrix of f with respect to the bases $\mathcal{X}' = \{\mathbf{x}'_1, \ldots, \mathbf{x}'_n\}$ and $\mathcal{Y}' = \{\mathbf{y}'_1, \ldots, \mathbf{y}'_m\}$ defined as follows. Let $\mathbf{G}^{-1} = ((q_{ij}))$. Then

$$\mathbf{x}'_j = \textstyle\sum_{i=1}^{n} h_{ij}\mathbf{x}_i \quad (j = 1, \ldots, n)$$
$$\mathbf{y}'_j = \textstyle\sum_{i=1}^{m} q_{ij}\mathbf{y}_i \quad (j = 1, \ldots, m)$$

This follows easily from *Theorems* 3.10.1, 3.10.2 and 3.10.3. Similarly if an $n \times n$ matrix $\mathbf{A}$ is the matrix of $f : V \to V$ with respect to some basis and $\mathbf{H}$ is a non-singular matrix of order n then $\mathbf{H}^{-1}\mathbf{AH}$ is the matrix of f with respect to another basis of V. Motivated by these, we give the following definitions.

Definition 3.10.5 Two $m \times n$ matrices $\mathbf{A}$ and $\mathbf{B}$ are said to be *equivalent* if there exist non-singular matrices $\mathbf{G}$ and $\mathbf{H}$ of orders m and n respectively such that $\mathbf{B} = \mathbf{GAH}$.

Definition 3.10.6 Two $n \times n$ matrices $\mathbf{A}$ and $\mathbf{B}$ are said to be *similar* if there exists a non-singular matrix $\mathbf{H}$ such that $\mathbf{B} = \mathbf{H}^{-1}\mathbf{AH}$.

Theorems 3.10.1, 3.10.2 and 3.10.3 show that *two $m \times n$ matrices* $\mathbf{A}$ *and* $\mathbf{B}$ *are equivalent iff they represent the same linear transformation from F^n to F^m* (with respect to two pairs of bases). We now check that '$\mathbf{A}$ is equivalent to $\mathbf{B}$' is an equivalence relation on the set of all $m \times n$ matrices. Reflexivity follows from $\mathbf{A} = \mathbf{I}_m\mathbf{A}\mathbf{I}_n$. If $\mathbf{B} = \mathbf{GAH}$ then $\mathbf{A} = \mathbf{G}^{-1}\mathbf{B}\mathbf{H}^{-1}$, so we have symmetry. If $\mathbf{B} = \mathbf{GAH}$ and $\mathbf{C} = \mathbf{KBL}$ then $\mathbf{C} = (\mathbf{KG})\mathbf{A}(\mathbf{HL})$ and we have transitivity since $\mathbf{KG}$ and $\mathbf{HL}$ are non-singular.

Similarly, *two $n \times n$ matrices* $\mathbf{A}$ *and* $\mathbf{B}$ *are similar iff they represent the same linear operator on F^n* (with respect to two bases). We leave it to the reader to check that similarity is an equivalence relation on the set of all $n \times n$ matrices.

We note that if $\mathbf{A}$ is an $m \times n$ matrix and $\mathbf{G}$ and $\mathbf{H}$ are non-singular matrices of orders m and n respectively, then $\mathbf{GAH}$ has two interpretations. One is that with respect to a fixed pair of bases (of F^n and F^m), $\mathbf{A}$ represents one linear transformation while $\mathbf{GAH}$ represents another with the same rank. The other is that if $\mathbf{A}$ represents a linear transformation f with respect to one pair of bases then $\mathbf{GAH}$ represents the same transformation f with respect to another pair of bases.

We illustrate *Theorems* 3.10.1 and 3.10.3 with

Example 3.10.7 Let f be the differentiation transformation from the vector space $\mathcal{P}_3$ to the vector space $\mathcal{P}_2$. Now $\mathcal{X} = \{1, t, t^2\}$ and $\mathcal{Y} = \{1, t\}$ are bases of $\mathcal{P}_3$ and $\mathcal{P}_2$ respectively. It is easy to check that the matrix of f with respect to $\mathcal{X}$ and $\mathcal{Y}$ is

$$\mathbf{A} = \begin{bmatrix} 0 & 1 & 0 \\ 0 & 0 & 2 \end{bmatrix} \tag{3.10.4}$$

Now let $\mathcal{X}' = \{\mathbf{x}'_1, \mathbf{x}'_2, \mathbf{x}'_3\}$, where $\mathbf{x}'_1(t) = 1 + t$, $\mathbf{x}'_2(t) = 2 + t$ and $\mathbf{x}'_3(t) = 1 + t + t^2$. Then it is easy to check that $\mathcal{X}'$ is also a basis of $\mathcal{P}_3$ and the matrix of transition from $\mathcal{X}$ to $\mathcal{X}'$ is

$$\mathbf{P} = \begin{bmatrix} 1 & 2 & 1 \\ 1 & 1 & 1 \\ 0 & 0 & 1 \end{bmatrix} \tag{3.10.5}$$

Similarly, let $\mathcal{Y}' = \{\mathbf{y}'_1, \mathbf{y}'_2\}$, where $\mathbf{y}'_1(t) = t$ and $\mathbf{y}'_2(t) = 3 + t$. Then it is easy to see that $\mathcal{Y}'$ is also a basis of $\mathcal{P}_2$ and the matrix of transition from $\mathcal{Y}$ to $\mathcal{Y}'$ is

$$\mathbf{Q} = \begin{bmatrix} 0 & 3 \\ 1 & 1 \end{bmatrix}$$

Clearly, $\mathbf{Q}$ is non-singular and

$$\mathbf{Q}^{-1} = \frac{1}{3}\begin{bmatrix} -1 & 3 \\ 1 & 0 \end{bmatrix} \tag{3.10.6}$$

is the matrix of transition from $\mathcal{Y}'$ to $\mathcal{Y}$ since

$$\begin{aligned} 1 &= -\tfrac{1}{3}t + \tfrac{1}{3}(3 + t) = -\tfrac{1}{3}\mathbf{y}'_1 + \tfrac{1}{3}\mathbf{y}'_2 \\ t &= 1 \cdot t + 0 \cdot (3 + t) = 1 \cdot \mathbf{y}'_1 + 0 \cdot \mathbf{y}'_2 \end{aligned}$$

Now let $\mathbf{B}$ be the matrix of f with respect to $\mathcal{X}'$ and $\mathcal{Y}'$. Then

$$\mathbf{B} = \frac{1}{3}\begin{bmatrix} -1 & -1 & 5 \\ 1 & 1 & 1 \end{bmatrix} \tag{3.10.7}$$

since

$$\begin{aligned} f(\mathbf{x}_1') &= f(1+t) = 1 = -\tfrac{1}{3}\mathbf{y}_1' + \tfrac{1}{3}\mathbf{y}_2' \\ f(\mathbf{x}_2') &= f(2+t) = 1 = -\tfrac{1}{3}\mathbf{y}_1' + \tfrac{1}{3}\mathbf{y}_2' \\ f(\mathbf{x}_3') &= f(1+t+t^2) = 1+2t = \tfrac{5}{3}\mathbf{y}_1' + \tfrac{1}{3}\mathbf{y}_2' \end{aligned}$$

Now it can easily be checked from (3.10.4)–(3.10.7) that $\mathbf{B} = \mathbf{Q}^{-1}\mathbf{AP}$. ∎

Exercises

1. If $\mathbf{P}$ and $\mathbf{R}$ are the matrices of transition from $\mathcal{X}$ to $\mathcal{X}'$ and from $\mathcal{X}'$ to $\mathcal{X}''$ respectively, what is the matrix of transition from $\mathcal{X}$ to $\mathcal{X}''$?
2. (a) Let $\mathcal{X} = \{\mathbf{x}_1, \ldots, \mathbf{x}_n\}$ and $\mathcal{X}' = \{\mathbf{x}_1', \ldots, \mathbf{x}_n'\}$ be two bases of F^n. Let $\mathbf{A} = [\,\mathbf{x}_1 : \cdots : \mathbf{x}_n\,]$ and $\mathbf{B} = [\,\mathbf{x}_1' : \cdots : \mathbf{x}_n'\,]$. Show that the matrix of transition from $\mathcal{X}$ to $\mathcal{X}'$ is $\mathbf{A}^{-1}\mathbf{B}$.

   (b) Find the matrix $\mathbf{P}$ of transition from the basis $\mathcal{X} = \{(1,-1),(4,0)\}$ of $\mathbb{R}^2$ to the basis $\mathcal{X}' = \{(1,1),(1,-1)\}$.
3. Consider the bases $\mathcal{Y} = \{(1,1,0),(1,0,1),(0,1,1)\}$ and $\mathcal{Y}' = \{(1,-1,0), (1,1,1),(1,1,-2)\}$ of $\mathbb{R}^3$.

   (a) Find the matrix $\mathbf{Q}$ of transition from $\mathcal{Y}$ to $\mathcal{Y}'$.

   (b) Find the functional form of the linear transformation σ taking the vectors of $\mathcal{Y}$ to the vectors of $\mathcal{Y}'$ in the same order. Also find the functional form of σ^{-1}.

   (c) Using definitions, find the matrix of σ with respect to $\mathcal{Y}$ as well as the matrix of σ with respect to $\mathcal{Y}'$. Also find the matrices of σ^{-1} with respect to $\mathcal{Y}$ and with respect to $\mathcal{Y}'$. How are these related?
4. Find the matrix of the linear transformation $f : (x_1, x_2) \mapsto (2x_1 - 3x_2, x_1, x_2 + 5x_1)$ from $\mathbb{R}^2$ to $\mathbb{R}^3$ with respect to $\mathcal{X}'$ given in *Exercise* 3.10.2(b) and $\mathcal{Y}'$ given in *Exercise* 3.10.3. Using the result of *Exercise* 2.2.6(b), verify *Theorem* 3.10.3 for this example.
5. Let f be the linear transformation in *Exercise* 2.2.14. If $\mathbf{P} = \begin{bmatrix} 4 & 1 \\ 7 & 2 \end{bmatrix}$, find the basis of $\mathbb{R}^2$ with respect to which the matrix of f is $\mathbf{P}^{-1}\begin{bmatrix} 2 & -1 \\ 3 & 5 \end{bmatrix}\mathbf{P}$. Also find this latter matrix and verify your answer.
6. Show that similarity is an equivalence relation on the set of all $n \times n$ matrices.

*7. If f is a linear operator on V, define the *trace of* f to be $\text{tr}(\mathbf{A})$ where $\mathbf{A}$ is the matrix of f with respect to any basis of V. Show that trace of f is well-defined, $\text{tr}(\alpha f) = \alpha\text{tr}(f)$, $\text{tr}(f+g) = \text{tr}(f) + \text{tr}(g)$ and $\text{tr}(h \circ f) = \text{tr}(f \circ h)$.

*8. Let $\mathcal{X}$ and $\mathcal{X}'$ be bases of V and let $\mathbf{P}$ be the matrix of transition from $\mathcal{X}$ to $\mathcal{X}'$. Let π be the linear transformation taking the vectors of $\mathcal{X}$ to the vectors of $\mathcal{X}'$ in the same order. Show that $\mathbf{P}$ is the matrix of π with respect to $\mathcal{X}$ as well as the matrix of π with respect to $\mathcal{X}'$. Deduce that $\mathbf{P}^{-1}$ is the matrix of π^{-1} with respect to $\mathcal{X}$ as well as the matrix of π^{-1} with respect to $\mathcal{X}'$.

3.11 Rank and nullity of linear transformations*

In this section we discuss briefly the rank, nullity and inverse of linear transformations.

If f is a linear transformation from V_1 to V_2 then the *range of* f is $\{f(\mathbf{x}) : \mathbf{x} \in V_1\}$. The range of f is clearly a subspace of V_2 and its dimension is called the *rank* of f, denoted by $\rho(f)$. Let $\mathbf{A}$ be the matrix of f with respect to any basis $\mathcal{X} = \{\mathbf{x}_1, \ldots, \mathbf{x}_n\}$ of V_1 and any basis $\mathcal{Y} = \{\mathbf{y}_1, \ldots, \mathbf{y}_m\}$ of V_2. Then we show that $\rho(f) = \rho(\mathbf{A})$. By *Theorem* 2.3.2,

$$f\Big(\sum_{j=1}^{n} \alpha_j \mathbf{x}_j\Big) = \sum_{i=1}^{m} \beta_i \mathbf{y}_i$$

where $\boldsymbol{\beta} = \mathbf{A}\boldsymbol{\alpha}$. Thus the range of f is $\left\{\sum_{i=1}^{m} \beta_i \mathbf{y}_i : \boldsymbol{\beta} \in \mathcal{C}(\mathbf{A})\right\}$. By the proof of *Theorem* 1.8.3, the map $\boldsymbol{\beta} \mapsto \sum_{i=1}^{m} \beta_i \mathbf{y}_i$ is an isomorphism from F^m onto V_2, so

$$\rho(\mathbf{A}) = d(\mathcal{C}(\mathbf{A})) = d\Big(\Big\{\sum_{i=1}^{m} \beta_i \mathbf{y}_i : \boldsymbol{\beta} \in \mathcal{C}(\mathbf{A})\Big\}\Big) = \rho(f)$$

We may thus say that the range of f corresponds to the column space of any matrix $\mathbf{A}$ representing f and $\rho(f) = \rho(\mathbf{A})$. Unfortunately, the row space of $\mathbf{A}$ is not so easy to explain in terms of linear transformations (see, however, Halmos (1958)).

The *kernel* $\mathcal{K}(f)$ of f is defined to be $\{\mathbf{x} \in V_1 : f(\mathbf{x}) = \mathbf{0}\}$. The kernel of f is a subspace of V_1 and its dimension is called the *nullity* of f, denoted $\nu(f)$. We now show that $\nu(f) = n - \rho(f)$, where n is the dimension of V_1. We leave it to the reader to prove this by imitating the proof of *Theorem* 3.5.9. We will instead prove it by using *Theorem* 3.5.9 itself. Fix bases $\mathcal{X}$ and $\mathcal{Y}$ as above and let $\mathbf{A}$ be the matrix of f with respect to $\mathcal{X}$ and $\mathcal{Y}$. Then the kernel of f is $\left\{\sum_{j=1}^{n} \alpha_j \mathbf{x}_j : \mathbf{A}\boldsymbol{\alpha} = \mathbf{0}\right\}$. Since $\boldsymbol{\alpha} \mapsto \sum_{j=1}^{n} \alpha_j \mathbf{x}_j$ is an isomorphism from F^n onto V_1, it follows that

$$\nu(f) = d\Big(\Big\{\sum_{j=1}^{n} \alpha_j \mathbf{x}_j : \mathbf{A}\boldsymbol{\alpha} = \mathbf{0}\Big\}\Big) = d(\{\boldsymbol{\alpha} : \mathbf{A}\boldsymbol{\alpha} = \mathbf{0}\}) = \nu(\mathbf{A})$$

Hence $\nu(f) = n - \rho(\mathbf{A}) = n - \rho(f)$.

We next prove the analogues of *Theorems* 3.3.3, 3.3.5 and 3.3.7 for linear transformations. A *left* (resp. *right*) *inverse* of a linear transformation $f : V_1 \to V_2$ is defined to be a linear transformation (not just a map) g from V_2 to V_1 such that $g \circ f = 1_{V_1}$ (resp. $f \circ g = 1_{V_2}$).

Theorem 3.11.1 Let f be a linear transformation from V_1 to V_2, where $d(V_2) = m$. Then the following statements are equivalent:

(i) f has a right inverse,

(ii) range of $f = V_2$,

(iii) $\rho(f) = m$.

Proof (i) $\Rightarrow$ (ii) Let g be a right inverse of f. If $\mathbf{y} \in V_2$ then $\mathbf{y} = f(g(\mathbf{y}))$ belongs to the range of f.

(ii) $\Rightarrow$ (i) Let $\mathcal{Y} = \{\mathbf{y}_1, \ldots, \mathbf{y}_m\}$ be a basis of V_2. For $i = 1, \ldots, m$, choose and fix an $\mathbf{x}_i$ such that $f(\mathbf{x}_i) = \mathbf{y}_i$ (note that $\mathbf{x}_i$ exists since range of f is V_2). Now it is easy to verify that $g : V_2 \to V_1$ defined by

$$g\Big(\sum_{i=1}^{m} \beta_i \mathbf{y}_i\Big) = \sum_{i=1}^{m} \beta_i \mathbf{x}_i$$

is a linear transformation and that $f \circ g = 1_{V_2}$. Thus g is a right inverse of f.

That (ii) and (iii) are equivalent is trivial. ∎

Theorem 3.11.2 Let f be a linear transformation from V_1 to V_2, where $d(V_1) = n$. Then the following statements are equivalent:

(i) f has a left inverse,

(ii) f is 1-1,

(iii) $\nu(f) = 0$ (i.e., $\rho(f) = n$).

Proof (i) $\Rightarrow$ (ii) Let g be a left inverse of f. If $f(\mathbf{x}) = f(\mathbf{z})$, then

$$\mathbf{x} = (g \circ f)(\mathbf{x}) = (g \circ f)(\mathbf{z}) = \mathbf{z}$$

(ii) $\Rightarrow$ (iii) If $f(\mathbf{x}) = \mathbf{0}$, then $f(\mathbf{x}) = f(\mathbf{0})$, so $\mathbf{x} = \mathbf{0}$ by hypothesis. Thus $\mathcal{K}(f) = \{\mathbf{0}\}$ and (iii) follows.

(iii) $\Rightarrow$ (i) The range S of f is a subspace of V_2 with dimension n. Let $\{\mathbf{y}_1, \ldots, \mathbf{y}_n\}$ be a basis of S. Extend this to a basis $\{\mathbf{y}_1, \ldots, \mathbf{y}_n, \mathbf{y}_{n+1}, \ldots, \mathbf{y}_m\}$ of V_2. For $i = 1, 2, \ldots, n$, take an $\mathbf{x}_i$ such that $f(\mathbf{x}_i) = \mathbf{y}_i$ (it exists since $\mathbf{y}_i \in S$). Since $\mathbf{y}_1, \ldots, \mathbf{y}_n$ are linearly independent it follows from *Exercise* 2.2.16 that $\mathbf{x}_1, \ldots, \mathbf{x}_n$ are also linearly independent and so form a basis of V_1. Let g be the linear transformation from V_2 to V_1 (see *Exercise* 2.2.17) taking $\mathbf{y}_i$ to $\mathbf{x}_i$ for $1 \leq i \leq n$ and to $\mathbf{0}$ for $n+1 \leq i \leq m$. Then $(g \circ f)(\mathbf{x}_i) = g(\mathbf{y}_i) = \mathbf{x}_i$ for $i = 1, \ldots, n$, so $g \circ f = 1_{V_1}$. ∎

The *inverse* of a linear operator f on V is a linear operator g on V which is both a left inverse and a right inverse of f. A linear operator with an inverse is said to be *non-singular*.

Theorem 3.11.3 For a linear operator f on V, the following statements are equivalent.

(i) f is non-singular, i.e., f has an inverse,

(ii) f has a left inverse,

(iii) f has a right inverse,

(iv) f is 1-1,

(v) f is onto V,

(vi) $\rho(f) = d(V)$ (or $\nu(f) = 0$).

This theorem follows easily from the two preceding theorems and the fact that if g is a left inverse of f and h is a right inverse then

$$g = g \circ 1_V = g \circ f \circ h = 1_V \circ h = h$$

Theorem 3.11.4 Let f be a non-singular linear transformation from V to itself. Let $\mathcal{X}$ and $\mathcal{Y}$ be two bases of V and let $\mathbf{A}$ be the matrix of f with respect to $\mathcal{X}$ and $\mathcal{Y}$. Then the matrix of f^{-1} with respect to $\mathcal{Y}$ and $\mathcal{X}$ is $\mathbf{A}^{-1}$.

Proof Clearly, $\mathbf{A}$ is a square matrix. Let $\mathbf{B}$ be the matrix of f^{-1} with respect to $\mathcal{Y}$ and $\mathcal{X}$. Then by *Theorem* 2.2.17, the matrix of $f \circ f^{-1}$ with respect to $\mathcal{X}$ is $\mathbf{AB}$. But $f \circ f^{-1}$ is the identity map, so $\mathbf{AB} = \mathbf{I}$ and $\mathbf{B} = \mathbf{A}^{-1}$. ∎

Corollary If $\mathbf{A}$ is the matrix of f with respect to $\mathcal{X}$ then $\mathbf{A}^{-1}$ is the matrix of f^{-1} with respect to $\mathcal{X}$.

It is easy to see that if f and g are non-singular linear operators on V, then so are f^{-1} and $f \circ g$. In fact, $(f^{-1})^{-1} = f$ and $(f \circ g)^{-1} = g^{-1} \circ f^{-1}$.
We now show that

$$\rho(f \circ g) \leq \min(\rho(f), \rho(g))$$

Clearly the range of $f \circ g$ is contained in the range of f, so $\rho(f \circ g) \leq \rho(f)$. To prove the other inequality, let S be the range of g. Then

$$\rho(f \circ g) = d(\text{range of } f \circ g) = d(f(S)) \leq d(S) = \rho(g)$$

The analogue of *Theorem* 3.5.10 for linear transformations is the following. Let g be a linear transformation from V_1 to V_2 and let T be a subspace of V_1. Then $\{g(\mathbf{x}) : \mathbf{x} \in T\}$ is a subspace of V_2 with dimension $d(T) - d(T \cap \mathcal{K}(g))$. The analogue of *Theorem* 3.5.11 is:

$$d\big(\text{range}(g) \cap \mathcal{K}(f)\big) = \rho(g) - \rho(f \circ g)$$

These results as well as the analogues of Sylvester's inequality and Frobenius inequality for linear transformations can be proved by imitating the proofs of the corresponding theorems for matrices.

Exercises

1. Prove the analogues of *Theorems* 3.5.10 to 3.5.13 for linear transformations.
2. If S is a subspace of a vector space V and T is a complement of S, the map η taking any $\mathbf{x}$ in V to its projection into S along T is called the *projector* or *projection into S along T*. *A projector* or *projection operator* is a projector into S along T for some S and T.
   (a) Show that if η is a linear operator on V and $\mathbf{A}$ is the matrix of η with respect to some basis $\mathcal{X}$, then η is a projector iff $\mathbf{A}^2 = \mathbf{A}$. Also then show that η is the projector into the range of η along the kernel of η.
   (b) State and prove the analogues of *Theorems* 3.6.3, 3.7.4 and 3.7.5 for linear operators on a vector space V.
3. Let f be a linear transformation from V_1 to V_2. Let $\rho(f) = r$ and $K = \mathcal{K}(f)$.

(a) Let $\{\mathbf{y}_1, \ldots, \mathbf{y}_r\}$ be a basis of the range of f and let $\mathbf{y}_i = f(\mathbf{x}_i)$ for $i = 1, \ldots, r$. If $S = \text{Sp}(\{\mathbf{x}_1, \ldots, \mathbf{x}_r\})$, show that $d(S) = r$ and $f(S) = f(V_1)$.

(b) If S is a subspace of V_1 such that $d(S) = r$ and $f(S) = f(V_1)$, show that S can be obtained as in (a), $S \oplus K = V_1$ and $S \simeq V_1/K$.

Chapter 4
Elementary operations and reduced forms

4.1 Introduction

In this chapter we shall study some simple operations called elementary operations which can be used to reduce any given matrix to one with a simple form thereby facilitating the solution of some problems to be solved for the original matrix.

Suppose we want to find the rank, a row basis and a column basis of $\mathbf{A}$ and solve the system $\mathbf{Ax} = \mathbf{b}$. If

$$\mathbf{A} = \begin{bmatrix} \mathbf{I}_r & \mathbf{B} \\ \mathbf{0} & \mathbf{0} \end{bmatrix} \text{ and } \mathbf{b} = \begin{bmatrix} \mathbf{c} \\ \mathbf{d} \end{bmatrix},$$

it is easy to see that $\rho(\mathbf{A}) = r$, the first r rows of $\mathbf{A}$ form a row basis and the first r columns form a column basis. Suppose now the order of $\mathbf{A}$ is $m \times n$, the order of $\mathbf{b}$ is $m \times 1$ and the order of $\mathbf{c}$ is $r \times 1$. Then $\mathbf{Ax} = \mathbf{b}$ has a solution iff $\mathbf{d} = \mathbf{0}$ and, when this happens, the $n \times 1$ vector $\left[\begin{smallmatrix} \mathbf{c} \\ \mathbf{0} \end{smallmatrix}\right]$ is a solution. Note that linear independence of vectors in F^n amounts to a suitable matrix being of full row rank and expressing a vector $\mathbf{b}$ in F^m as a linear combination of n given vectors amounts to solving a system of linear equations. We will see later how reduction to simple forms can be achieved without destroying what we want to study.

4.2 Elementary operations

Definition 4.2.1 An *elementary row operation* on a matrix is any one of the following:

(i) interchanging two rows,

(ii) multiplying a row by a non-zero scalar, and

(iii) replacing a row by the sum of that row and a scalar multiple of another row.

An *elementary column operation* is defined similarly. By an *elementary operation* we mean an elementary row operation or an elementary column operation. We shall use the notation R_{ik} for interchanging the i-th and k-th rows, $R_i(\alpha)$ for multiplying the i-th row by α and $R_{ik}(\beta)$ for adding β times the k-th row to the i-th row. The corresponding column operations are denoted by C_{ik}, $C_i(\alpha)$ and $C_{ik}(\beta)$ respectively. Clearly an elementary operation does not alter the order of a matrix.

To give an example, let

$$\mathbf{A} = \begin{bmatrix} 4 & 3 & 1 \\ 1 & 3 & 7 \\ -2 & 8 & 1 \\ 7 & 4 & 1 \end{bmatrix}$$

Then

$$\mathbf{B} = \begin{bmatrix} 4 & 3 & 1 \\ 7 & 4 & 1 \\ -2 & 8 & 1 \\ 1 & 3 & 7 \end{bmatrix}, \quad \mathbf{C} = \begin{bmatrix} 4 & 3 & 1 \\ 1 & 3 & 7 \\ -6 & 24 & 3 \\ 7 & 4 & 1 \end{bmatrix}, \quad \mathbf{D} = \begin{bmatrix} 4 & 3 & 1 \\ 0 & \frac{9}{4} & \frac{27}{4} \\ -2 & 8 & 1 \\ 7 & 4 & 1 \end{bmatrix}$$

are obtained from $\mathbf{A}$ by the elementary row operations R_{24}, $R_3(3)$ and $R_{21}(-\frac{1}{4})$.

We now show that an elementary row (resp. column) operation has the same effect as pre- (resp. post-) multiplication by a suitable matrix. For this we give

Definition 4.2.2 An *elementary matrix* is a matrix obtained from an identity matrix by a single elementary row operation.

We shall use the following notations for elementary matrices: $\mathbf{E}_{ik}$, $\mathbf{E}_i(\alpha)$ and $\mathbf{E}_{ik}(\beta)$ will denote, respectively, the matrices obtained from $\mathbf{I}$ by the elementary row operations R_{ik}, $R_i(\alpha)$ and $R_{ik}(\beta)$. Notice that, the order of an elementary matrix is omitted in the notation as it is usually known from the context.

Theorem 4.2.3 Making an elementary row operation on a matrix $\mathbf{A}$ is equivalent to premultiplying $\mathbf{A}$ by the corresponding elementary matrix.

Proof Let $\mathbf{A}$ be a matrix of order $m \times n$ and let $\mathbf{B}$, $\mathbf{C}$ and $\mathbf{D}$ be matrices obtained from $\mathbf{A}$ by the elementary row operations R_{ik}, $R_i(\alpha)$ and $R_{ik}(\beta)$ respectively. Then the theorem follows by direct verification as shown below:

$$\mathbf{B} = \begin{bmatrix} \mathbf{A}_{1*} \\ \cdots \\ \mathbf{A}_{k*} \\ \cdots \\ \mathbf{A}_{i*} \\ \cdots \\ \mathbf{A}_{m*} \end{bmatrix} \begin{array}{l} \\ \\ i\text{-th} \\ \\ k\text{-th} \\ \\ \\ \end{array} = \begin{array}{c} \begin{array}{ccccccc} & & i\text{-th} & & k\text{-th} & & \end{array} \\ \begin{bmatrix} 1 & \cdots & 0 & \cdots & 0 & \cdots & 0 \\ \cdots & \cdots & \cdots & \cdots & \cdots & \cdots & \cdots \\ 0 & \cdots & 0 & \cdots & 1 & \cdots & 0 \\ \cdots & \cdots & \cdots & \cdots & \cdots & \cdots & \cdots \\ 0 & \cdots & 1 & \cdots & 0 & \cdots & 0 \\ \cdots & \cdots & \cdots & \cdots & \cdots & \cdots & \cdots \\ 0 & \cdots & 0 & \cdots & 0 & \cdots & 1 \end{bmatrix} \end{array} \begin{bmatrix} \mathbf{A}_{1*} \\ \cdots \\ \mathbf{A}_{i*} \\ \cdots \\ \mathbf{A}_{k*} \\ \cdots \\ \mathbf{A}_{m*} \end{bmatrix} = \mathbf{E}_{ik}\mathbf{A}$$

$$\mathbf{C} = \begin{bmatrix} \mathbf{A}_{1*} \\ \cdots \\ \alpha\mathbf{A}_{i*} \\ \cdots \\ \mathbf{A}_{m*} \end{bmatrix} \begin{array}{l} \\ \\ i\text{-th} \\ \\ \\ \end{array} = \begin{array}{c} \begin{array}{ccccc} & & i\text{-th} & & \end{array} \\ \begin{bmatrix} 1 & \cdots & 0 & \cdots & 0 \\ \cdots & \cdots & \cdots & \cdots & \cdots \\ 0 & \cdots & \alpha & \cdots & 0 \\ \cdots & \cdots & \cdots & \cdots & \cdots \\ 0 & \cdots & 0 & \cdots & 1 \end{bmatrix} \end{array} \begin{bmatrix} \mathbf{A}_{1*} \\ \cdots \\ \mathbf{A}_{i*} \\ \cdots \\ \mathbf{A}_{m*} \end{bmatrix} = \mathbf{E}_i(\alpha)\mathbf{A}$$

$$\mathbf{D} = \begin{bmatrix} \mathbf{A}_{1*} \\ \cdots \\ \mathbf{A}_{i*} + \beta\mathbf{A}_{k*} \\ \cdots \\ \mathbf{A}_{k*} \\ \cdots \\ \mathbf{A}_{m*} \end{bmatrix} \begin{array}{l} \\ \\ i\text{-th} \\ \\ k\text{-th} \\ \\ \\ \end{array} = \begin{array}{c} \begin{array}{ccccccc} & & i\text{-th} & & k\text{-th} & & \end{array} \\ \begin{bmatrix} 1 & \cdots & 0 & \cdots & 0 & \cdots & 0 \\ \cdots & \cdots & \cdots & \cdots & \cdots & \cdots & \cdots \\ 0 & \cdots & 1 & \cdots & \beta & \cdots & 0 \\ \cdots & \cdots & \cdots & \cdots & \cdots & \cdots & \cdots \\ 0 & \cdots & 0 & \cdots & 1 & \cdots & 0 \\ \cdots & \cdots & \cdots & \cdots & \cdots & \cdots & \cdots \\ 0 & \cdots & 0 & \cdots & 0 & \cdots & 1 \end{bmatrix} \end{array} \begin{bmatrix} \mathbf{A}_{1*} \\ \cdots \\ \mathbf{A}_{i*} \\ \cdots \\ \mathbf{A}_{k*} \\ \cdots \\ \mathbf{A}_{m*} \end{bmatrix}$$

$$= \mathbf{E}_{ik}(\beta)\mathbf{A}.$$ ∎

Theorem 4.2.4 Making the column operations C_{jk}, $C_j(\alpha)$ and $C_{jk}(\beta)$ on a matrix $\mathbf{A}$ is equivalent to postmultiplying $\mathbf{A}$ by the elementary matrixs $\mathbf{E}_{jk}$, $\mathbf{E}_j(\alpha)$ and $\mathbf{E}_{kj}(\beta)$ respectively.

The proof of this theorem is by simple verification and is omitted.

We draw attention to the fact that the elementary matrix postmultiplying $\mathbf{A}$ *to describe the operation* $C_{jk}(\beta)$ *on* $\mathbf{A}$ *is* $\mathbf{E}_{kj}(\beta)$ *and not* $\mathbf{E}_{jk}(\beta)$.

Clearly, $\mathbf{E}_{ik}$ and $\mathbf{E}_i(\alpha)$ are symmetric and $\left(\mathbf{E}_{ik}(\beta)\right)^{\mathrm{T}} = \mathbf{E}_{ki}(\beta)$. Hence the next theorem follows from the preceding theorem.

Theorem 4.2.5 Let $\mathbf{E}$ be the elementary matrix required to premultiply a square matrix $\mathbf{A}$ for making an elementary row operation on $\mathbf{A}$. Then $\mathbf{E}^{\mathrm{T}}$ is the elementary matrix postmultiplying $\mathbf{A}$ for making the corresponding elementary column operation.

We make an interesting observation. If we perform an elementary row operation and an elementary column operation on $\mathbf{A}$, the order in which these are performed is immaterial since $(\mathbf{PA})\mathbf{Q} = \mathbf{P}(\mathbf{AQ})$ where $\mathbf{P}$ and $\mathbf{Q}$ are the elementary matrices corresponding to the row and column operations. However, when several row (or several column) operations are performed, the order in which they are performed is important.

Theorem 4.2.6 The elementary matrices $\mathbf{E}_{ik}$, $\mathbf{E}_i(\alpha)$ and $\mathbf{E}_{ik}(\beta)$ are non-singular with inverses $\mathbf{E}_{ik}$, $\mathbf{E}_i(\frac{1}{\alpha})$ and $\mathbf{E}_{ik}(-\beta)$ respectively.

Proof We will prove the result only for $\mathbf{E}_{ik}(\beta)$ as the others are proved similarly. By *Theorem* 4.2.3, $\mathbf{E}_{ik}(-\beta)\mathbf{E}_{ik}(\beta)\mathbf{I}$ is obtained from $\mathbf{I}$ by first adding β times the k-th row to the i-th row and then subtracting β times the k-th row from the i-th row and is thus $\mathbf{I}$ itself. Hence $\mathbf{E}_{ik}(-\beta)$ is the inverse of $\mathbf{E}_{ik}(\beta)$. ∎

Since pre- (resp. post-) multiplication by a non-singular matrix does not alter the row (resp. column) space, we have

Theorem 4.2.7 An elementary operation does not alter the rank of a matrix. Further, an elementary row operation does not alter the row space and an elementary column operation does not alter the column space.

Consider a finite sequence $\mathcal{S}$ of elementary row operations that can be performed on a matrix with m rows. If $\mathbf{E}_1, \mathbf{E}_2, \ldots, \mathbf{E}_q$ are the elementary matrices corresponding to $\mathcal{S}$, we call $\mathbf{P} := \mathbf{E}_q\mathbf{E}_{q-1}\cdots\mathbf{E}_1$ *the transforming matrix* corresponding to $\mathcal{S}$. The following theorem is trivial to prove:

Theorem 4.2.8 If a finite sequence $\mathcal{S}$ of elementary row operations is performed on any matrix $\mathbf{A}$, the resulting matrix is $\mathbf{PA}$ where $\mathbf{P}$ is the transforming matrix corresponding to $\mathcal{S}$.

We can similarly define *the transforming matrix* corresponding to a sequence $\mathcal{T}$ of elementary column operations as the matrix $\mathbf{Q}$ such that whenever $\mathcal{T}$ is performed on $\mathbf{A}$, the resulting matrix is $\mathbf{AQ}$.

Given any matrix **A**, by a finite sequence of elementary operations, we can reduce **A** to a matrix **B** with a simple structure such that the problem to be solved for **A** (like determining the rank, finding a row basis or a column basis, finding a rank-factorization, finding the inverse and solving linear equations) is much easier for **B**. Knowing the solution for **B**, it is also easy to get the solution of the problem for **A** because of the simple nature of elementary operations. The matrix **B** usually has many 0's and is called a *reduced form* of **A**. In *Sections* 4.4, 4.5 and 4.6, we will give various reduced forms and indicate their uses. The exact reduced form to be used depends on the purpose at hand.

Exercises

1. Consider the matrix

$$\mathbf{A} = \begin{bmatrix} 2 & 3 & -1 & 0 \\ 3 & 0 & 2 & -1 \\ 1 & 5 & 0 & 8 \end{bmatrix}$$

   (a) Write down the matrices obtained by performing each of the following elementary operations on **A**: (i) R_{23}, (ii) $R_1(\frac{1}{2})$, (iii) $R_{21}(-\frac{3}{2})$, (iv) $\mathbf{C}_3(-1)$ and (v) $\mathbf{C}_{31}(\frac{1}{2})$.

   (b) Which elementary row operation on **A** will convert the $(3,2)$-th element to 0? Write down the corresponding elementary matrix and the resulting matrix.

   (c) Is it possible to transform **A** by elementary operations to a matrix with the last two columns null? Why?

2. Consider the matrix **A** of the preceding exercise. By subtracting a multiple of the first row from the second, reduce the $(2,1)$-th element to 0. Next reduce the $(3,1)$-th element to 0 similarly. Find the transforming matrix and the resulting matrix. Repeat these interchanging the order of the two operations.

3. Show that interchanging two rows can be effected by elementary row operations of the other two types.

4. Show that each of $\mathbf{E}_{ij}$, $\mathbf{E}_i(\alpha)$ and $\mathbf{E}_{ij}(\beta)$ can also be obtained from **I** by an elementary column operation. (Be careful about the last matrix.)

*5. If $\alpha \neq 0$ and $\beta \neq 0$, show that $\mathbf{E}_{ij}(\alpha)$ and $\mathbf{E}_{k\ell}(\beta)$ commute iff $i \neq \ell$ and $j \neq k$. Which pairs of elementary matrices commute?

6. Verify *Theorem* 4.2.4.

7. Prove directly that an elementary row operation does not alter the row space.

8. Does it follow from *Theorem* 4.2.7 that if both elementary row operations and elementary column operations are performed on a matrix, neither the row space nor the column space changes? Why?
9. If $\mathcal{S}$ is a sequence of elementary row operations, what is the matrix obtained by performing $\mathcal{S}$ on $\mathbf{I}$?
10. Let $\mathbf{B}$ be obtained from $\mathbf{A}$ by making the elementary column operations $\mathbf{C}_{23}, \mathbf{C}_{25}(4), \mathbf{C}_{34}, \mathbf{C}_4(-2)$ and $\mathbf{C}_{53}(-3)$ in that order. Write down the transforming matrix $\mathbf{Q}$ (where $\mathbf{B} = \mathbf{AQ}$) as a product of matrices of the forms $\mathbf{E}_{ij}, \mathbf{E}_i(\alpha)$ and $\mathbf{E}_{ij}(\beta)$.
11. If $\mathbf{B}$ can be obtained from $\mathbf{A}$ by a sequence of elementary row operations, show that the transforming matrix $\mathbf{P}$ is not uniquely determined by $\mathbf{A}$ and $\mathbf{B}$ in general. Show, however, that $\mathbf{P}$ is unique (though the sequence of elementary row operations is not) if $\mathbf{A}$ is of full row rank.
12. Let $\mathbf{M}$ be an $m \times n$ matrix and $1 \leq k < m$. Show that the transforming matrix of a sequence of elementary row operations on $\mathbf{M}$ of the type $R_{ij}(\beta)$ with $1 \leq j \leq k$ and $k+1 \leq i \leq m$ is of the form $\begin{bmatrix} \mathbf{I} & \mathbf{0} \\ \mathbf{Q} & \mathbf{I} \end{bmatrix}$ for some $(m-k) \times k$ matrix $\mathbf{Q}$ and conversely.
13. Let
$$\mathbf{M} = \begin{bmatrix} \mathbf{A} & \mathbf{B} \\ \mathbf{C} & \mathbf{D} \end{bmatrix} \quad \text{and} \quad \mathbf{N} = \begin{bmatrix} \mathbf{A} & \mathbf{B} \\ \mathbf{C} - \mathbf{HA} & \mathbf{D} - \mathbf{HB} \end{bmatrix}$$
where $\mathbf{M}$ is of order $m \times n$ and $\mathbf{A}$ is of order $k \times \ell$.
    (a) Prove that $\mathbf{N}$ can be obtained from $\mathbf{M}$ by a sequence of elementary row operations of the type considered in the preceding exercise, for any $(m-k) \times k$ matrix $\mathbf{H}$.
    (b) Show that the submatrix $\mathbf{C}$ of $\mathbf{M}$ can be converted to $\mathbf{0}$ by such elementary row operations iff $\mathcal{R}(\mathbf{C}) \subseteq \mathcal{R}(\mathbf{A})$. If $k = \ell$ and $\mathbf{A}$ is non-singular, show that when $\mathbf{C}$ is thus converted to $\mathbf{0}$, $\mathbf{D}$ gets converted to $\mathbf{D} - \mathbf{CA}^{-1}\mathbf{B}$.
    (c) Work out the analogues of (a) and (b) for column operations.
14. Let $\mathbf{M}$ be as in the preceding exercise.
    (a) If $\mathbf{A}$ is non-singular, prove that $\rho(\mathbf{M}) = \rho(\mathbf{A}) + \rho(\mathbf{D} - \mathbf{CA}^{-1}\mathbf{B})$.
    (b) Deduce from (a) that if $\mathbf{A}$ is non-singular and if $\mathbf{B} = \mathbf{0}$ or $\mathbf{C} = \mathbf{0}$ then $\rho(\mathbf{M}) = \rho(\mathbf{A}) + \rho(\mathbf{D})$. Show that the hypothesis '$\mathbf{A}$ is non-singular' cannot be dropped here.

*15. Let $\mathbf{A}$ be the matrix of the linear transformation f with respect to the bases $\mathcal{X} = \{\mathbf{x}_1, \mathbf{x}_2, \ldots, \mathbf{x}_n\}$ and $\mathcal{Y} = \{\mathbf{y}_1, \mathbf{y}_2, \ldots, \mathbf{y}_m\}$. In each of the following cases, find the basis $\mathcal{Y}'$ such that $\mathbf{B}$ is the matrix of f with respect to $\mathcal{X}$ and $\mathcal{Y}'$: (i) $\mathbf{B} = \mathbf{E}_{ij}\mathbf{A}$, (ii) $\mathbf{B} = \mathbf{E}_i(\alpha)\mathbf{A}$ and

(iii) $\mathbf{B} = \mathbf{E}_{ij}(\beta)\mathbf{A}$. What are the corresponding results for column operations on $\mathbf{A}$?

4.3 Sweeping out a row or a column

In this section we will show how to reduce the ℓ-th column of $\mathbf{A}$ to $\mathbf{e}_k$ by elementary row operations and the k-th row to $\mathbf{e}_\ell^T$ by elementary column operations, if $a_{k\ell} \neq 0$. These are the basic steps and will be used repeatedly in the reduction procedures in later sections.

Algorithm 4.3.1 (*Sweeping out the ℓ-th column with the (k,ℓ)-th element as pivot*) Given: a matrix $\mathbf{A}$ with m rows and with $a_{k\ell} \neq 0$.

Step 1 Perform $R_k(\frac{1}{a_{k\ell}})$ on $\mathbf{A}$. (*We use* $\mathbf{A}$ *to denote the matrix at every stage, for convenience.*)

Step 2 Perform $R_{ik}(-a_{i\ell})$ on $\mathbf{A}$ for $i = 1, \ldots, k-1, k+1, \ldots, m$ and stop.

It is easy to see that the above algorithm converts $\mathbf{A}_{*\ell}$ to $\mathbf{e}_k$. Note that in *Step* 2, the k-th row is not altered.

In the above sweep-out, the k-th row is known as the *pivotal row*. The *sweep-out of the k-th row using the (k,ℓ)-th element as the pivot* is defined analogously. *Note that we use row operations for sweeping out a column and column operations for sweeping out a row.*

It is easy to check that the matrix $\mathbf{B}$ obtained by sweeping out the ℓ-th column of $\mathbf{A}$ using the (k,ℓ)-th element as the pivot is given by

$$b_{ij} = \begin{cases} a_{ij}/a_{k\ell} & \text{if } i = k \\ a_{ij} - (a_{i\ell}a_{kj})/a_{k\ell} & \text{if } i \neq k \end{cases} \tag{4.3.1}$$

Note the positions of the elements occurring in the formula (4.3.1):

$$\mathbf{A} = \begin{bmatrix} & \vdots & & \vdots & \\ \cdots & \boxed{a_{k\ell}} & \cdots & a_{kj} & \cdots \\ & \vdots & & \vdots & \\ \cdots & a_{i\ell} & \cdots & a_{ij} & \cdots \\ & \vdots & & \vdots & \end{bmatrix}$$

Here we have enclosed the pivot in a box.

Example 4.3.2 We will sweep out the fourth column of the matrix

$$\mathbf{A} = \begin{bmatrix} 2 & -3 & 4 & 3 & 6 \\ 0 & 1 & 3 & 0 & 2 \\ -1 & 0 & 1 & 2 & 4 \\ 5 & 7 & 0 & -2 & 0 \end{bmatrix} \tag{4.3.2}$$

with the $(3,4)$-th element as the pivot. We first divide the third row by the pivot. Then the third row becomes $(-\frac{1}{2}, 0, \frac{1}{2}, 1, 2)$. Next we subtract 3 times the third row from the first row and add 2 times the third row to the fourth row. Notice that since $a_{24} = 0$, we do not have to perform any operation for the second row. The final matrix is

$$\begin{bmatrix} \frac{7}{2} & -3 & \frac{5}{2} & 0 & 0 \\ 0 & 1 & 3 & 0 & 2 \\ -\frac{1}{2} & 0 & \frac{1}{2} & 1 & 2 \\ 4 & 7 & 1 & 0 & 4 \end{bmatrix} \tag{4.3.3}$$

Performing the same row operations (in the same order) on $\mathbf{I}_4$ we get

$$\begin{bmatrix} 1 & 0 & -\frac{3}{2} & 0 \\ 0 & 1 & 0 & 0 \\ 0 & 0 & \frac{1}{2} & 0 \\ 0 & 0 & 1 & 1 \end{bmatrix}$$

and this is the transforming matrix. (See also *Exercise* 4.3.7.)

Notice that to sweep out the third *row* of the matrix (4.3.3) with the $(3,4)$-th element as the pivot, we just have to replace all elements in the third row, except the $(3,4)$-th, by 0. ∎

Exercises

1. Consider the matrix **A** of *Exercise* 4.2.1.
   (a) Sweep out the first column using the $(3,1)$-th element as the pivot. Obtain the resulting matrix and the transforming matrix.
   (b) How do you sweep out the third row of the matrix obtained in (a) with the $(3,1)$-th element as the pivot? What is the corresponding transforming matrix?
   (c) Repeat (a) and (b) by performing the row sweep-out first and then the column sweep-out.

2. Write down the matrix obtained from $\mathbf{A}$ by sweeping out the k-th row and the ℓ-th column (in any order) with $a_{k\ell}$ as the pivot.
3. Let an $n \times n$ symmetric matrix $\mathbf{A}$ with $a_{11} \neq 0$ be reduced to
$$\mathbf{B} = \begin{bmatrix} a_{11} & \mathbf{x}^{\mathrm{T}} \\ \mathbf{0} & \mathbf{C} \end{bmatrix}$$
by elementary row operations of the type $R_{i1}(\beta)$. Let $\mathbf{P}$ be the transforming matrix. Show that $\mathbf{PAP}^{\mathrm{T}}$ is $\mathbf{B}$ with $\mathbf{x}^{\mathrm{T}}$ replaced by $\mathbf{0}$. Hence or otherwise show that $\mathbf{C}$ is symmetric.
4. Let $\mathbf{B}$ be obtained from $\mathbf{A}$ by sweeping out the ℓ-th column with $a_{k\ell}$ as the pivot followed by a permutation π of the rows not involving the k-th row. If the same permutation π of the rows is performed first and then the ℓ-th column is swept out with $a_{k\ell}$ as the pivot, show that the same matrix $\mathbf{B}$ is obtained.
5. Let $\mathbf{A}$ be a 3×2 matrix of rank 1 with $a_{21} \neq 0$. If the first column is swept out with a_{21} as the pivot show that the second column of the resulting matrix is of the form $(0, \alpha, 0)^{\mathrm{T}}$.

*6. Let $\mathbf{A}$ be an $m \times n$ matrix and let $\mathbf{B}$ be obtained from $\mathbf{A}$ by sweeping out the ℓ-th column with $a_{k\ell}$ as pivot. Then prove that $\mathbf{B} = \mathbf{UA}$ where $\mathbf{U}$ is obtained from $\mathbf{I}_m$ by replacing the k-th column by
$$-\left(a_{1\ell}, \ldots, a_{k-1,\ell}, -1, a_{k+1,\ell}, \ldots, a_{m\ell}\right)^{\mathrm{T}} / \, a_{k\ell}$$
Show also that $\mathbf{U}^{-1}$ is obtained from $\mathbf{I}_m$ by replacing the k-th column by $\mathbf{A}_{*\ell}$. Similarly prove that sweeping out the k-th row of $\mathbf{A}$ with $a_{k\ell}$ as pivot is equivalent to postmultiplying $\mathbf{A}$ by $\mathbf{V}$ where $\mathbf{V}$ is obtained from $\mathbf{I}_n$ by replacing the ℓ-th row by
$$-\left(a_{k1}, \ldots, a_{k,\ell-1}, -1, a_{k,\ell+1}, \ldots, a_{kn}\right) / \, a_{k\ell}$$
Also show that $\mathbf{V}^{-1}$ is obtained from $\mathbf{I}_n$ by replacing the ℓ-th row by $\mathbf{A}_{k*}$.

4.4 Echelon form

In this section we will show how elementary row operations can be used for finding (i) the rank of $\mathbf{A}$, (ii) a row basis and a column basis of $\mathbf{A}$, (iii) a left or right inverse of $\mathbf{A}$ whenever it exists and (iv) a solution of $\mathbf{Ax} = \mathbf{b}$ whenever it exists. Some other uses can be found in the exercises.

Definition 4.4.1 Let $\mathbf{A}$ be an $m \times n$ matrix with r non-null rows where $0 \leq r \leq m$. Then $\mathbf{A}$ is said to be in *echelon form* if

(i) the first r rows of $\mathbf{A}$ are non-null (and the last $m-r$ rows are null),

(ii) let the first non-zero element in the i-th row occur in the p_i-th position for $i = 1, \ldots, r$; then

$$p_1 < p_2 < \cdots < p_r \tag{4.4.1}$$

(iii) $a_{ip_i} = 1$ for $i = 1, \ldots, r$.

A typical example of a matrix in echelon form is

$$\begin{bmatrix} 0 & 1 & * & * & * & * \\ 0 & 0 & 0 & 1 & * & * \\ 0 & 0 & 0 & 0 & 1 & * \\ 0 & 0 & 0 & 0 & 0 & 0 \end{bmatrix} \tag{4.4.2}$$

where the elements denoted by $*$ are arbitrary. Note that, here, $r = 3$, $p_1 = 2$, $p_2 = 4$ and $p_3 = 5$.

Theorem 4.4.2 The rank of a matrix in echelon form is the number of non-null rows in it.

Proof Let $\mathbf{A}$ be as in *Definition* 4.4.1. We will show that the first r rows of $\mathbf{A}$ are linearly independent. Let $\alpha_1 \mathbf{A}_{1*} + \cdots + \alpha_r \mathbf{A}_{r*} = \mathbf{0}$. Comparing the p_1-th elements we get $\alpha_1 = 0$. Next comparing the p_2-th elements we get $\alpha_2 = 0$. Proceeding like this we see that all α_i's are 0, so the first r rows of $\mathbf{A}$ are linearly independent. ∎

Theorem 4.4.3 Every matrix can be reduced to a matrix in echelon form by a finite sequence of elementary row operations.

Proof Let $\mathbf{A}$ be an $m \times n$ matrix. If $\mathbf{A} = \mathbf{0}$, $\mathbf{A}$ itself is in echelon form. So let $\mathbf{A} \neq \mathbf{0}$.

Let the first non-null column of $\mathbf{A}$ be the p_1-th. Then by interchanging the first row and some other row, if necessary, make a_{1p_1} non-zero. (We will use $\mathbf{A}$ to denote the matrix at every stage, for convenience.) Now sweep out the p_1-th column using a_{1p_1} as the pivot. Then the matrix is of the form

$$\mathbf{A} = \begin{bmatrix} \mathbf{0} & 1 & \mathbf{x}^{\mathrm{T}} \\ \mathbf{0} & \mathbf{0} & \mathbf{B} \end{bmatrix}$$

where the $\mathbf{0}$ at top left is of order $1 \times (p_1 - 1)$. If $\mathbf{B} = \mathbf{0}$ then the current matrix is in echelon form. If $\mathbf{B} \neq \mathbf{0}$, let the first non-null column of the $[\mathbf{0} \quad \mathbf{0} \quad \mathbf{B}]$ be the p_2-th. Clearly, $p_1 < p_2$. Make $a_{2p_2} \neq 0$ by nterchanging the second row with some row below it, if necessary, and then sweep out the p_2-th column of $\mathbf{A}$ using a_{2p_2} as the pivot. Note that these operations do not alter the first $p_2 - 1$ columns of the matrix. Now the matrix is of the form

$$\mathbf{A} = \begin{bmatrix} \mathbf{0} & 1 & \mathbf{y}^{\mathrm{T}} & 0 & \mathbf{v}^{\mathrm{T}} \\ \mathbf{0} & 0 & \mathbf{0} & 1 & \mathbf{w}^{\mathrm{T}} \\ \mathbf{0} & \mathbf{0} & \mathbf{0} & \mathbf{0} & \mathbf{C} \end{bmatrix}$$

Proceeding similarly we reduce $\mathbf{A}$ to a matrix in echelon form in at most m steps. ∎

We now formulate the reduction procedure given in the proof of the preceding theorem as an algorithm. *In all such algorithms, we use* $\mathbf{A}$ *to denote the matrix obtained at every stage of the algorithm.*

Algorithm 4.4.4 (*Reduction to echelon form*) Let $\mathbf{A}$ be an $m \times n$ matrix.

Step 1 Set $i = 1$, $j = 1$ and $\ell = 1$.

Step 2 Check if $a_{ij} \neq 0$. If yes, go to *Step* 3. If no, go to *Step* 4.

Step 3 Interchange the ℓ-th and i-th rows and sweep out the j-th column with the (ℓ, j)-th element as the pivot. Increase ℓ by 1 and go to *Step* 5.

Step 4 If $i = m$, go to *Step* 5. Otherwise increase i by 1 and go to *Step* 2.

Step 5 If $\ell = m + 1$ or $j = n$ go to *Step* 6. Otherwise increase j by 1, put $i = \ell$ and go to *Step* 2.

Step 6 Stop. The matrix is in echelon form.

At any stage of this algorithm, $\ell - 1$ is the number of sweep-outs performed till then. Thus the final value of $\ell - 1$ gives the rank of the matrix.

Example 4.4.5 We apply the preceding algorithm to the matrix

$$\begin{bmatrix} 0 & 0 & 0 & 0 & 5 \\ 0 & 2 & -1 & 3 & -2 \\ 0 & -4 & 2 & -6 & 4 \\ 0 & 6 & -3 & 8 & 1 \end{bmatrix}$$

to reduce it to echelon form. We start with $i = j = \ell = 1$. The first time we get $a_{ij} \neq 0$ is when $i = j = 2$. So we interchange the first and second rows and then sweep out the second column with the $(1,2)$-th element as the pivot. The matrix becomes

$$\begin{bmatrix} 0 & 1 & -\frac{1}{2} & \frac{3}{2} & -1 \\ 0 & 0 & 0 & 0 & 5 \\ 0 & 0 & 0 & 0 & 0 \\ 0 & 0 & 0 & -1 & 7 \end{bmatrix}$$

We now increase ℓ to 2, j to 3 and put $i = 2$. The next time we get $a_{ij} \neq 0$ is when $i = 4$ and $j = 4$. So we interchange the second and fourth rows and sweep out the fourth column with the $(2,4)$-th element as the pivot. The matrix becomes

$$\begin{bmatrix} 0 & 1 & -\frac{1}{2} & 0 & \frac{19}{2} \\ 0 & 0 & 0 & 1 & -7 \\ 0 & 0 & 0 & 0 & 0 \\ 0 & 0 & 0 & 0 & 5 \end{bmatrix}$$

We now increase ℓ to 3, j to 5 and put $i = 3$. The next time we get $a_{ij} \neq 0$ is when $i = 4$ and $j = 5$. So we interchange the third and fourth rows and sweep out the fifth column with the $(3,5)$-th element as the pivot. The matrix now becomes

$$\begin{bmatrix} 0 & 1 & -\frac{1}{2} & 0 & 0 \\ 0 & 0 & 0 & 1 & 0 \\ 0 & 0 & 0 & 0 & 1 \\ 0 & 0 & 0 & 0 & 0 \end{bmatrix}$$

We now increase ℓ to 4. Since $j = n$, the algorithm stops and the final matrix is in echelon form. Since the final value of ℓ is 4, the rank of $\mathbf{A}$ is 3. ∎

If we apply *Algorithm* 4.4.4 to a square matrix $\mathbf{A}$, we get a matrix in echelon form which is clearly upper triangular. However, if we are interested only in reducing $\mathbf{A}$ to an upper triangular matrix by elementary row operations, we do not have to perform the sweep-outs completely and *Algorithm* 4.4.4 may be simplified as

Algorithm 4.4.6 (*Reduction to upper triangular form*) Let $\mathbf{A}$ be a square matrix of order n.

Step 1 Set $i = 1$ and $j = 1$.

Step 2 If $a_{ij} \neq 0$, interchange the i-th and j-th rows and go to *Step* 4. Otherwise go to *Step* 3.

Step 3 If $i < n$, increase i by 1 and go to *Step* 2. Otherwise go to *Step* 5.

Step 4 Subtract a_{kj}/a_{jj} times the j-th row from the k-th row for $k = j+1, \ldots, n$.

Step 5 If $j < n-1$, increase j by 1, then set $i = j$ and go to *Step* 2. Otherwise, stop since the matrix is in upper triangular form.

We note that the above algorithm uses only row operations of the types: interchanging two rows and adding a scalar multiple of one row to another.

We now show how reduction to echelon form can be used to find a row basis and a column basis.

Let $\mathbf{A}$ be an $m \times n$ matrix and $\mathbf{B}$ a matrix in echelon form obtained from $\mathbf{A}$ by elementary row operations. Then clearly the non-null rows of $\mathbf{B}$ form a basis of $\mathcal{R}(\mathbf{A})$. We can also find a row basis of $\mathbf{A}$ by maintaining an array with m positions which is initially set to $(1, 2, \ldots, m)$. Whenever two rows are interchanged (while converting $\mathbf{A}$ to $\mathbf{B}$ using *Algorithm* 4.4.4), the entries in the corresponding positions of the array are also interchanged. If the final array is $(s_1, s_2, \ldots, s_m)$ and r is the rank of $\mathbf{A}$, then the s_1-th, s_2-th, $\ldots, s_r$-th rows of the original matrix $\mathbf{A}$ form a basis of $\mathcal{R}(\mathbf{A})$. For a proof of this, see *Exercise* 4.4.10.

We can find a column basis of $\mathbf{A}$ as follows. Let $r = \rho(\mathbf{A})$ and let the first non-zero entry in the i-th row of $\mathbf{B}$ occur in the p_i-th position for $i = 1, \ldots, r$. Let $\mathbf{P}$ be the transforming matrix and let $\mathbf{C}$ and $\mathbf{D}$ be the submatrices of $\mathbf{A}$ and $\mathbf{B}$ formed by the p_1-th, p_2-th, $\ldots, p_r$-th columns. Then $\rho(\mathbf{D}) = \rho(\mathbf{PC}) = \rho(\mathbf{C})$. It is easy to see that the columns of $\mathbf{D}$ are linearly independent, so the same holds for $\mathbf{C}$. Thus the p_1-th, p_2-th, $\ldots, p_r$-th columns of $\mathbf{A}$ form a column basis of $\mathbf{A}$. However, the p_1-th, p_2-th, $\ldots, p_r$-th columns of $\mathbf{B}$ do not even belong to $\mathcal{C}(\mathbf{A})$.

Let us apply the above procedures to *Example* 4.4.5. Here the array changes from $(1, 2, 3, 4)$ to $(2, 1, 3, 4)$, then to $(2, 4, 3, 1)$ and finally to $(2, 4, 1, 3)$. Since $r = 3$ it follows that the second, fourth and first rows of the initial matrix $\mathbf{A}$ form a row basis of $\mathbf{A}$. Since $p_1 = 2$, $p_2 = 4$ and $p_3 = 5$, the second, fourth and fifth columns form a column basis for the initial $\mathbf{A}$.

It is sometimes possible to reduce a matrix in echelon form further by elementary row operations. For example, in the matrix (4.4.2), we can make the $(1, 4)$-th, $(1, 5)$-th and $(2, 5)$-th entries 0 by elementary row operations, without altering the other elements. Motivated by this, we give

Definition 4.4.7 A matrix $\mathbf{A}$ is said to be in *reduced echelon form*[†] if it is in echelon form and $\mathbf{A}_{*p_i} = \mathbf{e}_i$ for $i = 1, \ldots, r$, with the notation of *Definition* 4.4.1.

Thus a matrix in echelon form is in reduced echelon form iff the elements *above* the first non-zero entry of any row are all 0 (the elements *below* are already 0). *Algorithm* 4.4.4 actually converts any matrix to one in reduced echelon form. We thus have

Theorem 4.4.8 Any matrix can be reduced to a matrix in reduced echelon form by elementary row operations.

Suppose $\mathbf{A}$ is an $n \times n$ non-singular matrix in reduced echelon form. With the notation of *Definition* 4.4.1, we then have $r = n$. Moreover, (4.4.1) gives $p_i = i$ for all i and it follows that $\mathbf{A} = \mathbf{I}$. We thus have

Theorem 4.4.9 Any $n \times n$ non-singular matrix can be reduced to $\mathbf{I}_n$ by elementary row operations.

Corollary Any non-singular matrix is a product of elementary matrices.

Proof By the theorem, $\mathbf{E}_k\mathbf{E}_{k-1}\cdots\mathbf{E}_1\mathbf{A} = \mathbf{I}$ for some elementary matrices $\mathbf{E}_1, \mathbf{E}_2, \ldots, \mathbf{E}_k$. Clearly then $\mathbf{A} = \mathbf{E}_1^{-1}\mathbf{E}_2^{-1}\cdots\mathbf{E}_k^{-1}$ and each $\mathbf{E}_i^{-1}$ is an elementary matrix. ∎

When applied to a non-singular matrix $\mathbf{A}$, *Algorithm* 4.4.4 can be simplified because, then, $p_j = j$ for all j and so we may take $\ell = j$. Thus, for a non-singular matrix, *Algorithm* 4.4.4 can be simplified to

Algorithm 4.4.10 (*Reduction of an $n \times n$ non-singular matrix $\mathbf{A}$ to $\mathbf{I}_n$*)

Step 1 Set $i = 1$ and $j = 1$.

Step 2 Check whether $a_{ij} \neq 0$. If yes, go to *Step* 3. Otherwise increase i by 1 and go to *Step* 2.

Step 3 Interchange the j-th and i-th rows and sweep out the j-th column with the (j, j)-th element as the pivot. Go to *Step* 4.

[†]Some authors refer to our 'reduced echelon form' as 'echelon form'.

Step 4 If $j = n$, stop since the matrix is reduced to $\mathbf{I}$. Otherwise increase j by 1, then put $i = j$ and go to *Step* 2.

We note that the above algorithm can in fact be applied to any square matrix $\mathbf{A}$ of order n. If, at any stage, *Step* 2 cannot be carried out properly (because $a_{ij} = 0$ and $i = n$) then $\mathbf{A}$ is singular; otherwise $\mathbf{A}$ is non-singular.

We will now explain how the above algorithm can be used to find the inverse of a non-singular matrix and to solve linear equations. Suppose a sequence $\mathcal{S}$ of elementary row operations reduces a non-singular matrix $\mathbf{A}$ of order n to $\mathbf{I}$. Then the transforming matrix $\mathbf{P}$ corresponding to $\mathcal{S}$ is $\mathbf{A}^{-1}$ since $\mathbf{PA} = \mathbf{I}$. Further, $\mathcal{S}$ transforms $\mathbf{I}$ to $\mathbf{PI} = \mathbf{P} = \mathbf{A}^{-1}$. $\mathcal{S}$ also transforms $\mathbf{b}$ to $\mathbf{A}^{-1}\mathbf{b}$ which is the solution of the system of equations $\mathbf{Ax} = \mathbf{b}$.

We can now give a nice computational procedure for finding $\mathbf{A}^{-1}$ and for solving $\mathbf{Ax} = \mathbf{b}$, where $\mathbf{A}$ is a non-singular matrix of order n. Append $\mathbf{I}_n$ and $\mathbf{b}$ to the right of $\mathbf{A}$ as shown below:

$$[\mathbf{A} : \mathbf{I}_n : \mathbf{b}] \tag{4.4.3}$$

Then reduce $\mathbf{A}$ to $\mathbf{I}$ by elementary row operations but perform all the operations on the entire matrix (4.4.3). This can be done by applying *Algorithm* 4.4.10 to $\mathbf{A}$ or *Algorithm* 4.4.4 to the entire matrix (4.4.3). The final matrix obtained will be $[\mathbf{I} : \mathbf{A}^{-1} : \mathbf{u}]$ where $\mathbf{u}$ is the solution of $\mathbf{Ax} = \mathbf{b}$. For a numerical illustration, see *Example* 5.5.2.

We next show how reduced echelon form directly yields a rank-factorization. Let $\mathbf{A}$ be an $m \times n$ matrix with rank r and $\mathbf{B}$ a matrix in reduced echelon form obtained from $\mathbf{A}$ by elementary row operations. Let the first non-zero entry in the i-th row of $\mathbf{B}$ occur in the p_i-th position for $i = 1, \ldots, r$. Let $\mathbf{C} = [\,\mathbf{A}_{*p_1} : \mathbf{A}_{*p_2} : \cdots : \mathbf{A}_{*p_r}\,]$ and let $\mathbf{D}$ be the submatrix of $\mathbf{B}$ formed by the first r rows. Let $\mathbf{P}$ be the transforming matrix. Then $\mathbf{PA}_{*p_i} = \mathbf{e}_i$ for $i = 1, \ldots, r$. So $\mathbf{P}^{-1} = [\,\mathbf{C} : \mathbf{T}\,]$ for some matrix $\mathbf{T}$. Hence

$$\mathbf{A} = \mathbf{P}^{-1}\mathbf{B} = [\,\mathbf{C} : \mathbf{T}\,]\begin{bmatrix}\mathbf{D}\\ \mathbf{0}\end{bmatrix} = \mathbf{CD}$$

Thus $(\mathbf{C}, \mathbf{D})$ is a rank-factorization of $\mathbf{A}$. Note that we had earlier proved that the columns of $\mathbf{C}$ form a column basis of $\mathbf{A}$. What we have just shown is that if $\mathbf{B}$ is in reduced echelon form then the second matrix in a rank-factorization of $\mathbf{A}$ with $\mathbf{C}$ as the first matrix is readily available.

It can be proved that two $m \times n$ matrices $\mathbf{C}$ and $\mathbf{D}$ in reduced echelon form have the same row space iff $\mathbf{C} = \mathbf{D}$, see *Exercise* 4.4.11. Since row space is not altered by elementary row operations, it follows that any given matrix $\mathbf{A}$ can be converted by elementary row operations to one and only one matrix $\mathbf{C}$ in reduced echelon form, called *the reduced echelon form of* $\mathbf{A}$. Now the next theorem follows easily.

***Theorem 4.4.11** For any two $m \times n$ matrices $\mathbf{A}$ and $\mathbf{B}$, the following statements are equivalent:

(i) $\mathcal{R}(\mathbf{A}) = \mathcal{R}(\mathbf{B})$,

(ii) the reduced echelon forms of $\mathbf{A}$ and $\mathbf{B}$ are equal,

(iii) $\mathbf{B}$ can be obtained from $\mathbf{A}$ by a finite sequence of elementary row operations,

(iv) $\mathbf{B} = \mathbf{PA}$ for some non-singular matrix $\mathbf{P}$,

(v) if $\mathbf{A}$ represents a linear transformation f from V_1 to V_2 with respect to bases $\mathcal{X}$ and $\mathcal{Y}$, $\mathbf{B}$ represents f with respect to $\mathcal{X}$ and $\mathcal{Y}'$ for some basis $\mathcal{Y}'$ of V_2.

Two $m \times n$ matrices $\mathbf{A}$ and $\mathbf{B}$ are said to be *row-equivalent* if they satisfy (i) of the preceding theorem. Clearly row-equivalence is an equivalence relation. We express the fact that any given matrix is row-equivalent to one and only one matrix in reduced echelon form by saying that reduced echelon form is a *canonical form under row-equivalence.*

All the concepts and results given in this section deal with operations on the rows of a matrix, its row space etc. For this reason, a matrix satisfying *Definition* 4.4.1 is sometimes said to be in *row-echelon form.* We leave it to the reader to work out the concepts and results for columns.

Exercises

1. Reduce each of the following to a matrix in reduced echelon form by elementary row operations and find the rank, a row basis, a column basis and a rank-factorization.

$$(a) \begin{bmatrix} 2 & 1 & 0 & 0 & 1 \\ 3 & 0 & 3 & 0 & 2 \\ 5 & 7 & -9 & 2 & 5 \end{bmatrix}, \quad (b) \begin{bmatrix} 0 & 2 & 4 & 3.0 & 0 \\ 0 & 5 & 10 & 7.5 & 0 \\ 0 & 1 & 2 & 1.5 & 4 \\ 0 & 2 & 4 & 3.0 & 2 \end{bmatrix}$$

2. Reduce each of the following to an upper triangular matrix using *Algorithm* 4.4.6. Also reduce each of them to a matrix in reduced echelon

form by *Algorithm* 4.4.4. Find the transforming matrix in each case. Express the first matrix as a product of elementary matrices.

$$(a)\begin{bmatrix} 3 & 1 & 0 \\ 7 & 5 & 2 \\ 2 & 4 & 3 \end{bmatrix}, \quad (b)\begin{bmatrix} 0 & 1 & 2 \\ 1 & 0 & 2 \\ 2 & 2 & 0 \end{bmatrix}, \quad (c)\begin{bmatrix} 3 & 1 & 2 & 4 \\ 9 & 3 & 6 & 12 \\ 2 & 0 & 0 & 2 \\ 5 & 1 & 2 & 6 \end{bmatrix}$$

3. Show that every matrix with full row rank can be reduced to $[\mathbf{I} : \mathbf{0}]$ by elementary column operations but not by elementary row operations.
4. What is the reduced echelon form of a matrix with full column rank?
5. Prove that every square matrix is a product of triangular matrices.
6. (a) Find a basis of the null space of a matrix in reduced echelon form.
   (b) Show that elementary row operations do not alter the null space.
   (c) Find a basis of the null space for each of the matrices in *Exercise* 4.4.1.
7. For each of the following non-singular matrices $\mathbf{A}$, find $\mathbf{A}^{-1}$ and solve $\mathbf{Ax} = \mathbf{b}$, where $\mathbf{b} = (2 \ 1 \ 0 \ 4)^{\mathrm{T}}$:

$$\text{(a)}\begin{bmatrix} 0 & 1 & 1 & 2 \\ 1 & 0 & 2 & 1 \\ 1 & 2 & 0 & 2 \\ 2 & 1 & 2 & 0 \end{bmatrix}, \quad \text{(b)}\begin{bmatrix} 2 & 1 & 3 & 0 \\ 3 & 0 & 2 & 5 \\ 1 & 1 & 1 & 1 \\ 2 & 1 & 1 & 3 \end{bmatrix}$$

8. Use *Algorithm* 4.4.10 to check whether the last matrix of *Exercise* 4.4.2 is non-singular.
9. Using the procedure mentioned after *Algorithm* 4.4.10, prove that the inverse of a non-singular upper triangular matrix is upper triangular.

*10. (a) Let $\mathbf{B}$ be obtained from $\mathbf{A}$ by sweeping out a column. Then prove that a set of rows including the pivotal row is linearly independent in $\mathbf{B}$ iff it is linearly independent in $\mathbf{A}$.
   (b) Using the result in (a) and *Exercise* 4.3.4, show that the procedure given after *Algorithm* 4.4.6 indeed gives a row basis.

*11. (a) Let $\mathbf{C}$ and $\mathbf{D}$ be $m \times n$ matrices in reduced echelon form. Prove that $\mathbf{C}$ and $\mathbf{D}$ have the same row space iff $\mathbf{C} = \mathbf{D}$.
   (b) Prove *Theorem* 4.4.11.

12. Let $\mathbf{A}$ be an $m \times n$ matrix with reduced echelon form $\mathbf{B}$. If $m < n$, what is the reduced echelon form of the square matrix $\left[\begin{smallmatrix}\mathbf{A}\\ \mathbf{0}\end{smallmatrix}\right]$? If $m > n$, what is the reduced echelon form of the square matrix $[\mathbf{A} : \mathbf{0}]$?

*13. Show that whether $\mathcal{R}(\mathbf{B}) \subseteq \mathcal{R}(\mathbf{A})$ can be found as follows: form $\left[\begin{smallmatrix}\mathbf{A}\\ \mathbf{B}\end{smallmatrix}\right]$ and reduce $\mathbf{A}$ to its reduced echelon form using *Algorithm* 4.4.4 but perform the column sweep-outs on $\mathbf{B}$ also. Then $\mathcal{R}(\mathbf{B}) \subseteq \mathcal{R}(\mathbf{A})$ iff $\mathbf{B}$ gets converted to $\mathbf{0}$ by the time $\mathbf{A}$ is reduced to its reduced echelon form.

*14. (a) If a matrix $\mathbf{A}$, with the first k columns linearly independent, is reduced to $\mathbf{B}$ in reduced echelon form by elementary row operations, show that $p_1 = 1,\ p_2 = 2, \ldots, p_k = k$ with usual notation. Show how this can be used to extend a linearly independent set contained in a subspace S of F^m to a basis of S.

(b) Do *Exercise* 1.5.11 using the procedure indicated in (a).

*15. Show that a matrix with m rows and rank r can be converted to one in reduced echelon form by $(m+1)r$ or fewer elementary row operations.

*16. Work out the analogues of all the results of this section for columns.

4.5 Normal form

In this section we will derive the simplest form to which a given matrix can be reduced by using both elementary row and column operations. This is a powerful theoretical tool.

Definition 4.5.1 A matrix is said to be in *normal form* if it is

$$\begin{bmatrix} \mathbf{I}_r & \mathbf{0} \\ \mathbf{0} & \mathbf{0} \end{bmatrix} \tag{4.5.1}$$

for some r. We use the convention that (4.5.1) denotes a null matrix if $r = 0$.

Clearly, the rank of the matrix (4.5.1) is r. We will now prove that any matrix of rank r can be reduced to such a matrix by elementary operations.

Theorem 4.5.2 Every matrix can be reduced to a matrix in normal form by elementary (row and column) operations.

Proof If $\mathbf{A} = \mathbf{0}$ then $\mathbf{A}$ itself is in normal form. So let $\mathbf{A} \neq \mathbf{0}$. Then by interchanging the first row with another row and the first column with another column, if necessary, we make $a_{11} \neq 0$. Then we sweep out the first column and the first row using a_{11} as the pivot. At this stage the matrix is of the form

$$\mathbf{A} = \begin{bmatrix} 1 & \mathbf{0} \\ \mathbf{0} & \mathbf{B} \end{bmatrix}$$

If $\mathbf{B} = \mathbf{0}$, the current matrix is in normal form. If $\mathbf{B} \neq \mathbf{0}$, by interchanging the second row with some later row and the second column with some later column, if necessary, we make $a_{22} \neq 0$ and sweep out the second column and the second row with a_{22} as the pivot. Note that these operations do not disturb the first row and the first column. Now the matrix is of the form

$$\mathbf{A} = \begin{bmatrix} \mathbf{I}_2 & \mathbf{0} \\ \mathbf{0} & \mathbf{C} \end{bmatrix}$$

If $\mathbf{C} = \mathbf{0}$ we are done; otherwise we proceed as before until we arrive at a matrix in normal form. ∎

Since row operations amount to pre-multiplication and column operations amount to post-multiplication by non-singular matrices, we have

Theorem 4.5.3 Let $\mathbf{A}$ be an $m \times n$ matrix of rank r. Then there exist non-singular matrices $\mathbf{P}$ of order m and $\mathbf{Q}$ of order n such that

$$\mathbf{PAQ} = \begin{bmatrix} \mathbf{I}_r & \mathbf{0} \\ \mathbf{0} & \mathbf{0} \end{bmatrix} \tag{4.5.2}$$

By taking $\mathbf{R} = \mathbf{P}^{-1}$ and $\mathbf{S} = \mathbf{Q}^{-1}$ we can rewrite (4.5.2) as

$$\mathbf{A} = \mathbf{R} \begin{bmatrix} \mathbf{I}_r & \mathbf{0} \\ \mathbf{0} & \mathbf{0} \end{bmatrix} \mathbf{S} \tag{4.5.3}$$

This representation is often more useful than (4.5.2). For instance, we can readily get a rank-factorization of $\mathbf{A}$ from (4.5.3). To see this, partition $\mathbf{R}$ and $\mathbf{S}$ as

$$\mathbf{R} = [\,\mathbf{R}_1 : \mathbf{R}_2\,] \quad \text{and} \quad \mathbf{S} = \begin{bmatrix} \mathbf{S}_1 \\ \mathbf{S}_2 \end{bmatrix}$$

where $\mathbf{R}_1$ is of order $m \times r$ and $\mathbf{S}_1$ is of order $r \times n$. Then (4.5.3) gives $\mathbf{A} = \mathbf{R}_1\mathbf{S}_1$, so $(\mathbf{R}_1, \mathbf{S}_1)$ is a rank-factorization of $\mathbf{A}$.

We shall now give a procedure for finding a representation of a given matrix $\mathbf{A}$ in the form (4.5.3). Using the procedure given in the proof of the preceding theorem, we reduce $\mathbf{A}$ to the matrix (4.5.1), where $r = \rho(\mathbf{A})$. Note that the reduction is done in r stages. At the k-th stage, we perform (i) an interchange of two rows, (ii) an interchange of two columns, (iii) sweep-out of the k-th column with a_{kk} as the pivot and (iv) sweep-out of the k-th row with a_{kk} as the pivot. Let $\mathbf{E}_k$, $\mathbf{F}_k$, $\mathbf{U}_k$ and $\mathbf{V}_k$ be the corresponding transforming matrices. (If any of the four operations are not needed, we take the corresponding transforming matrices to be $\mathbf{I}$.) Then (4.5.2) holds with

$$\mathbf{P} = \mathbf{U}_r\mathbf{E}_r\mathbf{U}_{r-1}\mathbf{E}_{r-1}\cdots\mathbf{U}_1\mathbf{E}_1 \quad \text{and} \quad \mathbf{Q} = \mathbf{F}_1\mathbf{V}_1\mathbf{F}_2\mathbf{V}_2\cdots\mathbf{F}_r\mathbf{V}_r$$

Clearly then,

$$\mathbf{P}^{-1} = \mathbf{I}_m \mathbf{E}_1^{-1}\mathbf{U}_1^{-1} \cdots \mathbf{E}_r^{-1}\mathbf{U}_r^{-1} \quad \text{and} \quad \mathbf{Q}^{-1} = \mathbf{V}_r^{-1}\mathbf{F}_r^{-1} \cdots \mathbf{V}_1^{-1}\mathbf{F}_1^{-1}\mathbf{I}_n$$

Thus $\mathbf{P}^{-1}$ is obtained by performing the *column operations* corresponding to the transforming matrices $\mathbf{E}_1^{-1}, \mathbf{U}_1^{-1}, \ldots, \mathbf{E}_r^{-1}, \mathbf{U}_r^{-1}$, in that order, on $\mathbf{I}_m$. Similarly $\mathbf{Q}^{-1}$ is obtained by performing the *row operations* corresponding to the transforming matrices $\mathbf{F}_1^{-1}, \mathbf{V}_1^{-1}, \ldots, \mathbf{F}_r^{-1}, \mathbf{V}_r^{-1}$, in that order, on $\mathbf{I}_n$. We thus maintain two matrices $\mathbf{R}$ and $\mathbf{S}$ besides $\mathbf{A}$. We initially put $\mathbf{R} = \mathbf{I}_m$ and $\mathbf{S} = \mathbf{I}_n$. Whenever we interchange two rows (resp. columns) of $\mathbf{A}$, we interchange the corresponding columns (resp. rows) of $\mathbf{R}$ (resp. $\mathbf{S}$). When we sweep out the k-th column of $\mathbf{A}$, we have to postmultiply $\mathbf{R}$ by $\mathbf{U}_k^{-1}$ or, as seen from *Exercise* 4.3.6, replace the k-th column of $\mathbf{R}$ by $\mathbf{R}\mathbf{A}_{*k}$. Similarly, when we sweep out the k-th row of $\mathbf{A}$, we premultiply $\mathbf{S}$ by $\mathbf{V}_k^{-1}$ or, equivalently, we replace the k-th row of $\mathbf{S}$ by $\mathbf{A}_{k*}\mathbf{S}$. If $\mathbf{R}$ and $\mathbf{S}$ are the matrices obtained at the end and $\mathbf{A}$ is the initial matrix, then (4.5.3) holds. We present the above procedure as

Algorithm 4.5.4 (*Reduction to normal form)* Given: an $m \times n$ matrix.

Step 1 Set $\mathbf{A}$ = the given matrix, $\mathbf{R} = \mathbf{I}_m$ and $\mathbf{S} = \mathbf{I}_n$. Also set $i = 1$, $j = 1$ and $k = 0$.

Step 2 Check whether $a_{ij} = 0$. If yes, go to *Step* 3. Otherwise go to *Step* 5.

Step 3 If $i < m$, increase i by 1 and go to *Step* 2. Otherwise go to *Step* 4.

Step 4 If $j < n$, increase j by 1, set $i = k + 1$ and go to *Step* 2. Otherwise go to *Step* 10.

Step 5 Increase k by 1. Interchange the k-th and i-th rows of $\mathbf{A}$ and the k-th and j-th columns of $\mathbf{A}$.

Step 6 Interchange the k-th and i-th columns of $\mathbf{R}$ and the k-th and j-th rows of $\mathbf{S}$.

Step 7 Replace the k-th column of $\mathbf{R}$ by $\mathbf{R}\mathbf{A}_{*k}$ and then sweep out the k-th column of $\mathbf{A}$ with a_{kk} as the pivot.

Step 8 Replace the k-th row of $\mathbf{S}$ by $\mathbf{A}_{k*}\mathbf{S}$ and then sweep out the k-th row of $\mathbf{A}$ with a_{kk} as the pivot.

Step 9 If $k = \min(m, n)$ go to *Step* 10. Otherwise set $i = k + 1$, $j = k + 1$ and go to *Step* 2.

Step 10 Stop. Equation (4.5.3) holds for the final $\mathbf{R}$ and $\mathbf{S}$ and the initial $\mathbf{A}$, with r equal to the final k.

We illustrate the use of the preceding algorithm with a numerical example. Let

$$\mathbf{A} = \begin{bmatrix} 0 & 2 & 0 & 4 & -2 \\ 0 & 1 & 2 & 1 & 2 \\ 0 & 2 & -4 & 6 & -8 \\ 0 & 4 & 6 & 5 & 4 \end{bmatrix} \tag{4.5.4}$$

We start the algorithm with $\mathbf{I}_4$ as $\mathbf{R}$ and $\mathbf{I}_5$ as $\mathbf{S}$. We also put $i = 1$, $j = 1$ and $k = 0$. We get a non-zero element when $i = 1$ and $j = 2$. So we increase k to 1, interchange the first and second columns of $\mathbf{A}$ (no row interchange is needed since $k = i$) and interchange the first and second rows of $\mathbf{S}$. We then replace the first column of $\mathbf{R}$ by $\mathbf{RA}_{*1} = (2, 1, 2, 4)^{\mathrm{T}}$ and sweep out the first column of $\mathbf{A}$ with a_{11} as the pivot. At this stage, we have

$$\mathbf{A} = \begin{bmatrix} 1 & 0 & 0 & 2 & -1 \\ 0 & 0 & 2 & -1 & 3 \\ 0 & 0 & -4 & 2 & -6 \\ 0 & 0 & 6 & -3 & 8 \end{bmatrix}, \quad \mathbf{R} = \begin{bmatrix} 2 & 0 & 0 & 0 \\ 1 & 1 & 0 & 0 \\ 2 & 0 & 1 & 0 \\ 4 & 0 & 0 & 1 \end{bmatrix}, \quad \mathbf{S} = \begin{bmatrix} 0 & 1 & 0 & 0 & 0 \\ 1 & 0 & 0 & 0 & 0 \\ 0 & 0 & 1 & 0 & 0 \\ 0 & 0 & 0 & 1 & 0 \\ 0 & 0 & 0 & 0 & 1 \end{bmatrix}$$

Now we replace the first row of $\mathbf{S}$ by $\mathbf{A}_{1*}\mathbf{S} = (0, 1, 0, 2, -1)$ and sweep out the first row of $\mathbf{A}$ with a_{11} as the pivot (note that the only change in $\mathbf{A}$ is that a_{1j} is replaced by 0 for all $j \neq 1$). This is the end of Stage 1. Then we set $i = 2$, $j = 2$ and go to *Step* 2. We get a non-zero element when $i = 2$ and $j = 3$. We then increase k to 2, interchange the second and third columns of $\mathbf{A}$ and interchange the second and third rows of $\mathbf{S}$. We then replace the second column of $\mathbf{R}$ by $\mathbf{RA}_{*2} = (0, 2, -4, 6)^{\mathrm{T}}$ and sweep out the second column of $\mathbf{A}$ with a_{22} as the pivot. Next we replace the second row of $\mathbf{S}$ by $\mathbf{A}_{2*}\mathbf{S} = (0, 0, 1, -\frac{1}{2}, \frac{3}{2})$ and sweep out the second row of $\mathbf{A}$ with a_{22} as the pivot. This completes the second stage and we have

$$\mathbf{A} = \begin{bmatrix} 1 & 0 & 0 & 0 & 0 \\ 0 & 1 & 0 & 0 & 0 \\ 0 & 0 & 0 & 0 & 0 \\ 0 & 0 & 0 & 0 & -1 \end{bmatrix}, \quad \mathbf{R} = \begin{bmatrix} 2 & 0 & 0 & 0 \\ 1 & 2 & 0 & 0 \\ 2 & -4 & 1 & 0 \\ 4 & 6 & 0 & 1 \end{bmatrix}, \quad \mathbf{S} = \begin{bmatrix} 0 & 1 & 0 & 2 & -1 \\ 0 & 0 & 1 & -\frac{1}{2} & \frac{3}{2} \\ 1 & 0 & 0 & 0 & 0 \\ 0 & 0 & 0 & 1 & 0 \\ 0 & 0 & 0 & 0 & 1 \end{bmatrix}$$

We now set $i = 3$, $j = 3$ and go to *Step* 2. We get a non-zero element when $i = 4$ and $j = 5$. We increase k to 3, interchange the third and fourth rows of $\mathbf{A}$, the third and fifth columns of $\mathbf{A}$, the third and fourth columns of $\mathbf{R}$ and the third and fifth rows of $\mathbf{S}$. We then replace the third column of $\mathbf{R}$ by $\mathbf{R}\mathbf{A}_{*3} = (0, 0, 0, -1)^{\mathrm{T}}$ and sweep out the third column of $\mathbf{A}$ with a_{33} as the pivot. The algorithm now stops with

$$\mathbf{A} = \begin{bmatrix} 1 & 0 & 0 & 0 & 0 \\ 0 & 1 & 0 & 0 & 0 \\ 0 & 0 & 1 & 0 & 0 \\ 0 & 0 & 0 & 0 & 0 \end{bmatrix}, \ \mathbf{R} = \begin{bmatrix} 2 & 0 & 0 & 0 \\ 1 & 2 & 0 & 0 \\ 2 & -4 & 0 & 1 \\ 4 & 6 & -1 & 0 \end{bmatrix}, \ \mathbf{S} = \begin{bmatrix} 0 & 1 & 0 & 2 & -1 \\ 0 & 0 & 1 & -\frac{1}{2} & \frac{3}{2} \\ 0 & 0 & 0 & 0 & 1 \\ 0 & 0 & 0 & 1 & 0 \\ 1 & 0 & 0 & 0 & 0 \end{bmatrix}$$

We now have (4.5.3) with $\mathbf{R}$ and $\mathbf{S}$ as above, $\mathbf{A}$ given by (4.5.4) and $r = 3$. Notice that $(\mathbf{R}_1, \mathbf{S}_1)$ is a rank-factorization of the matrix (4.5.4), where $\mathbf{R}_1$ is the submatrix of $\mathbf{R}$ formed by the first three columns and $\mathbf{S}_1$ is the submatrix of $\mathbf{S}$ formed by the first three rows.

We have studied in *Theorem* 4.4.11 when two matrices can be obtained from each other by elementary row operations. We now give the corresponding result when both row and column operations are allowed.

Theorem 4.5.5 Let $\mathbf{A}$ and $\mathbf{B}$ be $m \times n$ matrices. Then the following statements are equivalent:

(i) $\mathbf{B}$ can be obtained from $\mathbf{A}$ by elementary operations,

(ii) $\mathbf{B} = \mathbf{PAQ}$ for some non-singular matrices $\mathbf{P}$ and $\mathbf{Q}$,

(iii) $\rho(\mathbf{A}) = \rho(\mathbf{B})$,

(iv) if $\mathbf{A}$ represents a linear transformation f from V_1 to V_2 with respect to some bases $\mathcal{X}$ and $\mathcal{Y}$ then $\mathbf{B}$ represents f with respect to some bases $\mathcal{X}'$ of V_1 and $\mathcal{Y}'$ of V_2.

Proof That (i) implies (ii) and (ii) implies (iii) are easy to prove. Given (iii), we can go from $\mathbf{A}$ to $\mathbf{B}$ *via* the matrix (4.5.1), where $r = \rho(\mathbf{A})$. Hence (iii) implies (i). That (ii) is equivalent to (iv) follows from *Theorem* 3.10.3. ∎

We recall that $\mathbf{A}$ and $\mathbf{B}$ are said to be equivalent if (ii) of the preceding theorem holds. Clearly normal form is a canonical form under equivalence in the sense that any given matrix is equivalent to one and only one matrix in normal form. Its significance for linear transformations is clear from (iv).

Exercises

1. Reduce each of the following matrices to its normal form and hence find the rank and a rank-factorization.

$$\text{(a)} \begin{bmatrix} 2 & 1 & 1 & 3 & 4 \\ 0 & 1 & -1 & 1 & 2 \\ 0 & 1 & -1 & 1 & 2 \\ 2 & 1 & 1 & 3 & 4 \end{bmatrix}, \quad \text{(b)} \begin{bmatrix} 0 & 0 & 2 & 0 & 5 \\ 0 & 0 & 0 & 1 & 3 \\ 2 & 0 & 5 & 1 & 0 \\ 0 & 2 & 2 & 4 & 1 \end{bmatrix}$$

2. Explain how normal form can be obtained from reduced echelon form.

3. Prove *Theorem* 4.5.3 as follows: consider $f : F^n \to F^m$ defined by $f(\mathbf{x}) = \mathbf{Ax}$. Extend a basis $\{\mathbf{y}_1, \mathbf{y}_2, \ldots, \mathbf{y}_r\}$ of the range of f to a basis $\mathcal{Y} = \{\mathbf{y}_1, \ldots, \mathbf{y}_r, \mathbf{y}_{r+1}, \ldots, \mathbf{y}_m\}$ of F^m. Let $f(\mathbf{x}_i) = \mathbf{y}_i$ for $i = 1, \ldots, r$ and let $\mathbf{x}_{r+1}, \ldots, \mathbf{x}_n$ form a basis of $\mathcal{K}(f)$. Show that the matrix of f with respect to $\mathcal{X} := \{\mathbf{x}_1, \ldots, \mathbf{x}_r, \mathbf{x}_{r+1}, \ldots, \mathbf{x}_n\}$ and $\mathcal{Y}$ is (4.5.1).

*4. Let $\mathbf{A}$ and $\mathbf{B}$ be square matrices of the same order such that $\rho(\mathbf{A}+\mathbf{B}) = \rho(\mathbf{A})+\rho(\mathbf{B})$. Show that there exist non-singular matrices $\mathbf{P}$ and $\mathbf{Q}$ such that $\mathbf{PAQ}$ and $\mathbf{PBQ}$ are both diagonal matrices with each diagonal entry 0 or 1. Show further that $\mathbf{PAQ} \cdot \mathbf{PBQ} = \mathbf{0}$. (Hint: use rank-factorizations for $\mathbf{A}$ and $\mathbf{B}$.)

*5. Let $\mathbf{A}$ and $\mathbf{B}$ be matrices of the same order such that $\mathcal{R}(\mathbf{B}) \subseteq \mathcal{R}(\mathbf{A})$ and $\mathcal{C}(\mathbf{B}) \subseteq \mathcal{C}(\mathbf{A})$. Let $\rho(\mathbf{A}) = r > 0$. Reduce $\mathbf{A}$ to its normal form by elementary row and column operations. Show that if the same sequence of operations is performed on $\mathbf{B}$, it gets reduced to a matrix of the form $\begin{bmatrix} \mathbf{D} & \mathbf{0} \\ \mathbf{0} & \mathbf{0} \end{bmatrix}$ where $\mathbf{D}$ is an $r \times r$ matrix. Is it necessarily true that $\mathbf{D}$ is diagonal? Justify your answer.

*6. Consider matrices over the integral domain $\mathbb{Z}$ (or $F[x]$). Show that any matrix $\mathbf{A}$ can be reduced by elementary operations of types I and III to $\mathbf{D} = \begin{bmatrix} \boldsymbol{\Delta} & \mathbf{0} \\ \mathbf{0} & \mathbf{0} \end{bmatrix}$ where $\boldsymbol{\Delta}$ is a diagonal matrix with non-zero diagonal entries. Deduce that $\mathbf{A} = \mathbf{PDQ}$ where $\mathbf{P}$ and $\mathbf{Q}$ have inverses over $\mathbb{Z}$ (or $F[x]$). (See *Theorem* 5 in *Chapter* 3 of Jacobson (1953) for a stronger result.)

4.6 Hermite canonical form

In *Section* 4.4 we saw that a non-singular matrix can be reduced to $\mathbf{I}$ by elementary row operations and that such a reduction is useful in solving linear equations. In this section we show that a square singular matrix

$\mathbf{A}$ can be reduced to a simple form known as Hermite canonical form which virtually serves the same purpose in solving linear equations as we shall see in the next chapter.

Definition 4.6.1 A square matrix $\mathbf{H}$ of order n is said to be in *Hermite canonical form* (HCF for short) if

(i) $\mathbf{H}$ is upper triangular,
(ii) each diagonal entry of $\mathbf{H}$ is 0 or 1,
(iii) the i-th row of $\mathbf{H}$ is null if $h_{ii} = 0$ and
(iv) the i-th column of $\mathbf{H}$ is $\mathbf{e}_i$ if $h_{ii} = 1$.

A typical example of a matrix in HCF is

$$\begin{bmatrix} 1 & * & 0 & 0 & * \\ 0 & 0 & 0 & 0 & 0 \\ 0 & 0 & 1 & 0 & * \\ 0 & 0 & 0 & 1 & * \\ 0 & 0 & 0 & 0 & 0 \end{bmatrix} \tag{4.6.1}$$

where the elements denoted by $*$ are arbitrary.

Theorem 4.6.2 Let $\mathbf{H}$ be an $n \times n$ matrix in HCF. Then the rank of $\mathbf{H}$ is the number of 1's on the diagonal of $\mathbf{H}$. Further, if the 1's on the diagonal occur in the p_1-th, $\ldots, p_r$-th positions then the p_1-th, $\ldots, p_r$-th columns of $\mathbf{H}$ form a column basis and

$$\mathcal{C}(\mathbf{H}) = \{(x_1, \ldots, x_n)^{\mathrm{T}} : x_i = 0 \text{ whenever } i \neq p_1, \ldots, p_r\} \tag{4.6.2}$$

Proof Since $\mathbf{H}$ has $n-r$ null rows, $\rho(\mathbf{H}) \leq r$. Since the p_i-th column of $\mathbf{H}$ is $\mathbf{e}_{p_i}$ for $i = 1, \ldots, r$, the p_1-th, $\ldots, p_r$-th columns of $\mathbf{H}$ are linearly independent and $\rho(\mathbf{H}) = r$ follows. Since the LHS of (4.6.2) contains the RHS and they have the same dimension, equality follows. ∎

Lemma Let $\mathbf{H}$ be an $n \times n$ matrix in HCF with rank r. Then there exists a permutation matrix $\mathbf{P}$ such that $\mathbf{P}^{\mathrm{T}}\mathbf{H}\mathbf{P}$ is of the form

$$\mathbf{P}^{\mathrm{T}}\mathbf{H}\mathbf{P} = \begin{bmatrix} \mathbf{I}_r & \mathbf{B} \\ \mathbf{0} & \mathbf{0} \end{bmatrix} \tag{4.6.3}$$

Proof Let the 1's on the diagonal of $\mathbf{H}$ occur in the p_1-th, $\ldots, p_r$-th positions. Let $\mathbf{P}$ be a permutation matrix with $\mathbf{e}_{p_1}, \ldots, \mathbf{e}_{p_r}$ as the first r columns. Then it is easy to see that the last $n-r$ rows of $\mathbf{P}^{\mathrm{T}}\mathbf{H}$ are null, so (4.6.3) holds. (The reader may try this out on the matrix (4.6.1).) ∎

Theorem 4.6.3 A matrix in HCF is idempotent.

Proof Let $\mathbf{H}$ be as in (4.6.2). Since $\mathbf{P}^{-1} = \mathbf{P}^{\mathrm{T}}$, we get

$$\mathbf{H}^2 = \mathbf{P}\begin{bmatrix} \mathbf{I}_r & \mathbf{B} \\ 0 & 0 \end{bmatrix}\mathbf{P}^{\mathrm{T}}\mathbf{P}\begin{bmatrix} \mathbf{I}_r & \mathbf{B} \\ 0 & 0 \end{bmatrix}\mathbf{P}^{\mathrm{T}} = \mathbf{P}\begin{bmatrix} \mathbf{I}_r & \mathbf{B} \\ 0 & 0 \end{bmatrix}\mathbf{P}^{\mathrm{T}} = \mathbf{H}. \quad \blacksquare$$

Theorem 4.6.4 Any square matrix can be reduced to a matrix in HCF by elementary row operations.

Proof Let $\mathbf{A}$ be an $n \times n$ matrix. For the purpose of this proof we shall say that the first k columns of $\mathbf{A}$ have the Hermite property if they are the first k columns of some matrix in HCF, i.e.,

(i) $a_{ij} = 0$ for $i = j+1, \ldots, n$ if $j \leq k$,

(ii) $a_{ii} = 0$ or 1 for $i = 1, 2, \ldots, k$,

(iii) $a_{ij} = 0$ for $j = 1, \ldots, k$ if $a_{ii} = 0$ and $i \leq k$,

(iv) $a_{ij} = 0$ for $i = 1, \ldots, j-1, j+1, \ldots, n$ if $a_{jj} = 1$ and $j \leq k$.

If the first column of $\mathbf{A}$ is non-null, we make $a_{11} \neq 0$ by interchanging the first row and some other row, if necessary, and sweep out the first column with a_{11} as the pivot. If the first column is null, we leave it unchanged. At this stage the first column of $\mathbf{A}$ has the Hermite property. Suppose by elementary row operations we have reduced the matrix to a stage where the first $j-1$ columns have the Hermite property. We then consider three cases.

Case 1 There exists an $i \geq j$ such that $a_{ij} \neq 0$. Then we choose such an i and interchange the j-th and i-th rows, if necessary, and sweep out the j-th column with a_{jj} as the pivot.

Case 2 $a_{ij} = 0$ for all $i \geq j$ and there exists an $\ell < j$ such that $a_{\ell j} \neq 0$ and $a_{\ell\ell} = 0$. Then we interchange the ℓ-th and j-th rows and sweep out the j-th column with a_{jj} as the pivot.

Case 3 $a_{ij} = 0$ for all $i \geq j$ and for each $\ell < j$, either $a_{\ell j} = 0$ or $a_{\ell\ell} \neq 0$. Then we leave the j-th column unchanged.

Notice that in all the three cases, the operations performed do not alter the first $j-1$ columns. Also, now, the first j columns of the matrix have the Hermite property. Proceeding thus, we reduce the given matrix to one in HCF in n stages. $\blacksquare$

We present the above procedure as

Algorithm 4.6.5 (*Reduction to HCF*) Let $\mathbf{A}$ be an $n \times n$ matrix.

Step 1 Set $i = 1$ and $j = 1$.

Step 2 If $a_{ij} = 0$, go to *Step* 3. Otherwise go to *Step* 8.

Step 3 If $i < n$, increase i by 1 and go to *Step* 2. Otherwise go to *Step* 4.

Step 4 If $j = 1$, go to *Step* 9. Otherwise set $i = 1$ and go to *Step* 5.

Step 5 If $a_{ij} = 0$, go to *Step* 7. Otherwise go to *Step* 6.

Step 6 If $a_{ii} = 0$, go to *Step* 8. Otherwise go to *Step* 7.

Step 7 If $i < j - 1$, increase i by 1 and go to *Step* 5. Otherwise go to *Step* 9.

Step 8 Interchange the i-th and j-th rows and sweep out the j-th column with a_{jj} as the pivot.

Step 9 If $j < n$, increase j by 1, set $i = j$ and go to *Step 2*. Otherwise stop, for, the current matrix is in HCF.

We mention that if $\mathbf{A}$ is non-singular, then the *Steps* 4 through 7 of the above algorithm will never be used and the algorithm reduces to *Algorithm* 4.4.10.

Example 4.6.6 We will reduce the matrix

$$\mathbf{A} = \begin{bmatrix} 0 & 2 & 4 & 3 \\ 0 & 1 & 2 & 4 \\ 2 & 3 & 8 & 2 \\ 0 & 4 & 8 & 16 \end{bmatrix}$$

to a matrix in HCF using the preceding algorithm. Setting $j = 1$, we get $a_{ij} \neq 0$ when $i = 3$. So we interchange the first and third rows and sweep out the first column with a_{11} as the pivot. The resulting matrix is

$$\begin{bmatrix} 1 & \frac{3}{2} & 4 & 1 \\ 0 & 1 & 2 & 4 \\ 0 & 2 & 4 & 3 \\ 0 & 4 & 8 & 16 \end{bmatrix}$$

We now increase j to 2 and sweep out the second column with a_{22} as the pivot to get

$$\begin{bmatrix} 1 & 0 & 1 & -5 \\ 0 & 1 & 2 & 4 \\ 0 & 0 & 0 & -5 \\ 0 & 0 & 0 & 0 \end{bmatrix}$$

We next increase j to 3. We find $a_{33} = a_{43} = 0$. Moreover, $a_{13} \neq 0$ but a_{11} is also non-zero. Also, $a_{23} \neq 0$ but a_{22} is also non-zero. So we leave the third column unchanged and increase j to 4.

We find $a_{44} = 0$. Moreover, $a_{14} \neq 0$ but a_{11} is also non-zero; similarly $a_{24} \neq 0$ but a_{22} is also non-zero. Finally we find $a_{34} \neq 0$ and $a_{33} = 0$. So we interchange the third and fourth rows and sweep out the fourth column with a_{44} as the pivot to get

$$\begin{bmatrix} 1 & 0 & 1 & 0 \\ 0 & 1 & 2 & 0 \\ 0 & 0 & 0 & 0 \\ 0 & 0 & 0 & 1 \end{bmatrix}$$

Since $j = 4$, the algorithm stops and the last matrix obtained is in HCF. ∎

Theorem 4.6.7 Any given square matrix **A** can be reduced to one and only one matrix **H** in HCF by elementary row operations.

Proof Suppose that an $n \times n$ matrix **A** can be reduced to **G** as well as to **H** by elementary row operations, where **G** and **H** are in HCF. Then **G** can be obtained from **H** by elementary row operations (*via* **A**), so there exists a non-singular matrix **P** such that **G** = **PH**. Postmultiplying by **H**, we get **GH** = **G** since **H** is idempotent. Interchanging the roles of **G** and **H**, we get **HG** = **H**. Since **G** and **H** are upper triangular, it follows that

$$g_{ii} = g_{ii}h_{ii} = h_{ii}g_{ii} = h_{ii} \quad \text{for } i = 1, \ldots, n$$

Hence by *Theorem* 4.6.2, **G** and **H** have the same column space. So there exists a matrix **Q** such that **G** = **HQ**. Premultiplying by **H**, we get **HG** = **G**. But **HG** = **H**, so **G** = **H**. ∎

There is a close connection between square matrices in reduced echelon form and matrices in HCF. If **A** is an $n \times n$ matrix in HCF with the 1's in the diagonal occurring in the p_1-th, ..., p_r-th positions then it is easy to check that the matrix obtained by permuting the rows in such a way that the p_i-th row goes to the i-th position for $i = 1, 2, \ldots, r$, is in reduced echelon form. This process can also be reversed. Using this observation, uniqueness of HCF and uniqueness of reduced echelon form can be deduced from each other, see *Exercise* 4.6.7.

The matrix **H** in *Theorem* 4.6.7 is called *the Hermite canonical form of* **A**. Clearly, Hermite canonical form is a canonical form under row-equivalence for square matrices.

Hermite canonical form as given in *Definition* 4.6.1 is with respect to rows. One can define an analogous concept with respect to columns and show that any given square matrix can be reduced to one and only one such matrix by elementary column operations.

Exercises

1. Reduce each of the following square matrices to its Hermite canonical form and hence determine the rank:

$$\text{(a)} \begin{bmatrix} 2 & 4 & 3 & 1 \\ 1 & 2 & 5 & 0 \\ 3 & 6 & 0 & 5 \\ 4 & 8 & 1 & 2 \end{bmatrix}, \quad \text{(b)} \begin{bmatrix} 1 & 2 & 3 & 2 \\ 2 & 4 & 10 & 2 \\ 4 & 8 & 16 & 6 \\ 1 & 2 & 3 & 1 \end{bmatrix}$$

2. Let $\mathbf{B}$ be the Hermite canonical form of $\mathbf{A}$. Show that $\rho(\mathbf{A})$ is the number of non-zero diagonal elements in $\mathbf{B}$. Is this conclusion true if $\mathbf{B}$ is simply an upper triangular matrix obtained from $\mathbf{A}$ by elementary row operations?
3. If $\mathbf{H}$ is a matrix in HCF, show that the non-null columns of $\mathbf{I}-\mathbf{H}$ form a basis for $\mathcal{N}(\mathbf{H})$.
4. Prove or disprove: if $\mathbf{H}_1$ and $\mathbf{H}_2$ are $n \times n$ matrices in HCF, $\mathbf{H}_1\mathbf{H}_2$ is in HCF.
5. Let $\mathbf{A}$ and $\mathbf{B}$ be real $n \times n$ matrices in HCF. Further let $a_{ii} = b_{ii}$ for all i. Show that $\frac{1}{2}(\mathbf{A}+\mathbf{B})$ is in HCF.
6. Let $\mathbf{H} \neq \mathbf{0}$ be the Hermite canonical form of $\mathbf{A}$. Let the 1's on the diagonal of $\mathbf{H}$ occur in the i_1-th,i_2-th,...,i_r-th positions. Then show that $(\mathbf{P}, \mathbf{Q})$ is a rank-factorization of $\mathbf{A}$ where $\mathbf{P}_{*k} = \mathbf{A}_{*i_k}$ and $\mathbf{Q}_{k*} = \mathbf{H}_{i_k*}$ for $k = 1, \ldots, r$. Hence write down rank-factorizations of the matrices in *Exercise* 4.6.1.
7. Deduce the results: (i) the reduced echelon form of a matrix is unique, and (ii) the HCF of a square matrix is unique, from each other. (Hint: use *Exercise* 4.4.12.)

Chapter 5
Linear equations

5.1 Introduction

Systems of linear equations occur in every branch of knowledge. We mention a few examples within Linear Algebra. Expressing $\mathbf{b}$ as a linear combination of $\mathbf{a}_1, \mathbf{a}_2, \ldots, \mathbf{a}_k$ is the same as solving $\mathbf{Ax} = \mathbf{b}$ where $\mathbf{A} = [\mathbf{a}_1 : \mathbf{a}_2 : \cdots : \mathbf{a}_k]$. The vectors $\mathbf{a}_1, \mathbf{a}_2, \ldots, \mathbf{a}_k$ are linearly dependent iff $\mathbf{Ax} = \mathbf{0}$ has a non-null solution. Solution of linear equations also plays an important role in obtaining approximate solutions of non-linear equations. In this chapter, we make a systematic study of the theoretical aspects of the solution of linear equations and give some computational procedures.

By a system of linear equations we mean a system of the form

$$\begin{array}{c} a_{11}x_1 + a_{12}x_2 + \cdots + a_{1n}x_n = b_1 \\ a_{21}x_1 + a_{22}x_2 + \cdots + a_{2n}x_n = b_2 \\ \cdots\cdots\cdots\cdots\cdots\cdots\cdots\cdots \\ a_{m1}x_1 + a_{m2}x_2 + \cdots + a_{mn}x_n = b_m \end{array} \tag{5.1.1}$$

Here the a_{ij}'s and the b_i's are given elements of the base field and x_j's are the unknowns whose values (belonging to the base field) have to be found satisfying (5.1.1). The system (5.1.1) can be written as $\mathbf{Ax} = \mathbf{b}$ where $\mathbf{A}$ is the $m \times n$ matrix $((a_{ij}))$, $\mathbf{b} = (b_1, b_2, \ldots, b_m)^{\mathrm{T}}$ and $\mathbf{x} = (x_1, x_2, \ldots, x_n)^{\mathrm{T}}$. Note that a_{ij} is the coefficient of x_j in the i-th equation and b_i is the 'right hand side' of the i-th equation. $\mathbf{A}$ is called the *matrix of coefficients*. Any vector $\mathbf{u}$ such that $\mathbf{Au} = \mathbf{b}$ is said to be a *solution* of $\mathbf{Ax} = \mathbf{b}$.

Note that the system

$$\begin{aligned} 4x_1 + 3x_2 &= 11 \\ 4x_1 - 3x_2 &= 5 \end{aligned} \tag{5.1.2}$$

has a unique solution, viz., $x_1 = 2$ and $x_2 = 1$. The system

$$\begin{aligned} 4x_1 + 3x_2 &= 11 \\ 8x_1 + 6x_2 &= 22 \end{aligned} \tag{5.1.3}$$

has more than one solution, e.g., $(2,1)^{\mathrm{T}}$ and $(0, \frac{11}{3})^{\mathrm{T}}$. In fact, for every α, $(\alpha, \frac{1}{3}(11-4\alpha))^{\mathrm{T}}$ is a solution. Finally the system

$$\begin{aligned} 4x_1 + 3x_2 &= 11 \\ 8x_1 + 6x_2 &= 20 \end{aligned} \tag{5.1.4}$$

has no solution at all. Thus the number of solutions of a system may be 0, 1 or more than 1.

The set of all solutions of $a_1x_1 + a_2x_2 + \cdots + a_nx_n = b$, with at least one a_i non-zero, is called a *hyperplane*. It is a line in F^2 when $n = 2$ and a plane in F^3 when $n = 3$. Clearly the solution set of (5.1.1) is the intersection of the hyperplanes corresponding to the m equations. We note that the two lines in (5.1.2) intersect, the two lines in (5.1.3) are coincident and the two lines in (5.1.4) are parallel.

Definition 5.1.1 A system of linear equations $\mathbf{Ax} = \mathbf{b}$ is said to be *consistent* if it has at least one solution and *inconsistent* otherwise. The system $\mathbf{Ax} = \mathbf{b}$ is said to be *homogeneous* if $\mathbf{b} = \mathbf{0}$ and *non-homogeneous* otherwise.

If a system $\mathbf{Ax} = \mathbf{b}$ is consistent, by a *general solution* we mean an expression which gives all possible solutions (and no others). Any specific solution is also called a *particular solution*. For example, $(2,1)^{\mathrm{T}}$ is a particular solution of (5.1.3) and $(\alpha, \frac{1}{3}(11-4\alpha))^{\mathrm{T}}$, where α is arbitrary, is a general solution.

Exercises

1. If a system $\mathbf{Ax} = \mathbf{b}$ of linear equations over $\mathbb{R}$ has two different solutions $\mathbf{u}$ and $\mathbf{v}$, show that there exist infinitely many solutions.
2. Let $\mathbf{A}$ be an $m \times n$ matrix. Are the following statements true?
   (a) $\mathbf{Ax} = \mathbf{b}$ has a solution for all $\mathbf{b}$ if $m < n$.
   (b) $\mathbf{Ax} = \mathbf{b}$ has at most one solution if $m > n$.
3. Show that the two equations $2x + 4y = 3$ and $x + 2y = 4$ are consistent over GF(5) but inconsistent over $\mathbb{R}$.
4. Show that the equations $2x + 4y = 1$ and $4x + 3y = 2$ have a unique solution over $\mathbb{R}$. Do they have a unique solution over GF(5)? If yes, prove it. If no, find all the solutions.

5. Let

$$\mathbf{L} = \begin{bmatrix} 1 & 0 & 0 & 0 \\ 2 & 1 & 0 & 0 \\ -3 & -1 & 1 & 0 \\ 2 & 0 & 2 & 1 \end{bmatrix}, \quad \mathbf{U} = \begin{bmatrix} 2 & 3 & 0 & 0 \\ 0 & 1 & 2 & 0 \\ 0 & 0 & 2 & 1 \\ 0 & 0 & 0 & 3 \end{bmatrix}, \quad \text{and } \mathbf{c} = \begin{bmatrix} 1 \\ 2 \\ 1 \\ 0 \end{bmatrix}$$

Find the solution of $\mathbf{Ly} = \mathbf{c}$ and, using it, the solution of $\mathbf{Ax} = \mathbf{c}$ where $\mathbf{A} = \mathbf{LU}$.

5.2 Homogeneous systems

Any homogeneous system $\mathbf{Ax} = \mathbf{0}$ is consistent since $\mathbf{0}$ is a solution. This solution is called the *trivial solution* and any other solution is called a *non-trivial solution*. Notice that a non-trivial solution has at least one component non-zero. The question 'when does a homogeneous system have a non-trivial solution?' is easily answered:

Theorem 5.2.1 Let $\mathbf{A}$ be a matrix with n columns. Then the system $\mathbf{Ax} = \mathbf{0}$ has a non-trivial solution iff $\rho(\mathbf{A}) < n$.

Proof Since $\mathbf{Ax} = x_1\mathbf{A}_{*1} + x_2\mathbf{A}_{*2} + \cdots + x_n\mathbf{A}_{*n}$, the system $\mathbf{Ax} = \mathbf{0}$ has a non-trivial solution iff the columns of $\mathbf{A}$ are linearly dependent. ∎

Theorem 5.2.2 Let $\mathbf{A}$ be a matrix with n columns. Then the set S of all solutions of the system $\mathbf{Ax} = \mathbf{0}$ is a subspace of F^n with dimension $\nu(\mathbf{A}) = n - \rho(\mathbf{A})$.

This theorem follows from *Theorem* 3.5.9 since $S = \mathcal{N}(\mathbf{A})$. The second proof of that theorem can be used to find a general solution of the system $\mathbf{Ax} = \mathbf{0}$ as shown in the following example.

Example 5.2.3 Consider the homogeneous system $\mathbf{Ax} = \mathbf{0}$ given below:

$$\begin{aligned} 2x_1 + 6x_2 + x_3 + 5x_4 &= 0 \\ x_1 + 3x_2 \qquad + 2x_4 &= 0 \\ x_1 + 3x_2 - x_3 + \ x_4 &= 0 \end{aligned} \tag{5.2.1}$$

The first and third columns of $\mathbf{A}$ form a column basis of $\mathbf{A}$ since $\mathbf{A}_{*2} = 3\mathbf{A}_{*1}$ and $\mathbf{A}_{*4} = 2\mathbf{A}_{*1} + \mathbf{A}_{*3}$. Hence $(-3, 1, 0, 0)^{\mathrm{T}}$ and $(-2, 0, -1, 1)^{\mathrm{T}}$ form a basis of the solution set of (5.2.1) and so $(-3\alpha - 2\beta, \alpha, -\beta, \beta)^{\mathrm{T}}$ is a general solution of (5.2.1) where α and β are arbitrary scalars. Notice that the third and fourth columns of $\mathbf{A}$ also form a basis of $\mathcal{C}(\mathbf{A})$ and

$\mathbf{A}_{*1} = \frac{1}{2}\mathbf{A}_{*4} - \frac{1}{2}\mathbf{A}_{*3}$, $\mathbf{A}_{*2} = \frac{3}{2}\mathbf{A}_{*4} - \frac{3}{2}\mathbf{A}_{*3}$. Hence $(1, 0, \frac{1}{2}, -\frac{1}{2})^{\mathrm{T}}$ and $(0, 1, \frac{3}{2}, -\frac{3}{2})^{\mathrm{T}}$ form another basis of the solution set of (5.2.1) and $(\gamma, \delta, \frac{\gamma}{2} + \frac{3\delta}{2}, -\frac{\gamma}{2} - \frac{3\delta}{2})^{\mathrm{T}}$ is also a general solution.

Exercises

1. Let $\mathbf{A}$ be an $m \times n$ matrix.
   (a) If $m < n$, show that $\mathbf{Ax} = \mathbf{0}$ has a non-trivial solution.
   (b) Show that $\mathbf{Ax} = \mathbf{0}$ has no non-trivial solution iff $\mathcal{R}(\mathbf{A}) = F^n$.
2. Prove or disprove: if $\mathbf{A}$ is an $m \times n$ matrix such that $m > n$, then $\mathbf{Ax} = \mathbf{0}$ has no non-trivial solution.
3. Prove or disprove: every system $\mathbf{Ax} = \mathbf{0}$, where $\mathbf{A}$ is a 2×3 matrix, has a solution $\mathbf{u}$ with $u_1 \neq 0$.
4. Obtain a basis of the solution space of the system of two linear equations: $2x_1 - 3x_2 + 5x_3 = 0$ and $x_1 - x_2 + x_3 = 0$.
5. Find a general solution of the system

$$\begin{aligned} x_1 + x_2 + 2x_3 + 3x_4 &= 0 \\ 2x_1 \qquad + 4x_3 + 4x_4 &= 0 \\ x_1 - x_2 + 2x_3 + x_4 &= 0 \\ x_1 - 2x_2 + 2x_3 \qquad &= 0 \end{aligned}$$

6. If $\mathbf{A}$ is a square matrix, show that $\mathbf{Ax} = \mathbf{0}$ has a non-trivial solution iff $\mathbf{y}^{\mathrm{T}}\mathbf{A} = \mathbf{0}$ has a non-trivial solution.
7. Let $\mathbf{A}$ and $\mathbf{B}$ be matrices with the same number of columns. Show that $\mathbf{Ax} = \mathbf{0}$ and $\begin{bmatrix}\mathbf{A}\\ \mathbf{B}\end{bmatrix}\mathbf{x} = \mathbf{0}$ have the same solution space iff $\mathcal{R}(\mathbf{B}) \subseteq \mathcal{R}(\mathbf{A})$.
8. Let $\mathbf{A}$ be an $m \times n$ matrix of rank $r < m$. Let $\mathbf{B}$ be a matrix in echelon form obtained from $\mathbf{A}$ by elementary row operations and let $\mathbf{P}$ be the transforming matrix. Show that the last $m - r$ rows of $\mathbf{P}$ form a basis of the solution space of $\mathbf{x}^{\mathrm{T}}\mathbf{A} = \mathbf{0}$.

5.3 General linear systems

In this section we shall investigate the conditions for consistency and uniqueness of a solution of a general (i.e., not necessarily homogeneous) system and establish a connection between the solution sets of non-homogeneous systems and those of homogeneous systems.

Theorem 5.3.1 The following statements about a system $\mathbf{Ax} = \mathbf{b}$ are equivalent:

(i) $\mathbf{Ax} = \mathbf{b}$ is consistent,

(ii) $\mathbf{b} \in \mathcal{C}(\mathbf{A})$,

(iii) $\rho(\mathbf{A} : \mathbf{b}) = \rho(\mathbf{A})$.

Proof The equivalence of (i) and (ii) is easy to see. Now, $\mathcal{C}(\mathbf{A}) \subseteq \mathcal{C}(\mathbf{A} : \mathbf{b})$. Given (ii), the reverse inclusion and so equality follow. Thus (ii) implies (iii). Given (iii), equality follows, so $\mathbf{b} \in \mathcal{C}(\mathbf{A})$. ∎

The matrix $[\,\mathbf{A} : \mathbf{b}\,]$ appearing in (iii) of the preceding theorem is called the *augmented matrix* of the system $\mathbf{Ax} = \mathbf{b}$. We next give an entirely different kind of characterization of consistent systems which is in the spirit of the *Duality Theorem of Linear Programming*.

Theorem 5.3.2 The system $\mathbf{Ax} = \mathbf{b}$ is consistent iff

$$\mathbf{A}^{\mathrm{T}}\mathbf{u} = \mathbf{0} \Longrightarrow \mathbf{b}^{\mathrm{T}}\mathbf{u} = 0 \tag{5.3.1}$$

Proof The *only if part* is trivial since if $\mathbf{Ay} = \mathbf{b}$ and $\mathbf{A}^{\mathrm{T}}\mathbf{u} = \mathbf{0}$ then $\mathbf{b}^{\mathrm{T}}\mathbf{u} = \mathbf{y}^{\mathrm{T}}\mathbf{A}^{\mathrm{T}}\mathbf{u} = 0$. To prove the *if part*, let (5.3.1) hold. Then clearly

$$\mathcal{N}(\mathbf{A}^{\mathrm{T}}) \subseteq \mathcal{N}\begin{bmatrix} \mathbf{A}^{\mathrm{T}} \\ \mathbf{b}^{\mathrm{T}} \end{bmatrix}$$

Since the reverse inclusion is trivial, we have equality. Taking dimensions we get $\rho(\mathbf{A}) = \rho(\mathbf{A} : \mathbf{b})$. Now the current theorem follows from the preceding theorem. ∎

The preceding theorem can be restated as: $\mathbf{Ax} = \mathbf{b}$ is inconsistent iff there exists $\mathbf{u}$ such that $\mathbf{A}^{\mathrm{T}}\mathbf{u} = \mathbf{0}$ and $\mathbf{b}^{\mathrm{T}}\mathbf{u} \neq 0$. Clearly such a $\mathbf{u}$ exists iff there exists a $\mathbf{v}$ such that $\mathbf{A}^{\mathrm{T}}\mathbf{v} = \mathbf{0}$ and $\mathbf{b}^{\mathrm{T}}\mathbf{v} = 1$. We may thus restate the theorem as:

$$\text{Exactly one of: } \mathbf{Ax} = \mathbf{b} \text{ and } \begin{bmatrix} \mathbf{A}^{\mathrm{T}} \\ \mathbf{b}^{\mathrm{T}} \end{bmatrix}\mathbf{v} = \begin{bmatrix} 0 \\ 1 \end{bmatrix} \text{ is consistent}$$

We next study the solution sets of consistent systems. Note that the solution set S of a non-homogeneous consistent system $\mathbf{Ax} = \mathbf{b}$ is not a subspace since it is not closed under addition and scalar multiplication. But we can still give a nice characterization of S.

Theorem 5.3.3 Let S be the set of all solutions of a consistent system $\mathbf{Ax} = \mathbf{b}$ and let $\mathbf{u}$ be any particular solution. Then $S = \mathbf{u} + \mathcal{N}(\mathbf{A})$.

Proof Suppose $\mathbf{v} \in S$. Then $\mathbf{A}(\mathbf{v}-\mathbf{u}) = \mathbf{b}-\mathbf{b} = \mathbf{0}$. Hence $\mathbf{v}-\mathbf{u} \in \mathcal{N}(\mathbf{A})$ and $\mathbf{v} = \mathbf{u}+(\mathbf{v}-\mathbf{u}) \in \mathbf{u}+\mathcal{N}(\mathbf{A})$.

Conversely let $\mathbf{v} \in \mathbf{u}+\mathcal{N}(\mathbf{A})$. Then $\mathbf{v} = \mathbf{u}+\mathbf{w}$ for some $\mathbf{w} \in \mathcal{N}(\mathbf{A})$. So $\mathbf{Av} = \mathbf{Au}+\mathbf{Aw} = \mathbf{b}+\mathbf{0} = \mathbf{b}$ and $\mathbf{v} \in S$. ∎

Motivated by the preceding theorem, we call $\mathbf{Ax} = \mathbf{0}$ the *homogeneous system corresponding to* (or *associated with*) $\mathbf{Ax} = \mathbf{b}$. The preceding theorem states that if $\mathbf{Ax} = \mathbf{b}$ is consistent, then adding a general solution of the corresponding homogeneous system to any particular solution of $\mathbf{Ax} = \mathbf{b}$ gives a general solution of $\mathbf{Ax} = \mathbf{b}$. For example, $(2,1)^{\mathrm{T}}$ is a particular solution of the system (5.1.3) and $(\beta, -\frac{4}{3}\beta)^{\mathrm{T}}$ is a general solution of the corresponding homogeneous system, so $(\beta+2, 1-\frac{4}{3}\beta)^{\mathrm{T}}$ is a general solution of (5.1.3).

The preceding theorem has a nice geometric interpretation. Recall that $\emptyset$, points, lines, planes etc., not necessarily containing the origin, are called *flats*. If W is a non-empty flat and $\mathbf{u}$ is a fixed vector then the *translate of* W *by* $\mathbf{u}$ is $\mathbf{u}+W$ and is a flat (of the same kind as W) parallel to W, see *Figure* 5.3.1.

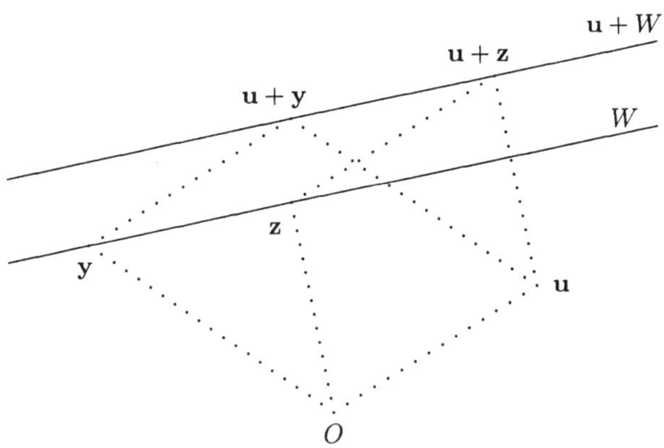

Figure 5.3.1

Clearly subspaces are the flats through the origin, so from the preceding theorem it follows that the solution set of any system $\mathbf{Ax} = \mathbf{b}$ is a flat. The converse is also true. If X is any non-empty flat and $\mathbf{u}$ is any fixed element of X then $W = -\mathbf{u}+X$ is a flat containing the origin and so is a subspace. By *Theorem* 3.7.4, W is the null space of a matrix $\mathbf{A}$ and $X = \mathbf{u} + W$ is the solution set of $\mathbf{Ax} = \mathbf{b}$ where $\mathbf{b} = \mathbf{Au}$.

Theorem 5.3.4 A consistent system $\mathbf{Ax} = \mathbf{b}$ has a unique solution iff $\mathbf{A}$ is of full column rank.

This theorem follows from the preceding theorem and can also deduced directly from definitions. The next theorem follows easily from *Theorems* 5.3.1 and 5.3.4.

Theorem 5.3.5 If $\mathbf{A}$ has n columns then the system $\mathbf{Ax} = \mathbf{b}$ has

(i) no solution iff $\rho(\mathbf{A}) < \rho(\mathbf{A} : \mathbf{b})$,

(ii) a unique solution iff $\rho(\mathbf{A}) = \rho(\mathbf{A} : \mathbf{b}) = n$,

(iii) at least two solutions iff $\rho(\mathbf{A}) = \rho(\mathbf{A} : \mathbf{b}) < n$.

We finally show that, to solve a consistent system $\mathbf{Ax} = \mathbf{b}$, it is enough to solve some r of the equations in it, where $r = \rho(\mathbf{A})$. For this, we may, without any loss of generality, assume that the submatrix of $\mathbf{A}$ formed by the first r rows and the first r columns is non-singular.

Theorem 5.3.6 Let $\mathbf{Ax} = \mathbf{b}$ be consistent, where $\mathbf{A}$ is an $m \times n$ matrix of rank r. Suppose that the submatrix of $\mathbf{A}$ formed by the first r rows and the first r columns is non-singular. Then the last $m - r$ equations in $\mathbf{Ax} = \mathbf{b}$ are redundant in the sense that the solution set does not change even if these are dropped. Moreover, all solutions of $\mathbf{Ax} = \mathbf{b}$ can be obtained by fixing $x_{r+1}, \ldots, x_n$ arbitrarily and solving the first r equations for $x_1, \ldots, x_r$.

Proof By *Theorem* 5.3.1, $\rho(\mathbf{A} : \mathbf{b}) = \rho(\mathbf{A}) = r$. Also the first r rows of $\mathbf{A}$ and so of $(\mathbf{A} : \mathbf{b})$ are linearly independent. Thus the first r rows of $(\mathbf{A} : \mathbf{b})$ form a row basis. So there exists a matrix $\mathbf{D}$ such that

$$\mathbf{A} = \begin{bmatrix} \mathbf{A}_1 \\ \mathbf{DA}_1 \end{bmatrix} \quad \text{and} \quad \mathbf{b} = \begin{bmatrix} \mathbf{b}_1 \\ \mathbf{Db}_1 \end{bmatrix}$$

where $\mathbf{A}_1$ and $\mathbf{b}_1$ have r rows each. Hence $\mathbf{Au} = \mathbf{b}$ iff $\mathbf{A}_1\mathbf{u} = \mathbf{b}_1$. This proves that the last $m-r$ equations in $\mathbf{Ax} = \mathbf{b}$ are redundant. The other conclusion follows on observing that, after fixing $x_{r+1}, \ldots, x_n$ arbitrarily, we can solve $\mathbf{A}_1\mathbf{x} = \mathbf{b}_1$ uniquely for $x_1, \ldots, x_r$ since the first r columns of $\mathbf{A}_1$ form a non-singular matrix. ∎

Example 5.3.7 Consider the system $\mathbf{Ax} = \mathbf{b}$ obtained from (5.2.1) by replacing the RHS vector by $(0, -1, -1)^{\mathrm{T}}$. Then $\rho(\mathbf{A}) = \rho[\mathbf{A} : \mathbf{b}] = 2$

and the submatrix of $\mathbf{A}$ formed by the first and second rows and the first and third columns is non-singular. So the third equation can be dropped (it is obtained by subtracting the first equation from 3 times the second). Also a general solution is obtained by fixing x_2 and x_4 arbitrarily and solving the two equations

$$\begin{aligned} 2x_1 + x_3 &= 0 - 6x_2 - 5x_4 \\ x_1 \quad &= -1 - 3x_2 - 2x_4 \end{aligned}$$

for x_1 and x_3. Thus we get the general solution $(-1 - 3x_2 - 2x_4, x_2, 2 - x_4, x_4)^{\mathrm{T}} = (-1, 0, 2, 0)^{\mathrm{T}} + x_2(-3, 1, 0, 0)^{\mathrm{T}} + x_4(-2, 0, -1, 1)^{\mathrm{T}}$, where x_2 and x_4 are arbitrary. Note that $(-1, 0, 2, 0)^{\mathrm{T}}$ is a particular solution and as x_2 and x_4 vary arbitrarily, $x_2(-3, 1, 0, 0)^{\mathrm{T}} + x_4(-2, 0, -1, 1)^{\mathrm{T}}$ is a general solution of $\mathbf{Ax} = \mathbf{0}$.

Exercises

1. Show that $\mathbf{Ax} = \mathbf{b}$ is consistent for all $\mathbf{b}$ iff $\mathbf{A}$ is of full row rank.
2. For which matrices $\mathbf{A}$ is it true that there does not exist any vector $\mathbf{b}$ such that $\mathbf{Ax} = \mathbf{b}$ has a unique solution?
3. Consider the system obtained from that in *Exercise* 5.2.4 by replacing the right hand sides of the two equations by 5 and 7 in that order. Find out whether this system is consistent and if so, find a general solution.
4. Obtain a system $\mathbf{Ax} = \mathbf{b}$ for which

$$\begin{bmatrix} 1 + 4\alpha + 3\beta \\ 2 + 3\alpha \\ 1 + 8\beta \\ \alpha + 5\beta \end{bmatrix}$$

is a general solution, where α and β are arbitrary scalars. (Hint: First get $\mathbf{A}$ using *Exercise* 3.5.12 and then get $\mathbf{b}$.)

5. (a) Let $\mathbf{Ax} = \mathbf{b}$ be consistent. Fix j. If the j-th column of $\mathbf{A}$ is a linear combination of the other columns of $\mathbf{A}$, show that for each $\alpha \in F$ there exists a solution of $\mathbf{Ax} = \mathbf{b}$ in which $x_j = \alpha$. Otherwise, show that x_j has the same value in all solutions of $\mathbf{Ax} = \mathbf{b}$. In the latter case, does it follow that $\mathbf{Ax} = \mathbf{b}$ has a unique solution?

   (b) Show that the following is true over $\mathbb{R}$ and false over GF(2): If for each j, $\mathbf{Ax} = \mathbf{0}$ has a solution $\mathbf{x}^{(j)}$ with the j-th component non-zero, then $\mathbf{Ax} = \mathbf{0}$ has a solution with all components non-zero.

   (c) Let $\mathbf{A}$ be an $m \times n$ matrix with rank $\leq n - 1$. Then show that every non-null solution of $\mathbf{Ax} = \mathbf{0}$ has all components non-zero iff the rank of every $m \times (n - 1)$ submatrix of $\mathbf{A}$ is $n - 1$.

6. Let $\mathbf{A}$ be a real $m \times n$ matrix. Show that $\mathbf{A}^T\mathbf{A}\mathbf{x} = \mathbf{A}^T\mathbf{b}$ is consistent for all $\mathbf{b} \in \mathbb{R}^m$. Show also that if $\mathbf{A}\mathbf{x} = \mathbf{b}$ is consistent, then the solution sets of the two systems are the same.
7. Consider the lines $a_1x + b_1y + c_1 = 0$ and $a_2x + b_2y + c_2 = 0$ in $\mathbb{R}^2$. Let $\mathbf{A}$ be the 2×2 matrix with (a_i, b_i) as the i-th row, $i = 1, 2$. Let $\mathbf{B}$ be the 2×3 matrix with (a_i, b_i, c_i) as the i-th row, $i = 1, 2$. Show the following:

   (a) The lines are identical iff $\rho(\mathbf{B}) = 1$.

   (b) The lines are parallel but not identical iff $\rho(\mathbf{A}) = 1$ and $\rho(\mathbf{B}) = 2$.

   (c) The lines intersect but are not identical iff $\rho(\mathbf{A}) = 2$.

8. Consider the three distinct lines $a_ix + b_iy + c_i = 0,\ i = 1, 2, 3$, in $\mathbb{R}^2$. Let $\mathbf{A}$ be the 3×3 matrix with (a_i, b_i, c_i) as the i-th row, $i = 1, 2, 3$. Show that the three lines are concurrent iff $\rho(\mathbf{A}) = 2$ and every 2×2 submatrix contained in the first two columns is non-singular.
9. If a and b are real numbers, not both zero, show that the three lines

$$\begin{aligned} ax + by - a - b &= 0 \\ bx - (a+b)y + a &= 0 \\ (a+b)x - ay - b &= 0 \end{aligned}$$

   in $\mathbb{R}^2$ are distinct and concurrent.

10. (a) Let $P_1, P_2, \ldots, P_k$ be distinct points in $\mathbb{R}^n$, P_i corresponding to $\mathbf{x}_i$, $i = 1, 2, \ldots, k$. Show that the following are equivalent:

    i. $P_1, P_2, \ldots, P_k$ are collinear (i.e., they are solutions of a system $\mathbf{A}\mathbf{x} = \mathbf{b}$ where $\mathbf{A}$ is an $(n-1) \times n$ matrix with rank $n-1$),

    ii. $\rho(\mathbf{x}_2 - \mathbf{x}_1 : \mathbf{x}_3 - x_1 : \cdots : \mathbf{x}_k - \mathbf{x}_1) \leq 1$,

    iii. $\rho \begin{bmatrix} \mathbf{x}_1 & \mathbf{x}_2 & \cdots & \mathbf{x}_k \\ 1 & 1 & \cdots & 1 \end{bmatrix} \leq 2$.

    (b) Obtain the analogous result for coplanarity.

11. Let $\mathbf{A}$ be an $m \times n$ matrix.

    (a) Show that $\mathbf{A}\mathbf{x} = \mathbf{b}$ has a solution belonging to $\mathcal{C}(\mathbf{B})$ iff $\mathbf{A}\mathbf{B}\mathbf{u} = \mathbf{b}$ is consistent.

    (b) Show that if $\mathbf{A}$ is real and $\mathbf{b} \in \mathcal{C}(\mathbf{A})$, then $\mathbf{A}\mathbf{x} = \mathbf{b}$ has a unique solution belonging to $\mathcal{C}(\mathbf{A}^T)$.

    (c) Show that for every $\mathbf{b} \in \mathcal{C}(\mathbf{A})$, $\mathbf{A}\mathbf{x} = \mathbf{b}$ has a solution belonging to $\mathcal{C}(\mathbf{A})$ iff (i) $\mathbf{A}$ is square and (ii) $\rho(\mathbf{A}) = \rho(\mathbf{A}^2)$.

12. Show that $\mathbf{c}^T \in \mathcal{R}(\mathbf{A})$ iff $\mathbf{c}^T\mathbf{u} = 0$ for all $\mathbf{u} \in \mathcal{N}(\mathbf{A})$. (Hint: Use *Theorem* 5.3.2.)

13. Let $\mathbf{Ax} = \mathbf{b}$ be consistent. Consider the new system obtained by including another equation $\mathbf{u}^{\mathrm{T}}\mathbf{x} = \beta$. If $\mathbf{u}^{\mathrm{T}} \notin \mathcal{R}(\mathbf{A})$, show that for any β, the new system is consistent but the solution sets of the old and new systems are different. If $\mathbf{u}^{\mathrm{T}} \in \mathcal{R}(\mathbf{A})$, then show that the new system is consistent iff the solution sets of the old and new systems are equal.
14. Let $\mathbf{Ax} = \mathbf{b}$ be consistent and $\mathbf{v}^{\mathrm{T}} \notin \mathcal{R}(\mathbf{A})$. Then show that the system $(\mathbf{A} + \mathbf{uv}^{\mathrm{T}})\mathbf{x} = \mathbf{b}$ is consistent. (Hint: use *Theorem* 5.3.2.)
15. Show that the system $\mathbf{AXB} = \mathbf{C}$ can be rewritten as $(\mathbf{A}\otimes\mathbf{B}^{\mathrm{T}})\mathrm{vec}(\mathbf{X}) = \mathrm{vec}(\mathbf{C})$ where $\mathrm{vec}(\mathbf{D}_{p\times q}) := (d_{11}, d_{12}, \ldots, d_{1q}, d_{21}, d_{22}, \ldots, d_{2q}, \ldots, d_{p1}, d_{p2}, \ldots, d_{pq})^{\mathrm{T}}$. (See *Exercise* 2.6.11.)

5.4 Generalized inverse of á matrix

The inverse of a non-singular matrix $\mathbf{A}$ has the property that once we know $\mathbf{A}^{-1}$, the solution to $\mathbf{Ax} = \mathbf{b}$ can simply be obtained as $\mathbf{A}^{-1}\mathbf{b}$ for any $\mathbf{b}$. Notice that $\mathbf{A}^{-1}$ does not depend on $\mathbf{b}$.

However, if $\mathbf{A}$ is a singular square matrix or a non-square matrix then $\mathbf{A}$ does not have inverse. In this section we show that given any $m \times n$ matrix $\mathbf{A}$, there exists a matrix $\mathbf{G}$ (which does not depend on $\mathbf{b}$) such that whenever $\mathbf{Ax} = \mathbf{b}$ is consistent, $\mathbf{Gb}$ is a solution to it. $\mathbf{G}$ serves virtually the same purpose as inverse in solving consistent systems and is called a generalized inverse. Generalized inverses help to give elegant solutions for many problems involving linear equations and projections.

Definition 5.4.1 Let $\mathbf{A}$ be an $m \times n$ matrix. Then $\mathbf{G}$ is said to be a *generalized inverse* of $\mathbf{A}$ (*g-inverse* for short) if $\mathbf{Gb}$ is a solution to $\mathbf{Ax} = \mathbf{b}$ whenever $\mathbf{Ax} = \mathbf{b}$ is consistent (that is, for all $\mathbf{b} \in \mathcal{C}(\mathbf{A})$). We denote a g-inverse of $\mathbf{A}$ by $\mathbf{A}^{-}$.

If $\mathbf{A}$ is an $m \times n$ matrix and $\mathbf{G}$ is a g-inverse of $\mathbf{A}$ then $\mathbf{AGb} = \mathbf{b}$ for all $\mathbf{b} \in \mathcal{C}(\mathbf{A})$, so $\mathbf{G}$ is an $n \times m$ matrix. We now prove that a g-inverse exists for every matrix.

Theorem 5.4.2 Every matrix has a g-inverse.

Proof Let $\mathbf{A}$ be an $m \times n$ matrix. If $\mathbf{A} = \mathbf{0}$ then $\mathbf{Ax} = \mathbf{b}$ is consistent iff $\mathbf{b} = \mathbf{0}$, so the null matrix of order $n \times m$ is a g-inverse of $\mathbf{A}$ (what are the other g-inverses?) Next let $\mathbf{A} \neq \mathbf{0}$. Then $\mathbf{A}$ has a rank-factorization $(\mathbf{P}, \mathbf{Q})$. Let $\mathbf{C}$ be a left inverse of $\mathbf{P}$ and $\mathbf{D}$ a right inverse of $\mathbf{Q}$ and $\mathbf{G} = \mathbf{DC}$. Now let $\mathbf{b} \in \mathcal{C}(\mathbf{A})$. Then $\mathbf{b} = \mathbf{Ay}$ for some $\mathbf{y}$. Notice that $\mathbf{A} = \mathbf{PQ}$, $\mathbf{QD} = \mathbf{I}$ and $\mathbf{CP} = \mathbf{I}$. So

$$\mathbf{AGb} = \mathbf{PQDCPQy} = \mathbf{PQy} = \mathbf{b}$$

Thus $\mathbf{Gb}$ is a solution of $\mathbf{Ax} = \mathbf{b}$ for all $\mathbf{b} \in \mathcal{C}(\mathbf{A})$, so $\mathbf{G}$ is a g-inverse of $\mathbf{A}$. ∎

We now give several useful characterizations of g-inverse.

Theorem 5.4.3 For any two matrices $\mathbf{A}$ and $\mathbf{G}$ the following statements are equivalent:

(i) $\mathbf{G}$ is a g-inverse of $\mathbf{A}$,

(ii) $\mathbf{AGA} = \mathbf{A}$,

(iii) $\mathbf{AG}$ is idempotent and $\rho(\mathbf{AG}) = \rho(\mathbf{A})$,

(iv) $\mathbf{GA}$ is idempotent and $\rho(\mathbf{GA}) = \rho(\mathbf{A})$.

Proof Let $\mathbf{A}$ be an $m \times n$ matrix. If any one of (i), (ii), (iii) and (iv) holds then it is easy to see that $\mathbf{G}$ is an $n \times m$ matrix.

We first establish the equivalence of (i) and (ii). Given (i), the system $\mathbf{Ax} = \mathbf{A}_{*j}$ is consistent, so $\mathbf{AGA}_{*j} = \mathbf{A}_{*j}$ for $j = 1, \ldots, n$ and (ii) follows. Given (ii), let $\mathbf{Ax} = \mathbf{b}$ be consistent and let $\mathbf{y}$ be a solution. Then $\mathbf{AGb} = \mathbf{AGAy} = \mathbf{Ay} = \mathbf{b}$ and (i) follows. This proves the equivalence of (i) and (ii).

We next prove that (ii) and (iii) are equivalent. Given (ii), $\mathbf{AGAG} = \mathbf{AG}$, so $\mathbf{AG}$ is idempotent. Also

$$\rho(\mathbf{AG}) \leq \rho(\mathbf{A}) = \rho(\mathbf{AGA}) \leq \rho(\mathbf{AG})$$

So $\rho(\mathbf{AG}) = \rho(\mathbf{A})$. Thus (ii) implies (iii). Given (iii), by *Theorem* 3.5.6 there exists $\mathbf{C}$ such that $\mathbf{AGC} = \mathbf{A}$. So $\mathbf{A} = \mathbf{AGC} = \mathbf{AGAGC} = \mathbf{AGA}$. Thus (ii) and (iii) are equivalent. That (ii) and (iv) are equivalent can be proved similarly. ∎

Corollary If $\mathbf{G}$ is a g-inverse of $\mathbf{A}$ then $\mathbf{AG}$ is the projector into $\mathcal{C}(\mathbf{A})$ along $\mathcal{N}(\mathbf{AG})$.

We now show that a g-inverse can be used not only to get a solution of a consistent system but also to check consistency and to get all solutions if the system is consistent.

Theorem 5.4.4 Let $\mathbf{G}$ be a g-inverse of $\mathbf{A}$. Then the system $\mathbf{Ax} = \mathbf{b}$ is consistent iff $\mathbf{AGb} = \mathbf{b}$.

Proof If $\mathbf{AGb} = \mathbf{b}$ then the system $\mathbf{Ax} = \mathbf{b}$ is clearly consistent. The converse follows from the definition of g-inverse. ∎

Corollary Let $\mathbf{G}$ be a g-inverse of $\mathbf{A}$. Then $\mathcal{C}(\mathbf{A}) \supseteq \mathcal{C}(\mathbf{B})$ iff $\mathbf{AGB} = \mathbf{B}$.

This corollary follows from the fact that $\mathcal{C}(\mathbf{A}) \supseteq \mathcal{C}(\mathbf{B})$ iff $\mathbf{Ax} = \mathbf{B}_{*j}$ is consistent for all j. Notice that the condition given in *Theorem* 3.5.1, namely, $\mathbf{B} = \mathbf{AC}$ for some matrix $\mathbf{C}$, is not easily verifiable. However, the condition in the preceding corollary specifies a choice of $\mathbf{C}$ so that the condition can be easily verified.

Theorem 5.4.5 Let $\mathbf{A}$ be an $m \times n$ matrix and let $\mathbf{G}$ be a g-inverse of $\mathbf{A}$. Then

(i) a general solution of $\mathbf{Ax} = \mathbf{0}$ is $(\mathbf{I} - \mathbf{GA})\mathbf{z}$ where $\mathbf{z}$ is an arbitrary vector in F^n,

(ii) if $\mathbf{Ax} = \mathbf{b}$ is consistent then $\mathbf{Gb} + (\mathbf{I} - \mathbf{GA})\mathbf{z}$ is a general solution of $\mathbf{Ax} = \mathbf{b}$, where $\mathbf{z}$ is an arbitrary vector in F^n,

(iii) if $\mathbf{y}^{\mathrm{T}}\mathbf{A} = \mathbf{c}^{\mathrm{T}}$ is consistent then $\mathbf{c}^{\mathrm{T}}\mathbf{G} + \mathbf{w}^{\mathrm{T}}(\mathbf{I} - \mathbf{AG})$ is a general solution, where $\mathbf{w}$ is an arbitrary vector in F^m.

Proof Since $\mathbf{AGA} = \mathbf{A}$ it easily follows that $\mathcal{N}(\mathbf{A}) = \mathcal{N}(\mathbf{GA})$. But $\mathbf{GA}$ is idempotent, so by *Theorem* 3.7.4, $\mathcal{N}(\mathbf{GA}) = \mathcal{C}(\mathbf{I} - \mathbf{GA})$. Thus $\mathcal{N}(\mathbf{A}) = \mathcal{C}(\mathbf{I} - \mathbf{GA})$ and (i) is proved. Now (ii) and (iii) follow from *Theorem* 5.3.3. ∎

We now identify the g-inverses of full row rank and full column rank matrices.

Theorem 5.4.6 If $\mathbf{A}$ is of full column rank, $\mathbf{G}$ is a g-inverse of $\mathbf{A}$ iff $\mathbf{G}$ is a left inverse of $\mathbf{A}$. If $\mathbf{A}$ is of full row rank then $\mathbf{G}$ is a g-inverse of $\mathbf{A}$ iff $\mathbf{G}$ is a right inverse of $\mathbf{A}$. If $\mathbf{A}$ is non-singular then $\mathbf{A}^{-1}$ is the unique g-inverse of $\mathbf{A}$.

Proof We will prove only the first statement since the proof of the second is similar. If $\mathbf{G}$ is a left inverse of $\mathbf{A}$ then $\mathbf{GA} = \mathbf{I}$, so $\mathbf{AGA} = \mathbf{A}$. Conversely if $\mathbf{A}$ is of full column rank and $\mathbf{G}$ is a g-inverse of $\mathbf{A}$ then $\mathbf{A}$ has a left inverse $\mathbf{C}$ and $\mathbf{AGA} = \mathbf{A}$, so $\mathbf{CAGA} = \mathbf{CA}$, i.e., $\mathbf{GA} = \mathbf{I}$ and $\mathbf{G}$ is a left inverse of $\mathbf{A}$. ∎

The next result shows how we can reduce the problem of finding a g-inverse of an $m \times n$ real or complex matrix to that of finding a g-inverse of a $k \times k$ matrix where $k = \min(m, n)$.

Theorem 5.4.7 If $\mathbf{A}$ is a complex matrix, $\mathbf{A}^*(\mathbf{AA}^*)^-$ and $(\mathbf{A}^*\mathbf{A})^-\mathbf{A}^*$ are g-inverses of $\mathbf{A}$.

Proof Clearly $\mathbf{AA}^*(\mathbf{AA}^*)^-\mathbf{AA}^* = \mathbf{AA}^*$. Since $\rho(\mathbf{AA}^*) = \rho(\mathbf{A})$, we can cancel $\mathbf{A}^*$ on the right by rank-cancellation. This proves the first conclusion of the theorem; the second is proved similarly. ∎

We now extend the results that the inverse of a symmetric matrix is symmetric and the inverse of a hermitian matrix is hermitian, to g-inverse. Recall that $\mathbf{A}$ is hermitian if $\mathbf{A}^* = \mathbf{A}$.

Theorem 5.4.8 If $1 + 1 \neq 0$ then every symmetric matrix has a symmetric g-inverse. Every (complex) hermitian matrix has a hermitian g-inverse.

Proof Let $\mathbf{G}$ be a g-inverse of $\mathbf{A}$ so that $\mathbf{AGA} = \mathbf{A}$. If $1 + 1 \neq 0$ and $\mathbf{A}$ is symmetric it follows that $\mathbf{AG}^{\mathrm{T}}\mathbf{A} = \mathbf{A}$, so $\frac{1}{2}(\mathbf{G} + \mathbf{G}^{\mathrm{T}})$ is a symmetric g-inverse of $\mathbf{A}$. If $\mathbf{A}$ is hermitian, $\mathbf{AG}^*\mathbf{A} = \mathbf{A}$ and $\frac{1}{2}(\mathbf{G}+\mathbf{G}^*)$ is a hermitian g-inverse of $\mathbf{A}$. ∎

We now prove a result which will be used in *Section* 7.5.

Theorem 5.4.9 If $\mathcal{R}(\mathbf{A}) \subseteq \mathcal{R}(\mathbf{B})$ and $\mathcal{C}(\mathbf{C}) \subseteq \mathcal{C}(\mathbf{B})$ then $\mathbf{AB}^-\mathbf{C}$ is invariant under different choices of $\mathbf{B}^-$.

Proof By hypothesis, $\mathbf{A} = \mathbf{DB}$ and $\mathbf{C} = \mathbf{BE}$ for some matrices $\mathbf{D}$ and $\mathbf{E}$. So $\mathbf{AB}^-\mathbf{C} = \mathbf{DBB}^-\mathbf{BE} = \mathbf{DBE}$ for any g-inverse $\mathbf{B}^-$ of $\mathbf{B}$. ∎

Corollary If $\mathbf{A}$ is a complex matrix, $\mathbf{A}(\mathbf{A}^*\mathbf{A})^-\mathbf{A}^*$ is invariant under different choices of $(\mathbf{A}^*\mathbf{A})^-$ and is hermitian.

Proof The first conclusion follows from the theorem since $\mathcal{R}(\mathbf{A}) = \mathcal{R}(\mathbf{A}^*\mathbf{A})$ and $\mathcal{C}(\mathbf{A}^*) = \mathcal{C}(\mathbf{A}^*\mathbf{A})$. Now $\mathbf{A}^*\mathbf{A}$ is hermitian and, by *Theorem* 5.4.8, has a hermitian g-inverse, so the second conclusion follows. ∎

If $\mathbf{G}$ is a g-inverse of an $m \times n$ matrix $\mathbf{A}$ then $\mathbf{AGA} = \mathbf{A}$ and $\mathbf{G}$ is an $n \times m$ matrix, so

$$\rho(\mathbf{A}) \leq \rho(\mathbf{G}) \leq \min(m, n)$$

It can be shown that if $\rho(\mathbf{A}) \leq s \leq \min(m, n)$, there exists a g-inverse $\mathbf{G}$ of $\mathbf{A}$ with rank s, see *Exercise* 5.4.9.

In the next section we will give a computational procedure for finding a g-inverse of a matrix, see *Algorithm* 5.5.5.

Exercises

1. If $\mathbf{G}_1$ and $\mathbf{G}_2$ are two g-inverses of $\mathbf{A}$, show that $\alpha\mathbf{G}_1 + (1-\alpha)\mathbf{G}_2$ is a g-inverse of $\mathbf{A}$ for all $\alpha \in F$.
2. Prove or disprove: if $\mathbf{G}$ is a g-inverse of a square matrix $\mathbf{A}$ then $\mathbf{G}^2$ is a g-inverse of $\mathbf{A}^2$.
3. Show that $\rho(\mathbf{A}) = \operatorname{tr}(\mathbf{GA})$ if $\mathbf{G}$ is a g-inverse of $\mathbf{A}$.
4. (a) Show that a g-inverse $\mathbf{G}$ of $\mathbf{A}$ has the property $\mathbf{GAG} = \mathbf{G}$ iff $\rho(\mathbf{G}) = \rho(\mathbf{A})$. Such a g-inverse is called a *reflexive g-inverse*. If $\mathbf{H}$ is any g-inverse of $\mathbf{A}$, show that $\mathbf{HAH}$ is a reflexive g-inverse of $\mathbf{A}$.

   (b) Let $(\mathbf{P}, \mathbf{Q})$ be a rank-factorization of a matrix $\mathbf{A}$. Then show that $\mathbf{G}$ is a reflexive g-inverse of $\mathbf{A}$ iff $\mathbf{G} = \mathbf{Q}_R^{-1}\mathbf{P}_L^{-1}$ for some choices of $\mathbf{Q}_R^{-1}$ and $\mathbf{P}_L^{-1}$.
5. Let $\mathbf{x}$ be a non-null vector and let $x_i \neq 0$. Show that $\mathbf{y}^{\mathrm{T}}$ is a g-inverse of $\mathbf{x}$ where $\mathbf{y}$ is the vector with $1/x_i$ in the i-th place and 0's elsewhere.
6. Let $\mathbf{A}$ be an $m \times n$ matrix. Show that an $n \times m$ matrix $\mathbf{G}$ is a g-inverse of $\mathbf{A}$ iff $\rho(\mathbf{I} - \mathbf{GA}) = n - \rho(\mathbf{A})$.
7. Let $\mathbf{G}$ be a g-inverse of $\mathbf{A}$. Then prove the following:

   (a) $\mathcal{R}(\mathbf{A}) \supseteq \mathcal{R}(\mathbf{C})$ iff $\mathbf{CGA} = \mathbf{C}$.

   (b) $\{\mathbf{u}^{\mathrm{T}} : \mathbf{u}^{\mathrm{T}}\mathbf{A} = \mathbf{0}\} = \mathcal{R}(\mathbf{I} - \mathbf{AG})$.
8. Show that $\mathbf{B}^-\mathbf{A}^-$ need not, in general, be a g-inverse of $\mathbf{AB}$. However, if $\rho(\mathbf{AB}) = \rho(\mathbf{A})$, show that $\mathbf{G} := \mathbf{B}(\mathbf{AB})^-$ is a g-inverse of $\mathbf{A}$ and that $\mathbf{B}^-\mathbf{G}$ is a g-inverse of $\mathbf{AB}$.
9. (a) Consider the $m \times n$ matrix $\mathbf{M}$ and the $n \times m$ matrix $\mathbf{G}$ given below:

   $$\mathbf{M} = \begin{bmatrix} \mathbf{A} & \mathbf{0} \\ \mathbf{0} & \mathbf{0} \end{bmatrix} \quad \text{and} \quad \mathbf{G} = \begin{bmatrix} \mathbf{A}^{-1} & \mathbf{B} \\ \mathbf{C} & \mathbf{D} \end{bmatrix}$$

   Show that $\mathbf{G}$ is a g-inverse of $\mathbf{M}$ and every g-inverse of $\mathbf{M}$ is of the form $\mathbf{G}$ for some $\mathbf{B}$, $\mathbf{C}$ and $\mathbf{D}$. (It is assumed that $\mathbf{A}$ is non-singular.)

   (b) Show that a matrix of order $m \times n$ with rank r has a g-inverse with rank s for every s such that $r \leq s \leq \min(m, n)$.

   (c) Show that a symmetric matrix has a symmetric g-inverse even when $1 + 1 = 0$. (Hint: use *Exercise* 3.3.11.)
10. Let $\mathbf{G}$ be a g-inverse of $\mathbf{A}$. Then prove that

    $$\{\mathbf{G} + (\mathbf{I} - \mathbf{GA})\mathbf{U} + \mathbf{V}(\mathbf{I} - \mathbf{AG}) : \mathbf{U}, \mathbf{V} \text{ arbitrary }\}$$

    is the class of all g-inverses of $\mathbf{A}$. (Hint: if $\mathbf{H}$ is any g-inverse, take $\mathbf{U} = \mathbf{H} - \mathbf{G}$.)

11. If $\mathbf{BA}^{-}\mathbf{C}$ is invariant for all choices of $\mathbf{A}^{-}$ and is non-null, prove that $\mathcal{R}(\mathbf{B}) \subseteq \mathcal{R}(\mathbf{A})$ and $\mathcal{C}(\mathbf{C}) \subseteq \mathcal{C}(\mathbf{A})$. (Hint: use the preceding exercise and *Exercise* 2.6.12.)
12. Show that there exists a matrix $\mathbf{X}$ such that $\mathbf{AXB} = \mathbf{C}$ iff $\mathbf{AA}^{-}\mathbf{CB}^{-}\mathbf{B} = \mathbf{C}$, where $\mathbf{A}^{-}$ and $\mathbf{B}^{-}$ are any given g-inverses of $\mathbf{A}$ and $\mathbf{B}$ respectively. Also then show that a general solution of $\mathbf{AXB} = \mathbf{C}$ is
$$\mathbf{A}^{-}\mathbf{CB}^{-} + (\mathbf{I} - \mathbf{A}^{-}\mathbf{A})\mathbf{U} + \mathbf{V}(\mathbf{I} - \mathbf{BB}^{-})$$
where $\mathbf{U}$ and $\mathbf{V}$ are arbitrary.
13. Let $\mathbf{Ax} = \mathbf{b}$ be a consistent system. Show that x_j has the same value in all solutions iff $(\mathbf{A}^{-}\mathbf{A})_{j*} = \mathbf{e}_j^{\mathrm{T}}$ for any $\mathbf{A}^{-}$. If $\mathbf{A}$ is a square matrix in HCF, show that the condition reduces to $\mathbf{A}_{j*} = \mathbf{e}_j^{\mathrm{T}}$. (See *Exercise* 5.3.5(a).)
14. Let $\rho(\mathbf{A}+\mathbf{B}) = \rho(\mathbf{A})+\rho(\mathbf{B})$. Then for every g-inverse $\mathbf{G}$ of $\mathbf{A}+\mathbf{B}$, show that $\mathbf{AGA} = \mathbf{A}$ and $\mathbf{AGB} = \mathbf{0}$. (Hint: Show that $\mathbf{AG}(\mathbf{A}+\mathbf{B}) = \mathbf{A}$ using *Exercise* 3.7.8.) Generalize this result to k matrices.
15. Let $\mathbf{M}$ be as in (3.8.1), where $\mathbf{A}$ and $\mathbf{D}$ may not be square. Let $\mathcal{C}(\mathbf{B}) \subseteq \mathcal{C}(\mathbf{A})$ and $\mathcal{R}(\mathbf{C}) \subseteq \mathcal{R}(\mathbf{A})$. Then prove (3.8.2) with inverse replaced by g-inverse throughout.
16. (a) Let $\mathbf{M}$ be as in *Exercise* 4.2.13 with $\mathcal{C}(\mathbf{B}) \subseteq \mathcal{C}(\mathbf{A})$ and $\mathcal{R}(\mathbf{C}) \subseteq \mathcal{R}(\mathbf{A})$. Show that when $\mathbf{C}$ is converted to $\mathbf{0}$ as in *Exercise* 4.2.12, $\mathbf{D}$ becomes $\mathbf{D} - \mathbf{CA}^{-}\mathbf{B}$. Deduce that $\rho(\mathbf{M}) = \rho(\mathbf{A}) + \rho(\mathbf{D} - \mathbf{CA}^{-}\mathbf{B})$.

    (b) Let $\mathbf{A}^{-}$ be a g-inverse of an $m \times n$ matrix $\mathbf{A}$, $\mathbf{u} \in \mathcal{C}(\mathbf{A})$ and $\mathbf{v}^{\mathrm{T}} \in \mathcal{R}(\mathbf{A})$. Write $\delta = 1 + \mathbf{v}^{\mathrm{T}}\mathbf{A}^{-}\mathbf{u}$. If $\delta = 0$, show that the rank of $\mathbf{A} + \mathbf{uv}^{\mathrm{T}}$ is $\rho(\mathbf{A}) - 1$ and $\mathbf{A}^{-}$ is a g-inverse of $\mathbf{A} + \mathbf{uv}^{\mathrm{T}}$. If $\delta \neq 0$, show that $\rho(\mathbf{A} + \mathbf{uv}^{\mathrm{T}}) = \rho(\mathbf{A})$ and $\mathbf{A}^{-} - \frac{1}{\delta}(\mathbf{A}^{-}\mathbf{uv}^{\mathrm{T}}\mathbf{A}^{-})$ is a g-inverse of $\mathbf{A} + \mathbf{uv}^{\mathrm{T}}$. (See *Exercise* 3.8.9.)
17. If $\mathcal{R}(\mathbf{A}) \subseteq \mathcal{R}(\mathbf{X})$, show that $\rho(\mathbf{X}) = \rho(\mathbf{XA}^{-}\mathbf{A}) + \rho(\mathbf{X}(\mathbf{I} - \mathbf{A}^{-}\mathbf{A})) = \rho(\mathbf{A}) + \rho(\mathbf{X}(\mathbf{I} - \mathbf{A}^{-}\mathbf{A}))$. (Hint: Use *Theorem* 3.7.1.)
18. Let $\mathbf{Ax} = \mathbf{b}$ be consistent, $\mathbf{u}^{\mathrm{T}} \notin \mathcal{R}(\mathbf{A})$ and let β be an arbitrary scalar. Fix a g-inverse $\mathbf{A}^{-}$ of $\mathbf{A}$. Then show that the solution $\mathbf{y} := \mathbf{A}^{-}\mathbf{b} + (\mathbf{I} - \mathbf{A}^{-}\mathbf{A})\mathbf{z}$ of $\mathbf{Ax} = \mathbf{b}$ is also a solution of $\mathbf{u}^{\mathrm{T}}\mathbf{x} = \beta$ iff
$$\mathbf{u}^{\mathrm{T}}(\mathbf{I} - \mathbf{A}^{-}\mathbf{A})\mathbf{z} = \beta - \mathbf{u}^{\mathrm{T}}\mathbf{A}^{-}\mathbf{b}$$
Prove that such a vector $\mathbf{z}$ exists for all β.
19. Let f be a linear transformation from V_1 to V_2. Let $\{\mathbf{y}_1, \ldots, \mathbf{y}_r, \mathbf{y}_{r+1}, \ldots, \mathbf{y}_m\}$ be a basis of V_2 where the first r vectors form a basis of the range of f. Let $\mathbf{x}_1, \ldots, \mathbf{x}_m \in V_1$ where $f(\mathbf{x}_i) = \mathbf{y}_i$ for $i = 1, \ldots, r$.

    (a) Show that there exists a linear transformation g from V_2 to V_1 which takes $\mathbf{y}_i$ to $\mathbf{x}_i$, $i = 1, \ldots, r$.

(b) Show that $f \circ g \circ f = f$ or, equivalently, whenever $\mathbf{y}$ belongs to the range of f, $f(g(\mathbf{y})) = \mathbf{y}$. Such a g may be called a g-*inverse* of f.

(c) If $f(\mathbf{u}) = \mathbf{b} \neq \mathbf{0}$, show that there exists a g-inverse g of f such that $\mathbf{u} = g(\mathbf{b})$.

20. Show that if the system $\mathbf{Ax} = \mathbf{b}$ is consistent and $\mathbf{b} \neq \mathbf{0}$ then the set of all solutions is $\{\mathbf{Gb} : \mathbf{G}$ is a g-inverse of $\mathbf{A}\}$. (Hint: use the preceding exercise.) Show also that '$\mathbf{b} \neq \mathbf{0}$' cannot be dropped.

5.5 Sweep-out method for solving $\mathbf{Ax} = \mathbf{b}$

In this section we shall give a unified computational method for checking $\mathbf{Ax} = \mathbf{b}$ for consistency, finding a general solution when it is consistent and finding a g-inverse of $\mathbf{A}$. We call this the sweep-out method since it uses sweep-out of the columns. Two other methods, which can be used only when certain conditions are satisfied, will be given in the next section and *Section* 6.6.

We first consider the case when $\mathbf{A}$ is a square matrix and then reduce the general case to this. The basic technique used is to reduce the given system by elementary row operations to an equivalent system which is easy to solve.

Theorem 5.5.1 The systems $\mathbf{Ax} = \mathbf{b}$ and $\mathbf{Cx} = \mathbf{d}$ are equivalent in the sense that they have the same set of solutions if $(\mathbf{C} : \mathbf{d})$ is obtained from $(\mathbf{A} : \mathbf{b})$ by elementary row operations.

Proof The transforming matrix $\mathbf{P}$ is non-singular, so $\mathbf{Au} = \mathbf{b}$ iff $\mathbf{PAu} = \mathbf{Pb}$. ∎

We choose the elementary row operations so that $\mathbf{C}$ has a nice structure and $\mathbf{Cx} = \mathbf{d}$ can be solved easily. Reducing $\mathbf{A}$ to an upper triangular matrix will be used in the next section. For a general solution when $\mathbf{C}$ is in reduced echelon form, see *Exercise* 5.5.9. The procedure when $\mathbf{C}$ is in Hermite Canonical Form is given below and will be used to develop a procedure for solving an arbitrary system. We start with the case of a square matrix.

Case 1. Square coefficient matrix

Consider the system $\mathbf{Ax} = \mathbf{b}$ where $\mathbf{A}$ is an $n \times n$ matrix. Suppose first that $\mathbf{A}$ is non-singular. Then we have seen after *Algorithm* 4.4.10

that if we start with the matrix $[\mathbf{A} : \mathbf{I}_n : \mathbf{b},]$ and reduce $\mathbf{A}$ to $\mathbf{I}$ by elementary row operations but perform all the operations on the entire matrix $[\mathbf{A} : \mathbf{I}_n : \mathbf{b}]$, then at the end we get $\mathbf{A}^{-1}$ in the place of $\mathbf{I}$ and the (unique) solution of $\mathbf{Ax} = \mathbf{b}$ in the place of $\mathbf{b}$.

Example 5.5.2 Suppose we want to solve the system of equations

$$\begin{aligned} 2x + 6y + z &= -1 \\ 3x + 9y + 2z &= -1 \\ -\ y + 3z &= 4 \end{aligned} \tag{5.5.1}$$

and find the inverse of the matrix of coefficients. We start by forming

$$[\mathbf{A} : \mathbf{I} : \mathbf{b}] = \left[\begin{array}{ccc|ccc|c} 2 & 6 & 1 & 1 & 0 & 0 & -1 \\ 3 & 9 & 2 & 0 & 1 & 0 & -1 \\ 0 & -1 & 3 & 0 & 0 & 1 & 4 \end{array}\right]$$

We then apply *Algorithm* 4.4.10 to $\mathbf{A}$ but perform all the row operations on the entire matrix. Sweeping out the first column with a_{11} as the pivot, we get

$$\left[\begin{array}{ccc|ccc|c} 1 & 3 & \frac{1}{2} & \frac{1}{2} & 0 & 0 & -\frac{1}{2} \\ 0 & 0 & \frac{1}{2} & -\frac{3}{2} & 1 & 0 & \frac{1}{2} \\ 0 & -1 & 3 & 0 & 0 & 1 & 4 \end{array}\right]$$

Now, we interchange the second and third rows and then sweep out the second column with the (new) a_{22} as the pivot to get

$$\left[\begin{array}{ccc|ccc|c} 1 & 0 & \frac{19}{2} & \frac{1}{2} & 0 & 3 & \frac{23}{2} \\ 0 & 1 & -3 & 0 & 0 & -1 & -4 \\ 0 & 0 & \frac{1}{2} & -\frac{3}{2} & 1 & 0 & \frac{1}{2} \end{array}\right]$$

Finally we sweep out the third column with a_{33} as the pivot to obtain

$$\left[\begin{array}{ccc|ccc|c} 1 & 0 & 0 & 29 & -19 & 3 & 2 \\ 0 & 1 & 0 & -9 & 6 & -1 & -1 \\ 0 & 0 & 1 & -3 & 2 & 0 & 1 \end{array}\right]$$

Thus the solution of (5.5.1) is $x = 2$, $y = -1$ and $z = 1$ and $\mathbf{A}^{-1}$ is the submatrix of the final matrix formed by the fourth, fifth and sixth columns. ∎

We now generalize the method to the case when $\mathbf{A}$ may be singular. As before, form the matrix $[\mathbf{A} : \mathbf{I}_n : \mathbf{b}]$. Then using *Algorithm* 4.6.5 reduce $\mathbf{A}$ to a matrix in Hermite canonical form but perform all the row operations on the entire matrix. Let the final matrix obtained be $[\mathbf{H} : \mathbf{G} : \mathbf{d}]$ where $\mathbf{H}$ and $\mathbf{G}$ are square matrices of order n. We then have

Theorem 5.5.3 With the notations as above, the system $\mathbf{Ax} = \mathbf{b}$ is consistent iff $d_i = 0$ whenever $h_{ii} = 0$. Also then $\mathbf{d}$ is a particular solution and $\mathbf{d} + (\mathbf{I} - \mathbf{H})\mathbf{z}$ is a general solution, where $\mathbf{z}$ is an arbitrary vector in F^n. The non-null columns of $\mathbf{I}-\mathbf{H}$ form a basis of $\mathcal{N}(\mathbf{A})$. $\mathbf{G}$ is a g-inverse of $\mathbf{A}$. If the 1's in the diagonal of $\mathbf{H}$ occur in the i_1-th,...,i_r-th positions, $(\mathbf{B}, \mathbf{C})$ is a rank-factorization of $\mathbf{A}$ where

$$\mathbf{B} = [\mathbf{A}_{*i_1} : \cdots : \mathbf{A}_{*i_r}] \text{ and } \mathbf{C} = \begin{bmatrix} \mathbf{H}_{i_1*} \\ \vdots \\ \mathbf{H}_{i_r*} \end{bmatrix}$$

Proof If $\mathbf{P}$ is the transforming matrix, we have $\mathbf{PA} = \mathbf{H}$, $\mathbf{PI} = \mathbf{G}$ and $\mathbf{Pb} = \mathbf{d}$. Thus $\mathbf{P} = \mathbf{G}$, $\mathbf{GA} = \mathbf{H}$ is idempotent and $\rho(\mathbf{GA}) = \rho(\mathbf{A})$. Hence by *Theorem* 5.4.3, $\mathbf{G}$ is a g-inverse of $\mathbf{A}$.

The systems $\mathbf{Ax} = \mathbf{b}$ and $\mathbf{Hx} = \mathbf{d}$ are equivalent by the preceding theorem. By *Theorem* 4.6.2, $\mathbf{Hx} = \mathbf{d}$ is consistent iff $d_i = 0$ whenever $h_{ii} = 0$. Suppose next the system $\mathbf{Hx} = \mathbf{d}$ is consistent. Since $\mathbf{H}^2 = \mathbf{H}$, $\mathbf{I}$ is a g-inverse of $\mathbf{H}$. So by *Theorem* 5.4.5, a general solution of $\mathbf{Hx} = \mathbf{d}$ is $\mathbf{d}+(\mathbf{I}-\mathbf{H})\mathbf{z}$ where $\mathbf{z}$ is arbitrary. Since $\mathbf{H}$ is upper triangular, it is easy to see that the $n - r$ non-null columns of $\mathbf{I} - \mathbf{H}$ are linearly independent and so form a basis of $\mathcal{C}(\mathbf{I} - \mathbf{H}) = \mathcal{N}(\mathbf{H}) = \mathcal{N}(\mathbf{A})$, The last statement of the theorem follows from *Exercise* 4.6.7. ∎

Case 2. Coefficient matrix $\mathbf{A}$ of order $m \times n$ with $m > n$.

The study of such a system $\mathbf{Ax} = \mathbf{b}$ can be reduced to that of a system with square coefficient matrix as follows. Consider the system

$$[\mathbf{A} : \mathbf{0}]\begin{bmatrix} \mathbf{x} \\ \mathbf{y} \end{bmatrix} = \mathbf{b} \tag{5.5.2}$$

where the null matrix is of order $m \times (m - n)$. Clearly, $\mathbf{Ax} = \mathbf{b}$ is consistent iff (5.5.2) is consistent. Also then $\mathbf{u}$ is a solution of $\mathbf{Ax} = \mathbf{b}$ iff $\left[\begin{smallmatrix} \mathbf{u} \\ \mathbf{y} \end{smallmatrix}\right]$ is a solution of (5.5.2) for some $\mathbf{y} \in F^{m-n}$. Thus a general solution of $\mathbf{Ax} = \mathbf{b}$ is obtained by taking the first n components of a general solution of (5.5.2). Since the coefficient matrix of the system (5.5.2) is square we can apply the method of *Case* 1 to it.

Hence we form the matrix $[\mathbf{A} : \mathbf{0} : \mathbf{I}_m : \mathbf{b}]$ and apply *Algorithm* 4.6.5 to $[\mathbf{A} : \mathbf{0}]$ but perform all the row operations on the entire matrix. Note that the null columns are not altered by the row operations. So $[\mathbf{A} : \mathbf{0}]$ is reduced to a matrix $\mathbf{H} = [\mathbf{H}_1 : \mathbf{0}]$ in HCF and the final matrix obtained is of the form $[\mathbf{H}_1 : \mathbf{0} : \mathbf{G} : \mathbf{d}]$. Partition this final matrix further as

$$\begin{bmatrix} \mathbf{H}_{11} & \mathbf{0} & \mathbf{G}_1 & \mathbf{d}_1 \\ \mathbf{0} & \mathbf{0} & \mathbf{G}_2 & \mathbf{d}_2 \end{bmatrix} \tag{5.5.3}$$

where $\mathbf{H}_{11}$ has n rows. We then have

Theorem 5.5.4 With the notations as above, the system $\mathbf{Ax} = \mathbf{b}$ is consistent iff '$\mathbf{d}_2 = \mathbf{0}$ and the i-th component of $\mathbf{d}_1$ is 0 whenever the (i, i)-th entry of $\mathbf{H}_{11}$ is 0'. Also then $\mathbf{d}_1$ is a particular solution and $\mathbf{d}_1 + (\mathbf{I} - \mathbf{H}_{11})\mathbf{z}_1$ is a general solution, where $\mathbf{z}_1$ is an arbitrary vector in F^n. $\mathbf{G}_1$ is a g-inverse of $\mathbf{A}$. A basis for $\mathcal{N}(\mathbf{A})$ and a rank-factorization of $\mathbf{A}$ can be obtained as in the preceding theorem by replacing $\mathbf{H}$ by $\mathbf{H}_{11}$.

Proof By the preceding theorem, $\begin{bmatrix} \mathbf{G}_1 \\ \mathbf{G}_2 \end{bmatrix}$ is a g-inverse of $[\mathbf{A} : \mathbf{0}]$, so $\mathbf{AG}_1\mathbf{A} = \mathbf{A}$ and $\mathbf{G}_1$ is a g-inverse of $\mathbf{A}$. The condition for consistency also follows from the same theorem on noting that, here, the last $m - n$ diagonal elements of the matrix $\mathbf{H}$ are 0. Moreover, if the system is consistent, a general solution of (5.5.2) is

$$\begin{bmatrix} \mathbf{d}_1 \\ \mathbf{0} \end{bmatrix} + \left(\begin{bmatrix} \mathbf{I} & \mathbf{0} \\ \mathbf{0} & \mathbf{I} \end{bmatrix} - \begin{bmatrix} \mathbf{H}_{11} & \mathbf{0} \\ \mathbf{0} & \mathbf{0} \end{bmatrix} \right) \begin{bmatrix} \mathbf{z}_1 \\ \mathbf{z}_2 \end{bmatrix} = \begin{bmatrix} \mathbf{d}_1 + (\mathbf{I} - \mathbf{H}_{11})\mathbf{z}_1 \\ \mathbf{z}_2 \end{bmatrix}$$

where $\mathbf{z}_1$ and $\mathbf{z}_2$ are arbitrary vectors in F^n and F^{m-n} respectively. Hence $\mathbf{d}_1 + (\mathbf{I} - \mathbf{H}_{11})\mathbf{z}_1$ is a general solution of $\mathbf{Ax} = \mathbf{b}$. The rest follows as in the preceding theorem. ∎

We finally note that it is not necessary to include the matrix $\mathbf{0}$ in $[\mathbf{A} : \mathbf{0} : \mathbf{I}_m : \mathbf{b}]$. We start with $[\mathbf{A} : \mathbf{I}_m : \mathbf{b}]$ and apply *Algorithm* 4.6.5 (with n of the algorithm replaced by m) to $\mathbf{A}$ until j reaches the value n (= the number of columns of $\mathbf{A}$) but perform all the row operations on the entire matrix. The final matrix obtained will be

$$\begin{bmatrix} \mathbf{H}_{11} & \mathbf{G}_1 & \mathbf{d}_1 \\ \mathbf{0} & \mathbf{G}_2 & \mathbf{d}_2 \end{bmatrix}$$

Then we proceed as already stated.

Case 3. Coefficient matrix **A** of order $m \times n$ with $m < n$.

We again convert the system $\mathbf{Ax} = \mathbf{b}$ into one with a square coefficient matrix. Clearly $\mathbf{Ax} = \mathbf{b}$ is equivalent to the system

$$\begin{bmatrix} \mathbf{A} \\ \mathbf{0} \end{bmatrix} \mathbf{x} = \begin{bmatrix} \mathbf{b} \\ \mathbf{0} \end{bmatrix} \tag{5.5.4}$$

where the null matrices have $n - m$ rows. Further if $\mathbf{G} = [\,\mathbf{G}_1 : \mathbf{G}_2\,]$ is a g-inverse of $\left[\begin{smallmatrix} \mathbf{A} \\ \mathbf{0} \end{smallmatrix}\right]$ where $\mathbf{G}_1$ has m columns then it is easy to see that $\mathbf{G}_1$ is a g-inverse of **A**.

We thus form the matrix

$$\left[\begin{array}{c|c|c} \mathbf{A} & & \mathbf{b} \\ & \mathbf{I}_n & \\ \mathbf{0} & & \mathbf{0} \end{array}\right]$$

and apply *Algorithm* 4.6.5 to $\left[\begin{smallmatrix} \mathbf{A} \\ \mathbf{0} \end{smallmatrix}\right]$ but perform all the row operations on the entire matrix. Let the final matrix obtained be $[\,\mathbf{H} : \mathbf{G} : \mathbf{d}\,]$. Then the submatrix of **G** formed by the first m columns is a g-inverse of **A**. The system $\mathbf{Ax} = \mathbf{b}$ is consistent iff $d_i = 0$ whenever $h_{ii} = 0$. When the system is consistent, **d** is a particular solution and $\mathbf{d} + (\mathbf{I} - \mathbf{H})\mathbf{z}$ is a general solution where **z** is an arbitrary vector in F^n. A basis for $\mathcal{N}(\mathbf{A})$ and a rank-factorization of **A** can be obtained exactly as in *Theorem* 5.5.3.

General System

We shall now present a unified algorithm for solving a general system $\mathbf{Ax} = \mathbf{b}$, combining the methods in the three cases given above.

Algorithm 5.5.5 (*Sweep-out method*) Given: an $m \times n$ matrix **A** and a vector **b** with m components.

Step 1 Form the matrix $\mathbf{T} = [\,\mathbf{T}_1 : \mathbf{T}_2 : \mathbf{T}_3\,]$ thus: If $m \geq n$ take $\mathbf{T}_1 = \mathbf{A}$, $\mathbf{T}_2 = \mathbf{I}_m$ and $\mathbf{T}_3 = \mathbf{b}$. If $m < n$ take $\mathbf{T}_1 = \left[\begin{smallmatrix} \mathbf{A} \\ \mathbf{0} \end{smallmatrix}\right]$, $\mathbf{T}_2 = \mathbf{I}_n$ and $\mathbf{T}_3 = \left[\begin{smallmatrix} \mathbf{b} \\ \mathbf{0} \end{smallmatrix}\right]$, where the null matrix in $\mathbf{T}_1$ and the null vector in $\mathbf{T}_3$ have $n - m$ rows each.

Step 2 Apply *Algorithm* 4.6.5 (with n replaced by the number of rows of $\mathbf{T}_1$) to $\mathbf{T}_1$ until j reaches the value n (= number of columns of **A**) but perform all the row operations on the entire matrix **T**. Let the final matrix obtained be $[\,\mathbf{H} : \mathbf{G} : \mathbf{d}\,]$ where **H**, **G** and **d** are of the same orders as $\mathbf{T}_1$, $\mathbf{T}_2$ and $\mathbf{T}_3$ respectively.

Step 3 Declare $\rho(\mathbf{A}) = \sum_{i=1}^{n} h_{ii}$.

Step 4 If $m = n$ go to *Step* 5. Otherwise go to *Step* 8.

Step 5 If $\rho(\mathbf{A}) < n$, go to *Step* 6. Otherwise declare $\mathbf{G}$ to be $\mathbf{A}^{-1}$ and $\mathbf{d}$ to be the unique solution to $\mathbf{Ax} = \mathbf{b}$ and stop.

Step 6 Declare $\mathbf{G}$ to be a g-inverse of $\mathbf{A}$.

Step 7 Check whether $d_i = 0$ whenever $h_{ii} = 0$. If yes, declare that $\mathbf{d}$ is a particular solution and $\mathbf{d} + (\mathbf{I} - \mathbf{H})\mathbf{z}$ is a general solution of $\mathbf{Ax} = \mathbf{b}$, where $\mathbf{z}$ is arbitrary, and stop. If no, declare that $\mathbf{Ax} = \mathbf{b}$ is inconsistent and stop.

Step 8 If $m > n$ go to *Step* 9. If $m < n$ declare that the submatrix of $\mathbf{G}$ formed by the first m columns of $\mathbf{G}$ is a g-inverse of $\mathbf{A}$ and go to *Step* 7.

Step 9 Partition $[\mathbf{H} : \mathbf{G} : \mathbf{d}]$ as $\begin{bmatrix} \mathbf{H}_{11} & \mathbf{G}_1 & \mathbf{d}_1 \\ \mathbf{H}_{21} & \mathbf{G}_2 & \mathbf{d}_2 \end{bmatrix}$ where $\mathbf{H}_{11}$ has n rows.

Step 10 Declare $\mathbf{G}_1$ to be a g-inverse of $\mathbf{A}$.

Step 11 Check whether $\mathbf{d}_2 = \mathbf{0}$ and the i-th component of $\mathbf{d}_1$ is 0 whenever the i-th component of $\mathbf{H}_{11}$ is 0 $(1 \le i \le n)$. If yes, declare that $\mathbf{d}_1$ is a particular solution and $\mathbf{d}_1 + (\mathbf{I} - \mathbf{H}_{11})\mathbf{z}$ is a general solution of $\mathbf{Ax} = \mathbf{b}$, where $\mathbf{z}$ is arbitrary, and stop. If no, declare that $\mathbf{Ax} = \mathbf{b}$ is inconsistent and stop.

We illustrate the use of the above algorithm with

Example 5.5.6 Consider the systems $\mathbf{Ax} = \mathbf{b}$ and $\mathbf{Ax} = \mathbf{c}$ where

$$\mathbf{A} = \begin{bmatrix} 0 & 0 & 0 & 1 \\ 2 & 4 & 8 & 6 \\ 1 & 2 & 4 & 5 \\ 3 & 5 & 11 & 8 \\ 4 & 6 & 14 & 0 \end{bmatrix}, \ \mathbf{b} = \begin{bmatrix} -1 \\ 10 \\ 3 \\ 15 \\ 28 \end{bmatrix} \text{ and } \mathbf{c} = \begin{bmatrix} 2 \\ 6 \\ 7 \\ 8 \\ -10 \end{bmatrix}$$

Since $m > n$ we form the matrix $\mathbf{T} = [\mathbf{A} : \mathbf{I}_5 : \mathbf{b} : \mathbf{c}]$ thus:

$$\mathbf{T} = \left[\begin{array}{cccc|ccccc|c|c} 0 & 0 & 0 & 1 & 1 & 0 & 0 & 0 & 0 & -1 & 2 \\ 2 & 4 & 8 & 6 & 0 & 1 & 0 & 0 & 0 & 10 & 6 \\ 1 & 2 & 4 & 5 & 0 & 0 & 1 & 0 & 0 & 3 & 7 \\ 3 & 5 & 11 & 8 & 0 & 0 & 0 & 1 & 0 & 15 & 8 \\ 4 & 6 & 14 & 0 & 0 & 0 & 0 & 0 & 1 & 28 & -10 \end{array}\right]$$

We now apply *Algorithm* 4.6.5. Setting $j = 1$, we interchange the first and second rows and perform the sweep-out on the first column with the $(1, 1)$-th element as the pivot to get

$$\left[\begin{array}{rrrr|rrrrr|r|r} 1 & 2 & 4 & 3 & 0 & 0.5 & 0 & 0 & 0 & 5 & 3 \\ 0 & 0 & 0 & 1 & 1 & 0 & 0 & 0 & 0 & -1 & 2 \\ 0 & 0 & 0 & 2 & 0 & -0.5 & 1 & 0 & 0 & -2 & 4 \\ 0 & -1 & -1 & -1 & 0 & -1.5 & 0 & 1 & 0 & 0 & -1 \\ 0 & -2 & -2 & -12 & 0 & -2 & 0 & 0 & 1 & 8 & -22 \end{array}\right]$$

Setting $j = 2$, we interchange the second and fourth rows and perform the sweep-out on the second column with the $(2, 2)$-th element as the pivot to get

$$\left[\begin{array}{rrrr|rrrrr|r|r} 1 & 0 & 2 & 1 & 0 & -2.5 & 0 & 2 & 0 & 5 & 1 \\ 0 & 1 & 1 & 1 & 0 & 1.5 & 0 & -1 & 0 & 0 & 1 \\ 0 & 0 & 0 & 2 & 0 & -0.5 & 1 & 0 & 0 & -2 & 4 \\ 0 & 0 & 0 & 1 & 1 & 0 & 0 & 0 & 0 & -1 & 2 \\ 0 & 0 & 0 & -10 & 0 & 1 & 0 & -2 & 1 & 18 & -20 \end{array}\right]$$

Now setting $j = 3$, we notice that $a_{i3} = 0$ for all $i \geq 3$. Further, even though a_{13} and a_{23} are non-zero, a_{11} and a_{22} are equal to 1. So we increase j to 4 and perform the sweep-out on the fourth column with the $(4, 4)$-element as the pivot to get

$$\left[\begin{array}{rrrr|rrrrr|r|r} 1 & 0 & 2 & 0 & -1 & -2.5 & 0 & 2 & 0 & 6 & -1 \\ 0 & 1 & 1 & 0 & -1 & 1.5 & 0 & -1 & 0 & 1 & -1 \\ 0 & 0 & 0 & 0 & -2 & -0.5 & 1 & 0 & 0 & 0 & 0 \\ 0 & 0 & 0 & 1 & 1 & 0 & 0 & 0 & 0 & -1 & 2 \\ 0 & 0 & 0 & 0 & 10 & 1 & 0 & -2 & 1 & 8 & 0 \end{array}\right]$$

We halt *Algorithm* 4.6.5 here. The rank of $\mathbf{A}$ is 3. We partition the final matrix as mentioned in *Step* 9 of *Algorithm* 5.5.5 and get

$$\begin{bmatrix} -1 & -2.5 & 0 & 2 & 0 \\ -1 & 1.5 & 0 & -1 & 0 \\ -2 & -0.5 & 1 & 0 & 0 \\ 1 & 0 & 0 & 0 & 0 \end{bmatrix}$$

as a g-inverse of $\mathbf{A}$. Corresponding to the system $\mathbf{Ax} = \mathbf{b}$, $\mathbf{d}_2 = 8 \neq \mathbf{0}$, so the system $\mathbf{Ax} = \mathbf{b}$ is not consistent. However, corresponding to the system $\mathbf{Ax} = \mathbf{c}$, $\mathbf{d}_2 = \mathbf{0}$ and '$d_i = 0$ if the i-th diagonal element of $\mathbf{H}_{11}$ is 0' holds since only the third diagonal element of $\mathbf{H}_{11}$ is 0 and the third component of $\mathbf{d}_1$ is 0. Hence $\mathbf{Ax} = \mathbf{c}$ is consistent and a general solution is $\mathbf{d}_1 + (\mathbf{I} - \mathbf{H}_{11})\mathbf{z}$, that is

$$\begin{bmatrix} -1 \\ -1 \\ 0 \\ 2 \end{bmatrix} + \begin{bmatrix} 0 & 0 & -2 & 0 \\ 0 & 0 & -1 & 0 \\ 0 & 0 & 1 & 0 \\ 0 & 0 & 0 & 0 \end{bmatrix} \mathbf{z} = \begin{bmatrix} -1-2\alpha \\ -1-\alpha \\ \alpha \\ 2 \end{bmatrix}$$

where $\alpha = z_3$ is arbitrary. ∎

Example 5.5.7 Using the sweep-out technique we will find (i) when the system $\mathbf{Ax} = \mathbf{b}$ is consistent and (ii) a general solution of $\mathbf{Ax} = \mathbf{b}$ whenever it is consistent, where

$$\mathbf{A} = \begin{bmatrix} 1 & 1 & 1 \\ \alpha & \beta & \gamma \\ \alpha^2 & \beta^2 & \gamma^2 \end{bmatrix} \quad \text{and} \quad \mathbf{b} = \begin{bmatrix} 1 \\ \epsilon \\ \epsilon^2 \end{bmatrix}$$

We start with the matrix $[\mathbf{A} : \mathbf{b}]$. By sweeping out the first column with a_{11} as the pivot we obtain

$$\begin{bmatrix} 1 & 1 & 1 & 1 \\ 0 & \beta-\alpha & \gamma-\alpha & \epsilon-\alpha \\ 0 & \beta^2-\alpha^2 & \gamma^2-\alpha^2 & \epsilon^2-\alpha^2 \end{bmatrix} \tag{5.5.5}$$

We now consider two cases.

Case 1. $\alpha \neq \beta$. Then we sweep out the second column of (5.5.5) with the $(2,2)$-element as the pivot to obtain

$$\begin{bmatrix} 1 & 0 & (\beta-\gamma)/(\beta-\alpha) & (\beta-\epsilon)/(\beta-\alpha) \\ 0 & 1 & (\gamma-\alpha)/(\beta-\alpha) & (\epsilon-\alpha)/(\beta-\alpha) \\ 0 & 0 & (\gamma-\alpha)(\gamma-\beta) & (\epsilon-\alpha)(\epsilon-\beta) \end{bmatrix}$$

If, now, $\gamma \neq \alpha$ and $\gamma \neq \beta$ then $\mathbf{A}$ is non-singular and the value of x_3 in the solution of $\mathbf{Ax} = \mathbf{b}$ is $(\epsilon-\alpha)(\epsilon-\beta)/(\gamma-\alpha)(\gamma-\beta)$. The values of x_1 and x_2 can be written down by symmetry. Next let $\gamma = \alpha$. Then $\mathbf{Ax} = \mathbf{b}$ is consistent iff $\epsilon = \alpha$ or $\epsilon = \beta$. Also then a general solution of $\mathbf{Ax} = \mathbf{b}$ is $((\beta-\epsilon)/(\beta-\alpha)-x_3, (\epsilon-\alpha)/(\beta-\alpha), x_3)^{\mathrm{T}}$ where x_3 is arbitrary. Finally let $\gamma = \beta$. Then $\mathbf{Ax} = \mathbf{b}$ is consistent iff $\epsilon = \alpha$ or $\epsilon = \beta$ and then a general solution is $((\beta-\epsilon)/(\beta-\alpha), (\epsilon-\alpha)/(\beta-\alpha) - x_3, x_3)^{\mathrm{T}}$ where x_3 is arbitrary.

Case 2. $\alpha = \beta$. If $\alpha \neq \gamma$ the situation is similar to *Case* 1. So we may take $\alpha = \beta = \gamma$. Then $\mathbf{Ax} = \mathbf{b}$ is consistent iff $\epsilon = \alpha$ and then a general solution of $\mathbf{Ax} = \mathbf{b}$ is $(1 - x_2 - x_3, x_2, x_3)^{\mathrm{T}}$ where x_2 and x_3 are arbitrary. ∎

Exercises

1. Consider the system of linear equations $\mathbf{Hx} = \mathbf{b}$ where

$$\mathbf{H} = \begin{bmatrix} 1 & -4 & 0 & -3 \\ 0 & 0 & 0 & 0 \\ 0 & 0 & 1 & 4 \\ 0 & 0 & 0 & 0 \end{bmatrix} \text{ and } \mathbf{b} = \begin{bmatrix} 2 \\ 0 \\ 5 \\ 0 \end{bmatrix}$$

   (a) Show that $\mathbf{Hx} = \mathbf{b}$ is consistent and obtain a general solution.

   (b) Does there exist a solution with the second component negative? What about the first component?

   (c) If one more equation $2x_1 + 3x_2 + 4x_3 + 5x_4 = 7$ is added to the system $\mathbf{Hx} = \mathbf{b}$, show that the new system is consistent and obtain a solution of the new system (Hint: use *Exercise* 5.4.18).

2. In the context of solving $\mathbf{Ax} = \mathbf{b}$, show that sweeping out the first column of $[\mathbf{A} : \mathbf{b}]$ with a_{11} as the pivot amounts to expressing x_1 in terms of the other x_j's using the first equation and substituting this in the other equations.

3. In each of the following cases, solve the system $\mathbf{Ax} = \mathbf{b}$ (i.e., find whether it is consistent and a general solution if it is consistent). Also find for $\mathbf{A}$, a g-inverse, the rank, a rank-factorization and a basis of the null space.

   (a) $\mathbf{A} = \begin{bmatrix} 2 & 1 & 3 \\ 4 & 0 & -1 \\ 3 & 1 & 1 \end{bmatrix}, \quad \mathbf{b} = \begin{bmatrix} 1 \\ 5 \\ 4 \end{bmatrix}$

   (b) $\mathbf{A} = \begin{bmatrix} 2 & 1 & 1 \\ 1 & 2 & -4 \\ 0 & 1 & -3 \\ -1 & 0 & -2 \end{bmatrix}, \quad \mathbf{b} = \begin{bmatrix} 1 \\ -1 \\ -1 \\ -1 \end{bmatrix}$

   (c) $\mathbf{A}$ as in (b) above and $\mathbf{b} = (3\ \ 3\ \ 1\ \ 1)^{\mathrm{T}}$

   (d) $\mathbf{A} = \begin{bmatrix} 1 & 2 & -1 & 1 \\ 2 & 1 & 4 & 1 \\ 1 & 1 & -1 & -1 \end{bmatrix}, \quad \mathbf{b} = \begin{bmatrix} 1 \\ 1 \\ 1 \end{bmatrix}$

(e) $\mathbf{A} = \begin{bmatrix} 2 & -2 & 0 & 4 \\ -1 & 0 & 3 & 1 \\ 6 & -6 & 1 & 8 \\ 1 & 2 & -7 & -16 \end{bmatrix}, \quad \mathbf{b} = \begin{bmatrix} 2 \\ 6 \\ 12 \\ -7 \end{bmatrix}$

(f) $\mathbf{A} = \begin{bmatrix} 2 & 1 & 1 & 1 \\ 1 & 2 & 1 & 1 \\ 1 & 1 & 2 & 1 \\ 1 & 1 & 1 & 2 \end{bmatrix}, \quad \mathbf{b} = \begin{bmatrix} 1 \\ 2 \\ 3 \\ 4 \end{bmatrix}$

4. Let

$$\mathbf{A} = \begin{bmatrix} 8 & 1 & 0 & 0 \\ 0 & 0 & 1 & 3 \\ 4 & 0.5 & 2 & 4 \\ 16 & 2 & 1 & 2 \end{bmatrix} \quad \text{and} \quad \mathbf{b} = \begin{bmatrix} b_1 \\ b_2 \\ b_3 \\ b_4 \end{bmatrix}$$

Using *Algorithm* 5.5.5, find out when the system $\mathbf{Ax} = \mathbf{b}$ is consistent. When it is, find a general solution. Also find a g-inverse of $\mathbf{A}$ and $\rho(\mathbf{A})$.

5. Consider the matrix $\begin{bmatrix} \mathbf{A} & \mathbf{b} \\ \mathbf{c}^{\mathrm{T}} & 0 \end{bmatrix}$ where $\mathbf{A}$ is an $m \times n$ matrix, $\mathbf{c}^{\mathrm{T}} \in \mathcal{R}(\mathbf{A})$ and $\mathbf{b} \in \mathcal{C}(\mathbf{A})$.

   (a) Show that the value γ of $\mathbf{c}^{\mathrm{T}}\mathbf{x}$ is the same at all solutions of $\mathbf{Ax} = \mathbf{b}$ and is $\mathbf{c}^{\mathrm{T}}\mathbf{A}^{-}\mathbf{b}$ for any g-inverse $\mathbf{A}^{-}$ of $\mathbf{A}$.

   (b) Show that γ can be obtained as follows: convert $\mathbf{c}^{\mathrm{T}}$ to $\mathbf{0}^{\mathrm{T}}$ by elementary row operations where the only operations involving the last row are those of the type $R_{m+1,i}(\beta)$. Then we get $-\gamma$ in the $(m+1, n+1)$-th position.

   (c) Using (b), find the value of $2x_1 + 3x_2 - 5x_3$ at the solution to $\mathbf{Ax} = \mathbf{b}$ in *Exercise* 5.5.3(b).

6. Find, for what values of α and β, the system

$$\begin{aligned} 2x + 4y + (\alpha+3)z &= 2 \\ x + 3y + \qquad z &= 2 \\ (\alpha-2)x + 2y + \qquad 3z &= \beta \end{aligned}$$

is consistent and find a general solution whenever it is consistent.

7. Find when the system

$$\begin{aligned} x + y + z &= 1 \\ \alpha x + \beta y + \gamma z &= \epsilon \\ \alpha^3 x + \beta^3 y + \gamma^3 z &= \epsilon^3 \end{aligned}$$

is consistent and find a general solution whenever it is consistent.

8. Find when the following system over $\mathbb{R}$ is consistent and find a general solution whenever it is consistent.

$$\begin{aligned} x + \ \ y + \ \ z &= 0 \\ \alpha x + \beta y + \gamma z &= 0 \\ \beta x + \gamma y + \alpha z &= \delta \end{aligned}$$

9. (*Solution of linear equations using reduced echelon form*) Consider the system $\mathbf{Bx} = \mathbf{d}$ where $\mathbf{B}$ is an $m \times n$ matrix in reduced echelon form with rank r. Let the leading 1 in the i-th row of $\mathbf{B}$ occur in the p_i-th column, $i = 1, \ldots, r$.

   (a) Show that $\mathbf{Bx} = \mathbf{d}$ is consistent iff $d_i = 0$ for all $i > r$.

   (b) If $\mathbf{Bx} = \mathbf{d}$ is consistent, show that $\mathbf{u}$ is a solution, where

$$u_j = \begin{cases} d_i & \text{if } j = p_i,\ 1 \leq i \leq r \\ 0 & \text{otherwise} \end{cases}$$

   (c) Show that

$$(\mathbf{KB})_{j*} = \begin{cases} \mathbf{B}_{i*} & \text{if } j = p_i,\ 1 \leq i \leq r \\ \mathbf{0} & \text{otherwise} \end{cases}$$

   where $\mathbf{K}$ is the $n \times m$ matrix $[\mathbf{e}_{p_1} : \mathbf{e}_{p_2} : \cdots : \mathbf{e}_{p_r} : * : \cdots : *]$ and $*$ denotes an arbitrary column vector.

   (d) With $\mathbf{K}$ as in (c), show that $\mathbf{K}$ is a g-inverse of $\mathbf{B}$ and obtain a general solution of $\mathbf{Bx} = \mathbf{d}$ when it is consistent.

   (e) Suppose $(\mathbf{B} : \mathbf{P} : \mathbf{d})$ is obtained from $(\mathbf{A} : \mathbf{I} : \mathbf{b})$ by elementary row operations. Show that the $n \times m$ matrix $\mathbf{G}$ is a g-inverse of $\mathbf{A}$ where

$$\mathbf{G}_{j*} = \begin{cases} \mathbf{P}_{i*} & \text{if } j = p_i,\ 1 \leq i \leq r \\ \mathbf{0} & \text{otherwise} \end{cases}$$

5.6 Compact form of Gauss-Doolittle method

The method we have given in the preceding section for solving a system of linear equations is suitable for use on a computer. However it is not convenient for hand computations since it involves writing down many intermediate matrices. In this section we give a method which can be used if a calculator with a single memory is available, provided the coefficient matrix satisfies a certain condition. This method is also of theoretical importance.

We consider a system $\mathbf{Ax} = \mathbf{b}$ where $\mathbf{A}$ is an $n \times n$ matrix and make the following

Assumption All leading principal submatrices of $\mathbf{A}$ are non-singular.

We incidentally note that if $\mathbf{A}$ is non-singular, the system $\mathbf{Ax} = \mathbf{b}$ can be converted to one satisfying the above assumption by a renaming of the variables (i.e., a permutation of the columns of $\mathbf{A}$).

The method consists in reducing the system to one in which $\mathbf{A}$ is upper triangular with diagonal elements 1. Such a system is very easy to solve by the method of 'back substitution'. For example, if $n = 3$, the system is:

$$\begin{aligned} x_1 + a_{12}x_2 + a_{13}x_3 &= b_1 \\ x_2 + a_{23}x_3 &= b_2 \\ x_3 &= b_3 \end{aligned}$$

Its solution is obtained by first solving for x_3 from the third equation, then solving for x_2 from the second equation and finally solving for x_1 from the first equation. Thus we get $x_3 = b_3$, then $x_2 = b_2 - a_{23}b_3$ and finally $x_1 = b_1 - a_{13}b_3 - a_{12}b_2 + a_{12}a_{23}b_3$.

We note that the system $\mathbf{Ax} = \mathbf{b}$ is consistent and has a unique solution since, by our assumption, $\mathbf{A}$ is non-singular. Also, $\mathbf{A}^{-1}$ can be obtained by solving each of the systems $\mathbf{Ax} = \mathbf{e}_1, \ldots, \mathbf{Ax} = \mathbf{e}_n$.

We thus form the matrix $[\mathbf{A} : \mathbf{I}_n : \mathbf{b}]$. Our task is to convert this, by elementary row operations, to a matrix with the first n columns forming an upper triangular matrix. This can be done by *Algorithm* 4.4.6 but this involves noting down the intermediate matrices. We derive below a compact method by which the final matrix can be obtained in a single step with the help of an auxiliary matrix. For convenience we denote the bigger matrix $[\mathbf{A} : \mathbf{I}_n : \mathbf{b}]$ by $\widetilde{\mathbf{A}}$ and use a_{ij} to denote its (i, j)-th element for $i = 1, \ldots, n$ and $j = 1, \ldots, p$ where p denotes the number of columns of $\widetilde{\mathbf{A}}$.

Theorem 5.6.1 Let $\widetilde{\mathbf{A}} = ((a_{ij}))$ be an $n \times p$ matrix such that $n \leq p$ and the n leading principal submatrices of $\widetilde{\mathbf{A}}$ are non-singular. Then there exist unique matrices $\mathbf{L}$ and $\widetilde{\mathbf{U}}$ such that

(i) $\mathbf{L}$ is an $n \times n$ lower triangular (non-singular) matrix,

(ii) $\widetilde{\mathbf{U}}$ is an $n \times p$ matrix such that the first n columns form an upper triangular matrix with diagonal entries 1 and

(iii) $\widetilde{\mathbf{A}} = \mathbf{L}\widetilde{\mathbf{U}}$.

Proof We have to prove the existence and uniqueness of ℓ_{ij} for $1 \leq j \leq i \leq n$ and u_{ij} for $i+1 \leq j \leq p$ and $1 \leq i \leq n$ such that

$$\begin{bmatrix} a_{11} & a_{12} & a_{13} & \cdots & a_{1n} & \cdots & a_{1p} \\ a_{21} & a_{22} & a_{23} & \cdots & a_{2n} & \cdots & a_{2p} \\ \cdots & \cdots & \cdots & \cdots & \cdots & \cdots & \cdots \\ a_{n1} & a_{n2} & a_{n3} & \cdots & a_{nn} & \cdots & a_{np} \end{bmatrix} =$$

$$\begin{bmatrix} \ell_{11} & 0 & \cdots & 0 \\ \ell_{21} & \ell_{22} & \cdots & 0 \\ \cdots & \cdots & \cdots & \cdots \\ \ell_{n1} & \ell_{n2} & \cdots & \ell_{nn} \end{bmatrix} \begin{bmatrix} 1 & u_{12} & u_{13} & \cdots & u_{1n} & \cdots & u_{1p} \\ 0 & 1 & u_{23} & \cdots & u_{2n} & \cdots & u_{2p} \\ \cdots & \cdots & \cdots & \cdots & \cdots & \cdots & \cdots \\ 0 & 0 & 0 & \cdots & 1 & \cdots & u_{np} \end{bmatrix} \tag{5.6.1}$$

It is easy to see that for $i \geq k$, the (i,k)-th elements of the two sides of (5.6.1) are equal iff

$$\ell_{ik} = a_{ik} - \sum_{s=1}^{k-1} \ell_{is} u_{sk} \tag{5.6.2}$$

Similarly for $k < j$, the (k,j)-th elements of the two sides of (5.6.1) are equal iff

$$u_{kj} = \frac{1}{\ell_{kk}} \left(a_{kj} - \sum_{s=1}^{k-1} \ell_{ks} u_{sj} \right) \tag{5.6.3}$$

provided $\ell_{kk} \neq 0$. Notice that the RHS of (5.6.2) involves only the first $k-1$ columns of $\mathbf{L}$ and the first $k-1$ rows of $\widetilde{\mathbf{U}}$ besides a_{ik}. Similarly the RHS of (5.6.3) involves only the first k columns of $\mathbf{L}$ and the first $k-1$ rows of $\widetilde{\mathbf{U}}$. We can now give the procedure for finding ℓ's and u's satisfying (5.6.2) and (5.6.3).

We start by finding the first column of $\mathbf{L}$ by taking $k = 1$ in (5.6.2). Then we find the first row of $\widetilde{\mathbf{U}}$ (to the right of the diagonal element) by taking $k = 1$ in (5.6.3). Notice that this can be done since $\ell_{11} = a_{11} \neq 0$ by hypothesis. Next we find the second column of $\mathbf{L}$ (from the diagonal element downwards) by taking $k = 2$ in (5.6.2). At this stage we have

$$\begin{bmatrix} a_{11} & a_{12} \\ a_{21} & a_{22} \end{bmatrix} = \begin{bmatrix} \ell_{11} & 0 \\ \ell_{21} & \ell_{22} \end{bmatrix} \begin{bmatrix} 1 & u_{12} \\ 0 & 1 \end{bmatrix}$$

Since the matrix on the left is non-singular it follows that $\ell_{22} \neq 0$. So we find the second row of $\widetilde{\mathbf{U}}$ (to the right of the diagonal element) by taking $k = 2$ in (5.6.3). Proceeding thus we find ℓ's and u's satisfying (5.6.2) whenever $i \geq k$ and (5.6.3) whenever $k < j$. $\mathbf{L}$ and $\widetilde{\mathbf{U}}$ thus obtained satisfy (5.6.1) and so (i), (ii) and (iii) of the theorem.

To prove the uniqueness, let $\mathbf{L}$ and $\widetilde{\mathbf{U}}$ satisfy (i), (ii) and (iii) so that (5.6.1) holds. Then (5.6.2) shows that the first column of $\mathbf{L}$ is uniquely determined. Since the first n columns of $\widetilde{\mathbf{A}}$ form a non-singular matrix it follows from (5.6.1) that $\ell_{ii} \neq 0$ for all i. Now (5.6.3) shows that the first row of $\widetilde{\mathbf{U}}$ is unique. Then (5.6.2) shows that the second column of $\mathbf{L}$ is unique. Next (5.6.3) shows that the second row of $\widetilde{\mathbf{U}}$ is unique. Proceeding thus we see that $\mathbf{L}$ and $\widetilde{\mathbf{U}}$ are uniquely determined by $\widetilde{\mathbf{A}}$. ∎

Since $\mathbf{L}$ is non-singular it follows from the corollary to *Theorem* 4.4.9 that the matrix $\widetilde{\mathbf{U}}$ can be obtained from $\widetilde{\mathbf{A}}$ by a sequence of elementary row operations. However we have shown in the proof of the preceding theorem how $\widetilde{\mathbf{U}}$ can be obtained directly. Note that the transforming matrix is $\mathbf{L}^{-1}$. If $\widetilde{\mathbf{A}}$ contains $\mathbf{I}_n$ as a submatrix, the corresponding submatrix of $\widetilde{\mathbf{U}}$ is $\mathbf{L}^{-1}$. We now present the procedure given above as

Algorithm 5.6.2 (*Gauss-Doolittle method*) Given: an $n \times n$ matrix $\mathbf{A}$ such that all leading principal submatrices of $\mathbf{A}$ are non-singular and an $n \times 1$ vector $\mathbf{b}$.

Step 1 Form $\widetilde{\mathbf{A}} = [\mathbf{A} : \mathbf{I}_n : \mathbf{b}]$. Denote the (i, j)-th element of $\widetilde{\mathbf{A}}$ by a_{ij} for $i = 1, \ldots, n$ and $j = 1, \ldots, p$, where $p = 2n + 1$. Write 0 in the cells above the diagonal in the matrix $\mathbf{L}$. Write 0 in the cells below the diagonal and 1's in the cells on the diagonal in the $n \times p$ matrix $\widetilde{\mathbf{U}}$. Set $k = 1$.

Step 2 Find $\ell_{kk}, \ell_{k+1,k}, \ldots, \ell_{nk}$ using (5.6.2) and enter them in the appropriate places in $\mathbf{L}$.

Step 3 Find $u_{k,k+1}, u_{k,k+2}, \ldots, u_{kp}$ using (5.6.3) and enter them in the appropriate places in $\widetilde{\mathbf{U}}$.

Step 4 If $k = n$ go to *Step* 5. Otherwise increase k by 1 and go to *Step* 2.

Step 5 Let $\mathbf{U}$ be the submatrix of $\widetilde{\mathbf{U}}$ formed by the first n columns. Get the solution of $\mathbf{Ux} = \widetilde{\mathbf{U}}_{*p}$ by back substitution. Declare it to be the solution of $\mathbf{Ax} = \mathbf{b}$.

Step 6 Find $\mathbf{A}^{-1}$ thus: $(\mathbf{A}^{-1})_{*i}$ is the solution of $\mathbf{Ux} = \widetilde{\mathbf{U}}_{*,n+i}$ obtained by back substitution. Stop.

We now illustrate the use of this algorithm with an example.

Example 5.6.3 We will solve the system

$$\begin{aligned} 2x_1 + 6x_2 + 5x_3 - x_4 &= -3 \\ 3x_1 + 8x_2 + 7x_3 + 2x_4 &= 0 \\ x_2 + x_3 + 2x_4 &= 1 \\ -x_1 + x_2 - 2x_3 - 5x_4 &= -8 \end{aligned} \tag{5.6.4}$$

and find the inverse of the matrix of coefficients by the Gauss-Doolittle method. So we form the matrix

$$\widetilde{\mathbf{A}} = \left[\begin{array}{rrrr|rrrr|r|r} 2 & 6 & 5 & -1 & 1 & 0 & 0 & 0 & -3 & 10 \\ 3 & 8 & 7 & 2 & 0 & 1 & 0 & 0 & 0 & 21 \\ 0 & 1 & 1 & 2 & 0 & 0 & 1 & 0 & 1 & 6 \\ -1 & 1 & -2 & -5 & 0 & 0 & 0 & 1 & -8 & -14 \end{array}\right] \tag{5.6.5}$$

Here we have included an extra column consisting of the row sums which can be used as computational checks.

We set $k = 1$ and find the first column of $\mathbf{L}$. Note that this coincides with the first column of $\mathbf{A}$. (It is convenient to note down $\widetilde{\mathbf{U}}$ below $\widetilde{\mathbf{A}}$ and $\mathbf{L}$ to the left of $\widetilde{\mathbf{U}}$.) We next compute the first row of $\widetilde{\mathbf{U}}$ using $u_{1j} = a_{1j}/a_{11}$ for $j = 2, 3, \ldots, 10$. At this stage we check that $u_{1,10}$ is the sum of $u_{11}, u_{12}, \ldots, u_{19}$. We next increase k to 2 and compute the second column of $\mathbf{L}$ using (5.6.2). For example,

$$\ell_{22} = a_{22} - \ell_{21}u_{12} = 8 - 3 \times 3 = -1$$

$$\ell_{32} = a_{32} - \ell_{31}u_{12} = 1 - 0 \times 3 = 1$$

etc. Then we compute the second row of $\widetilde{\mathbf{U}}$ using (5.6.3). For example,

$$u_{24} = \frac{1}{\ell_{22}}(a_{24} - \ell_{21}u_{14}) = \frac{1}{(-1)}(2 - 3 \times (-0.5)) = -3.5$$

After completing the second row of $\widetilde{\mathbf{U}}$ we check that $u_{2,10}$ is the sum of $u_{22}, u_{23}, \ldots, u_{29}$ (note that $u_{21} = 0$). We then increase k to 3 and compute the third column of $\mathbf{L}$ using (5.6.2). For example,

$$\ell_{33} = a_{33} - \ell_{31}u_{13} - \ell_{32}u_{23} = 1 - 0 \times 2.5 - 1 \times 0.5 = 0.5$$

We then compute the third row of $\widetilde{\mathbf{U}}$ using (5.6.3). Thus

$$u_{34} = \frac{1}{\ell_{33}}(a_{34} - \ell_{31}u_{14} - \ell_{32}u_{24}) = \frac{1}{(0.5)}(2 - 0 \times (-0.5) - 1 \times (-3.5)) = 11$$

etc. We then check that $u_{3,10}$ is the sum of $u_{33}, u_{34}, \ldots, u_{39}$. We next increase k to 4 and compute ℓ_{44} as

$$\begin{aligned}\ell_{44} &= a_{44} - \ell_{41}u_{14} - \ell_{42}u_{24} - \ell_{43}u_{34} \\ &= (-5) - (-1) \times (-0.5) - 4 \times (-3.5) - (-1.5) \times 11 = 25\end{aligned}$$

We then compute the fourth row of $\widetilde{\mathbf{U}}$ and check that $u_{4,10}$ is the sum of $u_{44}, u_{45}, \ldots, u_{49}$. For example,

$$\begin{aligned}u_{49} &= (a_{49} - \ell_{41}u_{19} - \ell_{42}u_{29} - \ell_{43}u_{39})/\ell_{44} \\ &= ((-8) - (-1) \times (-1.5) - 4 \times (-4.5) - (-1.5) \times 11)/25 = 1\end{aligned}$$

We thus have

$$\mathbf{L} = \begin{bmatrix} 2 & 0 & 0 & 0 \\ 3 & -1 & 0 & 0 \\ 0 & 1 & 0.5 & 0 \\ -1 & 4 & -1.5 & 25 \end{bmatrix}$$

and

$$\widetilde{\mathbf{U}} = \left[\begin{array}{cccc|cccc|c|c} 1 & 3 & 2.5 & -0.5 & 0.5 & 0 & 0 & 0 & -1.5 & 5 \\ 0 & 1 & 0.5 & -3.5 & 1.5 & -1 & 0 & 0 & -4.5 & -6 \\ 0 & 0 & 1 & 11 & -3 & 2 & 2 & 0 & 11 & 24 \\ 0 & 0 & 0 & 1 & -0.4 & 0.28 & 0.12 & 0.04 & 1 & 2.04 \end{array}\right]$$

Taking $\mathbf{U}$ as the submatrix of $\widetilde{\mathbf{U}}$ formed by the first four columns, we get the solution of $\mathbf{Ax} = \mathbf{b}$ by solving $\mathbf{Ux} = \widetilde{\mathbf{U}}_{*9}$ by the method of back substitution:

$$\begin{aligned}x_4 &= 1 \\ x_3 &= 11 - 11 \times 1 = 0 \\ x_2 &= (-4.5) - (-3.5) \times 1 - 0.5 \times 0 = -1 \\ x_1 &= (-1.5) - (-0.5) \times 1 - 2.5 \times 0 - 3 \times (-1) = 2\end{aligned}$$

Similarly by solving the systems $\mathbf{Ux} = \widetilde{\mathbf{U}}_{*i}$ for $i = 5, 6, 7$ and 8 we get the four columns of $\mathbf{A}^{-1}$. The reader may verify that

$$\mathbf{A}^{-1} = \begin{bmatrix} -1.4 & 1.28 & -1.88 & 0.04 \\ -0.6 & 0.52 & 0.08 & 0.36 \\ 1.4 & -1.08 & 0.68 & -0.44 \\ -0.4 & 0.28 & 0.12 & 0.04 \end{bmatrix}. \quad \blacksquare$$

We now show that the diagonal elements of $\mathbf{L}$ have a special significance.

Theorem 5.6.4 Let $\mathbf{A}$ be an $n \times n$ matrix with the first $n-1$ leading principal submatrices non-singular. Then there exist unique matrices $\mathbf{L}$ and $\mathbf{U}$ such that $\mathbf{L}$ is an $n \times n$ lower triangular matrix, $\mathbf{U}$ is an $n \times n$ upper triangular matrix with 1's on the diagonal and $\mathbf{A} = \mathbf{LU}$. Further,

$$\ell_{nn} = a_{nn} - \mathbf{c}^{\mathrm{T}}\mathbf{B}^{-1}\mathbf{d} \tag{5.6.6}$$

where

$$\mathbf{A} = \begin{bmatrix} \mathbf{B} & \mathbf{d} \\ \mathbf{c}^{\mathrm{T}} & a_{nn} \end{bmatrix} \tag{5.6.7}$$

Proof The first statement follows from the proof of *Theorem* 5.6.1 (note that ℓ_{nn} may be 0 but $\ell_{11}, \ell_{22}, \ldots, \ell_{n-1,n-1}$ are non-zero). To prove the second statement, partition $\mathbf{L}$ and $\mathbf{U}$ as

$$\mathbf{L} = \begin{bmatrix} \mathbf{L}_1 & \mathbf{0} \\ \mathbf{w}^{\mathrm{T}} & \ell_{nn} \end{bmatrix} \text{ and } \mathbf{U} = \begin{bmatrix} \mathbf{U}_1 & \mathbf{u} \\ \mathbf{0} & 1 \end{bmatrix} \tag{5.6.8}$$

Then $\mathbf{A} = \mathbf{LU}$ gives $\mathbf{B} = \mathbf{L}_1\mathbf{U}_1$, $\mathbf{d} = \mathbf{L}_1\mathbf{u}$, $\mathbf{c}^{\mathrm{T}} = \mathbf{w}^{\mathrm{T}}\mathbf{U}_1$ and $a_{nn} = \mathbf{w}^{\mathrm{T}}\mathbf{u} + \ell_{nn}$. Noting that $\mathbf{L}_1$ and $\mathbf{U}_1$ are non-singular, we have

$$a_{nn} - \mathbf{c}^{\mathrm{T}}\mathbf{B}^{-1}\mathbf{d} = \mathbf{w}^{\mathrm{T}}\mathbf{u} + \ell_{nn} - \mathbf{w}^{\mathrm{T}}\mathbf{U}_1\mathbf{U}_1^{-1}\mathbf{L}_1^{-1}\mathbf{L}_1\mathbf{u} = \ell_{nn}. \qquad \blacksquare$$

The preceding theorem can be used to find the value of the linear form $\mathbf{c}^{\mathrm{T}}\mathbf{x}$ at the solution of $\mathbf{Bx} = \mathbf{d}$ without actually finding the solution. For example, suppose we want the value of $2x_1 + 3x_2 + x_3 - 2x_4$ at the solution of the system (5.6.4) of *Example* 5.6.3. We then treat the ninth column of (5.6.5) as the fifth column and attach another row $(2, 3, 1, -2, 0)$ at the bottom. We then compute $\ell_{51}, \ldots, \ell_{55}$. It may be checked that $\ell_{51} = 2$, $\ell_{52} = -3$, $\ell_{53} = -2.5$, $\ell_{54} = 16$ and $\ell_{55} = 1$. Thus the required value of the linear form is -1 and this can be verified from the solution.

For those knowing Statistics: It follows from *Theorem* 5.6.4 that if $\mathbf{A}$ is the variance-covariance matrix of random variables $X_1, X_2, \ldots, X_n$ then ℓ_{nn} is the variance of the residual $X_{n\cdot 12\ldots(n-1)}$. It can also be proved that $(\ell_{kk}, \ell_{k+1,k}, \ldots, \ell_{nk})^{\mathrm{T}}$ is the first column of the variance-covariance matrix of the residuals $X_{k\cdot 12\ldots(k-1)}, \ldots, X_{n\cdot 12\ldots(k-1)}$.

Exercises

1. In each of the following cases, solve the system $\mathbf{Ax} = \mathbf{b}$ and find $\mathbf{A}^{-1}$ by the compact form of the Gauss-Doolittle method. Also find the sum of the components of the solution vector using *Theorem* 5.6.4. Verify your answers.

(a) $\mathbf{A}$ and $\mathbf{b}$ as in *Exercise* 5.5.3 (a).

(b) $\mathbf{A}$ and $\mathbf{b}$ as in *Exercise* 5.5.3 (e).

(c) $\mathbf{A}$ and $\mathbf{b}$ as in *Exercise* 5.5.3 (f).

(d) $\mathbf{A} = \begin{bmatrix} 0.12 & 0.02 & 0.05 \\ -0.06 & 0.34 & 0.10 \\ 0.20 & 0.08 & 0.26 \end{bmatrix}$ and $\mathbf{b} = \begin{bmatrix} -0.01 \\ 1.48 \\ 0.50 \end{bmatrix}$.

2. Deduce the second conclusion of *Theorem* 5.6.4 from *Theorems* 3.8.1 and 3.8.2 assuming that $\mathbf{A}$ is also non-singular.
3. Deduce the uniqueness of $\mathbf{L}$ and $\mathbf{U}$ in *Theorem* 5.6.1 from *Theorem* 3.8.2.
4. Let $\widetilde{\mathbf{A}}$ be as in *Theorem* 5.6.1. Then prove that there exist $\mathbf{L}$ and $\widetilde{\mathbf{U}}$ such that

   (a) $\mathbf{L}$ is an $n \times n$ lower triangular matrix,

   (b) $\widetilde{\mathbf{U}}$ is an $n \times p$ matrix such that the submatrix formed by the first n columns is upper triangular,

   (c) $\widetilde{\mathbf{A}} = \mathbf{L}\widetilde{\mathbf{U}}$ and

   (d) *either* $\ell_{11}, \ell_{22}, \ldots, \ell_{nn}$ are preassigned non-zero numbers *or* $u_{11}, u_{22}, \ldots, u_{nn}$ are preassigned non-zero numbers.
5. Let $\widetilde{\mathbf{A}}$ be as in *Theorem* 5.6.1. Then show that there exist unique matrices $\mathbf{L}$, $\widetilde{\mathbf{U}}$ and $\mathbf{D}$ such that the following hold: (i) and (ii) of the theorem, $\ell_{ii} = 1$ for $i = 1, \ldots, n$, $\mathbf{D}$ is a diagonal matrix and $\widetilde{\mathbf{A}} = \mathbf{L}\mathbf{D}\widetilde{\mathbf{U}}$.
6. If $\mathbf{A} = \mathbf{L}\mathbf{U}$, where $\mathbf{L}$ and $\mathbf{U}$ are non-singular lower and upper triangular matrices, show that the leading principal minors of $\mathbf{A}$ are non-singular.
7. Show that an $n \times n$ matrix $\mathbf{A}$ can be written as $\mathbf{U}\mathbf{L}$ where $\mathbf{U}$ and $\mathbf{L}$ are non-singular upper and lower triangular matrices respectively iff the principal submatrix of $\mathbf{A}$ formed by the last k rows is non-singular for $k = 1, 2, \ldots, n$. (Hint: use *Exercise* 2.6.14.)

Chapter 6
Determinants

6.1 Introduction

Determinant is a scalar associated with a square matrix in a particular way. One of the most important uses of determinants within Linear Algebra is in the study of eigenvalues (*Chapter* 8). They also occur in Cramer's rule for solving linear equations and can be used to give a formula for the inverse of a non-singular matrix. In the Calculus of several variables, the Jacobian used in transforming a multiple integral uses determinant. This use arises from the fact that determinant is the volume of a certain parallelopiped. Determinants are also useful in various other subjects like Physics, Astronomy and Statistics.

We continue to study matrices over a general field. In fact the results in this chapter which do not involve inverse and rank hold for matrices over any integral domain. This comment will be used in *Chapter* 8.

Exercise

1. Let $P = (\alpha_1, \alpha_2)$ and $Q = (\beta_1, \beta_2)$ be two points in $\mathbb{R}^2$. Show that the area of the parallelogram $OPRQ$, where O is the origin, is $\alpha_1\beta_2 - \alpha_2\beta_1$. (By convention, the area is positive or negative according as the direction $OPRQ$ is counter-clockwise or clockwise.) If $A = (\gamma_1, \gamma_2)$, show that the area of the parallelogram $APSQ$ is
$$(\alpha_1 - \gamma_1)(\beta_2 - \gamma_2) - (\alpha_2 - \gamma_2)(\beta_1 - \gamma_1)$$
What is the area of the triangle APQ?

6.2 Permutations

In this section we give some elementary properties of permutations which will be needed in the definition of determinant. The reader may skip the proofs if he is willing to assume the results.

Definition 6.2.1 A *permutation* is a 1–1 map from a finite non-empty set S onto itself.

If π is a permutation of $\{s_1, s_2, \ldots, s_n\}$, we sometimes denote $\pi(s_i)$ by π_{s_i} for convenience and represent π as

$$\pi = \begin{pmatrix} s_1 & s_2 & \cdots & s_n \\ \pi_{s_1} & \pi_{s_2} & \cdots & \pi_{s_n} \end{pmatrix}$$

We treat $\left(\begin{smallmatrix} 1 & 2 & 3 & 4 \\ 2 & 4 & 3 & 1 \end{smallmatrix}\right)$ and $\left(\begin{smallmatrix} 1 & 2 & 4 \\ 2 & 4 & 1 \end{smallmatrix}\right)$ as the same permutation since fixed points do not matter for our purpose.

The *resultant* or *product* $\pi\theta$ of two permutations π and θ is again a permutation. In general, $\pi\theta \neq \theta\pi$. For example, if

$$\pi = \begin{pmatrix} 1 & 2 & 3 & 4 \\ 1 & 3 & 4 & 2 \end{pmatrix} \text{ and } \theta = \begin{pmatrix} 3 & 4 & 5 \\ 5 & 3 & 4 \end{pmatrix}, \tag{6.2.1}$$

$(\pi\theta)(3) = \pi(\theta(3)) = \pi(5) = 5$ whereas $(\theta\pi)(3) = \theta(\pi(3)) = \theta(4) = 3$. However, permutations of disjoint subsets commute. Also, multiplication of permutations is associative. The *identity permutation* leaves every element fixed and is denoted by 1. Every permutation π has a unique inverse π^{-1} such that $\pi\pi^{-1} = \pi^{-1}\pi = 1$. For example, the inverse of the π in (6.2.1) is $\left(\begin{smallmatrix} 1 & 2 & 3 & 4 \\ 1 & 4 & 2 & 3 \end{smallmatrix}\right)$. It is easy to see that $(\pi\theta)^{-1} = \theta^{-1}\pi^{-1}$ for any permutations π and θ.

Definition 6.2.2 A *cyclic permutation* is a permutation of the form

$$[t_1\ t_2\ \cdots\ t_k]^{\dagger} := \begin{pmatrix} t_1 & t_2 & \cdots & t_{k-1} & t_k \\ t_2 & t_3 & \cdots & t_k & t_1 \end{pmatrix} \tag{6.2.2}$$

where $t_1, \ldots, t_k$ are distinct and $k \geq 1$.

We call $\{t_1, \ldots, t_k\}$ the *support* and k the *length* of the cycle (6.2.2). Note that (6.2.2) is the identity permutation when $k = 1$. The cycle θ of (6.2.1) is [3 5 4]. We say that two cyclic permutations are *disjoint* if their supports are disjoint. Similarly we say that the cycles $\theta_1, \theta_2, \ldots, \theta_p$ *cover* S if the union of their supports contains S.

Theorem 6.2.3 Any permutation π of S is a product of pair-wise disjoint cycles (covering S) which are uniquely determined by π.

---

† The usual notation is $(t_1\ t_2\ \cdots\ t_k)$ but we avoid this as parentheses are used for many things.

Proof We first prove that π is a product of disjoint cycles covering S. Choose any element s of S. Consider $s, \pi(s), \pi^2(s), \ldots$. Since S is finite, there exist non-negative integers i and k with $i < k$ such that $\pi^k(s) = \pi^i(s)$. Choose such a pair with k smallest possible. If $i \geq 1$ then $\pi^{k-1}(s) = \pi^{i-1}(s)$ since π is one-to-one, a contradiction to the minimality of k. Thus $i = 0$ and $\theta_1 := [s, \pi(s), \ldots, \pi^{k-1}(s)]$ is a cycle of π. If $\pi = \theta_1$, we are done. Otherwise we start with an element t not in the support of θ_1 and obtain a cycle θ_2 as above. Clearly the cycles θ_1 and θ_2 are disjoint. Proceeding thus, we can express π as a product $\theta_1\theta_2 \cdots \theta_p$ of disjoint cycles which cover S.

To prove uniqueness, let $\pi = \theta_1\theta_2 \cdots \theta_p$ where the θ's are disjoint cycles covering S. Let s_i be any element of the support S_i of θ_i. Then S_i is the set of all distinct elements among $s_i, \pi(s_i), \pi^2(s_i), \ldots$ and θ_i is the restriction of π to S_i. Thus the θ_i's are uniquely determined by π. ∎

The representation of a permutation π of S as a product of pair-wise disjoint cycles covering S is called the *disjoint cycle decomposition* of π. For example, the disjoint cycle decomposition of the permutation

$$\begin{pmatrix} 1 & 2 & 3 & 4 & 5 & 6 & 7 & 8 \\ 3 & 2 & 6 & 1 & 8 & 4 & 7 & 5 \end{pmatrix} \tag{6.2.3}$$

of the set $\{1, 2, \ldots, 8\}$ is $[1\ 3\ 6\ 4]\,[2]\,[5\ 8]\,[7]$.

A cycle of length 2 interchanges two elements and is called a *transposition*. Since

$$[t_1\ t_2\ \ldots\ t_k] = [t_1\ t_k]\,[t_1\ t_{k-1}] \cdots [t_1\ t_2],$$

every cycle and, so, every permutation π is a product of (not necessarily disjoint) transpositions. Note that such a representation is not unique. For example, $[1\ 2\ 3\ 4] = [1\ 4]\,[1\ 3]\,[1\ 2] = [1\ 4]\,[3\ 4]\,[2\ 4]\,[3\ 4]\,[1\ 3]$. However, we show below that the number of transpositions has the same parity in all such representations of π. For this we need the following

Definition 6.2.4 Let π be a permutation of a set S of size n and p the number of cycles (including cycles of length 1) in the disjoint cycle decomposition of π. Then $(-1)^{n-p}$ is called the *sign* of π and is denoted by $\epsilon(\pi)$. Also π is said to be *even* or *odd* according as $\epsilon(\pi)$ is $+1$ or -1.

Note that $\epsilon(\pi)$ does not change if π is considered as a permutation of a superset of S since then n and p increase by the same number. The identity permutation is even since $p = n$ for it. A cycle is even or odd according as its length is odd or even. In particular, a transposition is odd.

Theorem 6.2.5 If π is a permutation and σ is a transposition then

$$\epsilon(\pi\sigma) = \epsilon(\sigma\pi) = -\epsilon(\pi)$$

Proof We will prove $\epsilon(\pi\sigma) = -\epsilon(\pi)$; the proof of $\epsilon(\sigma\pi) = -\epsilon(\pi)$ is similar. Let $\theta_1\theta_2\cdots\theta_p$ be the disjoint cycle decomposition of π and let $\sigma = [uv]$. Let S_i denote the support of θ_i. We consider two cases.

Case (i). $u, v \in S_i$ (for some i). Let $\theta_i = [ux_1\cdots x_k v y_1\cdots y_\ell]$, where k and ℓ are non-negative integers. Now

$$[ux_1\cdots x_k v y_1\cdots y_\ell]\,[uv] = [uy_1\cdots y_\ell]\,[vx_1\cdots x_k] \tag{6.2.4}$$

Thus the cycle θ_i in the disjoint cycle decomposition of π splits into two in that of $\pi\sigma$, so $\epsilon(\pi\sigma) = -\epsilon(\pi)$.

Case (ii). $u \in S_i$ and $v \in S_j$ with $i \neq j$. Let $\theta_i = [uy_1\cdots y_\ell]$ and $\theta_j = [vx_1\cdots x_k]$ where k and ℓ are non-negative integers. Then post-multiplying both sides of (6.2.4) by $[uv]$ we get

$$\theta_i\theta_j[uv] = [ux_1\cdots x_k v y_1\cdots y_\ell]$$

Thus the two cycles θ_i and θ_j of π are combined into one cycle in $\pi\sigma$, hence $\epsilon(\pi\sigma) = -\epsilon(\pi)$. ∎

Theorem 6.2.6 For all permutations π and θ, $\epsilon(\pi\theta) = \epsilon(\pi)\epsilon(\theta)$.

Proof We can express θ as a product of transpositions, say, $\theta = \sigma_1\sigma_2\cdots\sigma_t$. Then by the preceding theorem we have

$$\epsilon(\pi\theta) = \epsilon(\pi\sigma_1\sigma_2\cdots\sigma_t) = -\epsilon(\pi\sigma_1\sigma_2\cdots\sigma_{t-1}) = \cdots = (-1)^t\epsilon(\pi)$$

Since this is true for all π, we get $\epsilon(\theta) = (-1)^t$ by taking $\pi = 1$. ∎

The preceding theorem shows that the product of two even as well as two odd permutations is even and the product of an even permutation and an odd permutation (in either order) is odd. It follows that the product of k transpositions is even or odd according as k is even or odd. Taking $\theta = \pi^{-1}$ in the preceding theorem, we get

Theorem 6.2.7 For any permutation π, $\epsilon(\pi^{-1}) = \epsilon(\pi)$.

In the next few sections we will be interested in permutations of the set $S = \{1, 2, \ldots, n\}$. *For convenience we will denote the permutation*

$$\begin{pmatrix} 1 & 2 & \cdots & n \\ j_1 & j_2 & \cdots & j_n \end{pmatrix} \tag{6.2.5}$$

by $(j_1\ j_2\ \dots\ j_n)$. Note that $(1\ 2\ \dots\ n)$ is the identity permutation. It is easy to see that

$$(j_1\ j_2\ \dots\ j_n)(k_1\ k_2\ \dots\ k_n) = (j_{k_1}\ j_{k_2}\ \cdots\ j_{k_n})$$

Postmultiplication of $(j_1\ j_2\ \cdots\ j_n)$ by $[k\ \ell]$ interchanges j_k and j_ℓ while premultiplication interchanges k and ℓ wherever they may occur. For example,

$$(2\ 4\ 5\ 3\ 1)[2\ 5] = (2\ 1\ 5\ 3\ 4) \quad \text{and} \quad [2\ 5](2\ 4\ 5\ 3\ 1) = (5\ 4\ 2\ 3\ 1)$$

The inverse of $(2\ 4\ 5\ 3\ 1)$ is $(5\ 1\ 4\ 2\ 3)$.

The next result gives the number of interchanges needed to bring specified rows or columns of a matrix to the first few positions without disturbing the order of the others.

Theorem 6.2.8 Let $1 \le i_1 < i_2 < \cdots < i_k \le n$ and let $i_{k+1}, i_{k+2}, \dots, i_n$ be the integers in $\{1, 2, \dots, n\} - \{i_1, i_2, \dots, i_k\}$ written in increasing order. Then $\pi = (i_1, i_2, \dots, i_n)$ can be obtained from $(1, 2, \dots, n)$ by

$$t(\pi) = i_1 + i_2 + \cdots + i_k - \frac{k(k+1)}{2}$$

transpositions and $\epsilon(\pi) = (-1)^{t(\pi)}$.

Proof We first note that for any p, the element in the p-th position can be brought to the first position without disturbing the order of the other elements by $p-1$ transpositions. Now start with $(1, 2, \dots, n)$ and bring the i_1-th element to the first position by $i_1 - 1$ transpositions. Next bring the i_2-th element (note that its position has not changed) to the second position by $i_2 - 2$ transpositions. Proceeding thus we arrive at π after

$$(i_1 - 1) + (i_2 - 2) + \cdots + (i_k - k) = i_1 + i_2 + \cdots + i_k - \frac{k(k+1)}{2}$$

transpositions and the theorem follows. ∎

Exercises

1. If $\pi = \begin{pmatrix} 1 & 2 & 3 & 4 \\ 2 & 4 & 3 & 1 \end{pmatrix}$, $\theta = \begin{pmatrix} 2 & 3 & 5 & 7 \\ 3 & 7 & 2 & 5 \end{pmatrix}$ and $\eta = \begin{pmatrix} 1 & 3 & 5 & 8 & 9 \\ 9 & 3 & 8 & 1 & 5 \end{pmatrix}$, find $\pi\theta$, $\theta\eta$, $\pi\theta\eta$, $\theta\eta\pi$, π^{-1}, θ^{-1}, $\pi^{-1}\theta^{-1}$, $\theta^{-1}\pi^{-1}$ and $(\pi\theta)^{-1}$.

2. Find the disjoint cycle decomposition of each of the following permutations:
(a) $\begin{pmatrix} 1 & 2 \\ 1 & 2 \end{pmatrix}$, (b) $\begin{pmatrix} 1 & 2 \\ 2 & 1 \end{pmatrix}$ and (c) $\begin{pmatrix} 1 & 2 & 3 & 4 & 5 & 6 & 7 & 8 & 9 \\ 2 & 7 & 8 & 4 & 9 & 5 & 1 & 3 & 6 \end{pmatrix}$.
3. What is the permutation $[s_1\ s_2]\,[s_1\ s_3]\cdots[s_1\ s_k]$?
4. If σ is a cycle of length k, show that $\sigma^p = 1$ iff k divides p. If $n_1, n_2, \ldots, n_t$ are the lengths of the cycles in the disjoint cycle decomposition of π, when is $\pi^p = 1$?
5. Prove *Theorem* 6.2.7 from definitions.
6. Show that the set $\mathcal{S}$ of all permutation matrices of order n forms a group under matrix multiplication and the set $\mathcal{T}$ of all permutations of $\{1, \ldots, n\}$ written in the form $(j_1, j_2, \ldots, j_n)$ forms a group under composition. Show also that $\varphi : \mathbf{P} \mapsto (1, 2, \ldots, n)\mathbf{P}$ is an isomorphism from $\mathcal{S}$ to $\mathcal{T}$.
7. (a) Find [3 6](2 4 3 6 5 1) and (2 4 3 6 5 1)[3 6].
   (b) Find [6 5 1](2 4 3 6 5 1).
   (c) Express the inverse of (2 4 3 6 5 1) in the same form.
8. If $\pi = (j_1, j_2, \ldots, j_n)$ is a permutation of $\{1, \ldots, n\}$ and θ is obtained from π by interchanging two j's, show that $\epsilon(\theta) = -\epsilon(\pi)$. Deduce that if π can be converted to $(1\ 2\ \ldots\ n)$ by k interchanges then $\epsilon(\pi) = (-1)^k$.
9. If $\pi = (j_1, j_2, \ldots, j_n)$ is a permutation of $\{1, 2, \ldots, n\}$, let $s(\pi)$ denote the number of *inversions* in π, i.e., the number of ordered pairs (k, ℓ) such that $1 \leq k < \ell \leq n$ and $j_k > j_\ell$. Show that π is even or odd according as $s(\pi)$ is even or odd. (Hint: If $j_k = n$ then by premultiplying π by a cycle of length $n-k+1$, n can be taken to the last position without altering the order of the other terms of π, see *Exercise* 6.2.7(b).)
10. (a) Show that the permutation $(n, n-1, \ldots, 1)$ is even iff n is of the form $4k$ or $4k+1$.
    (b) Deduce from the result in (a) that it is not possible to reverse the order of the numbers $1, 2, \ldots, 15$ in the '15-squares puzzle'.

6.3 Determinant and its elementary properties

Definition 6.3.1 If $\mathbf{A} = ((a_{ij}))$ is a square matrix of order n, the *determinant* of $\mathbf{A}$ is

$$|\mathbf{A}| = \sum \epsilon(j_1\ j_2\ \ldots\ j_n) a_{1j_1} a_{2j_2} \cdots a_{nj_n} \tag{6.3.1}$$

where $(j_1\ j_2\ \ldots\ j_n)$ runs over the $n!$ permutations of $(1, 2, \ldots, n)$.

The determinant of (a_{11}) is a_{11}. Since $\epsilon(1\ 2) = 1$ and $\epsilon(2\ 1) = -1$,

$$\begin{vmatrix} a_{11} & a_{12} \\ a_{21} & a_{22} \end{vmatrix} = a_{11}a_{22} - a_{12}a_{21} \tag{6.3.2}$$

Note that we drop the brackets when we use vertical bars. When $n = 3$, it can be verified that

$$\epsilon(1\ 2\ 3) = \epsilon(2\ 3\ 1) = \epsilon(3\ 1\ 2) = +1$$

and

$$\epsilon(1\ 3\ 2) = \epsilon(2\ 1\ 3) = \epsilon(3\ 2\ 1) = -1,$$

so

$$\begin{vmatrix} a_{11} & a_{12} & a_{13} \\ a_{21} & a_{22} & a_{23} \\ a_{31} & a_{32} & a_{33} \end{vmatrix} = \begin{aligned} &a_{11}a_{22}a_{33} - a_{11}a_{23}a_{32} + a_{12}a_{23}a_{31} \\ &- a_{12}a_{21}a_{33} + a_{13}a_{21}a_{32} - a_{13}a_{22}a_{31} \end{aligned} \tag{6.3.3}$$

For higher order determinants, evaluation from the definition is cumbersome; a more efficient method is given after *Theorem* 6.3.8.

Each term in the expression (6.3.1) contains exactly one element from each row and exactly one element from each column of the matrix. Hence it follows that the determinant of $\mathbf{A}$ is 0 if $\mathbf{A}$ has either a null row or a null column. Noting that the coefficient of $a_{11}a_{22}\cdots a_{nn}$ is $+1$, it also follows that the determinant of $\mathbf{I}$ is 1 and the determinant of a diagonal matrix is the product of the diagonal elements. More generally we have

Theorem 6.3.2 The determinant of a triangular matrix is the product of the diagonal elements.

Proof We prove the result for an upper triangular matrix $\mathbf{A}$ of order n. The proof for lower triangular matrices is similar (see also the next theorem). Suppose the term $a_{1j_1}a_{2j_2}\cdots a_{nj_n}$ in (6.3.1) is nonzero. Then $j_n = n$. Noting that $j_1, j_2, \ldots, j_n$ are distinct it follows that $j_{n-1} = n-1, \ldots, j_1 = 1$. Thus all terms other than $a_{11}a_{22}\cdots a_{nn}$ are zero and the theorem follows. ∎

Theorem 6.3.3 For any square matrix $\mathbf{A}$, $|\mathbf{A}^T| = |\mathbf{A}|$.

Proof Let $\mathbf{A}$ be an $n \times n$ matrix. Then

$$|\mathbf{A}| = \textstyle\sum_\pi \epsilon(\pi)\, a_{1,\pi(1)}a_{2,\pi(2)}\cdots a_{n,\pi(n)} \tag{6.3.4}$$

where π runs over all permutations of $S = \{1, 2, \ldots, n\}$. Since $(\pi^{-1})^{-1} = \pi$, it follows that as π runs over the permutations of S, so does π^{-1}.

Writing θ for π^{-1} it is easy to see that

$$\epsilon(\pi)\, a_{1,\pi(1)} a_{2,\pi(2)} \cdots a_{n,\pi(n)} = \epsilon(\theta)\, a_{\theta(1),1} a_{\theta(2),2} \cdots a_{\theta(n),n} \qquad (6.3.5)$$

For example, if $n = 4$ and $\pi = (2\ 4\ 3\ 1)$ then $\pi^{-1} = (4\ 1\ 3\ 2)$ and (6.3.5) becomes $a_{12}a_{24}a_{33}a_{41} = a_{41}a_{12}a_{33}a_{24}$. Thus

$$|\mathbf{A}| = \textstyle\sum_\theta \epsilon(\theta)\, a_{\theta_1,1} a_{\theta_2,2} \cdots a_{\theta_n,n} \qquad (6.3.6)$$

where $\theta = (\theta_1\ \theta_2\ \ldots\ \theta_n)$ runs over all permutations of S. Clearly, the RHS of (6.3.6) is the determinant of $\mathbf{A}^{\mathrm{T}}$. ∎

The preceding theorem shows that results on determinant proved with respect to rows can also be proved with respect to columns and vice versa.

Theorem 6.3.4 If $\mathbf{B}$ is obtained from $\mathbf{A}$ by multiplying any one row or column by α, then $|\mathbf{B}| = \alpha|\mathbf{A}|$.

Proof Let $\mathbf{B}$ be obtained from $\mathbf{A}$ by $R_i(\alpha)$. Then

$$\begin{aligned} |\mathbf{B}| &= \textstyle\sum_\pi \epsilon(\pi)\, b_{1,\pi(1)} b_{2,\pi(2)} \cdots b_{n,\pi(n)} \\ &= \textstyle\sum_\pi \epsilon(\pi)\, a_{1,\pi_1} \cdots a_{k-1,\pi_{k-1}} (\alpha a_{k,\pi_k}) a_{k+1,\pi_{k+1}} \cdots a_{n,\pi_n} \\ &= \alpha \textstyle\sum_\pi \epsilon(\pi)\, a_{1,\pi_1} a_{2,\pi_2} \cdots a_{n,\pi_n} \\ &= \alpha|\mathbf{A}|. \end{aligned}$$

∎

Theorem 6.3.5 If $\mathbf{B}$ is obtained from $\mathbf{A}$ by interchanging two rows or by interchanging two columns, $|\mathbf{B}| = -|\mathbf{A}|$.

Proof We will prove the theorem when $\mathbf{B}$ is obtained from $\mathbf{A}$ by $R_{k\ell}$, where $k < \ell$. Clearly

$$\begin{aligned} |\mathbf{B}| &= \textstyle\sum_\pi \epsilon(\pi)\, b_{1,\pi_1} \cdots b_{k,\pi_k} \cdots b_{\ell,\pi_\ell} \cdots b_{n,\pi_n} \\ &= \textstyle\sum_\pi \epsilon(\pi)\, a_{1,\pi_1} \cdots a_{\ell,\pi_k} \cdots a_{k,\pi_\ell} \cdots a_{n,\pi_n} \qquad (6.3.7) \end{aligned}$$

Now, as π runs over the distinct permutations of $\{1, 2, \ldots, n\}$, so does $\theta := \pi[k\ \ell]$ since any permutation η is $\pi[k\ \ell]$ where $\pi = \eta[k\ \ell]$. Also $\epsilon(\theta) = -\epsilon(\pi)$. Since θ is $(\pi_1\ \pi_2\ \ldots\ \pi_n)$ with π_k and π_ℓ interchanged, (6.3.7) can be rewritten as

$$|\mathbf{B}| = -\sum_\theta \epsilon(\theta)\, a_{1,\theta_1} a_{2,\theta_2} \cdots a_{n,\theta_n} = -|\mathbf{A}|.$$

∎

Theorem 6.3.6 If two rows or two columns of $\mathbf{A}$ are equal, $|\mathbf{A}| = 0$.

Proof Suppose the k-th and the ℓ-th rows of $\mathbf{A}$ are equal. Interchanging these rows will not alter the matrix. So $|\mathbf{A}| = -|\mathbf{A}|$ or $|\mathbf{A}|(1+1) = 0$. If $1+1 \neq 0$, we immediately have $|\mathbf{A}| = 0$.

Suppose now $1+1 = 0$. Then $\alpha + \alpha = 0$ for every scalar α. Since $-1 = +1$, we have

$$|\mathbf{A}| = \textstyle\sum_\pi a_{1,\pi_1} a_{2,\pi_2} \cdots a_{n,\pi_n}$$

We can pair off each term $a_{1,\pi_1} a_{2,\pi_2} \cdots a_{n,\pi_n}$ here with the term

$$a_{1,\pi_1} \cdots a_{k-1,\pi_{k-1}} a_{k,\pi_\ell} a_{k+1,\pi_{k+1}} \cdots a_{\ell-1,\pi_{\ell-1}} a_{\ell,\pi_k} a_{\ell+1,\pi_{\ell+1}} \cdots a_{n,\pi_n}$$

Observe that these two terms have the same value since the k-th and ℓ-th rows of $\mathbf{A}$ are equal. Thus their sum is 0 and it follows that $|\mathbf{A}| = 0$. ∎

Theorem 6.3.7 For a fixed k, let the k-th row of $\mathbf{A}$ be the sum of two row vectors $\mathbf{x}^{\mathrm{T}}$ and $\mathbf{y}^{\mathrm{T}}$. Then $|\mathbf{A}| = |\mathbf{B}| + |\mathbf{C}|$ where $\mathbf{B}$ (resp. $\mathbf{C}$) is obtained from $\mathbf{A}$ by replacing the k-th row by $\mathbf{x}^{\mathrm{T}}$ (resp. $\mathbf{y}^{\mathrm{T}}$).

Proof We have

$$\begin{aligned}
|\mathbf{A}| &= \textstyle\sum_\pi \epsilon(\pi) a_{1,\pi_1} \cdots a_{k-1,\pi_{k-1}} (x_{\pi_k} + y_{\pi_k}) a_{k+1,\pi_{k+1}} \cdots a_{n,\pi_n} \\
&= \textstyle\sum_\pi \epsilon(\pi) a_{1,\pi_1} \cdots a_{k-1,\pi_{k-1}} x_{\pi_k} a_{k+1,\pi_{k+1}} \cdots a_{n,\pi_n} \\
&\quad + \textstyle\sum_\pi \epsilon(\pi) a_{1,\pi_1} \cdots a_{k-1,\pi_{k-1}} y_{\pi_k} a_{k+1,\pi_{k+1}} \cdots a_{n,\pi_n} \\
&= |\mathbf{B}| + |\mathbf{C}|.
\end{aligned}$$ ∎

An analogous result can be proved for columns. *Theorems* 6.3.4 and 6.3.7 show that the determinant is a linear function of the k-th row when the other rows are kept fixed. The next theorem is an immediate consequence of *Theorems* 6.3.4, 6.3.6 and 6.3.7.

Theorem 6.3.8 If a scalar multiple of one row (resp. column) is added to another row (resp. column) of a matrix, the determinant of the matrix is not altered.

The results proved above give an efficient method for evaluating $|\mathbf{A}|$: we reduce $\mathbf{A}$ to an upper triangular matrix $\mathbf{B}$ using elementary row operations. Suppose $\alpha_1, \alpha_2, \ldots, \alpha_p$ are the scalars used in the row operations of the type 'multiplying a row by a non-zero scalar' and suppose q interchanges of rows are used. Then clearly

$$|\mathbf{A}| = (-1)^q \frac{b_{11} b_{22} \cdots b_{nn}}{\alpha_1 \alpha_2 \cdots \alpha_p} \tag{6.3.8}$$

The above equation has an important consequence. Since the α_i's are non-zero, it shows that $|\mathbf{A}| \neq 0$ iff all the diagonal elements of $\mathbf{B}$ are non-zero, i.e., iff $\mathbf{B}$ and so $\mathbf{A}$ are non-singular. We thus have

Theorem 6.3.9 A square matrix $\mathbf{A}$ is non-singular iff $|\mathbf{A}| \neq 0$.

Let us illustrate the above procedure with the matrix

$$\mathbf{A} = \begin{bmatrix} 0 & 0 & 1 & 2 \\ 2 & 4 & 6 & 2 \\ 1 & 3 & 5 & 1 \\ 4 & 5 & 9 & 6 \end{bmatrix}$$

We reduce $\mathbf{A}$ to an upper triangular matrix by using *Algorithm* 4.4.6. It can be checked that the sequence of row operations used is: R_{12}, $R_{31}(-\frac{1}{2})$, $R_{41}(-2)$, R_{23}, $R_{42}(3)$, $R_{43}(-3)$ and the final matrix is

$$\mathbf{B} = \begin{bmatrix} 2 & 4 & 6 & 2 \\ 0 & 1 & 2 & 0 \\ 0 & 0 & 1 & 2 \\ 0 & 0 & 0 & -4 \end{bmatrix}$$

Since exactly two interchanges have been used in this process and no multiplications of rows by non-zero scalars are used, we get

$$|\mathbf{A}| = (-1)^2(2)(1)(1)(-4) = -8$$

Sometimes the determinant of a matrix can be evaluated more efficiently by utilizing its structure to reduce it to a triangular matrix instead of applying *Algorithm* 4.4.6. We illustrate this with

Example 6.3.10 Consider the $n \times n$ matrix

$$\mathbf{A} = \begin{bmatrix} a & b & b & \cdots & b \\ b & a & b & \cdots & b \\ \cdots & \cdots & \cdots & \cdots & \cdots \\ b & b & b & \cdots & a \end{bmatrix}$$

We first add the second, third, ..., n-th columns to the first column and then subtract the first row from each of the second, third, ..., n-th rows. Then we obtain the upper triangular matrix

$$\mathbf{B} = \begin{bmatrix} a+(n-1)b & b & b & \cdots & b \\ 0 & a-b & 0 & \cdots & 0 \\ \cdots & \cdots & \cdots & \cdots & \cdots \\ 0 & 0 & 0 & \cdots & a-b \end{bmatrix}$$

Hence $|\mathbf{A}| = |\mathbf{B}| = \big(a+(n-1)b\big)(a-b)^{n-1}$. ∎

Exercises

1. If $\mathbf{A}$ is of order 7, find the signs of the terms: $a_{13}a_{26}a_{32}a_{45}a_{54}a_{67}a_{71}$, $a_{41}a_{52}a_{73}a_{34}a_{15}a_{66}a_{27}$ and $a_{17}a_{24}a_{35}a_{41}a_{53}a_{66}a_{72}$ in the expansion of $|\mathbf{A}|$.
2. What is $|\alpha\mathbf{A}|$ if $\mathbf{A}$ is of order n?
3. Show that the determinant of a real skew-symmetric matrix of odd order is 0.
4. If $\mathbf{A}$ is a real skew-symmetric matrix of order 2 or 4, express $|\mathbf{A}|$ as the square of a polynomial in the elements above the diagonal. (This result is, in fact, true for any even order.)
5. Show that $|\mathbf{E}_{ij}| = -1$, $|\mathbf{E}_i(\alpha)| = \alpha$ and $|\mathbf{E}_{ij}(\beta)| = 1$.
6. Prove that
$$\begin{vmatrix} 1+a_1 & a_2 & \cdots & a_n \\ a_1 & 1+a_2 & \cdots & a_n \\ \cdots & \cdots & \cdots & \cdots \\ a_1 & a_2 & \cdots & 1+a_n \end{vmatrix} = 1 + a_1 + \cdots + a_n$$
7. Let $x_1 \le x_2 \le \cdots \le x_n$ be real numbers and $\mathbf{A}$ the $n\times n$ matrix $((a_{ij}))$ with $a_{ij} = |x_i - x_j|$. Show that
$$|\mathbf{A}| = 2^{n-2}(x_n - x_1)\prod_{i=1}^{n-1}(x_i - x_{i+1})$$
(Hint: subtract the $(i+1)$-th column from the i-th column for $i = 1, 2, \ldots, n-1$ in that order and then add the first row to the others.)
8. If $\mathbf{A}$ is the real $n\times n$ matrix $((a_{ij}))$ where $a_{ij} = \rho^{|i-j|}$, show that $|\mathbf{A}| = (1-\rho^2)^{n-1}$.
9. Show that if the permutation matrix $\mathbf{P}$ corresponds to π under the map in *Exercise* 6.2.6, then $|\mathbf{P}| = 1$ or -1 according as π is even or odd.
10. Deduce *Theorem* 6.3.5 from *Theorems* 6.3.6 and 6.3.7. (Hint: Consider $|\mathbf{D}|$ with $\mathbf{D}_{k*} = \mathbf{D}_{\ell*} = \mathbf{A}_{k*} + \mathbf{A}_{\ell*}$ and $\mathbf{D}_{i*} = \mathbf{A}_{i*}$ for all other i.)
11. Let $\mathbf{A}$ be an $n\times n$ matrix whose elements are (differentiable) functions of a real variable x. Then show that
$$\frac{d}{dx}|\mathbf{A}| = \sum_{k=1}^{n} |\mathbf{A}_k|$$
where $\mathbf{A}_k$ is the matrix obtained from $\mathbf{A}$ by differentiating the elements of the k-th row with respect to x.

12. Let f be a map from $F^n \times \cdots \times F^n$($n$ copies) to F. Consider the following conditions:
    (i) $f(\mathbf{x}_1, \mathbf{x}_2, \ldots, \mathbf{x}_n) = 0$ whenever two of the $\mathbf{x}_i$'s are equal. (Such an f is said to be *alternating*.)
    (ii) $f(\mathbf{x}_1, \mathbf{x}_2, \ldots, \mathbf{x}_n)$ is linear in each $\mathbf{x}_i$ when all the other $\mathbf{x}_j$'s are kept fixed. (Such an f is said to be *multilinear*.)
    (iii) $f(\mathbf{x}_1, \mathbf{x}_2, \ldots, \mathbf{x}_n)$ is multiplied by α if any $\mathbf{x}_i$ is multiplied by α.
    (iv) $f(\mathbf{x}_1, \mathbf{x}_2, \ldots, \mathbf{x}_n)$ is not altered if a scalar multiple of $\mathbf{x}_j$ is added to $\mathbf{x}_i$, $i \neq j$.
    (v) $f(\mathbf{e}_1, \mathbf{e}_2, \ldots, \mathbf{e}_n) = 1$.

    (a) If f satisfies (i) and (ii), show that $f(\mathbf{x}_1, \mathbf{x}_2, \ldots, \mathbf{x}_n) = c|\mathbf{A}|$ for some constant $c \in F$, where $\mathbf{A} = [\mathbf{x}_1 : \cdots : \mathbf{x}_n]^{\mathrm{T}}$. (Hint: Use *Exercise* 6.3.10.) If, moreover, f satisfies (v), show that $f(\mathbf{x}_1, \mathbf{x}_2, \ldots, \mathbf{x}_n) = |\mathbf{A}|$.
    (b) Do (a) when f satisfies (iii) and (iv) instead of (i) and (ii).

    Thus determinant can be defined as a function of the rows (or columns) satisfying either '(i), (ii) and (v)' or '(iii), (iv) and (v)'. Then some of our theorems follow trivially but, to prove the existence of determinant, one has to again consider the RHS of (6.3.1) and obtain its properties.
13. Solve *Exercise* 6.1.1 using (b) of the preceding exercise.
14. For a 3×3 matrix $\mathbf{A}$, let $f(\mathbf{A})$ be the volume of the parallelopiped formed with OP, OQ and OR as three edges, where O is the origin and P, Q and R are the points of $\mathbb{R}^3$ corresponding to the three rows of $\mathbf{A}$. By convention, the volume is positive iff OP, OQ and OR form a right-handed system, i.e., the following happens: imagine a right-handed screw placed perpendicular to the plane OPQ with its tip at the origin. If the screw is rotated from OP to OQ by the smaller angle, the line OR lies on the side of the plane OPQ into which the screw advances. Using *Exercise* 6.3.12(b), show that $f(\mathbf{A}) = |\mathbf{A}|$.
15. Let F be a subfield of a field G and $\mathbf{A}$ a matrix over F. Using *Theorem* 6.3.9 show that $\mathbf{A}$ has an inverse over F iff $\mathbf{A}$ has an inverse over G. Deduce that the rank of $\mathbf{A}$ over F is the same as the rank of $\mathbf{A}$ over G.

6.4 Cofactors and expansion theorems

In this section we give a recursive method for evaluating determinants by expressing a determinant of order n in terms of determinants of order

$n-1$. This will be used in *Section* 6.6 to obtain a formula for the inverse of a non-singular matrix in terms of determinants.

Definition 6.4.1 Let $\mathbf{A}$ be a matrix of order $n \geq 2$. Then the *cofactor of* a_{ij} *in* $\mathbf{A}$ is

$$A_{ij} = (-1)^{i+j}|\mathbf{M}_{ij}| \tag{6.4.1}$$

where $\mathbf{M}_{ij}$ is the matrix obtained from $\mathbf{A}$ by deleting the i-th row and the j-th column.

We note that the cofactor of a_{ij} is a scalar which depends on i and j but not on the value of a_{ij}. Notice that we use A_{ij} to denote the cofactor of a_{ij} in $\mathbf{A}$, $\mathbf{A}_{ij}$ to denote the (i,j)-th block of a partitioned matrix $\mathbf{A}$ and $(\mathbf{A})_{ij}$ to denote the (i,j)-th element of an ordinary matrix $\mathbf{A}$.

Regarding the sign $(-1)^{i+j}$ in (6.4.1), we note that it is $+1$ if $i = j$. It changes (from $+1$ to -1 and from -1 to $+1$) whenever we move from one cell to an adjacent cell along a row or a column. For example, the cofactors of the $(2,3)$-th and $(2,4)$-th elements in

$$\begin{bmatrix} 1 & 2 & 5 & 2 \\ -2 & 3 & 0 & -4 \\ 1 & -1 & 0 & 2 \\ 0 & 1 & 4 & 2 \end{bmatrix} \tag{6.4.2}$$

are

$$(-1)\begin{vmatrix} 1 & 2 & 2 \\ 1 & -1 & 2 \\ 0 & 1 & 2 \end{vmatrix} = 6 \text{ and } +\begin{vmatrix} 1 & 2 & 5 \\ 1 & -1 & 0 \\ 0 & 1 & 4 \end{vmatrix} = -7$$

We will now show that a determinant can be 'expanded by any row'. Using this we can find the determinant of an $n \times n$ matrix by evaluating n determinants of order $n-1$. We will first consider a special case.

Lemma 6.4.2 Fix k and j. If $a_{k\ell} = 0$ for all $\ell \neq j$, $|\mathbf{A}| = a_{kj}A_{kj}$.

Proof Let $\mathbf{A}$ be of order n. We first consider the case $k = j = n$. Then

$$|\mathbf{A}| = \textstyle\sum_\pi \epsilon(\pi) a_{1,\pi(1)} a_{2,\pi(2)} \cdots a_{n,\pi(n)} \tag{6.4.3}$$

where π runs over all permutations of $\{1, 2, \ldots, n\}$. We may take $\pi(n) = n$ as, otherwise, the term corresponding to π is 0. Any such permutation π can be looked upon as a permutation θ of $T := \{1, 2, \ldots, n-1\}$ and $\epsilon(\pi) = \epsilon(\theta)$. Thus (6.4.3) can be rewritten as

$$|\mathbf{A}| = a_{nn} \textstyle\sum_\theta \epsilon(\theta) a_{1,\theta(1)} a_{2,\theta(2)} \cdots a_{n-1,\theta(n-1)} \tag{6.4.4}$$

where θ runs over all permutations of T. Clearly the sum on the RHS of (6.4.4) is the determinant of the matrix $\mathbf{M}_{nn}$ obtained from $\mathbf{A}$ by deleting the n-th row and the n-th column. Since $A_{nn} = (-1)^{n+n}|\mathbf{M}_{nn}| = |\mathbf{M}_{nn}|$, the lemma follows in this case.

We now consider the general case. It is easy to see that by $n-k$ interchanges of rows and $n-j$ interchanges of columns, we can take the k-th row and the j-th column of $\mathbf{A}$ to the last positions, without disturbing the others. The final matrix thus obtained is

$$\mathbf{B} = \begin{bmatrix} \mathbf{M}_{kj} & * \\ \mathbf{0} & a_{kj} \end{bmatrix}$$

where $\mathbf{M}_{kj}$ is the matrix obtained from $\mathbf{A}$ by deleting the k-th row and the j-th column. By what was proved in the preceding paragraph, it follows that $|\mathbf{B}| = a_{kj}|\mathbf{M}_{kj}|$. But $|\mathbf{B}| = (-1)^{n-k+n-j}|\mathbf{A}| = (-1)^{k+j}|\mathbf{A}|$, so the lemma follows. ∎

Theorem 6.4.3 Let $\mathbf{A}$ be a square matrix of order n and k an integer such that $1 \leq k \leq n$. Then

$$|\mathbf{A}| = \sum_{j=1}^{n} a_{kj}A_{kj} \tag{6.4.5}$$

Proof Clearly $\mathbf{A}_{k*}$ can be written as $\mathbf{x}_1^{\mathrm{T}} + \mathbf{x}_2^{\mathrm{T}} + \cdots + \mathbf{x}_n^{\mathrm{T}}$ where

$$\mathbf{x}_j^{\mathrm{T}} = (0, \ldots, 0, a_{kj}, 0, \ldots, 0),$$

a_{kj} occurring in the j-th position. Then by *Theorem* 6.3.7 we have $|\mathbf{A}| = \sum_j |\mathbf{B}_j|$ where $\mathbf{B}_j$ is obtained from $\mathbf{A}$ by replacing the k-th row by $\mathbf{x}_j^{\mathrm{T}}$. Now, the (k,j)-th element and its cofactor in $\mathbf{B}_j$ are the same as those in $\mathbf{A}$, so $|\mathbf{B}_j| = a_{kj}A_{kj}$ by the lemma and the theorem follows. ∎

The expression (6.4.5) is called the *expansion of* $|\mathbf{A}|$ *by the* k-th *row*. We can similarly expand $|\mathbf{A}|$ by the k-th column as follows:

$$|\mathbf{A}| = \sum_{i=1}^{n} a_{ik}A_{ik}$$

for any k. Clearly it is best to expand a determinant by a row or a column having the maximum number of zeros. Expanding the determinant of the matrix in (6.4.2) by the third column, we see that it is equal to

$$5\begin{vmatrix} -2 & 3 & -4 \\ 1 & -1 & 2 \\ 0 & 1 & 2 \end{vmatrix} - 4\begin{vmatrix} 1 & 2 & 2 \\ -2 & 3 & -4 \\ 1 & -1 & 2 \end{vmatrix} = -10 - 0 = -10$$

Example 6.4.4 We will show that the determinant of the *Vandermonde matrix*

$$\mathbf{A}_n = \begin{bmatrix} a_1^{n-1} & a_1^{n-2} & \cdots & a_1 & 1 \\ a_2^{n-1} & a_2^{n-2} & \cdots & a_2 & 1 \\ \cdots & \cdots & \cdots & \cdots & \cdots \\ a_n^{n-1} & a_n^{n-2} & \cdots & a_n & 1 \end{bmatrix} \tag{6.4.6}$$

is $\prod_{1\leq i<j\leq n}(a_i - a_j)$. We prove this by induction on n. The result is trivial if $n = 1$ or 2. So assume the result for matrices of order $n-1$ and let $n \geq 3$. If any two a_i's are equal, then $|\mathbf{A}_n| = 0$ and the result follows. So let $a_1, \ldots, a_n$ be distinct. Now form a new matrix $\mathbf{B}_n$ by replacing a_1 in $\mathbf{A}_n$ by an indeterminate x. (To be rigorous, $\mathbf{B}_n$ is a matrix not over F but over the integral domain $F[x]$ of polynomials in x with coefficients from F.) Expanding $|\mathbf{B}_n|$ by the first row we see that it is a polynomial $f(x)$ with degree $n-1$ and with the coefficient of x^{n-1} equal to

$$\begin{vmatrix} a_2^{n-2} & \cdots & a_2 & 1 \\ a_3^{n-2} & \cdots & a_3 & 1 \\ \cdots & \cdots & \cdots & \cdots \\ a_n^{n-2} & \cdots & a_n & 1 \end{vmatrix}$$

which is $\prod_{2\leq i<j\leq n}(a_i-a_j)$ by induction hypothesis. Clearly, $a_2, a_3, \ldots, a_n$ are roots of $f(x)$, so $(x-a_2)\cdots(x-a_n)$ divides $f(x)$. Since $f(x)$ is also of degree $n-1$, it follows that

$$f(x) = c(x-a_2)\cdots(x-a_n) \tag{6.4.7}$$

where c is a scalar. Comparing the coefficients of x^{n-1} on the two sides, we get

$$c = \prod_{2\leq i<j\leq n}(a_i - a_j)$$

Substituting this value of c in (6.4.7) and noting that $|\mathbf{A}_n| = f(a_1)$, the required result follows. ∎

The Vandermonde matrix occurs in many contexts like polynomial interpolation, regression, weighing designs and Coding Theory. Clearly $\mathbf{A}_n$ is non-singular iff $a_1, a_2, \ldots, a_n$ are distinct.

Sometimes we can evaluate a determinant more quickly by expanding it by several rows simultaneously instead of by a single row. To show how this can be done, we need some further notation and definitions.

Let $\mathbf{A} = ((a_{ij}))$ be an $n \times n$ matrix and let $i_1, i_2, \ldots, i_k$ and $j_1, j_2, \ldots, j_k$ be numbers between 1 and n (inclusive). Then

$$\mathbf{A}(i_1, i_2, \ldots, i_k \,|\, j_1, j_2, \ldots, j_k) = \begin{bmatrix} a_{i_1 j_1} & a_{i_1 j_2} & \cdots & a_{i_1 j_k} \\ a_{i_2 j_1} & a_{i_2 j_2} & \cdots & a_{i_2 j_k} \\ \cdots & \cdots & \cdots & \cdots \\ a_{i_k j_1} & a_{i_k j_2} & \cdots & a_{i_k j_k} \end{bmatrix} \tag{6.4.8}$$

Suppose now

$$I = \{i_1, i_2, \ldots, i_k\} \text{ where } 1 \le i_1 < i_2 < \cdots < i_k \le n \tag{6.4.9}$$

and

$$J = \{j_1, j_2, \ldots, j_k\} \text{ where } 1 \le j_1 < j_2 < \cdots < j_k \le n \tag{6.4.10}$$

Then (6.4.8) is a submatrix of $\mathbf{A}$ and we will denote it by $\mathbf{A}(I \,|\, J)$. Its determinant is called a *minor of* $\mathbf{A}$ or a *k-rowed minor of* $\mathbf{A}$ (these terms are used even when $\mathbf{A}$ is not square).

Definition 6.4.5 Let $\mathbf{A}$ be an $n \times n$ matrix and let I and J be given by (6.4.9) and (6.4.10). Then the *cofactor of* $\mathbf{A}(I \,|\, J)$ *in* $\mathbf{A}$ is

$$A_{IJ} = (-1)^{i_1 + \cdots + i_k + j_1 + \cdots + j_k} |\mathbf{A}(\bar{I} \,|\, \bar{J})|$$

where $\bar{I}$ and $\bar{J}$ are the complements of I and J respectively in $\{1, 2, \ldots, n\}$.

For example, the cofactor of $\mathbf{A}(1, 3 \,|\, 2, 3)$ in the matrix $\mathbf{A}$ given in (6.4.2) is

$$(-1)^{1+3+2+3} \begin{vmatrix} -2 & -4 \\ 0 & 2 \end{vmatrix} = (-1)(-4) = 4$$

When $k = 1$, the above definition reduces to *Definition* 6.4.1.

Theorem 6.4.6 (*Laplace expansion*) Let $\mathbf{A}$ be an $n \times n$ matrix and let I be as in (6.4.9). Then

$$|\mathbf{A}| = \textstyle\sum_J |\mathbf{A}(I \,|\, J)| \, A_{IJ} \tag{6.4.11}$$

where J runs over all subsets of $\{1, 2, \ldots, n\}$ with size k.

[†]**Proof** Let $\bar{I} = \{i_{k+1}, i_{k+2}, \ldots, i_n\}$ where $i_{k+1} < i_{k+2} < \cdots < i_n$ and let $\mu = (i_1, i_2, \ldots, i_n)$. Now

$$|\mathbf{A}| = \textstyle\sum_\pi \epsilon(\pi) \, a_{1\pi_1} a_{2\pi_2} \cdots a_{n\pi_n} = \sum_\pi \epsilon(\pi) \, a_{i_1 \pi(i_1)} a_{i_2 \pi(i_2)} \cdots a_{i_n \pi(i_n)} \tag{6.4.12}$$

Suppose now π is a permutation of $(1, 2, \ldots, n)$. Let $J = \{j_1, j_2, \ldots, j_k\}$

[†]The proof may be omitted in the first reading.

where $j_1, j_2, \ldots, j_k$ are $\pi(i_1), \pi(i_2), \ldots, \pi(i_k)$ written in increasing order. Also let $\bar{J} = \{j_{k+1}, j_{k+2}, \ldots, j_n\}$ where $j_{k+1} < j_{k+2} < \cdots < j_n$ and $\nu = (j_1, j_2, \ldots, j_n)$. Let θ be the permutation of $\{1, 2, \ldots, k\}$ defined by

$$\pi(i_s) = \nu(\theta(s)) \quad \text{for } s = 1, 2, \ldots, k$$

and τ the permutation of $\{k+1, k+2, \ldots, n\}$ defined by

$$\pi(i_t) = \nu(\tau(t)) \quad \text{for } t = k+1, k+2, \ldots, n$$

Then it is easy to check that $\pi = \nu\theta\tau\mu^{-1}$. For example, if $1 \leq s \leq k$, then $(\nu\theta\tau\mu^{-1})(i_s) = (\nu\theta\tau)(s) = (\nu\theta)(s) = \pi(i_s)$. So $\epsilon(\pi) = \epsilon(\mu)\epsilon(\nu)\epsilon(\theta)\epsilon(\tau)$.

It is also easy to check that the map $\pi \mapsto (J, \theta, \tau)$ is a 1-1 correspondence between permutations of $(1, 2, \ldots, n)$ and triples (J, θ, τ) where J is as in (6.4.10), θ is a permutation of $\{1, 2, \ldots, k\}$ and τ is a permutation of $\{k+1, k+2, \ldots, n\}$. Hence we can rewrite (6.4.12) as

$$\begin{aligned}|\mathbf{A}| = \textstyle\sum_J \epsilon(\mu)\epsilon(\nu)\big(\sum_\theta \epsilon(\theta)\, a_{i_1 j_{\theta(1)}} a_{i_2 j_{\theta(2)}} \cdots a_{i_k j_{\theta(k)}}\big) \times \\ \big(\textstyle\sum_\tau \epsilon(\tau)\, a_{i_{k+1} j_{\tau(k+1)}} a_{i_{k+2} j_{\tau(k+2)}} \cdots a_{i_n j_{\tau(n)}}\big) \qquad (6.4.13)\end{aligned}$$

The first sum in parentheses here is $|\mathbf{A}(I \,|\, J)|$ (this is easily seen by calling $\mathbf{A}(I \,|\, J)$ as $\mathbf{B}$ and using the definition of determinant). Similarly the second sum in parentheses is $|\mathbf{A}(\bar{I} \,|\, \bar{J})|$. Also, by *Theorem* 6.2.8,

$$\epsilon(\mu)\epsilon(\nu) = (-1)^{i_1 + \cdots + i_k + j_1 + \cdots + j_k - k(k+1)}$$

Since $k(k+1)$ is even, it can be dropped and $\epsilon(\mu)\epsilon(\nu)\,|\mathbf{A}(\bar{I} \,|\, \bar{J})| = A_{IJ}$. Hence the theorem follows from (6.4.13). ∎

Corollary If $\mathbf{A}$ and $\mathbf{D}$ are square matrices (of possibly different orders),

$$\begin{vmatrix} \mathbf{A} & \mathbf{B} \\ \mathbf{0} & \mathbf{D} \end{vmatrix} = \begin{vmatrix} \mathbf{A} & \mathbf{0} \\ \mathbf{C} & \mathbf{D} \end{vmatrix} = |\mathbf{A}| \cdot |\mathbf{D}|$$

The expression (6.4.11) is called the expansion of $|\mathbf{A}|$ by the rows numbered $i_1, i_2, \ldots, i_k$. We can similarly expand a determinant by any set of columns.

We note that by taking I to be a singleton set in the preceding theorem, we can get an alternative proof of *Theorem* 6.4.3.

We shall illustrate the use of Laplace expansion by expanding the determinant of the matrix in (6.4.2) by the second and third rows. We thus take $I = \{2, 3\}$. It is easy to see that $|\mathbf{A}(I \,|\, J)| = 0$ unless $J = \{1, 2\}$

or $J = \{2, 4\}$. Thus

$$\begin{aligned}|\mathbf{A}| &= \begin{vmatrix} -2 & 3 \\ 1 & -1 \end{vmatrix} (-1)^{2+3+1+2} \begin{vmatrix} 5 & 2 \\ 4 & 2 \end{vmatrix} + \begin{vmatrix} 3 & -4 \\ -1 & 2 \end{vmatrix} (-1)^{2+3+2+4} \begin{vmatrix} 1 & 5 \\ 0 & 4 \end{vmatrix} \\ &= -2 - 8 = -10.\end{aligned}$$

Exercises

1. Verify (6.3.3) by expanding the determinant by the first row.
2. Using *Exercise* 6.1.1, prove that the area of the triangle formed by the three points $A = (x_1, y_1), P = (x_2, y_2)$ and $Q = (x_3, y_3)$ in $\mathbb{R}^2$ is

$$\frac{1}{2}\begin{vmatrix} x_1 & y_1 & 1 \\ x_2 & y_2 & 1 \\ x_3 & y_3 & 1 \end{vmatrix}$$

   Deduce that A, P and Q are collinear iff the determinant is 0. (This can also be deduced from *Exercise* 5.3.12.)
3. (a) Let $P = (x_1, y_1)$ and $Q = (x_2, y_2)$ be two distinct points in $\mathbb{R}^2$. Show that the equation of the line passing through P and Q is

$$\begin{vmatrix} x & y & 1 \\ x_1 & y_1 & 1 \\ x_2 & y_2 & 1 \end{vmatrix} = 0$$

   by showing that this equation represents a line, i.e., the coefficients of x and y are not both zero, and that it contains P and Q.

   (b) State and prove an analogous result for the plane containing three non-collinear points in $\mathbb{R}^3$.

   (c) Let $P_i = (u_i, v_i, w_i)$, $i = 1, 2, 3$, be points in $\mathbb{R}^3$ such that O, P_1, P_2 are not collinear. Show that the equation of the plane containing P_3 and parallel to the plane OP_1P_2 is

$$\begin{vmatrix} x - u_3 & y - v_3 & z - w_3 \\ u_1 & v_1 & w_1 \\ u_2 & v_2 & w_2 \end{vmatrix} = 0$$

4. If $\mathbf{A}$ is the $n \times n$ matrix with $a_{ij} = i + j - 2$ for all i and j, show that $|\mathbf{A}| = 0$ whenever $n \geq 4$.
5. (a) Evaluate the determinant

$$\begin{vmatrix} x & -1 & 0 \\ 0 & x & -1 \\ c & b & x + a \end{vmatrix}$$

   where x is a variable and a, b, c are constants.

(b) Show that any monic polynomial in x with degree ≥ 1 can be written as $|x\mathbf{I} + \mathbf{A}|$ for some matrix $\mathbf{A}$.

6. Evaluate

$$\begin{vmatrix} 3 & 1 & 4 & -2 \\ 0 & -2 & 0 & 0 \\ 3 & 4 & 0 & 0 \\ 7 & 1 & 0 & -5 \end{vmatrix}$$

in three ways: (i) using expansion by one row or column at each stage, (ii) using Laplace expansion and (iii) by transforming it to a triangular matrix by a suitable permutations of rows and columns.

7. Using Laplace expansion by the first and second rows show that

$$\begin{vmatrix} a & -b & -a & b \\ b & a & -b & -a \\ c & -d & c & -d \\ d & c & d & c \end{vmatrix} = 4(a^2 + b^2)(c^2 + d^2)$$

8. Let $\mathbf{A}_n$ be the following $n \times n$ *tridiagonal* matrix

$$\begin{bmatrix} a & b & 0 & \cdots & 0 & 0 \\ c & a & b & \cdots & 0 & 0 \\ 0 & c & a & \cdots & 0 & 0 \\ \cdots & \cdots & \cdots & \cdots & \cdots & \cdots \\ 0 & 0 & 0 & \cdots & a & b \\ 0 & 0 & 0 & \cdots & c & a \end{bmatrix}$$

Show that $|\mathbf{A}_n| = a|\mathbf{A}_{n-1}| - bc|\mathbf{A}_{n-2}|$ for $n \geq 3$. If $a = 1 + bc$, show that

$$|\mathbf{A}_n| = 1 + bc + (bc)^2 + \cdots + (bc)^n$$

If $a = 2\cos\theta$ with $0 < \theta < \pi$ and $b = c = 1$ then show that

$$|\mathbf{A}_n| = \frac{\sin{(n+1)\theta}}{\sin\theta}$$

9. If $A_{ij} \neq 0$, show that by changing the single element a_{ij}, we can change $|\mathbf{A}|$ to any specified scalar.

10. Using *Example* 6.4.4, show that a polynomial with degree at most n and with $n + 1$ distinct roots must be the zero polynomial.

11. Prove that if $x_1, x_2, \ldots, x_n$ are distinct real numbers and $y_1, y_2, \ldots, y_n$ are arbitrary real numbers then there exists a unique polynomial $p(t) \in \mathcal{P}_n$ such that $p(x_i) = y_i$ for $i = 1, \ldots, n$.

*12. Show that the determinant of the matrix (6.4.6) with a_i^{n-1} replaced by a_i^n in the first column for all i, is $\pm(a_1 + \cdots + a_n)\prod_{1\leq i<j\leq n}(a_j - a_i)$ according as $n(n-1)$ is divisible by 4 or not.

13. Without expanding the determinants, prove that

$$\begin{vmatrix} 1 & 1 & 1 \\ x & y & z \\ yz & xz & xy \end{vmatrix} = \begin{vmatrix} 1 & 1 & 1 \\ x & y & z \\ x^2 & y^2 & z^2 \end{vmatrix}$$

14. Prove the following without expanding the determinant:

$$\begin{vmatrix} a & a & a & a \\ a & b & b & b \\ a & b & c & c \\ a & b & c & d \end{vmatrix} = a(b-a)(c-b)(d-c)$$

*15. Prove the corollary to *Theorem* 6.4.6 by imitating the proof of *Lemma* 6.4.2.

*16. If $\mathbf{A}$ is an $n \times n$ matrix over $\mathbb{C}$, show that $|\det \mathbf{A}| \leq \prod_{i=1}^{n} \left(\sum_{j=1}^{n} |a_{ij}| \right)$.

*17. (a) Let $\mathbf{A}$ be an $n \times n$ strictly diagonally dominated complex matrix (see *Definition* 3.3.8). Prove by induction on n that

$$\prod_{i=1}^{n} (|a_{ii}| - t_i) \leq |\det \mathbf{A}| \leq \prod_{i=1}^{n} (|a_{ii}| + t_i)$$

where $t_i = \sum_{j=i+1}^{n} |a_{ij}|$ for $i = 1, \ldots, n-1$ and $t_n = 0$. (Hint: replace a_{11} by $a_{11} - (\det \mathbf{A})/A_{11}$ in $\mathbf{A}$ and use *Theorem* 3.3.9.) Show also that the conclusion may not hold if it is assumed only that $|a_{ii}| > t_i$ for $i = 1, \ldots, n$.

(b) Let $\mathbf{A}$ be a real $n \times n$ matrix with positive diagonal entries which is strictly dominated by its diagonal entries. Show that

$$\prod_{i=1}^{n} (a_{ii} - t_i) \leq \det \mathbf{A} \leq \prod_{i=1}^{n} (a_{ii} + t_i)$$

where t_i is as defined in (a). Deduce that $\det \mathbf{A} > 0$.

6.5 Determinant of a product

In this section we study the determinant of the product of two matrices $\mathbf{A}$ and $\mathbf{B}$. We first study the case when $\mathbf{A}$ and $\mathbf{B}$ are square and later extend it to the case when $\mathbf{AB}$ is square but $\mathbf{A}$ and $\mathbf{B}$ are rectangular.

Theorem 6.5.1 If $\mathbf{A}$ and $\mathbf{B}$ are $n \times n$ matrices, $|\mathbf{AB}| = |\mathbf{A}| \cdot |\mathbf{B}|$.

Proof Suppose first that $\mathbf{A}$ is an elementary matrix. Then the present theorem follows from *Exercise* 6.3.5 since premultiplication of

$\mathbf{B}$ by an elementary matrix amounts to performing the corresponding elementary row operation on $\mathbf{B}$. Next let $\mathbf{A}$ be any non-singular matrix. Then $\mathbf{A}$ is a product of elementary matrices, say $\mathbf{A} = \mathbf{E}_1\mathbf{E}_2 \cdots \mathbf{E}_k$. By repeatedly using the result already proved, we get

$$|\mathbf{AB}| = |\mathbf{E}_1| \cdot |\mathbf{E}_2 \cdots \mathbf{E}_k\mathbf{B}| = \cdots = |\mathbf{E}_1| \cdot |\mathbf{E}_2| \cdots |\mathbf{E}_k| \cdot |\mathbf{B}| \qquad (6.5.1)$$

for any matrix $\mathbf{B}$. Taking $\mathbf{B} = \mathbf{I}$, we get $|\mathbf{A}| = |\mathbf{E}_1| \cdot |\mathbf{E}_2| \cdots |\mathbf{E}_k|$. Substituting this in (6.5.1) the theorem follows when $\mathbf{A}$ is non-singular. Finally let $\mathbf{A}$ be singular. Then $\mathbf{AB}$ is also singular and the present theorem follows from *Theorem* 6.3.9. ∎

Corollary 1 If $\mathbf{A}$ is non-singular, $|\mathbf{A}| \neq 0$ and $|\mathbf{A}^{-1}| = 1/|\mathbf{A}|$.

Corollary 2 Similar matrices have the same determinant.

The second corollary follows thus: $|\mathbf{P}^{-1}\mathbf{AP}| = |\mathbf{P}^{-1}| \cdot |\mathbf{A}| \cdot |\mathbf{P}| = |\mathbf{A}|$. It shows that (see the corollary to *Theorem* 3.9.8) we can define the determinant of a linear operator φ on a vector space V as the determinant of the matrix of φ with respect to any basis of V.

We now extend the preceding theorem to a square matrix written as the product of two rectangular matrices.

Theorem 6.5.2 (*Cauchy, Binet*) Let $\mathbf{C} = \mathbf{AB}$ where $\mathbf{A}$ is an $n \times p$ matrix and $\mathbf{B}$ is a $p \times n$ matrix. If $n > p$, $|\mathbf{C}| = 0$. If $n \leq p$,

$$|\mathbf{C}| = \sum_{1 \leq j_1 < \cdots < j_n \leq p} |\mathbf{A}(1, \ldots, n \,|\, j_1, \ldots, j_n)| \, |\mathbf{B}(j_1, \ldots, j_n \,|\, 1, \ldots, n)| \qquad (6.5.2)$$

Proof The first conclusion follows easily from: $\rho(\mathbf{C}) \leq \rho(\mathbf{A}) \leq p$. To prove the second, let $n \leq p$. Then we have

$$\begin{aligned} |\mathbf{C}| &= \textstyle\sum_\pi \epsilon(\pi) c_{1\pi_1} \cdots c_{n\pi_n} \\ &= \textstyle\sum_\pi \epsilon(\pi) \left(\sum_{\mu_1=1}^p a_{1\mu_1} b_{\mu_1\pi_1}\right) \cdots \left(\sum_{\mu_n=1}^p a_{n\mu_n} b_{\mu_n\pi_n}\right) \\ &= \textstyle\sum_{\mu_1} \cdots \sum_{\mu_n} a_{1\mu_1} \cdots a_{n\mu_n} \sum_\pi \epsilon(\pi) b_{\mu_1\pi_1} \cdots b_{\mu_n\pi_n} \\ &= \textstyle\sum_{\mu_1} \cdots \sum_{\mu_n} a_{1\mu_1} \cdots a_{n\mu_n} |\mathbf{B}(\mu_1, \ldots, \mu_n \,|\, 1, \ldots, n)| \qquad (6.5.3) \end{aligned}$$

Now if any two μ's are equal then $|\mathbf{B}(\mu_1, \ldots, \mu_n \,|\, 1, \ldots, n)| = 0$, so it is enough to take the summations in (6.5.3) over distinct values of $\mu_1, \ldots, \mu_n$. All such choices of $\mu_1, \ldots, \mu_n$ can be obtained by first choosing $j_1, \ldots, j_n$ such that $1 \leq j_1 < \cdots < j_n \leq p$ and then permuting them.

Thus (6.5.3) can be written as

$$\begin{aligned}
|\mathbf{C}| &= \sum_{1\le j_1<\cdots<j_n\le p} \sum_{\theta} a_{1j_{\theta(1)}} \cdots a_{nj_{\theta(n)}} |\mathbf{B}(j_{\theta(1)},\ldots,j_{\theta(n)}\,|\,1,\ldots,n)| \\
&= \sum_{1\le j_1<\cdots<j_n\le p} \sum_{\theta} a_{1j_{\theta(1)}} \cdots a_{nj_{\theta(n)}} \epsilon(\theta) |\mathbf{B}(j_1,\ldots,j_n\,|\,1,\ldots,n)| \\
&= \sum_{1\le j_1<\cdots<j_n\le p} |\mathbf{A}(1,\ldots,n\,|\,j_1,\ldots,j_n)|\cdot|\mathbf{B}(j_1,\ldots,j_n\,|\,1,\ldots,n)|
\end{aligned}$$

and the theorem follows. ∎

We make two observations regarding the proof of the preceding theorem. First: even when $n > p$, (6.5.3) holds and shows that $|\mathbf{C}| = 0$ since, then, $\mu_1,\ldots,\mu_n$ cannot be chosen to be distinct. Second: it gives an alternative proof of *Theorem* 6.5.1 which is valid over any commutative ring with identity.

Exercises

1. (a) Show that $\begin{vmatrix} b & c & 0 \\ a & 0 & c \\ 0 & a & b \end{vmatrix} = -2abc.$

   (b) Deduce from the result in (a) that

$$\begin{vmatrix} b^2+c^2 & ab & ac \\ ab & a^2+c^2 & bc \\ ac & bc & a^2+b^2 \end{vmatrix} = 4a^2b^2c^2$$

   and

$$\begin{vmatrix} b^2+ac & bc & c^2 \\ ab & 2ac & bc \\ a^2 & ab & b^2+ac \end{vmatrix} = 4a^2b^2c^2$$

2. Prove that

$$\begin{vmatrix} a & b & c & d \\ -b & a & -d & c \\ -c & d & a & -b \\ -d & -c & b & a \end{vmatrix} = (a^2+b^2+c^2+d^2)^2$$

3. If $\mathbf{A}$ is the matrix in *Exercise* 6.4.7, compute $|\mathbf{A}|^2$ as $|\mathbf{A}\mathbf{A}^{\mathrm{T}}|$.

4. Verify (6.5.2) for the matrices

$$\mathbf{A} = \begin{bmatrix} 1 & 2 & 0 & -1 \\ 3 & 4 & 0 & 1 \end{bmatrix} \text{ and } \mathbf{B} = \begin{bmatrix} 1 & -1 \\ 2 & -2 \\ 3 & 1 \\ -1 & 4 \end{bmatrix}.$$

5. Let $\mathbf{M} = \begin{bmatrix} \mathbf{A} & \mathbf{0} \\ -\mathbf{I} & \mathbf{B} \end{bmatrix}$. Prove *Theorem* 6.5.1 by evaluating $|\mathbf{M}|$ in two ways: (i) directly and (ii) after reducing $\mathbf{B}$ to $\mathbf{0}$ by adding suitable linear combinations of the first n columns to the last n.

*6. Prove *Theorem* 6.5.2 also as in the preceding exercise.

*7. Show that for fixed $\mathbf{B}$, $|\mathbf{AB}|$ is an alternating multilinear function of the rows of $\mathbf{A}$. Hence by *Exercise* 6.3.12, there exists a scalar c such that $|\mathbf{AB}| = c|\mathbf{A}|$. Show that $c = |\mathbf{B}|$ and deduce *Theorem* 6.5.1.

8. Show that the linear operator $\mathbf{x} \mapsto \mathbf{Ax}$ on $\mathbb{R}^2$ takes a parallelogram in $\mathbb{R}^2$ to another in $\mathbb{R}^2$ whose area is $|\mathbf{A}|$ times the area of the former.

9. Show that the set of all $n \times n$ matrices with determinant 1 forms a group under multiplication. This group is known as the *Special Linear group* and is denoted $SL(n)$. (See also *Exercise* 3.4.11.)

6.6 Classical adjoint and inverse

In this section we derive an explicit expression for the inverse of a non-singular matrix using determinants and prove Cramer's rule for solving linear equations. In the process, we define classical adjoint and study some of its properties. We start by characterizing rank in terms of determinants.

Theorem 6.6.1 The rank of a (non-null) matrix $\mathbf{A}$ is the largest integer k for which $\mathbf{A}$ has a non-vanishing minor of order k.

Proof Let $r = \rho(\mathbf{A})$. By *Theorem* 3.5.5, $\mathbf{A}$ has a non-singular submatrix $\mathbf{B}$ of order r and $|\mathbf{B}| \neq 0$. If $\mathbf{C}$ is any submatrix of $\mathbf{A}$ with order k and $|\mathbf{C}| \neq 0$, then $\mathbf{C}$ is non-singular, so $k = \rho(\mathbf{C}) \leq \rho(\mathbf{A}) = r$. ∎

To give a formula for the inverse of a non-singular matrix, we need

Theorem 6.6.2 Let $\mathbf{A}$ be an $n \times n$ matrix. Then

$$\sum_{j=1}^{n} a_{ij} A_{kj} = \begin{cases} |\mathbf{A}| & \text{if } i = k \\ 0 & \text{otherwise} \end{cases} \tag{6.6.1}$$

Proof When $i = k$, (6.6.1) follows from *Theorem* 6.4.3. So let $i \neq k$. Let $\mathbf{B}$ be the matrix obtained from $\mathbf{A}$ by replacing the k-th row by the i-th row. Expanding $|\mathbf{B}|$ by the k-th row, we see that it equals the LHS of (6.6.1). But $|\mathbf{B}| = 0$ since $\mathbf{B}$ has two equal rows. ∎

Since the LHS of (6.6.1) is the (i,k)-th element of $\mathbf{A}\,((A_{ij}))^{\mathrm{T}}$, (6.6.1) can be rewritten as

$$\mathbf{A}\,((A_{ij}))^{\mathrm{T}} = |\mathbf{A}| \cdot \mathbf{I} \tag{6.6.2}$$

Using the expansion of a determinant by a column it follows similarly that $((A_{ij}))^{\mathrm{T}}\,\mathbf{A} = |\mathbf{A}| \cdot \mathbf{I}$.

Theorem 6.6.3 If $\mathbf{A}$ is non-singular,

$$\mathbf{A}^{-1} = \frac{1}{|\mathbf{A}|}((A_{ij}))^{\mathrm{T}}$$

This theorem follows immediately from (6.6.2). It is an important theoretical tool and can be used to find the inverses of non-singular matrices of small orders. When $n = 2$, it gives (3.3.1). When $n = 3$ it gives

$$\begin{bmatrix} a_{11} & a_{12} & a_{13} \\ a_{21} & a_{22} & a_{23} \\ a_{31} & a_{32} & a_{33} \end{bmatrix}^{-1}$$

$$= \frac{1}{|\mathbf{A}|}\begin{bmatrix} a_{22}a_{33} - a_{23}a_{32} & a_{32}a_{13} - a_{12}a_{33} & a_{12}a_{23} - a_{13}a_{22} \\ a_{31}a_{23} - a_{21}a_{33} & a_{11}a_{33} - a_{13}a_{31} & a_{21}a_{13} - a_{11}a_{23} \\ a_{21}a_{32} - a_{22}a_{31} & a_{31}a_{12} - a_{11}a_{32} & a_{11}a_{22} - a_{12}a_{21} \end{bmatrix} \tag{6.6.3}$$

where $|\mathbf{A}|$ is given by (6.3.3).

Theorem 6.6.4 (*Cramer's rule*) Let $\mathbf{A}$ be a non-singular matrix of order n. Then the solution to the system $\mathbf{Ax} = \mathbf{b}$ is given by

$$x_j = |\mathbf{A}_j| \,/\, |\mathbf{A}| \quad \text{for } j = 1, \ldots, n \tag{6.6.4}$$

where $\mathbf{A}_j$ denotes the matrix obtained from $\mathbf{A}$ by replacing $\mathbf{A}_{*j}$ by $\mathbf{b}$.

Proof Since the solution is $\mathbf{A}^{-1}\mathbf{b}$, we have

$$x_j = (\mathbf{A}^{-1})_{j*}\,\mathbf{b} = \frac{1}{|\mathbf{A}|}\sum_{i=1}^{n} A_{ij} b_i = \frac{|\mathbf{A}_j|}{|\mathbf{A}|}.$$ ∎

If $|\mathbf{A}_j| \neq 0$ for all j, Cramer's rule can also be stated as

$$\frac{x_1}{|\mathbf{A}_1|} = \cdots = \frac{x_n}{|\mathbf{A}_n|} = \frac{1}{|\mathbf{A}|}$$

We illustrate the use of Cramer's rule by solving the system

$$\begin{aligned} 2x_1 + 3x_2 - x_3 &= 4 \\ x_1 + 4x_2 + x_3 &= -1 \\ 5x_1 - x_2 + 2x_3 &= 1 \end{aligned} \tag{6.6.5}$$

Let $\mathbf{A}$ be the matrix of coefficients. Sweeping out the first column of $\mathbf{A}$ with $a_{21} = 1$ as the pivot, we get

$$|\mathbf{A}| = \begin{vmatrix} 2 & 3 & -1 \\ 1 & 4 & 1 \\ 5 & -1 & 2 \end{vmatrix} = \begin{vmatrix} 0 & -5 & -3 \\ 1 & 4 & 1 \\ 0 & -21 & -3 \end{vmatrix} = -\begin{vmatrix} -5 & -3 \\ -21 & -3 \end{vmatrix} = 48$$

Also

$$|\mathbf{A}_1| = \begin{vmatrix} 4 & 3 & -1 \\ -1 & 4 & 1 \\ 1 & -1 & 2 \end{vmatrix} = \begin{vmatrix} 0 & 7 & -9 \\ 0 & 3 & 3 \\ 1 & -1 & 2 \end{vmatrix} = +\begin{vmatrix} 7 & -9 \\ 3 & 3 \end{vmatrix} = 48$$

We leave it to the reader to verify that $|\mathbf{A}_2| = 0$ and $|\mathbf{A}_3| = -96$. Thus by Cramer's rule, the solution to (6.6.5) is $(1, 0, -2)^{\mathrm{T}}$.

Motivated by *Theorem* 6.6.3 we give a name to the matrix $((A_{ij}))^{\mathrm{T}}$ and study some of its properties.

Definition 6.6.5 Let $\mathbf{A}$ be a square matrix of order $n \geq 2$. Then $((A_{ij}))^{\mathrm{T}}$ is called the *(classical) adjoint of* $\mathbf{A}$ or the *adjugate matrix of* $\mathbf{A}$. We will denote it by $\mathbf{A}^{\circledast}$.[†]

We mention that the (modern) adjoint $\mathbf{A}^*$ defined as $\overline{\mathbf{A}}^{\mathrm{T}}$ for a complex matrix $\mathbf{A}$ has nothing to do with the classical adjoint defined above. *We use only the classical adjoint in this chapter and only the modern adjoint in the remaining chapters* except where explicitly stated otherwise. We start the study of the classical adjoint with a rather surprising result.

Theorem 6.6.6 Let $\mathbf{A}$ be a square matrix of order $n \geq 2$. Then

$$\rho(\mathbf{A}^{\circledast}) = \begin{cases} 0 & \text{if } \rho(\mathbf{A}) \leq n-2 \\ 1 & \text{if } \rho(\mathbf{A}) = n-1 \\ n & \text{otherwise} \end{cases}$$

Proof Suppose first that $\rho(\mathbf{A}) \leq n - 2$. Then by *Theorem* 6.6.1, $A_{ij} = 0$ for all i and j and so $\mathbf{A}^{\circledast} = \mathbf{0}$.

[†]The properties of the classical adjoint proved below will not be needed in the sequel.

Next let $\rho(\mathbf{A}) = n - 1$. Then some A_{ij} is non-zero, so $\rho(\mathbf{A}^{\circledast}) \geq 1$. Since $|\mathbf{A}| = 0$, we have $\mathbf{A}\mathbf{A}^{\circledast} = \mathbf{0}$ from (6.6.2). Hence $\mathcal{C}(\mathbf{A}^{\circledast}) \subseteq \mathcal{N}(\mathbf{A})$. Taking dimensions, we get $\rho(\mathbf{A}^{\circledast}) \leq 1$ and equality follows.

Finally let $\rho(\mathbf{A}) = n$. Then $|\mathbf{A}| \neq 0$ and $\mathbf{A}^{\circledast} = |\mathbf{A}|\,\mathbf{A}^{-1}$ is non-singular. ∎

Theorem 6.6.7 For a square matrix $\mathbf{A}$ of order n, $|\mathbf{A}^{\circledast}| = |\mathbf{A}|^{n-1}$.

Proof Taking determinants of both sides of (6.6.2), we get

$$|\mathbf{A}| \cdot |\mathbf{A}^{\circledast}| = |\mathbf{A}|^{n} \tag{6.6.6}$$

If $|\mathbf{A}| \neq 0$, we can divide both sides of (6.6.6) by $|\mathbf{A}|$ and the theorem follows. If $|\mathbf{A}| = 0$ then $|\mathbf{A}^{\circledast}| = 0$ by the preceding theorem. ∎

Theorem 6.6.8 (*Jacobi*) Let $\mathbf{A}$ be an $n \times n$ matrix. Denote $\mathbf{A}^{\circledast}$ by $\mathbf{B}$ and let I and J be subsets of $\{1, 2, \ldots, n\}$ with size k. Then

$$|\mathbf{B}(I \,|\, J)| = |\mathbf{A}|^{k-1} A_{JI} \tag{6.6.7}$$

Proof Let $\mathbf{C}$ be the matrix obtained from $\mathbf{B}$ by replacing $\mathbf{B}(I \,|\, \bar{J})$ by $\mathbf{0}$ and $\mathbf{B}(\bar{I} \,|\, \bar{J})$ by the identity matrix of order $n - k$. Expanding $|\mathbf{C}|$ by the columns in $\bar{J}$ we get

$$|\mathbf{C}| = |\mathbf{C}(\bar{I} \,|\, \bar{J})| \cdot C_{\bar{I}\bar{J}} = (-1)^{s(\bar{I})+s(\bar{J})}|\mathbf{B}(I \,|\, J)|$$

where $s(\bar{I})$ denotes the sum of the elements of $\bar{I}$. Let us denote the matrix $\mathbf{AB} = |\mathbf{A}| \cdot \mathbf{I}$ by $\mathbf{G}$ and let $\mathbf{H} = \mathbf{AC}$. Then it is easy to see that $\mathbf{H}$ is obtained from $\mathbf{G}$ by replacing $\mathbf{G}(1, \ldots, n | \bar{J})$ by $\mathbf{A}(1, \ldots, n | \bar{I})$. Expanding $|\mathbf{H}|$ by the columns in J, we get

$$|\mathbf{H}| = |\mathbf{H}(J \,|\, J)| \cdot H_{JJ} = |\mathbf{A}|^{k}|\mathbf{A}(\bar{J} \,|\, \bar{I})|$$

But $|\mathbf{H}| = |\mathbf{A}| \cdot |\mathbf{C}|$, so

$$|\mathbf{A}|(-1)^{s(\bar{I})+s(\bar{J})}|\mathbf{B}(I \,|\, J)| = |\mathbf{A}|^{k}|\mathbf{A}(\bar{J} \,|\, \bar{I})| \tag{6.6.8}$$

Since $s(\bar{I}) + s(\bar{J}) + s(I) + s(J) = n(n+1)$ is even, (6.6.8) gives

$$|\mathbf{A}| \cdot |\mathbf{B}(I \,|\, J)| = |\mathbf{A}|^{k} A_{JI} \tag{6.6.9}$$

If $|\mathbf{A}| \neq 0$, we can divide both sides of (6.6.9) by $|\mathbf{A}|$ and the theorem follows. So let $|\mathbf{A}| = 0$. Then by *Theorem* 6.6.6, $\rho(\mathbf{B}) \leq 1$. Now if $k = 1$ then (6.6.7) holds trivially (here 0^0 is taken to be 1). So let $k \geq 2$. Then $|\mathbf{B}(I \,|\, J)| = 0$ and the theorem follows. ∎

Theorem 6.6.9 For any $n \times n$ matrix $\mathbf{A}$, $(\mathbf{A}^{\circledast})^{\circledast} = |\mathbf{A}|^{n-2}\mathbf{A}$.

Proof Clearly

$$\mathbf{A}^{\circledast}\left(\mathbf{A}^{\circledast}\right)^{\circledast} = |\mathbf{A}^{\circledast}| \cdot \mathbf{I}$$

Premultiplying both sides by $\mathbf{A}$ we get

$$|\mathbf{A}|\left(\mathbf{A}^{\circledast}\right)^{\circledast} = |\mathbf{A}|^{n-1}\mathbf{A} \tag{6.6.10}$$

If $|\mathbf{A}| \neq 0$, we can divide both sides by $|\mathbf{A}|$ and the theorem follows. So let $|\mathbf{A}| = 0$. Then $\rho(\mathbf{A}^{\circledast}) \leq 1$. If $n = 2$, the theorem can be verified directly. So let $n \geq 3$. Then by *Theorem* 6.6.6, $\rho\big((\mathbf{A}^{\circledast})^{\circledast}\big) = 0$ and the present theorem follows. ∎

Theorem 6.6.10 If $\mathbf{A}$ and $\mathbf{B}$ are square matrices of the same order, $(\mathbf{AB})^{\circledast} = \mathbf{B}^{\circledast}\mathbf{A}^{\circledast}$.

Proof Using (6.6.2) and (6.6.3) we get

$$\begin{aligned} |\mathbf{A}| \cdot |\mathbf{B}| \cdot \mathbf{B}^{\circledast}\mathbf{A}^{\circledast} &= |\mathbf{AB}| \cdot \mathbf{I} \cdot \mathbf{B}^{\circledast}\mathbf{A}^{\circledast} \\ &= (\mathbf{AB})^{\circledast}\mathbf{ABB}^{\circledast}\mathbf{A}^{\circledast} \\ &= (\mathbf{AB})^{\circledast}\mathbf{A}(|\mathbf{B}| \cdot \mathbf{I})\mathbf{A}^{\circledast} \\ &= |\mathbf{A}| \cdot |\mathbf{B}| \cdot (\mathbf{AB})^{\circledast} \end{aligned} \tag{6.6.11}$$

If $\mathbf{A}$ and $\mathbf{B}$ are non-singular, we can divide both sides by $|\mathbf{A}| \cdot |\mathbf{B}|$. If $\mathbf{A}$ or $\mathbf{B}$ is singular we proceed as follows.

Consider the integral domain G of all polynomials in the $2n^2$ variables x_{ij}'s and y_{ij}'s (i, j run from 1 to n) with coefficients from the base field F. Clearly $\mathbf{X} = ((x_{ij}))$ and $\mathbf{Y} = ((y_{ij}))$ are matrices over G and (6.6.11) holds with $\mathbf{A}$ and $\mathbf{B}$ replaced by $\mathbf{X}$ and $\mathbf{Y}$ respectively. Comparing the (i, j)-th elements of the two sides we get

$$|\mathbf{X}| \cdot |\mathbf{Y}| \cdot \left(\mathbf{Y}^{\circledast}\mathbf{X}^{\circledast}\right)_{ij} = |\mathbf{X}| \cdot |\mathbf{Y}| \cdot \left((\mathbf{XY})^{\circledast}\right)_{ij} \tag{6.6.12}$$

Since $|\mathbf{X}| \cdot |\mathbf{Y}|$ is a non-zero polynomial over F, it can be cancelled in (6.6.12). Then substituting a_{ij}'s for the variables x_{ij}'s and b_{ij}'s for the variables y_{ij}'s, the theorem follows. ∎

We mention that the technique used in the preceding theorem can be used to cancel $|\mathbf{A}|$ in (6.6.6), (6.6.9) and (6.6.10) also (after considering the (i, j)-th elements in the last case). Thus one can avoid the use of *Theorem* 6.6.6 if one wants to.

Exercises

1. Deduce *Theorem* 6.3.9 from *Theorems* 6.5.1 and 6.6.2.
2. Write the solution of the system

$$\begin{aligned} a_{11}x_1 + a_{12}x_2 &= b_1 \\ a_{21}x_1 + a_{22}x_2 &= b_2 \end{aligned}$$

   using Cramer's rule (assuming the matrix of coefficients is non-singular).
3. Using Cramer's rule and *Example* 6.4.4, write down the solution of the system in *Example* 5.5.7 when α, β and γ are distinct.
4. Find the inverses of the following matrices using *Theorem* 6.6.3:

$$\text{(a)} \begin{bmatrix} 2 & 1 \\ 8 & -3 \end{bmatrix}, \text{(b)} \begin{bmatrix} 2 & -1 & 4 \\ 0 & 1 & 0 \\ 0 & 0 & 3 \end{bmatrix}, \text{(c)} \begin{bmatrix} 1 & 0 & 0 & 0 \\ 3 & 2 & 0 & 0 \\ 0 & 1 & -1 & 0 \\ -2 & 3 & 1 & 3 \end{bmatrix} \text{ and}$$

   (d) the matrix in *Example* 6.3.10 when it is non-singular.
5. Solve the following system using Cramer's rule:

$$\begin{aligned} 2x_1 - x_2 + x_3 &= -3 \\ x_1 + x_2 - 3x_3 &= 17 \\ 5x_1 - 2x_2 - 4x_3 &= 20 \end{aligned}$$

6. Using Cramer's rule, find the value of x_4 in the solution of $\mathbf{Ax} = \mathbf{b}$ where $\mathbf{A}$ is the matrix in *Exercise* 6.6.4(c) and $\mathbf{b} = (2, 4, -1, -4)^{\mathrm{T}}$.
7. Show that the intersection of the two distinct planes $a_1x + b_1y + c_1z = 0$ and $a_2x + b_2y + c_2z = 0$ in $\mathbb{R}^3$ is

$$\frac{x}{b_1c_2 - b_2c_1} = \frac{y}{c_1a_2 - c_2a_1} = \frac{z}{a_1b_2 - a_2b_1}$$

   provided the denominators are non-zero.
8. What is $(\alpha\mathbf{A})^{\circledast}$? Show that $(\mathbf{A}^{\mathrm{T}})^{\circledast} = (\mathbf{A}^{\circledast})^{\mathrm{T}}$ and, if $\mathbf{A}$ is non-singular, then $(\mathbf{A}^{-1})^{\circledast} = (\mathbf{A}^{\circledast})^{-1}$.
9. Solve *Exercise* 3.6.7 using minors.
10. Let $\mathbf{A}$ be an $n \times n$ integral matrix (i.e., a matrix over $\mathbb{Z}$, the ring of integers). Show that the following are equivalent:

    (a) for each $\mathbf{b} \in \mathbb{Z}^n$, the system $\mathbf{Ax} = \mathbf{b}$ has a unique solution over $\mathbb{R}$ and the solution belongs to $\mathbb{Z}^n$,

    (b) $\mathbf{A}^{-1}$ exists and is an integral matrix,

    (c) $|\mathbf{A}| = +1$ or -1.

    A matrix $\mathbf{A}$ satisfying (c) is said to be *unimodular*.

*11. Let $\mathbf{A}$ be an $m \times n$ integral matrix. Consider the statements:

(a) for every integral $\mathbf{b}$ such that $\mathbf{Ax} = \mathbf{b}$ is consistent (over $\mathbb{R}$), there is an integral solution,

(b) $\mathbf{A}$ has an integral g-inverse,

(c) every minor of $\mathbf{A}$ is 0, 1 or -1.

Show that (a) and (b) are equivalent. (Hint: use *Exercise* 4.5.6.) Show also that (c) implies (b) and (b) does not imply (c). If (c) holds, $\mathbf{A}$ is said to be *totally unimodular*.

6.7 Determinant of a partitioned matrix

In this section we obtain the determinant of a partitioned matrix.

Theorem 6.7.1 If $\mathbf{A}$ and $\mathbf{D}$ are square and $\mathbf{A}$ non-singular,

$$\begin{vmatrix} \mathbf{A} & \mathbf{B} \\ \mathbf{C} & \mathbf{D} \end{vmatrix} = |\mathbf{A}| \cdot |\mathbf{D} - \mathbf{C}\mathbf{A}^{-1}\mathbf{B}| \tag{6.7.1}$$

Proof The determinant of the matrix on the LHS of (6.7.1) is not altered by adding $-\mathbf{C}\mathbf{A}^{-1}[\,\mathbf{A} : \mathbf{B}\,]$ to $[\,\mathbf{C} : \mathbf{D}\,]$ since this amounts to adding linear combinations of certain rows to some other rows. Thus the determinant equals

$$\begin{vmatrix} \mathbf{A} & \mathbf{B} \\ \mathbf{0} & \mathbf{D} - \mathbf{C}\mathbf{A}^{-1}\mathbf{B} \end{vmatrix} = |\mathbf{A}| \cdot |\mathbf{D} - \mathbf{C}\mathbf{A}^{-1}\mathbf{B}|. \qquad \blacksquare$$

The case when $\mathbf{D}$ is a scalar is particularly important. We can then write (6.7.1) as

$$\begin{vmatrix} \mathbf{A} & \mathbf{b} \\ \mathbf{c}^{\mathrm{T}} & \alpha \end{vmatrix} = |\mathbf{A}| \cdot (\alpha - \mathbf{c}^{\mathrm{T}}\mathbf{A}^{-1}\mathbf{b}) \tag{6.7.2}$$

We can use this to give an alternative proof of (5.6.6) in *Theorem* 5.6.4. With the notations of that theorem, we have $|\mathbf{A}| = |\mathbf{L}| \cdot |\mathbf{U}| = \ell_{11}\ell_{22}\cdots\ell_{nn}$ and $|\mathbf{B}| = |\mathbf{L}_1||\mathbf{U}_1| = \ell_{11}\ell_{22}\cdots\ell_{n-1,n-1}$. Hence

$$\ell_{nn} = \frac{|\mathbf{A}|}{|\mathbf{B}|} = a_{nn} - \mathbf{c}^{\mathrm{T}}\mathbf{B}^{-1}\mathbf{d}$$

by (6.7.2).

Exercises

1. If $\mathbf{D}$ is non-singular, show that

$$\begin{vmatrix} \mathbf{A} & \mathbf{B} \\ \mathbf{C} & \mathbf{D} \end{vmatrix} = |\mathbf{D}| \cdot |\mathbf{A} - \mathbf{B}\mathbf{D}^{-1}\mathbf{C}|$$

Write down the corresponding result when $\mathbf{B}$ is non-singular.

2. (a) If $\mathbf{A}$ is non-singular, show that $|\mathbf{A} + \mathbf{b}\mathbf{c}^{\mathrm{T}}| = |\mathbf{A}|(1 + \mathbf{c}^{\mathrm{T}}\mathbf{A}^{-1}\mathbf{b})$. Deduce that $|\mathbf{I} + \mathbf{u}\mathbf{v}^{\mathrm{T}}| = 1 + \mathbf{v}^{\mathrm{T}}\mathbf{u}$ and solve *Exercise* 6.3.6 using this.

   (b) If $\mathbf{A}$, $\mathbf{B}$ and $\mathbf{C}$ are of orders $n \times n, n \times k$ and $k \times n$ respectively and $\mathbf{A}$ is non-singular, show that $|\mathbf{A} + \mathbf{B}\mathbf{C}| = |\mathbf{A}| \cdot |\mathbf{I}_k + \mathbf{C}\mathbf{A}^{-1}\mathbf{B}|$.

*3. Prove that

$$\begin{vmatrix} \mathbf{A} & \mathbf{b} \\ \mathbf{c}^{\mathrm{T}} & \alpha \end{vmatrix} = \alpha|\mathbf{A}| - \mathbf{c}^{\mathrm{T}}\mathbf{A}^{\circledast}\mathbf{b}$$

(Hint: expand the determinant on the LHS by the last column and each of the minors thus obtained by the last row.)

*4. Show that $|\mathbf{A} + \mathbf{b}\mathbf{c}^{\mathrm{T}}| = |\mathbf{A}| + \mathbf{c}^{\mathrm{T}}\mathbf{A}^{\circledast}\mathbf{b}$. Hence deduce that $|\mathbf{A} + \alpha\mathbf{J}| = |\mathbf{A}| + \alpha\,\mathbf{1}^{\mathrm{T}}\mathbf{A}^{\circledast}\mathbf{1}$ where $\mathbf{J} = ((1))$.

5. Solve $\det(\mathrm{diag}(a, b, c) + x\mathbf{1}\mathbf{1}^{\mathrm{T}}) = 0$ for x where a, b, c are arbitrary real numbers.

6. Show that

$$\begin{vmatrix} 1+x_1 & 1 & \cdots & 1 \\ 1 & 1+x_2 & \cdots & 1 \\ \cdots & \cdots & \cdots & \cdots \\ 1 & 1 & \cdots & 1+x_n \end{vmatrix} = \left(1 + \sum_{i=1}^{n} \frac{1}{x_i}\right)\left(\prod_{i=1}^{n} x_i\right)$$

if $x_1, x_2, \ldots, x_n$ are non-zero. Also find the value of the determinant when (i) exactly one of the x_i's is 0, (ii) two or more of the x_i's are 0.

7. Let $\mathbf{A}, \mathbf{B}, \mathbf{C}$ and $\mathbf{D}$ be $m \times m$ matrices.

   (a) If $\mathbf{A}$ is non-singular and $\mathbf{A}$ commutes with $\mathbf{C}$, show that

$$\begin{vmatrix} \mathbf{A} & \mathbf{B} \\ \mathbf{C} & \mathbf{D} \end{vmatrix} = |\mathbf{A}\mathbf{D} - \mathbf{C}\mathbf{B}|$$

   (b) Find similar expressions for the determinant on the left hand side in each of the cases: (i) $\mathbf{D}$ is non-singular and $\mathbf{D}$ commutes with $\mathbf{B}$ and (ii) $\mathbf{A}$ is non-singular and $\mathbf{A}$ commutes with $\mathbf{B}$.

   (c) Show that the result in (a) is false even if $\mathbf{A}, \mathbf{B}, \mathbf{C}$ and $\mathbf{D}$ are non-singular and $m = 2$ if $\mathbf{A}$ and $\mathbf{C}$ do not commute.

Chapter 7
Inner product and orthogonality

7.1 Introduction

Till now we have studied concepts like subspace, dimension, linear transformations and their representations, linear equations, inverse and determinant, which are valid over any field F. We have also given the geometric interpretation of some of these over $\mathbb{R}$. In the Euclidean spaces $\mathbb{R}^2$ and $\mathbb{R}^3$ there are two other concepts, viz., length (or distance) and angle which have no analogues over a general field. In this chapter we study these two concepts.

Fortuitously there is a single concept usually known as scalar product or dot product which covers both the concepts of length and angle. Let $\mathbf{x} = (x_1, x_2)$ and $\mathbf{z} = (z_1, z_2)$ be vectors in $\mathbb{R}^2$ represented by the points P and Q as in *Figure* 7.1.1.

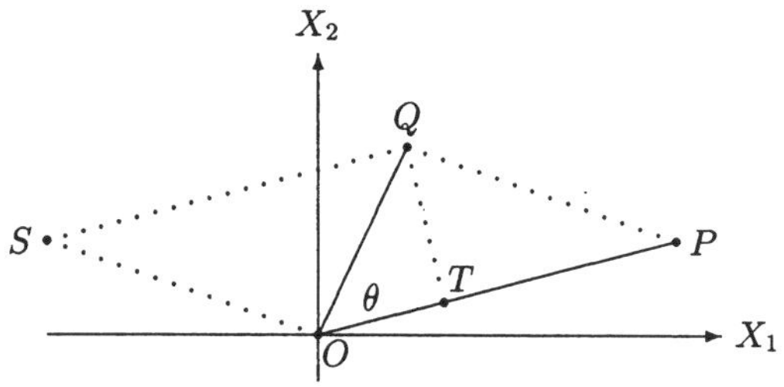

Figure 7.1.1

Then the scalar product of $\mathbf{x}$ and $\mathbf{z}$ is defined to be

$$\langle \mathbf{x}, \mathbf{z} \rangle = \ell_1 \ell_2 \cos\theta \tag{7.1.1}$$

where ℓ_1 is the length of OP, ℓ_2 is the length of OQ and θ is the angle between OP and OQ. It can be shown using trigonometry that $\ell_1 \ell_2 \cos\theta = x_1 z_1 + x_2 z_2$, so

$$\langle \mathbf{x}, \mathbf{z} \rangle = x_1 z_1 + x_2 z_2 \tag{7.1.2}$$

In Linear Algebra, scalar product is called inner product.

Note that the length OP can be defined in terms of the inner product (7.1.1) since $OP^2 = \langle \mathbf{x}, \mathbf{x} \rangle$. If $OPQS$ is a parallelogram, the distance $PQ = OS = \sqrt{\langle \mathbf{z}-\mathbf{x}, \mathbf{z}-\mathbf{x} \rangle}$ since $S = \mathbf{z} - \mathbf{x}$. The angle θ can be obtained as

$$\theta = \cos^{-1}\left(\frac{\langle \mathbf{x}, \mathbf{z} \rangle}{\sqrt{\langle \mathbf{x}, \mathbf{x} \rangle \cdot \langle \mathbf{z}, \mathbf{z} \rangle}}\right) \tag{7.1.3}$$

The above concepts and results have obvious analogues in $\mathbb{R}^3$. In this chapter we shall extend them to arbitrary (finite-dimensional) vector spaces over $\mathbb{R}$ and $\mathbb{C}$. One does not extend inner product to vector spaces over a general field mainly because $\langle \mathbf{x}, \mathbf{x} \rangle \geq 0$ has no meaning in a general field.

Throughout this chapter we take F to be $\mathbb{R}$ or $\mathbb{C}$.

We recall a few definitions and results. For a complex number α, $\overline{\alpha}$ denotes the complex conjugate of α. For a complex matrix $\mathbf{A} = ((a_{ij}))$, $\overline{\mathbf{A}} = ((\overline{a}_{ij}))$ and $\mathbf{A}^* = \overline{\mathbf{A}}^{\mathrm{T}}$. We have $\overline{\alpha+\beta} = \overline{\alpha} + \overline{\beta}$, $\overline{\alpha\beta} = \overline{\alpha} \cdot \overline{\beta}$ and $\overline{\overline{\alpha}} = \alpha$. Also $|\alpha|^2 = \alpha\overline{\alpha} \geq 0$ and $|\alpha| = 0$ iff $\alpha = 0$. Finally, $|\alpha + \beta| \leq |\alpha| + |\beta|$ and $|\alpha\beta| = |\alpha| \cdot |\beta|$. For matrices we have

$$\overline{\mathbf{A}+\mathbf{B}} = \overline{\mathbf{A}} + \overline{\mathbf{B}}, \quad \overline{\mathbf{AB}} = \overline{\mathbf{A}} \cdot \overline{\mathbf{B}}, \quad \overline{\overline{\mathbf{A}}} = \mathbf{A}$$
$$(\mathbf{A}+\mathbf{B})^* = \mathbf{A}^* + \mathbf{B}^*, \ (\mathbf{AB})^* = \mathbf{B}^*\mathbf{A}^* \text{ and } (\mathbf{A}^*)^* = \mathbf{A}$$

If $\mathbf{A}$ is non-singular, then $(\mathbf{A}^*)^{-1} = (\mathbf{A}^{-1})^*$. It is worth noting that for a 1×1 matrix (α), $(\alpha)^* = (\overline{\alpha})$.

Exercises

1. Deduce (7.1.2) from (7.1.1).
2. When are OP and OQ perpendicular to each other? Give the answer in terms of $\langle \mathbf{x}, \mathbf{y} \rangle$, where $\mathbf{x}$ and $\mathbf{y}$ are the vectors representing P and Q.
3. Let $\mathbf{z}$ be a fixed nonnull vector in the plane. What is the locus of the point $\mathbf{x}$ such that $\langle \mathbf{x}, \mathbf{z} \rangle = 0$? What happens if 0 is replaced by a non-zero scalar?
4. If x_1, x_2, y_1, y_2 are real numbers, show that
$$(x_1x_2 + y_1y_2)^2 \leq (x_1^2 + y_1^2)(x_2^2 + y_2^2)$$
Hence deduce that $PQ + QR \geq PR$ for any three points P, Q and R in the plane.

7.2 Inner product

Motivated by the usual inner product (7.1.2) on $\mathbb{R}^2$ we now give the axiomatic definition of inner product on a vector space over F, where F is either $\mathbb{R}$ or $\mathbb{C}$.

Definition 7.2.1 An inner product on a vector space V over F is a map $(\mathbf{x}, \mathbf{y}) \mapsto \langle \mathbf{x}, \mathbf{y} \rangle$ from $V \times V$ to F satisfying the following three conditions:

(i) $\langle \mathbf{x}, \mathbf{y} \rangle = \overline{\langle \mathbf{y}, \mathbf{x} \rangle}$

(ii) $\langle \alpha_1 \mathbf{x}_1 + \alpha_2 \mathbf{x}_2, \mathbf{y} \rangle = \alpha_1 \langle \mathbf{x}_1, \mathbf{y} \rangle + \alpha_2 \langle \mathbf{x}_2, \mathbf{y} \rangle$

(iii) $\langle \mathbf{x}, \mathbf{x} \rangle \geq 0$; $\langle \mathbf{x}, \mathbf{x} \rangle = 0 \Rightarrow \mathbf{x} = \mathbf{0}$.

A vector space together with an inner product is called an *inner product space*. A real inner product space is also called a *Euclidean space* and a complex inner product space a *unitary space*.

We note a few points about *Definition* 7.2.1. By (i), $\langle \mathbf{x}, \mathbf{x} \rangle$ is real (even if $F = \mathbb{C}$), so (iii) is meaningful. Condition (ii) says that when the second argument is held fixed, inner product is linear in the first. Finally, the definition is applicable even when $F = \mathbb{R}$; condition (i) then reduces to $\langle \mathbf{x}, \mathbf{y} \rangle = \langle \mathbf{y}, \mathbf{x} \rangle$. We now give several examples of inner product.

Example 7.2.2 The inner product

$$\langle \mathbf{x}, \mathbf{y} \rangle = \sum_{i=1}^{n} x_i y_i = \mathbf{y}^{\mathrm{T}} \mathbf{x} \tag{7.2.1}$$

on $\mathbb{R}^n$ is called the *canonical inner product* on $\mathbb{R}^n$.

Example 7.2.3 On $\mathbb{C}^n$, define

$$\langle \mathbf{x}, \mathbf{y} \rangle = \sum_{i=1}^{n} x_i \bar{y}_i = \mathbf{y}^* \mathbf{x} \tag{7.2.2}$$

Then conditions (i) and (ii) are easy to verify. Condition (iii) follows from $\langle \mathbf{x}, \mathbf{x} \rangle = \mathbf{x}^* \mathbf{x} = \sum_i |x_i|^2$. We call (7.2.2) the *canonical inner product* on $\mathbb{C}^n$. Note that (7.2.2) reduces to (7.2.1) if all components of $\mathbf{x}$ and $\mathbf{y}$ are real.

Example 7.2.4 On $\mathbb{R}^2$, $\langle \mathbf{x}, \mathbf{y} \rangle = 2x_1 y_1 - x_1 y_2 - x_2 y_1 + 3x_2 y_2$ is an inner product. To see this, note that $\langle \mathbf{x}, \mathbf{y} \rangle = \mathbf{y}^{\mathrm{T}} \mathbf{A} \mathbf{x}$ where

$$\mathbf{A} = \begin{bmatrix} 2 & -1 \\ -1 & 3 \end{bmatrix}$$

Condition (i) is satisfied since $\mathbf{A}$ is symmetric. Condition (ii) is easily verified. Condition (iii) follows from

$$\langle \mathbf{x}, \mathbf{x} \rangle = \mathbf{x}^{\mathrm{T}} \mathbf{A} \mathbf{x} = 2x_1^2 - 2x_1x_2 + 3x_2^2 = 2\left(x_1 - \frac{x_2}{2}\right)^2 + \frac{5}{2}x_2^2$$

Example 7.2.5 $V = \mathbb{C}^n$ and $\langle \mathbf{x}, \mathbf{y} \rangle = \mathbf{y}^* \mathbf{B}^* \mathbf{B} \mathbf{x}$ where $\mathbf{B}$ is a matrix with n columns and rank n. Now condition (i) follows from

$$\overline{\langle \mathbf{y}, \mathbf{x} \rangle} = \overline{\mathbf{x}^* \mathbf{B}^* \mathbf{B} \mathbf{y}} = (\mathbf{x}^* \mathbf{B}^* \mathbf{B} \mathbf{y})^* = \mathbf{y}^* \mathbf{B}^* \mathbf{B} \mathbf{x} = \langle \mathbf{x}, \mathbf{y} \rangle$$

Condition (ii) is easy to prove. Clearly $\langle \mathbf{x}, \mathbf{x} \rangle = (\mathbf{B}\mathbf{x})^*(\mathbf{B}\mathbf{x}) \geq 0$. If $\langle \mathbf{x}, \mathbf{x} \rangle = 0$ then $\mathbf{B}\mathbf{x} = \mathbf{0}$ and so $\mathbf{x} = \mathbf{0}$ and (iii) follows.

***Example 7.2.6** Let V be the vector space of all real-valued random variables with mean 0 and finite variance, defined on a fixed probability space. Let $F = \mathbb{R}$ and define $\langle \mathbf{x}, \mathbf{y} \rangle$ to be the covariance between $\mathbf{x}$ and $\mathbf{y}$. Then conditions (i) and (ii) follow from well known properties of expectation. Also, $\langle \mathbf{x}, \mathbf{x} \rangle = \mathrm{var}(\mathbf{x}) \geq 0$. If $\mathrm{var}(\mathbf{x}) = 0$, then $\mathbf{x} = \mathbf{0}$ with probability 1 since the mean of $\mathbf{x}$ is 0. If we agree to treat two random variables which are equal with probability 1 as equal, then it follows that $\mathbf{x} = \mathbf{0}$ and covariance is an inner product.

Clearly, the restriction of an inner product to a subspace is an inner product. We now derive two simple properties of an inner product. More will be given in later sections.

Theorem 7.2.7 In any inner product space, we have

(i) $\langle \mathbf{x}, \beta_1\mathbf{y}_1 + \beta_2\mathbf{y}_2 \rangle = \overline{\beta}_1 \langle \mathbf{x}, \mathbf{y}_1 \rangle + \overline{\beta}_2 \langle \mathbf{x}, \mathbf{y}_2 \rangle$,

(ii) $\langle \mathbf{0}, \mathbf{y} \rangle = \langle \mathbf{x}, \mathbf{0} \rangle = 0$.

Proof We have from (i) and (ii) of *Definition* 7.2.1,

$$\begin{aligned} \langle \mathbf{x}, \beta_1\mathbf{y}_1 + \beta_2\mathbf{y}_2 \rangle &= \overline{\langle \beta_1\mathbf{y}_1 + \beta_2\mathbf{y}_2, \mathbf{x} \rangle} \\ &= \overline{\beta_1 \langle \mathbf{y}_1, \mathbf{x} \rangle + \beta_2 \langle \mathbf{y}_2, \mathbf{x} \rangle} \\ &= \overline{\beta}_1 \overline{\langle \mathbf{y}_1, \mathbf{x} \rangle} + \overline{\beta}_2 \overline{\langle \mathbf{y}_2, \mathbf{x} \rangle} \\ &= \overline{\beta}_1 \langle \mathbf{x}, \mathbf{y}_1 \rangle + \overline{\beta}_2 \langle \mathbf{x}, \mathbf{y}_2 \rangle \end{aligned}$$

This proves (i). To prove (ii) we have

$$\langle \mathbf{0}, \mathbf{y} \rangle + \langle \mathbf{0}, \mathbf{y} \rangle = \langle \mathbf{0} + \mathbf{0}, \mathbf{y} \rangle = \langle \mathbf{0}, \mathbf{y} \rangle$$

So $\langle \mathbf{0}, \mathbf{y}\rangle = 0$. Now $\langle \mathbf{x}, \mathbf{0}\rangle = \overline{\langle \mathbf{0}, \mathbf{x}\rangle} = \overline{0} = 0$. ∎

Corollary $\left\langle \sum_{i=1}^{k} \alpha_i \mathbf{x}_i, \sum_{j=1}^{\ell} \beta_j \mathbf{y}_j \right\rangle = \sum_{i=1}^{k} \sum_{j=1}^{\ell} \alpha_i \overline{\beta}_j \langle \mathbf{x}_i, \mathbf{y}_j\rangle$

We express the result (i) of *Theorem* 7.2.7 by saying that an inner product is *conjugate-linear* in the second argument.

Let V be an inner product space and $\mathcal{U} = \{\mathbf{u}_1, \mathbf{u}_2, \ldots, \mathbf{u}_n\}$ a basis of V. Let $\boldsymbol{\alpha} = (\alpha_1, \ldots, \alpha_n)^{\mathrm{T}}$ and $\boldsymbol{\beta} = (\beta_1, \ldots, \beta_n)^{\mathrm{T}}$ be the coordinate vectors of $\mathbf{x}$ and $\mathbf{y}$ respectively w.r.t. $\mathcal{U}$ and let $\mathbf{A} = ((a_{ij}))$ where $a_{ij} = \langle \mathbf{u}_j, \mathbf{u}_i\rangle$. Then, by the preceding corollary,

$$\left\langle \sum \alpha_i \mathbf{u}_i, \sum \beta_j \mathbf{u}_j \right\rangle = \sum\sum \overline{\beta}_j\, a_{ji}\, \alpha_i = \boldsymbol{\beta}^* \mathbf{A} \boldsymbol{\alpha} \qquad (7.2.3)$$

Note that $\mathbf{A}$ cannot be any matrix. Condition (i) of *Definition* 7.2.1 gives (a) $\mathbf{A} = \mathbf{A}^*$. Condition (iii) gives (b) $\boldsymbol{\alpha}^* \mathbf{A} \boldsymbol{\alpha} \geq 0$ for all $\boldsymbol{\alpha} \in F^n$ and (c) if $\boldsymbol{\alpha}^* \mathbf{A} \boldsymbol{\alpha} = 0$ then $\boldsymbol{\alpha} = \mathbf{0}$. Conversely if $\mathbf{A}$ is a matrix satisfying (a), (b) and (c), then $\langle \cdot, \cdot \rangle$ defined by (7.2.3) is an inner product on V. We will show in *Chapter* 9 that such an $\mathbf{A}$ can be written as $\mathbf{B}^*\mathbf{B}$ for some non-singular $\mathbf{B}$. If $F = \mathbb{R}$, $\mathbf{A}$ and $\mathbf{B}$ will be real and adjoint is to be replaced by transpose.

Exercises

1. Find a necessary and sufficient condition for $\langle \mathbf{x}, \mathbf{y}\rangle = \sum_{i=1}^{n} \alpha_i x_i y_i$ to be an inner product on $\mathbb{R}^n$. Prove your statement.
2. Fix any finite subset A of $\mathbb{R}$ with size $\geq n$. Let $V = \mathcal{P}_n$ over $\mathbb{R}$. Show that $\langle p, q\rangle := \sum_{a \in A} p(a) q(a)$ is an inner product on V.
3. Show that

$$\langle f, g\rangle = \int_a^b f(t)\, g(t)\, dt$$

is an inner product on the vector space V of all real-valued continuous functions on an interval $[a, b]$. (Incidentally, V is not finite-dimensional.) If $h \in V$ is such that $h(t) > 0$ for all $t \in [a, b]$, show that $\langle f, g\rangle = \int_a^b h(t)\, f(t)\, g(t)\, dt$ is also an inner product.
4. (a) Show that $\langle \mathbf{A}, \mathbf{B}\rangle = \mathrm{tr}(\mathbf{B}^* \mathbf{A})$ is an inner product on $\mathbb{C}^{m \times n}$.
   (b) Show that $\langle \mathbf{A}, \mathbf{B}\rangle = \sum_{i=1}^{n} a_{ii} \overline{b}_{ii}$ is not an inner product on $\mathbb{C}^{n \times n}$. What are all the axioms which are violated?
5. Show that $\langle \mathbf{x}, \mathbf{y}\rangle := \mathbf{y}^{\mathrm{T}} \mathbf{x}$ and $\langle \mathbf{x}, \mathbf{y}\rangle := \mathbf{x}^* \mathbf{y}$ are not inner products on $\mathbb{C}^n$.

6. Show that $\langle \mathbf{x}, \mathbf{y} \rangle = 0$ for all $\mathbf{y}$ iff $\mathbf{x} = \mathbf{0}$.
7. Show that $\langle \mathbf{x} + i\mathbf{y}, \mathbf{u} + i\mathbf{v} \rangle := \mathbf{u}^{\mathrm{T}}\mathbf{x} + \mathbf{v}^{\mathrm{T}}\mathbf{y}$ is an inner product on the vector space $\mathbb{C}^n$ over $\mathbb{R}$, where $\mathbf{x}, \mathbf{y}, \mathbf{u}$ and $\mathbf{v}$ belong to $\mathbb{R}^n$. What is its connection with the canonical inner product on $\mathbb{C}^n$?

7.3 Norm

In the following axiomatic definition of norm we capture the concept of length in real and complex vector spaces.

Definition 7.3.1 A *norm* on a (real or complex) vector space V is a map $\mathbf{x} \mapsto ||\mathbf{x}||$ from V to $\mathbb{R}$ satisfying the following three conditions:

(i) $||\mathbf{x}|| \geq 0$; $\mathbf{x} = \mathbf{0}$ if $||\mathbf{x}|| = 0$

(ii) $||\alpha\mathbf{x}|| = |\alpha| \cdot ||\mathbf{x}||$

(iii) $||\mathbf{x} + \mathbf{y}|| \leq ||\mathbf{x}|| + ||\mathbf{y}||$.

A vector space together with a norm on it is called a *normed vector space* or *normed linear space*.

Condition (iii) above is called the *triangle inequality* for the following reason. If P and S represent $\mathbf{x}$ and $\mathbf{y}$ in *Figure* 7.1.1, then Q represents $\mathbf{x} + \mathbf{y}$ and (iii) says: $OQ \leq OP + PQ$.

We leave it to the reader to prove the following simple properties of a norm:

(a) $||\mathbf{0}|| = 0$,

(b) $||-\mathbf{x}|| = ||\mathbf{x}||$,

(c) $|\,||\mathbf{x}|| - ||\mathbf{y}||\,| \leq ||\mathbf{x} - \mathbf{y}|| \leq ||\mathbf{x}|| + ||\mathbf{y}||$.

We will prove later that every inner product induces a norm. We now give a family of norms which are not induced by inner products. For this we need some famous inequalities which are also of independent interest.

Lemma 7.3.2 Let $0 < \alpha < 1$ and c, d be non-negative (real) numbers. Then

$$c^\alpha d^{1-\alpha} \leq c\alpha + d(1-\alpha) \tag{7.3.1}$$

Proof The result is trivial if $c = 0$ or $d = 0$ or $c = d$. So let $c > d > 0$. Let $y = c/d$. Then (7.3.1) reduces to $f(y) \geq 0$ where $f(x) = \alpha x + (1-\alpha) - x^\alpha$. Now $f(1) = 0$ and $f'(x) = \alpha - \alpha x^{\alpha-1} \geq 0$ for $x \geq 1$. So $f(x) \geq 0$ for all $x \geq 1$ and the lemma follows. ∎

Theorem 7.3.3 (*Hölder's inequality*) Let p be a real number > 1 and let $q = p/(p-1)$ so that $(1/p) + (1/q) = 1$. If a_i and b_i are non-negative real numbers for $i = 1, 2, \ldots, n$, then

$$\sum_{i=1}^{n} a_i b_i \leq \left(\sum_{i=1}^{n} a_i^p\right)^{1/p} \left(\sum_{i=1}^{n} b_i^q\right)^{1/q} \tag{7.3.2}$$

Proof Let $a = \sum a_i^p$ and $b = \sum b_i^q$. If either a or b is 0 then (7.3.2) holds trivially. So let $a \neq 0$ and $b \neq 0$. Define $c_i = a_i^p/a$ and $d_i = b_i^q/b$. Then

$$\begin{aligned} \sum a_i b_i &= \sum (ac_i)^{1/p}(bd_i)^{1/q} \\ &\leq a^{1/p}b^{1/q} \sum \left(c_i \cdot \tfrac{1}{p} + d_i \cdot \tfrac{1}{q}\right) \quad \text{by (7.3.1)} \\ &= a^{1/p}b^{1/q} \quad \text{since } \sum c_i = \sum d_i = 1. \end{aligned}$$

∎

Theorem 7.3.4 (*Minkowski's inequality*) Let $p > 1$ and let a_i, b_i be non-negative numbers for $i = 1, 2, \ldots, n$. Then

$$\left(\sum_{i=1}^{n} (a_i + b_i)^p\right)^{1/p} \leq \left(\sum_{i=1}^{n} a_i^p\right)^{1/p} + \left(\sum_{i=1}^{n} b_i^p\right)^{1/p} \tag{7.3.3}$$

Proof Let q be defined as in the preceding theorem and let $S = \sum (a_i + b_i)^p$. Then

$$\begin{aligned} S &= \sum (a_i + b_i)(a_i + b_i)^{p-1} \\ &= \sum a_i (a_i + b_i)^{p-1} + \sum b_i (a_i + b_i)^{p-1} \\ &\leq (\sum a_i^p)^{1/p} \cdot \left\{\sum (a_i + b_i)^{(p-1)q}\right\}^{1/q} \\ &\quad + (\sum b_i^p)^{1/p} \cdot \left\{\sum (a_i + b_i)^{(p-1)q}\right\}^{1/q} \quad \text{by (7.3.2)} \\ &= \left\{(\sum a_i^p)^{1/p} + \sum (b_i^p)^{1/p}\right\} S^{1/q} \quad \text{since } (p-1)q = p. \end{aligned}$$

Now (7.3.3) follows easily. ∎

There are continuous versions of Hölder's and Minkowski's inequalities obtained essentially by replacing sum by integral but we will not need them. We are now ready to give a family of norms on $\mathbb{R}^n$ and $\mathbb{C}^n$.

Example 7.3.5 Let p be a real number > 1. Then the L_p-norm on $\mathbb{R}^n$ as well as on $\mathbb{C}^n$ is defined by

$$\|\mathbf{x}\|_p = \left(\sum_{i=1}^{n} |x_i|^p\right)^{1/p} \tag{7.3.4}$$

It is easy to see that this satisfies conditions (i) and (ii) of *Definition* 7.3.1. Condition (iii) follows from Minkowski's inequality thus:

$$\begin{aligned} \|\mathbf{x}+\mathbf{y}\|_p &= (\sum |x_i + y_i|^p)^{1/p} \leq \{\sum(|x_i| + |y_i|)^p\}^{1/p} \\ &\leq (\sum |x_i|^p)^{1/p} + (\sum |y_i|^p)^{1/p} \quad \text{by (7.3.3)} \\ &= \|\mathbf{x}\|_p + \|\mathbf{y}\|_p \end{aligned}$$

Thus (7.3.4) is a norm. Note that $\|\mathbf{x}\|_2$ is the usual length of $O\mathbf{x}$ and arises from the canonical inner product as $\sqrt{\langle \mathbf{x}, \mathbf{x}\rangle}$. However, if $p \neq 2$, the L_p-norm does not arise from any inner product, see *Exercise* 7.3.9.

Example 7.3.6 On $\mathbb{R}^n$ as well as on $\mathbb{C}^n$, each of the following is a norm:

$$\|\mathbf{x}\|_\infty = \max_{1\leq i\leq n} |x_i| \tag{7.3.5}$$

$$\|\mathbf{x}\|_1 = \textstyle\sum_{i=1}^n |x_i| \tag{7.3.6}$$

We leave it to the reader to verify that these are norms and that the RHS of (7.3.5) is the limit of the RHS of (7.3.4) as $p \to \infty$. Clearly, (7.3.6) is obtained by taking $p = 1$ in (7.3.4). The norm (7.3.5) is called the L_∞-*norm* and the norm (7.3.6) is called the L_1-*norm*.

Example 7.3.7 On $\mathbb{R}^n$ as well as on $\mathbb{C}^n$, $\|\mathbf{x}\| = \sum_{j=1}^n j|x_j|$ is a norm.

There is a continuous version of the L_p-norm but we will not need it. The L_p-norm on the vector space of random variables with mean 0 and finite p-th moment is defined as $\{E(|\mathbf{x}|^p)\}^{1/p}$. The L_2-norm here is the standard deviation.

We now study some simple properties of norms and establish the connection between inner products and norms. In a normed vector space, we can *normalize* any non-null vector $\mathbf{x}$ by dividing it by $\|\mathbf{x}\|$. If $\mathbf{y} = \frac{1}{\|\mathbf{x}\|}\mathbf{x}$, then $\|\mathbf{y}\| = 1$. Such a vector $\mathbf{y}$ is said to be a *normalized vector*.

Theorem 7.3.8 In a normed vector space, $d(\mathbf{x}, \mathbf{y}) = \|\mathbf{x} - \mathbf{y}\|$ is a metric which is invariant under translations.

Proof That $d(\mathbf{x}, \mathbf{y}) = d(\mathbf{y}, \mathbf{x})$, $d(\mathbf{x}, \mathbf{y}) \geq 0$, $d(\mathbf{x}, \mathbf{y}) = 0$ iff $\mathbf{x} = \mathbf{y}$, and $d(\mathbf{x}, \mathbf{z}) \leq d(\mathbf{x}, \mathbf{y}) + d(\mathbf{y}, \mathbf{z})$ are easy to verify. Further, $d(\mathbf{x}+\mathbf{u}, \mathbf{y}+\mathbf{u}) = d(\mathbf{x}, \mathbf{y})$, so d is invariant under translations. ∎

The metric induced by the L_2-norm on $\mathbb{R}^n$ is the usual distance function. We next show that every inner product induces a norm. For this we need the following famous

Theorem 7.3.9 (*Cauchy-Schwarz inequality*) In any inner product space

$$|\langle \mathbf{x}, \mathbf{y} \rangle|^2 \leq \langle \mathbf{x}, \mathbf{x} \rangle \cdot \langle \mathbf{y}, \mathbf{y} \rangle \tag{7.3.7}$$

Further, equality holds iff $\mathbf{x}, \mathbf{y}$ are linearly dependent.

Proof If $\mathbf{x} = \mathbf{0}$ then both sides of (7.3.7) are 0 and $\mathbf{x}, \mathbf{y}$ are linearly dependent. So let $\mathbf{x} \neq \mathbf{0}$. Then $\alpha := \langle \mathbf{x}, \mathbf{x} \rangle \neq 0$. Let $\beta = \langle \mathbf{y}, \mathbf{x} \rangle / \langle \mathbf{x}, \mathbf{x} \rangle$ and let $\mathbf{z} = \mathbf{y} - \beta \mathbf{x}$. Then

$$\langle \mathbf{z}, \mathbf{z} \rangle = \langle \mathbf{y}, \mathbf{y} \rangle - \overline{\beta} \langle \mathbf{y}, \mathbf{x} \rangle - \beta \langle \mathbf{x}, \mathbf{y} \rangle + \beta \overline{\beta} \langle \mathbf{x}, \mathbf{x} \rangle \tag{7.3.8}$$

Since $\langle \mathbf{x}, \mathbf{x} \rangle$ is real and $\overline{\langle \mathbf{y}, \mathbf{x} \rangle} = \langle \mathbf{x}, \mathbf{y} \rangle$, it is easily seen that each of the last three terms on the RHS of (7.3.8) is, ignoring the sign, equal to $|\langle \mathbf{x}, \mathbf{y} \rangle|^2 / \langle \mathbf{x}, \mathbf{x} \rangle$. Thus

$$\langle \mathbf{z}, \mathbf{z} \rangle = \langle \mathbf{y}, \mathbf{y} \rangle - \frac{|\langle \mathbf{x}, \mathbf{y} \rangle|^2}{\langle \mathbf{x}, \mathbf{x} \rangle} \tag{7.3.9}$$

Now $\langle \mathbf{z}, \mathbf{z} \rangle \geq 0$ gives (7.3.7) since $\langle \mathbf{x}, \mathbf{x} \rangle > 0$.

Suppose now $\mathbf{x}, \mathbf{y}$ are linearly dependent. Then $\mathbf{y} = \gamma \mathbf{x}$ for some scalar γ (note $\mathbf{x} \neq \mathbf{0}$) and it is easy to verify that $\beta = \gamma$, so $\mathbf{z} = \mathbf{0}$ and equality in (7.3.7) follows from (7.3.9). Conversely, suppose equality holds in (7.3.7). Then (7.3.9) gives $\mathbf{z} = \mathbf{0}$, so $\mathbf{y} = \beta \mathbf{x}$ and $\mathbf{x}, \mathbf{y}$ are linearly dependent. ∎

We caution the reader that $\beta = \langle \mathbf{x}, \mathbf{y} \rangle / \langle \mathbf{x}, \mathbf{x} \rangle$ will not work in the above proof when $F = \mathbb{C}$. Also, several concepts and proofs in this chapter have analogues in Statistics. We will be pointing them but they can be ignored by readers who do not know Statistics. For example, in the inner product space of random variables, the β considered in the above proof is nothing but the regression coefficient of $\mathbf{y}$ on $\mathbf{x}$, $\mathbf{z}$ is the residual of $\mathbf{y}$ on $\mathbf{x}$, (7.3.9) says that the residual variance is $V(\mathbf{y})(1 - \rho^2)$ and (7.3.7) says that $\rho^2 \leq 1$ where ρ is the correlation coefficient between $\mathbf{x}$ and $\mathbf{y}$.

Note that Cauchy-Schwarz inequality for the usual inner product in $\mathbb{R}^2$ says that $\cos^2 \theta \leq 1$ where θ is the angle between $O\mathbf{x}$ and $O\mathbf{y}$.

Theorem 7.3.10 In any inner product space, $||\mathbf{x}|| = +\sqrt{\langle \mathbf{x}, \mathbf{x} \rangle}$ is a norm.

Proof Condition (i) of *Definition* 7.3.1 follows from (iii) of *Definition* 7.2.1. Condition (ii) follows thus:

$$||\alpha\mathbf{x}|| = \sqrt{\langle\alpha\mathbf{x}, \alpha\mathbf{x}\rangle} = \sqrt{\alpha\overline{\alpha}\langle\mathbf{x}, \mathbf{x}\rangle} = \sqrt{|\alpha|^2\langle\mathbf{x}, \mathbf{x}\rangle} = |\alpha| \cdot ||\mathbf{x}||$$

To prove condition (iii), we have

$$\begin{aligned} ||\mathbf{x}+\mathbf{y}||^2 &= \langle\mathbf{x}+\mathbf{y}, \mathbf{x}+\mathbf{y}\rangle \\ &= \langle\mathbf{x}, \mathbf{x}\rangle + \langle\mathbf{x}, \mathbf{y}\rangle + \langle\mathbf{y}, \mathbf{x}\rangle + \langle\mathbf{y}, \mathbf{y}\rangle \\ &= ||\mathbf{x}||^2 + 2\text{Re}\langle\mathbf{x}, \mathbf{y}\rangle + ||\mathbf{y}||^2 \\ &\leq ||\mathbf{x}||^2 + 2|\langle\mathbf{x}, \mathbf{y}\rangle| + ||\mathbf{y}||^2 \\ &\leq ||\mathbf{x}||^2 + 2||\mathbf{x}|| \cdot ||\mathbf{y}|| + ||\mathbf{y}||^2 \\ &= (||\mathbf{x}|| + ||\mathbf{y}||)^2 \end{aligned}$$

Taking non-negative square root, (iii) follows. ∎

We say that $||\mathbf{x}|| = \sqrt{\langle\mathbf{x}, \mathbf{x}\rangle}$ is the *norm induced by* the inner product $\langle\mathbf{x}, \mathbf{y}\rangle$. The norm induced by the canonical inner product on $\mathbb{R}^n$, viz. $\sqrt{\sum x_i^2}$ is the usual length of the line segment $O\mathbf{x}$. The norm induced by the inner product of *Example* 7.2.6 is the standard deviation. We next prove that the norm induced by an inner product has a property not possessed by norms in general.

Theorem 7.3.11 (*Parallelogram law*) The norm induced by an inner product on a real vector space has the property

$$||\mathbf{x}+\mathbf{y}||^2 + ||\mathbf{x}-\mathbf{y}||^2 = 2||\mathbf{x}||^2 + 2||\mathbf{y}||^2 \tag{7.3.10}$$

The proof of this theorem is simple and is left to the reader. This theorem shows that the norm induced by an inner product is essentially quadratic in nature. It is called the parallelogram law because it says that the sum of the squares of the diagonals of a parallelogram equals the sum of the squares of the four sides and is a restatement of the Apollonius theorem.

Exercises

1. Deduce the properties (a), (b) and (c) of a norm stated before *Lemma* 7.3.2, from the axioms.
2. Find all possible norms on $\mathbb{R}^1$.
3. When is $||\mathbf{x}|| = \sum_{j=1}^{n} \alpha_j |x_j|$ a norm on $\mathbb{R}^n$?

4. Is a norm determined by its values for the vectors in a basis?
5. Prove that (7.3.4) is not a norm if $0 < p < 1$.
6. Prove that (7.3.5) and (7.3.6) are norms and that for fixed $\mathbf{x}$, the RHS of (7.3.4) converges to the RHS of (7.3.5) as $p \to \infty$.
7. Verify that the norm given in *Example* 7.3.7 is indeed a norm.
8. Show that $\|\mathbf{A}\| = \sqrt{\sum_i \sum_j |a_{ij}|^2}$ is a norm on the vector space of all complex $n \times n$ matrices and that it is induced by an inner product.
9. (a) Prove *Theorem* 7.3.11.

   *(b) Let $1 \leq p \leq \infty$. Verify that the L_p-norm satisfies (7.3.10) iff $p = 2$.
10. (a) Show that an inner product on a real vector space is uniquely determined by the induced norm.

    *(b) In a complex inner product space, prove the polarization identity:
$$\langle \mathbf{x}, \mathbf{y} \rangle = \tfrac{1}{4}\{\|\mathbf{x}+\mathbf{y}\|^2 - \|\mathbf{x}-\mathbf{y}\|^2 + i\|\mathbf{x}+i\mathbf{y}\|^2 - i\|\mathbf{x}-i\mathbf{y}\|^2\}$$
11. Show that d defined by: $d(x, y) = 0$ if $x = y$ and 1 otherwise, is a metric on $\mathbb{R}$ which is invariant under translations and that it does not arise from a norm as in *Theorem* 7.3.8.
12. Let $\|\cdot\|$ be a norm on $V \neq \{\mathbf{0}\}$. Given any positive number α, show that there exists a vector $\mathbf{x} \in V$ such that $\|\mathbf{x}\| = \alpha$.
13. Let $N_1(\cdot)$ and $N_2(\cdot)$ be two norms on $\mathbb{R}^n$. Show that $\alpha N_1 + \beta N_2$ is a norm if α and β are non-negative and not both are zero.
14. (a) Find the distance between $(2, 5)$ and $(3, -4)$ under each of the norms L_1, L_2 and L_∞.

    *(b) Show that
$$n^{-1/2}\|\mathbf{x}\|_1 \leq \|\mathbf{x}\|_2 \leq \|\mathbf{x}\|_1$$
    and
$$\|\mathbf{x}\|_\infty \leq \|\mathbf{x}\|_2 \leq n^{1/2}\|\mathbf{x}\|_\infty$$
    for all $\mathbf{x} \in \mathbb{C}^n$. (Thus the norms L_1, L_2 and L_∞ are topologically equivalent.)

*15. Let $N(\cdot)$ be a norm on $\mathbb{R}^n$. For an $n \times n$ matrix $\mathbf{A}$, define
$$\|\mathbf{A}\| = \sup_{\mathbf{x} \neq 0} \frac{N(\mathbf{A}\mathbf{x})}{N(\mathbf{x})}$$
   (a) Show that $\|\cdot\|$ is a norm on $\mathbb{R}^{n \times n}$. This is called *the matrix norm induced by the vector norm N*.
   (b) Show that $\|\mathbf{I}\| = 1$.
   (c) Show that $N(\mathbf{A}\mathbf{x}) \leq \|\mathbf{A}\|\, N(\mathbf{x})$ for all $\mathbf{x}$.

(d) Show that $||\mathbf{AB}|| \le ||\mathbf{A}|| \cdot ||\mathbf{B}||$, where $\mathbf{A}$ and $\mathbf{B}$ are $n \times n$ matrices.

(e) Show that the matrix norm induced by the L_1-norm on $\mathbb{R}^n$ is

$$||\mathbf{A}|| = \max_j \sum_{i=1}^n |a_{ij}|$$

(f) Show that the matrix norm induced by the L_∞-norm on $\mathbb{R}^n$ is

$$||\mathbf{A}|| = \max_i \sum_{j=1}^n |a_{ij}|$$

(g) Let $||\mathbf{A}|| < 1$, where the matrix norm is induced by a vector norm $N(\cdot)$. Then show that $\mathbf{I} - \mathbf{A}$ and $\mathbf{I} + \mathbf{A}$ are non-singular. Show also that

$$\frac{1}{1 + ||\mathbf{A}||} \le ||(\mathbf{I} - \mathbf{A})^{-1}|| \le \frac{1}{1 - ||\mathbf{A}||}$$

What would be the bounds for $||(\mathbf{I} + \mathbf{A})^{-1}||$?

(h) Let $\mathbf{u}$ and $\mathbf{u} + \mathbf{h}$ be solutions of $\mathbf{Ax} = \mathbf{b}$ and $(\mathbf{A} + \mathbf{E})\mathbf{x} = \mathbf{b} + \mathbf{g}$ respectively, where $\mathbf{A}$ is non-singular. Let the relative errors in $\mathbf{A}$, $\mathbf{b}$ and $\mathbf{u}$ be defined as $r_\mathbf{A} = ||\mathbf{E}||/||\mathbf{A}||$, $r_\mathbf{b} = N(\mathbf{g})/N(\mathbf{b})$ and $r_\mathbf{u} = N(\mathbf{h})/N(\mathbf{u})$ respectively. Define the condition number $c(\mathbf{A})$ of $\mathbf{A}$ as $||\mathbf{A}|| \cdot ||\mathbf{A}^{-1}||$. If $||\mathbf{A}^{-1}|| \cdot ||\mathbf{E}|| < 1$, show that

$$r_\mathbf{u} \le \frac{c(\mathbf{A})}{1 - c(\mathbf{A})r_\mathbf{A}}(r_\mathbf{A} + r_\mathbf{b})$$

(See *Exercise* 9.5.6 for more on matrix norms.)

*16. Show that any norm on a real vector space satisfying (7.3.11) is induced by an inner product.

7.4 Orthogonality and orthonormal basis

We saw in the *Introduction* to this chapter that inner product combines the concepts of length and angle. We have discussed the first concept (length) in the preceding section. In the present section and the next we shall discuss an important special case of the second, viz., the angle between two vectors being 90°. It follows from (7.1.3) that the lines $O\mathbf{x}$ and $O\mathbf{y}$ are at right angles iff the usual inner product $\langle \mathbf{x}, \mathbf{y} \rangle$ is 0. Motivated by this we give the following

Definition 7.4.1 Vectors $\mathbf{x}$ and $\mathbf{y}$ in an inner product space V are said to be *orthogonal* (to each other) if $\langle \mathbf{x}, \mathbf{y} \rangle = 0$. We then write $\mathbf{x} \perp \mathbf{y}$. A set $A \subseteq V$ is said to be *orthogonal* if $\mathbf{x} \perp \mathbf{y}$ for every pair of distinct vectors $\mathbf{x}$ and $\mathbf{y}$ in A. A is said to be *orthonormal* if A is orthogonal and every vector in A has norm 1. If A and B are subsets of V, we say that A is *orthogonal to* B if every vector in A is orthogonal to every vector in B. Sets $A_1, A_2, \ldots, A_k$ are said to be *orthogonal to one another* if A_i is orthogonal to A_j whenever $i \neq j$.

Note that $\langle \mathbf{x}, \mathbf{y} \rangle = 0 \Rightarrow \langle \mathbf{y}, \mathbf{x} \rangle = 0$. Clearly $\mathbf{0} \perp \mathbf{x}$ for all $\mathbf{x}$. Also $\mathbf{x} \perp \mathbf{x}$ iff $\mathbf{x} = \mathbf{0}$. If $\mathbf{x} \perp \mathbf{y}$ and $\mathbf{x} \perp \mathbf{z}$ then it is easy to see that $\mathbf{x} \perp (\mathbf{y}+\mathbf{z})$ and $\mathbf{x} \perp (\alpha \mathbf{y})$ for all scalars α.

We emphasize that a set of vectors is orthogonal iff its elements are pair-wise orthogonal. The corresponding statement for linear independence is false; linear independence is a property of the entire set whereas orthogonality is a property of pairs. The empty set is orthonormal (in a vacuous sense). Any orthogonal set not containing the null vector can be converted to an orthonormal set by normalizing each vector in it.

Example 7.4.2 Consider $V = \mathbb{R}^4$ with $\langle \mathbf{x}, \mathbf{y} \rangle = \mathbf{y}^{\mathrm{T}}\mathbf{x}$. Then the canonical basis $\{\mathbf{e}_1, \mathbf{e}_2, \mathbf{e}_3, \mathbf{e}_4\}$ is an orthonormal set. The three vectors $\mathbf{x} = (1,1,1,1)^{\mathrm{T}}$, $\mathbf{y} = (1,-1,1,-1)^{\mathrm{T}}$ and $\mathbf{z} = (1,0,-1,0)^{\mathrm{T}}$ form an orthogonal set. The set $\{\frac{1}{2}\mathbf{x}, \frac{1}{2}\mathbf{y}, \frac{1}{\sqrt{2}}\mathbf{z}\}$ is orthonormal. The sets

$$\begin{aligned} A &= \{(\xi_1, 0, \xi_3, 0)^{\mathrm{T}} : \xi_1, \xi_3 \in \mathbb{R}\} \\ B &= \{(0, \xi_2, 0, 0)^{\mathrm{T}} : \xi_2 \in \mathbb{R}\} \\ C &= \{(0, 0, 0, \xi_4)^{\mathrm{T}} : \xi_4 \in \mathbb{R}\} \end{aligned}$$

are orthogonal to one another but none of them is an orthogonal set. ∎

In the inner product space of random variables, an orthogonal set is a set of pairwise uncorrelated random variables. They form an orthonormal set if, further, each of them has unit variance.

It is easy to verify that the vectors $(1,0)^{\mathrm{T}}$ and $(1,2)^{\mathrm{T}}$ are orthogonal in the inner product space of *Example* 7.2.5 but not with respect to the canonical inner product on $\mathbb{R}^2$.

Theorem 7.4.3 (*Pythagoras theorem*) In a real inner product space, $||\mathbf{x}+\mathbf{y}||^2 = ||\mathbf{x}||^2 + ||\mathbf{y}||^2$ if $\mathbf{x} \perp \mathbf{y}$. More generally, $||\sum_{i=1}^k \mathbf{x}_i||^2 = \sum_{i=1}^k ||\mathbf{x}_i||^2$ if $\{\mathbf{x}_1, \mathbf{x}_2, \ldots, \mathbf{x}_k\}$ is orthogonal.

We leave the proof of this theorem to the reader, see *Exercise* 7.4.4.

Theorem 7.4.4 Any orthogonal set A not containing the null vector is linearly independent.

Proof Let $\mathbf{x}_1, \mathbf{x}_2, \ldots, \mathbf{x}_k$ be distinct vectors belonging to A such that $\alpha_1\mathbf{x}_1 + \cdots + \alpha_k\mathbf{x}_k = \mathbf{0}$. Since $\langle \mathbf{x}_i, \mathbf{x}_j \rangle = 0$ whenever $i \neq j$, we have

$$0 = \langle \alpha_1\mathbf{x}_1 + \cdots + \alpha_k\mathbf{x}_k, \mathbf{x}_j \rangle = \alpha_j \langle \mathbf{x}_j, \mathbf{x}_j \rangle$$

Now $\langle \mathbf{x}_j, \mathbf{x}_j \rangle \neq 0$, hence $\alpha_j = 0$ for all j. ∎

Corollary Any orthonormal set is linearly independent.

Theorem 7.4.5 If the subspaces $S_1, S_2, \ldots, S_k$ are orthogonal to one another then $S_1 + S_2 + \cdots + S_k$ is direct.

Proof We will verify (iii) of *Theorem* 1.7.8. So let $\mathbf{x}_1 + \cdots + \mathbf{x}_k = \mathbf{0}$ where $\mathbf{x}_i \in S_i$ for $i = 1, \ldots, k$. Suppose at least one $\mathbf{x}_i$ is non-null. Then the non-null vectors among $\mathbf{x}_1, \ldots, \mathbf{x}_k$ are linearly independent by the preceding theorem, a contradiction since their sum is $\mathbf{0}$. So $\mathbf{x}_i = \mathbf{0}$ for all i and the present theorem follows from *Theorem* 1.7.8. ∎

Let S be a subspace of an inner product space. We then say that B is an *orthogonal basis* (resp. an *orthonormal basis*) of S if B is a basis of S and B is an orthogonal (resp. an orthonormal) set. If r is the dimension of S, then by *Theorem* 7.4.4, any r non-null vectors in S which are pair-wise orthogonal form an orthogonal basis for S.

We have seen in *Section* 1.5 that a basis corresponds to a coordinate system. An orthonormal basis corresponds to a system of rectangular coordinates where the reference point on each axis is at unit distance from the origin. Finding the coordinates with respect to such a coordinate system is easy as shown in the following

Theorem 7.4.6 Let $B = \{\mathbf{x}_1, \mathbf{x}_2, \ldots, \mathbf{x}_n\}$ be an orthonormal basis of an inner product space V. Then for any $\mathbf{x} \in V$, we have

$$\mathbf{x} = \textstyle\sum_{j=1}^{n} \langle \mathbf{x}, \mathbf{x}_j \rangle \mathbf{x}_j \tag{7.4.2}$$

Proof Since B is a basis, $\mathbf{x} = \sum_{i=1}^{n} \alpha_i\mathbf{x}_i$ for unique scalars $\alpha_1, \ldots, \alpha_n$. Since B is orthonormal, we have $\langle \mathbf{x}, \mathbf{x}_j \rangle = \langle \alpha_j\mathbf{x}_j, \mathbf{x}_j \rangle = \alpha_j$. ∎

If $\{\mathbf{x}_1, \mathbf{x}_2, \ldots, \mathbf{x}_n\}$ is only an orthogonal basis, then it is easy to see that (7.4.2) has to be modified by replacing $\langle \mathbf{x}, \mathbf{x}_j \rangle$ by $\langle \mathbf{x}, \mathbf{x}_j \rangle / \langle \mathbf{x}_j, \mathbf{x}_j \rangle$.

Suppose next $A = \{\mathbf{x}_1, \mathbf{x}_2, \ldots, \mathbf{x}_k\}$ is any orthogonal set of non-null vectors in V. Then for any $\mathbf{x} \in V$, we call

$$\mathbf{z} := \mathbf{x} - \sum_{j=1}^{k} \frac{\langle \mathbf{x}, \mathbf{x}_j \rangle}{\langle \mathbf{x}_j, \mathbf{x}_j \rangle} \mathbf{x}_j$$

the *residual of* $\mathbf{x}$ *with respect to* A.

It is easy to see that the residual is orthogonal to each $\mathbf{x}_i$. Notice that in the inner product space of random variables, the sum on the right is the linear regression of $\mathbf{x}$ on $\mathbf{x}_1, \ldots, \mathbf{x}_k$.

Using the observation that the residual is orthogonal to each $\mathbf{x}_i$, we can obtain an orthogonal basis $\{\mathbf{z}_1, \mathbf{z}_2, \ldots, \mathbf{z}_r\}$ of a subspace S starting from any basis $\{\mathbf{x}_1, \mathbf{x}_2, \ldots, \mathbf{x}_r\}$ of S, as follows. We take $\mathbf{z}_1 = \mathbf{x}_1$ and define $\mathbf{z}_2$ to be the residual of $\mathbf{x}_2$ with respect to $\{\mathbf{z}_1\}$. Then $\{\mathbf{z}_1, \mathbf{z}_2\}$ is orthogonal. Also $\mathbf{z}_2 \neq \mathbf{0}$ since $\mathbf{x}_2 \notin \mathrm{Sp}(\{\mathbf{x}_1\}) = \mathrm{Sp}(\{\mathbf{z}_1\})$. Moreover, $\mathbf{z}_2$ is a linear combination of $\mathbf{x}_1$ and $\mathbf{x}_2$, and $\mathbf{x}_2$ is a linear combination of $\mathbf{z}_1$ and $\mathbf{z}_2$, so $\{\mathbf{z}_1, \mathbf{z}_2\}$ and $\{\mathbf{x}_1, \mathbf{x}_2\}$ have the same span. We next take $\mathbf{z}_3$ to be the residual of $\mathbf{x}_3$ with respect to $\{\mathbf{z}_1, \mathbf{z}_2\}$. Then it can be proved similarly that $\mathbf{z}_3 \neq \mathbf{0}$, the set $\{\mathbf{z}_1, \mathbf{z}_2, \mathbf{z}_3\}$ is orthogonal and $\{\mathbf{z}_1, \mathbf{z}_2, \mathbf{z}_3\}$ and $\{\mathbf{x}_1, \mathbf{x}_2, \mathbf{x}_3\}$ have the same span. Proceeding like this, we get non-null vectors $\mathbf{z}_1, \mathbf{z}_2, \ldots, \mathbf{z}_r$ which form an orthogonal set and so an orthogonal basis of S. We have thus proved

Theorem 7.4.7 (*Gram-Schmidt orthogonalization process*) Let $\{\mathbf{x}_1, \mathbf{x}_2, \ldots, \mathbf{x}_r\}$ be a basis of S. Define $\mathbf{z}_1, \mathbf{z}_2, \ldots, \mathbf{z}_r$ inductively by:

$$\mathbf{z}_k = \mathbf{x}_k - \sum_{j=1}^{k-1} \frac{\langle \mathbf{x}_k, \mathbf{z}_j \rangle}{\langle \mathbf{z}_j, \mathbf{z}_j \rangle} \mathbf{z}_j \quad (k = 1, \ldots, r) \tag{7.4.3}$$

(Note that $\mathbf{z}_1 = \mathbf{x}_1$.) Then $\{\mathbf{z}_1, \mathbf{z}_2, \ldots, \mathbf{z}_r\}$ is an orthogonal basis of S.

An orthonormal basis of S can be obtained by normalizing the $\mathbf{z}_k$'s. We now extend the above procedure to the case where we start with a finite generating set $\{\mathbf{x}_1, \mathbf{x}_2, \ldots, \mathbf{x}_s\}$ of S instead of a basis.

Algorithm 7.4.8 (*Generalized Gram-Schmidt Process*) Given: vectors $\mathbf{x}_1, \mathbf{x}_2, \ldots, \mathbf{x}_s$ in an inner product space.

Step 1 Set $k = 1$.

Step 2 Compute

$$\mathbf{z}_k = \mathbf{x}_k - \textstyle\sum_{j=1}^{k-1} \langle \mathbf{x}_k, \mathbf{y}_j \rangle \mathbf{y}_j \tag{7.4.4}$$

Step 3 Compute $\mathbf{y}_k := \frac{1}{\|\mathbf{z}_k\|}\mathbf{z}_k$ or $\mathbf{0}$ according as $\mathbf{z}_k \neq \mathbf{0}$ or $\mathbf{z}_k = \mathbf{0}$.

Step 4 If $k < s$, increase k by 1 and go to *Step* 2. Otherwise go to *Step* 5.

Step 5 Stop. For $i = 1, \ldots, s$, the set B_i of all non-null vectors among $\mathbf{y}_1, \ldots, \mathbf{y}_i$ is an orthonormal basis of the span S_i of $\{\mathbf{x}_1, \ldots, \mathbf{x}_i\}$.

The following statements can easily be proved for $k = 1, \ldots, s$ by using induction on k. In fact, the proof is similar to that of the preceding theorem and is left as an exercise to the reader.

(a) $\mathbf{z}_k \in S_k$ and so $\mathbf{y}_k \in S_k$,

(b) B_k is an orthonormal set, and

(c) $\mathrm{Sp}(B_k) = S_k$

Remark 1 In the preceding algorithm, $\mathbf{y}_k = \mathbf{0}$ iff $\mathbf{x}_k$ is a linear combination of $\mathbf{x}_1, \ldots, \mathbf{x}_{k-1}$.

The *only if part* of this remark is trivial. To prove the *if part*, let $\mathbf{x}_k$ be a linear combination of $\mathbf{x}_1, \ldots, \mathbf{x}_{k-1}$. Then $S_k = S_{k-1}$. Since $B_{k-1} \subseteq B_k$ it follows that $B_k = B_{k-1}$ and $\mathbf{y}_k = \mathbf{0}$.

Remark 2 In the preceding algorithm, if $\mathbf{x}_1, \mathbf{x}_2, \ldots, \mathbf{x}_\ell$ form an orthonormal set then $\mathbf{y}_j = \mathbf{x}_j$ for $j = 1, 2, \ldots, \ell$.

This remark can be proved easily by proving $\mathbf{y}_1 = \mathbf{x}_1$, $\mathbf{y}_2 = \mathbf{x}_2$, $\ldots, \mathbf{y}_\ell = \mathbf{x}_\ell$ in that order. We next derive an important consequence of the preceding algorithm and its proof.

Theorem 7.4.9 Every subspace S of a finite-dimensional inner product space has an orthonormal basis. Moreover, any orthonormal subset of S can be extended to an orthonormal basis of S.

Proof Starting from any basis of S we can construct an orthonormal basis by the Gram-Schmidt process. This proves the first statement. To prove the second, let $A = \{\mathbf{x}_1, \ldots, \mathbf{x}_\ell\}$ be an orthonormal subset of S. Extend A to a generating set $\{\mathbf{x}_1, \ldots, \mathbf{x}_\ell, \mathbf{x}_{\ell+1}, \ldots, \mathbf{x}_s\}$ of S by appending a basis. Applying the generalized Gram-Schmidt process to $\mathbf{x}_1, \ldots, \mathbf{x}_s$, get $\mathbf{y}_1, \ldots, \mathbf{y}_s$. Then the non-null vectors among $\mathbf{y}_1, \ldots, \mathbf{y}_s$ form an orthonormal basis of S and contain $\mathbf{x}_1, \ldots, \mathbf{x}_\ell$ by *Remark* 2 above. ∎

We note that the orthonormal basis obtained by the Gram-Schmidt process from $\mathbf{x}_1, \ldots, \mathbf{x}_s$ may be quite different from that obtained from a rearrangement of $\mathbf{x}_1, \ldots, \mathbf{x}_s$.

Example 7.4.10 Consider $\mathbb{R}^4$ with the usual inner product. Let

$$\mathbf{x}_1 = \frac{1}{\sqrt{3}}(1, 0, 1, -1)^{\mathrm{T}} \text{ and } \mathbf{x}_2 = \frac{1}{\sqrt{7}}(-2, 1, 1, -1)^{\mathrm{T}}$$

We will extend $\{\mathbf{x}_1, \mathbf{x}_2\}$ to an orthonormal basis by the method of the preceding theorem. We append the canonical basis so that $\mathbf{x}_3 = \mathbf{e}_1$, $\mathbf{x}_4 = \mathbf{e}_2$, $\mathbf{x}_5 = \mathbf{e}_3$ and $\mathbf{x}_6 = \mathbf{e}_4$ and apply the Generalized Gram-Schmidt Process to $\mathbf{x}_1, \ldots, \mathbf{x}_6$. We get $\mathbf{y}_1 = \mathbf{z}_1 = \mathbf{x}_1$ and $\mathbf{y}_2 = \mathbf{z}_2 = \mathbf{x}_2$. We next compute

$$\begin{aligned}\mathbf{z}_3 &= \mathbf{x}_3 - \langle \mathbf{x}_3, \mathbf{y}_1 \rangle \mathbf{y}_1 - \langle \mathbf{x}_3, \mathbf{y}_2 \rangle \mathbf{y}_2 \\ &= (1,0,0,0)^{\mathrm{T}} - \frac{1}{\sqrt{3}} \cdot \frac{1}{\sqrt{3}}(1,0,1,-1)^{\mathrm{T}} - \frac{1}{\sqrt{7}}(-2)\frac{1}{\sqrt{7}}(-2,1,1,-1)^{\mathrm{T}} \\ &= \frac{1}{21}(2,6,-1,1)^{\mathrm{T}}\end{aligned}$$

Notice that to get $\mathbf{y}_3$ we may as well normalize $(2,6,-1,1)^{\mathrm{T}}$. Thus $\mathbf{y}_3 = \frac{1}{\sqrt{42}}(2,6,-1,1)^{\mathrm{T}}$. We leave it to the reader to verify that $\mathbf{z}_4$ becomes $\mathbf{0}$, so $\mathbf{y}_4 = \mathbf{0}$. We next find $\mathbf{z}_5$ to be $\frac{1}{2}(0,0,1,1)^{\mathrm{T}}$, so $\mathbf{y}_5 = \frac{1}{\sqrt{2}}(0,0,1,1)^{\mathrm{T}}$. At this stage we have already got an orthonormal set of size 4, viz., $B = \{\mathbf{y}_1, \mathbf{y}_2, \mathbf{y}_3, \mathbf{y}_5\}$. Since $d(\mathbb{R}^4) = 4$ we can stop here and B is an extension of $\{\mathbf{x}_1, \mathbf{x}_2\}$ to an orthonormal basis of $\mathbb{R}^4$. One may verify that $\mathbf{z}_6$ is indeed $\mathbf{0}$. ∎

Exercises

1. Show that sets $A_1, A_2, \ldots, A_k$ of vectors are orthogonal to one another iff their spans are orthogonal to one another.
2. (a) Prove *Theorem* 7.4.3. Also prove the converse of the first statement and disprove the converse of the second.

   (b) Prove *Theorem* 7.4.3 for a complex inner product space. Show that now even the converse of the first statement is false.
*3. Let $\mathbf{A}$ and $\mathbf{B}$ be diagonal matrices with positive diagonal entries. Let $\perp_{\mathbf{A}}$ denote orthogonality with respect to the inner product $\langle \mathbf{x}, \mathbf{y} \rangle = \mathbf{y}^{\mathrm{T}} \mathbf{A} \mathbf{x}$. Then show that '$\mathbf{x} \perp_{\mathbf{A}} \mathbf{y} \Leftrightarrow \mathbf{x} \perp_{\mathbf{B}} \mathbf{y}$' holds iff $\mathbf{B}$ is a scalar multiple of $\mathbf{A}$.
4. Let $\mathbf{x}_1, \mathbf{x}_2, \ldots, \mathbf{x}_k$ form an orthonormal set.

   (a) Show that $\|\sum_{i=1}^k \alpha_i \mathbf{x}_i\|^2 = \sum_{i=1}^k |\alpha_i|^2$.

   (b) If $\mathbf{z}$ is the residual of $\mathbf{x}$ on $\{\mathbf{x}_1, \ldots, \mathbf{x}_k\}$, show that

   $$\|\mathbf{z}\|^2 = \|\mathbf{x}\|^2 - \|\sum_{i=1}^k \langle \mathbf{x}, \mathbf{x}_i \rangle \mathbf{x}_i\|^2 = \|\mathbf{x}\|^2 - \sum_{i=1}^k |\langle \mathbf{x}, \mathbf{x}_i \rangle|^2.$$

   (c) Prove *Bessel's inequality*:

   $$\|\mathbf{x}\|^2 \geq \sum_{i=1}^k |\langle \mathbf{x}, \mathbf{x}_i \rangle|^2$$

   for any $\mathbf{x}$. Show also that equality holds iff $\mathbf{x} \in \mathrm{Sp}(\{\mathbf{x}_1, \ldots, \mathbf{x}_k\})$.

5. Let $B = \{\mathbf{x}_1, \mathbf{x}_2, \ldots, \mathbf{x}_k\}$ be an orthonormal set in a finite-dimensional inner product space V. Show that the following statements are equivalent:
   (a) B is maximal,
   (b) $\langle \mathbf{x}, \mathbf{x}_i \rangle = 0$ for $i = 1, \ldots, k \Rightarrow \mathbf{x} = \mathbf{0}$,
   (c) B generates V,
   (d) if $\mathbf{x} \in V$ then $\mathbf{x} = \sum_{i=1}^{k} \langle \mathbf{x}, \mathbf{x}_i \rangle \mathbf{x}_i$,
   (e) if $\mathbf{x}, \mathbf{y} \in V$ then $\langle \mathbf{x}, \mathbf{y} \rangle = \sum_{i=1}^{k} \langle \mathbf{x}, \mathbf{x}_i \rangle \cdot \langle \mathbf{x}_i, \mathbf{y} \rangle$,
   (f) if $\mathbf{x} \in V$ then $\|\mathbf{x}\|^2 = \sum_{i=1}^{k} |\langle \mathbf{x}, \mathbf{x}_i \rangle|^2$.
6. In (7.4.2), can $\langle \mathbf{x}, \mathbf{x}_j \rangle$ be replaced by $\langle \mathbf{x}_j, \mathbf{x} \rangle$?
7. Show that $\mathbf{z} := \mathbf{x} + \sum_{i=1}^{k} \alpha_i \mathbf{x}_i$ is the residual of $\mathbf{x}$ with respect to the orthogonal set $\{\mathbf{x}_1, \ldots, \mathbf{x}_k\}$ iff $\langle \mathbf{z}, \mathbf{x}_j \rangle = 0$ for $j = 1, \ldots, k$.
8. Prove the statements (a)–(c) given after *Algorithm* 7.4.8.
9. Find an orthonormal basis of the subspace of $\mathbb{R}^4$ spanned by $(2, -1, 0, 1)$, $(6, 1, 4, -5)$ and $(4, 1, 3, -4)$.
10. Let $\mathbf{x}_1 = (1, 1, 1, 1)$, $\mathbf{x}_2 = (0, 1, 1, 1)$, $\mathbf{x}_3 = (0, 0, 1, 1)$ and $\mathbf{x}_4 = (0, 0, 0, 1)$ in $\mathbb{R}^4$. Starting from $\{\mathbf{x}_1, \mathbf{x}_2, \mathbf{x}_3, \mathbf{x}_4\}$ obtain an orthonormal basis of $\mathbb{R}^4$. If you use $\{\mathbf{x}_4, \mathbf{x}_3, \mathbf{x}_2, \mathbf{x}_1\}$ what is the orthonormal basis obtained?

*11. With respect to the inner product of *Example* 7.2.4, find an orthonormal basis of $\mathbb{R}^2$ using Gram-Schmidt orthogonalization process on the canonical basis.

*12. Consider the inner product $\langle \mathbf{x}, \mathbf{y} \rangle = \mathbf{y}^{\mathrm{T}} \mathbf{A} \mathbf{x}$ on $\mathbb{R}^3$ where

$$\mathbf{A} = \begin{bmatrix} 2 & 1 & -1 \\ 1 & 1 & 0 \\ -1 & 0 & 3 \end{bmatrix}$$

Find an orthonormal basis B of $S := \{(x_1, x_2, x_3) : x_1 + x_2 + x_3 = 0\}$ and then extend it to an orthonormal basis C of $\mathbb{R}^3$.

*13. (**QR**-*decomposition*) Let $\mathbf{A}$ be an $n \times s$ matrix with rank p. Let $\mathbf{y}_1, \mathbf{y}_2, \ldots, \mathbf{y}_s$ be the vectors obtained when *Algorithm* 7.4.8 is applied to the columns of $\mathbf{A}$. Let $\mathbf{P} = [\,\mathbf{y}_1 : \mathbf{y}_2 : \cdots : \mathbf{y}_s\,]$ and let $\mathbf{U}$ be the $s \times s$ upper triangular matrix $((u_{ik}))$ where

$$u_{ik} = \begin{cases} \langle \mathbf{A}_{*k}, \mathbf{y}_i \rangle & \text{if } i < k \\ \|\mathbf{z}_k\| & \text{if } i = k \\ 0 & \text{otherwise} \end{cases}$$

Then show that $\mathbf{A} = \mathbf{PU}$. Also show that if $\mathbf{Q}$ is the submatrix of $\mathbf{P}$ formed by the non-null columns and $\mathbf{R}$ the submatrix of $\mathbf{U}$ formed by the corresponding rows, then $(\mathbf{Q}, \mathbf{R})$ is a rank-factorization of $\mathbf{A}$ and $\mathbf{Q}^*\mathbf{Q} = \mathbf{I}_p$. When $\mathbf{A}$ is of full column rank, $(\mathbf{Q}, \mathbf{R}) = (\mathbf{P}, \mathbf{U})$ is known as a $\mathbf{QR}$-decomposition of $\mathbf{A}$.

*14. Using the preceding exercise, find a rank factorization for each of the matrices in *Exercise* 4.5.1.

*15. Show that QR-decomposition (see *Exercise* 7.4.12) is unique if we insist that the diagonal elements of $\mathbf{R}$ are real and positive, i.e., if $\mathbf{A}$ is of full column rank, show that there exist unique matrices $\mathbf{Q}$ and $\mathbf{R}$ such that $\mathbf{A} = \mathbf{QR}$, $\mathbf{Q}^*\mathbf{Q} = \mathbf{I}$, $\mathbf{R}$ is upper triangular and $r_{ii} > 0$ for all i.

7.5 Orthogonal complement and orthogonal projector

In this section we study the orthogonal complement of a subspace. We also study the concept of orthogonal projection and give an explicit expression for the orthogonal projector into the column space of a matrix.

Definition 7.5.1 For any set A of vectors in an inner product space V, we define

$$A^{\perp} = \{\mathbf{y} \in V : \mathbf{y} \perp \mathbf{x} \text{ for every } \mathbf{x} \in A\}$$

This may be read 'A perpendicular'. Note that $\emptyset^{\perp} = V$.

We now show that $A^{\perp}$ is a subspace of V for any set $A \subseteq V$. Clearly $\mathbf{0} \in A^{\perp}$. If $\mathbf{y}, \mathbf{z} \in A^{\perp}$ and $\alpha \in F$ then

$$\langle \alpha\mathbf{y} + \mathbf{z}, \mathbf{x}\rangle = \alpha\langle \mathbf{y}, \mathbf{x}\rangle + \langle \mathbf{z}, \mathbf{x}\rangle = 0$$

for all $\mathbf{x} \in A$, so $\alpha\mathbf{y} + \mathbf{z} \in A^{\perp}$. Thus $A^{\perp}$ is a subspace.

Theorem 7.5.2 Let $B = \{\mathbf{x}_1, \mathbf{x}_2, \ldots, \mathbf{x}_k\}$ be any orthonormal basis of a subspace S and let $\{\mathbf{x}_1, \mathbf{x}_2, \ldots, \mathbf{x}_n\}$ be any extension of B to an orthonormal basis of V. Then $S^{\perp}$ is the span of $\{\mathbf{x}_{k+1}, \ldots, \mathbf{x}_n\}$.

Proof Clearly $\mathbf{x}_{k+1}, \ldots, \mathbf{x}_n$ belong to $S^{\perp}$. Suppose now $\mathbf{x} \in S^{\perp}$. By *Theorem* 7.4.6, $\mathbf{x} = \sum_{i=1}^{n} \langle \mathbf{x}, \mathbf{x}_i\rangle \mathbf{x}_i$. Since $\mathbf{x} \in S^{\perp}$, we have $\langle \mathbf{x}, \mathbf{x}_i\rangle = 0$ for $i = 1, \ldots, k$. So $\mathbf{x} \in \mathrm{Sp}(\{\mathbf{x}_{k+1}, \ldots, \mathbf{x}_n\})$ and the theorem follows. ∎

Theorem 7.5.3 If S is a subspace of V then $S^{\perp}$ is a complement of S, $d(S^{\perp}) = d(V) - d(S)$ and $(S^{\perp})^{\perp} = S$.

This theorem follows immediately from the preceding theorem. The first conclusion of this theorem justifies the following

Definition 7.5.4 If S is a subspace, $S^{\perp}$ is called the *orthogonal complement* of S.

We leave it to the reader to check that in $\mathbb{R}^2$, the orthogonal complement of the x-axis is the y-axis and the orthogonal complement of the line $y = x$ is the line $y = -x$. In $\mathbb{R}^3$, the orthogonal complement of the x-axis is the y-z plane and vice versa.

We now study some simple properties of orthogonal complements. We first note that if W is a complement of S and is orthogonal to S then $W = S^\perp$. This follows on noting that the union of an orthonormal basis of S and an orthonormal basis of W is an orthonormal basis of V.

Suppose $S_1, S_2, \ldots, S_k$ are subspaces which are orthogonal to one another and $S_1 + \cdots + S_k = V$. Then by *Theorem* 7.4.5, $S_1 + \cdots + S_k$ is direct. Now, for any fixed i $(1 \le i \le k)$, $\sum_{j \ne i} S_j$ is a complement of S_i and is orthogonal to S_i, so it is the orthogonal complement of S_i.

Theorem 7.5.5 If $S \subseteq T$ then $S^\perp \supseteq T^\perp$.

This theorem follows easily from definitions. Note the reversal of inclusion.

Theorem 7.5.6 If S and T are subspaces then

$$(S+T)^\perp = S^\perp \cap T^\perp \tag{7.5.1}$$

and

$$(S \cap T)^\perp = S^\perp + T^\perp \tag{7.5.2}$$

Proof It is enough to prove the first conclusion since the second follows from it. Since $S + T$ contains S and T, it follows that $(S+T)^\perp$ is contained in $S^\perp$ and $T^\perp$. So LHS $\subseteq$ RHS in (7.5.1). To prove the reverse inclusion, let $\mathbf{x} \in S^\perp \cap T^\perp$. Consider an arbitrary $\mathbf{y} \in S + T$. Then $\mathbf{y} = \mathbf{u} + \mathbf{v}$ for some $\mathbf{u} \in S$ and $\mathbf{v} \in T$. So

$$\langle \mathbf{x}, \mathbf{y} \rangle = \langle \mathbf{x}, \mathbf{u} + \mathbf{v} \rangle = \langle \mathbf{x}, \mathbf{u} \rangle + \langle \mathbf{x}, \mathbf{v} \rangle = 0 + 0 = 0$$

and $\mathbf{x} \in (S+T)^\perp$. Thus LHS $\supseteq$ RHS in (7.5.1) and equality follows. ∎

The result $S = (S^\perp)^\perp$ is quite powerful and is closely related to a result known as Farkas Lemma which is equivalent to the Duality Theorem of Linear Programming. Notice that when $F = \mathbb{R}$, *Theorem* 5.3.2 follows easily from *Theorems* 7.5.3 and 7.5.5.

Definition 7.5.7 If S is a subspace of V and $\mathbf{x} \in V$, the projection of $\mathbf{x}$ into S along $S^\perp$ is called the *orthogonal projection of* $\mathbf{x}$ *into* S.

Geometrically, the orthogonal projection of $\mathbf{x}$ into S is the foot of the perpendicular drawn from $\mathbf{x}$ to S. Since $(S^\perp)^\perp = S$, it follows that if $\mathbf{y}$ is the orthogonal projection of $\mathbf{x}$ into S then $\mathbf{x}-\mathbf{y}$ is the orthogonal projection of $\mathbf{x}$ into $S^{\perp}$.

Suppose $\{\mathbf{x}_1, \mathbf{x}_2, \ldots, \mathbf{x}_k\}$ is an orthonormal basis of S. For any $\mathbf{x} \in V$, let

$$\mathbf{y} = \sum_{i=1}^{k} \langle \mathbf{x}, \mathbf{x}_i \rangle \mathbf{x}_i \tag{7.5.3}$$

Then, clearly, $\mathbf{y} \in S$. Moreover, the residual $\mathbf{x}-\mathbf{y}$ is orthogonal to each of $\mathbf{x}_1, \mathbf{x}_2, \ldots, \mathbf{x}_k$, so $\mathbf{x}-\mathbf{y} \in S^{\perp}$. We thus have

Theorem 7.5.8 Let $\{\mathbf{x}_1, \mathbf{x}_2, \ldots, \mathbf{x}_k\}$ be an orthonormal basis of S. Then for any $\mathbf{x} \in V$, $\mathbf{y}$ defined by (7.5.3) is the orthogonal projection of $\mathbf{x}$ into S and $\mathbf{x}-\mathbf{y}$ is the orthogonal projection of $\mathbf{x}$ into $S^{\perp}$.

The preceding theorem shows that the residual of $\mathbf{x}$ with respect to an orthonormal basis of S does not depend on the choice of the basis. Residual is, really, with respect to the subspace S.

We now study orthogonal projection into a flat (see *Sections* 1.6 and 5.3).

Theorem 7.5.9 Let S be a subspace and $W = \mathbf{u} + S$ a flat. Then any vector $\mathbf{x}$ can be written uniquely as $\mathbf{w}+\mathbf{y}$ where $\mathbf{w} \in W$ and $\mathbf{y} \in S^{\perp}$.

Proof We first show the existence of $\mathbf{w}$ and $\mathbf{y}$. Let $\mathbf{x}-\mathbf{u} = \mathbf{s}+\mathbf{t}$ where $\mathbf{s} \in S$ and $\mathbf{t} \in S^{\perp}$. Then $\mathbf{x} = (\mathbf{u}+\mathbf{s})+\mathbf{t}$, so we may take $\mathbf{u}+\mathbf{s}$ as $\mathbf{w}$ and $\mathbf{t}$ as $\mathbf{y}$.

We next prove the uniqueness of $\mathbf{w}$ and $\mathbf{y}$. Suppose $\mathbf{x}$ can also be written as $\mathbf{w}'+\mathbf{y}'$ where $\mathbf{w}' \in W$ and $\mathbf{y}' \in S^{\perp}$. Then $\mathbf{w}-\mathbf{w}' = \mathbf{y}'-\mathbf{y}$. Since $\mathbf{w}-\mathbf{w}' \in S$ and $\mathbf{y}'-\mathbf{y} \in S^{\perp}$, it follows that $\mathbf{w} = \mathbf{w}'$ and $\mathbf{y} = \mathbf{y}'$. ∎

We shall call the vector $\mathbf{w}$ of the preceding theorem *the orthogonal projection of* $\mathbf{x}$ *into* W. Geometrically, it is the foot of the perpendicular from $\mathbf{x}$ to W. It follows from the proof of the theorem that $\mathbf{w} = \mathbf{u} + \mathbf{P}(\mathbf{x}-\mathbf{u})$ where $\mathbf{P}$ is the orthogonal projector into S. If $\mathbf{u} = \mathbf{0}$ and $W = S$, this definition reduces to the earlier definition of orthogonal projection into a subspace. We next deduce an important property of orthogonal projection from Pythagoras theorem.

Theorem 7.5.10 Let $\mathbf{w}$ be the orthogonal projection of $\mathbf{x}$ into a flat W. Then

$$\min_{\mathbf{z} \in W} \|\mathbf{x}-\mathbf{z}\|$$

is attained at $\mathbf{w}$ and only at $\mathbf{w}$.

Proof Let $W = \mathbf{u} + S$ where S is a subspace. Take any $\mathbf{z} \in W$. Then $\mathbf{w} - \mathbf{z} \in S$ and $\mathbf{x} - \mathbf{w} \in S^{\perp}$, so $(\mathbf{x} - \mathbf{w}) \perp (\mathbf{w} - \mathbf{z})$. Hence

$$||\mathbf{x} - \mathbf{z}||^2 = ||(\mathbf{x} - \mathbf{w}) + (\mathbf{w} - \mathbf{z})||^2 = ||\mathbf{x} - \mathbf{w}||^2 + ||\mathbf{w} - \mathbf{z}||^2$$

Since $||\mathbf{w} - \mathbf{z}||^2 \geq 0$ and equality holds iff $\mathbf{z} = \mathbf{w}$, the theorem follows. ∎

In the remaining part of this section, we consider only the coordinate vector spaces $\mathbb{R}^n$ and $\mathbb{C}^n$ equipped with the canonical inner product.

Theorem 7.5.11 For any real matrix $\mathbf{A}$, $\mathcal{N}(\mathbf{A}) = (\mathcal{R}(\mathbf{A}))^{\perp}$. For a complex matrix $\mathbf{A}$, $\mathcal{N}(\mathbf{A}) = (\mathcal{C}(\mathbf{A}^*))^{\perp}$.

Proof We give the proof for complex matrices; the proof in the real case is easier (note that by $(\mathcal{R}(\mathbf{A}))^{\perp}$ we really mean $(\mathcal{C}(\mathbf{A}^{\mathrm{T}}))^{\perp}$).

$$\begin{aligned} \mathbf{x} \perp \mathcal{C}(\mathbf{A}^*) &\Leftrightarrow \langle \mathbf{x}, \mathbf{A}^*\mathbf{z}\rangle = 0 \text{ for all } \mathbf{z} \\ &\Leftrightarrow \mathbf{z}^*\mathbf{A}\mathbf{x} = 0 \text{ for all } \mathbf{z} \Leftrightarrow \mathbf{A}\mathbf{x} = \mathbf{0} \Leftrightarrow \mathbf{x} \in \mathcal{N}(\mathbf{A}) \end{aligned}$$ ∎

We next study orthogonal projectors. If S is a subspace of F^n, the *orthogonal projector into* S is the $n \times n$ matrix $\mathbf{P}$ such that for every $\mathbf{x} \in F^n$, $\mathbf{Px}$ is the orthogonal projection of $\mathbf{x}$ into S. We say that an $n \times n$ matrix $\mathbf{Q}$ is *an orthogonal projector* if it is the orthogonal projector into some subspace S of F^n. It is then clear (see *Section* 3.7) that $S = \mathcal{C}(\mathbf{Q})$, $\mathbf{Q}$ is the projector into S along $S^{\perp}$ and $\mathbf{I} - \mathbf{Q}$ is the orthogonal projector into $S^{\perp}$ (which equals $\mathcal{C}(\mathbf{I} - \mathbf{Q})$). We now characterize orthogonal projectors (compare with *Theorem* 3.7.5).

Theorem 7.5.12 The following statements about an $n \times n$ matrix $\mathbf{Q}$ are equivalent:

(i) $\mathbf{Q}$ is an orthogonal projector

(ii) $\mathbf{Q}^*\mathbf{Q} = \mathbf{Q}$

(iii) $\mathbf{Q}^* = \mathbf{Q}$ and $\mathbf{Q}^2 = \mathbf{Q}$.

Proof We first prove that (i) and (ii) are equivalent. $\mathbf{Q}$ is an orthogonal projector iff $\mathbf{Qx}$ is the orthogonal projection of $\mathbf{x}$ into $\mathcal{C}(\mathbf{Q})$ for all $\mathbf{x} \in F^n$, i.e. iff

$$\mathbf{x} - \mathbf{Qx} \perp \mathcal{C}(\mathbf{Q}) \text{ for all } \mathbf{x} \in F^n \tag{7.5.4}$$

Now, (7.5.4) holds iff $\langle \mathbf{Qy}, (\mathbf{I} - \mathbf{Q})\mathbf{x}\rangle = 0$ for all $\mathbf{x}, \mathbf{y} \in F^n$, which is equivalent to $(\mathbf{I} - \mathbf{Q})^*\mathbf{Q} = \mathbf{0}$. Thus (i) $\Leftrightarrow$ (ii). Given (ii), taking adjoints we get $\mathbf{Q}^*\mathbf{Q} = \mathbf{Q}^*$ and (iii) follows. That (iii) implies (ii) is trivial. ∎

We next obtain an explicit formula for the orthogonal projector into the column space of an arbitrary matrix.

Theorem 7.5.13 The orthogonal projector $\mathbf{P_A}$ into $\mathcal{C}(\mathbf{A})$ is given by

$$\mathbf{P_A} = \mathbf{A}(\mathbf{A}^*\mathbf{A})^-\mathbf{A}^* \tag{7.5.5}$$

Proof By the corollary to *Theorem* 5.4.9, the matrix $\mathbf{P}$ on the RHS of (7.5.5) is independent of the choice of the g-inverse of $\mathbf{A}^*\mathbf{A}$ and is hermitian. Also, by *Theorems* 5.4.7 and 5.4.3, $\mathbf{P}$ is idempotent and $\mathcal{C}(\mathbf{P}) = \mathcal{C}(\mathbf{A})$. Hence, by the preceding theorem, $\mathbf{P}$ is the orthogonal projector into $\mathcal{C}(\mathbf{A})$. ∎

The properties $\mathbf{P_A^*} = \mathbf{P_A}$, $\mathbf{P_A^2} = \mathbf{P_A}$, $\mathbf{P_A A} = \mathbf{A}$ and $\mathbf{A^*P_A} = \mathbf{A}^*$ which follow directly from the definition of $\mathbf{P_A}$ are very useful. The expression (7.5.5) itself can be derived from these (see *Exercise* 7.5.7). Clearly, the orthogonal projector into $(\mathcal{C}(\mathbf{A}))^\perp$ is $\mathbf{I} - \mathbf{P_A}$.

Note that $\mathbf{P_A}$ can be computed thus: Apply the generalized Gram-Schmidt orthogonalization process to the columns of $\mathbf{A}$. If $\mathbf{D}$ is the matrix formed with the resulting non-null vectors as columns, then $\mathbf{D^*D} = \mathbf{I}$ and $\mathcal{C}(\mathbf{D}) = \mathcal{C}(\mathbf{A})$, so $\mathbf{P_A} = \mathbf{P_D} = \mathbf{DD^*}$ by (7.5.5).

Example 7.5.14 We will find the orthogonal projector into the column space of

$$\mathbf{A} = \begin{bmatrix} 1 & 0 & 1 \\ 0 & 1 & -1 \\ 1 & 3 & -2 \\ -1 & -3 & 2 \end{bmatrix}$$

Taking the columns of $\mathbf{A}$ as $\mathbf{x}_1, \mathbf{x}_2$ and $\mathbf{x}_3$ we apply generalized Gram-Schmidt process. We get $\mathbf{z}_1 = \mathbf{x}_1$ and $\mathbf{y}_1 = \frac{1}{\sqrt{3}}(1,0,1,-1)^{\mathrm{T}}$. We next get

$$\mathbf{z}_2 = (0,1,3,-3)^{\mathrm{T}} - \frac{6}{\sqrt{3}} \cdot \frac{1}{\sqrt{3}}(1,0,1,-1)^{\mathrm{T}} = (-2,1,1,-1)^{\mathrm{T}}$$

so $\mathbf{y}_2 = \frac{1}{\sqrt{7}}(-2,1,1,-1)^{\mathrm{T}}$. Finally we get

$$\mathbf{z}_3 = (1,-1,-2,2)^{\mathrm{T}} - \frac{(-3)}{\sqrt{3}} \cdot \frac{1}{\sqrt{3}}(1,0,1,-1)^{\mathrm{T}} + \frac{(-7)}{7}(-2,1,1,-1)^{\mathrm{T}} = \mathbf{0}$$

Thus $\mathbf{y}_3 = \mathbf{0}$ and $\mathbf{D} = [\,\mathbf{y}_1 : \mathbf{y}_2\,]$. Hence

$$\mathbf{P_A} = \frac{1}{21}\begin{bmatrix} 19 & -6 & 1 & -1 \\ -6 & 3 & 3 & -3 \\ 1 & 3 & 10 & -10 \\ -1 & -3 & -10 & 10 \end{bmatrix}$$

We may also proceed thus: form the matrix $[\,\mathbf{A}^*\mathbf{A} : \mathbf{I}\,]$ and reduce $\mathbf{A}^*\mathbf{A}$ to a matrix in HCF but perform the elementary row operations on the entire matrix. Then $\mathbf{I}$ gets converted to a g-inverse of $\mathbf{A}^*\mathbf{A}$. We leave it to the reader to verify that the g-inverse of $\mathbf{A}^*\mathbf{A}$ thus obtained is

$$\mathbf{C} = \begin{bmatrix} \frac{19}{21} & -\frac{2}{7} & 0 \\ -\frac{2}{7} & \frac{1}{7} & 0 \\ -1 & 1 & 1 \end{bmatrix}$$

and $\mathbf{ACA}^*$ is the matrix $\mathbf{P_A}$ obtained above. The orthogonal projection of $(1,1,1,1)^{\mathrm{T}}$ into $\mathscr{C}(\mathbf{A})$ is $\left(\frac{13}{21}, -\frac{1}{7}, \frac{4}{21}, -\frac{4}{21}\right)^{\mathrm{T}}$. ∎

Exercises

Note: In *Exercises* 3, 4 and 6 through 14 below, assume the inner product to be the canonical inner product.

1. Let A be a set of vectors. Show that $A^\perp = (\mathrm{Sp}(A))^\perp$. What is $(A^\perp)^\perp$? Prove that $(A^\perp)^\perp = A$ iff A is a subspace.
2. In a real inner product space, show that $(\mathbf{x}+\mathbf{y}) \perp (\mathbf{x}-\mathbf{y})$ iff $\|\mathbf{x}\| = \|\mathbf{y}\|$.
3. Consider the subspaces $S = \{(\xi_1, \cdots, \xi_4) : \xi_1 = \xi_2 = \xi_3\}$ and $T = \{(\xi_1, \cdots, \xi_4) : \xi_1 = \xi_2 \text{ and } \xi_4 = 0\}$ of $\mathbb{R}^4$. Find $S+T, S^\perp, T^\perp$ and verify that $(S+T)^\perp = S^\perp \cap T^\perp$.
4. Let $\{\mathbf{u}, \mathbf{v}\}$ be a basis of a plane S in $\mathbb{R}^3$. Let $\mathbf{A}$ be any 3×3 matrix with $\mathbf{A}_{2*} = \mathbf{u}^{\mathrm{T}}$ and $\mathbf{A}_{3*} = \mathbf{v}^{\mathrm{T}}$. Show that $\{(A_{11}, A_{12}, A_{13})^{\mathrm{T}}\}$ is a basis of $S^\perp$ and deduce that $S = \{(x,y,z)^{\mathrm{T}} : A_{11}x + A_{12}y + A_{13}z = 0\}$.
5. Deduce (7.5.2) from (7.5.1).
6. Deduce *Theorem* 7.5.8 from the fact that $\mathbf{P_A} = \mathbf{AA}^*$ if the columns of $\mathbf{A}$ form an orthonormal set.
7. Derive (7.5.5) thus: First show that $\mathbf{P_A} = \mathbf{AB}$ for some $\mathbf{B}$. Then show that $\mathbf{P_A} = \mathbf{ABB}^*\mathbf{A}^*$ and finally show that $\mathbf{BB}^* = (\mathbf{A}^*\mathbf{A})^-$.
8. Show that $\mathbf{Au} = \mathbf{P_A b}$ iff $\mathbf{A}^*\mathbf{Au} = \mathbf{A}^*\mathbf{b}$. Show also that, given $\mathbf{A}$ and $\mathbf{b}$, $\mathbf{Au}$ satisfying $\mathbf{A}^*\mathbf{Au} = \mathbf{A}^*\mathbf{b}$ is unique though $\mathbf{u}$ is not.

9. Obtain $\mathbf{P_A}$ where $\mathbf{A} = \begin{bmatrix} 3 & 2 & 1 \\ 1 & 3 & -2 \\ -2 & 1 & -3 \end{bmatrix}$.

10. Let $\mathbf{x}^{\mathrm{T}} = (\frac{1}{2}, \frac{1}{2}, \frac{1}{2}, \frac{1}{2})$, $\mathbf{y}^{\mathrm{T}} = (\frac{1}{2}, -\frac{1}{2}, \frac{1}{2}, -\frac{1}{2})$ and $\mathbf{A} = \mathbf{I} - \mathbf{x}\mathbf{x}^{\mathrm{T}} - \mathbf{y}\mathbf{y}^{\mathrm{T}}$. Show that $\mathbf{A}$ is an orthogonal projector and find an orthonormal basis of $\mathcal{C}(\mathbf{A})$.

11. Let $\mathbf{P}$ be an orthogonal projector over $\mathbb{R}$. Then show that
    (a) $0 \leq p_{jj} \leq 1$ for all j,
    (b) $-1/2 \leq p_{ij} \leq 1/2$ whenever $i \neq j$.

12. (a) If $\mathbf{P}$ and $\mathbf{Q}$ are orthogonal projectors of order n, show that $\mathbf{P}+\mathbf{Q}$ is an orthogonal projector iff $\mathcal{C}(\mathbf{P}) \perp \mathcal{C}(\mathbf{Q})$ and, when this condition holds, $\mathbf{P}+\mathbf{Q}$ is the orthogonal projector into $\mathcal{C}(\mathbf{P}) + \mathcal{C}(\mathbf{Q})$.
    (b) Show that $\mathbf{P}_{(\mathbf{X}:\mathbf{Y})} = \mathbf{P_X} + \mathbf{P_Z}$ where $\mathbf{Z} = (\mathbf{I} - \mathbf{P_X})\mathbf{Y}$.

13. Let $W = \{(\alpha, 1, 1) : \alpha \in \mathbb{R}\}$ and S the subspace of $\mathbb{R}^3$ which is a translate of W. Find the orthogonal projections of $(1, 2, 3)$ into W and into S.

*14. Let $W = \mathbf{u} + S$ where S is a subspace and $\mathbf{u} \notin S$. Show that the orthogonal projections $\mathbf{w}$ and $\mathbf{s}$ of $\mathbf{x}$ into W and into S respectively, are collinear with $\mathbf{x}$ (for all $\mathbf{x}$) iff S is a hyperplane (i.e., $d(S) = d(V) - 1$).

15. Let S be a subspace of V and $\mathbf{y} \in V$. Then show that $\|\mathbf{x}\| \leq \|\mathbf{x}+\mathbf{z}\|$ for all $\mathbf{z} \in S$ iff $\mathbf{x} \in S^{\perp}$. Deduce that, with respect to the canonical inner product, $\|\mathbf{Au}\| \leq \|\mathbf{Au} + \mathbf{Bv}\|$ for all $\mathbf{u}$ and $\mathbf{v}$ iff $\mathbf{A}^*\mathbf{B} = \mathbf{0}$.

*16. (a) Let $\mathbf{u}_1, \mathbf{u}_2, \ldots, \mathbf{u}_n$ form a basis of $\mathbb{C}^n$.
        i. Show that $\mathbf{A} = \left(\sum \mathbf{u}_i\mathbf{u}_i^*\right)^{-1}$ exists.
        ii. Show that $\mathbf{A}$ is a g-inverse of $\mathbf{u}_i\mathbf{u}_i^*$ for $i = 1, \ldots, n$.
        iii. Show that $\langle \mathbf{x}, \mathbf{y} \rangle = \mathbf{y}^*\mathbf{A}\mathbf{x}$ is an inner product on $\mathbb{C}^n$ and that $\mathbf{u}_1, \mathbf{u}_2, \ldots, \mathbf{u}_n$ form an orthonormal basis with respect to it. (Hint: use *Exercise* 5.4.14.)
    (b) Let $\mathbf{y}$ and $\mathbf{z}$ be any two linearly independent vectors in $\mathbb{C}^n$. Show that there exists a norm $\|\cdot\|$ induced by an inner product such that $\|\mathbf{y}\| < \|\mathbf{z}\|$.

*17. What is the orthogonal complement of the subspace of even polynomials in $\mathcal{P}_n(\mathbb{R})$ with respect to the inner product $\langle p, q \rangle = \int_{-1}^{1} p(t)q(t)\,dt$?

*18. (a) In *Exercise* 7.4.11, find the orthogonal projections of $(1, 1, 1)$ and $(1, 1, 2)$ into S. Also find an orthonormal basis of $S^{\perp}$.
    (b) Determine the orthogonal complement of the x_1-axis in *Example* 7.2.4.

(c) In the inner product space of *Exercise* 7.2.4(a), determine the orthogonal complement of the subspace of all diagonal matrices.

*19. (a) Show that $\mathbf{A}$ is an orthogonal projector with respect to the inner product $\langle \mathbf{x}, \mathbf{y} \rangle = \mathbf{y}^* \Sigma \mathbf{x}$ iff $\mathbf{A}$ is idempotent and $\mathbf{A}^* \Sigma (\mathbf{I} - \mathbf{A}) = \mathbf{0}$.

(b) If $\mathbf{A}$ is idempotent and $\Sigma = (\mathbf{A}\mathbf{A}^* + (\mathbf{I} - \mathbf{A})(\mathbf{I} - \mathbf{A})^*)^{-1}$, show that $\mathbf{A}$ is an orthogonal projector with respect to $\langle \mathbf{x}, \mathbf{y} \rangle = \mathbf{y}^* \Sigma \mathbf{x}$.

7.6 Orthogonal and unitary matrices

Transformations which preserve distances are called *isometries* and are important. For example, the motion of a rigid body is an isometry. Linear transformations which preserve inner product (and so distances and angles) are isometries. In this section we study their matrices.

Throughout this section we consider only the canonical inner product on $\mathbb{C}^n$ and $\mathbb{R}^n$.

Definition 7.6.1 A *unitary* matrix is a (complex) square matrix $\mathbf{A}$ such that $\mathbf{A}\mathbf{A}^* = \mathbf{I}$. An *orthogonal* matrix is a real square matrix $\mathbf{A}$ such that $\mathbf{A}\mathbf{A}^{\mathrm{T}} = \mathbf{I}$.

Note that a square matrix $\mathbf{A}$ is unitary iff $\mathbf{A}^*\mathbf{A} = \mathbf{I}$ since this, as well as the defining condition, is equivalent to $\mathbf{A}^* = \mathbf{A}^{-1}$. Similarly a real square matrix $\mathbf{A}$ is orthogonal iff $\mathbf{A}^{\mathrm{T}}\mathbf{A} = \mathbf{I}$.

It is easy to see that $(\mathbf{A}\mathbf{A}^*)_{ij} = \langle \mathbf{A}_{i*}, \mathbf{A}_{j*} \rangle$ and $(\mathbf{A}^*\mathbf{A})_{ij} = \langle \mathbf{A}_{*j}, \mathbf{A}_{*i} \rangle$. Hence $\mathbf{A}$ is unitary iff the rows as well as the columns of $\mathbf{A}$ form orthonormal bases of F^n. If $\mathbf{A}$ is real, the word 'unitary' can be replaced by 'orthogonal'. Before we proceed to study orthogonal and unitary matrices let us give some examples.

Clearly $\mathbf{I}$ and all permutation matrices are orthogonal (and so unitary). If $\mathbf{A}$ is unitary, the matrix obtained from $\mathbf{A}$ by any permutation of rows or columns is also unitary. The matrix obtained by multiplying any row or column of a unitary matrix by a scalar of unit modulus is also unitary. The unitary matrices of order 1 are $\exp(i\theta)$, $0 \le \theta < 2\pi$. The orthogonal matrices of order 1 are 1 and -1. We next give some non-trivial examples.

Example 7.6.2 It is easy to see that

$$\mathbf{A}_\theta = \begin{bmatrix} \cos\theta & -\sin\theta \\ \sin\theta & \cos\theta \end{bmatrix} \quad \text{and} \quad \mathbf{B}_\theta = \begin{bmatrix} \cos\theta & \sin\theta \\ \sin\theta & -\cos\theta \end{bmatrix}$$

are orthogonal matrices. We will now show that any 2×2 orthogonal matrix $\mathbf{A}$ is $\mathbf{A}_\theta$ or $\mathbf{B}_\theta$ for some θ. Let P and Q be the points in $\mathbb{R}^2$ corresponding to the two columns of $\mathbf{A}$ and let θ be the angle between the x-axis and OP. Then length $OP = 1$, so $P = (\cos\theta, \sin\theta)^{\mathrm{T}}$. Length OQ is also 1 and OQ is perpendicular to OP, so $Q = (\cos\varphi, \sin\varphi)^{\mathrm{T}}$ where φ is $\theta + \frac{\pi}{2}$ or $\theta - \frac{\pi}{2}$. Hence $\mathbf{A}$ is $\mathbf{A}_\theta$ or $\mathbf{B}_\theta$. We incidentally note that the map $\mathbf{u} \mapsto \mathbf{A}\mathbf{u}$ is the rotation of the plane about the origin by the angle θ in the counter-clockwise direction or the reflection of the plane in the line $y = \left(\tan\frac{\theta}{2}\right)x$ according as $\mathbf{A} = \mathbf{A}_\theta$ or $\mathbf{B}_\theta$. ■.

Example 7.6.3 Let $\mathbf{u}$ be any vector in $\mathbb{C}^n$ with $||\mathbf{u}|| = 1$ and let $\mathbf{A} = \mathbf{I} - 2\mathbf{u}\mathbf{u}^*$. Then $\mathbf{A}$ is hermitian and

$$\mathbf{A}\mathbf{A}^* = \mathbf{A}^2 = \mathbf{I} - 4\mathbf{u}\mathbf{u}^* + 4\mathbf{u}\mathbf{u}^*\mathbf{u}\mathbf{u}^* = \mathbf{I}$$

since $\mathbf{u}^*\mathbf{u} = 1$. Thus $\mathbf{A}$ is unitary. If $\mathbf{u}$ is real then $\mathbf{A}$ is symmetric and orthogonal. ■

We shall now study some properties of orthogonal and unitary matrices. The next theorem is proved by straightforward verification.

Theorem 7.6.4 If $\mathbf{A}$ and $\mathbf{B}$ are unitary (or orthogonal) then so are $\mathbf{AB}$, $\mathbf{A}^{\mathrm{T}}$, $\overline{\mathbf{A}}$ and $\mathbf{A}^{-1}$. If $\mathbf{A}$ and $\mathbf{C}$ are unitary (or orthogonal) matrices of possibly different orders then so is

$$\begin{bmatrix} \mathbf{A} & \mathbf{0} \\ \mathbf{0} & \mathbf{C} \end{bmatrix}$$

Theorem 7.6.5 The determinant of a unitary matrix has modulus 1. The determinant of an orthogonal matrix is 1 or -1.

This theorem follows from $\mathbf{A}\mathbf{A}^* = \mathbf{I}$ since $\det(\mathbf{A}^*) = \overline{\det(\mathbf{A})}$. An orthogonal matrix is said to be *proper* or *improper* according as its determinant is 1 or -1. Note that $\mathbf{A}_\theta$ of *Example* 7.6.2 is proper and $\mathbf{B}_\theta$ improper.

Theorem 7.6.6 Let $\mathbf{A}$ be an $n \times n$ matrix. Then the following statements are equivalent:

(i) $\mathbf{A}$ is unitary
(ii) $\langle \mathbf{Ax}, \mathbf{Ay} \rangle = \langle \mathbf{x}, \mathbf{y} \rangle$ for all $\mathbf{x}, \mathbf{y} \in \mathbb{C}^n$
(iii) $||\mathbf{Ax}|| = ||\mathbf{x}||$ for all $\mathbf{x} \in \mathbb{C}^n$
(iv) $||\mathbf{Ax}|| = 1$ whenever $||\mathbf{x}|| = 1$ and $\mathbf{x} \in \mathbb{C}^n$
(v) $||\mathbf{Ax} - \mathbf{Ay}|| = ||\mathbf{x} - \mathbf{y}||$ for all $\mathbf{x}, \mathbf{y} \in \mathbb{C}^n$

(vi) $\{\mathbf{Ax}_1, \ldots, \mathbf{Ax}_n\}$ is an orthonormal basis of $\mathbb{C}^n$ whenever $\{\mathbf{x}_1, \ldots, \mathbf{x}_n\}$ is an orthonormal basis of $\mathbb{C}^n$.

Further, if $\mathbf{A}$ is real, then 'unitary' can be replaced by 'orthogonal' in (i) and $\mathbb{C}^n$ by $\mathbb{R}^n$ in (ii) through (vi).

Proof Given (i), $\mathbf{A}^*\mathbf{A} = \mathbf{I}$, so $\langle \mathbf{Ax}, \mathbf{Ay} \rangle = \mathbf{y}^*\mathbf{A}^*\mathbf{Ax} = \mathbf{y}^*\mathbf{x} = \langle \mathbf{x}, \mathbf{y} \rangle$. Thus (i) implies (ii). That (ii) implies (iii) follows by taking $\mathbf{y} = \mathbf{x}$.

We next prove that (iii) implies (i). Denote $\mathbf{A}^*\mathbf{A}$ by $\mathbf{B}$. Taking $\mathbf{x} = \mathbf{e}_i$, the i-th vector in the canonical basis, we have

$$||\mathbf{Ae}_i||^2 = \langle \mathbf{Ae}_i, \mathbf{Ae}_i \rangle = \mathbf{e}_i^* \mathbf{A}^* \mathbf{Ae}_i = \mathbf{e}_i^{\mathrm{T}} \mathbf{Be}_i = b_{ii}$$

Since $||\mathbf{e}_i||^2 = 1$, (iii) gives $b_{ii} = 1$ for all i. Next taking $\mathbf{x} = \mathbf{e}_i + \mathbf{e}_j$ we have

$$||\mathbf{Ax}||^2 = (\mathbf{e}_i + \mathbf{e}_j)^{\mathrm{T}}\mathbf{B}(\mathbf{e}_i + \mathbf{e}_j) = b_{ii} + b_{ij} + b_{ji} + b_{jj}$$

Since $||\mathbf{x}||^2 = 2$, it follows that $b_{ij} + b_{ji} = 0$ whenever $i \neq j$. If $\mathbf{A}$ is real then $\mathbf{B}$ is symmetric, so it follows that $\mathbf{B} = \mathbf{I}$ and $\mathbf{A}$ is orthogonal. If $\mathbf{A}$ is not real, taking $\mathbf{x} = \sqrt{-1}\mathbf{e}_i + \mathbf{e}_j$, we get

$$||\mathbf{Ax}||^2 = \mathbf{x}^*\mathbf{Bx} = b_{ii} - \sqrt{-1}\, b_{ij} + \sqrt{-1}\, b_{ji} + b_{jj}$$

Since $||\mathbf{x}||^2 = 2$ it follows that $b_{ij} - b_{ji} = 0$. Since we have already proved that $b_{ij} + b_{ji} = 0$, we get $b_{ij} = b_{ji} = 0$. Thus $\mathbf{B} = \mathbf{I}$ and $\mathbf{A}$ is unitary. Thus (i), (ii) and (iii) are equivalent.

That (iii) implies (iv) is trivial. Conversely, given (iv), we have $||\mathbf{A}(\mathbf{x}/||\mathbf{x}||)|| = 1$ for all $\mathbf{x} \neq \mathbf{0}$ which is the same as (iii). That (iii) implies (v) follows by replacing $\mathbf{x}$ by $\mathbf{x} - \mathbf{y}$. The converse implication follows by taking $\mathbf{y} = \mathbf{0}$. Clearly (ii) and (iv) together imply (vi). Conversely (vi) implies (iv) since if $||\mathbf{x}|| = 1$ then $\{\mathbf{x}\}$ can be extended to an orthonormal basis. This completes the proof of the equivalence of (i) through (vi). The statement about real $\mathbf{A}$ follows easily from the above proof. ∎

The equivalence of (i)–(iii) of the preceding theorem shows that if $\mathbf{A}$ is orthogonal then the map $\mathbf{x} \mapsto \mathbf{Ax}$ preserves angles. Statements (iii) and (v) say that the map preserves length and distance. Note that such a map leaves the surface of a sphere, with centre at the origin, invariant. Written out in full, (iii) says: if we make a change of variables from $x_1, x_2, \ldots, x_n$ to $y_1, y_2, \ldots, y_n$ by $\mathbf{y} = \mathbf{Ax}$, where $\mathbf{A}$ is orthogonal, then

$$y_1^2 + y_2^2 + \cdots + y_n^2 = x_1^2 + x_2^2 + \cdots + x_n^2 \tag{7.6.1}$$

This is what makes orthogonal transformations useful in many subjects. For example, this is used in Statistics to show that the sample mean and sample variance are independently distributed if the population is normal.

Suppose $\mathbf{A}$ is unitary. Then the map $\mathbf{x} \mapsto \mathbf{A}\mathbf{x}+\mathbf{c}$, known as an *affine transformation*, also preserves distances. We prove a strong form of the converse in the following theorem where we do not assume that the map is an affine transformation.

Theorem 7.6.7 Let f be any map from $\mathbb{R}^n$ to itself such that

$$\|f(\mathbf{x}) - f(\mathbf{y})\| = \|\mathbf{x} - \mathbf{y}\| \quad \text{for all } \mathbf{x}, \mathbf{y} \in \mathbb{R}^n \tag{7.6.2}$$

Then there exists an orthogonal matrix $\mathbf{A}$ and a vector $\mathbf{c} \in \mathbb{R}^n$ such that $f(\mathbf{x}) = \mathbf{A}\mathbf{x} + \mathbf{c}$ for all $\mathbf{x} \in \mathbb{R}^n$.

Proof Define a map $g : \mathbb{R}^n \to \mathbb{R}^n$ by $g(\mathbf{x}) - f(\mathbf{x}) - f(\mathbf{0})$. We first prove that g preserves inner products and then show that $g(\mathbf{x}) = \mathbf{A}\mathbf{x}$ for some orthogonal matrix $\mathbf{A}$.

Clearly $g(\mathbf{x}) - g(\mathbf{y}) = f(\mathbf{x}) - f(\mathbf{y})$, so (7.6.2) gives

$$\|g(\mathbf{x}) - g(\mathbf{y})\| = \|\mathbf{x} - \mathbf{y}\| \quad \text{for all } \mathbf{x}, \mathbf{y} \in \mathbb{R}^n \tag{7.6.3}$$

Squaring both sides of (7.6.3) we get

$$\begin{aligned} &\langle g(\mathbf{x}), g(\mathbf{x})\rangle - 2\langle g(\mathbf{x}), g(\mathbf{y})\rangle + \langle g(\mathbf{y}), g(\mathbf{y})\rangle \\ &\quad = \langle \mathbf{x}, \mathbf{x}\rangle - 2\langle \mathbf{x}, \mathbf{y}\rangle + \langle \mathbf{y}, \mathbf{y}\rangle \quad \text{for all } \mathbf{x}, \mathbf{y} \in \mathbb{R}^n \end{aligned} \tag{7.6.4}$$

Now $g(\mathbf{0}) = \mathbf{0}$, so taking $\mathbf{y} = \mathbf{0}$ in (7.6.4), we see that $\langle g(\mathbf{x}), g(\mathbf{x})\rangle = \langle \mathbf{x}, \mathbf{x}\rangle$. Similarly $\langle g(\mathbf{y}), g(\mathbf{y})\rangle = \langle \mathbf{y}, \mathbf{y}\rangle$, so (7.6.4) gives

$$\langle g(\mathbf{x}), g(\mathbf{y})\rangle = \langle \mathbf{x}, \mathbf{y}\rangle \quad \text{for all } \mathbf{x}, \mathbf{y} \in \mathbb{R}^n \tag{7.6.5}$$

Thus g preserves inner products. Let $\mathbf{A} = [\, g(\mathbf{e}_1) : g(\mathbf{e}_2) : \cdots : g(\mathbf{e}_n)\,]$ where $\{\mathbf{e}_1, \mathbf{e}_2, \ldots, \mathbf{e}_n\}$ is the canonical basis of $\mathbb{R}^n$. Then by (7.6.5) the columns of $\mathbf{A}$ form an orthonormal basis of $\mathbb{R}^n$, so $\mathbf{A}$ is orthogonal. Also, for any $\mathbf{x} \in \mathbb{R}^n$,

$$\begin{aligned} g(\mathbf{x}) &= \textstyle\sum_{i=1}^n \langle g(\mathbf{x}), g(\mathbf{e}_i)\rangle g(\mathbf{e}_i) \\ &= \textstyle\sum_{i=1}^n \langle \mathbf{x}, \mathbf{e}_i\rangle g(\mathbf{e}_i) \\ &= \textstyle\sum_{i=1}^n \langle \mathbf{x}, \mathbf{e}_i\rangle \mathbf{A}\mathbf{e}_i \\ &= \mathbf{A}\big(\textstyle\sum_{i=1}^n \langle \mathbf{x}, \mathbf{e}_i\rangle \mathbf{e}_i\big) \\ &= \mathbf{A}\mathbf{x} \end{aligned}$$

So $f(\mathbf{x}) = \mathbf{A}\mathbf{x} + \mathbf{c}$ where $\mathbf{c} = f(\mathbf{0})$. ∎

In *Example* 7.6.2 we saw that every 2×2 orthogonal matrix corresponds to either a rotation or a reflection of the plane depending upon whether it is proper or improper. We will prove in *Section* 8.7 that every orthogonal matrix of order 3 corresponds to either a rotation of $\mathbb{R}^3$ about a line through the origin or such a rotation followed by a reflection in the origin, depending upon whether it is proper or improper.

In *Section* 3.10 we studied transition matrices and the effect of a change of bases on the matrix of a linear transformation. We will now obtain corresponding results for orthonormal bases.

Theorem 7.6.8 The matrix of transition $\mathbf{P}$ from an orthonormal basis $\mathcal{X} = \{\mathbf{x}_1, \mathbf{x}_2, \ldots, \mathbf{x}_n\}$ of an inner product space to another orthonormal basis $\mathcal{X}' = \{\mathbf{x}'_1, \mathbf{x}'_2, \ldots, \mathbf{x}'_n\}$ is a unitary matrix. Conversely, if $\mathcal{X}$ is an orthonormal basis and $\mathcal{X}'$ is defined by (3.10.3), where $\mathbf{P} = ((p_{ij}))$ is unitary, then $\mathcal{X}'$ is also an orthonormal basis.

Proof The theorem follows from:

$$\langle \mathbf{x}'_k, \mathbf{x}'_\ell \rangle = \left\langle \sum_{i=1}^{n} p_{ik}\mathbf{x}_i, \sum_{j=1}^{n} p_{j\ell}\mathbf{x}_j \right\rangle = \sum_{i=1}^{n} p_{ik}\overline{p}_{i\ell} = (\mathbf{P}^*\mathbf{P})_{\ell k}$$ ∎

Corollary Let f be a linear operator on an inner product space V. If $\mathcal{X}$ and $\mathcal{X}'$ are two orthonormal bases of V and $\mathbf{A}$ and $\mathbf{B}$ are the matrices of f with respect to $\mathcal{X}$ and $\mathcal{X}'$ respectively, $\mathbf{B} = \mathbf{P}^{-1}\mathbf{A}\mathbf{P}$ for some unitary matrix $\mathbf{P}$.

This corollary follows from the preceding theorem and *Theorem* 3.10.4. When $\mathbf{A}$ and $\mathbf{B}$ are related thus, we say that they are *unitarily similar to each other.* It can easily be checked that unitary similarity is an equivalence relation.

We end this section by mentioning a simple way of constructing a large class of unitary matrices provided one can invert a matrix.

Definition 7.6.9 A *skew-hermitian matrix* is a square matrix $\mathbf{S}$ such that $\mathbf{S}^* = -\mathbf{S}$. A real skew-hermitian matrix is said to be *skew-symmetric.*

We leave the proof of the following theorem as an exercise.

Theorem 7.6.10 Let $\mathbf{A}$ be a square matrix such that $\mathbf{I} + \mathbf{A}$ is non-singular and let $\widetilde{\mathbf{A}} = (\mathbf{I} - \mathbf{A})(\mathbf{I} + \mathbf{A})^{-1}$. Then (i) $\mathbf{I} + \widetilde{\mathbf{A}}$ is also non-singular, (ii) $\widetilde{\widetilde{\mathbf{A}}} = \mathbf{A}$ and (iii) $\widetilde{\mathbf{A}}$ is unitary iff $\mathbf{A}$ is skew-hermitian. If $\mathbf{A}$ is real, 'unitary' and 'skew-hermitian' can be replaced by 'orthogonal' and 'skew-symmetric' respectively in (iii).

We will prove later (see the *Corollary* to *Theorem* 8.7.1) that if $\mathbf{S}$ is skew-hermitian then $\mathbf{I} + \mathbf{S}$ is non-singular. Thus *Theorem* 7.6.10 shows that $\mathbf{S} \leftrightarrow \widetilde{\mathbf{S}}$ is a 1-1 correspondence between skew-hermitian matrices and unitary matrices $\mathbf{U}$ such that $\mathbf{I} + \mathbf{U}$ is non-singular. Note that it is easy to generate the skew-hermitian matrices thus: put arbitrary purely imaginary numbers on the diagonal, arbitrary complex numbers above the diagonal and then fill the cells below the diagonal by using $s_{ij} = -\bar{s}_{ji}$. Now taking $\widetilde{\mathbf{S}}$ we get all unitary matrices $\mathbf{U}$ such that $\mathbf{I} + \mathbf{U}$ is non-singular.

Exercises

1. Characterize upper triangular unitary matrices.
2. Obtain an orthogonal matrix with $\left(\frac{1}{2}, \frac{1}{2}, \frac{1}{2}, \frac{1}{2}\right)^{\mathrm{T}}$ as the first column.
3. Show that the following matrices are unitary:

$$\text{(i)}\ \frac{1}{\sqrt{2}}\begin{bmatrix} 1 & -1 \\ 1 & 1 \end{bmatrix},\ \text{(ii)}\ \frac{1}{\sqrt{2}}\begin{bmatrix} -1 & -i \\ i & 1 \end{bmatrix},\ \text{(iii)}\ \frac{1}{2}\begin{bmatrix} 1-i & 1+i \\ 1-i & -1-i \end{bmatrix}$$

4. Consider the three statements: (i) $\mathbf{A}$ is unitary, (ii) $\mathbf{A}$ is hermitian, and (iii) $\mathbf{A}^2 = \mathbf{I}$, about a square matrix $\mathbf{A}$. Show that none of them implies the others but any two of them imply the third.
5. If $\mathbf{A} = \frac{1}{2}\begin{bmatrix} \sqrt{3} & 0 \\ \sqrt{-1} & 2 \end{bmatrix}$, show that $\|\mathbf{Ax}\| = \|\mathbf{x}\|$ does not hold for some $\mathbf{x} \in \mathbb{C}^2$ but holds whenever the components of $\mathbf{x}$ are real.
6. Let $\mathbf{A}$ be an $n \times n$ matrix. Show that the following statements are equivalent:
   (a) $\mathbf{Ax} \perp \mathbf{Ay}$ iff $\mathbf{x} \perp \mathbf{y}$,
   (b) $\mathbf{A}$ is a non-zero scalar times a unitary matrix,
   (c) the columns of $\mathbf{A}$ are orthogonal and have equal norms,
   (d) the rows of $\mathbf{A}$ are orthogonal and have equal norms.
7. Prove *Theorem* 7.6.6 thus: (i) $\Rightarrow$ (ii) $\Rightarrow$ (iii) $\Rightarrow$ (vi) $\Rightarrow$ (i), (iii) $\Leftrightarrow$ (iv) and (iii) $\Leftrightarrow$ (v). To prove (iii) $\Rightarrow$ (vi), use *Exercise* 7.3.10 and to prove (vi) $\Rightarrow$ (i), take $\mathbf{x}_i = \mathbf{e}_i$ for $i = 1, \ldots, n$.

8. Let $\mathbf{A}$ be an $m \times n$ matrix and $\mathbf{P}, \mathbf{Q}$ be orthogonal matrices of orders m and n respectively. If $\mathbf{B} = \mathbf{PAQ}$, show that

$$\sum_{i=1}^{m} \sum_{j=1}^{n} a_{ij}^2 = \sum_{i=1}^{m} \sum_{j=1}^{n} b_{ij}^2$$

9. Show that if $\mathbf{A}$ is unitary, then $\mathcal{C}(\mathbf{I} - \mathbf{A})$ and $\mathcal{N}(\mathbf{I} - \mathbf{A})$ are orthogonal complements.
10. (a) If $\mathbf{A}$ is an orthogonal matrix, prove that $|a_{11}| = |A_{11}|$.
    (b) If $\mathbf{A}$ is an orthogonal matrix, show that $|\det \mathbf{A}(1, \ldots, k \mid 1, \ldots, k)| = |\det \mathbf{A}(k+1, \ldots, n \mid k+1, \ldots, n)|$.
11. Prove *Theorem* 7.6.10. (Hint: note that $\mathbf{I} - \mathbf{A}$ and $(\mathbf{I} + \mathbf{A})^{-1}$ commute.)
12. If $\mathbf{A}$ is skew-hermitian, $\mathbf{B}$ is symmetric, $\mathbf{AB} = \mathbf{BA}$ and $\mathbf{B} + \mathbf{A}$ is non-singular, show that $(\mathbf{B} - \mathbf{A})(\mathbf{B} + \mathbf{A})^{-1}$ is unitary.
13. If $\mathbf{A}$ and $\mathbf{B}$ are commuting orthogonal matrices such that $\mathbf{I} + \mathbf{A}$ and $\mathbf{I} + \mathbf{B}$ are non-singular, show that $\mathbf{C} := (\mathbf{I} - \mathbf{AB})(\mathbf{I} + \mathbf{A} + \mathbf{B} + \mathbf{AB})^{-1}$ is skew-symmetric.
14. Which of the matrices in *Example* 7.6.2 has the property $\mathbf{I} + \mathbf{A}$ is non-singular? For such $\mathbf{A}$, find $\widetilde{\mathbf{A}}$ and verify *Theorem* 7.6.10.
15. With $\mathbf{A}_\theta$ and $\mathbf{B}_\theta$ as in *Example* 7.6.2, find $\mathbf{A}_\theta \mathbf{A}_\eta$, $\mathbf{A}_\theta \mathbf{B}_\eta$, $\mathbf{B}_\eta \mathbf{A}_\theta$ and $\mathbf{B}_\theta \mathbf{B}_\eta$.
16. Show that the set of all $n \times n$ orthogonal matrices forms a group under multiplication. This group is denoted O_n. Show also that the set of all proper orthogonal matrices of order n forms a subgroup of O_n. This subgroup is denoted SO_n. (See *Exercise* 6.5.9.)
17. Show that for an $m \times n$ matrix $\mathbf{A}$, $\|\mathbf{Ax}\| = \|\mathbf{x}\|$ for all $\mathbf{x} \in \mathbb{C}^n$ iff $\mathbf{A}^*\mathbf{A} = \mathbf{I}$. (Such a rectangular matrix is called a *semi-unitary matrix*.)
18. Given $\mathbf{u}$ and $\mathbf{v}$ in $\mathbb{R}^n$ with $\|\mathbf{u}\| = \|\mathbf{v}\|$, explain how an orthogonal matrix $\mathbf{C}$ can be obtained so that $\mathbf{Cu} = \mathbf{v}$.
19. Let $\mathbf{u}$ and $\mathbf{x}$ be fixed vectors in $\mathbb{R}^n$. Find the maximum and the minimum values of $(\mathbf{x}^{\mathrm{T}}\mathbf{C}\mathbf{u})^2$ as $\mathbf{C}$ varies over all $n \times n$ orthogonal matrices.

Chapter 8
Eigenvalues

8.1 Introduction

In this chapter we study eigenvalues and eigenvectors associated with a complex square matrix. These are useful in the study of canonical forms of a matrix under similarity and in the study of quadratic forms. They have applications in many subjects like Geometry, Mechanics, Astronomy, Engineering, Economics and Statistics.

Throughout this chapter we take the base field to be $\mathbb{C}$ except in a few places where we take it to be $\mathbb{R}$. $\mathbf{A}$ will denote an $n \times n$ matrix unless specified otherwise.

8.2 Characteristic roots

For any $n \times n$ matrix $\mathbf{A}$, consider the polynomial

$$\chi_{\mathbf{A}}(\lambda) := |\lambda\mathbf{I} - \mathbf{A}| = \begin{vmatrix} \lambda - a_{11} & -a_{12} & \cdots & -a_{1n} \\ -a_{21} & \lambda - a_{22} & \cdots & -a_{2n} \\ \cdots & \cdots & \cdots & \cdots \\ -a_{n1} & -a_{n2} & \cdots & \lambda - a_{nn} \end{vmatrix}$$

Clearly this is a polynomial of degree n and is monic since λ^n occurs only in the term $\prod_{i=1}^{n}(\lambda - a_{ii})$ and has coefficient 1. So, by the *fundamental theorem of algebra*, $\chi(\lambda)$ has exactly n (not necessarily distinct) roots, usually denoted by $\lambda_1, \lambda_2, \ldots, \lambda_n$. Clearly

$$\chi_{\mathbf{A}}(\lambda) = (\lambda - \lambda_1)(\lambda - \lambda_2)\cdots(\lambda - \lambda_n) \tag{8.2.1}$$

Strictly speaking, $\lambda\mathbf{I} - \mathbf{A}$ is a matrix not over $\mathbb{C}$ but over the integral domain $\mathbb{C}[\lambda]$ of all polynomials in the indeterminate λ with coefficients from $\mathbb{C}$. As mentioned in *Section* 6.1, most of the results on determinants are valid for such matrices.

Definition 8.2.1 $\chi_{\mathbf{A}}(\lambda)$ is called the *characteristic polynomial* of $\mathbf{A}$ and $\chi_{\mathbf{A}}(\lambda) = 0$ is called the *characteristic equation* of $\mathbf{A}$. The n roots of $\chi_{\mathbf{A}}(\lambda)$ are called the *characteristic roots* of **A**. The *spectrum* of **A** is the set of distinct characteristic roots of **A**.

Clearly, a complex number α is a characteristic root of **A** iff $\alpha\mathbf{I} - \mathbf{A}$ is singular. In particular, 0 is a characteristic root of **A** iff **A** is singular.

Example 8.2.2 Let

$$\mathbf{A} = \begin{bmatrix} 1 & 1 \\ -1 & 3 \end{bmatrix}, \ \mathbf{B} = \begin{bmatrix} 1 & 0 & 2 \\ 0 & -1 & 1 \\ 0 & -1 & 0 \end{bmatrix} \text{ and } \mathbf{C} = \begin{bmatrix} 1 & 1 \\ 0 & i \end{bmatrix}$$

Then

$$\chi_{\mathbf{A}}(\lambda) = \begin{vmatrix} \lambda - 1 & -1 \\ 1 & \lambda - 3 \end{vmatrix} = (\lambda - 2)^2$$

So the characteristic roots of **A** are 2, 2 and the spectrum of **A** is $\{2\}$. It can be verified that $\chi_{\mathbf{B}}(\lambda) = (\lambda - 1)(\lambda^2 + \lambda + 1)$, so the characteristic roots of **B** are 1, $(-1 - \sqrt{3}\,i)/2$ and $(-1 + \sqrt{3}\,i)/2$. Finally, $\chi_{\mathbf{C}}(\lambda) = (\lambda - 1)(\lambda - i)$ and the spectrum of **C** is $\{1, i\}$. ∎

Lemma 8.2.3 The constant term and the coefficient of λ^{n-1} in $\chi_{\mathbf{A}}(\lambda)$ are $(-1)^n|\mathbf{A}|$ and $-\mathrm{tr}(\mathbf{A})$ respectively.

Proof The first statement follows since the constant term is $\chi(0) = |-\mathbf{A}|$. The second statement follows from the fact that λ^{n-1} occurs only in the term $\prod_{i=1}^{n}(\lambda - a_{ii})$. ∎

Theorem 8.2.4 The sum of the characteristic roots of **A** is $\mathrm{tr}(\mathbf{A})$ and the product of the characteristic roots of **A** is $|\mathbf{A}|$.

This theorem follows easily from the preceding lemma and (8.2.1). Finding the characteristic roots of a matrix is not easy in general, since there is no easy way of finding the roots of a polynomial of degree greater than 3. See *Section* 8.10 for some computational methods.

Theorem 8.2.5 If **A** is (upper or lower) triangular then $\chi_{\mathbf{A}}(\lambda) = \prod_{i=1}^{n}(\lambda - a_{ii})$ and the characteristic roots of **A** are the diagonal entries of **A**.

This theorem follows from the fact that $\lambda\mathbf{I} - \mathbf{A}$ is also triangular with $\lambda - a_{11}, \ldots, \lambda - a_{nn}$ on the diagonal.

Since $\lambda\mathbf{I} - \mathbf{A}^{\mathrm{T}} = (\lambda\mathbf{I} - \mathbf{A})^{\mathrm{T}}$ it follows from *Theorem* 6.3.3 that characteristic polynomials of **A** and $\mathbf{A}^{\mathrm{T}}$ are the same.

Theorem 8.2.6 Similar matrices have the same characteristic polynomial.

This theorem follows from *Corollary* 2 to *Theorem* 6.5.1 on observing that $\lambda\mathbf{I}-\mathbf{P}^{-1}\mathbf{AP}=\mathbf{P}^{-1}(\lambda\mathbf{I}-\mathbf{A})\mathbf{P}$. It shows that just like determinant, characteristic polynomial can be defined for a linear operator φ on a vector space V as the characteristic polynomial of the matrix of φ with respect to any basis of V. It also follows from the theorem that if $\mathbf{B}$ is obtained from $\mathbf{A}$ by applying the same permutation to the rows and columns, then $\mathbf{A}$ and $\mathbf{B}$ have the same characteristic polynomial.

Theorem 8.2.7 Let $\mathbf{A}$ and $\mathbf{B}$ be matrices of orders $m\times n$ and $n\times m$ respectively, where $m\leq n$. Then $\chi_{\mathbf{BA}}(\lambda)=\lambda^{n-m}\chi_{\mathbf{AB}}(\lambda)$.

Proof By *Theorem* 4.5.3 there exist non-singular matrices $\mathbf{P}$ and $\mathbf{Q}$ such that

$$\mathbf{PAQ}=\begin{bmatrix}\mathbf{I}_r & \mathbf{0}\\ \mathbf{0} & \mathbf{0}\end{bmatrix}$$

where r is the rank of $\mathbf{A}$. Partition $\mathbf{Q}^{-1}\mathbf{BP}^{-1}$ as

$$\mathbf{Q}^{-1}\mathbf{BP}^{-1}=\begin{bmatrix}\mathbf{C} & \mathbf{D}\\ \mathbf{E} & \mathbf{G}\end{bmatrix}$$

where $\mathbf{C}$ is of order $r\times r$. Then

$$\mathbf{PABP}^{-1}=\mathbf{PAQQ}^{-1}\mathbf{BP}^{-1}=\begin{bmatrix}\mathbf{C} & \mathbf{D}\\ \mathbf{0} & \mathbf{0}\end{bmatrix} \tag{8.2.3}$$

and

$$\mathbf{Q}^{-1}\mathbf{BAQ}=\mathbf{Q}^{-1}\mathbf{BP}^{-1}\mathbf{PAQ}=\begin{bmatrix}\mathbf{C} & \mathbf{0}\\ \mathbf{E} & \mathbf{0}\end{bmatrix} \tag{8.2.4}$$

From (8.2.3) we get

$$\chi_{\mathbf{AB}}(\lambda)=\chi_{\mathbf{PABP}^{-1}}(\lambda)=\begin{vmatrix}\lambda\mathbf{I}_r-\mathbf{C} & -\mathbf{D}\\ \mathbf{0} & \lambda\mathbf{I}_{m-r}\end{vmatrix}=|\lambda\mathbf{I}_r-\mathbf{C}|\lambda^{m-r}$$

It follows similarly from (8.2.4) that $\chi_{\mathbf{BA}}(\lambda)$ is $|\lambda\mathbf{I}_r-\mathbf{C}|\lambda^{n-r}$. Hence the theorem follows. ∎

Corollary For any two $n\times n$ matrices $\mathbf{A}$ and $\mathbf{B}$ the characteristic polynomials of $\mathbf{AB}$ and $\mathbf{BA}$ are the same.

If $\mathbf{AB}$ is square, *Theorem* 8.2.7 shows that the non-zero characteristic roots of $\mathbf{AB}$ are the same as those of $\mathbf{BA}$.

Exercises

1. Find the characteristic roots of each of the following matrices:

$$(a) \begin{bmatrix} \cos\theta & -\sin\theta \\ \sin\theta & \cos\theta \end{bmatrix}, \ (b) \begin{bmatrix} 3 & 5 \\ 1 & 4 \end{bmatrix}, \ (c) \begin{bmatrix} 0 & 1 & 2 \\ -1 & 0 & -3 \\ -2 & 3 & 0 \end{bmatrix},$$

$$(d) \begin{bmatrix} 0 & -1 & 2 \\ -1 & 0 & 2 \\ 2 & 2 & -3 \end{bmatrix}, \ (e) \frac{1}{25}\begin{bmatrix} 34 & -26(1+i) & 14(i-1) \\ 26(i-1) & 3 & 8i \\ -14(1+i) & -8i & 63 \end{bmatrix}$$

2. The characteristic roots of a 3×3 matrix $\mathbf{A}$ are known to be in arithmetic progression. Determine them given $\mathrm{tr}(\mathbf{A}) = 15$ and $|\mathbf{A}| = 80$.
3. (a) Show that the characteristic polynomial of the $n \times n$ matrix

$$\begin{bmatrix} 0 & 1 & 0 & \cdots & 0 \\ 0 & 0 & 1 & \cdots & 0 \\ \cdots & \cdots & \cdots & \cdots & \cdots \\ 0 & 0 & 0 & \cdots & 1 \\ -a_0 & -a_1 & -a_2 & \cdots & -a_{n-1} \end{bmatrix}$$

is $f(\lambda) := a_0 + a_1\lambda + \cdots + a_{n-1}\lambda^{n-1} + \lambda^n$. (The matrix is called the *companion matrix* of $f(\lambda)$.)

   (b) Given complex numbers $\lambda_1, \lambda_2, \ldots, \lambda_n$, not necessarily distinct, show that there is a matrix with characteristic roots $\lambda_1, \lambda_2, \ldots, \lambda_n$. Prove this in two ways: (i) directly and (ii) using (a).
4. Show that the characteristic roots of the $n \times n$ permutation matrix $\mathbf{P}$ with $p_{i,i+1} = 1$ for $i = 1, \ldots, n-1$ and $p_{n1} = 1$, are the n-th roots of unity.
5. If $\mathbf{A}$ is non-singular, deduce the corollary to *Theorem* 8.2.7 by showing that $\mathbf{BA}$ is similar to $\mathbf{AB}$.
6. Verify that $\chi_{\mathbf{D}}(\mathbf{D}) = 0$ for $\mathbf{D} = \mathbf{A}$, $\mathbf{B}$ and $\mathbf{C}$ of *Example* 8.2.2.
7. Find a rank-factorization of the matrix

$$\mathbf{C} = \begin{bmatrix} 2 & 4 & 2 & 4 & 4 \\ 1 & 2 & 1 & 2 & 2 \\ 3 & 0 & 3 & 3 & 0 \\ 0 & -4 & 0 & -2 & -4 \\ 5 & 2 & 5 & 6 & 2 \end{bmatrix}$$

and hence the characteristic roots of $\mathbf{C}$.
8. Express the characteristic polynomial of $\alpha\mathbf{I} + \beta\mathbf{A}$ in terms of that of $\mathbf{A}$. Hence find the characteristic roots of $\alpha\mathbf{I} + \beta\mathbf{A}$. What are the characteristic roots of $-\mathbf{A}$? (See *Theorem* 8.3.11 for a more general result.)

9. Show that if β is a characteristic root of $\mathbf{A}$ and $\mathbf{A}$ is non-singular, $1/\beta$ is a characteristic root of $\mathbf{A}^{-1}$.
10. Show that the characteristic roots of a matrix do not determine the rank (except when zero occurs as a characteristic root at most once; see *Theorem* 8.5.6).
11. If $\mathbf{A}$ be non-singular, when is $\mathbf{A} + \beta\mathbf{B}$ is singular?
12. Let $\mathbf{A}$ be a 2×2 matrix. Then show that $|\mathbf{I}+\mathbf{A}| = \mathbf{1}+|\mathbf{A}|$ iff $\mathrm{tr}(\mathbf{A}) = 0$.
*13. Show that the coefficient of λ^k in the characteristic polynomial of $\mathbf{A}$ is $(-1)^{n-k}$ times the sum of all the $(n-k)$-rowed principal minors of $\mathbf{A}$ and deduce *Lemma* 8.2.3.

8.3 Eigenvectors and eigenspaces

The importance of characteristic roots arises from the fact that if α is a characteristic root of $\mathbf{A}$, there exists a non-null vector $\mathbf{x}$ such that $\mathbf{Ax} = \alpha\mathbf{x}$. For such an $\mathbf{x}$, known as an eigenvector corresponding to the eigenvalue α, the line $\{\beta\mathbf{x} : \beta \in \mathbb{C}\}$ is invariant under the map $\mathbf{x} \mapsto \mathbf{Ax}$. In this section we study some elementary properties of eigenvectors and deduce some results about eigenvalues from these.

Definition 8.3.1 A complex number α is said to be an *eigenvalue* of $\mathbf{A}$ if there exists a non-null vector $\mathbf{x} \in \mathbb{C}^n$ such that $\mathbf{Ax} = \alpha\mathbf{x}$. Any such (non-null) $\mathbf{x}$ is called an *eigenvector of* $\mathbf{A}$ *corresponding to the eigenvalue* α. By *an eigenvector* of $\mathbf{A}$ we mean an eigenvector of $\mathbf{A}$ corresponding to some eigenvalue of $\mathbf{A}$.

Theorem 8.3.2 A number α is an eigenvalue of $\mathbf{A}$ iff α is a characteristic root of $\mathbf{A}$.

This theorem follows from the two observations: (i) α is an eigenvalue of $\mathbf{A}$ iff the system $(\alpha\mathbf{I} - \mathbf{A})\mathbf{x} = \mathbf{0}$ has a non-trivial solution and (ii) α is a characteristic root of $\mathbf{A}$ iff $\alpha\mathbf{I} - \mathbf{A}$ is singular.

If $\mathbf{A}$ is a real matrix and α is a real eigenvalue of $\mathbf{A}$ then the system $(\alpha\mathbf{I} - \mathbf{A})\mathbf{x} = \mathbf{0}$ has a non-trivial solution over $\mathbb{R}$ and so there exists a *real* eigenvector of $\mathbf{A}$ corresponding to α. There will, of course, be non-real eigenvectors of $\mathbf{A}$ corresponding to α, for example: $\sqrt{-1}\,\mathbf{x}$.

The preceding theorem shows that eigenvalues are the same as characteristic roots. However, *by 'the characteristic roots of* $\mathbf{A}$*' we mean the* n *roots of the characteristic polynomial of* $\mathbf{A}$ *whereas 'the eigenvalues of* **A**' *would mean the distinct characteristic roots of* **A**. Eigenvalues are also known as proper values, latent roots etc. and eigenvectors are also called characteristic vectors, latent vectors etc.

To illustrate the use of eigenvectors, let $f(\lambda)$ be a polynomial and β an eigenvalue of **A**. Let **x** be an eigenvector of **A** corresponding to β. Then $\mathbf{Ax} = \beta\mathbf{x}$. Premultiplying by **A**, we get $\mathbf{A}^2\mathbf{x} = \beta\mathbf{Ax} = \beta^2\mathbf{x}$. Again premultiplying by **x**, we get $\mathbf{A}^3\mathbf{x} = \beta^3\mathbf{x}$. Proceeding like this we see that $\mathbf{A}^k\mathbf{x} = \beta^k\mathbf{x}$ for all $k \geq 0$, so $f(\mathbf{A})\mathbf{x} = f(\beta)\mathbf{x}$. Since $\mathbf{x} \neq \mathbf{0}$, it follows that $f(\beta)$ is an eigenvalue of $f(\mathbf{A})$. Thus we have proved

Theorem 8.3.3 Let $f(\lambda)$ be a polynomial and β an eigenvalue of **A**. Then $f(\beta)$ is an eigenvalue of $f(\mathbf{A})$.

Corollary Each eigenvalue of an idempotent matrix **A** is 0 or 1.

Proof Let β be an eigenvalue of **A** and let $f(\lambda) = \lambda^2 - \lambda$. Then $f(\mathbf{A}) = \mathbf{0}$, so by the theorem, $f(\beta) = 0$. Hence β is 0 or 1. ∎

More generally, it follows from the preceding theorem that if β is an eigenvalue of a matrix **A** and $f(\lambda)$ is any polynomial such that $f(\mathbf{A}) = \mathbf{0}$ then $f(\beta) = 0$.

If α is an eigenvalue of **A**, the set of all eigenvectors of **A** corresponding to α, together with **0**, forms $\mathcal{N}(\alpha\mathbf{I} - \mathbf{A})$. This justifies

Definition 8.3.4 Let α be an eigenvalue of **A**. Then the subspace of $\mathbb{C}^n$ consisting of all eigenvectors of **A** corresponding to α together with **0** is called the *eigenspace of* **A** *corresponding to* α and is denoted $\text{ES}(\mathbf{A}, \alpha)$. Its dimension is called the *geometric multiplicity of* α *with respect to* **A**.

We note that $\text{ES}(\mathbf{A}, 0) = \mathcal{N}(\mathbf{A})$. It is also easy to see that $\text{ES}(\mathbf{A}, \alpha) \subseteq \mathcal{C}(\mathbf{A})$ if $\alpha \neq 0$.

Clearly the geometric multiplicity of an eigenvalue α of **A** is the nullity of $\alpha\mathbf{I} - \mathbf{A}$ and is at least 1. Suppose now α is a real eigenvalue of a real matrix **A** and S is the corresponding eigenspace. Then S has a basis consisting of real vectors which can even be chosen to be orthogonal. This is because the nullity of $\alpha\mathbf{I} - \mathbf{A}$ over $\mathbb{C}$ is the same as that over $\mathbb{R}$ by *Exercise* 6.3.15 and the Gram-Schmidt orthogonalization process gives a real orthonormal basis if we start with a real basis.

There is another type of multiplicity of an eigenvalue α of **A**, viz., the number of times α appears as a root of the characteristic equation of $\mathbf{A}$. This is called the *algebraic multiplicity of α with respect to* $\mathbf{A}$. We now establish an inequality between the two multiplicities.

Theorem 8.3.5 For any eigenvalue α of $\mathbf{A}$, the algebraic multiplicity of α with respect to $\mathbf{A}$ is not less than the geometric multiplicity of α with respect to $\mathbf{A}$.

Proof Let $\{\mathbf{x}_1, \mathbf{x}_2, \ldots, \mathbf{x}_k\}$ be a basis of $\mathrm{ES}(\mathbf{A}, \alpha)$ and $\{\mathbf{x}_1, \mathbf{x}_2, \ldots, \mathbf{x}_n\}$ an extension to a basis of $\mathbb{C}^n$. Then $\mathbf{P} := [\,\mathbf{x}_1 : \mathbf{x}_2 : \cdots : \mathbf{x}_n\,]$ is non-singular and

$$\begin{aligned}\mathbf{P}^{-1}\mathbf{AP} &= \mathbf{P}^{-1}[\,\mathbf{Ax}_1 : \mathbf{Ax}_2 : \cdots : \mathbf{Ax}_n\,] \\ &= \mathbf{P}^{-1}[\,\alpha\mathbf{x}_1 : \cdots : \alpha\mathbf{x}_k : \mathbf{Ax}_{k+1} : \cdots : \mathbf{Ax}_n\,]\end{aligned}$$

Now, $\mathbf{P}^{-1}(\alpha\mathbf{x}_j) = \alpha\mathbf{P}^{-1}\mathbf{P}_{*j} = \alpha\mathbf{e}_j$ for $j = 1, \ldots, k$, so

$$\mathbf{P}^{-1}\mathbf{AP} = \begin{bmatrix} \alpha\mathbf{I}_k & \mathbf{B} \\ \mathbf{0} & \mathbf{C} \end{bmatrix}$$

for some matrices $\mathbf{B}$ and $\mathbf{C}$. Hence

$$\chi_{\mathbf{A}}(\lambda) = \chi_{\mathbf{P}^{-1}\mathbf{AP}}(\lambda) = (\lambda - \alpha)^k \chi_{\mathbf{C}}(\lambda)$$

Thus the algebraic multiplicity of α with respect to $\mathbf{A}$ is at least k and the theorem follows. ∎

The technique used in the proof of the preceding theorem is very important and will be used several times in the remainder of this chapter.

We note that the algebraic multiplicity of an eigenvalue can be strictly greater than the geometric multiplicity. For example if

$$\mathbf{A} = \begin{bmatrix} 1 & 2 & 0 \\ 0 & 1 & 3 \\ 0 & 0 & 1 \end{bmatrix}$$

then the algebraic multiplicity of 1 with respect to $\mathbf{A}$ is 3 and the geometric multiplicity of 1 with respect to $\mathbf{A}$ is

$$\nu(\mathbf{I} - \mathbf{A}) = 3 - \rho(\mathbf{I} - \mathbf{A}) = 1$$

Definition 8.3.6 An eigenvalue α of $\mathbf{A}$ is said to be *regular* if the algebraic and the geometric multiplicities of α with respect to $\mathbf{A}$ are equal. α is said to be a *simple* eigenvalue of $\mathbf{A}$ if the algebraic multiplicity of α with respect to $\mathbf{A}$ is 1.

If α is a simple eigenvalue of $\mathbf{A}$ it follows immediately from *Theorem* 8.3.5 that α is regular and there exists a unique (upto multiplication by a non-zero scalar) eigenvector of **A** corresponding to α. We next study eigenvectors corresponding to distinct eigenvalues.

Theorem 8.3.7 Let $\alpha_1, \alpha_2, \ldots, \alpha_k$ be distinct eigenvalues of **A** and let $\mathbf{x}_1, \mathbf{x}_2, \ldots, \mathbf{x}_k$ be corresponding eigenvectors. Then $\mathbf{x}_1, \mathbf{x}_2, \ldots, \mathbf{x}_k$ are linearly independent.

Proof Suppose the conclusion is false. Let j be the smallest positive integer such that $\mathbf{x}_1, \mathbf{x}_2, \ldots, \mathbf{x}_j$ are linearly dependent. Then $j \geq 2$ and

$$\mathbf{x}_j = \beta_1 \mathbf{x}_1 + \cdots + \beta_{j-1} \mathbf{x}_{j-1} \tag{8.3.1}$$

for some $\beta_1, \ldots, \beta_{j-1}$. Premultiplying by **A**, we get

$$\alpha_j \mathbf{x}_j = \beta_1 \alpha_1 \mathbf{x}_1 + \cdots + \beta_{j-1} \alpha_{j-1} \mathbf{x}_{j-1}$$

Suppose now $\alpha_j = 0$. Then $\beta_1 \alpha_1 = \cdots = \beta_{j-1} \alpha_{j-1} = 0$. Since $\alpha_1, \ldots, \alpha_j$ are distinct, we get $\beta_1 = \cdots = \beta_{j-1} = 0$, so $\mathbf{x}_j = \mathbf{0}$, a contradiction. So $\alpha_j \neq 0$ and

$$\mathbf{x}_j = \beta_1 \frac{\alpha_1}{\alpha_j} \mathbf{x}_1 + \cdots + \beta_{j-1} \frac{\alpha_{j-1}}{\alpha_j} \mathbf{x}_{j-1}$$

Comparing this with (8.3.1) we get $\beta_i \alpha_i / \alpha_j = \beta_i$ for $i = 1, \ldots, j-1$ since $\mathbf{x}_1, \ldots, \mathbf{x}_{j-1}$ are linearly independent. Now for at least one i, β_i is non-zero and so $\alpha_i = \alpha_j$, a contradiction which proves the theorem. ∎

Corollary If $S_1, S_2, \ldots, S_k$ are the eigenspaces corresponding to distinct eigenvalues $\alpha_1, \alpha_2, \ldots, \alpha_k$ of a matrix **A**, then $S_1 + S_2 + \cdots + S_k$ is direct.

We have seen in *Theorem* 8.2.7 that if **AB** is a square matrix then every non-zero eigenvalue of **AB** is also an eigenvalue of **BA** with the same algebraic multiplicity. We now show that the geometric multiplicity also remains the same.

Theorem 8.3.8 Let α be a non-zero eigenvalue of a square matrix **AB**, where **A** and **B** need not be square. Then α is an eigenvalue of **BA** with the same geometric multiplicity. If $\mathbf{x}_1, \ldots, \mathbf{x}_r$ are linearly independent eigenvectors of **AB** corresponding to α then $\mathbf{Bx}_1, \ldots, \mathbf{Bx}_r$ are linearly independent eigenvectors of **BA** corresponding to α.

Proof Let $\{\mathbf{x}_1, \mathbf{x}_2, \ldots, \mathbf{x}_r\}$ be a basis of $\mathrm{ES}(\mathbf{AB}, \alpha)$. Then $\mathbf{ABx}_i = \alpha \mathbf{x}_i$ and so $\mathbf{BABx}_i = \alpha \mathbf{Bx}_i$ for $i = 1, \ldots, r$. Suppose $\sum_{i=1}^{r} \beta_i \mathbf{Bx}_i = \mathbf{0}$. Premultiplying by **A** we get $\alpha \sum \beta_i \mathbf{x}_i = \mathbf{0}$. Since $\alpha \neq 0$ and $\mathbf{x}_1, \ldots, \mathbf{x}_r$

are linearly independent it follows that $\beta_i = 0$ for $i = 1, \ldots, r$. Thus $\mathbf{Bx}_1, \ldots, \mathbf{Bx}_r$ are linearly independent eigenvectors of $\mathbf{BA}$ corresponding to α. Hence the geometric multiplicity of α with respect to $\mathbf{BA} \geq$ the geometric multiplicity of α with respect to $\mathbf{AB}$. By symmetry the reverse inequality holds and equality follows. ∎

The above theorem can be used effectively to find eigenvectors of $\mathbf{BA}$ when $\mathbf{AB}$ is of smaller order than $\mathbf{BA}$, for example if $(\mathbf{B}, \mathbf{A})$ is a rank-factorization of a singular matrix.

***Theorem 8.3.9** Let $\mathbf{x}$ be a non-null vector. Then there exists an eigenvector $\mathbf{y}$ of $\mathbf{A}$ belonging to the span of $\{\mathbf{x}, \mathbf{Ax}, \mathbf{A}^2\mathbf{x}, \ldots\}$.

Proof Let k be the smallest positive integer such that $\mathbf{x}, \mathbf{Ax}, \mathbf{A}^2\mathbf{x}, \ldots, \mathbf{A}^k\mathbf{x}$ are linearly dependent. (Why does it exist?) Let

$$\sum_{i=0}^{k} c_i \mathbf{A}^i \mathbf{x} = \mathbf{0} \tag{8.3.2}$$

where $c_k \neq 0$. Now let $\beta_1, \beta_2, \ldots, \beta_k$ be the roots of the polynomial $g(t) = \sum_{i=0}^{k} c_i t^i$. Then $g(t) = c_k(t - \beta_1)(t - \beta_2) \cdots (t - \beta_k)$. By *Theorem* 2.4.4, $\sum_{i=1}^{k} c_i \mathbf{A}^i = g(\mathbf{A}) = c_k \prod_{i=1}^{k} (\mathbf{A} - \beta_i \mathbf{I})$. Taking $\mathbf{y} = \big(\prod_{i=2}^{k} (\mathbf{A} - \beta_i \mathbf{I})\big)\mathbf{x}$, it is easy to see that $\mathbf{y} \neq \mathbf{0}$ by the minimality of k and $(\mathbf{A} - \beta_1 \mathbf{I})\mathbf{y} = \mathbf{0}$ by (8.3.2), so the theorem follows. ∎

Remark In the proof of the preceding theorem, β_1 and, by symmetry, each β_i is an eigenvalue of $\mathbf{A}$. So if $\mathbf{A}$ is a real matrix with real eigenvalues and if $\mathbf{x}$ is real, then the $\mathbf{y}$ obtained is real.

Theorem 8.3.10 Every $n \times n$ matrix $\mathbf{A}$ is similar to an upper triangular matrix over $\mathbb{C}$.

Proof We prove the theorem by induction on n. If $n = 1$ the result holds trivially. So assume it for matrices of order $n - 1$ and let $\mathbf{A}$ be of order n. Let α be an eigenvalue of $\mathbf{A}$ and let $\mathbf{x}$ be an eigenvector of $\mathbf{A}$ corresponding to α. Let $\mathbf{P}$ be a non-singular matrix with $\mathbf{x}$ as the first column. Then as in the proof of *Theorem* 8.3.5 we have

$$\mathbf{P}^{-1}\mathbf{AP} = \begin{bmatrix} \alpha & \mathbf{y}^{\mathrm{T}} \\ \mathbf{0} & \mathbf{C} \end{bmatrix}$$

for some $1 \times (n - 1)$ vector $\mathbf{y}^{\mathrm{T}}$ and some $(n - 1) \times (n - 1)$ matrix $\mathbf{C}$. By induction hypothesis there exists a non-singular matrix $\mathbf{W}$ of order

$n-1$ such that $\mathbf{T} := \mathbf{W}^{-1}\mathbf{C}\mathbf{W}$ is upper triangular. It is easy to see that $\mathbf{Q} := \begin{bmatrix} 1 & 0 \\ 0 & \mathbf{W} \end{bmatrix}$ is non-singular, so $\mathbf{PQ}$ is also non-singular and

$$(\mathbf{PQ})^{-1}\mathbf{A}(\mathbf{PQ}) = \mathbf{Q}^{-1}\mathbf{P}^{-1}\mathbf{APQ} =$$
$$\begin{bmatrix} 1 & 0 \\ 0 & \mathbf{W}^{-1} \end{bmatrix}\begin{bmatrix} \alpha & \mathbf{y}^{\mathrm{T}} \\ 0 & \mathbf{C} \end{bmatrix}\begin{bmatrix} 1 & 0 \\ 0 & \mathbf{W} \end{bmatrix} = \begin{bmatrix} \alpha & \mathbf{y}^{\mathrm{T}}\mathbf{W} \\ 0 & \mathbf{T} \end{bmatrix}$$

is upper triangular. ∎

Note The preceding theorem is a powerful tool. Using it we will now prove a stronger form of *Theorem* 8.3.3. Note, however, that the preceding theorem does not hold over $\mathbb{R}$ since a real matrix may not have real eigenvalues.

Theorem 8.3.11 Let $\lambda_1, \lambda_2, \ldots, \lambda_n$ be the characteristic roots of $\mathbf{A}$ and let $f(\lambda)$ be a polynomial. Then $f(\lambda_1), f(\lambda_2), \ldots, f(\lambda_n)$ are the characteristic roots of $f(\mathbf{A})$.

Proof By the preceding theorem there exists a non-singular matrix $\mathbf{P}$ such that $\mathbf{T} := \mathbf{P}^{-1}\mathbf{AP}$ is upper triangular. Since $\mathbf{A}$ and $\mathbf{T}$ have the same characteristic roots, we may take $t_{ii} = \lambda_i$ for $i = 1, \ldots, n$. Now it is easy to prove by induction on k that $\mathbf{T}^k = \mathbf{P}^{-1}\mathbf{A}^k\mathbf{P}$ for all $k \geq 0$. Hence if $f(\lambda) = \alpha_0 + \alpha_1\lambda + \cdots + \alpha_s\lambda^s$, we have

$$\begin{aligned} f(\mathbf{T}) &= \alpha_0\mathbf{I} + \alpha_1\mathbf{T} + \cdots + \alpha_s\mathbf{T}^s \\ &= \alpha_0\mathbf{P}^{-1}\mathbf{P} + \alpha_1\mathbf{P}^{-1}\mathbf{AP} + \cdots + \alpha_s\mathbf{P}^{-1}\mathbf{A}^s\mathbf{P} \\ &= \mathbf{P}^{-1}(\alpha_0\mathbf{I} + \alpha_1\mathbf{A} + \cdots + \alpha_s\mathbf{A}^s)\mathbf{P} \\ &= \mathbf{P}^{-1}f(\mathbf{A})\,\mathbf{P} \end{aligned}$$

But, from *Theorem* 2.5.3 it easily follows that $f(\mathbf{T})$ is upper triangular with $f(t_{11}), f(t_{22}), \ldots, f(t_{nn})$ as the diagonal entries, hence the characteristic roots of $f(\mathbf{A})$ are $f(\lambda_1), f(\lambda_2), \ldots, f(\lambda_n)$. ∎

Corollary If $\mathbf{A}$ is singular, the algebraic multiplicities of 0 with respect to $\mathbf{A}^\ell$ and with respect to $\mathbf{A}$ are equal for any positive integer ℓ.

Exercises

1. Determine all the eigenvectors of the matrices $\mathbf{A}$ and $\mathbf{C}$ in *Example* 8.2.2 corresponding to each eigenvalue.
2. For the matrix $\mathbf{C}$ of *Exercise* 8.2.7, find a basis of the eigenspace corresponding to each eigenvalue.

3. Let α be an eigenvalue of $\mathbf{A}$. Then show that $\mathrm{ES}(\mathbf{A}^k, \alpha^k) \supseteq \mathrm{ES}(\mathbf{A}, \alpha)$ if $k \geq 1$. Extend the result to $k = -1$ if $\mathbf{A}$ is non-singular. Show also that proper inclusion is possible.
4. If k, ℓ and n are integers such that $1 \leq k \leq \ell \leq n$, show that there exists an $n \times n$ matrix $\mathbf{A}$ and an eigenvalue α of $\mathbf{A}$ such that k and ℓ are the geometric and algebraic multiplicities of α with respect to $\mathbf{A}$.
5. If $\alpha_1, \ldots, \alpha_k$ are the distinct eigenvalues of an $n \times n$ matrix $\mathbf{A}$ with geometric multiplicities $n_1, \ldots, n_k$ respectively, then $n_1 + \cdots + n_k \leq n$. Deduce this from *Theorem* 8.3.5 and also from the corollary to *Theorem* 8.3.7.
6. (a) Let δ be an eigenvalue of $\mathbf{A}$ with algebraic multiplicity a and let $\beta \neq 0$. Then show that $\alpha + \beta\delta$ is an eigenvalue of $\alpha\mathbf{I} + \beta\mathbf{A}$ with algebraic multiplicity a and $\mathrm{ES}(\alpha\mathbf{I} + \beta\mathbf{A}, \alpha + \beta\delta) = \mathrm{ES}(\mathbf{A}, \delta)$.

   (b) Prove or disprove: if δ is an eigenvalue of $\mathbf{A}$, the algebraic and geometric multiplicities of $f(\delta)$ with respect to $f(\mathbf{A})$ are the same as those of δ with respect to $\mathbf{A}$ for any polynomial f.

*7. Consider the circulant

$$\mathbf{A} = \begin{bmatrix} a_1 & a_2 & \cdots & a_n \\ a_n & a_1 & \cdots & a_{n-1} \\ \cdots & \cdots & \cdots & \cdots \\ a_2 & a_3 & \cdots & a_1 \end{bmatrix}$$

Write $\mathbf{A}$ as a polynomial in a suitable permutation matrix $\mathbf{P}$. Deduce that the characteristic roots of $\mathbf{A}$ are $a_1 + a_2\omega_i + a_3\omega_i^2 + \cdots + a_n\omega_i^{n-1}$, $i = 1, \ldots, n$, where $\omega_1, \omega_2, \ldots, \omega_n$ are the n-th roots of unity.

8. If $\mathbf{A}$ is an $n \times n$ singular matrix with k distinct eigenvalues, show that $k - 1 \leq \rho(\mathbf{A}) \leq n - 1$. Also show by construction that $\rho(\mathbf{A})$ can take any value ℓ between $k - 1$ and $n - 1$.
9. Deduce *Theorem* 8.3.7 from *Example* 6.4.4 thus: given $\sum_{i=1}^{k} \beta_i \mathbf{x}_i = \mathbf{0}$, premultiply this equation by $\mathbf{A}, \mathbf{A}^2, \ldots, \mathbf{A}^{k-1}$.
10. Let $\mathbf{A} = \mathbf{u}\mathbf{u}^*$ where $\mathbf{u}$ is a non-null vector.

    (a) Show that the eigenvalues of $\mathbf{A}$ are 0 and $\mathbf{u}^*\mathbf{u}$.

    (b) Show that $\mathbf{u}^*\mathbf{u}$ is a simple eigenvalue of $\mathbf{A}$.

    (c) Identify $\mathrm{ES}(\mathbf{A}, 0)$ and $\mathrm{ES}(\mathbf{A}, \mathbf{u}^*\mathbf{u})$ and deduce the result in (b).

    (d) Show that $\mathbf{A}$ is similar to a diagonal matrix.

11. Find the eigenvalues and their algebraic and geometric multiplicities for each of the real $n \times n$ matrices (i) $(\alpha - \beta)\mathbf{I} + \beta\mathbf{1}\mathbf{1}^{\mathrm{T}}$ and (ii) $\alpha\mathbf{I} + \mathbf{u}\mathbf{1}^{\mathrm{T}} + \mathbf{1}\mathbf{u}^{\mathrm{T}}$. Here $\mathbf{1}$ denotes a vector with all entries 1 and $\mathbf{u}$ is an arbitrary vector.

12. If α is an eigenvalue of $\mathbf{A}$, then it is an eigenvalue of $\mathbf{A}^{\mathrm{T}}$ also. An eigenvector of $\mathbf{A}^{\mathrm{T}}$ corresponding to α, i.e., a vector $\mathbf{x} \neq \mathbf{0}$ such that $\mathbf{x}^{\mathrm{T}}\mathbf{A} = \alpha\mathbf{x}^{\mathrm{T}}$, is called a *left eigenvector of* $\mathbf{A}$ *corresponding to* α. Viewed in the same spirit, eigenvectors of $\mathbf{A}$ as defined in *Definition* 8.3.1 are called *right eigenvectors*. Let λ_1 and λ_2 be distinct eigenvalues of $\mathbf{A}$. If $\mathbf{x}$ is a left eigenvector of $\mathbf{A}$ corresponding to λ_1 and $\mathbf{y}$ is a right eigenvector of $\mathbf{A}$ corresponding to λ_2, then show that $\mathbf{x}^{\mathrm{T}}\mathbf{y} = 0$.
13. Let λ be an eigenvalue of $\mathbf{A}$. Let r be the geometric multiplicity of λ. Show that the dimension of the space of the left eigenvectors of $\mathbf{A}$ corresponding to λ is also r.
14. Deduce *Theorem* 8.2.4 from *Theorem* 8.3.10.

*15. Let $\mathbf{A}$ be an $n \times n$ matrix and let

$$\rho_i = \sum_{\substack{j=1 \\ j \neq i}}^{n} |a_{ij}| \quad (i = 1, \ldots, n)$$

(a) If α is an eigenvalue of $\mathbf{A}$, show that $|\alpha - a_{ii}| \leq \rho_i$ for at least one i.

(b) If $\mathbf{A}$ is strictly diagonally dominant, deduce that

$$|\det \mathbf{A}| \geq (|a_{11}| - \rho_1)(|a_{22}| - \rho_2) \cdots (|a_{nn}| - \rho_n)$$

For a sharper bound, see *Exercise* 6.4.17(a).

16. Let α be an eigenvalue of $\mathbf{A}$. Then show that $|\alpha| \leq \|\mathbf{A}\|$ where $\|\cdot\|$ is the matrix norm induced by any vector norm (see *Exercise* 7.3.15).

17. (a) Let $\mathbf{A}$ be an $n \times n$ idempotent matrix. Then show that $\mathcal{C}(\mathbf{A}) = \mathrm{ES}(\mathbf{A}, 1)$ and $\mathcal{C}(\mathbf{I} - \mathbf{A}) = \mathrm{ES}(\mathbf{A}, 0)$ and that $\mathbf{A}$ has n linearly independent eigenvectors.

(b) If each eigenvalue of $\mathbf{A}$ is 0 or 1, does it follow that $\mathbf{A}$ is idempotent?

*18. Let f be a linear operator on a complex vector space V.

(a) Show that there exists a subspace S of V with $d(S) = 1$ such that $f(S) \subseteq S$. An S satisfying the latter condition is said to be *invariant under* f.

*(b) Show using (a) and *Exercise* 2.1.15 that there exists a basis B of V with respect to which the matrix of f is upper triangular. Hence deduce *Theorem* 8.3.10.

19. Let $\mathbf{A}$ be an $n \times n$ matrix and let $\mathbf{D}$ be the $n \times n$ matrix with (i, j)-th element $\mathrm{tr}(\mathbf{A}^{i+j-2})$. Show that the characteristic roots of $\mathbf{A}$ are distinct iff $\mathbf{D}$ is non-singular.

8.4 Cayley-Hamilton theorem and minimal polynomial

In this section we prove that for any matrix $\mathbf{A}$, $\chi_{\mathbf{A}}(\mathbf{A}) = \mathbf{0}$ and show how this can be used to evaluate large powers of $\mathbf{A}$ or the value of a polynomial with large degree in $\mathbf{A}$.

Definition 8.4.1 A polynomial $f(\lambda)$ is said to *annihilate* $\mathbf{A}$ if $f(\mathbf{A}) = \mathbf{0}$.

Theorem 8.4.2 (*Cayley-Hamilton Theorem*) For every matrix $\mathbf{A}$, the characteristic polynomial of $\mathbf{A}$ annihilates $\mathbf{A}$.

Proof By *Theorem* 8.3.10, there exists a non-singular matrix $\mathbf{P}$ such that $\mathbf{T} := \mathbf{P}^{-1}\mathbf{A}\mathbf{P}$ is upper triangular. Let $\mathbf{W}_k = \prod_{i=1}^{k}(\mathbf{T} - t_{ii}\mathbf{I})$. We prove by induction on k that the first k columns of $\mathbf{W}_k$ are null. This is trivial for $k = 1$, so assume it for $k-1$. Let $\ell \leq k$, $\mathbf{B} = \mathbf{W}_{k-1}$ and $\mathbf{C} = \mathbf{T} - t_{kk}\mathbf{I}$. Then

$$(\mathbf{W}_k)_{i\ell} = \sum_{j=1}^{n} b_{ij}c_{j\ell} = 0$$

since $b_{ij} = 0$ if $j \leq k-1$ and $c_{j\ell} = 0$ if $j \geq k$. Thus the first k columns of $\mathbf{W}_k$ are null for $k = 1, \ldots, n$. Hence $\mathbf{W}_n = \mathbf{0}$. Thus $f(\mathbf{T}) = \mathbf{0}$ where $f(\lambda) = \prod_{i=1}^{n}(\lambda - t_{ii})$. But $f(\mathbf{T}) = \mathbf{P}^{-1}f(\mathbf{A})\mathbf{P}$, so $f(\mathbf{A}) = \mathbf{0}$. Since $\mathbf{A}$ and $\mathbf{T}$ are similar, $\chi_{\mathbf{A}}(\lambda) = \chi_{\mathbf{T}}(\lambda) = \prod_{i=1}^{n}(\lambda - t_{ii}) = f(\lambda)$. ∎

We note that the preceding theorem cannot be proved simply by replacing λ by $\mathbf{A}$ in $\chi_{\mathbf{A}}(\lambda) = |\lambda\mathbf{I} - \mathbf{A}|$. To see this, observe that if we replace λ by $\mathbf{A}$ in the RHS what we get is the scalar 0 and not the null matrix as required.

The Cayley-Hamilton theorem is usually stated thus: every matrix satisfies its own characteristic equation. We illustrate two uses of this theorem with

Example 8.4.3 Consider the matrix $\mathbf{A}$ of *Example* 8.2.2. Then $\chi_{\mathbf{A}}(\lambda) = 4 - 4\lambda + \lambda^2$ and the reader may verify that $4\mathbf{I} - 4\mathbf{A} + \mathbf{A}^2 = \mathbf{0}$, so $\mathbf{A}^{-1} = \mathbf{I} - \frac{1}{4}\mathbf{A}$. If we want $\mathbf{A}^6$, we divide λ^6 by $\lambda^2 - 4\lambda + 4$ to get

$$\lambda^6 = (\lambda^2 - 4\lambda + 4)(\lambda^4 + 4\lambda^3 + 12\lambda^2 + 32\lambda + 80) + 192\lambda - 320$$

Hence

$$\mathbf{A}^6 = 192\mathbf{A} - 320\,\mathbf{I} = \begin{bmatrix} -128 & 192 \\ -192 & 256 \end{bmatrix}$$ ∎

We note that, in view of *Lemma* 8.2.3, the method of the preceding example can be used to express $\mathbf{A}^{-1}$ as a polynomial in $\mathbf{A}$ whenever $\mathbf{A}$ is non-singular. The method can also be used to evaluate a polynomial in $\mathbf{A}$ with large degree even if $\mathbf{A}$ is singular.

In what follows, we use the convention that the degree of the zero polynomial is $-\infty$.

Cayley-Hamilton theorem shows that for any square matrix $\mathbf{A}$, there exists a non-zero annihilating polynomial. (This also follows from the fact that $\mathbf{I}, \mathbf{A}, \ldots, \mathbf{A}^{n^2}$ are linearly dependent in $F^{n\times n}$.) If f annihilates $\mathbf{A}$, αf also annihilates $\mathbf{A}$, so there exists a monic polynomial annihilating $\mathbf{A}$. Suppose k is the minimum degree of a non-zero polynomial annihilating $\mathbf{A}$ and f and g are two monic polynomials of degree k annihilating $\mathbf{A}$. Then $h = f - g$ also annihilates $\mathbf{A}$ and has degree less than k, so $h = 0$ and $f = g$. This justifies the following

Definition 8.4.4 The monic polynomial of the least degree which annihilates $\mathbf{A}$ is called the *minimal polynomial* of $\mathbf{A}$.

By Cayley-Hamilton theorem, the degree of the minimal polynomial of an $n \times n$ matrix $\mathbf{A}$ is at most n. We now show that the minimal polynomial not only has the least degree among the non-zero polynomials annihilating $\mathbf{A}$ but also divides each of them.

Theorem 8.4.5 The minimal polynomial of $\mathbf{A}$ divides every polynomial which annihilates $\mathbf{A}$.

Proof Let $f(\lambda)$ be the minimal polynomial of $\mathbf{A}$ and let $g(\mathbf{A}) = \mathbf{0}$. Since $f \neq 0$, there exist polynomials $q(\lambda)$ and $r(\lambda)$ such that

$$g(\lambda) = f(\lambda)q(\lambda) + r(\lambda)$$

where the degree of $r(\lambda)$ is less than that of $f(\lambda)$. Then

$$\mathbf{0} = g(\mathbf{A}) = f(\mathbf{A})q(\mathbf{A}) + r(\mathbf{A}) = r(\mathbf{A})$$

Thus $r(\lambda)$ annihilates $\mathbf{A}$. By the minimality of f, it follows that $r(\lambda) = 0$, so f divides g. ∎

Corollary The minimal polynomial of $\mathbf{A}$ divides the characteristic polynomial of $\mathbf{A}$.

The preceding theorem shows that once an annihilating polynomial $g(\lambda)$ is known, the search for the minimal polynomial can be restricted to the factors of $g(\lambda)$. Thus if $\mathbf{A}$ is idempotent then $\lambda^2 - \lambda$ annihilates $\mathbf{A}$, so the minimal polynomial of $\mathbf{A}$ is λ, $\lambda - 1$ or $\lambda^2 - \lambda$. If $\mathbf{A}$ is neither $\mathbf{0}$ nor $\mathbf{I}$ it follows that the minimal polynomial of $\mathbf{A}$ is $\lambda^2 - \lambda$.

Theorem 8.4.6 A complex number α is a root of the minimal polynomial of $\mathbf{A}$ iff α is a characteristic root of $\mathbf{A}$.

Proof The *only if part* follows from the preceding corollary. The *if part* follows from *Theorem* 8.3.3. ∎

The preceding theorem shows that the distinct roots of the minimal polynomial coincide with those of the characteristic polynomial. It therefore follows that if the n characteristic roots of $\mathbf{A}$ are distinct then the minimal polynomial of $\mathbf{A}$ coincides with the characteristic polynomial of $\mathbf{A}$. A matrix $\mathbf{A}$ with the latter property is said to be *non-derogatory*.

Theorem 8.4.7 The minimal polynomial of a diagonal matrix $\mathbf{A}$ is $\prod_{i=1}^{k}(\lambda - d_i)$ where $d_1, d_2, \ldots, d_k$ are the distinct diagonal entries of $\mathbf{A}$.

Proof Without any loss of generality we take $\mathbf{A} = \text{diag}(d_1\mathbf{I}, \ldots, d_k\mathbf{I})$ where the $\mathbf{I}$'s are identity matrices of suitable orders. Then

$$(\mathbf{A} - d_1\mathbf{I})(\mathbf{A} - d_2\mathbf{I}) \cdots (\mathbf{A} - d_k\mathbf{I}) = \mathbf{0}$$

So by *Theorem* 8.4.2, the minimal polynomial divides the polynomial $f(\lambda) = \prod_{i=1}^{k}(\lambda - d_i)$. But, by the preceding theorem, each d_i is a root of the minimal polynomial of $\mathbf{A}$. Since the minimal polynomial is monic, it follows that the minimal polynomial is $f(\lambda)$. ∎

Caution The minimal polynomial of a matrix need not be a product of distinct linear factors (see *Theorem* 8.5.2).

Theorem 8.4.8 Similar matrices have the same minimal polynomial.

Proof Let $\mathbf{B} = \mathbf{P}^{-1}\mathbf{A}\mathbf{P}$. Then as in the proof of *Theorem* 8.3.11, $f(\mathbf{B}) = \mathbf{P}^{-1}f(\mathbf{A})\mathbf{P}$ for any polynomial f. Thus $f(\mathbf{B}) = \mathbf{0}$ iff $f(\mathbf{A}) = \mathbf{0}$, so $\mathbf{A}$ and $\mathbf{B}$ have the same minimal polynomial. ∎

In view of the preceding theorem we can define the minimal polynomial of a linear operator φ on a vector space V as the minimal polynomial of the matrix of φ with respect to any basis of V. Now, if f is any polynomial and $\mathbf{A}$ is the matrix of φ with respect to $\mathcal{X}$ then by *Theorem* 2.2.17, $f(\mathbf{A})$ is the matrix of $f(\varphi)$ with respect to $\mathcal{X}$. Thus $f(\mathbf{A}) = \mathbf{0}$ iff

$f(\varphi) = 0$ and the minimal polynomial of φ is the monic polynomial of the least degree which annihilates φ.

Exercises

1. Verify Cayley-Hamilton theorem for the matrix in *Exercise* 8.2.1(b).
2. Find the inverse of the matrix in *Exercise* 8.2.1(d) using Cayley-Hamilton theorem.
3. Find $\mathbf{A}^{10}$ for the matrix $\mathbf{A}$ in *Exercise* 8.2.1(c) using Cayley-Hamilton theorem.
4. Using Cayley-Hamilton theorem, find the inverse of the matrix $\mathbf{B}$ of *Example* 8.2.2. If $h(\lambda) = (\lambda^3 - \lambda^2 + 2\lambda + 1)/(\lambda^2 - \lambda + 1)$, find $h(\mathbf{B})$.
5. If h is a rational function such that $h(\mathbf{A})$ is defined, show that $h(\mathbf{A})$ equals a polynomial in $\mathbf{A}$. (See *Exercise* 3.4.5(b).)
6. Prove Cayley-Hamilton theorem thus: let $\mathbf{H} := (\lambda\mathbf{I} - \mathbf{A})^{\circledast} = \mathbf{H}_0 + \lambda\mathbf{H}_1 + \cdots + \lambda^{n-1}\mathbf{H}_{n-1}$. Then $(\lambda\mathbf{I} - \mathbf{A})\mathbf{H} = \chi_{\mathbf{A}}(\lambda)\,\mathbf{I}$. Multiply by $\mathbf{A}^i$ the equation obtained by comparing the coefficients of λ^i on the two sides and sum up to get $\chi_{\mathbf{A}}(\mathbf{A}) = \mathbf{0}$.
7. Find the minimal polynomials of the following, where α, β and γ are arbitrary scalars: $\mathbf{A} = \begin{bmatrix} 1 & \alpha \\ 0 & 1 \end{bmatrix}, \mathbf{B} = \begin{bmatrix} 1 & \alpha & \beta \\ 0 & 1 & \gamma \\ 0 & 0 & 2 \end{bmatrix}$ and $\mathbf{C} = \begin{bmatrix} 1 & \alpha & \beta \\ 0 & 1 & \gamma \\ 0 & 0 & 1 \end{bmatrix}$.
8. $\mathbf{A}$ is said to be nilpotent if $\mathbf{A}^k = \mathbf{0}$ for some positive integer k. Show that $\mathbf{A}$ is nilpotent iff all the characteristic roots of $\mathbf{A}$ are 0.
9. Let $\mathbf{A}$ be nilpotent.
   (a) If $\mathbf{A} \neq \mathbf{0}$, show that $\mathbf{A}$ cannot be similar to a diagonal matrix.
   (b) What can you say about the minimal polynomial of $\mathbf{A}$?
10. Show that the constant term in the minimal polynomial of $\mathbf{A}$ is non-zero iff $\mathbf{A}$ is non-singular.
11. Show that the minimal polynomial coincides with the characteristic polynomial for the matrix in *Exercise* 8.2.3.
12. Determine the minimal polynomials of $\begin{bmatrix} 1 & 0 & 0 \\ 0 & 0 & 0 \\ 0 & 0 & 0 \end{bmatrix}$ and $\begin{bmatrix} 1 & 0 & 0 \\ 0 & 0 & 1 \\ 0 & 0 & 0 \end{bmatrix}$.
    Notice that these two matrices have the same characteristic polynomial but different minimal polynomials. Also, the minimal polynomial of the latter matrix has a repeated root.
13. Let $\mathbf{A}$ and $\mathbf{B}$ be square matrices of the same order and let $\mathbf{C} = \mathbf{AB} - \mathbf{BA}$. Show that $\mathbf{I} - \mathbf{C}$ is not nilpotent.

14. What is the minimal polynomial of $\alpha\mathbf{A}$?
15. Find the minimal polynomial of the $n \times n$ matrix $\mathbf{J} = \mathbf{1}\mathbf{1}^{\mathrm{T}}$.
16. Prove that the minimal polynomial of $\mathrm{diag}(\mathbf{A}, \mathbf{B})$ is the L.C.M. of the minimal polynomials of $\mathbf{A}$ and $\mathbf{B}$.

8.5 Spectral representation of a semi-simple matrix*

We saw in the preceding section that every matrix is similar to an upper triangular matrix. But not every matrix is similar to a diagonal matrix. For example, if $\begin{bmatrix} 0 & 1 \\ 0 & 0 \end{bmatrix}$ is similar to a diagonal matrix $\mathbf{D}$, then both the characteristic roots of $\mathbf{D}$ are 0 and so $\mathbf{D} = \mathbf{0}$, an impossibility. In this section we give some necessary and sufficient conditions for a matrix to be similar to a diagonal matrix and study some nice representations of such matrices.

Definition 8.5.1 A matrix is said to be *semi-simple* or *diagonable* if it is similar to a diagonal matrix.

If $\mathbf{A}$ is the matrix of a linear operator φ on V with respect to some basis, then $\mathbf{A}$ is semi-simple iff there is a coordinate system (with the same origin) each of whose coordinate axes is left invariant by φ.

There is a nice connection between semi-simple matrices and eigenvectors. Suppose $\mathbf{A}$ is semi-simple and $\mathbf{P}^{-1}\mathbf{A}\mathbf{P} = \mathbf{D} := \mathrm{diag}(d_1, \ldots, d_n)$. Then $\mathbf{A}\mathbf{P} = \mathbf{P}\mathbf{D}$, so $\mathbf{A}\mathbf{P}_{*j} = d_j\mathbf{P}_{*j}$. Thus the columns of $\mathbf{P}$ are linearly independent eigenvectors of $\mathbf{A}$ (corresponding to the diagonal entries of $\mathbf{D}$ in the same order). Conversely, if $\mathbf{A}$ has n linearly independent eigenvectors and $\mathbf{P}$ is the matrix formed with these vectors as the columns then $\mathbf{P}^{-1}\mathbf{A}\mathbf{P}$ is diagonal. In the following theorem we give some more characterizations of a semi-simple matrix.

Theorem 8.5.2 The following statements about an $n \times n$ matrix $\mathbf{A}$ are equivalent:

(i) $\mathbf{A}$ is semi-simple,

(ii) the minimal polynomial of $\mathbf{A}$ is a product of distinct linear factors or, equivalently, there exists an annihilating polynomial of $\mathbf{A}$ which is a product of distinct linear factors,

(iii) all the eigenvalues of $\mathbf{A}$ are regular,

(iv) the sum of the eigenspaces of **A** is $\mathbb{C}^n$,

(v) **A** has n linearly independent eigenvectors.

Proof That (i) implies (ii) follows from *Theorems* 8.4.7 and 8.4.8.

(ii) $\Rightarrow$ (iii) Let the minimal polynomial of **A** be $(\lambda-\alpha_1)\cdots(\lambda-\alpha_k)$. Then $(\mathbf{A}-\alpha_1\mathbf{I})\cdots(\mathbf{A}-\alpha_k\mathbf{I}) = \mathbf{0}$. So by Sylvester's law of nullity,

$$\nu(\mathbf{A}-\alpha_1\mathbf{I}) + \cdots + \nu(\mathbf{A}-\alpha_k\mathbf{I}) \geq n \tag{8.5.1}$$

Now by *Theorem* 8.4.6, $\alpha_1, \ldots, \alpha_k$ are the distinct eigenvalues of **A**. If n_i and r_i denote the algebraic and geometric multiplicities of α_i with respect to **A**, then $\nu(\mathbf{A}-\alpha_i\mathbf{I}) = r_i \leq n_i$ and $\sum n_i = n$, so (8.5.1) gives $r_i = n_i$ for $i = 1, \ldots, k$.

We incidentally note that (i) $\Rightarrow$ (iii) follows easily from definitions.

(iii) $\Rightarrow$ (iv) This follows from the corollary to *Theorem* 8.3.7 and the fact that the algebraic multiplicities add up to n.

(iv) $\Rightarrow$ (v) This follows from the corollary to *Theorem* 8.3.7.

(v) $\Rightarrow$ (i) This follows easily as noted in the paragraph preceding this theorem. ∎

Corollary 1 An $n \times n$ matrix with n distinct eigenvalues is semi-simple.

This corollary follows from the fact that if all the characteristic roots of **A** are distinct then each is simple and so regular.

Corollary 2 An idempotent matrix is semi-simple.

This corollary follows from the fact that $\lambda(\lambda-1)$ annihilates an idempotent matrix.

In the next theorem we give several useful ways of representing a semi-simple matrix.

Theorem 8.5.3 The following statements about an $n \times n$ matrix **A** are equivalent:

(i) **A** is semi-simple and has rank r,

(ii) There exists a non-singular matrix **P** of order n and a diagonal non-singular matrix $\mathbf{\Delta}$ of order r such that

$$\mathbf{A} = \mathbf{P}\begin{bmatrix} \mathbf{\Delta} & \mathbf{0} \\ \mathbf{0} & \mathbf{0} \end{bmatrix}\mathbf{P}^{-1} \tag{8.5.2}$$

(iii) There exist non-zero scalars $\delta_1, \ldots, \delta_r$ and vectors $\mathbf{u}_1, \ldots, \mathbf{u}_r$ and $\mathbf{v}_1, \ldots, \mathbf{v}_r \in \mathbb{C}^n$ such that $\mathbf{v}_i^{\mathrm{T}}\mathbf{u}_j = \delta_{ij}$ for all i, j and

$$\mathbf{A} = \sum_{i=1}^{r} \delta_i \, \mathbf{u}_i \mathbf{v}_i^{\mathrm{T}} \tag{8.5.3}$$

(iv) There exist matrices $\mathbf{R}$, $\mathbf{S}$ and $\boldsymbol{\Delta}$ of orders $n \times r$, $r \times n$ and $r \times r$ respectively such that $\boldsymbol{\Delta}$ is diagonal and non-singular, $\mathbf{SR} = \mathbf{I}$ and

$$\mathbf{A} = \mathbf{R\Delta S} \tag{8.5.4}$$

Proof (i) $\Rightarrow$ (ii) This follows on observing that the non-zero diagonal entries of a diagonal matrix can be brought to the first few positions by applying the same permutation to the rows and columns, i.e., by postmultiplying by some permutation matrix $\mathbf{Q}$ and premultiplying by $\mathbf{Q}^{-1}$ (see *Theorems* 3.4.6 and 4.2.5).

To prove (ii) $\Rightarrow$ (iii), take δ_i to be the i-th diagonal entry of $\boldsymbol{\Delta}$, $\mathbf{u}_i$ to be the i-th column of $\mathbf{P}$ and $\mathbf{v}_i^{\mathrm{T}}$ to be the i-th row of $\mathbf{P}^{-1}$.

To prove (iii) $\Rightarrow$ (iv), take $\mathbf{R} = [\,\mathbf{u}_1 : \cdots : \mathbf{u}_r\,]$, $\mathbf{S}^{\mathrm{T}} = [\,\mathbf{v}_1 : \cdots : \mathbf{v}_r\,]$ and $\boldsymbol{\Delta} = \operatorname{diag}(\delta_1, \ldots, \delta_r)$.

(iv) $\Rightarrow$ (ii) Since $\mathbf{SR} = \mathbf{I}_r$, so $\rho(\mathbf{R}) = \rho(\mathbf{S}) = r$, $\mathbf{RS}$ and $\mathbf{I} - \mathbf{RS}$ are idempotent and $\rho(\mathbf{I} - \mathbf{RS}) = n - r$. Let $(\mathbf{X}, \mathbf{Y})$ be a rank-factorization of $(\mathbf{I} - \mathbf{RS})$. Then $\mathbf{RS} + \mathbf{XY} = \mathbf{I}_n$. So $[\,\mathbf{R} : \mathbf{X}\,]\left[\begin{smallmatrix}\mathbf{S}\\ \mathbf{Y}\end{smallmatrix}\right] = \mathbf{I}_n$. Note that $[\,\mathbf{R} : \mathbf{X}\,]$ is a square matrix. Taking $\mathbf{P} = [\,\mathbf{R} : \mathbf{X}\,]$, (8.5.2) follows.

That (ii) $\Rightarrow$ (i) is trivial. ∎

We call each of the representations (8.5.2), (8.5.3) and (8.5.4) a *spectral decomposition* of the semi-simple matrix $\mathbf{A}$. Note that they are not unique since there can be different sets of n linearly independent eigenvectors of $\mathbf{A}$.

Example 8.5.4 We shall find a spectral decomposition of the matrix

$$\mathbf{A} = \begin{bmatrix} 7 & -6 & 6 \\ 2 & 0 & 4 \\ 1 & -2 & 6 \end{bmatrix}$$

For this we first find the eigenvalues of $\mathbf{A}$. It can be checked that $\chi_{\mathbf{A}}(\lambda) = (\lambda - 4)^2(\lambda - 5)$, so the eigenvalues are 4 and 5. We next find a basis for each eigenspace.

ES($\mathbf{A}$, 4) is the set of all solutions of $(\mathbf{A} - 4\mathbf{I})\mathbf{x} = \mathbf{0}$, i.e.,

$$\begin{bmatrix} 3 & -6 & 6 \\ 2 & -4 & 4 \\ 1 & -2 & 2 \end{bmatrix} \begin{bmatrix} x_1 \\ x_2 \\ x_3 \end{bmatrix} = \begin{bmatrix} 0 \\ 0 \\ 0 \end{bmatrix} \tag{8.5.5}$$

Clearly the matrix of coefficients has rank 1, so the set of all solutions has dimension 2. Thus 4 is a regular eigenvalue. It can be checked that $(2,1,0)^{\mathrm{T}}$ and $(-2,0,1)^{\mathrm{T}}$ form a basis for the solution space of (8.5.5).

We leave it to the reader to check that the geometric multiplicity of 5 is 1 and that $\{(3,2,1)^{\mathrm{T}}\}$ is a basis of the corresponding eigenspace.

Since all the eigenvalues of $\mathbf{A}$ are regular, $\mathbf{A}$ is semi-simple. To get a spectral decomposition, we form the matrix $\mathbf{P}$ with $(2,1,0)^{\mathrm{T}}$, $(-2,0,1)^{\mathrm{T}}$ and $(3,2,1)^{\mathrm{T}}$ as the columns in that order. By the corollary to *Theorem* 8.3.7, $\mathbf{P}$ is non-singular. It can be checked that

$$\mathbf{P}^{-1} = \begin{bmatrix} -2 & 5 & -4 \\ -1 & 2 & -1 \\ 1 & -2 & 2 \end{bmatrix}$$

Hence

$$\mathbf{A} = \begin{bmatrix} 2 & -2 & 3 \\ 1 & 0 & 2 \\ 0 & 1 & 1 \end{bmatrix} \begin{bmatrix} 4 & 0 & 0 \\ 0 & 4 & 0 \\ 0 & 0 & 5 \end{bmatrix} \begin{bmatrix} -2 & 5 & -4 \\ -1 & 2 & -1 \\ 1 & -2 & 2 \end{bmatrix}$$

is a spectral decomposition of $\mathbf{A}$ in the form (8.5.2). The corresponding spectral decomposition in the form (8.5.3) is

$$\mathbf{A} = 4\begin{bmatrix} 2 \\ 1 \\ 0 \end{bmatrix}(-2,5,-4) + 4\begin{bmatrix} -2 \\ 0 \\ 1 \end{bmatrix}(-1,2,-1) + 5\begin{bmatrix} 3 \\ 2 \\ 1 \end{bmatrix}(1,-2,2)$$

In this example, a spectral decomposition of $\mathbf{A}$ in the form (8.5.4) coincides with that in the form (8.5.2) since $r = n$. We also note that a different spectral decomposition in the form (8.5.2) can be obtained by taking some other bases of the eigenspaces. ∎

We next study another representation of a semi-simple matrix which is unique unlike spectral decomposition.

Theorem 8.5.5 Let $\mathbf{A}$ be an $n \times n$ matrix. Then $\mathbf{A}$ is semi-simple iff for some k $(1 \le k \le n)$ there exist (complex) numbers $\alpha_1, \ldots, \alpha_k$ and $n \times n$ matrices $\mathbf{E}_1, \ldots, \mathbf{E}_k$ such that the following four conditions are satisfied:

(i) $\mathbf{A} = \alpha_1\mathbf{E}_1 + \cdots + \alpha_k\mathbf{E}_k$,

(ii) $\alpha_1, \ldots, \alpha_k$ are distinct and $\mathbf{E}_1, \ldots, \mathbf{E}_k$ are non-null,

(iii) $\mathbf{E}_1 + \cdots + \mathbf{E}_k = \mathbf{I}$,

(iv) $\mathbf{E}_i^2 = \mathbf{E}_i$ for $i = 1, \ldots, k$.

Further, α's and $\mathbf{E}$'s are uniquely determined by $\mathbf{A}$ thus: $\alpha_1, \ldots, \alpha_k$ are the distinct eigenvalues of $\mathbf{A}$ and if S_i denotes the eigen subspace of $\mathbf{A}$ corresponding to α_i then $\mathbf{E}_i$ is the projector into S_i along $\sum_{j\neq i} S_j$.

Proof *Only if part* Without any loss of generality we may take

$$\mathbf{A} = \mathbf{P}\ \mathrm{diag}(\alpha_1\mathbf{I}_{n_1}, \ldots, \alpha_k\mathbf{I}_{n_k})\,\mathbf{P}^{-1} \tag{8.5.6}$$

where $\alpha_1, \ldots, \alpha_k$ are distinct. Partition $\mathbf{P}$ and $\mathbf{P}^{-1}$ as

$$[\,\mathbf{P}_1 : \cdots : \mathbf{P}_k\,] \ \text{ and } \begin{bmatrix} \mathbf{Q}_1 \\ \vdots \\ \mathbf{Q}_k \end{bmatrix} \tag{8.5.7}$$

respectively, where $\mathbf{P}_i$ has n_i columns and $\mathbf{Q}_i$ has n_i rows for $i = 1, \ldots, k$. Define $\mathbf{E}_i = \mathbf{P}_i\mathbf{Q}_i$. Then (i) follows from (8.5.6), (ii) follows on observing that $\mathbf{P}_i$ is of full column rank and $\mathbf{Q}_i$ is of full row rank, (iii) follows from $\mathbf{P}\mathbf{P}^{-1} = \mathbf{I}$ and (iv) follows from $\mathbf{P}^{-1}\mathbf{P} = \mathbf{I}$.

If part Let $\rho(\mathbf{E}_i) = n_i$ and $(\mathbf{P}_i, \mathbf{Q}_i)$ be a rank-factorization of $\mathbf{E}_i$ for $i = 1, \ldots, k$. Then $\sum n_i = \sum \mathrm{tr}(\mathbf{E}_i) = \mathrm{tr}(\mathbf{I}) = n$, so the matrices $\mathbf{P}$ and $\mathbf{Q}$ in (8.5.7) are $n \times n$ matrices. By (iii), $\mathbf{PQ} = \mathbf{I}$, so $\mathbf{Q} = \mathbf{P}^{-1}$ and (8.5.6) follows from (i). Hence $\mathbf{A}$ is semi-simple and $\alpha_1, \ldots, \alpha_k$ are the distinct eigenvalues of $\mathbf{A}$. Now $\mathbf{QP} = \mathbf{I}$ gives $\mathbf{E}_i\mathbf{E}_j = \mathbf{0}$ whenever $i \neq j$. So postmultiplying both sides of (i) by $\mathbf{E}_i$ we get $\mathbf{AE}_i = \alpha_i\mathbf{E}_i$ and $S_i \supseteq \mathcal{C}(\mathbf{E}_i)$. By (iii), any $\mathbf{x} \in \mathbb{C}^n$ can be written as

$$\mathbf{x} = \mathbf{E}_1\mathbf{x} + \cdots + \mathbf{E}_k\mathbf{x} \tag{8.5.8}$$

Hence $\mathcal{C}(\mathbf{E}_1) + \cdots + \mathcal{C}(\mathbf{E}_k) = \mathbb{C}^n$. Since $S_i \supseteq \mathcal{C}(\mathbf{E}_i)$ it follows that $S_1 + \cdots + S_k = \mathbb{C}^n$. Since α's are distinct, $S_1 + \cdots + S_k$ is direct and so $\mathcal{C}(\mathbf{E}_1) + \cdots + \mathcal{C}(\mathbf{E}_k)$ is also direct. Since both these are $\mathbb{C}^n$, it follows that $\mathcal{C}(\mathbf{E}_i)$ and S_i have the same dimension and so are equal. Now (8.5.8) shows that $\mathbf{E}_i$ is the projector into S_i along $\sum_{j\neq i} S_j$. ∎

The unique representation of a semi-simple matrix $\mathbf{A}$ in the form (i) of the preceding theorem, where (ii), (iii) and (iv) hold, is called the *spectral form* or *spectral representation* of $\mathbf{A}$.

Remark By *Theorem* 3.7.7, it follows that in the preceding theorem, (iv) can be replaced by any one of the following:

(v) $\mathbf{E}_i\mathbf{E}_j = \mathbf{0}$ whenever $i \neq j$,

(vi) $\rho(\mathbf{E}_1) + \rho(\mathbf{E}_2) + \cdots + \rho(\mathbf{E}_k) = n$.

Note that given $\mathbf{E}_i$'s, we can get a $\mathbf{P}$ such that (8.5.6) holds by taking $\mathbf{P} = [\,\mathbf{P}_1; \cdots : \mathbf{P}_k\,]$ where the columns of $\mathbf{P}_i$ form a basis of $\mathcal{C}(\mathbf{E}_i)$. This follows from the proof of the if part of the preceding theorem. We can give an explicit expression for the matrix $\mathbf{E}_i$ in the spectral form of $\mathbf{A}$:

$$\mathbf{E}_i = \prod_{\substack{j=1 \\ j \neq i}}^{k} \frac{(\alpha_j\mathbf{I} - \mathbf{A})}{(\alpha_j - \alpha_i)} \tag{8.5.9}$$

To prove this, let $\mathbf{G}_i$ denote the matrix on the RHS. Then it is easy to see that $\mathbf{G}_i\mathbf{x}$ is $\mathbf{x}$ if $\mathbf{x} \in S_i$ and $\mathbf{0}$ if $\mathbf{x} \in S_j$ for any $j \neq i$. Hence $\mathbf{G}_i$ is the projector into S_i along $\sum_{j \neq i} S_j$ and so equals $\mathbf{E}_i$.

We now give one use of spectral form (or spectral decomposition) of $\mathbf{A}$. Using (iv) and (v), it is easy to prove by induction on p that $\mathbf{A}^p = \sum_i \alpha_i^p \mathbf{E}_i$ for all positive integers p. Hence it follows that $f(\mathbf{A}) = \sum_i f(\alpha_i)\mathbf{E}_i$ for any polynomial f. If $\mathbf{A}$ is non-singular then $\mathbf{A}^{-1} = \sum_i \frac{1}{\alpha_i}\mathbf{E}_i$ since $\left(\sum \alpha_i\mathbf{E}_i\right)\left(\sum \frac{1}{\alpha_j}\mathbf{E}_j\right) = \sum \mathbf{E}_i = \mathbf{I}$. If $\mathbf{A}$ is singular and $\alpha_k = 0$, it is easy to verify that $\sum_{i=1}^{k-1} \frac{1}{\alpha_i}\mathbf{E}_i$ is a g-inverse of $\mathbf{A}$.

We have seen in *Theorem* 8.5.2 that all the eigenvalues of $\mathbf{A}$ are regular iff $\mathbf{A}$ is similar to a diagonal matrix. In the following theorem, we characterise when 0 is a regular eigenvalue of $\mathbf{A}$.

***Theorem 8.5.6** For any $n \times n$ matrix $\mathbf{A}$ with rank $r < n$, the following statements are equivalent:

(i) 0 is a regular eigenvalue of $\mathbf{A}$,

(ii) $\mathbf{A}$ is similar to $\mathbf{G} := \begin{bmatrix} \mathbf{0} & \mathbf{0} \\ \mathbf{0} & \mathbf{C} \end{bmatrix}$ for some $r \times r$ matrix $\mathbf{C}$,

(iii) $\rho(\mathbf{A}) = \rho(\mathbf{A}^2)$,

(iv) $\mathcal{N}(\mathbf{A})$ is a complement of $\mathcal{C}(\mathbf{A})$.

Proof Since 0 is an eigenvalue with geometric multiplicity $n - r$, it follows from the proof of *Theorem* 8.3.5 that

$$\mathbf{P}^{-1}\mathbf{A}\mathbf{P} = \begin{bmatrix} \mathbf{0} & \mathbf{B} \\ \mathbf{0} & \mathbf{C} \end{bmatrix} \tag{8.5.10}$$

for some matrices $\mathbf{P}$, $\mathbf{B}$ and $\mathbf{C}$, where $\mathbf{C}$ is of order $r \times r$.

(i) $\Rightarrow$ (ii) Since the characteristic roots of $\mathbf{A}$ and $\mathbf{P}^{-1}\mathbf{AP}$ are the same, it follows from (8.5.10) that $\mathbf{C}$ is non-singular. Now it is easy to verify that

$$\begin{bmatrix} \mathbf{I} & \mathbf{BC}^{-1} \\ \mathbf{0} & \mathbf{I} \end{bmatrix}^{-1} \begin{bmatrix} \mathbf{0} & \mathbf{B} \\ \mathbf{0} & \mathbf{C} \end{bmatrix} \begin{bmatrix} \mathbf{I} & \mathbf{BC}^{-1} \\ \mathbf{0} & \mathbf{I} \end{bmatrix} = \begin{bmatrix} \mathbf{0} & \mathbf{0} \\ \mathbf{0} & \mathbf{C} \end{bmatrix}$$

so (ii) follows.

(ii) $\Rightarrow$ (iii) The matrix $\mathbf{C}$ appearing in $\mathbf{G}$ is non-singular, so (iii) follows easily.

(iii) $\Rightarrow$ (i) Squaring the two sides of (8.5.10), we get

$$\rho(\mathbf{A}^2) = \rho\begin{bmatrix} \mathbf{0} & \mathbf{BC} \\ \mathbf{0} & \mathbf{C}^2 \end{bmatrix} = \rho\begin{bmatrix} \mathbf{BC} \\ \mathbf{C}^2 \end{bmatrix} = \rho\left(\begin{bmatrix} \mathbf{B} \\ \mathbf{C} \end{bmatrix}\mathbf{C}\right) \leq \rho(\mathbf{C}) \qquad (8.5.11)$$

so $\mathbf{C}$ is non-singular. Hence (i) follows from (8.5.10).

By *Theorem* 3.5.11, (iii) is equivalent to: $\mathcal{C}(\mathbf{A}) \cap \mathcal{N}(\mathbf{A}) = \{\mathbf{0}\}$ which is equivalent to (iv) since the dimensions of $\mathcal{C}(\mathbf{A})$ and $\mathcal{N}(\mathbf{A})$ add up to n. This proves the theorem. ∎

Suppose $\mathbf{A}$ is a singular matrix with index p (see *Exercise* 3.5.17). Let a be the algebraic multiplicity of 0 with respect to $\mathbf{A}$. By the corollary to *Theorem* 8.3.11, the algebraic multiplicity of 0 with respect to $\mathbf{A}^\ell$ remains a for all ℓ. But the geometric multiplicity of 0 with respect to $\mathbf{A}^\ell$, viz., $\nu(\mathbf{A}^\ell)$ increases with ℓ until $\ell = p$. We now show that the geometric multiplicity equals the algebraic multiplicity when $\ell = p$. Let $\mathbf{H} = \mathbf{A}^p$. Then clearly $\rho(\mathbf{H}) = \rho(\mathbf{H}^2)$. So by the preceding theorem, 0 is a regular eigenvalue of $\mathbf{H}$. We have thus proved

***Theorem 8.5.7**· Let $\mathbf{A}$ be a singular matrix with index p and let a be the algebraic multiplicity of 0 with respect to $\mathbf{A}$. Then $a = \nu(\mathbf{A}^p)$.

Exercises

1. If $\mathbf{A}$ is a 2×2 matrix such that $\mathbf{A}^2 = \mathbf{0}$, show that either $\mathbf{A} = \mathbf{0}$ or $\mathbf{A}$ is similar to $\begin{bmatrix} 0 & 1 \\ 0 & 0 \end{bmatrix}$.

2. Show that $\mathbf{A} = \begin{bmatrix} a & b \\ c & d \end{bmatrix}$ is semi-simple iff either $\mathbf{A}$ is a scalar matrix or $(a-d)^2 + 4bc \neq 0$.

3. If $\mathbf{A}^k = \mathbf{I}$ for some positive integer k, show that $\mathbf{A}$ is semi-simple.
4. Show that $\mathbf{A}$ is idempotent iff each eigenvalue of $\mathbf{A}$ is 0 or 1 and $\mathbf{A}$ is semi-simple.
5. If $\mathbf{A}$ is a semi-simple matrix such that $\mathbf{A}^2 = \mathbf{A}^3$, show that $\mathbf{A}$ is idempotent. Show also that the condition that $\mathbf{A}$ is semi-simple cannot be dropped.
6. Prove that $\mathbf{A}$ is a g-inverse of itself iff $\mathbf{A}$ is semi-simple and the spectrum of $\mathbf{A}$ is a subset of $\{-1, 0, 1\}$.
7. If $\mathbf{A}$ is semi-simple, show that any polynomial in $\mathbf{A}$ is also semi-simple.
8. Let $\mathbf{A} = \begin{bmatrix} 2 & 1 \\ 1 & 2 \end{bmatrix}$. Obtain a spectral decomposition of $\mathbf{A}$. Hence find $\mathbf{A}^{10}$ and write down a spectral decomposition of $\mathbf{A}^{-1}$.
9. A matrix $\mathbf{A}$ is said to be *stochastic* if $a_{ij} \geq 0$ for all i and j and $\sum_j a_{ij} = 1$ for all i. Let $\mathbf{A}$ be a 2×2 stochastic matrix $\neq \mathbf{I}$.
   (a) Find a spectral decomposition of $\mathbf{A}$.
   (b) Obtain an expression for $\mathbf{A}^k$ where k is an arbitrary positive integer.
   (c) Show that there exists a 3×3 stochastic (upper triangular) matrix which is not semi-simple.
10. Find a spectral decomposition of

$$\begin{bmatrix} 2 & -1 & 1 & 3 \\ 1 & 0 & 1 & 1 \\ 0 & 2 & 2 & -2 \\ -1 & 1 & 0 & -2 \end{bmatrix}$$

in each of the forms (8.5.2), (8.5.3) and (8.5.4). (Hint: The first two columns form a column basis. Find a rank-factorization and use *Theorem* 8.3.8.)

11. If $\sum_{i=1}^{k} \alpha_i \mathbf{E}_i$ is the spectral form of $\mathbf{A}$ and $\alpha_k = 0$, show that $\sum_{i=1}^{k-1} \frac{1}{\alpha_i} \mathbf{E}_i$ is a g-inverse of $\mathbf{A}$.
12. Let $\mathbf{P}$ be a non-singular matrix, the columns of which are eigen vectors of $\mathbf{A}$. Then show that the rows of $\mathbf{P}^{-1}$ are left eigen vectors of $\mathbf{A}$ (see *Exercise* 8.3.12).
13. Let $\mathbf{A}$ be semi-simple and let $\mathbf{A} = \sum_{i=1}^{k} \alpha_i \mathbf{E}_i$ be the spectral form of $\mathbf{A}$. Then prove that $\mathbf{B}$ commutes with $\mathbf{A}$ iff $\mathbf{B}$ commutes with $\mathbf{E}_i$ for $i = 1, \ldots, k$.
14. Let $\mathbf{A}$ be a square matrix with real eigenvalues such that $\rho(\mathbf{A}) = \rho(\mathbf{A}^2)$ and $\mathrm{tr}(\mathbf{A}^2) \neq 0$. Then show that

$$\rho(\mathbf{A}) \geq \frac{(\mathrm{tr}(\mathbf{A}))^2}{\mathrm{tr}(\mathbf{A}^2)}$$

and that equality holds iff all the non-zero characteristic roots of $\mathbf{A}$ are equal.

15. In *Theorem* 8.5.5, show that the sum of the $\mathbf{E}_i$'s corresponding to the i's for which $\alpha_i \neq 0$, is the projector into $\mathcal{C}(\mathbf{A})$ along $\mathcal{N}(\mathbf{A})$.
16. Let $\mathbf{A}$ and $\mathbf{B}$ be $n \times n$ semi-simple matrices. Show that the following statements are equivalent:
    (a) $\mathbf{AB} = \mathbf{BA}$,
    (b) $\mathbf{A}$ and $\mathbf{B}$ are simultaneously diagonable (i.e., there exists a non-singular matrix $\mathbf{P}$ such that $\mathbf{P}^{-1}\mathbf{AP}$ and $\mathbf{P}^{-1}\mathbf{BP}$ are diagonal),
    (c) $\mathbf{A}$ and $\mathbf{B}$ are polynomials in a common semi-simple matrix.
17. For the matrix $\mathbf{A}$ in *Example* 8.5.4, find a different spectral decomposition. Compute the spectral form of $\mathbf{A}$ from each of these decompositions and check that it remains the same.

8.6 Jordan canonical form*

In *Sections* 8.3 and 8.5 we saw that every matrix is similar to an upper triangular matrix and *not* every matrix is similar to a diagonal matrix. In this section we will prove that every matrix is similar to a unique upper triangular matrix of a particular type which is almost diagonal. We will give a quick proof (which is due to P. S. S. N. V. P. Rao) of this result without much discussion. The results of this section will not be used in the sequel.

We need some notation. Let $\mathbf{J}_\alpha(n)$ denote the $n \times n$ matrix

$$\mathbf{J}_\alpha(n) = \begin{bmatrix} \alpha & 1 & 0 & \cdots & 0 & 0 \\ 0 & \alpha & 1 & \cdots & 0 & 0 \\ \cdots & \cdots & \cdots & \cdots & \cdots & \cdots \\ 0 & 0 & 0 & \cdots & \alpha & 1 \\ 0 & 0 & 0 & \cdots & 0 & \alpha \end{bmatrix}$$

Note that each diagonal element of $\mathbf{J}_\alpha(n)$ is α, each element immediately above the diagonal (some authors call such elements *superdiagonal elements*) is 1 and all other elements are 0. By a $\mathbf{J}_\alpha$*-block* we mean $\mathbf{J}_\alpha(n)$ for some n and by a *Jordan block* we mean $\mathbf{J}_\alpha(n)$ for some α and some n.

If $\mathbf{T}_1, \ldots, \mathbf{T}_k$ are square matrices of possibly different orders, we call the matrix $\text{diag}(\mathbf{T}_1, \ldots, \mathbf{T}_k)$ the *direct sum* of $\mathbf{T}_1, \ldots, \mathbf{T}_k$ in that order and denote it by $\mathbf{T}_1 \oplus \cdots \oplus \mathbf{T}_k$.

Lemma 8.6.1 Let $\mathbf{B}$ be a singular matrix of order n and let a be the algebraic multiplicity of 0 with respect to $\mathbf{B}$. Then there exists an $n \times a$ matrix $\mathbf{P}$ with full column rank such that $\mathbf{BP} = \mathbf{PT}$, where $\mathbf{T}$ is an $a \times a$ matrix which is a direct sum of $\mathbf{J}_0$-blocks.

Proof By *Theorem* 3.5.11 we have

$$\rho(\mathbf{B}^i) - \rho(\mathbf{B}^{i+1}) = d(\mathcal{C}(\mathbf{B}^i) \cap \mathcal{N}(\mathbf{B})) \quad \text{for } i = 0, 1, \ldots \tag{8.6.1}$$

Let us denote $\mathcal{C}(\mathbf{B}^i) \cap \mathcal{N}(\mathbf{B})$ by S_i and let p be the index of $\mathbf{B}$. Then

$$\mathcal{N}(\mathbf{B}) = S_0 \supseteq S_1 \supseteq \cdots \supseteq S_{p-1}$$

Adding (8.6.1) over $i = 0, 1, \ldots, p-1$ and using *Theorem* 8.5.7 we get

$$a = \nu(\mathbf{B}^p) = \rho(\mathbf{B}^0) - \rho(\mathbf{B}^p) = \sum_{i=0}^{p-1} d(S_i) \tag{8.6.2}$$

We now construct a basis of S_0 as follows: we start with a basis of S_{p-1}, say $\{\mathbf{B}^{p-1}\mathbf{x}_{p-1,1}, \ldots, \mathbf{B}^{p-1}\mathbf{x}_{p-1,d_{p-1}}\}$. (Note that $\mathbf{x}_{p-1,j}$ is a vector for each j.) Extend this to a basis of S_{p-2} and let the new vectors included be $\mathbf{B}^{p-2}\mathbf{x}_{p-2,1}, \ldots, \mathbf{B}^{p-2}\mathbf{x}_{p-2,d_{p-2}}$. We then extend this basis of S_{p-2} to a basis of S_{p-3} and denote the new vectors included by $\mathbf{B}^{p-3}\mathbf{x}_{p-3,1}, \ldots, \mathbf{B}^{p-3}\mathbf{x}_{p-3,d_{p-3}}$. We proceed like this until we get a basis of S_0. We note down the vectors of this basis in the first column of a table as shown in *Figure* 8.6.1 (note that the vectors in the first i blocks of this column together form a basis of S_{p-i}). We then complete the table by noting down the vectors in the other columns as shown in the table. By (8.6.2) there are exactly a vectors in the table since there are $d(S_0)$ vectors in the first column, $d(S_1)$ vectors in the second column, ... ,$d(S_{p-1})$ vectors in the last column. Call these vectors $\mathbf{y}_1, \mathbf{y}_2, \ldots, \mathbf{y}_a$, where the vectors in the table are read-off row-wise. Since the vectors in the first column belong to $S_0 = \mathcal{N}(\mathbf{B})$, it follows that

$$\mathbf{B}^k \mathbf{x}_{ij} = \mathbf{0} \quad \text{whenever } k \geq i+1 \tag{8.6.3}$$

We now claim that all the a vectors in the table are linearly independent. To prove this let

$$\alpha_1 \mathbf{y}_1 + \cdots + \alpha_a \mathbf{y}_a = \mathbf{0} \tag{8.6.4}$$

Premultiplying by $\mathbf{B}^{p-1}$ we get

$$\alpha_1 \mathbf{B}^{p-1}\mathbf{y}_1 + \cdots + \alpha_a \mathbf{B}^{p-1}\mathbf{y}_a = \mathbf{0} \tag{8.6.5}$$

$\mathbf{B}^{p-1}\mathbf{x}_{p-1,1}$ $\vdots$ $\mathbf{B}^{p-1}\mathbf{x}_{p-1,d_{p-1}}$	$\mathbf{B}^{p-2}\mathbf{x}_{p-1,1}$ $\vdots$ $\mathbf{B}^{p-2}\mathbf{x}_{p-1,d_{p-1}}$	$\cdots$	$\mathbf{B}\mathbf{x}_{p-1,1}$ $\vdots$ $\mathbf{B}\mathbf{x}_{p-1,d_{p-1}}$	$\mathbf{x}_{p-1,1}$ $\vdots$ $\mathbf{x}_{p-1,d_{p-1}}$
$\mathbf{B}^{p-2}\mathbf{x}_{p-2,1}$ $\vdots$ $\mathbf{B}^{p-2}\mathbf{x}_{p-2,d_{p-2}}$	$\mathbf{B}^{p-3}\mathbf{x}_{p-2,1}$ $\vdots$ $\mathbf{B}^{p-3}\mathbf{x}_{p-2,d_{p-2}}$	$\cdots$	$\mathbf{x}_{p-2,1}$ $\vdots$ $\mathbf{x}_{p-2,d_{p-2}}$	
$\vdots$	$\vdots$	$\iddots$		
$\mathbf{B}\mathbf{x}_{11}$ $\vdots$ $\mathbf{B}\mathbf{x}_{1d_1}$	$\mathbf{x}_{11}$ $\vdots$ $\mathbf{x}_{1d_1}$			
$\mathbf{x}_{01}$ $\vdots$ $\mathbf{x}_{0d_0}$				

Figure 8.6.1

Using (8.6.3) we see that all vectors in (8.6.5) except those corresponding to the vectors in the last column of the table are $\mathbf{0}$. So (8.6.5) becomes: a linear combination of $\mathbf{B}^{p-1}\mathbf{x}_{p-1,1}, \ldots, \mathbf{B}^{p-1}\mathbf{x}_{p-1,d_{p-1}}$ is $\mathbf{0}$. But these vectors are linearly independent (note that they occur in the first column of the table), so the corresponding α's are 0. Now drop these terms from (8.6.4) and premultiply by $\mathbf{B}^{p-2}$. We can show as above that the α's corresponding to the vectors in the penultimate column of the table are 0. Proceeding like this we can show that all α's are 0, so $\mathbf{y}_1, \mathbf{y}_2, \ldots, \mathbf{y}_a$ are linearly independent. Now let $\mathbf{P}$ be the $n \times a$ matrix $[\,\mathbf{y}_1 : \mathbf{y}_2 : \cdots : \mathbf{y}_a\,]$. Clearly $\mathbf{P}$ is of full column rank. Partition $\mathbf{P}$ as $[\,\mathbf{P}_1 : \mathbf{P}_2 : \cdots : \mathbf{P}_s\,]$ where the number of columns of $\mathbf{P}_j$ is n_j, the number of vectors in the j-th row of the table, for $j = 1, 2, \ldots, s = \sum d_i$. Since $\mathbf{B}\mathbf{y}_i = \mathbf{0}$ if $\mathbf{y}_i$

occurs in the first column of the table and $\mathbf{B}\mathbf{y}_i = \mathbf{y}_{i-1}$ otherwise, it follows that $\mathbf{B}\mathbf{P}_j = \mathbf{P}_j\mathbf{J}_0(n_j)$. Thus $\mathbf{BP} = \mathbf{PT}$ where $\mathbf{T}$ is the $a \times a$ matrix $\mathbf{J}_0(n_1) \oplus \cdots \oplus \mathbf{J}_0(n_s)$. ∎

Remark If p is the index of $\mathbf{B}$, the matrix $\mathbf{P}$ constructed in the proof of the preceding lemma has the property $\mathcal{C}(\mathbf{P}) = \mathcal{N}(\mathbf{B}^p)$ since $\mathbf{B}^p\mathbf{y}_i = \mathbf{0}$ for all i and $a = \nu(\mathbf{B}^p)$.

We now make a few simple observations about Jordan blocks. Clearly $\mathbf{J}_\alpha(n) + \beta\mathbf{I} = \mathbf{J}_{\alpha+\beta}(n)$. $\mathbf{J}_\alpha(n)$ is non-singular if $\alpha \neq 0$. For any $\ell \geq 0$, $\rho(\mathbf{J}_0(n))^\ell = \max(n - \ell, 0)$.

Theorem 8.6.2 Every square matrix is similar to a direct sum of Jordan blocks.

Proof Let $\mathbf{A}$ be an $n \times n$ matrix and let $\alpha_1, \ldots, \alpha_k$ be the distinct eigenvalues of $\mathbf{A}$. Denote $\mathbf{A} - \alpha_i\mathbf{I}$ by $\mathbf{B}_i$ and let a_i be the algebraic multiplicity of 0 with respect to $\mathbf{B}_i$, i.e., the algebraic multiplicity of α_i with respect to $\mathbf{A}$. Also let p_i be the index of $\mathbf{B}_i$. Then by the proof of the preceding lemma there exists an $n \times a_i$ matrix $\mathbf{P}_i$ with full column rank and an $a_i \times a_i$ matrix $\mathbf{T}_i$ which is a direct sum of $\mathbf{J}_0$-blocks such that $\mathbf{B}_i\mathbf{P}_i = \mathbf{P}_i\mathbf{T}_i$ and $\mathcal{C}(\mathbf{P}_i) = \mathcal{N}(\mathbf{B}_i^{p_i})$. So $\mathbf{AP}_i = \mathbf{P}_i(\mathbf{T}_i + \alpha_i\mathbf{I})$. Taking $\mathbf{P} = [\,\mathbf{P}_1 : \cdots : \mathbf{P}_k\,]$ we have $\mathbf{AP} = \mathbf{PJ}$ where

$$\mathbf{J} = (\mathbf{T}_1 + \alpha_1\mathbf{I}) \oplus \cdots \oplus (\mathbf{T}_k + \alpha_k\mathbf{I})$$

Clearly $\mathbf{T}_i + \alpha_i\mathbf{I}$ is a direct sum of $\mathbf{J}_{\alpha_i}$-blocks. Since $\sum a_i = n$, $\mathbf{P}$ is an $n \times n$ matrix.

It remains only to prove that $\mathbf{P}$ is non-singular. Since the columns of each $\mathbf{P}_i$ are linearly independent it is enough to prove that $\mathcal{C}(\mathbf{P}_1) + \cdots + \mathcal{C}(\mathbf{P}_k)$ is direct. Let $\mathbf{Q}_i$ denote the matrix obtained from $\mathbf{P}$ by deleting the columns of $\mathbf{P}_i$ and let $\mathbf{L}_i$ be the matrix obtained from $\mathbf{J}$ by deleting the corresponding rows and columns. Then to prove that $\mathcal{C}(\mathbf{P}_1) + \cdots + \mathcal{C}(\mathbf{P}_k)$ is direct, it is enough to show that $\mathcal{C}(\mathbf{P}_i) \cap \mathcal{C}(\mathbf{Q}_i) = \{\mathbf{0}\}$ for $i = 1, \ldots, k$. Now $\mathbf{AQ}_i = \mathbf{Q}_i\mathbf{L}_i$, so $\mathbf{B}_i\mathbf{Q}_i = \mathbf{Q}_i(\mathbf{L}_i - \alpha_i\mathbf{I})$. Using this repeatedly we get $\mathbf{B}_i^{p_i}\mathbf{Q}_i = \mathbf{Q}_i(\mathbf{L}_i - \alpha_i\mathbf{I})^{p_i}$. Now $\mathbf{L}_i - \alpha_i\mathbf{I}$ is non-singular since the α's are distinct, so we get $\rho(\mathbf{B}_i^{p_i}\mathbf{Q}_i) = \rho(\mathbf{Q}_i)$. Hence by *Theorem* 3.5.11, $\mathcal{N}(\mathbf{B}_i^{p_i}) \cap \mathcal{C}(\mathbf{Q}_i) = \{\mathbf{0}\}$. But $\mathcal{N}(\mathbf{B}_i^{p_i}) = \mathcal{C}(\mathbf{P}_i)$, so $\mathcal{C}(\mathbf{P}_i) \cap \mathcal{C}(\mathbf{Q}_i) = \{\mathbf{0}\}$ follows. ∎

A matrix which is a direct sum of Jordan blocks is said to be in *Jordan canonical form* or, simply, *Jordan form*. One may also insist that the

Jordan blocks corresponding to the same α_i occur consecutively and their orders are decreasing as one reads from left to right. Clearly every matrix is similar to a matrix in such a Jordan form.

We now show that any given matrix is similar to a matrix in Jordan form which is *unique* except for a rearrangement of the Jordan blocks in it. Suppose $\mathbf{P}^{-1}\mathbf{AP} = \mathbf{J}$ where $\mathbf{J}$ is the direct sum of the matrices $\mathbf{J}_{\alpha_i}(n_{ij}) : j = 1, \ldots, g_i$ and $i = 1, \ldots, k$, with $\alpha_1, \ldots, \alpha_k$ distinct and $n_{i1} \geq n_{i2} \geq \cdots \geq n_{ig_i}$ for each i. Then, clearly, $\alpha_1, \ldots, \alpha_k$ are the distinct eigenvalues of $\mathbf{A}$ and the algebraic multiplicity a_i of α_i with respect to $\mathbf{A}$ is $\sum_{j=1}^{g_i} n_{ij}$. Let us denote $\mathbf{J} - \alpha_i\mathbf{I}$ by $\mathbf{J}_i$. Then we have $\mathbf{P}^{-1}(\mathbf{A} - \alpha_i\mathbf{I})^\ell\mathbf{P} = \mathbf{J}_i^\ell$ and so $\rho((\mathbf{A} - \alpha_i\mathbf{I})^\ell) = \rho(\mathbf{J}_i^\ell)$ for all $\ell \geq 0$. Now *fix any* i. Then for all $\ell \geq 1$, we have

$$\rho(\mathbf{J}_i^{\ell-1}) = \sum_{j=1}^{g_i} \max(n_{ij} - \ell + 1, 0) + (n - a_i) \tag{8.6.6}$$

and

$$\rho(\mathbf{J}_i^{\ell}) = \sum_{j=1}^{g_i} \max(n_{ij} - \ell, 0) + (n - a_i) \tag{8.6.7}$$

Now $\max(n_{ij} - \ell + 1, 0) - \max(n_{ij} - \ell, 0)$ is 1 or 0 according as $n_{ij} \geq \ell$ or not. So, subtracting (8.6.7) from (8.6.6) we see that the number m_ℓ of n_{ij}'s which are $\geq \ell$ is $\rho((\mathbf{A} - \alpha_i\mathbf{I})^{\ell-1}) - \rho((\mathbf{A} - \alpha_i\mathbf{I})^\ell)$. Thus the sequence $m_1, m_2, m_3, \ldots$ is uniquely determined. Since all n_{ij}'s are ≥ 1, it follows that $g_i = m_1$. Now $n_{i1}, n_{i2}, \ldots, n_{ig_i}$ are uniquely determined since the number of times ℓ appears among them is $m_\ell - m_{\ell+1}$ for $\ell = 1, 2, 3, \ldots$. This proves the uniqueness of $\mathbf{J}$. Incidentally, (8.6.7) with $\ell = 1$ gives $\rho(\mathbf{J}_i) = n - g_i$, so $g_i = \nu(\mathbf{J}_i) = \nu(\mathbf{A} - \alpha_i\mathbf{I})$ is the geometric multiplicity of α_i with respect to $\mathbf{A}$. We have thus proved

Theorem 8.6.3 Any given square matrix is similar to a unique (unique upto rearrangement of the Jordan blocks) matrix in Jordan form.

The matrix in Jordan form which is similar to $\mathbf{A}$ is called the *Jordan form of* $\mathbf{A}$ or the *classical canonical form of* $\mathbf{A}$. The next theorem follows easily from the proof of *Theorem* 8.6.3.

Theorem 8.6.4 If $\mathbf{A}$ and $\mathbf{B}$ are square matrices of the same order, the following statements are equivalent:

(i) $\mathbf{A}$ and $\mathbf{B}$ are similar,

(ii) $\rho((\mathbf{A} - \alpha_i\mathbf{I})^\ell) = \rho((\mathbf{B} - \alpha_i\mathbf{I})^\ell)$ for every complex number α and every positive integer ℓ,

(iii) **A** and **B** have the same Jordan form.

Corollary Every square matrix is similar to its transpose.

Theorem 8.6.5 If the Jordan form of **A** is the direct sum of $\mathbf{J}_{\alpha_i}(n_{ij}) : j = 1, \ldots, g_i$ and $i = 1, \ldots, k$, where $\alpha_1, \ldots, \alpha_k$ are distinct, then the characteristic polynomial of **A** is $\prod_i (\lambda - \alpha_i)^{\sum_j n_{ij}}$ and the minimal polynomial of **A** is $\prod_i (\lambda - \alpha_i)^{m_i}$ where $m_i = \max_j n_{ij}$.

We leave the proof of this theorem as a simple exercise to the reader. The polynomials $(\lambda - \alpha_i)^{n_{ij}}$ are called the *elementary divisors of* **A**.

Exercises

1. Show that $\rho(\mathbf{J}_0(n))^\ell = \max(n - \ell, 0)$.
2. Show that the vectors in the first i columns of the table in *Figure* 8.6.1 form a basis of $\mathcal{N}(\mathbf{B}^i)$.
3. If $\mathbf{B}, \mathbf{P}$ and $\mathbf{T}$ are as in *Lemma* 8.6.1, show that $\mathcal{C}(\mathbf{P}) = \mathcal{N}(\mathbf{B}^p)$ where p is the index of **B**.
4. In *Lemma* 8.6.1, show that the number of $\mathbf{J}_0$-blocks in **T** is the geometric multiplicity g of 0 with respect to **B**.
5. Find the Jordan canonical forms (and the transforming matrices) of (i) $\begin{bmatrix} 1 & 2 & 0 \\ 0 & 0 & 1 \\ 0 & 0 & 1 \end{bmatrix}$, (ii) $\begin{bmatrix} 1 & 0 & 2 \\ 0 & 0 & 0 \\ 0 & 0 & 1 \end{bmatrix}$ and (iii) $\begin{bmatrix} 1 & 0 & 0 \\ 0 & 0 & 2 \\ 0 & 0 & 1 \end{bmatrix}$.
6. Prove *Theorems* 8.6.4 and 8.6.5.
7. Prove that **A** is similar to $\mathbf{A}^*$.
8. Show that any square matrix **A** can be written uniquely as $\mathbf{A} = \mathbf{D} + \mathbf{N}$ where **D** is semi-simple, **N** is nilpotent and both **D** and **N** are polynomials in **A**.
9. For any square matrix **A**, show that there exist unique matrices **R** and **M** such that $\mathbf{A} = \mathbf{R} + \mathbf{M}$, $\rho(\mathbf{R}) = \rho(\mathbf{R}^2)$, **M** is nilpotent and $\mathbf{RM} = \mathbf{MR} = \mathbf{0}$.

8.7 Spectral theorem

In this section, we prove the result, usually known as spectral theorem, that any real symmetric matrix **A** is orthogonally similar to a diagonal matrix. We also characterize matrices which are unitarily similar to diagonal matrices.

We had stated in the Introduction to this chapter that we would take the base field to be $\mathbb{C}$. However, the results of the first six sections remain valid over any algebraically closed field (i.e., a field over which every polynomial can be expressed as a product of linear factors). *The results of this and the next section will specifically require the base field to be* $\mathbb{C}$ (or $\mathbb{R}$ sometimes). $\mathbf{A}$ will continue to denote an $n \times n$ complex matrix. The inner product used is the canonical inner product, viz., $\langle \mathbf{x}, \mathbf{y} \rangle = \mathbf{y}^*\mathbf{x}$ on $\mathbb{C}^n$ and $\langle \mathbf{x}, \mathbf{y} \rangle = \mathbf{y}^{\mathrm{T}}\mathbf{x}$ on $\mathbb{R}^n$. The facts

$$\langle \mathbf{A}\mathbf{x}, \mathbf{y} \rangle = \langle \mathbf{x}, \mathbf{A}^*\mathbf{y} \rangle \quad \text{and} \quad \langle \mathbf{x}, \mathbf{A}\mathbf{y} \rangle = \langle \mathbf{A}^*\mathbf{x}, \mathbf{y} \rangle$$

will be used repeatedly in this section.

Theorem 8.7.1 The eigenvalues of a hermitian matrix and hence of a real symmetric matrix are all real.

Proof Let α be an eigenvalue of a hermitian matrix $\mathbf{A}$ and let $\mathbf{x}$ be an eigenvector of $\mathbf{A}$ corresponding to α. Then

$$\langle \mathbf{A}\mathbf{x}, \mathbf{x} \rangle = \langle \alpha\mathbf{x}, \mathbf{x} \rangle = \alpha\langle \mathbf{x}, \mathbf{x} \rangle$$

But

$$\begin{aligned} \langle \mathbf{A}\mathbf{x}, \mathbf{x} \rangle &= \langle \mathbf{x}, \mathbf{A}^*\mathbf{x} \rangle = \langle \mathbf{x}, \mathbf{A}\mathbf{x} \rangle \quad \text{since } \mathbf{A} \text{ is hermitian} \\ &= \langle \mathbf{x}, \alpha\mathbf{x} \rangle = \overline{\alpha}\langle \mathbf{x}, \mathbf{x} \rangle \end{aligned}$$

Since $\langle \mathbf{x}, \mathbf{x} \rangle \neq 0$ it follows that $\alpha = \overline{\alpha}$ and α is real. ∎

Corollary The eigenvalues of a skew-hermitian (and so of a real skew-symmetric) matrix $\mathbf{A}$ are purely imaginary. Consequently $\mathbf{I} - \mathbf{A}$ and $\mathbf{I} + \mathbf{A}$ are non-singular.

Proof Since $\mathbf{A}$ is skew-hermitian we have

$$(i\mathbf{A})^* = \overline{i}\mathbf{A}^* = (-i)(-\mathbf{A}) = i\mathbf{A}$$

Thus $i\mathbf{A}$ is hermitian, so the eigenvalues of $i\mathbf{A}$ are real. Since β is an eigenvalue of $\mathbf{A}$ iff $i\beta$ is an eigenvalue of $i\mathbf{A}$, the corollary follows. ∎

Theorem 8.7.2 (*Spectral theorem for real symmetric matrices*) Any real symmetric matrix is orthogonally similar to a diagonal matrix.

Proof We imitate the proof of *Theorem* 8.3.10. The result holds trivially for a matrix of order 1. So assume it for matrices of order $n-1$ and let $\mathbf{A}$ be a real symmetric matrix of order n. Let α be an eigenvalue of $\mathbf{A}$. By the preceding theorem, α is real. Let $\mathbf{x}$ be a normalized real eigenvector of $\mathbf{A}$ corresponding to α. Then there exists an orthogonal matrix $\mathbf{P}$ with $\mathbf{x}$ as the first column. Then as in the proof of *Theorem* 8.3.10 we have

$$\mathbf{P}^{-1}\mathbf{A}\mathbf{P} = \begin{bmatrix} \alpha & \mathbf{y}^{\mathrm{T}} \\ \mathbf{0} & \mathbf{C} \end{bmatrix}$$

for some vector $\mathbf{y}^{\mathrm{T}}$ and some $(n-1) \times (n-1)$ matrix $\mathbf{C}$. Now $\mathbf{P}^{-1}\mathbf{A}\mathbf{P} = \mathbf{P}^{\mathrm{T}}\mathbf{A}\mathbf{P}$ is symmetric, so $\mathbf{C}$ is symmetric and $\mathbf{y}^{\mathrm{T}} = \mathbf{0}$. So by induction hypothesis there exists an orthogonal matrix $\mathbf{W}$ of order $n - 1$ such that $\mathbf{D} := \mathbf{W}^{-1}\mathbf{C}\mathbf{W}$ is diagonal. Now $\mathbf{Q} := \mathrm{diag}(1, \mathbf{W})$ and so $\mathbf{PQ}$ are orthogonal and

$$(\mathbf{PQ})^{-1}\mathbf{A}(\mathbf{PQ}) = \mathbf{Q}^{-1}\mathbf{P}^{-1}\mathbf{A}\mathbf{P}\mathbf{Q} = \begin{bmatrix} 1 & \mathbf{0} \\ \mathbf{0} & \mathbf{W}^{-1} \end{bmatrix}\begin{bmatrix} \alpha & \mathbf{0} \\ \mathbf{0} & \mathbf{C} \end{bmatrix}\begin{bmatrix} 1 & \mathbf{0} \\ \mathbf{0} & \mathbf{W} \end{bmatrix} = \begin{bmatrix} \alpha & \mathbf{0} \\ \mathbf{0} & \mathbf{D} \end{bmatrix}$$

is diagonal. ∎

It is easy to see that the converse of the preceding theorem also holds. The theorem itself says that if $\mathbf{A}$ is symmetric, then the matrix of the linear operator $\mathbf{x} \mapsto \mathbf{A}\mathbf{x}$ with respect to some othonormal basis is diagonal. Orthogonal similarity is very useful since $\mathbf{P}^{-1} = \mathbf{P}^{\mathrm{T}}$, as we will see in the next chapter. We next prove spectral theorem for hermitian matrices.

Theorem 8.7.3 (*Spectral theorem for hermitian matrices*) Any hermitian matrix is unitarily similar to a real diagonal matrix.

The proof of this theorem is similar to that of the preceding theorem. We only note that α would be real and $\mathbf{P}$ unitary, so $\mathbf{P}^{-1}\mathbf{A}\mathbf{P} = \mathbf{P}^{*}\mathbf{A}\mathbf{P}$ would be hermitian. The converse of this theorem also holds. We now describe the spectral decomposition of a hermitian matrix.

Theorem 8.7.4 Let $\mathbf{A}$ be an $n \times n$ hermitian matrix with rank r. Then $\mathbf{A}$ can be represented in each of the following equivalent forms:

(i) There exists a unitary matrix $\mathbf{P}$ of order n and a real diagonal non-singular matrix $\mathbf{\Delta}$ of order r such that

$$\mathbf{A} = \mathbf{P}\begin{bmatrix} \mathbf{\Delta} & \mathbf{0} \\ \mathbf{0} & \mathbf{0} \end{bmatrix}\mathbf{P}^{*} \tag{8.7.1}$$

(ii) There exist non-zero real numbers $\delta_1, \delta_2, \ldots, \delta_r$ (not necessarily distinct) and orthonormal vectors $\mathbf{u}_1, \mathbf{u}_2, \ldots, \mathbf{u}_r \in \mathbb{C}^n$ such that

$$\mathbf{A} = \textstyle\sum_{i=1}^{r} \delta_i\, \mathbf{u}_i \mathbf{u}_i^* \tag{8.7.2}$$

(iii) There exist matrices $\mathbf{R}$ and $\mathbf{\Delta}$ of orders $n \times r$ an $r \times r$ respectively such that $\mathbf{\Delta}$ is real, diagonal and non-singular, $\mathbf{R}^*\mathbf{R} = \mathbf{I}$ and

$$\mathbf{A} = \mathbf{R}\mathbf{\Delta}\mathbf{R}^* \tag{8.7.3}$$

(iv) There exist unique non-null orthogonal projectors $\mathbf{E}_1$, $\mathbf{E}_2, \ldots, \mathbf{E}_k$ and distinct real numbers $\alpha_1, \alpha_2, \ldots, \alpha_k$ such that

$$\mathbf{A} = \textstyle\sum_{i=1}^{k} \alpha_i \mathbf{E}_i \text{ and } \mathbf{I} = \sum_{i=1}^{k} \mathbf{E}_i \tag{8.7.4}$$

This theorem follows from the preceding theorem and *Theorems* 8.5.3 and 8.5.5. We note in passing that in (8.7.4), $\mathbf{E}_i\mathbf{E}_j = \mathbf{0}$ whenever $i \neq j$.

Remark The preceding theorem holds for a real symmetric matrix $\mathbf{A}$ if all the matrices and vectors are assumed to be over $\mathbb{R}$.

To prove spectral theorem for the largest possible class of complex matrices, we start with a stronger form of *Theorem* 8.3.10.

Theorem 8.7.5 (*Schur, Jacobi*) Every complex matrix $\mathbf{A}$ is unitarily similar to an upper triangular matrix.

The proof of this theorem imitates that of *Theorem* 8.3.10 (see the proof of *Theorem* 8.7.2 above) and is left to the reader. The corresponding result for real matrices, namely that every real matrix $\mathbf{A}$ is orthogonally similar to an upper triangular matrix, is false since $\mathbf{A}$ may not have real eigenvalues and real eigenvectors. However, it can be proved that every real matrix $\mathbf{A}$ with real eigenvalues is orthogonally similar to an upper triangular matrix.

We now determine the matrices which are unitarily similar to diagonal matrices.† Suppose $\mathbf{A} = \mathbf{U}^{-1}\mathbf{D}\mathbf{U}$ where $\mathbf{U}$ is unitary and $\mathbf{D}$ is diagonal. Then $\mathbf{U}^{-1} = \mathbf{U}^*$, so we have

$$\mathbf{A}\mathbf{A}^* = (\mathbf{U}^*\mathbf{D}\mathbf{U})(\mathbf{U}^*\overline{\mathbf{D}}\mathbf{U}) = \mathbf{U}^*\mathbf{D}\overline{\mathbf{D}}\mathbf{U} = \mathbf{U}^*\overline{\mathbf{D}}\mathbf{D}\mathbf{U} = \mathbf{A}^*\mathbf{A}$$

It turns out that this condition is also sufficient for $\mathbf{A}$ to be unitarily similar to a diagonal matrix. We give a name to such matrices in

†This and the remaining results in this section may be omitted in a first course.

Definition 8.7.6 A matrix $\mathbf{A}$ is said to be *normal* if $\mathbf{AA}^* = \mathbf{A}^*\mathbf{A}$.

We now show that every normal matrix is unitarily similar to a diagonal matrix. We will deduce this from *Theorem* 8.7.5 and the following

Lemma 8.7.7 An upper triangular matrix is normal iff it is diagonal.

Proof The *if part* is trivial. To prove the *only if part*, let $\mathbf{T}$ be an upper triangular normal matrix of order n. Then equating the k-th diagonal entries of $\mathbf{T}^*\mathbf{T}$ and $\mathbf{TT}^*$ we get

$$\sum_{i=1}^{k} |t_{ik}|^2 = \sum_{j=k}^{n} |t_{kj}|^2$$

Taking $k = 1$ we get $t_{1j} = 0$ for $j \geq 2$. Then taking $k = 2$ we get $t_{2j} = 0$ for $j \geq 3$. Proceeding thus we see that $\mathbf{T}$ is diagonal. ∎

Theorem 8.7.8 A matrix is unitarily similar to a diagonal matrix iff it is normal.

Proof The *only if part* has been proved above. To prove the *if part*, let $\mathbf{A}$ be normal. By *Theorem* 8.7.5, there exists a unitary matrix $\mathbf{U}$ such that $\mathbf{T} = \mathbf{U}^{-1}\mathbf{AU}$ is upper triangular. Now it is easy to see that $\mathbf{TT}^* = \mathbf{U}^*\mathbf{AA}^*\mathbf{U} = \mathbf{U}^*\mathbf{A}^*\mathbf{AU} = \mathbf{T}^*\mathbf{T}$. So by the preceding lemma, $\mathbf{T}$ is diagonal. ∎

We note that the class of normal matrices includes hermitian and so real symmetric matrices, skew-hermitian and so real skew-symmetric matrices, unitary and so orthogonal matrices and, of course, diagonal matrices. It follows that all these types of matrices are semi-simple.

Let $\mathbf{A}$ be an $n \times n$ normal matrix and let $\mathbf{A} = \mathbf{UDU}^*$ where $\mathbf{D}$ is diagonal and $\mathbf{U}$ is unitary. Without any loss of generality assume that $\mathbf{D} = \text{diag}(\delta_1\mathbf{I}_{n_1}, \ldots, \delta_k\mathbf{I}_{n_k})$ where $\delta_1, \ldots, \delta_k$ are distinct. Then

$$\mathbf{A} = \mathbf{UDU}^* = \delta_1\mathbf{U}_1\mathbf{U}_1^* + \delta_2\mathbf{U}_2\mathbf{U}_2^* + \cdots + \delta_k\mathbf{U}_k\mathbf{U}_k^*$$

where $\mathbf{U} = [\,\mathbf{U}_1 : \cdots : \mathbf{U}_k\,]$ and $\mathbf{U}_i$ is of order $n \times n_i$. Let $\mathbf{E}_i = \mathbf{U}_i\mathbf{U}_i^*$ for $i = 1, \ldots, k$. Then it is easy to see that $\mathbf{E}_i$'s are orthogonal projectors and $\sum_{i=1}^k \mathbf{E}_i = \mathbf{UU}^* = \mathbf{I}$. Thus $\sum_{i=1}^k \delta_i\mathbf{E}_i$ is the spectral form of $\mathbf{A}$. Clearly, $\mathbf{E}_i^* = \mathbf{E}_i = \mathbf{E}_i^2$ and $\mathbf{E}_i^*\mathbf{E}_j = \mathbf{E}_i\mathbf{E}_j = \mathbf{0}$ whenever $i \neq j$.

Theorem 8.7.9 Let $\sum_{i=1}^k \alpha_i\mathbf{E}_i$ be the spectral form of a semi-simple matrix $\mathbf{A}$ and let S_i be the eigenspace of $\mathbf{A}$ corresponding to α_i. Then the following statements are equivalent.

(i) $\mathbf{A}$ is normal,

(ii) $\mathbf{E}_i^*\mathbf{E}_j = \mathbf{0}$ or, equivalently, S_i is orthogonal to S_j whenever $i \neq j$,

(iii) $\mathbf{E}_i$ is hermitian or, equivalently, $\mathbf{E}_i$ is the orthogonal projector into S_i for all i.

Proof That (i) implies (ii) follows from the above discussion.

(ii) $\Rightarrow$ (iii) Taking adjoints of the two sides of $\mathbf{E}_1 + \cdots + \mathbf{E}_k = \mathbf{I}$ and postmultiplying by $\mathbf{E}_i$ we get $\mathbf{E}_i^*\mathbf{E}_i = \mathbf{E}_i$. Since $\mathbf{E}_i^*\mathbf{E}_i$ is hermitian, so is $\mathbf{E}_i$. The equivalence of the two statements in (iii) follows from *Theorem* 7.5.12. (That $\mathbf{E}_i$ is the orthogonal projector into S_i can also be deduced from the fact that $\sum_{j\neq i} S_j$ is the orthogonal complement of S_i).

(iii) $\Rightarrow$ (i) By hypothesis $\mathbf{E}_i^* = \mathbf{E}_i$, so $\mathbf{A}^* = \sum \overline{\alpha}_i\mathbf{E}_i$. Since $\mathbf{E}_i\mathbf{E}_j = \mathbf{0}$ whenever $i \neq j$ and $\mathbf{E}_i^2 = \mathbf{E}_i$, we have $\mathbf{A}\mathbf{A}^* = \sum \alpha_i\overline{\alpha}_i\mathbf{E}_i = \sum \overline{\alpha}_i\alpha_i\mathbf{E}_i = \mathbf{A}^*\mathbf{A}$. ∎

The preceding theorem, combined with the fact that a normal matrix is semi-simple, has the following important consequence. If $\alpha_1, \alpha_2, \ldots, \alpha_k$ are the distinct eigenvalues of a normal matrix $\mathbf{A}$ and if $\mathbf{x}_{i1}, \mathbf{x}_{i2}, \ldots, \mathbf{x}_{in_i}$ form an orthonormal basis of the eigenspace of $\mathbf{A}$ corresponding to α_i for $i = 1, \ldots, k$ then

$$\mathbf{U}^{-1}\mathbf{A}\mathbf{U} = \operatorname{diag}(\alpha_1\mathbf{I}_{n_1}, \alpha_2\mathbf{I}_{n_2}, \ldots, \alpha_k\mathbf{I}_{n_k})$$

where $\mathbf{U}$ is the unitary matrix

$$[\mathbf{x}_{11} \cdots \mathbf{x}_{1n_1}\mathbf{x}_{21} \cdots \mathbf{x}_{2n_2} \cdots \mathbf{x}_{k1} \cdots \mathbf{x}_{kn_k}]$$

If $\mathbf{A}$ is real symmetric then $\alpha_1, \ldots, \alpha_k$ are all real and the $\mathbf{x}_{ij}$'s can be chosen to be real; the matrix $\mathbf{U}$ will then be an orthogonal matrix.

Characterizations similar to those given in the preceding theorem can be given for various types of matrices included in the class of normal matrices. We leave it to the reader to prove the following: *Theorem* 8.7.9 remains true if 'normal' is replaced by 'hermitian' in (i) and the condition that α_i's are real is added to both (ii) and (iii). The corresponding result for unitary matrices requires the following

Theorem 8.7.10 Each eigenvalue of a unitary matrix has unit modulus.

Proof Let α be an eigenvalue of the unitary matrix $\mathbf{A}$ and let $\mathbf{x}$ be an eigenvector of $\mathbf{A}$ corresponding to α. Then

$$||\mathbf{x}|| = ||\mathbf{A}\mathbf{x}|| = ||\alpha\mathbf{x}|| = |\alpha| \cdot ||\mathbf{x}||$$

Since $\mathbf{x} \neq \mathbf{0}$ it follows that $|\alpha| = 1$. ■

It can now be proved easily that *Theorem* 8.7.9 remains true if 'normal' is replaced by 'unitary' in (i) and the condition that α_i's have unit modulus is added to both (ii) and (iii). If 'unitary' is replaced by 'orthogonal', the condition that $\mathbf{A}$ is real should also be added to (ii) and (iii). Note that α_i's and $\mathbf{E}_i$'s need not be real.

We now characterise geometrically the orthogonal transformations in $\mathbb{R}^3$ (see the discussion after *Theorem* 7.6.7).

Theorem 8.7.11 Let $\mathbf{A}$ be a 3×3 orthogonal matrix. Then $f : \mathbf{x} \mapsto \mathbf{Ax}$ is a rotation of space about some line through the origin by some angle if $|\mathbf{A}| = 1$ and a rotation about some line through the origin by some angle followed by inversion in the origin if $|\mathbf{A}| = -1$.

Proof: By inversion in the origin we mean $\iota : \mathbf{x} \mapsto -\mathbf{x}$. Note that ι commutes with every linear operator, so the inversion may follow or precede the rotation.

Since $\mathbf{A}$ is of order 3, it has a real eigenvalue. Since every eigenvalue of an orthogonal matrix has unit modulus, it follows that either 1 or -1 is an eigenvalue of $\mathbf{A}$.

Case 1. 1 is an eigenvalue of $\mathbf{A}$. Let $\mathbf{u}$ be an eigenvector of $\mathbf{A}$ with unit norm, corresponding to the eigenvalue 1. Let $\mathbf{P} = [\,\mathbf{u} : \mathbf{v} : \mathbf{w}\,]$ be an orthogonal matrix with $\mathbf{u}$ as the first column. Then $\mathbf{P}^{-1}\mathbf{AP}$ is

$$\mathbf{C} := \begin{bmatrix} 1 & \mathbf{0}^{\mathrm{T}} \\ \mathbf{0} & \mathbf{B} \end{bmatrix}$$

for some 2×2 orthogonal matrix $\mathbf{B}$ since $\mathbf{P}^{-1}\mathbf{AP}$ is orthogonal. Now let R be the point corresponding to $\mathbf{x}$ in $\mathbb{R}^3$ and let $\mathcal{B}$ be the orthonormal basis $\{\mathbf{u}, \mathbf{v}, \mathbf{w}\}$. Then the coordinate vector of R w.r.t. $\mathcal{B}$ is $\mathbf{P}^{-1}\mathbf{x}$.

By *Example* 7.6.2, $\mathbf{B} = \mathbf{A}_\theta$ or $\mathbf{B}_\theta$. Also, $\mathbf{z} \mapsto \mathbf{A}_\theta\,\mathbf{z}$ is rotation of the plane about the origin by angle θ and $\mathbf{z} \mapsto \mathbf{B}_\theta\,\mathbf{z}$ is reflection of the plane in the line passing through the origin and making an angle $\theta/2$ with the first axis. We now consider the cases $\mathbf{B} = \mathbf{A}_\theta$ and $\mathbf{B} = \mathbf{B}_\theta$ separately.

Case 1.1. $\mathbf{B} = \mathbf{A}_\theta$. Let S be the image of R when 3-dimensional space is rotated about the line $O\mathbf{u}$ by angle θ. Then the coordinate vector of S w.r.t. $\mathcal{B}$ (note that $O\mathbf{u}$ is the first axis of $\mathcal{B}$) is $\mathbf{CP}^{-1}\mathbf{x}$, so the coordinate vector of S w.r.t. the original coordinate system is $\mathbf{PCP}^{-1}\mathbf{x} = \mathbf{Ax}$. Thus $\mathbf{x} \mapsto \mathbf{Ax}$ is the rotation of space about $O\mathbf{u}$ by angle θ. Note that in this case $|\mathbf{A}| = |\mathbf{C}| = 1$.

Case 1.2. $\mathbf{B} = \mathbf{B}_\theta$. Then

$$\mathbf{P}^{-1}(-\mathbf{A})\mathbf{P} = \mathbf{D} := \begin{bmatrix} -1 & \mathbf{0}^{\mathrm{T}} \\ \mathbf{0} & -\mathbf{B}_\theta \end{bmatrix} = \begin{bmatrix} -1 & \mathbf{0}^{\mathrm{T}} \\ \mathbf{0} & \mathbf{B}_\eta \end{bmatrix}$$

where $\eta = \pi + \theta$. Now let T be the image of R when 3-dimensional space is rotated by 180° about the line L passing through O, lying in the plane $O\mathbf{vw}$ and making an angle $\eta/2$ with $O\mathbf{v}$. Let R' (resp. T') be the foot of the perpendicular from R (resp. T) to the plane $O\mathbf{vw}$. Then it is easy to see that T' is the reflection of R' in L and $R'R = -T'T$. (Note that the plane containing R, R', T and T' is perpendicular to L.) So the coordinate vector of T w.r.t. $\mathcal{B}$ is $\mathbf{DP}^{-1}\mathbf{x}$ and the coordinate vector of T w.r.t. the original coordinate system is $\mathbf{PDP}^{-1}\mathbf{x} = -\mathbf{Ax}$. Thus $\mathbf{x} \mapsto \mathbf{Ax}$ is now the rotation of space by 180° about L followed by inversion in the origin. Note that, now, $|\mathbf{A}| = -|\mathbf{D}| = -1$.

Case 2. 1 is not an eigenvalue of $\mathbf{A}$. Then -1 is an eigenvalue of $\mathbf{A}$ and 1 is an eigenvalue of $-\mathbf{A}$. Also, $-\mathbf{A}$ is orthogonal. So $\mathbf{x} \mapsto -\mathbf{Ax}$ is a rotation or a rotation followed by inversion in the origin according as $|-\mathbf{A}| = 1$ or -1. Since ι commutes with rotation and ι^2 is the identity map, it follows that $\mathbf{x} \mapsto \mathbf{Ax}$ is a rotation or a rotation followed by inversion in the origin according as $|\mathbf{A}| = 1$ or -1. ∎

Orthogonal transformations of $\mathbb{R}^3$ can also be characterized through reflections in planes, see *Exercise* 8.7.17.

Exercises

1. Find the spectral form of $\frac{1}{\sqrt{2}}\begin{bmatrix} 1 & -1 \\ 1 & 1 \end{bmatrix}$.

2. Find a unitary matrix (if possible, an orthogonal matrix) $\mathbf{P}$ such that $\mathbf{P}^*\mathbf{AP}$ is upper triangular, where

$$\mathbf{A} = \begin{bmatrix} -7 & -13 & -5 \\ 2 & 5 & -5 \\ -8 & -2 & 11 \end{bmatrix}$$

3. Find an orthogonal matrix $\mathbf{P}$ such that $\mathbf{P}^{\mathrm{T}}\mathbf{AP}$ is diagonal, where

$$\mathbf{A} = \begin{bmatrix} 5 & -4 & -2 \\ -4 & 5 & -2 \\ -2 & -2 & 8 \end{bmatrix}$$

4. Let $\mathbf{A}$ be a real skew-symmetric matrix of order n.
   (a) If n is odd, show that $|\mathbf{A}| = 0$.
   (b) If n is even, show that $|\mathbf{A}| \geq 0$.
   (c) For any n, show that $|\mathbf{I} + \mathbf{A}| \geq 1$.
5. (a) Let $\mathbf{A} = \mathbf{x}\mathbf{x}^{\mathrm{T}}$ where $\mathbf{x} = (1, 2, 1, 4)^{\mathrm{T}}$. Obtain a spectral decomposition of $\mathbf{A}$ in the form (8.7.1), where $\mathbf{P}$ is orthogonal.
   (b) Find the spectral form of the $(k+1)\times(k+1)$ matrix $\mathbf{B} = \begin{bmatrix} \mathbf{0} & \mathbf{1} \\ \mathbf{1}^{\mathrm{T}} & 0 \end{bmatrix}$.
   (Hint: use rank-factorization.)
6. (a) Given a spectral decomposition of a real symmetric matrix $\mathbf{A}$ in any of the forms (8.7.1) through (8.7.4), find the corresponding spectral decomposition of $a\mathbf{I} + b\mathbf{A}$ where a and b are real numbers.
   (b) Obtain a spectral decomposition of $(a - b)\mathbf{I} + b\mathbf{1}\mathbf{1}^{\mathrm{T}}$ in each of the forms (8.7.1) and (8.7.4).
7. Let $\mathbf{A} = \frac{1}{15}\begin{bmatrix} 11 & -2 & -2 & -6 \\ -2 & 14 & -1 & -3 \\ -2 & -1 & 14 & -3 \\ -6 & -3 & -3 & 6 \end{bmatrix}$.
   (a) Show that $\mathbf{A}$ is an orthogonal projector.
   (b) Obtain a spectral decomposition of $\mathbf{A}$ in the form (8.7.2).
   (c) Hence obtain a spectral decomposition of $\mathbf{I} - \mathbf{A}$.
8. Find a normal matrix which is none of: hermitian, skew-hermitian, unitary and diagonal.
9. Show that a normal matrix is unitary iff every eigenvalue has unit modulus.
10. Prove theorems similar to *Theorem* 8.7.8 characterizing hermitian, real symmetric, skew-hermitian, unitary and orthogonal matrices.
11. Prove or disprove: every complex symmetric matrix is normal.
12. Show that the $n \times n$ matrix $\mathbf{1}\mathbf{1}^{\mathrm{T}}$ is similar to the $n \times n$ matrix $\begin{bmatrix} n & \mathbf{0} \\ \mathbf{0} & \mathbf{0} \end{bmatrix}$.
13. Show that $\mathbf{A}$ is an orthogonal projector iff $\mathbf{A}$ is hermitian and each eigenvalue of $\mathbf{A}$ belongs to $\{0, 1\}$.
14. Prove that $\mathbf{A}$ is normal iff $||\mathbf{A}^*\mathbf{x}|| = ||\mathbf{A}\mathbf{x}||$ for all $\mathbf{x}$.
15. Let $\mathbf{A}$ be a real symmetric matrix.
   (a) If $\mathbf{A}^k = \mathbf{I}$ for some positive integer k, show that $\mathbf{A}^2 = \mathbf{I}$.
   (b) If the eigenvalues of $\mathbf{A}$ are all positive and if $\mathbf{A}^k = \mathbf{I}$ for some positive integer k then show that $\mathbf{A} = \mathbf{I}$.
   (c) If $\mathbf{A}^k = \mathbf{0}$ for some positive integer k, then show that $\mathbf{A} = \mathbf{0}$.

16. **A** is said to be *range-hermitian* if $\mathcal{C}(\mathbf{A}) = \mathcal{C}(\mathbf{A}^*)$.
    (a) Let **A** be range-hermitian. Show that $\mathcal{N}(\mathbf{A}) = \mathcal{N}(\mathbf{A}^*)$, $\rho(\mathbf{A}) = \rho(\mathbf{A}^2)$ and $\overline{\mathbf{A}}$ and $\mathbf{A}^{\mathrm{T}}$ are range-hermitian.
    (b) Show that every normal matrix is range-hermitian.
17. (a) Show that any translation by distance d in 3-dimensional space is the resultant of reflections in two planes perpendicular to the direction of the translation and separated by distance $d/2$.
    (b) Show that a rotation about a line through the origin is the resultant of reflections in two planes containing the line. (Hint: $\mathbf{A}_\theta = \mathbf{B}_\theta \mathbf{B}_0$ in the notation of *Example* 7.6.2.)
    (c) Show that inversion in the origin is the resultant of reflections in three planes through the origin.
    (d) If **A** is a 3×3 orthogonal matrix, show that $f : \mathbf{x} \mapsto \mathbf{A}\mathbf{x}$ is the resultant of three or less reflections in planes.

8.8 Singular value decomposition*

Unlike in the earlier sections **A** *will denote an* $m \times n$ *matrix throughout this section.* If $\rho(\mathbf{A}) = r$, we saw in *Section* 4.5 that the normal form $\begin{bmatrix} \mathbf{I}_r & \mathbf{0} \\ \mathbf{0} & \mathbf{0} \end{bmatrix}$ is a canonical form of **A** under equivalence. In this section we obtain a canonical form under what might be called unitary equivalence, viz., $\mathbf{B} = \mathbf{UAV}$ where **U** and **V** are unitary. The significance of such equivalence is that if **A** represents a linear transformation f from $\mathbb{C}^n$ to $\mathbb{C}^m$ with respect to orthonormal bases $\mathcal{X}$ and $\mathcal{Y}$, **B** represents f with respect to some orthonormal bases $\mathcal{X}'$ and $\mathcal{Y}'$. Note that the canonical form we are seeking cannot be the normal form because that would imply $|\det \mathbf{A}| = 1$ if **A** is non-singular.

Definition 8.8.1 A *singular value decomposition* of an $m \times n$ matrix **A** is a representation of **A** in the form

$$\mathbf{A} = \mathbf{U} \begin{bmatrix} \mathbf{\Delta} & \mathbf{0} \\ \mathbf{0} & \mathbf{0} \end{bmatrix} \mathbf{V}^* \tag{8.8.1}$$

where **U** and **V** are unitary matrices and $\mathbf{\Delta}$ is a diagonal matrix with (strictly) positive diagonal entries.

We note that, following common notation, we used $\mathbf{V}^*$ instead of **V** in (8.8.1). Before we prove its existence, we give two other forms of a singular value decomposition of $\mathbf{A}$ in the spirit of *Theorem* 8.5.3 and establish the equivalence.

Given a singular value decomposition of a matrix $\mathbf{A}$ in the form (8.8.1), let d_i be the i-th diagonal entry of $\mathbf{\Delta}$, $\mathbf{u}_i$ be the i-th column of $\mathbf{U}$ and $\mathbf{v}_i$ be the i-th column of $\mathbf{V}$ for $i = 1, \ldots, r$, where r is the order of $\mathbf{\Delta}$. Then

$$\mathbf{A} = \sum_{i=1}^{r} d_i\, \mathbf{u}_i \mathbf{v}_i^* \tag{8.8.2}$$

where $d_1, \ldots, d_r$ are (strictly) positive real numbers and $\{\mathbf{u}_1, \ldots, \mathbf{u}_r\}$ and $\{\mathbf{v}_1, \ldots, \mathbf{v}_r\}$ are orthonormal subsets of $\mathbb{C}^m$ and $\mathbb{C}^n$ respectively. We say that $d_1, \ldots, d_r$ are the *singular values of* $\mathbf{A}$ and $\mathbf{u}_i$ and $\mathbf{v}_i$ are *singular vectors of* $\mathbf{A}$ *corresponding to* d_i.

Given a singular value decomposition of $\mathbf{A}$ in the form (8.8.2), let $\mathbf{\Delta} = \text{diag}(d_1, \ldots, d_r)$, $\mathbf{R} = [\,\mathbf{u}_1 : \cdots : \mathbf{u}_r\,]$ and $\mathbf{S} = [\,\mathbf{v}_1 : \cdots : \mathbf{v}_r\,]$. Then

$$\mathbf{A} = \mathbf{R\Delta S}^* \tag{8.8.3}$$

where $\mathbf{\Delta}$ is an $r \times r$ diagonal matrix with positive diagonal entries and $\mathbf{R}$, $\mathbf{S}$ satisfy $\mathbf{R}^*\mathbf{R} = \mathbf{S}^*\mathbf{S} = \mathbf{I}_r$.

Finally, given a singular value decomposition of $\mathbf{A}$ in the form (8.8.3), let $\mathbf{U}$ and $\mathbf{V}$ be unitary matrices such that the submatrices of $\mathbf{U}$ and $\mathbf{V}$ formed by the first r columns are $\mathbf{R}$ and $\mathbf{S}$ respectively. Then clearly $\mathbf{U}\begin{bmatrix} \mathbf{\Delta} & \mathbf{0} \\ \mathbf{0} & \mathbf{0} \end{bmatrix}\mathbf{V}^*$ is a singular value decomposition of $\mathbf{A}$ in the form (8.8.1). This proves the equivalence of the three forms (8.8.1), (8.8.2) and (8.8.3).

It is easy to see that if $\mathbf{R\Delta S}^*$ is a singular value decomposition of $\mathbf{A}$ in the form (8.8.3), then $\mathbf{A} = \mathbf{RR}^*\mathbf{A}$ and $\mathbf{A} = \mathbf{ASS}^*$.

Theorem 8.8.2 Every matrix has a singular value decomposition.

Proof Let $\mathbf{A}$ be an $m \times n$ matrix with rank r. Then $\mathbf{AA}^*$ is hermitian and has rank r, so by *Theorem* 8.7.4 there exists a matrix $\mathbf{R}$ and a real non-singular diagonal matrix $\mathbf{\Lambda} = \text{diag}(\delta_1, \ldots, \delta_r)$ such that

$$\mathbf{AA}^* = \mathbf{R\Lambda R}^* \quad \text{and} \quad \mathbf{R}^*\mathbf{R} = \mathbf{I}$$

Clearly $\mathbf{\Lambda} = \mathbf{B}^*\mathbf{B}$ where $\mathbf{B} = \mathbf{A}^*\mathbf{R}$. Hence $\delta_j = \|\mathbf{B}_{*j}\|^2 > 0$ for $j = 1, \ldots, r$. Now define $\mathbf{S} = \mathbf{BG}$ where $\mathbf{G} = \text{diag}(1/\sqrt{\delta_1}, \ldots, 1/\sqrt{\delta_r})$. Then $\mathbf{S}^*\mathbf{S} = \mathbf{G\Lambda G} = \mathbf{I}$. Now let $\mathbf{H} = \mathbf{RG}^{-1}\mathbf{S}^*$. Then $\mathbf{H} = \mathbf{RG}^{-1}\mathbf{GR}^*\mathbf{A} = \mathbf{RR}^*\mathbf{A}$. So $\mathbf{HA}^* = \mathbf{RR}^*\mathbf{AA}^* = \mathbf{AA}^*$. Rank-cancelling $\mathbf{A}^*$ on the right, we get $\mathbf{H} = \mathbf{A}$. Clearly $\mathbf{R\Delta S}^*$ is a singular value decomposition of $\mathbf{A}$ in the form (8.8.3) where $\mathbf{\Delta} = \mathbf{G}^{-1}$. ∎

Remark If $\mathbf{A}$ is real, we can choose $\mathbf{R}$ and $\mathbf{S}$ and so $\mathbf{U}$, $\mathbf{V}$, $\mathbf{u}_i$'s and $\mathbf{v}_i$'s to be real.

We now identify the singular values and singular vectors. From (8.8.1) it is easy to see that

$$\mathbf{AA}^* = \mathbf{U}\begin{bmatrix} \mathbf{\Delta}^2 & \mathbf{0} \\ \mathbf{0} & \mathbf{0} \end{bmatrix}\mathbf{U}^* \quad \text{and} \quad \mathbf{A}^*\mathbf{A} = \mathbf{V}\begin{bmatrix} \mathbf{\Delta}^2 & \mathbf{0} \\ \mathbf{0} & \mathbf{0} \end{bmatrix}\mathbf{V}^* \tag{8.8.4}$$

Thus $d_1^2, \ldots, d_r^2$ are the non-zero characteristic roots of both $\mathbf{AA}^*$ and $\mathbf{A}^*\mathbf{A}$ and $\mathbf{u}_i$ (resp. $\mathbf{v}_i$) is an eigenvector of $\mathbf{AA}^*$ (resp. $\mathbf{A}^*\mathbf{A}$) corresponding to d_i^2.

It should be noted that if $\mathbf{U}$ and $\mathbf{V}$ are unitary matrices and $\mathbf{\Delta}$ is a diagonal matrix with positive diagonal entries such that (8.8.4) holds, it does *not* follow that (8.8.1) holds. However, from the proof of the preceding theorem we see that if $\mathbf{R\Lambda R}^*$ is a spectral decomposition of $\mathbf{AA}^*$ in the form (8.7.3) where $\mathbf{R}^*\mathbf{R} = \mathbf{I}$ and $\mathbf{S} = \mathbf{A}^*\mathbf{R\Lambda}^{-1/2}$, then $\mathbf{R\Lambda}^{1/2}\mathbf{S}^*$ is a singular value decomposition of $\mathbf{A}$ in the form (8.8.3). If $m > n$, one starts from $\mathbf{A}^*\mathbf{A}$ instead of $\mathbf{AA}^*$. We leave it to the reader to work out the details. It is easy to see that in a singular value decomposition of $\mathbf{A}$ in the form (8.8.3), given one of $\mathbf{R}$ and $\mathbf{S}$, the other is uniquely determined.

We now give a form of singular value decomposition which is similar to the spectral form of a semi-simple matrix and is unique.

Let $\mathbf{R\Delta S}^*$ be a singular value decomposition of an $m \times n$ matrix $\mathbf{A}$ in the form (8.8.3). Without any loss of generality we assume that $\mathbf{\Delta} = \text{diag}(\beta_1\mathbf{I}_{n_1}, \ldots, \beta_k\mathbf{I}_{n_k})$, where $\beta_1, \ldots, \beta_k$ are distinct. Let

$$\mathbf{R} = [\,\mathbf{R}_1 : \cdots : \mathbf{R}_k\,] \quad \text{and} \quad \mathbf{S}^* = \begin{bmatrix} \mathbf{S}_1^* \\ \vdots \\ \mathbf{S}_k^* \end{bmatrix}$$

where $\mathbf{R}_i$ has n_i columns and $\mathbf{S}_i^*$ has n_i rows. Write $\mathbf{T}_i = \mathbf{R}_i\mathbf{S}_i^*$ for $i = 1, \ldots, k$. We then prove the following:

(i) $\mathbf{A} = \sum_{i=1}^k \beta_i\mathbf{T}_i$,

(ii) $\beta_1, \ldots, \beta_k$ are distinct positive numbers and $\mathbf{T}_1, \ldots, \mathbf{T}_k$ are non-null matrices,

(iii) $\mathbf{T}_i\mathbf{T}_j^* = \delta_{ij}\mathbf{P}_{\mathbf{T}_i}$ and $\mathbf{T}_i^*\mathbf{T}_j = \delta_{ij}\mathbf{P}_{\mathbf{T}_i^*}$ for all i, j.

Property (i) follows from (8.8.3). To prove (ii), it is enough to note that

$(\mathbf{R}_i, \mathbf{S}_i^*)$ is a rank-factorization of $\mathbf{T}_i$. Also $\mathbf{S}_i^*\mathbf{S}_j = \delta_{ij}\mathbf{I}$ and $\mathbf{R}_i^*\mathbf{R}_j = \delta_{ij}\mathbf{I}$, hence $\mathbf{R}_i\mathbf{R}_i^*$ and $\mathbf{S}_i\mathbf{S}_i^*$ are orthogonal projectors and (iii) follows.

Conversely, suppose (i), (ii) and (iii) are satisfied. Denote $\mathcal{C}(\mathbf{T}_i) = \mathcal{C}(\mathbf{T}_i\mathbf{T}_i^*)$ by S_i and let $d(S_i) = n_i$. Let $\mathbf{R}_i$ be a matrix whose columns form an orthonormal basis of S_i. Then by (iii), $\mathbf{T}_i\mathbf{T}_i^* = \mathbf{R}_i\mathbf{R}_i^*$. Also, for $i \neq j$, $\mathbf{T}_i^*\mathbf{T}_j = \mathbf{0}$, so $S_i \perp S_j$. Let $\mathbf{R} = [\mathbf{R}_1 : \cdots : \mathbf{R}_k]$ and $\mathbf{\Lambda} = \text{diag}(\beta_1^2\mathbf{I}_{n_1}, \ldots, \beta_k^2\mathbf{I}_{n_k})$. Then $\mathbf{R}^*\mathbf{R} = \mathbf{I}$ and (i) gives

$$\mathbf{AA}^* = \textstyle\sum_i \beta_i^2\mathbf{T}_i\mathbf{T}_i^* = \mathbf{R\Lambda R}^*$$

Hence $\mathbf{R\Delta S}^*$ is a singular value decomposition of $\mathbf{A}$ in the form (8.8.3) where $\mathbf{\Delta} = \mathbf{\Lambda}^{1/2}$ and $\mathbf{S} = \mathbf{A}^*\mathbf{R\Lambda}^{-1/2}$.

We now prove the uniqueness of the β_i's and $\mathbf{T}_i$'s. Clearly $\mathbf{R\Lambda R}^*$ is a spectral decomposition of $\mathbf{AA}^*$ in the form (8.7.3). Thus $\beta_1^2, \ldots, \beta_k^2$ are the distinct non-zero eigenvalues of $\mathbf{AA}^*$ and hence are uniquely determined. Further, $\mathcal{C}(\mathbf{T}_i) = S_i = \mathcal{C}(\mathbf{R}_i)$ is the eigenspace of $\mathbf{AA}^*$ corresponding to β_i^2. Suppose now $\sum_{i=1}^k \beta_i\mathbf{W}_i$ is another representation of $\mathbf{A}$ satisfying (ii) and (iii). Then $\mathcal{C}(\mathbf{W}_i) = S_i$. Also $\beta_i(\mathbf{T}_i - \mathbf{W}_i) = \sum_{j\neq i}\beta_j(\mathbf{W}_j - \mathbf{T}_j)$. Since $\mathcal{C}(\text{LHS}) \subseteq S_i$ and $\mathcal{C}(\text{RHS}) \perp S_i$, it follows that $\beta_i(\mathbf{T}_i - \mathbf{W}_i) = \mathbf{0}$ and so $\mathbf{T}_i = \mathbf{W}_i$.

Since the singular values $d_1, \ldots, d_r$ of $\mathbf{A}$ are the positive square roots of the non-zero characteristic roots of $\mathbf{AA}^*$, it easily follows that two $m \times n$ matrices $\mathbf{A}$ and $\mathbf{C}$ are unitarily equivalent iff $\mathbf{AA}^*$ and $\mathbf{CC}^*$ have the same characteristic roots. Also, $\begin{bmatrix} \mathbf{\Delta} & \mathbf{0} \\ \mathbf{0} & \mathbf{0} \end{bmatrix}$ is a canonical form of $\mathbf{A}$ under unitary equivalence, where $\mathbf{\Delta} = \text{diag}(d_1, \ldots, d_r)$.

As an application of singular value decomposition, we show that the condition $\mathbf{AA}^* = \mathbf{CC}^*$ which is obviously necessary for the existence of a unitary $\mathbf{Z}$ such that $\mathbf{A} = \mathbf{CZ}$, is also sufficient.

Theorem 8.8.3 If $\mathbf{A}$ and $\mathbf{C}$ are $m \times n$ matrices such that $\mathbf{AA}^* = \mathbf{CC}^*$ then there exists a unitary matrix $\mathbf{Z}$ such that $\mathbf{A} = \mathbf{CZ}$.

Proof Let $\mathbf{R\Lambda R}^*$ be a spectral decomposition of $\mathbf{AA}^* = \mathbf{CC}^*$ in the form (8.7.3). Then from the proof of *Theorem* 8.8.2, it follows that $\mathbf{R\Lambda}^{1/2}\mathbf{S}^*$ and $\mathbf{R\Lambda}^{1/2}\mathbf{W}^*$ are singular value decompositions of $\mathbf{A}$ and $\mathbf{C}$ in the form (8.8.3) for some $\mathbf{S}$ and $\mathbf{W}$. Hence there exist unitary matrices $\mathbf{U}$, $\mathbf{V}$ and $\mathbf{X}$ such that

$$\mathbf{A} = \mathbf{U}\begin{bmatrix} \mathbf{\Lambda}^{1/2} & \mathbf{0} \\ \mathbf{0} & \mathbf{0} \end{bmatrix}\mathbf{V}^*$$

and

$$\mathbf{C} = \mathbf{U}\begin{bmatrix} \mathbf{\Lambda}^{1/2} & 0 \\ 0 & 0 \end{bmatrix}\mathbf{X}^*$$

Now $\mathbf{Z} = \mathbf{X}\mathbf{V}^*$ is unitary and $\mathbf{A} = \mathbf{C}\mathbf{Z}$. ∎

Exercises

1. Compute a singular value decomposition of

$$\mathbf{A} = \begin{bmatrix} 2 & 1 & 0 & 1 \\ -1 & 1 & 1 & -2 \\ 1 & 2 & 1 & -1 \end{bmatrix}$$

2. Give a proof of *Theorem* 8.8.2 starting from $\mathbf{A}^*\mathbf{A}$ instead of $\mathbf{A}\mathbf{A}^*$.
3. Prove or disprove: If d is a singular value of $\mathbf{A}$ then d^2 is a singular value of $\mathbf{A}^2$.
4. Show that $\rho(\mathbf{A})$ is the number of singular values of $\mathbf{A}$. (Recall that for a square matrix $\mathbf{A}$, $\rho(\mathbf{A})$ need not be equal to the number of non-zero characteristic roots of $\mathbf{A}$).
5. Let (8.8.1) be a singular value decomposition of $\mathbf{A}$ and $\mathbf{G} = \mathbf{V}\begin{bmatrix} \mathbf{E} & \mathbf{L} \\ \mathbf{M} & \mathbf{N} \end{bmatrix}\mathbf{U}^*$. Show that $\mathbf{G}$ is a g-inverse of $\mathbf{A}$ iff $\mathbf{E} = \mathbf{\Delta}^{-1}$ and that $\mathbf{G}$ is a reflexive g-inverse of $\mathbf{A}$ iff $\mathbf{E} = \mathbf{\Delta}^{-1}$ and $\mathbf{N} = \mathbf{MDL}$.
6. (a) If $\mathbf{A} = \sum_{i=1}^{k} \beta_i \mathbf{T}_i$ is the singular value decomposition of $\mathbf{A}$ in the unique form, show that $\sum_{i=1}^{k} \mathbf{T}_i\mathbf{T}_i^* = \mathbf{P}_{\mathbf{A}}$ and $\sum_{i=1}^{k} \mathbf{T}_i^*\mathbf{T}_i = \mathbf{P}_{\mathbf{A}^*}$.
   (b) Show that (i) $\mathbf{T}_i\mathbf{T}_i^*$ is idempotent implies $\mathbf{T}_i\mathbf{T}_i^* = \mathbf{P}_{\mathbf{T}_i}$ and $\mathbf{T}_i^*\mathbf{T}_i = \mathbf{P}_{\mathbf{T}_i^*}$ but (ii) $\mathbf{T}_i\mathbf{T}_j^* = \mathbf{0}$ whenever $i \neq j$ does not imply $\mathbf{T}_i^*\mathbf{T}_j = \mathbf{0}$.
7. Let $\lambda_1, \lambda_2, \ldots, \lambda_r$ be the non-zero characteristic roots of a normal matrix $\mathbf{A}$. Then show that the singular values of $\mathbf{A}$ are $|\lambda_1|, |\lambda_2|, \ldots, |\lambda_r|$.

*8. Let $\mathbf{A}$ be a square matrix with characteristic roots $\lambda_1, \lambda_2, \ldots, \lambda_n$ and singular values $d_1, d_2, \ldots, d_r$. Then prove that $|\sum_{i=1}^{n} \lambda_i^2| \leq \sum_{j=1}^{r} d_j^2$. Suppose now $\mathbf{A}$ is real. Then show that the above inequality holds as equality iff $\mathbf{A}$ is symmetric or skew-symmetric.

8.9 Computational methods

In this section we give two iterative methods for computing eigenvalues and eigenvectors. $\mathbf{A}$ will denote a real $n \times n$ matrix.

The first method is known as *Jacobi's method*. By this method, given a real symmetric matrix $\mathbf{A}$, we can find an orthogonal matrix $\mathbf{P}$ such that $\mathbf{\Delta} := \mathbf{P}^{\mathrm{T}}\mathbf{A}\mathbf{P}$ is approximately diagonal. Then the diagonal elements of $\mathbf{\Delta}$ are approximately the characteristic roots of $\mathbf{A}$ and the columns of $\mathbf{P}$ are corresponding approximate eigenvectors.

We start by showing that it is easy to make any off-diagonal element of $\mathbf{A}$ zero by an orthogonal transformation. For this we need some notation. Let $\mathbf{Q}_{k\ell}(\theta)$ denote the $n \times n$ matrix $((q_{ij}))$ where $q_{ij} = \delta_{ij}$ whenever $(i,j) \neq (k,k),(k,\ell),(\ell,k)$ and (ℓ,ℓ), $q_{kk} = q_{\ell\ell} = \cos\theta$ and $q_{k\ell} = -q_{\ell k} = \sin\theta$. It is easy to see that $\mathbf{Q}_{k\ell}(\theta)$ is an orthogonal matrix.

Theorem 8.9.1 Let $\mathbf{A}$ be a real symmetric matrix of order n and $1 \leq k < \ell \leq n$. Then there exists a θ such that the (k,ℓ)-th element of $\mathbf{B}$ is 0, where $\mathbf{B} = \mathbf{Q}^{\mathrm{T}}\mathbf{A}\mathbf{Q}$ and $\mathbf{Q} = \mathbf{Q}_{k\ell}(\theta)$. Moreover, if

$$|a_{k\ell}| = \max_{1\leq i<j\leq n} |a_{ij}| \tag{8.9.1}$$

then

$$\sum\sum_{i\neq j} b_{ij}^2 \leq \left(1 - \frac{2}{n(n-1)}\right) \sum\sum_{i\neq j} a_{ij}^2$$

Proof It is easy to verify that the (k,ℓ)-th element of $\mathbf{B}$ is

$$\begin{aligned}(a_{kk} - a_{\ell\ell})\cos\theta\sin\theta + a_{k\ell}\cos^2\theta - a_{\ell k}\sin^2\theta \\ = \tfrac{1}{2}(a_{kk} - a_{\ell\ell})\sin 2\theta + a_{k\ell}\cos 2\theta\end{aligned}$$

Clearly this can be made 0 by taking

$$\theta = \frac{1}{2}\tan^{-1}\mu \quad \text{where } \mu = \frac{2a_{k\ell}}{a_{\ell\ell} - a_{kk}} \tag{8.9.2}$$

If $a_{\ell\ell} - a_{kk} = 0$, we take $\theta = \pi/4$.

To prove the second part, we first note that

$$\begin{aligned}\textstyle\sum_{i=1}^n \sum_{j=1}^n b_{ij}^2 = \operatorname{tr}(\mathbf{B}^{\mathrm{T}}\mathbf{B}) &= \operatorname{tr}(\mathbf{Q}^{\mathrm{T}}\mathbf{A}^{\mathrm{T}}\mathbf{A}\mathbf{Q}) = \operatorname{tr}(\mathbf{A}^{\mathrm{T}}\mathbf{A}\mathbf{Q}\mathbf{Q}^{\mathrm{T}}) \\ &= \operatorname{tr}(\mathbf{A}^{\mathrm{T}}\mathbf{A}) = \textstyle\sum_{i=1}^n \sum_{j=1}^n a_{ij}^2\end{aligned} \tag{8.9.3}$$

Now, without any loss of generality, we may assume that $k = 1$ and $\ell = 2$. Then $\mathbf{Q} = \begin{bmatrix}\mathbf{S} & \mathbf{0}\\ \mathbf{0} & \mathbf{I}\end{bmatrix}$ where $\mathbf{S} = \begin{bmatrix}\cos\theta & \sin\theta\\ -\sin\theta & \cos\theta\end{bmatrix}$. Partition $\mathbf{A}$ as $\begin{bmatrix}\mathbf{A}_1 & \mathbf{A}_2\\ \mathbf{A}_2^{\mathrm{T}} & \mathbf{A}_3\end{bmatrix}$ and $\mathbf{B}$ as $\begin{bmatrix}\mathbf{B}_1 & \mathbf{B}_2\\ \mathbf{B}_2^{\mathrm{T}} & \mathbf{B}_3\end{bmatrix}$ where $\mathbf{A}_1$ and $\mathbf{B}_1$ are 2×2 matrices.

Then

$$\begin{bmatrix} \mathbf{B}_1 & \mathbf{B}_2 \\ \mathbf{B}_2^{\mathrm{T}} & \mathbf{B}_3 \end{bmatrix} = \begin{bmatrix} \mathbf{S}^{\mathrm{T}} & 0 \\ 0 & \mathbf{I} \end{bmatrix} \begin{bmatrix} \mathbf{A}_1 & \mathbf{A}_2 \\ \mathbf{A}_2^{\mathrm{T}} & \mathbf{A}_3 \end{bmatrix} \begin{bmatrix} \mathbf{S} & 0 \\ 0 & \mathbf{I} \end{bmatrix} = \begin{bmatrix} \mathbf{S}^{\mathrm{T}}\mathbf{A}_1\mathbf{S} & \mathbf{S}^{\mathrm{T}}\mathbf{A}_2 \\ \mathbf{A}_2^{\mathrm{T}}\mathbf{S} & \mathbf{A}_3 \end{bmatrix}$$

Thus $\mathbf{B}_1 = \mathbf{S}^{\mathrm{T}}\mathbf{A}_1\mathbf{S}$. Since $\mathbf{S}$ is orthogonal, it follows as in (8.9.3) that $b_{11}^2 + b_{22}^2 = a_{11}^2 + a_{22}^2 + 2a_{12}^2$ (note that $b_{12} = 0$). Also $\mathbf{B}_3 = \mathbf{A}_3$, so we have

$$\sum_{i=1}^{n} b_{ii}^2 = b_{11}^2 + b_{22}^2 + \sum_{i=3}^{n} b_{ii}^2 = \sum_{i=1}^{n} a_{ii}^2 + 2a_{12}^2$$

Hence (8.9.3) gives

$$\sum_{i \neq j}\sum b_{ij}^2 = \sum_{i \neq j}\sum a_{ij}^2 - 2a_{12}^2 \leq \sum_{i \neq j}\sum a_{ij}^2 - \frac{2}{n(n-1)} \sum_{i \neq j}\sum a_{ij}^2$$

where the last inequality follows from the fact that maximum cannot be less than the average. The theorem follows. ∎

We can now give Jacobi's method. Write $\mathbf{A}_1 = \mathbf{A}$ and $\mathbf{P}_1 = \mathbf{I}$. At the m-th stage ($m \geq 1$), find k and ℓ satisfying (8.9.2), where the a's are the elements of $\mathbf{A}_m$. Then find the matrix $\mathbf{Q}_m$ as in *Theorem* 8.9.1 with $\mathbf{A}_m$ in the place of $\mathbf{A}$. Compute $\mathbf{A}_{m+1} = \mathbf{Q}_m^{\mathrm{T}}\mathbf{A}_m\mathbf{Q}_m$ and $\mathbf{P}_{m+1} = \mathbf{P}_m\mathbf{Q}_m$. Clearly $\mathbf{P}_{m+1} = \mathbf{Q}_1\mathbf{Q}_2\cdots\mathbf{Q}_m$ is orthogonal and $\mathbf{A}_{m+1} = \mathbf{P}_{m+1}^{\mathrm{T}}\mathbf{A}\mathbf{P}_{m+1}$. By *Theorem* 8.9.1, the off-diagonal elements of $\mathbf{A}_m$ go to 0 as $m \to \infty$. Thus for large m, $\mathbf{A}_{m+1}$ is approximately diagonal. It can also be shown that $\mathbf{A}_m$ converges to a fixed diagonal matrix. For a proof of this, see Wilkinson (1965).

Before we give Jacobi's method as an algorithm we make a few observations. To find $\mathbf{Q}$ as in *Theorem* 8.9.1, we need to find only $\cos\theta$ and $\sin\theta$ and not θ itself. Clearly we may choose θ satisfying (8.9.2) so that $-\pi/2 \leq 2\theta \leq \pi/2$. Then $\cos\theta$ and $\cos 2\theta$ are positive. Now (8.9.2) gives $\sec^2 2\theta = 1 + \tan^2 2\theta = 1 + \mu^2$. Hence it is easy to see that

$$\cos\theta = \sqrt{\frac{1}{2}\left(1 + \sqrt{\frac{1}{1+\mu^2}}\right)} \text{ and } \sin\theta = \pm\sqrt{\frac{1}{2}\left(1 - \sqrt{\frac{1}{1+\mu^2}}\right)} \quad (8.9.4)$$

where $\sin\theta$ has the same sign as μ. It is also easy to compute $\mathbf{Q}^{\mathrm{T}}\mathbf{A}\mathbf{Q}$ and $\mathbf{P}\mathbf{Q}$ because of the simple structure of $\mathbf{Q}$. These are incorporated in the following algorithm. We also assume that a small positive number ϵ is given so that when the off-diagonal elements are less than ϵ in absolute value, the matrix is considered to be approximately diagonal.

Algorithm 8.9.2 (*Jacobi's method*) Given: a real symmetric matrix of order n and a small positive real number ϵ.

Step 1 Set $\mathbf{A}$ = the given matrix and $\mathbf{P} = \mathbf{I}_n$.

Step 2 Find k and ℓ such that $1 \le k < \ell \le n$ and (8.9.1) holds.

Step 3 If $|a_{k\ell}| < \epsilon$, stop and declare that $a_{11}, a_{22}, \ldots, a_{nn}$ are the characteristic roots of the given matrix and the columns of $\mathbf{P}$ are the corresponding eigenvectors. Otherwise go to *Step* 4.

Step 4 If $a_{kk} = a_{\ell\ell}$, set $\cos\theta = \sin\theta = 1/\sqrt{2}$ and go to *Step* 6. Otherwise go to *Step* 5.

Step 5 Compute μ defined in (8.9.2) and $\cos\theta$ and $\sin\theta$ from (8.9.4), where $\sin\theta$ has the same sign as μ.

Step 6 Replace the k-th and ℓ-th columns of $\mathbf{A}$ by $(\cos\theta)\mathbf{A}_{*k} - (\sin\theta)\mathbf{A}_{*\ell}$ and $(\sin\theta)\mathbf{A}_{*k} + (\cos\theta)\mathbf{A}_{*\ell}$ respectively. (The new matrix is again called $\mathbf{A}$.)

Step 7 Replace a_{kk} and $a_{\ell\ell}$ by $a_{kk}\cos\theta - a_{k\ell}\sin\theta$ and $a_{k\ell}\sin\theta + a_{\ell\ell}\cos\theta$ respectively. Then replace $a_{k\ell}$ and $a_{\ell k}$ by 0.

Step 8 Replace the k-th and ℓ-th rows of $\mathbf{A}$ by $\mathbf{A}_{*k}{}^{\mathrm{T}}$ and $\mathbf{A}_{*\ell}{}^{\mathrm{T}}$ respectively (notice that this makes $\mathbf{A}$ symmetric).

Step 9 Replace the k-th and ℓ-th columns of $\mathbf{P}$ by $(\cos\theta)\mathbf{P}_{*k} - (\sin\theta)\mathbf{P}_{*\ell}$ and $(\sin\theta)\mathbf{P}_{*k} + (\cos\theta)\mathbf{P}_{*\ell}$ respectively. Go to *Step* 2. ∎

A few remarks are in order. An off-diagonal element of $\mathbf{A}$ which is made zero in one iteration may become non-zero in a subsequent iteration. So the method may not terminate in $n(n-1)/2$ iterations. In fact, the matrix may not become diagonal in a finite number of steps. If n is large, the factor $1 - \frac{2}{n(n-1)}$ appearing in *Theorem* 8.9.1 is close to 1 and convergence to a diagonal matrix may be slow.

Power Method

We now give the second method known as the *power method*. To apply this method, we have to *assume that* $\mathbf{A}$ *is a real matrix which is similar to* $\operatorname{diag}(\lambda_1, \lambda_2, \ldots, \lambda_n)$, *where* λ*'s are real and*

$$|\lambda_1| > |\lambda_2| \ge \cdots \ge |\lambda_n|$$

The power method can then be used to compute approximately λ_1 and a corresponding eigenvector.

Let $\mathbf{x}_1, \mathbf{x}_2, \ldots, \mathbf{x}_n$ be normalized real eigenvectors of $\mathbf{A}$ corresponding to $\lambda_1, \lambda_2, \ldots, \lambda_n$ respectively, such that they form a basis of $\mathbb{R}^n$. The power method is based on the following

Theorem 8.9.3 Let $\mathbf{y} = \sum_{i=1}^n \alpha_i \mathbf{x}_i$ where $\alpha_1 > 0$. Write $\mathbf{y}_0 = \mathbf{y}$ and define $\mathbf{z}_k$ and $\mathbf{y}_k$ for $k = 1, 2, \ldots$ inductively by: $\mathbf{z}_k = \mathbf{A}\mathbf{y}_{k-1}$ and $\mathbf{y}_k = (1/\|\mathbf{z}_k\|)\mathbf{z}_k$. Then $\|\mathbf{z}_k\| \to \lambda_1$ as $k \to \infty$ and $\mathbf{y}_{2m} \to \mathbf{x}_1$ as $m \to \infty$. Moreover, $\mathbf{y}_{2m+1}$ converges to $\mathbf{x}_1$ or $-\mathbf{x}_1$ according as λ_1 is positive or negative.

Proof It is easy to see that $\mathbf{y}_k = (1/\|\mathbf{A}^k\mathbf{y}\|)\mathbf{A}^k\mathbf{y}$ for all $k \geq 1$. Now $\mathbf{A}^k\mathbf{y} = \sum_{i=1}^n \alpha_i \lambda_i^k \mathbf{x}_i = \lambda_1^k \Big(\sum_{i=1}^n \alpha_i \big(\frac{\lambda_i}{\lambda_1}\big)^k \mathbf{x}_i \Big)$, so

$$\mathbf{y}_k = \pm \frac{\alpha_1 \mathbf{x}_1 + \sum_{i=2}^n \alpha_i \big(\frac{\lambda_i}{\lambda_1}\big)^k \mathbf{x}_i}{\Big\| \alpha_1 \mathbf{x}_1 + \sum_{i=2}^n \alpha_i \big(\frac{\lambda_i}{\lambda_1}\big)^k \mathbf{x}_i \Big\|} \tag{8.9.5}$$

where the sign is $+$ or $-$ according as λ_1^k is positive or negative. Since $|\lambda_i| < |\lambda_1|$ for $i = 2, \ldots, n$, it follows that the numerator in the RHS of (8.9.5) goes to $\alpha_1 \mathbf{x}_1$ and the denominator to α_1 as $k \to \infty$. Hence $\mathbf{y}_{2m} \to \mathbf{x}_1$ and $\mathbf{y}_{2m+1}$ converges to $\mathbf{x}_1$ or $-\mathbf{x}_1$ according as λ_1 is positive or negative. Since $\mathbf{z}_k = \mathbf{A}\mathbf{y}_{k-1}$ it follows that $\mathbf{z}_{2m+1} \to \mathbf{A}\mathbf{x}_1$ and $\mathbf{z}_{2m}$ goes to $\mathbf{A}\mathbf{x}_1$ or $-\mathbf{A}\mathbf{x}_1$. Since $\|\mathbf{A}\mathbf{x}_1\| = \|\lambda_1 \mathbf{x}_1\| = |\lambda_1|$, $\|\mathbf{z}_k\| \to |\lambda_1|$. ∎

Clearly the preceding theorem remains true when $\alpha_1 < 0$ provided $\mathbf{x}_1$ is replaced by $-\mathbf{x}_1$ in the conclusion. We now give the power method as

Algorithm 8.9.4 (*Power method*)

Step 1 Choose a vector $\mathbf{y}$ with unit norm. Set $k = 1$ and $\mathbf{y}_0 = \mathbf{y}$.

Step 2 Compute $\mathbf{z}_k = \mathbf{A}\mathbf{y}_{k-1}$, $\|\mathbf{z}_k\|$ and $\mathbf{y}_k = \mathbf{z}_k/\|\mathbf{z}_k\|$.

Step 3 Check if $\mathbf{y}_k$ is approximately equal to $\mathbf{y}_{k-1}$. If so, declare that $\|\mathbf{z}_k\|$ is the maximum modulus eigenvalue of $\mathbf{A}$ and $\mathbf{y}_k$ is a corresponding eigenvector and stop. Otherwise go to *Step* 4.

Step 4 Check if $\mathbf{y}_k$ is approximately equal to $-\mathbf{y}_{k-1}$. If so, declare that $-\|\mathbf{z}_k\|$ is the maximum modulus eigenvalue of $\mathbf{A}$ and $\mathbf{y}_k$ is a corresponding eigenvector and stop. Otherwise increase k by 1 and go to *Step* 2. ∎

A few remarks are in order. This algorithm is applicable only when $\mathbf{A}$ satisfies the conditions stated at the beginning. Also, $\mathbf{y}$ should be such that $\alpha_1 \neq 0$, where $\mathbf{y} = \sum_{i=1}^{n} \alpha_i \mathbf{x}_i$. Usually, even if α_1 is zero, it becomes non-zero at some stage because of the rounding-off errors, and the algorithm works.

If $\mathbf{P}$ is any non-singular matrix with $\mathbf{x}_1$ as the first column, then $\mathbf{P}^{-1}\mathbf{AP} = \begin{bmatrix} \lambda_1 & \mathbf{u}^{\mathrm{T}} \\ 0 & \mathbf{B} \end{bmatrix}$ for some $\mathbf{u}$ and $\mathbf{B}$. Then the characteristic roots of $\mathbf{A}$ are those of $\mathbf{B}$ together with λ_1. If $\mu \neq \lambda_1$ is an eigenvalue of $\mathbf{B}$ with $\mathbf{v}$ as a corresponding eigenvector, then it is easy to check that $\mathbf{P}\left[\begin{smallmatrix} \mathbf{B} \\ \mathbf{v} \end{smallmatrix}\right]$ is an eigenvector of $\mathbf{A}$ corresponding to μ, where $\beta = (\mathbf{u}^{\mathrm{T}}\mathbf{v})/(\mu - \lambda_1)$. Thus the power method can be used to obtain the eigenvalues and the corresponding eigenvectors if $|\lambda_1| > |\lambda_2| > \cdots > |\lambda_n|$.

Exercises

1. Using Jacobi's method, find an orthogonal matrix $\mathbf{P}$ such that $\mathbf{P}^{\mathrm{T}}\mathbf{AP}$ is diagonal where

$$\mathbf{A} = \begin{bmatrix} 3 & 1 & 2 & -4 \\ 1 & 2 & -3 & 1 \\ 2 & -3 & 1 & 0 \\ -4 & 1 & 0 & 2 \end{bmatrix}$$

   Also find $\mathbf{P}^{\mathrm{T}}\mathbf{AP}$.

2. Using the power method, find the maximum modulus eigenvalue of the matrix $\mathbf{A}$ in the preceding exercise and a corresponding normalized eigenvector.

3. Let $\mathbf{P}^{-1}\mathbf{AP} = \begin{bmatrix} \lambda_1 & \mathbf{y}^{\mathrm{T}} \\ 0 & \mathbf{C} \end{bmatrix}$. Let λ_2 ($\neq \lambda_1$) be an eigenvalue of $\mathbf{C}$ with corresponding eigenvector $\mathbf{u}$. Then show that $\mathbf{P}\begin{bmatrix} \beta \\ \mathbf{u} \end{bmatrix}$ is an eigenvector of $\mathbf{A}$ with respect to the eigenvalue λ_2, where

$$\beta = \frac{\mathbf{y}^{\mathrm{T}}\mathbf{u}}{\lambda_2 - \lambda_1}$$

Chapter 9
Quadratic forms

9.1 Introduction

Quadratic forms are homogeneous polynomials of degree 2 in several variables like $2x^2 + y^2 - 3z^2 + 2xy + yz$. They occur in the study of conics in geometry, energy in Physics, and have wide applications in various subjects including Statistics. Within Linear Algebra they occur in the context of inner products.

In the first seven sections of this chapter we study real quadratic forms (i.e., the coefficients are real and the variables are real variables). The complex analogue of a real quadratic form, called a hermitian form, is relevant for the study of inner products over $\mathbb{C}$. We take a brief look at them in the last section of the chapter.

Unless otherwise stated, *we take the base field to be $\mathbb{R}$ throughout the first seven sections of this chapter.*

Definition 9.1.1 A *quadratic form* in the variables $x_1, x_2, \ldots, x_n$ is an expression of the form

$$q(x_1, \ldots, x_n) = \sum_{i=1}^{n} \alpha_i x_i^2 + \sum_{1 \le i < j \le n} \beta_{ij} x_i x_j \tag{9.1.1}$$

where the α's and β's belong to $\mathbb{R}$. A quadratic form in n variables is said to be n-ary.

Thus a general unary (1-ary) quadratic form is $\alpha_1 x_1^2$ and a general binary (2-ary) quadratic form is $\alpha_1 x_1^2 + \alpha_2 x_2^2 + \beta_{12} x_1 x_2$.

Notice that the RHS of (9.1.1) can be written as $\mathbf{x}^{\mathrm{T}}\mathbf{A}\mathbf{x}$ where $\mathbf{A}$ is the $n \times n$ symmetric matrix $((a_{ij}))$ with

$$a_{ij} = \begin{cases} \alpha_i & \text{if } i = j \\ \beta_{ij}/2 & \text{if } i < j \\ \beta_{ji}/2 & \text{if } i > j \end{cases}$$

Conversely if $\mathbf{A}$ is any symmetric matrix of order n then $\mathbf{x}^{\mathrm{T}}\mathbf{A}\mathbf{x}$ is the RHS of (9.1.1) where $\alpha_i = a_{ii}$ and $\beta_{ij} = 2a_{ij}$ for $i < j$. This shows that a quadratic form can be expressed uniquely as $\mathbf{x}^T\mathbf{A}\mathbf{x}$ where $\mathbf{A}$ is symmetric. (See also *Exercise* 2.6.8.)

From now on, we assume that $\mathbf{A}$ *is symmetric whenever we refer to the quadratic form* $\mathbf{x}^T\mathbf{A}\mathbf{x}$. We call $\mathbf{A}$ the *matrix of the quadratic form* $\mathbf{x}^T\mathbf{A}\mathbf{x}$.

Since $\mathbf{A} \leftrightarrow \mathbf{x}^T\mathbf{A}\mathbf{x}$ is a 1-1 correspondence between symmetric matrices of order n and n-ary quadratic forms, we sometimes attribute concepts defined for one, like rank and determinant defined for matrices and definiteness category (see *Section* 9.2) defined for quadratic forms, to the other also.

Exercises

1. Write down the matrices of the following 3-ary quadratic forms:

   (a) $x_1^2 + x_2^2 - 3x_3^2 + 2x_1x_2 - 6x_1x_3$, (b) $x_1^2 + 2x_3^2 - x_1x_2$,

   (c) x_2x_3, (d) $(2x_1 - x_2 + 3x_3)^2$ and (e) $(\mathbf{u}^T\mathbf{x})^2$.

2. If $\mathbf{A}$ is not symmetric, what is the matrix of $\mathbf{x}^T\mathbf{A}\mathbf{x}$ viewed as a quadratic form?

3. Show that $\mathbf{x}^T\mathbf{A}\mathbf{x} = 0\,\forall\,\mathbf{x} \Rightarrow \mathbf{A} = \mathbf{0}$ and $\mathbf{x}^T\mathbf{A}\mathbf{x} = \mathbf{x}^T\mathbf{B}\mathbf{x}\,\forall\,\mathbf{x} \Rightarrow \mathbf{A} = \mathbf{B}$.

4. If $\overline{x} = \frac{1}{n}(x_1 + \cdots + x_n)$, find the matrices of the quadratic forms $n\overline{x}^2$ and $\sum_{i=1}^n (x_i - \overline{x})^2$. Verify that they are idempotent and add up to $\mathbf{I}$.

5. A map $\psi : \mathbb{R}^m \times \mathbb{R}^n \to \mathbb{R}$ is said to be a *bilinear form* if $\psi(\alpha\mathbf{x}, \mathbf{y}) = \psi(\mathbf{x}, \alpha\mathbf{y}) = \alpha\psi(\mathbf{x}, \mathbf{y})$, $\psi(\mathbf{x}+\mathbf{z}, \mathbf{y}) = \psi(\mathbf{x}, \mathbf{y}) + \psi(\mathbf{z}, \mathbf{y})$ and $\psi(\mathbf{x}, \mathbf{y}+\mathbf{u}) = \psi(\mathbf{x}, \mathbf{y}) + \psi(\mathbf{x}, \mathbf{u})$.

   (a) Show that every bilinear form $\psi(\mathbf{x}, \mathbf{y})$ can be written as $\mathbf{x}^T\mathbf{A}\mathbf{y}$ for some $m \times n$ matrix $\mathbf{A}$.

   (b) When $m = n$ show that a bilinear form gives rise to a quadratic form if we put $\mathbf{x} = \mathbf{y}$.

   (c) If $m = n$ and $\mathbf{A}$ is symmetric, show that there is a unique bilinear form $\psi(\mathbf{x}, \mathbf{y})$ such that $\psi(\mathbf{x}, \mathbf{y}) = \psi(\mathbf{y}, \mathbf{x})$ which gives rise to the quadratic form $\mathbf{x}^T\mathbf{A}\mathbf{x}$ as in (b).

9.2 Classification of quadratic forms

We classify a quadratic form into one of several categories based on its range. Notice, however, that both the quadratic forms $x_1^2 + x_2^2$ and

$x_1^2 + x_2^2 - 2x_1x_2$ in x_1, x_2 have range equal to the set of all non-negative real numbers. But the former becomes 0 when and only when $x_1 = 0$ and $x_2 = 0$ while the latter can be 0 even when x_1 and x_2 are non-zero. We distinguish between such quadratic forms in the following

Definition 9.2.1 An n-ary quadratic form $\mathbf{x}^T\mathbf{A}\mathbf{x}$ is said to be
non-negative definite (*n.n.d.*) if $\mathbf{x}^T\mathbf{A}\mathbf{x} \geq 0$ for all $\mathbf{x} \in \mathbb{R}^n$,
positive definite (*p.d.*) if $\mathbf{x}^T\mathbf{A}\mathbf{x} > 0$ for all $\mathbf{x} \neq \mathbf{0}$,
non-positive definite (*n.p.d.*) if $\mathbf{x}^T\mathbf{A}\mathbf{x} \leq 0$ for all $\mathbf{x} \in \mathbb{R}^n$,
negative definite (*n.d.*) if $\mathbf{x}^T\mathbf{A}\mathbf{x} < 0$ for all $\mathbf{x} \neq \mathbf{0}$, and
indefinite if it is neither n.n.d. nor n.p.d.

We say that the real symmetric matrix $\mathbf{A}$ is *n.n.d., p.d.* etc. if the associated quadratic form $\mathbf{x}^T\mathbf{A}\mathbf{x}$ is n.n.d., p.d. etc.

It is obvious that every p.d. (resp. n.d.) quadratic form is n.n.d. (resp. n.p.d.). $\mathbf{x}^T\mathbf{A}\mathbf{x}$ is n.p.d. (resp. n.d.) iff $\mathbf{x}^T(-\mathbf{A})\mathbf{x}$ is n.n.d. (resp. p.d.). $\mathbf{x}^T\mathbf{A}\mathbf{x}$ is indefinite iff it takes both positive and negative values. An n.n.d. quadratic form which is not p.d. is said to be *positive semi-definite* (*p.s.d.*). *Negative semi-definiteness* (*n.s.d.*) is defined similarly.

Note that every non-zero quadratic form belongs to exactly one of the categories: p.d., p.s.d., n.d., n.s.d. and indefinite. It is easy to give examples of quadratic forms in two variables x_1 and x_2 belonging to each of these categories: x_1^2 and $9x_1^2 + x_2^2 + 6x_1x_2$ are p.s.d., $x_1^2 + x_2^2 + x_1x_2 = (x_1 + \frac{1}{2}x_2)^2 + \frac{3}{4}x_2^2$ is p.d. and $2x_1^2 - x_2^2$ and x_1x_2 are indefinite.

It is easy to see the importance of positive definiteness for inner products. We showed in the last paragraph of *Section* 7.2 that $\langle \mathbf{x}, \mathbf{y} \rangle = \mathbf{y}^T\mathbf{A}\mathbf{x}$ is an inner product on $\mathbb{R}^n$ iff $\mathbf{A}$ satisfies the conditions (a), (b) and (c) given there which we now recognize to be positive definiteness.

The study of quadratic forms is often simplified by a suitable change of variables. For example, the quadratic form $x_1^2 + x_2^2 + x_1x_2$ becomes $y_1^2 + \frac{3}{4}y_2^2$ on making the change of variables $y_1 = x_1 + \frac{1}{2}x_2$ and $y_2 = x_2$ as seen above. Hence it follows that the quadratic form is p.d. We now study the effect of a change of variables on a general quadratic form.

Suppose $\mathbf{P}$ is a non-singular matrix and we make a change of variables, from $x_1, x_2, \ldots, x_n$ to $y_1, y_2, \ldots, y_n$, where $\mathbf{x} = \mathbf{P}\mathbf{y}$ (or, equivalently, $\mathbf{y} = \mathbf{P}^{-1}\mathbf{x}$). Then the quadratic form $\mathbf{x}^T\mathbf{A}\mathbf{x}$ transforms to $\mathbf{y}^T\mathbf{B}\mathbf{y}$ where $\mathbf{B} = \mathbf{P}^T\mathbf{A}\mathbf{P}$. Note that $\mathbf{P}^T\mathbf{A}\mathbf{P}$ is symmetric. Since $\mathbf{P}\mathbf{y} \neq \mathbf{0}$ iff $\mathbf{y} \neq \mathbf{0}$, it follows that $\mathbf{x}^T\mathbf{A}\mathbf{x}$ and $\mathbf{y}^T\mathbf{B}\mathbf{y}$ have the same definiteness category. Thus a non-singular transformation of the variables changes a quadratic form into another with the same definiteness category.

Motivated by the discussion in the preceding paragraph we define two $n \times n$ matrices $\mathbf{A}$ and $\mathbf{B}$ to be *congruent* if there exists a non-singular matrix $\mathbf{P}$ such that $\mathbf{B} = \mathbf{P}^{\mathrm{T}}\mathbf{A}\mathbf{P}$. It is easy to check that this is an equivalence relation. By *Theorem* 4.2.5 and the corollary to *Theorem* 4.4.9, $\mathbf{B}$ is congruent to $\mathbf{A}$ iff $\mathbf{B}$ can be obtained from $\mathbf{A}$ by performing a sequence of elementary row operations and the corresponding sequence of elementary column operations. Clearly, congruent symmetric matrices correspond to quadratic forms which can be obtained from each other by non-singular transformations of variables.

Exercises

1. Prepare a table showing the possible types of definiteness (p.d., p.s.d., n.d., n.s.d. and indefinite) of $\mathbf{A} + \mathbf{B}$ given those of $\mathbf{A}$ and $\mathbf{B}$.
2. Find the quadratic form to which $x_1^2+2x_2^2-x_3^2+2x_1x_2+x_2x_3$ transforms by the change of variables $y_1 = x_1 - x_3, y_2 = x_2 - x_3, y_3 = x_3$ by actual substitution. Verify that the matrix of the resulting quadratic form is congruent to the matrix of the original quadratic form.
3. Prove that congruence is an equivalence relation on the set of all $n \times n$ symmetric matrices.
4. If $\mathbf{A}$ and $\mathbf{B}$ are n.n.d., then show that $\mathrm{diag}(\mathbf{A}, \mathbf{B})$ is n.n.d. If $\mathbf{A}$ is p.d. and $\mathbf{B}$ is n.d., what can be said about $\mathrm{diag}(\mathbf{A}, \mathbf{B})$?
5. (a) If $\mathbf{A}$ is an n.n.d. matrix of order n and $\mathbf{P}$ is an $m \times n$ matrix, show that $\mathbf{P}\mathbf{A}\mathbf{P}^{\mathrm{T}}$ is n.n.d. Deduce that $\mathbf{P}\mathbf{P}^{\mathrm{T}}$ is n.n.d. for any matrix $\mathbf{P}$.

   *(b) Let $\mathbf{A}$ be symmetric and $\mathbf{P}\mathbf{A}\mathbf{P}^{\mathrm{T}}$ be n.n.d. Show that $\mathbf{P} = \mathbf{A}$ implies $\mathbf{A}$ is n.n.d. and that the condition $\mathbf{P} = \mathbf{A}$ cannot be dropped.
6. Let $\mathbf{A}$ and $\mathbf{B}$ be n.n.d. Show that $\mathbf{A} + \mathbf{B} = \mathbf{0}$ iff $\mathbf{A} = \mathbf{B} = \mathbf{0}$. Deduce that if $\mathbf{C}$ and $\mathbf{D}$ are symmetric and $\mathbf{C}^2 + \mathbf{D}^2 = \mathbf{0}$, then $\mathbf{C} = \mathbf{D} = \mathbf{0}$.

9.3 Rank and signature

In this section we show that any given quadratic form can be converted to a diagonal form (a quadratic form of the type $\delta_1 y_1^2 + \cdots + \delta_n y_n^2$) by a non-singular transformation and that the numbers of positive and negative δ's are independent of the transformation used.

Theorem 9.3.1 Any quadratic form $\mathbf{x}^T\mathbf{A}\mathbf{x}$ can be reduced to a diagonal form by a non-singular transformation of the variables.

Proof We first give a short proof using Spectral theorem. Since $\mathbf{A}$ is real symmetric, there exists an orthogonal matrix $\mathbf{P}$ such that $\mathbf{P}^{-1}\mathbf{A}\mathbf{P}$ is diagonal. But $\mathbf{P}^{-1} = \mathbf{P}^T$, so the transformation $\mathbf{x} = \mathbf{P}\mathbf{y}$ converts $\mathbf{x}^T\mathbf{A}\mathbf{x}$ to a diagonal form.

The above proof, though short, is not convenient for actual reduction since the eigenvalues and orthogonal eigenvectors required are not easy to obtain. So we give below an elementary constructive proof using the idea of *completing the square.*

If $\mathbf{A} = \mathbf{0}$, there is nothing to be done. So let $\mathbf{A} \neq \mathbf{0}$. We now show that by a non-singular transformation of the variables, we can make a_{11} non-zero. If $a_{ii} \neq 0$, interchanging x_1 and x_i will make a_{11} non-zero. So let $a_{11} = a_{22} = \cdots = a_{nn} = 0$. Since $\mathbf{A} \neq \mathbf{0}$, $a_{ij} \neq 0$ for some i and j with $i \neq j$. Now we make the change of variables

$$x_k = \begin{cases} y_j + y_i & \text{if } k = j \\ y_k & \text{if } k \neq j \end{cases} \tag{9.3.1}$$

It is easy to see that this transformation is non-singular and converts $\mathbf{x}^T\mathbf{A}\mathbf{x}$ to $\mathbf{y}^T\mathbf{B}\mathbf{y}$ with $b_{ii} = 2a_{ij} \neq 0$. We can then make the $(1,1)$-element non-zero (note that the resultant of non-singular transformations is also non-singular). Thus we may take $a_{11} \neq 0$. So

$$\begin{aligned}
\mathbf{x}^T\mathbf{A}\mathbf{x} &= a_{11}x_1^2 + 2a_{12}x_1x_2 + \cdots + 2a_{1n}x_1x_n + \sum_{i=2}^{n}\sum_{j=2}^{n} a_{ij}x_ix_j \\
&= a_{11}\left(x_1 + \frac{a_{12}}{a_{11}}x_2 + \cdots + \frac{a_{1n}}{a_{11}}x_n\right)^2 + \sum_{i=2}^{n}\sum_{j=2}^{n}\left(a_{ij} - \frac{a_{i1}a_{1j}}{a_{11}}\right)x_ix_j
\end{aligned}$$

We now make the change of variables

$$z_i = \begin{cases} x_1 + \dfrac{a_{12}}{a_{11}}x_2 + \cdots + \dfrac{a_{1n}}{a_{11}}x_n & \text{if } i = 1 \\ x_i & \text{if } i \neq 1 \end{cases} \tag{9.3.2}$$

Clearly this transformation is non-singular and converts $\mathbf{x}^T\mathbf{A}\mathbf{x}$ to $a_{11}z_1^2 + q'(z_2, \ldots, z_n)$ where q' is an $(n-1)$-ary quadratic form. We next repeat the above procedure for the quadratic form $q'(z_2, \ldots, z_n)$ and $\mathbf{x}^T\mathbf{A}\mathbf{x}$ can be reduced to a diagonal form in $n-1$ stages. ∎

The procedure given in the second proof of *Theorem* 9.3.1 is known as *Lagrange's reduction*. It can also be described in terms of the matrix instead of the quadratic form. We give this procedure with a slight simplification as

Algorithm 9.3.2 (*Congruent reduction of an $n \times n$ real symmetric matrix* $\mathbf{A}$ *to diagonal form*)

Step 1 Form the matrix $[\mathbf{A} : \mathbf{I}]$ and set $i = 1$.

Step 2 If $i = n$, stop. Let $[\mathbf{B} : \mathbf{Q}]$ be the final matrix and $\mathbf{C}$ the diagonal matrix obtained from $\mathbf{B}$ by replacing the elements above the diagonal by 0's; then $\mathbf{Q}$ is non-singular and $\mathbf{QAQ}^{\mathrm{T}} = \mathbf{C}$ where $\mathbf{A}$ is the given matrix. If $i < n$, go to *Step* 3.

Step 3 If $a_{ii} = 0$, go to *Step* 4. Otherwise go to *Step* 6.

Step 4 If $a_{jj} \neq 0$ for some $j > i$, choose one such j, perform R_{ij} on $[\mathbf{A} : \mathbf{I}]$ and C_{ij} on $\mathbf{A}$ and go to *Step* 6. If there is no such j, go to *Step* 5.

Step 5 If $a_{ji} \neq 0$ for some $j > i$, choose one such j, perform $R_{ij}(1)$ on $[\mathbf{A} : \mathbf{I}]$ and $C_{ij}(1)$ on $\mathbf{A}$ and go to *Step* 6. If there is no such j, increase i by 1 and go to *Step* 2.

Step 6 Perform $R_{ki}(-a_{ki}/a_{ii})$ on $[\mathbf{A} : \mathbf{I}]$ for $k = i+1, \ldots, n$, then increase i by 1 and go to *Step* 2.

We note that in this algorithm, we do not explicitly perform on $\mathbf{A}$ the column operations corresponding to the row operations performed in *Step* 6. ∎

Example 9.3.3 We will reduce the 3-ary quadratic form $2x_1x_3 + x_2x_3$ to diagonal form using Lagrange's method.

Since the diagonal terms are absent and the coefficient of x_1x_3 is non-zero we make the change of variables $x_1 = y_1$, $x_2 = y_2$ and $x_3 = y_3 + y_1$. The quadratic form is transformed to

$$2y_1^2 + y_1y_2 + 2y_1y_3 + y_2y_3 = 2\left(y_1 + \tfrac{1}{4}y_2 + \tfrac{1}{2}y_3\right)^2 - \tfrac{1}{8}y_2^2 - \tfrac{1}{2}y_3^2 + \tfrac{1}{2}y_2y_3$$

With $z_1 = y_1 + \frac{1}{4}y_2 + \frac{1}{2}y_3$, $z_2 = y_2$ and $z_3 = y_3$, the quadratic form becomes

$$2z_1^2 - \tfrac{1}{8}z_2^2 - \tfrac{1}{2}z_3^2 + \tfrac{1}{2}z_2z_3 = 2z_1^2 - \tfrac{1}{8}(z_2 - 2z_3)^2$$

Finally, with the transformation $z_1 = u_1$, $z_2 - 2z_3 = u_2$ and $z_3 = u_3$, the quadratic form becomes

$$2u_1^2 - \tfrac{1}{8}u_2^2$$

It can be checked that the transformation from x's to u's is given by: $x_1 = u_1 - \frac{1}{4}u_2 - u_3$, $x_2 = u_2 + 2u_3$ and $x_3 = u_1 - \frac{1}{4}u_2$. The u's can be expressed in terms of x's as: $u_1 = \frac{1}{2}x_1 + \frac{1}{4}x_2 + \frac{1}{2}x_3$, $u_2 = 2x_1 + x_2 - 2x_3$ and $u_3 = x_3 - x_1$. Thus the given quadratic form can be written as

$$2\left(\tfrac{1}{2}x_1 + \tfrac{1}{4}x_2 + \tfrac{1}{2}x_3\right)^2 - \tfrac{1}{8}(2x_1 + x_2 - 2x_3)^2$$

which can also be written as

$$\left(\tfrac{1}{\sqrt{2}}x_1 + \tfrac{1}{2\sqrt{2}}x_2 + \tfrac{1}{\sqrt{2}}x_3\right)^2 - \left(\tfrac{1}{\sqrt{2}}x_1 + \tfrac{1}{2\sqrt{2}}x_2 - \tfrac{1}{\sqrt{2}}x_3\right)^2$$

with coefficients 1, -1 and 0. ∎

As in the preceding example, we can easily deduce the following result from *Theorem* 9.3.1.

Theorem 9.3.4 Any quadratic form $\mathbf{x}^{\mathrm{T}}\mathbf{A}\mathbf{x}$ can be reduced to a diagonal form of the type $z_1^2+\cdots+z_k^2-z_{k+1}^2-\cdots-z_{k+\ell}^2$ by a non-singular transformation. Hence there exist linear functionals $\mathbf{q}_1^{\mathrm{T}}\mathbf{x}, \ldots, \mathbf{q}_{k+\ell}^{\mathrm{T}}\mathbf{x}$ such that

$$\mathbf{x}^{\mathrm{T}}\mathbf{A}\mathbf{x} = (\mathbf{q}_1^{\mathrm{T}}\mathbf{x})^2 + \cdots + (\mathbf{q}_k^{\mathrm{T}}\mathbf{x})^2 - (\mathbf{q}_{k+1}^{\mathrm{T}}\mathbf{x})^2 - \cdots - (\mathbf{q}_{k+\ell}^{\mathrm{T}}\mathbf{x})^2$$

We now show that the k and ℓ of the preceding theorem are uniquely determined by $\mathbf{A}$.

Theorem 9.3.5 (*Sylvester's law of inertia*) Let a quadratic form $\mathbf{x}^{\mathrm{T}}\mathbf{A}\mathbf{x}$ be transformed to the diagonal forms $\mathbf{y}^{\mathrm{T}}\mathbf{B}\mathbf{y}$ and $\mathbf{z}^{\mathrm{T}}\mathbf{C}\mathbf{z}$ by two non-singular transformations. Then the numbers of positive, negative and zero diagonal entries in $\mathbf{B}$ are the same as those in $\mathbf{C}$.

Proof By hypothesis, $\mathbf{A}$ is congruent to both $\mathbf{B}$ and $\mathbf{C}$. So $\mathbf{C} = \mathbf{Q}^{\mathrm{T}}\mathbf{B}\mathbf{Q}$ for some non-singular matrix $\mathbf{Q}$. We assume without any loss of generality that the positive diagonal entries of $\mathbf{B}$ are $b_{11}, \ldots, b_{kk}$ and the positive diagonal entries of $\mathbf{C}$ are $c_{11}, \ldots, c_{mm}$. Suppose now $k > m$. Then there exists $\mathbf{z} \neq \mathbf{0}$ such that $z_i = 0$ for $i = 1, \ldots, m$ and $\mathbf{Q}_{i*}\mathbf{z} = 0$ for $i = k+1, \ldots, n$ since these are less than n homogeneous linear equations in n unknowns. Since $z_1, \ldots, z_m$ are 0, $\mathbf{z}^{\mathrm{T}}\mathbf{C}\mathbf{z} \leq 0$. Now $\mathbf{Q}\mathbf{z} \neq \mathbf{0}$ since $\mathbf{Q}$ is non-singular and $\mathbf{z} \neq \mathbf{0}$. Also the last $n-k$ components of $\mathbf{Q}\mathbf{z}$ are 0, so

$$\mathbf{z}^{\mathrm{T}}\mathbf{C}\mathbf{z} = \mathbf{z}^{\mathrm{T}}\mathbf{Q}^{\mathrm{T}}\mathbf{B}\mathbf{Q}\mathbf{z} = (\mathbf{Q}\mathbf{z})^{\mathrm{T}}\mathbf{B}(\mathbf{Q}\mathbf{z}) > 0,$$

a contradiction which proves that $k \leq m$. By symmetry the reverse inequality follows and $k = m$. Similarly (or by considering $-\mathbf{B}$ and

$-\mathbf{C}$) it can be shown that $\mathbf{B}$ and $\mathbf{C}$ have the same number of negative diagonal entries and the theorem follows. ∎

In view of the preceding theorem we give

Definition 9.3.6 We shall denote by $P(\mathbf{A})$ and $N(\mathbf{A})$ the numbers of positive and negative diagonal entries in any diagonal form to which $\mathbf{x}^{\mathrm{T}}\mathbf{A}\mathbf{x}$ can be reduced by a non-singular transformation. The number $P(\mathbf{A}) - N(\mathbf{A})$ is called the *signature* $\mathbf{x}^{\mathrm{T}}\mathbf{A}\mathbf{x}$. When the matrix is clear from the context we write P and N for $P(\mathbf{A})$ and $N(\mathbf{A})$ respectively.

Since $\mathbf{x}^{\mathrm{T}}\mathbf{A}\mathbf{x}$ can be transformed to $\lambda_1 y_1^2 + \cdots + \lambda_n y_n^2$ by an orthogonal transformation, where $\lambda_1, \ldots, \lambda_n$ are the characteristic roots of $\mathbf{A}$, it follows that P and N are the numbers of positive and negative characteristic roots of $\mathbf{A}$. Since $P + N$ is the rank of $\mathbf{A}$, it follows that rank and signature determine P and N (and vice versa). We next determine the definiteness category of $\mathbf{x}^{\mathrm{T}}\mathbf{A}\mathbf{x}$ in terms of P and N. We leave the proof of the following theorem as a simple exercise to the reader.

Theorem 9.3.7 Let $\mathbf{x}^{\mathrm{T}}\mathbf{A}\mathbf{x}$ be an n-ary quadratic form with rank r and signature s and let P and N be as defined above. Then

(i) $\mathbf{A}$ is p.d. iff $P = n$ (equivalently, $s = n$),

(ii) $\mathbf{A}$ is n.n.d. iff $N = 0$ (equivalently, $r = s$),

(iii) $\mathbf{A}$ is n.d. iff $N = n$ (equivalently, $s = -n$),

(iv) $\mathbf{A}$ is n.p.d. iff $P = 0$ (equivalently, $r = -s$),

(v) $\mathbf{A}$ is indefinite iff $P \geq 1$ and $N \geq 1$ (equivalently, $|s| < r$).

Corollary $\mathbf{A}$ is p.d. iff $\mathbf{A}$ is n.n.d. and non-singular.

For the quadratic form $2x_1x_3 + x_2x_3$, $P = 1$ and $N = 1$ as seen from *Example* 9.3.3, hence the rank is 2 and signature is 0 and the quadratic form is indefinite (the last assertion can also be proved directly from the definition). The next theorem is an immediate consequence of (i) and (ii) of the preceding theorem.

Theorem 9.3.8 A symmetric matrix $\mathbf{A}$ is p.d. (resp. n.n.d.) iff each eigenvalue of $\mathbf{A}$ is positive (resp. non-negative).

Theorem 9.3.9 Two n-ary quadratic forms can be obtained from each other by non-singular transformations (i.e., their matrices are congruent) iff they have the same rank and the same signature.

Proof The *only if part* follows from *Theorem* 9.3.5. The *if part* follows from *Theorem* 9.3.4 since we can go from one of the quadratic forms to the other *via* $z_1^2 + \cdots + z_P^2 - z_{P+1}^2 - \cdots - z_{P+N}^2$. ∎

Clearly $\text{diag}(\mathbf{I}, -\mathbf{I}, \mathbf{0})$ is a canonical form under congruence transformation in the sense that every real symmetric matrix is congruent to one and only one such matrix, where the orders of $\mathbf{I}$ and $-\mathbf{I}$ may be different.

Equation of second degree in the plane

We now study the locus of the point (x, y) in $\mathbb{R}^2$ satisfying a general equation of degree two (or less):

$$ax^2 + by^2 + 2hxy + 2gx + 2fy + c = 0 \tag{9.3.3}$$

To determine the locus, we change the coordinate axes in such a way that the equation takes a readily recognizable form. We do not want any distortion of distances and angles, so we allow only a rotation of both the axes about the origin by a common angle and a translation of the origin. Let the new coordinates of the point (x, y) be denoted by x' and y'.

From the proof of *Theorem* 8.9.1, it follows that there exists θ (given by (8.9.2)) such that $\mathbf{P}^{\mathrm{T}}\mathbf{A}\mathbf{P}$ is diagonal, where

$$\mathbf{A} = \begin{bmatrix} a & h \\ h & b \end{bmatrix} \quad \text{and} \quad \mathbf{P} = \begin{bmatrix} \cos\theta & -\sin\theta \\ \sin\theta & \cos\theta \end{bmatrix}$$

We rotate the coordinate axes around the origin by an angle θ so that $(x, y)^{\mathrm{T}} = \mathbf{P}(x', y')^{\mathrm{T}}$ and (9.3.3) gets transformed to an equation of the form $a'x'^2 + b'y'^2 + 2g'x' + 2f'y' + c = 0$. For the sake of convenience, we drop the primes ($'$) and write the equation as

$$ax^2 + by^2 + 2gx + 2fy + c = 0 \tag{9.3.4}$$

We now consider several cases.

Case 1. $a = b = 0$. If at least one of f and g is non-zero, (9.3.4) represents a line. If $f = g = 0$ and $c \neq 0$, (9.3.4) has no solution. If $f = g = c = 0$, (9.3.4) represents the whole of $\mathbb{R}^2$.

Case 2. $a = 0$ and $b \neq 0$. If we shift the origin to the point $(0, -f/b)$,' we have $x = x'$ and $y = y' - f/b$ where x' and y' are the new coordinates of the point (x, y) and the equation takes the form $b'y'^2 + 2g'x' + c' = 0$. We again drop the primes.

Subcase 2a. $g \neq 0$. Then on shifting the origin to $(-c/2g, 0)$, the equation takes the form $by^2 + 2gx = 0$, which represents a parabola.

Subcase 2b. $g = 0$. Then the equation represents a pair of parallel lines, a single line or the empty set according as bc is negative, 0 or positive.

Case 3. $a \neq 0$ and $b = 0$. This is similar to the preceding case.

Case 4. Both a and b are non-zero. Then on shifting the origin to $(-g/a, -f/b)$, the equation takes the form $ax^2 + by^2 + c = 0$.

Subcase 4a. a and b have the same sign. If c is of the opposite sign, the equation represents an ellipse (which is a circle in case $a = b$). If $c = 0$, the equation represents a single point. If c has the same sign as a and b, the equation has no solution.

Subcase 4b. a and b have opposite signs. If $c \neq 0$, the the equation represents a hyperbola. If $c = 0$, the equation represents a pair of intersecting lines.

This exhausts all possibilities. We note that if at least one of a and b is non-zero, (9.3.3) represents an ellipse, a parabola or a hyperbola according as the signature of the quadratic form $ax^2 + by^2 + 2hxy$ (with the original a, b and h) is 2, 1 or 0 in modulus. An ellipse may degenerate into a point or even the empty set; a parabola may degenerate into a pair of parallel or coincident lines or the empty set, and a hyperbola may degenerate into a pair of intersecting straight lines.

We also note that the rotation of the coordinate axes we used makes the axes parallel to the principal axes of the conic. If the origin is at the centre of the conic, the points in $\mathbb{R}^2$ corresponding to the columns of $\mathbf{P}$, which are eigenvectors of $\mathbf{A}$, lie on the principal axes. For this reason, the spectral theorem is sometimes known as the principal axes theorem.

Exercises

1. Prove that every orthogonal projector is an n.n.d. matrix.
2. If $\mathbf{B} = \mathbf{A}^{-1}$, show that $\mathbf{x}^{\mathrm{T}}\mathbf{A}\mathbf{x}$ and $\mathbf{x}^{\mathrm{T}}\mathbf{B}\mathbf{x}$ have the same signature.
3. Let $\mathbf{P}$ be of full column rank. Show that $\mathbf{A}$ and $\mathbf{P}\mathbf{A}\mathbf{P}^{\mathrm{T}}$ have the same rank and the same signature. What about the number of zero eigenvalues?
4. Prove that a quadratic form $\mathbf{x}^{\mathrm{T}}\mathbf{A}\mathbf{x}$ can be written as the product of two linearly independent linear forms in $\mathbf{x}$ iff $\mathbf{A}$ has rank 2 and signature 0.

5. Show that *Algorithm* 9.3.2 works.
6. Work out *Example* 9.3.3 using *Algorithm* 9.3.2.
7. Examine the definiteness category of each of the quadratic forms in *Exercise* 9.1.1 by reducing it to a diagonal form using *Algorithm* 9.3.2.
8. Find the rank and signature of each of the following quadratic forms:
   (a) $x_1^2-3x_2^2-8x_3^2-x_4^2+2x_1x_2-2x_1x_3+2x_1x_4-14x_2x_3+10x_2x_4+10x_3x_4$
   (b) $x_1x_2 + x_3x_4 + \cdots + x_{2k-1}x_{2k}$
   (c) $\sum_{i,j=1}^{n}(x_i - x_j)^2$
9. (a) Using Lagrange's method, reduce $Q = x_1x_2 + x_2x_3 + x_1x_3$ to a diagonal form with each coefficient 1,−1 or 0. Also obtain the transforming matrix.
   (b) Repeat (a) using *Algorithm* 9.3.2.
   (c) What are the rank and signature of the quadratic form Q in (a)?
   (d) What is the type of definiteness of Q? Prove your answer using (c) and also from definitions.
10. Prove *Theorem* 9.3.7 and its corollary.
11. If $\mathbf{A}$ is any real symmetric matrix, show that there exists a real number α such that $\alpha\mathbf{I} + \mathbf{A}$ is positive definite.
12. Show that every real symmetric matrix can be written as the difference of two p.d. matrices.
13. Show that the set $\{\mathbf{x} : \mathbf{x}^{\mathrm{T}}\mathbf{A}\mathbf{x} \le 1\}$ is bounded iff $\mathbf{A}$ is p.d.
14. Consider the locus of the point (x, y) in $\mathbb{R}^2$ satisfying $Q(x, y) = 1$, where $Q(x, y) = ax^2 + by^2 + 2hxy$. Let r and s be the rank and signature of Q. Show that the locus is
    (a) an ellipse iff $r = 2$ and $s = 2$,
    (b) a hyperbola iff $r = 2$ and $s = 0$,
    (c) a pair of parallel straight lines iff $r = 1$ and $s = 1$,
    (d) empty otherwise.

    Show also that the locus is a circle iff the matrix of Q is $\alpha\mathbf{I}$ for some $\alpha > 0$.
*15. (a) Let $\mathbf{M} = \begin{bmatrix} \mathbf{A} & \mathbf{B} \\ -\mathbf{B}^{\mathrm{T}} & \mathbf{C} \end{bmatrix}$ be skew-symmetric, where $\mathbf{A}$ is non-singular. Find a non-singular matrix $\mathbf{P}$ such that $\mathbf{P}^{\mathrm{T}}\mathbf{M}\mathbf{P}$ is of the form $\begin{bmatrix} \mathbf{A} & \mathbf{0} \\ \mathbf{0} & \mathbf{D} \end{bmatrix}$ for some $\mathbf{D}$.
    (b) Show that any real skew-symmetric matrix is congruent to a unique matrix of the form $\operatorname{diag}(\mathbf{S}_1, \mathbf{S}_2, \ldots, \mathbf{S}_k)$ where $\mathbf{S}_i = \begin{bmatrix} 0 & 1 \\ -1 & 0 \end{bmatrix}$ for $i = 1, \ldots, k-1$ and $\mathbf{S}_k$ is either $\mathbf{S}_1$ or $\mathbf{0}$.
    (c) Show that the rank of a real skew-symmetric matrix is even.

9.4 p.d. and n.n.d. matrices

In the preceding section we gave criteria for each definiteness category in terms of rank and signature. In this section we give a different type of characterization of p.d. and n.n.d. matrices which is much more useful in many applications. We shall also study several properties of n.n.d. and p.d. matrices and give a third characterization in terms of minors.

Theorem 9.4.1 A (real) matrix $\mathbf{A}$ is n.n.d. iff there exists a real matrix $\mathbf{B}$ such that $\mathbf{A} = \mathbf{B}^T\mathbf{B}$.

Proof If $\mathbf{A} = \mathbf{B}^T\mathbf{B}$ then $\mathbf{x}^T\mathbf{A}\mathbf{x} = \mathbf{x}^T\mathbf{B}^T\mathbf{B}\mathbf{x} = (\mathbf{B}\mathbf{x})^T\mathbf{B}\mathbf{x} \geq 0$ for all $\mathbf{x}$, so $\mathbf{A}$ is n.n.d. Conversely, let $\mathbf{A}$ be n.n.d. Then by *Theorem* 9.3.4, there exists a non-singular matrix $\mathbf{P}$ such that $\mathbf{P}^T\mathbf{A}\mathbf{P} = \text{diag}(\mathbf{I}_r, \mathbf{0})$. Taking $\mathbf{B} = \text{diag}(\mathbf{I}_r, \mathbf{0})\mathbf{P}^{-1}$, we have $\mathbf{A} = \mathbf{B}^T\mathbf{B}$. In *Section* 9.8 we will give a compact method to compute a $\mathbf{B}$ which is upper triangular. ∎

We note that for the *if part* of the preceding theorem, $\mathbf{B}$ need not be a square matrix. Also, $\mathbf{A} = \mathbf{B}^T\mathbf{B}$ means simply that $a_{ij} = \langle \mathbf{B}_{*i}, \mathbf{B}_{*j} \rangle$ for all i and j, where the inner product is the canonical inner product. The matrix $\mathbf{B}^T\mathbf{B}$ is called the *Gram matrix of* $\mathbf{B}$ and is said to be a *Grammian*. $\mathbf{B}$ itself is said to be a *square-root* of $\mathbf{B}^T\mathbf{B}$. (Note that $\mathbf{B}^2$ may not equal $\mathbf{B}$.)

It is worth noting that if $\mathbf{A}$ is n.n.d., then $\mathbf{P}^T\mathbf{A}\mathbf{P}$ is n.n.d. for any matrix $\mathbf{P}$ ($\mathbf{P}$ need not even be square).

Theorem 9.4.2 $\mathbf{A}$ is p.d. iff $\mathbf{A} = \mathbf{B}^T\mathbf{B}$ for some real non-singular matrix $\mathbf{B}$.

Proof The *if part* follows from definitions and the *only if part* follows from the proof of the preceding theorem. ∎

Corollary The inverse of a p.d. matrix is p.d.

Theorem 9.4.3 Let

$$\mathbf{A} = \begin{bmatrix} \mathbf{B} & \mathbf{C} \\ \mathbf{C}^T & \mathbf{D} \end{bmatrix} \tag{9.4.1}$$

be an n.n.d. matrix, where $\mathbf{B}$ and $\mathbf{D}$ are square. Then $\mathcal{C}(\mathbf{C}) \subseteq \mathcal{C}(\mathbf{B})$.

Proof By *Theorem* 9.4.1, there exists a matrix $\mathbf{S}$ such that $\mathbf{A} = \mathbf{S}^T\mathbf{S}$. Partition $\mathbf{S}$ as $\mathbf{S} = [\mathbf{S}_1 : \mathbf{S}_2]$ where $\mathbf{S}_1$ has the same number of columns as $\mathbf{B}$. Then $\mathbf{A} = \mathbf{S}^T\mathbf{S}$ gives $\mathbf{B} = \mathbf{S}_1^T\mathbf{S}_1$ and $\mathbf{C} = \mathbf{S}_1^T\mathbf{S}_2$. So $\mathcal{C}(\mathbf{C}) \subseteq \mathcal{C}(\mathbf{S}_1^T) = \mathcal{C}(\mathbf{S}_1^T\mathbf{S}_1) = \mathcal{C}(\mathbf{B})$. ∎

Since a permutation matrix $\mathbf{P}$ is non-singular, $\mathbf{P}^{\mathrm{T}}\mathbf{A}\mathbf{P}$ and $\mathbf{A}$ have the same definiteness category. Since any principal submatrix of $\mathbf{A}$ can be brought to the top left corner by applying the same permutation to the rows and columns, many results on leading principal submatrices of n.n.d. and p.d. matrices can easily be extended to all principal submatrices. The next theorem follows thus from the preceding theorem.

Theorem 9.4.4 Let $\mathbf{A}$ be an n.n.d. matrix of order n and let $1 \le i_1 < i_2 < \cdots < i_k \le n$ and $1 \le j \le n$. Then

$$\mathbf{A}(i_1, \ldots, i_k \,|\, j) \in \mathcal{C}(\mathbf{A}(i_1, \ldots, i_k \,|\, i_1, \ldots, i_k)) \tag{9.4.2}$$

Corollary 1 If $\mathbf{A}$ is n.n.d. and $a_{ii} = 0$ then $a_{ij} = a_{ji} = 0$ for all j.

Corollary 2 If $\mathbf{A}$ is n.n.d., the rank of the submatrix formed by any set of rows equals the rank of the principal submatrix formed by them.

Theorem 9.4.5 Every principal submatrix $\mathbf{B}$ of an n.n.d. (resp. p.d.) matrix $\mathbf{A}$ is n.n.d. (resp. p.d.).

Proof Without loss of generality, let $\mathbf{A}$ and $\mathbf{B}$ be as in (9.4.1). Then the theorem follows since $\mathbf{y}^{\mathrm{T}}\mathbf{B}\mathbf{y} = [\,\mathbf{y}^{\mathrm{T}} : \mathbf{0}^{\mathrm{T}}\,]\mathbf{A}\begin{bmatrix}\mathbf{y}\\ \mathbf{0}\end{bmatrix}$. ∎

Theorem 9.4.6 If $\mathbf{A}$ and $\mathbf{C}$ are n.n.d. matrices of the same order then $\mathbf{A}+\mathbf{C}$ is n.n.d. and $\mathcal{C}(\mathbf{A}+\mathbf{C}) = \mathcal{C}(\mathbf{A}) + \mathcal{C}(\mathbf{C})$. If, moreover, at least one of $\mathbf{A}$ and $\mathbf{C}$ is p.d. then $\mathbf{A}+\mathbf{C}$ is p.d.

Proof Since $\mathbf{x}^{\mathrm{T}}(\mathbf{A}+\mathbf{C})\mathbf{x} = \mathbf{x}^{\mathrm{T}}\mathbf{A}\mathbf{x} + \mathbf{x}^{\mathrm{T}}\mathbf{C}\mathbf{x}$, the first and the third statements follow easily. To prove the second statement, let $\mathbf{A} = \mathbf{B}^{\mathrm{T}}\mathbf{B}$ and $\mathbf{C} = \mathbf{D}^{\mathrm{T}}\mathbf{D}$. Then

$$\begin{aligned}\mathcal{C}(\mathbf{A}+\mathbf{C}) = \mathcal{C}\left([\,\mathbf{B}^{\mathrm{T}} : \mathbf{D}^{\mathrm{T}}\,]\begin{bmatrix}\mathbf{B}\\ \mathbf{D}\end{bmatrix}\right) &= \mathcal{C}(\mathbf{B}^{\mathrm{T}} : \mathbf{D}^{\mathrm{T}})\\ &= \mathcal{C}(\mathbf{B}^{\mathrm{T}}) + \mathcal{C}(\mathbf{D}^{\mathrm{T}}) = \mathcal{C}(\mathbf{A}) + \mathcal{C}(\mathbf{C})\end{aligned}$$ ∎

We now show the importance of n.n.d. matrices in Statistics. The next theorem and the paragraph following it may be omitted by those who do not know Statistics.

***Theorem 9.4.7** Any variance-covariance matrix is n.n.d. and every n.n.d. matrix is a variance-covariance matrix.

Proof Suppose first that $\boldsymbol{\Sigma} = ((\sigma_{ij}))$ is the variance-covariance matrix of the random variables $Z_1, \ldots, Z_n$. Then

$$\mathbf{x}^{\mathrm{T}}\boldsymbol{\Sigma}\mathbf{x} = \sum_{i=1}^{n}\sum_{j=1}^{n}\sigma_{ij}x_ix_j = \operatorname{Var}(x_1Z_1 + \cdots + x_nZ_n) \geq 0$$

for all $\mathbf{x} \in \mathbb{R}^n$, so $\boldsymbol{\Sigma}$ is n.n.d. To prove the converse, let $\mathbf{A}$ be n.n.d. Then $\mathbf{A} = \mathbf{B}^{\mathrm{T}}\mathbf{B}$ for some matrix $\mathbf{B}$. Let k be the number of rows of $\mathbf{B}$ and let $Y_1, \ldots, Y_k$ be k pair-wise uncorrelated random variables, each with unit variance. Define $\mathbf{Z} = \mathbf{B}^{\mathrm{T}}\mathbf{Y}$. Then it is easy to check that the variance-covariance matrix of $\mathbf{Z}$ is $\mathbf{B}^{\mathrm{T}}\mathbf{I}\mathbf{B} = \mathbf{A}$. ∎

The preceding theorem can also be used to deduce several results about n.n.d. matrices from corresponding results in Statistics. For example, *Corollary* 1 to *Theorem* 9.4.4 follows from the fact that if $\operatorname{var}(Z_i) = 0$ then Z_i is a constant, so $\operatorname{cov}(Z_i, Z_j) = 0$ for all j. *Theorem* 9.4.5 follows from the fact that a principal submatrix of a variance-covariance matrix is also a variance-covariance matrix. That the square of the coefficient of correlation between two random variables is at most 1 gives the result that any principal minor with order 2 of an n.n.d. matrix is non-negative. This last statement suggests that there may be a connection between the principal minors of $\mathbf{A}$ and the definiteness category of $\mathbf{A}$. We now proceed to establish this.

Theorem 9.4.8 If $\mathbf{A}$ is p.d. then all principal minors of $\mathbf{A}$ are positive. Conversely, if all the leading principal minors of a real symmetric matrix $\mathbf{A}$ are positive then $\mathbf{A}$ is p.d.

Proof Let $\mathbf{A}$ be p.d. Then $\mathbf{A} = \mathbf{B}^{\mathrm{T}}\mathbf{B}$ for some non-singular matrix $\mathbf{B}$, so $|\mathbf{A}| = |\mathbf{B}|^2 > 0$. Since every principal submatrix of $\mathbf{A}$ is also p.d., it follows that all principal minors of $\mathbf{A}$ are positive.

We prove the converse by induction on n. If $n = 1$ the result is trivial. So assume it for matrices of order $n - 1$ and let $\mathbf{A}$ be of order n. Let $\mathbf{A}_k$ denote the k-th order leading principal submatrix of $\mathbf{A}$. By hypothesis, $a_{11} > 0$. Make the $(i, 1)$-th element of $\mathbf{A}$ null by performing $R_{i1}(-a_{i1}/a_{11})$ and the $(1, i)$-th element 0 by $C_{i1}(-a_{1i}/a_{11})$ for $i = 2, \ldots, n$. Let $\mathbf{B}$ be the resulting matrix and let $\mathbf{C}$ be the matrix obtained from $\mathbf{B}$ by deleting the first row and the first column. Since $\mathbf{B}_k$ is obtained from $\mathbf{A}_k$ by elementary operations of the third type, it follows that $|\mathbf{A}_k| = |\mathbf{B}_k| = a_{11}|\mathbf{C}_{k-1}|$ for $k = 2, \ldots, n$. Hence the leading principal minors of $\mathbf{C}$ are positive and, by induction hypothesis, $\mathbf{C}$ is p.d. So $\mathbf{B} = \operatorname{diag}(a_{11}, \mathbf{C})$ is p.d. Since $\mathbf{A}$ is congruent to $\mathbf{B}$ it follows that $\mathbf{A}$ is p.d. ∎

Corollary $\mathbf{A}$ is n.d. iff all principal minors of $\mathbf{A}$ with even order are positive and all principal minors of $\mathbf{A}$ with odd order are negative.

It follows easily from the preceding theorem and *Exercise* 6.4.17(b) that every $n \times n$ matrix $\mathbf{A}$ with the property $a_{ii} > \sum_{j \neq i} |a_{ij}|$ for $i = 1, 2, \ldots, n$, is p.d.

Theorem 9.4.9 A real symmetric matrix $\mathbf{A}$ is n.n.d. iff all principal minors of $\mathbf{A}$ are non-negative.

Proof The *only if part* follows as in the preceding theorem since $\mathbf{A} = \mathbf{B}^{\mathrm{T}}\mathbf{B}$ for some square matrix $\mathbf{B}$.

We prove the *if part* by induction on n. If $n = 1$ the result is trivial. So assume it for matrices of order $n - 1$ and let $\mathbf{A}$ be of order n. If all diagonal entries of $\mathbf{A}$ are 0, then $|\mathbf{A}(i, j \,|\, i, j)| = -a_{ij}^2 \geq 0$, so $a_{ij} = 0$ for all i, j and $\mathbf{A} = \mathbf{0}$ is n.n.d. So let $\mathbf{A}$ have at least one positive diagonal entry, say the i-th. We can then make a_{11} positive by the operations R_{1i} and C_{1i}. Note that all principal minors of the new matrix would still be non-negative and the new matrix would have the same definiteness category as $\mathbf{A}$. We may thus assume without any loss of generality that $a_{11} > 0$. We now make a_{1i} and a_{i1} zero for $i = 2, \ldots, n$ as in the proof of the preceding theorem and consider $\mathbf{B}$ and $\mathbf{C}$ as defined there. Then

$$\begin{aligned} |\mathbf{A}(1, i_1, \ldots, i_k \,|\, 1, i_1, \ldots, i_k)| &= |\mathbf{B}(1, i_1, \ldots, i_k \,|\, 1, i_1, \ldots, i_k)| \\ &= a_{11}|\mathbf{C}(i_1 - 1, \ldots, i_k - 1 \,|\, i_1 - 1, \ldots, i_k - 1)| \end{aligned}$$

whenever $2 \leq i_1 < \cdots < i_k \leq n$. Hence all the principal minors of $\mathbf{C}$ are non-negative and, by induction hypothesis, $\mathbf{C}$ is n.n.d. So $\mathbf{B} = \operatorname{diag}(a_{11}, \mathbf{C})$ is n.n.d. Since $\mathbf{A}$ is congruent to $\mathbf{B}$, $\mathbf{A}$ is n.n.d. ∎

Corollary $\mathbf{A}$ is n.p.d. iff all principal minors of $\mathbf{A}$ with even order are non-negative and all principal minors of $\mathbf{A}$ with odd order are non-positive.

We point out that for the *if part* of *Theorem* 9.4.9, it is not enough to assume that the leading principal minors of $\mathbf{A}$ are non-negative. To see this consider the matrix $\operatorname{diag}(0, -1)$.

Theorem 9.4.10 Let $\mathbf{A}$ be an n.n.d. matrix of order n. Fix an i such that $1 \leq i \leq n$. Then $|\mathbf{A}| \leq a_{ii}A_{ii}$. Suppose now $\mathbf{A}$ is p.d. Then $|\mathbf{A}| = a_{ii}A_{ii}$ iff $a_{ij} = 0$ for all $j \neq i$.

Proof Without any loss of generality we may take $i = n$. If $|\mathbf{A}| = 0$ then $|\mathbf{A}| \leq a_{nn}A_{nn}$ follows from the preceding theorem. So let $|\mathbf{A}| \neq 0$. Then by the corollary to *Theorem* 9.3.7, $\mathbf{A}$ is p.d. Partition $\mathbf{A}$ as

$$\mathbf{A} = \begin{bmatrix} \mathbf{D} & \mathbf{u} \\ \mathbf{u}^{\mathrm{T}} & a_{nn} \end{bmatrix}$$

Then $\mathbf{D}$ is p.d. and, by the corollary to *Theorem* 9.4.2, $\mathbf{D}^{-1}$ is p.d. So $\mathbf{u}^{\mathrm{T}}\mathbf{D}^{-1}\mathbf{u} \geq 0$. Now by *Theorem* 6.7.1

$$|\mathbf{A}| = A_{nn}(a_{nn} - \mathbf{u}^{\mathrm{T}}\mathbf{D}^{-1}\mathbf{u}) \leq a_{nn}A_{nn}$$

Further, equality holds iff $\mathbf{u}^{\mathrm{T}}\mathbf{D}^{-1}\mathbf{u} = 0$ which happens iff $\mathbf{u} = \mathbf{0}$. ∎

The inequality of the preceding theorem can also be stated thus: if $\mathbf{A}$ is p.d., then $(\mathbf{A}^{-1})_{ii} \geq 1/a_{ii}$.

Theorem 9.4.11 If $\mathbf{A}$ is n.n.d., $|\mathbf{A}| \leq a_{11}a_{22}\cdots a_{nn}$ where n is the order of $\mathbf{A}$. If $\mathbf{A}$ is p.d. and equality holds then $\mathbf{A}$ is diagonal.

Proof By using the preceding theorem repeatedly we get

$$|\mathbf{A}| \leq a_{11}A_{11} \leq a_{11}a_{22}A_{\{1,2\},\{1,2\}} \leq \cdots \leq a_{11}a_{22}\cdots a_{nn}$$

Noting that $|\mathbf{A}| = a_{11}a_{22}\cdots a_{nn}$ iff equality holds throughout the above string of inequalities, the second conclusion of the theorem also follows from the preceding theorem. ∎

Corollary (Hadamard's inequality) For a real $n \times n$ matrix $\mathbf{B}$,

$$(\det \mathbf{B})^2 \leq \prod_{j=1}^{n}(b_{1j}^2 + b_{2j}^2 + \cdots + b_{nj}^2)$$

This corollary follows from the theorem by taking $\mathbf{A} = \mathbf{B}^{\mathrm{T}}\mathbf{B}$.

Exercises

1. If $\mathbf{A}$ is n.n.d., show that $\mathbf{x}^{\mathrm{T}}\mathbf{A}\mathbf{x} = 0$ iff $\mathbf{A}\mathbf{x} = \mathbf{0}$. Show also that $\mathbf{x}^{\mathrm{T}}\mathbf{A}\mathbf{x} = 0$ iff $\mathbf{y}^{\mathrm{T}}\mathbf{A}\mathbf{x} = 0$ for all $\mathbf{y}$. Hence deduce *Corollary* 1 to *Theorem* 9.4.4.
2. Let $\mathbf{A}$ be an $n \times n$ p.d. matrix and let $\mathbf{P}$ be an $n \times r$ matrix of rank r. Then show that $\mathbf{P}^{\mathrm{T}}\mathbf{A}\mathbf{P}$ is p.d.
3. If $\mathbf{A}$ is an n.n.d. matrix of order n with rank r and if $k \geq r$, prove that there exists an $n \times k$ matrix $\mathbf{C}$ such that $\mathbf{A} = \mathbf{C}\mathbf{C}^{\mathrm{T}}$. Note that if $k = r$, $(\mathbf{C}, \mathbf{C}^{\mathrm{T}})$ is a rank-factorization of $\mathbf{A}$.

4. Let $(\mathbf{C}, \mathbf{C}^{\mathrm{T}})$ be a rank-factorization of an n.n.d. matrix $\mathbf{A}$ of order n and let $\mathbf{C}_L^{-1}$ be a left inverse of $\mathbf{C}$.
   (a) Show that $([\mathbf{C} : \mathbf{u}], [\mathbf{C} : \mathbf{u}]^{\mathrm{T}})$ is a rank-factorization of $\mathbf{A} + \mathbf{u}\mathbf{u}^{\mathrm{T}}$ if $\mathbf{u} \notin \mathscr{C}(\mathbf{A})$.
   *(b) Let $\mathbf{u} \in \mathscr{C}(\mathbf{A})$, $\alpha = (\sqrt{1 \pm \mathbf{u}^{\mathrm{T}}\mathbf{A}^{-}\mathbf{u}} - 1)/(\mathbf{u}^{\mathrm{T}}\mathbf{A}^{-}\mathbf{u})$ and $\mathbf{T} = \mathbf{I} + \alpha\mathbf{C}_L^{-1}\mathbf{u}\mathbf{u}^{\mathrm{T}}(\mathbf{C}_L^{-1})^{\mathrm{T}}$. Show that $(\mathbf{CT}, (\mathbf{CT})^{\mathrm{T}})$ is a rank-factorization of $\mathbf{A} \pm \mathbf{u}\mathbf{u}^{\mathrm{T}}$. Assume that $\mathbf{u}^{\mathrm{T}}\mathbf{A}^{-}\mathbf{u} < 1$ while considering $\mathbf{A} - \mathbf{u}\mathbf{u}^{\mathrm{T}}$.
5. If $\mathbf{A}$ is a p.d. matrix and $\mathbf{W}^{\mathrm{T}}\mathbf{A}\mathbf{W}$ is defined then show that $\rho(\mathbf{W}^{\mathrm{T}}\mathbf{A}\mathbf{W}) = \rho(\mathbf{W})$, $\mathscr{R}(\mathbf{W}^{\mathrm{T}}\mathbf{A}\mathbf{W}) = \mathscr{R}(\mathbf{W})$ and $\mathbf{W}(\mathbf{W}^{\mathrm{T}}\mathbf{A}\mathbf{W})^{-}\mathbf{W}^{\mathrm{T}}$ is invariant under different choices of the g-inverse.
6. Let $\mathbf{A}$ be an n.n.d. matrix and let p be a positive integer. Show that there is a unique n.n.d. matrix $\mathbf{B}$ such that $\mathbf{B}^p = \mathbf{A}$. (Hint: use *Theorem* 8.7.4(iv)).
7. Show that $(1-\rho)\mathbf{I} + \rho\mathbf{1}\mathbf{1}^{\mathrm{T}}$ is p.d. iff $-\frac{1}{n-1} < \rho < 1$ where n is the order of the matrix and $\mathbf{1}^{\mathrm{T}} = (1, 1, \ldots, 1)$. Prove this using *Theorem* 9.4.8 and also by using *Theorem* 9.3.7.
8. Show that a symmetric matrix $\mathbf{A}$ is n.n.d. iff it has an n.n.d. g-inverse.
9. If $\mathbf{A}$ is a p.s.d. matrix of order n, show that there exists an n.n.d. matrix $\mathbf{B}$ of order n such that $\rho(\mathbf{A} + \mathbf{B}) = \rho(\mathbf{A}) + \rho(\mathbf{B}) = n$. Also then, prove that $(\mathbf{A} + \mathbf{B})^{-1}$ is a p.d. g-inverse of $\mathbf{A}$.
10. If $\mathbf{A}, \mathbf{B}$ are symmetric matrices of the same order, write $\mathbf{A} \geq \mathbf{B}$ if $\mathbf{A} - \mathbf{B}$ is n.n.d. Then prove the following:
    (a) $\mathbf{A} \geq \mathbf{B}$ and $\mathbf{B} \geq \mathbf{A} \Rightarrow \mathbf{A} = \mathbf{B}$
    (b) $\mathbf{A} \geq \mathbf{B}$ and $\mathbf{B} \geq \mathbf{C} \Rightarrow \mathbf{A} \geq \mathbf{C}$
    (c) $\mathbf{B}$ n.n.d. and $\mathbf{A} \geq \mathbf{B} \Rightarrow |\mathbf{A}| \geq |\mathbf{B}|$
    (d) $\mathbf{B}$ p.d., $\mathbf{A} \geq \mathbf{B}$ and $|\mathbf{A}| = |\mathbf{B}| \Rightarrow \mathbf{A} = \mathbf{B}$

    (Hint for (c) and (d): use *Theorem* 6.5.2.)
11. Let $\mathbf{A}$ be p.d. and $\mathbf{M} = \begin{bmatrix} \mathbf{A} & \mathbf{b} \\ \mathbf{b}^{\mathrm{T}} & d \end{bmatrix}$. Show that $\mathbf{M}$ is p.d., p.s.d. or indefinite according as $d - \mathbf{b}^{\mathrm{T}}\mathbf{A}^{-1}\mathbf{b}$ is positive, zero or negative.
12. Show that the matrix $\mathbf{A}_\theta$ of *Exercise* 3.8.13 is p.d. or p.s.d. according as $\theta^2 > \frac{1}{n-1}$ or $\theta^2 = \frac{1}{n-1}$.
13. Let $\mathbf{M} = \begin{bmatrix} \mathbf{A} & \mathbf{B} \\ \mathbf{B}^{\mathrm{T}} & \mathbf{D} \end{bmatrix}$ be symmetric, where $\mathbf{A}$ is square.
    (a) Prove that $\mathbf{M}$ is p.d. iff $\mathbf{A}$ and $\mathbf{D} - \mathbf{B}^{\mathrm{T}}\mathbf{A}^{-1}\mathbf{B}$ are p.d. Show also that $\mathbf{M}$ is p.d. iff $\mathbf{D}$ and $\mathbf{A} - \mathbf{B}\mathbf{D}^{-1}\mathbf{B}^{\mathrm{T}}$ are p.d.
    (b) Prove that $\mathbf{M}$ is n.n.d. iff $\mathbf{A}$ and $\mathbf{D} - \mathbf{B}^{\mathrm{T}}\mathbf{A}^{-}\mathbf{B}$ are n.n.d. and $\mathscr{C}(\mathbf{B}) \subseteq \mathscr{C}(\mathbf{A})$.

(c) If $\mathbf{M}$ is p.d. and $\mathbf{L}$ is the leading principal submatrix of $\mathbf{M}^{-1}$ with the same order as $\mathbf{A}$, prove that $\mathbf{L} - \mathbf{A}^{-1}$ is n.n.d.

(d) If $\mathbf{M}$ is n.n.d., prove that $|\mathbf{M}| \leq |\mathbf{A}| \cdot |\mathbf{D}|$. Suppose next $\mathbf{M}$ is p.d. Then prove that $|\mathbf{M}| = |\mathbf{A}| \cdot |\mathbf{D}|$ iff $\mathbf{B} = \mathbf{0}$.

14. If $\mathbf{A}$ is p.d., show that $\begin{bmatrix} \mathbf{A} & \mathbf{I} \\ \mathbf{I} & \mathbf{A}^{-1} \end{bmatrix}$ is p.s.d.

15. (a) Let $\mathbf{A}$ be p.d. Then show that $\mathbf{A} - \mathbf{b}\mathbf{b}^{\mathrm{T}}$ is p.d. iff $\mathbf{b}^{\mathrm{T}}\mathbf{A}^{-1}\mathbf{b} < 1$.

 (b) Let $\mathbf{A}$ be n.n.d. Then show that $\mathbf{A} - \mathbf{b}\mathbf{b}^{\mathrm{T}}$ is n.n.d. iff $\mathbf{b} \in \mathcal{C}(\mathbf{A})$ and $\mathbf{b}^{\mathrm{T}}\mathbf{A}^{-}\mathbf{b} \leq 1$.

16. If $\mathbf{E}$ is an $m \times n$ matrix with $m < n$, show that there exists an $(n-m) \times n$ matrix $\mathbf{H}$ such that $\left[\begin{smallmatrix} \mathbf{E} \\ \mathbf{H} \end{smallmatrix}\right]$ is p.d. iff the submatrix of $\mathbf{E}$ formed by the first m columns is p.d.

*17. If $\mathbf{A} = \mathbf{B} + \mathbf{C}$ where $\mathbf{B}$ is p.d. and $\mathbf{C}$ is (real and) skew-symmetric, prove that $|\mathbf{A}| \geq |\mathbf{B}|$. (Hint: use *Exercise* 8.7.4.)

*18. Show that a real matrix $\mathbf{A}$ is the product of a p.d. matrix and a real symmetric matrix iff $\mathbf{A}$ is semi-simple with real eigenvalues.

*19. Let $\mathbf{A}$ and $\mathbf{B}$ be n.n.d. Show that the eigenvalues of $\mathbf{AB}$ are non-negative. If $\mathbf{AB} \neq \mathbf{0}$, show that $\mathbf{AB}$ has a positive eigenvalue.

*20. Let $\mathbf{A}$ be p.d. and $\mathbf{B}$ real symmetric. Then prove that $\mathbf{A} + \mathbf{B}$ is p.d. iff each eigenvalue of $\mathbf{A}^{-1}\mathbf{B}$ is greater than -1.

21. (a) Let $\mathbf{A}$ be a real $n \times n$ matrix such that $|a_{ij}| \leq 1$ for all i, j.

 i. Prove that $|\det \mathbf{A}| \leq n^{n/2}$.

 ii. Show that $|\det \mathbf{A}| = n^{n/2}$ iff $a_{ij} = 1$ or -1 for all i, j and the rows (or columns) of $\mathbf{A}$ are pairwise orthogonal. Such a matrix $\mathbf{A}$ is called a *Hadamard matrix.*

 *(b) Prove that a necessary condition for the existence of a Hadamard matrix of order n is that $n = 2$ or n is a multiple of 4. (Whether this condition is also sufficient is an open problem.)

*22. If $\mathbf{A}$ and $\mathbf{B}$ are n.n.d. and if $\rho(\mathbf{A} - \mathbf{B}) = \rho(\mathbf{A}) - \rho(\mathbf{B})$, show that $\mathbf{A} - \mathbf{B}$ is n.n.d.

*23. Let $\mathbf{A}_{11}$, $\mathbf{B}_{11}$ and $\mathbf{C}_{11}$ be the $k \times k$ leading principal submatrices of $\mathbf{A}$, $\mathbf{B}$ and $\mathbf{C} := \mathbf{A} + \mathbf{B}$, where $\mathbf{A}$ and $\mathbf{B}$ are n.n.d. If $\rho(\mathbf{C}) = \rho(\mathbf{C}_{11})$, show that $\rho(\mathbf{A}) = \rho(\mathbf{A}_{11})$ and $\rho(\mathbf{B}) = \rho(\mathbf{B}_{11})$.

*24. Show that a $k \times \ell$ matrix $\mathbf{B}$ is a submatrix of some $n \times n$ orthogonal matrix iff no singular value of $\mathbf{B}$ is greater than 1 and the number of times 1 is a singular value is at least $k + \ell - n$.

*25. Let $\mathbf{V}$ be an n.n.d. matrix of order n and let $\mathbf{X}$ be a matrix with n rows.

 (a) Show that $\mathcal{C}(\mathbf{V} + \mathbf{X}\mathbf{X}^{\mathrm{T}}) = \mathcal{C}(\mathbf{X} : \mathbf{V}) = \mathcal{C}(\mathbf{X} : (\mathbf{I} - \mathbf{P}_{\mathbf{X}})\mathbf{V})$.

(b) Show that $\mathbf{V} + \mathbf{X}\mathbf{X}^{\mathrm{T}}$ is p.d. iff $\rho((\mathbf{I} - \mathbf{P_X})\mathbf{V}) = \rho(\mathbf{I} - \mathbf{P_X})$.

(c) Show that $\rho\begin{bmatrix} \mathbf{V} & \mathbf{X} \\ \mathbf{X}^{\mathrm{T}} & \mathbf{0} \end{bmatrix} = \rho(\mathbf{V} : \mathbf{X}) + \rho(\mathbf{X})$. (Hint: Show that $\mathcal{R}(\mathbf{V} : \mathbf{X}) \cap \mathcal{R}(\mathbf{X}^{\mathrm{T}} : \mathbf{0}) = \{\mathbf{0}\}$.)

(d) Show that $\mathcal{C}\begin{bmatrix} \mathbf{0} \\ \mathbf{X}^{\mathrm{T}} \end{bmatrix} \subseteq \mathcal{C}\begin{bmatrix} \mathbf{V} & \mathbf{X} \\ \mathbf{X}^{\mathrm{T}} & \mathbf{0} \end{bmatrix}$. (Hint: use *Theorem* 7.5.5.)

9.5 Extrema of quadratic forms

In this section we obtain some results on the extreme values of quadratic forms. It is easy to see that $\sup_{\mathbf{x}} \mathbf{x}^{\mathrm{T}}\mathbf{A}\mathbf{x}$ is ∞ if $P(\mathbf{A}) \geq 1$ and 0 otherwise, where $P(\mathbf{A})$ is the number of positive eigenvalues of $\mathbf{A}$. We next see what happens when $\mathbf{x}$ is restricted to have unit norm.

Theorem 9.5.1 For any quadratic form $\mathbf{x}^{\mathrm{T}}\mathbf{A}\mathbf{x}$,

$$\max_{\|\mathbf{x}\|=1} \mathbf{x}^{\mathrm{T}}\mathbf{A}\mathbf{x} = \max_{\mathbf{y}\neq\mathbf{0}} \frac{\mathbf{y}^{\mathrm{T}}\mathbf{A}\mathbf{y}}{\mathbf{y}^{\mathrm{T}}\mathbf{y}} = \lambda_1 \tag{9.5.1}$$

and

$$\min_{\|\mathbf{x}\|=1} \mathbf{x}^{\mathrm{T}}\mathbf{A}\mathbf{x} = \min_{\mathbf{y}\neq\mathbf{0}} \frac{\mathbf{y}^{\mathrm{T}}\mathbf{A}\mathbf{y}}{\mathbf{y}^{\mathrm{T}}\mathbf{y}} = \lambda_n \tag{9.5.2}$$

where λ_1 and λ_n are the largest and the smallest eigenvalues of $\mathbf{A}$. Moreover,

$$\frac{\mathbf{y}^{\mathrm{T}}\mathbf{A}\mathbf{y}}{\mathbf{y}^{\mathrm{T}}\mathbf{y}} = \lambda_1 \Leftrightarrow \mathbf{y} \text{ is an eigenvector of } \mathbf{A} \text{ corresponding to } \lambda_1 \tag{9.5.3}$$

and a similar statement holds for λ_n.

Proof Let $\lambda_1 \geq \lambda_2 \geq \cdots \geq \lambda_n$ be the characteristic roots of $\mathbf{A}$ and

$$\mathbf{A} = \lambda_1\mathbf{u}_1\mathbf{u}_1^{\mathrm{T}} + \lambda_2\mathbf{u}_2\mathbf{u}_2^{\mathrm{T}} + \cdots + \lambda_n\mathbf{u}_n\mathbf{u}_n^{\mathrm{T}}$$

be a spectral decomposition of $\mathbf{A}$. Since $\mathbf{u}_1, \ldots, \mathbf{u}_n$ form an orthonormal basis of $\mathbb{R}^n$, any $\mathbf{y}$ can be written as $\mathbf{y} = \alpha_1\mathbf{u}_1 + \cdots + \alpha_n\mathbf{u}_n$ for some α's and then $\mathbf{y}^{\mathrm{T}}\mathbf{A}\mathbf{y} = \lambda_1\alpha_1^2 + \cdots + \lambda_n\alpha_n^2$ and $\mathbf{y}^{\mathrm{T}}\mathbf{y} = \alpha_1^2 + \cdots + \alpha_n^2$. So $\mathbf{y}^{\mathrm{T}}\mathbf{A}\mathbf{y}/\mathbf{y}^{\mathrm{T}}\mathbf{y} \leq \lambda_1$. Now let $k = \max\{j : \lambda_1 = \lambda_j\}$. If $k = n$, $\mathbf{A} = \lambda_1\mathbf{I}$ and (9.5.3) is trivial. Next let $k < n$. Then each side of the equivalence in (9.5.3) holds iff $\alpha_{k+1} = \cdots = \alpha_n = 0$, so (9.5.3) follows. The remaining parts of the theorem are trivial. ∎

With λ's and $\mathbf{u}$'s as in the proof of the preceding theorem it is also easy to see that

$$\begin{aligned}\lambda_k &= \max\left\{\tfrac{\mathbf{y}^{\mathrm{T}}\mathbf{A}\mathbf{y}}{\mathbf{y}^{\mathrm{T}}\mathbf{y}} : \mathbf{u}_i^{\mathrm{T}}\mathbf{y} = 0 \text{ for } i = 1, \ldots, k-1\right\} \\ &= \min\left\{\tfrac{\mathbf{y}^{\mathrm{T}}\mathbf{A}\mathbf{y}}{\mathbf{y}^{\mathrm{T}}\mathbf{y}} : \mathbf{u}_i^{\mathrm{T}}\mathbf{y} = 0 \text{ for } i = k+1, \ldots, n\right\}\end{aligned}$$

Theorem 9.5.2 If $\mathbf{B}$ is a principal submatrix of a symmetric matrix $\mathbf{A}$, all eigenvalues of $\mathbf{B}$ lie between the smallest and the largest eigenvalues of $\mathbf{A}$.

This theorem follows easily from the preceding theorem. In *Theorem* 9.5.4 below, we give a refinement of this result. For this we need

Theorem 9.5.3 Let $\mathbf{A}$ be a real symmetric matrix of order n. Fix k, $2 \leq k \leq n$ and let $\mathbf{B}$ denote an $n \times (k-1)$ matrix. Then

$$\inf_{\mathbf{B}} \sup_{\mathbf{B}^{\mathrm{T}}\mathbf{y}=\mathbf{0}} \frac{\mathbf{y}^{\mathrm{T}}\mathbf{A}\mathbf{y}}{\mathbf{y}^{\mathrm{T}}\mathbf{y}} = \lambda_k \tag{9.5.4}$$

and

$$\sup_{\mathbf{B}} \inf_{\mathbf{B}^{\mathrm{T}}\mathbf{y}=\mathbf{0}} \frac{\mathbf{y}^{\mathrm{T}}\mathbf{A}\mathbf{y}}{\mathbf{y}^{\mathrm{T}}\mathbf{y}} = \lambda_{n-k+1} \tag{9.5.5}$$

where $\lambda_1 \geq \lambda_2 \geq \cdots \geq \lambda_n$ are the characteristic roots of $\mathbf{A}$.

Proof Let $\mathbf{u}_1, \mathbf{u}_2, \ldots, \mathbf{u}_n$ be real orthonormal eigenvectors of $\mathbf{A}$ corresponding to $\lambda_1, \lambda_2, \ldots, \lambda_n$ respectively and $\mathbf{U} = [\mathbf{u}_1 : \mathbf{u}_2 : \ldots : \mathbf{u}_n]$ so that $\mathbf{U}^{\mathrm{T}}\mathbf{A}\mathbf{U} = \operatorname{diag}(\lambda_1, \lambda_2, \ldots, \lambda_n)$. Let $\mathbf{B}$ be an $n \times (k-1)$ matrix. Writing $\mathbf{y} = \mathbf{U}\mathbf{z}$, we get

$$\sup_{\mathbf{B}^{\mathrm{T}}\mathbf{y}=\mathbf{0}} \frac{\mathbf{y}^{\mathrm{T}}\mathbf{A}\mathbf{y}}{\mathbf{y}^{\mathrm{T}}\mathbf{y}} = \sup_{\mathbf{B}^{\mathrm{T}}\mathbf{U}\mathbf{z}=\mathbf{0}} \frac{\mathbf{z}^{\mathrm{T}}\mathbf{U}^{\mathrm{T}}\mathbf{A}\mathbf{U}\mathbf{z}}{\mathbf{z}^{\mathrm{T}}\mathbf{U}^{\mathrm{T}}\mathbf{U}\mathbf{z}} = \sup_{\mathbf{B}^{\mathrm{T}}\mathbf{U}\mathbf{z}=\mathbf{0}} \frac{\sum_{i=1}^{n} \lambda_i z_i^2}{\sum_{i=1}^{n} z_i^2}$$

Since $\mathbf{B}^{\mathrm{T}}\mathbf{U}$ is of order $(k-1) \times n$, clearly there exists a non-null vector $\mathbf{w}$ such that $\mathbf{B}^{\mathrm{T}}\mathbf{U}\mathbf{w} = \mathbf{0}$ and $w_{k+1} = w_{k+2} = \cdots = w_n = 0$, so

$$\sup_{\mathbf{B}^{\mathrm{T}}\mathbf{U}\mathbf{z}=\mathbf{0}} \frac{\sum_{i=1}^{n} \lambda_i z_i^2}{\sum_{i=1}^{n} z_i^2} \geq \frac{\sum_{i=1}^{k} \lambda_i w_i^2}{\sum_{i=1}^{k} w_i^2} \geq \lambda_k$$

Since this is true for every $\mathbf{B}$, we get LHS $\geq$ RHS in (9.5.4). Taking $\mathbf{B} = [\,\mathbf{u}_1 : \ldots : \mathbf{u}_{k-1}\,]$, we get equality by the comment following *Theorem* 9.5.1. This proves (9.5.4). Equation (9.5.5) is proved similarly. ∎

Theorem 9.5.4 (*Interlacing Theorem or Sturmian Separation Theorem*) Let $\mathbf{A}$ be a real symmetric matrix of order n and let $\mathbf{A}_i$ denote the leading principal submatrix of $\mathbf{A}$ with order i, $i = 1, \ldots, n$. Let $\lambda_1(\mathbf{A}_i) \geq \lambda_2(\mathbf{A}_i) \geq \cdots \geq \lambda_i(\mathbf{A}_i)$ be the characteristic roots of $\mathbf{A}_i$ for each i. Then

$$\lambda_{k+1}(\mathbf{A}_{i+1}) \leq \lambda_k(\mathbf{A}_i) \leq \lambda_k(\mathbf{A}_{i+1})$$

Proof Fix k. Let $f(\mathbf{B}) = \sup\{\mathbf{y}^T\mathbf{A}\mathbf{y}/\mathbf{y}^T\mathbf{y} : \mathbf{B}^T\mathbf{y} = \mathbf{0}\}$ for any $n \times (k-1)$ matrix $\mathbf{B}$ and $g(\mathbf{D}) = \sup\{\mathbf{v}^T\mathbf{A}_{n-1}\mathbf{v}/\mathbf{v}^T\mathbf{v} : \mathbf{D}^T\mathbf{v} = \mathbf{0}\}$ for any $(n-1) \times (k-1)$ matrix $\mathbf{D}$. It is easy to see that if $\mathbf{D}$ is the submatrix of $\mathbf{B}$ formed by the first $n-1$ rows, then $f(\mathbf{B}) \geq g(\mathbf{D})$. So every $f(\mathbf{B}) \geq$ some $g(\mathbf{D})$ and hence $\inf_{\mathbf{B}} f(\mathbf{B}) \geq \inf_{\mathbf{D}} g(\mathbf{D})$. This proves $\lambda_k(\mathbf{A}_{n-1}) \leq \lambda_k(\mathbf{A})$. It follows similarly that $\lambda_k(\mathbf{A}_i) \leq \lambda_k(\mathbf{A}_{i+1})$ for all i.

To prove the other inequality, let $f(\mathbf{B}) = \sup\{\mathbf{y}^T\mathbf{A}\mathbf{y}/\mathbf{y}^T\mathbf{y} : \mathbf{B}^T\mathbf{y} = \mathbf{0}\}$ for any $n \times k$ matrix $\mathbf{B}$ and $g(\mathbf{D}) = \sup\{\mathbf{v}^T\mathbf{A}_{n-1}\mathbf{v}/\mathbf{v}^T\mathbf{v} : \mathbf{D}^T\mathbf{v} = \mathbf{0}\}$ for any $(n-1) \times (k-1)$ matrix $\mathbf{D}$. It is easy to see that if the matrix obtained from $\mathbf{B}$ by dropping the last row and the last column is $\mathbf{D}$ and if $\mathbf{B}_{*k} = \mathbf{e}_n$ then $f(\mathbf{B}) = g(\mathbf{D})$. So every $g(\mathbf{D})$ is an $f(\mathbf{B})$ and hence $\inf_{\mathbf{B}} f(\mathbf{B}) \leq \inf_{\mathbf{D}} g(\mathbf{D})$. This proves $\lambda_{k+1}(\mathbf{A}) \leq \lambda_k(\mathbf{A}_{n-1})$. It follows similarly that $\lambda_{k+1}(\mathbf{A}_{i+1}) \leq \lambda_k(\mathbf{A}_i)$ for all i. ∎

Theorem 9.5.5 (*Poincare Separation Theorem*) Let $\mathbf{A}$ be a real symmetric matrix of order n and $\mathbf{B}$ an $n \times k$ matrix such that $\mathbf{B}^T\mathbf{B} = \mathbf{I}$. Let $\lambda_i(\mathbf{A})$ denote the i-th largest characteristic root of $\mathbf{A}$. Then

$$\lambda_i(\mathbf{B}^T\mathbf{A}\mathbf{B}) \leq \lambda_i(\mathbf{A}) \quad \text{for } i = 1, \ldots, k$$
$$\lambda_{k-j}(\mathbf{B}^T\mathbf{A}\mathbf{B}) \geq \lambda_{n-j}(\mathbf{A}) \quad \text{for } j = 0, 1, \ldots, k-1$$

Proof Since $\mathbf{B}^T\mathbf{B} = \mathbf{I}$, there exists $\mathbf{C}$ such that $\mathbf{D} := [\mathbf{B} : \mathbf{C}]$ is orthogonal. Now $\mathbf{B}^T\mathbf{A}\mathbf{B}$ is the k-th order leading principal submatrix of $\mathbf{D}^T\mathbf{A}\mathbf{D} = \mathbf{D}^{-1}\mathbf{A}\mathbf{D}$. So the present theorem follows from the preceding theorem. ∎

Theorem 9.5.6 Let $\mathbf{B}$ be p.d. Then

$$\max_{\mathbf{x} \neq \mathbf{0}} \frac{\mathbf{x}^T\mathbf{A}\mathbf{x}}{\mathbf{x}^T\mathbf{B}\mathbf{x}} = \lambda_1(\mathbf{B}^{-1}\mathbf{A}) \tag{9.5.6}$$

where $\lambda_1(\mathbf{B}^{-1}\mathbf{A})$ is the largest eigenvalue of $\mathbf{B}^{-1}\mathbf{A}$. Also, the maximum is attained at $\mathbf{x}_0$ iff $\mathbf{x}_0$ is an eigenvector of $\mathbf{B}^{-1}\mathbf{A}$ corresponding to $\lambda_1(\mathbf{B}^{-1}\mathbf{A})$.

Proof Let $\mu = \lambda_1(\mathbf{B}^{-1}\mathbf{A})$ and $\mathbf{B} = \mathbf{C}^T\mathbf{C}$ where $\mathbf{C}$ is non-singular.

Writing $\mathbf{Cx} = \mathbf{y}$ we get

$$\max_{\mathbf{x}\neq\mathbf{0}} \frac{\mathbf{x}^{\mathrm{T}}\mathbf{A}\mathbf{x}}{\mathbf{x}^{\mathrm{T}}\mathbf{B}\mathbf{x}} = \max_{\mathbf{y}\neq\mathbf{0}} \frac{\mathbf{y}^{\mathrm{T}}(\mathbf{C}^{-1})^{\mathrm{T}}\mathbf{A}\mathbf{C}^{-1}\mathbf{y}}{\mathbf{y}^{\mathrm{T}}\mathbf{y}} = \lambda_1\left((\mathbf{C}^{-1})^{\mathrm{T}}\mathbf{A}\mathbf{C}^{-1}\right)$$

Now the characteristic roots of $(\mathbf{C}^{-1})^{\mathrm{T}}\mathbf{A}\mathbf{C}^{-1}$ are all real and, by *Theorem* 8.2.7, are the same as the characteristic roots of $\mathbf{C}^{-1}(\mathbf{C}^{-1})^{\mathrm{T}}\mathbf{A} = \mathbf{B}^{-1}\mathbf{A}$. Hence (9.5.6) follows. Also, $\mathbf{x}_0^{\mathrm{T}}\mathbf{A}\mathbf{x}_0/\mathbf{x}_0^{\mathrm{T}}\mathbf{B}\mathbf{x}_0 = \mu$ iff $\mathbf{C}\mathbf{x}_0$ is an eigenvector of $(\mathbf{C}^{-1})^{\mathrm{T}}\mathbf{A}\mathbf{C}^{-1}$ corresponding to μ which is equivalent to: $\mathbf{x}_0$ is an eigenvector of $\mathbf{B}^{-1}\mathbf{A}$ corresponding to μ. ∎

Theorem 9.5.7 Let $\mathbf{B}$ be p.d. and $\mathbf{u} \neq \mathbf{0}$. Then

$$\max_{\mathbf{x}\neq\mathbf{0}} \frac{(\mathbf{u}^{\mathrm{T}}\mathbf{x})^2}{\mathbf{x}^{\mathrm{T}}\mathbf{B}\mathbf{x}} = \mathbf{u}^{\mathrm{T}}\mathbf{B}^{-1}\mathbf{u}$$

and the maximum is attained at $\mathbf{x}_0$ iff $\mathbf{x}_0$ is a scalar multiple of $\mathbf{B}^{-1}\mathbf{u}$.

Proof Since $(\mathbf{u}^{\mathrm{T}}\mathbf{x})^2 = \mathbf{x}^{\mathrm{T}}\mathbf{u}\mathbf{u}^{\mathrm{T}}\mathbf{x}$ and $\mathbf{u}^{\mathrm{T}}\mathbf{B}^{-1}\mathbf{u} > 0$, it follows that the required maximum is $\lambda_1(\mathbf{B}^{-1}\mathbf{u}\mathbf{u}^{\mathrm{T}}) = \lambda_1(\mathbf{u}^{\mathrm{T}}\mathbf{B}^{-1}\mathbf{u}) = \mathbf{u}^{\mathrm{T}}\mathbf{B}^{-1}\mathbf{u}$. Moreover, the maximum is attained at $\mathbf{x}_0$ iff $\mathbf{B}^{-1}\mathbf{u}\mathbf{u}^{\mathrm{T}}\mathbf{x}_0 = (\mathbf{u}^{\mathrm{T}}\mathbf{B}^{-1}\mathbf{u})\mathbf{x}_0$. Since $\mathbf{u}^{\mathrm{T}}\mathbf{x}_0$ is a scalar, this is equivalent to: $\mathbf{x}_0$ is a scalar multiple of $\mathbf{B}^{-1}\mathbf{u}$. ∎

We next obtain the maximum of a *bilinear form* $\mathbf{y}^{\mathrm{T}}\mathbf{A}\mathbf{x}$.

Theorem 9.5.8 Let $\mathbf{A}$ be a real $m \times n$ matrix. Then

$$\max_{\|\mathbf{x}\|=1,\,\|\mathbf{y}\|=1} \mathbf{y}^{\mathrm{T}}\mathbf{A}\mathbf{x} = \max_{\mathbf{s}\neq\mathbf{0},\,\mathbf{t}\neq\mathbf{0}} \frac{\mathbf{t}^{\mathrm{T}}\mathbf{A}\mathbf{s}}{\sqrt{(\mathbf{t}^{\mathrm{T}}\mathbf{t})(\mathbf{s}^{\mathrm{T}}\mathbf{s})}} = d_1 \tag{9.5.7}$$

where d_1 is the largest singular value of $\mathbf{A}$. The maximum of $\mathbf{y}^{\mathrm{T}}\mathbf{A}\mathbf{x}$ in (9.5.7) is attained at $\mathbf{x}_0$ and $\mathbf{y}_0$, where $\mathbf{x}_0$ and $\mathbf{y}_0$ are normalized singular vectors of $\mathbf{A}$ corresponding to d_1.

Proof Let $d_1 \geq d_2 \geq \cdots \geq d_r$ be the singular values of $\mathbf{A}$ and $\sum_{k=1}^{r} d_k\mathbf{u}_k\mathbf{v}_k^{\mathrm{T}}$ a singular value decomposition of $\mathbf{A}$, where $\mathbf{u}_1, \ldots, \mathbf{u}_m$ form an orthonormal basis of $\mathbb{R}^m$ and $\mathbf{v}_1, \ldots, \mathbf{v}_n$ form an orthonormal basis of $\mathbb{R}^n$. Now let $\mathbf{t} = \sum_{i=1}^{m} \alpha_i\mathbf{u}_i$ and $\mathbf{s} = \sum_{j=1}^{n} \beta_j\mathbf{v}_j$. Then

$$\begin{aligned} |\mathbf{t}^{\mathrm{T}}\mathbf{A}\mathbf{s}| &= |\textstyle\sum_{k=1}^{r} d_k\alpha_k\beta_k| \leq d_1 \sum_{k=1}^{r} |\alpha_k\beta_k| \\ &\leq d_1\sqrt{(\textstyle\sum_{k=1}^{r} \alpha_k^2)(\sum_{k=1}^{r} \beta_k^2)} \leq d_1\sqrt{(\sum_{i=1}^{m} \alpha_i^2)(\sum_{j=1}^{n} \beta_j^2)} \\ &= d_1\sqrt{(\mathbf{t}^{\mathrm{T}}\mathbf{t})(\mathbf{s}^{\mathrm{T}}\mathbf{s})} \end{aligned}$$

Also, $\mathbf{t}^{\mathrm{T}}\mathbf{A}\mathbf{s} = d_1$ and $\mathbf{t}^{\mathrm{T}}\mathbf{t} = \mathbf{s}^{\mathrm{T}}\mathbf{s} = 1$ if $\mathbf{t} = \mathbf{u}_1$ and $\mathbf{s} = \mathbf{v}_1$. ∎

We next determine the minimum value of a p.d. quadratic form in $\mathbf{x}$ when $\mathbf{x}$ is subject to linear constraints.

Theorem 9.5.9 Let $\mathbf{x}^T\mathbf{\Lambda x}$ be a p.d. quadratic form and let $\mathbf{Ax} = \mathbf{b}$ be a consistent system. Then

$$\min_{\mathbf{Ax}=\mathbf{b}} \mathbf{x}^T\mathbf{\Lambda x} = \mathbf{b}^T\mathbf{S}^-\mathbf{b}$$

where $\mathbf{S} = \mathbf{A\Lambda}^{-1}\mathbf{A}^T$. Also the minimum is attained at the unique point $\mathbf{x}_0 = \mathbf{\Lambda}^{-1}\mathbf{A}^T\mathbf{S}^-\mathbf{b}$.

Proof We first consider the case $\mathbf{\Lambda} = \mathbf{I}$. By *Theorem* 5.4.7, $\mathbf{G} := \mathbf{A}^T(\mathbf{AA}^T)^-$ is a g-inverse of $\mathbf{A}$. Let W be the flat $\{\mathbf{x} : \mathbf{Ax} = \mathbf{b}\} = \mathbf{Gb} + \mathcal{N}(\mathbf{A})$. Then $\mathbf{Gb} \in W$ and $-\mathbf{Gb} \in \mathcal{C}(\mathbf{A}^T) = (\mathcal{N}(\mathbf{A}))^\perp$, so $\mathbf{Gb}$ is the orthogonal projection of $\mathbf{0}$ into W. Hence by *Theorem* 7.5.10, $\min\{\mathbf{x}^T\mathbf{x} : \mathbf{Ax} = \mathbf{b}\}$ is attained at $\mathbf{Gb}$ and at no other point. Since $(\mathbf{AA}^T)^-$ in $\mathbf{G}$ can be taken to be a reflexive g-inverse of $\mathbf{AA}^T$, we get $\mathbf{b}^T\mathbf{G}^T\mathbf{Gb} = \mathbf{b}^T(\mathbf{AA}^T)^-\mathbf{b}$. This proves the theorem when $\mathbf{\Lambda} = \mathbf{I}$. The general case can be reduced to the special case by writing $\mathbf{\Lambda} = \mathbf{CC}^T$ where $\mathbf{C}$ is non-singular. We leave it to the reader to work out the details. ∎

Exercises

1. Prove *Theorem* 9.5.2.
2. If $\mathbf{J}$ denotes the $n \times n$ matrix with each entry 1 and $\mathbf{x} \in \mathbb{R}^n$ show that $\mathbf{x}^T\mathbf{Jx} \leq n\ \mathbf{x}^T\mathbf{x}$. (Prove this in two ways: first directly from Cauchy-Schwarz inequality and second from *Theorem* 9.5.7).
3. (a) If α is an eigenvalue of an $n \times n$ real symmetric matrix $\mathbf{A}$, show that $|\alpha| \leq n(\max_{i,j}|a_{ij}|)$.
   (b) If $\lambda_1, \lambda_2, \ldots, \lambda_n$ are the characteristic roots of a matrix $\mathbf{A}$, prove using *Theorem* 8.7.5 that $\sum_{i=1}^n |\lambda_i|^2 \leq \sum_{i,j=1}^n |a_{ij}|^2$. Deduce using the inequality between A.M. and G.M. that $|\det \mathbf{A}|^2 \leq (\frac{1}{n}\sum_{i,j}|a_{ij}|^2)^n$.
4. Show that for any $m \times n$ matrix $\mathbf{A}$, $\max\{\|\mathbf{Ax}\| : \|\mathbf{x}\| = 1\}$ is the square-root of the largest eigenvalue of $\mathbf{A}^T\mathbf{A}$. If $\mathbf{A}$ is a real $n \times n$ normal matrix with eigenvalues $\lambda_1, \ldots, \lambda_n$, show that $\max\{\|\mathbf{Ax}\| : \|\mathbf{x}\| = 1\} = \max_i|\lambda_i|$.
5. Prove *Theorem* 9.5.7 by writing $\mathbf{B} = \mathbf{C}^T\mathbf{C}$ where $\mathbf{C}$ is non-singular and applying Cauchy-Schwarz inequality to $\mathbf{Cx}$ and $(\mathbf{C}^{-1})^T\mathbf{u}$.

6. (a) Show that the largest singular value of a square matrix $\mathbf{A}$ is $\|\mathbf{A}\|$ where $\|\cdot\|$ is the matrix norm induced by the Euclidean norm for vectors.

   (b) Deduce that the modulus of any eigenvalue of $\mathbf{A}$ cannot be greater than the largest singular value of $\mathbf{A}$.

   (c) Consider the norm $\|\mathbf{A}\|_E$ of *Exercise* 7.3.8 and the norm $\|\mathbf{A}\|_2$ induced by the L_2-norm on $\mathbb{R}^n$ (see *Exercise* 7.3.15) for an $n \times n$ matrix $\mathbf{A}$. Show that $\|\mathbf{A}\|_2 \leq \|\mathbf{A}\|_E \leq n^{1/2}\|\mathbf{A}\|_2$.

9.6 Simultaneous diagonalization

In this section we study the simultaneous reduction of two or more quadratic forms to diagonal forms by the same non-singular transformation. We note that this is not always possible since the quadratic forms $x_1^2 - x_2^2$ and $x_1^2 + x_2^2 + 2x_1x_2$ cannot be simultaneously diagonalized by a non-singular transformation (verify this!).

Theorem 9.6.1 If at least one of $\mathbf{A}$ and $\mathbf{B}$ is p.d. or n.d., then $\mathbf{x}^{\mathrm{T}}\mathbf{A}\mathbf{x}$ and $\mathbf{x}^{\mathrm{T}}\mathbf{B}\mathbf{x}$ can be simultaneously reduced to diagonal forms by a non-singular transformation.

Proof Without any loss of generality we may take $\mathbf{B}$ to be p.d. Then by *Theorem* 9.4.2 there exists a non-singular matrix $\mathbf{M}$ such that $\mathbf{M}^{\mathrm{T}}\mathbf{B}\mathbf{M} = \mathbf{I}$. Now $\mathbf{M}^{\mathrm{T}}\mathbf{A}\mathbf{M}$ is symmetric, so there exists an orthogonal matrix $\mathbf{P}$ such that $\mathbf{P}^{\mathrm{T}}\mathbf{M}^{\mathrm{T}}\mathbf{A}\mathbf{M}\mathbf{P}$ is diagonal. Clearly $\mathbf{P}^{\mathrm{T}}\mathbf{M}^{\mathrm{T}}\mathbf{B}\mathbf{M}\mathbf{P} = \mathbf{P}^{\mathrm{T}}\mathbf{P} = \mathbf{I}$ and $\mathbf{x} = \mathbf{M}\mathbf{P}\mathbf{y}$ is the required the transformation. ∎

Remark We have actually proved that if $\mathbf{B}$ is p.d. then there exists a non-singular matrix $\mathbf{S}$ such that $\mathbf{x} = \mathbf{S}\mathbf{y}$ transforms $\mathbf{x}^{\mathrm{T}}\mathbf{A}\mathbf{x}$ to a diagonal form $\mathbf{y}^{\mathrm{T}}\boldsymbol{\Delta}\mathbf{y}$ and $\mathbf{x}^{\mathrm{T}}\mathbf{B}\mathbf{x}$ to $\mathbf{y}^{\mathrm{T}}\mathbf{y}$.

We now identify the diagonal elements $\alpha_1, \alpha_2, \ldots, \alpha_n$ of $\boldsymbol{\Delta}$ and the columns $\mathbf{x}_1, \mathbf{x}_2, \ldots, \mathbf{x}_n$ of the matrix $\mathbf{S}$. Since $\mathbf{S}^{\mathrm{T}}\mathbf{B}\mathbf{S} = \mathbf{I}$ iff $\mathbf{S}^{\mathrm{T}} = \mathbf{S}^{-1}\mathbf{B}^{-1}$,

$$\begin{aligned}
\mathbf{S}^{\mathrm{T}}\mathbf{A}\mathbf{S} = \boldsymbol{\Delta} \text{ and } \mathbf{S}^{\mathrm{T}}\mathbf{B}\mathbf{S} = \mathbf{I} &\Leftrightarrow \mathbf{S}^{\mathrm{T}}\mathbf{B}\mathbf{S} = \mathbf{I} \text{ and } \mathbf{S}^{-1}\mathbf{B}^{-1}\mathbf{A}\mathbf{S} = \boldsymbol{\Delta} \\
&\Leftrightarrow \mathbf{x}_i^{\mathrm{T}}\mathbf{B}\mathbf{x}_j = \delta_{ij} \text{ and } \mathbf{B}^{-1}\mathbf{A}\mathbf{x}_i = \alpha_i\mathbf{x}_i \quad \text{for all } i \text{ and } j \\
&\Leftrightarrow \mathbf{x}_i^{\mathrm{T}}\mathbf{B}\mathbf{x}_j = \delta_{ij} \text{ and } (\mathbf{A} - \alpha_i\mathbf{B})\mathbf{x}_i = \mathbf{0} \quad \text{for all } i \text{ and } j \qquad (9.6.1)
\end{aligned}$$

$\mathbf{S}^{-1}\mathbf{B}^{-1}\mathbf{A}\mathbf{S} = \boldsymbol{\Delta}$ shows that $\alpha_1, \alpha_2, \ldots, \alpha_n$ are the characteristic roots of $\mathbf{B}^{-1}\mathbf{A}$ or, equivalently, the roots of $|\mathbf{A} - \lambda\mathbf{B}| = 0$. We now show that the condition $\mathbf{x}_i^{\mathrm{T}}\mathbf{B}\mathbf{x}_j = 0$ in (9.6.1) can be dropped whenever $\alpha_i \neq \alpha_j$.

Theorem 9.6.2 Let $(\mathbf{A} - \alpha\mathbf{B})\mathbf{x} = \mathbf{0}$ and $(\mathbf{A} - \beta\mathbf{B})\mathbf{y} = \mathbf{0}$, where $\mathbf{A}$ and $\mathbf{B}$ are symmetric and $\alpha \neq \beta$. Then $\mathbf{x}^{\mathrm{T}}\mathbf{B}\mathbf{y} = 0$.

Proof $\alpha\,\mathbf{x}^{\mathrm{T}}\mathbf{B}\mathbf{y} = \alpha\,\mathbf{y}^{\mathrm{T}}\mathbf{B}\mathbf{x} = \mathbf{y}^{\mathrm{T}}\mathbf{A}\mathbf{x} = \mathbf{x}^{\mathrm{T}}\mathbf{A}\mathbf{y} = \beta\,\mathbf{x}^{\mathrm{T}}\mathbf{B}\mathbf{y}$. Since $\alpha \neq \beta$, it follows that $\mathbf{x}^{\mathrm{T}}\mathbf{B}\mathbf{y} = \mathbf{0}$. ∎

We can now give a procedure to find $\mathbf{S}$: we first find the roots of $|\mathbf{A} - \lambda\mathbf{B}| = 0$. Let $\beta_1, \ldots, \beta_k$ be the distinct roots, β_i occurring n_i times. We then find n_i vectors $\mathbf{x}_{i1}, \mathbf{x}_{i2}, \ldots, \mathbf{x}_{in_i}$ belonging to the solution space of $(\mathbf{A} - \beta_i\mathbf{B})\mathbf{x} = \mathbf{0}$ which are orthonormal with respect to the inner product $\langle \mathbf{x}, \mathbf{y} \rangle = \mathbf{y}^{\mathrm{T}}\mathbf{B}\mathbf{x}$. (Such vectors are said to be **B**-*orthonormal* and the norm induced by this inner product is called **B**-*norm*.) Taking

$$\mathbf{S} = [\,\mathbf{x}_{11} : \cdots : \mathbf{x}_{1n_1} : \mathbf{x}_{21} : \cdots : \mathbf{x}_{2n_2} : \cdots : \mathbf{x}_{k1} : \cdots : \mathbf{x}_{kn_k}\,]$$

we have $\mathbf{S}^{\mathrm{T}}\mathbf{A}\mathbf{S} = \operatorname{diag}(\beta_1\mathbf{I}_{n_1}, \beta_2\mathbf{I}_{n_2}, \ldots, \beta_k\mathbf{I}_{n_k})$ and $\mathbf{S}^{\mathrm{T}}\mathbf{B}\mathbf{S} = \mathbf{I}$.

Example 9.6.3 We will simultaneously diagonalize the quadratic forms $\mathbf{x}^{\mathrm{T}}\mathbf{A}\mathbf{x}$ and $\mathbf{x}^{\mathrm{T}}\mathbf{B}\mathbf{x}$ where

$$\mathbf{A} = \begin{bmatrix} 10 & 2 & 6 \\ 2 & 3 & -8 \\ 6 & -8 & 14 \end{bmatrix} \quad \text{and} \quad \mathbf{B} = \begin{bmatrix} 5 & 1 & 3 \\ 1 & 3 & -1 \\ 3 & -1 & 13 \end{bmatrix}$$

Here it can be verified that

$$|\mathbf{A} - \lambda\mathbf{B}| = -144(\lambda - 2)^2(\lambda + 1)$$

(it will be convenient to apply the row operations $R_{12}(-5)$ and $R_{32}(-3)$ to evaluate this determinant). Thus the distinct roots of $|\mathbf{A} - \lambda\mathbf{B}| = 0$ are 2 and -1, the former occurring twice and the latter once.

So we have to find two solutions of $(\mathbf{A} - 2\mathbf{B})\mathbf{x} = \mathbf{0}$ which are **B**-orthonormal. The system $(\mathbf{A} - 2\mathbf{B})\mathbf{x} = \mathbf{0}$ is easily seen to be equivalent to the single equation $x_2 + 2x_3 = 0$. We may start with an arbitrary basis of the solution space and get a **B**-orthonormal basis by the Gram-Schmidt process. We shall take $\{(1,0,0)^{\mathrm{T}}, (0,2,-1)^{\mathrm{T}}\}$ as the initial basis. The **B**-norm of $(1,0,0)^{\mathrm{T}}$ is $\sqrt{5}$, so the first vector of the orthonormal basis is $(1/\sqrt{5}, 0, 0)^{\mathrm{T}}$. We next consider

$$\begin{aligned} &(0,2,-1)^{\mathrm{T}} - \left\langle (0,2,-1)^{\mathrm{T}}, \left(\tfrac{1}{\sqrt{5}}, 0, 0\right)^{\mathrm{T}} \right\rangle \left(\tfrac{1}{\sqrt{5}}, 0, 0\right)^{\mathrm{T}} \\ &\quad = (0,2,-1)^{\mathrm{T}} - \left(\tfrac{1}{\sqrt{5}}, 0, 0\right)\mathbf{B}\begin{bmatrix} 0 \\ 2 \\ -1 \end{bmatrix} \left(\tfrac{1}{\sqrt{5}}, 0, 0\right)^{\mathrm{T}} = \left(\tfrac{1}{5}, 2, -1\right)^{\mathrm{T}} \end{aligned}$$

It can be verified that the **B**-norm of this vector is $12/\sqrt{5}$, so the second vector of the **B**-orthonormal basis is $\left(\frac{\sqrt{5}}{60}, \frac{\sqrt{5}}{6}, -\frac{\sqrt{5}}{12}\right)^{\mathrm{T}}$.

We next find a **B**-normal solution of $(\mathbf{A}+\mathbf{B})\mathbf{x} = \mathbf{0}$. It can be checked that the dimension of the solution space is 1 and that $(-1,2,1)^{\mathrm{T}}$ is a solution. The **B**-norm of this vector is 4, so $\left(-\frac{1}{4}, \frac{1}{2}, \frac{1}{4}\right)^{\mathrm{T}}$ is a **B**-normal solution. We now form the matrix

$$\mathbf{S} = \begin{bmatrix} 1/\sqrt{5} & \sqrt{5}/60 & -1/4 \\ 0 & \sqrt{5}/6 & 1/2 \\ 0 & -\sqrt{5}/12 & 1/4 \end{bmatrix}$$

It can be verified that $\mathbf{S}^{\mathrm{T}}\mathbf{A}\mathbf{S} = \mathrm{diag}(2,2,-1)$ and $\mathbf{S}^{\mathrm{T}}\mathbf{B}\mathbf{S} = \mathbf{I}$. ∎

We now consider a generalization of *Theorem* 9.6.1 to the case when one of the matrices is n.n.d. (or n.p.d.). We prove

Theorem 9.6.4 Let **A** and **B** be $n \times n$ real symmetric matrices such that **A** is n.n.d. and $\mathcal{C}(\mathbf{B}) \subseteq \mathcal{C}(\mathbf{A})$. Then there exists a non-singular matrix **P** such that $\mathbf{P}^{\mathrm{T}}\mathbf{A}\mathbf{P}$ and $\mathbf{P}^{\mathrm{T}}\mathbf{B}\mathbf{P}$ are diagonal.

Proof By *Exercise* 9.4.3, $\mathbf{A} = \mathbf{C}^{\mathrm{T}}\mathbf{C}$ for some $r \times n$ matrix **C**, where $r = \rho(\mathbf{A})$. Then $\mathcal{C}(\mathbf{B}) \subseteq \mathcal{C}(\mathbf{A}) = \mathcal{C}(\mathbf{C}^{\mathrm{T}})$, so by *Exercise* 3.6.6(b), $\mathbf{B} = \mathbf{C}^{\mathrm{T}}\mathbf{D}\mathbf{C}$ for some $r \times r$ real symmetric matrix **D**. Now $\mathbf{D} = \mathbf{S}^{\mathrm{T}}\mathbf{E}\mathbf{S}$ for some orthogonal matrix **S** and some diagonal matrix **E**. So $\mathbf{A} = \mathbf{C}^{\mathrm{T}}\mathbf{S}^{\mathrm{T}}\mathbf{S}\mathbf{C}$ and $\mathbf{B} = \mathbf{C}^{\mathrm{T}}\mathbf{S}^{\mathrm{T}}\mathbf{E}\mathbf{S}\mathbf{C}$. Since **SC** is an $r \times n$ matrix with rank r, there exists a non-singular matrix **P** such that $\mathbf{SCP} = [\,\mathbf{I}_r : \mathbf{0}\,]$. Then

$$\mathbf{P}^{\mathrm{T}}\mathbf{A}\mathbf{P} = \mathbf{P}^{\mathrm{T}}\mathbf{C}^{\mathrm{T}}\mathbf{S}^{\mathrm{T}}\mathbf{S}\mathbf{C}\mathbf{P} = \mathrm{diag}(\mathbf{I}_r : \mathbf{0})$$

and

$$\mathbf{P}^{\mathrm{T}}\mathbf{B}\mathbf{P} = \mathbf{P}^{\mathrm{T}}\mathbf{C}^{\mathrm{T}}\mathbf{S}^{\mathrm{T}}\mathbf{E}\mathbf{S}\mathbf{C}\mathbf{P} = \begin{bmatrix} \mathbf{I}_r \\ \mathbf{0} \end{bmatrix} \mathbf{E}\,[\,\mathbf{I}_r : \mathbf{0}\,] = \begin{bmatrix} \mathbf{E} & \mathbf{0} \\ \mathbf{0} & \mathbf{0} \end{bmatrix}.$$ ∎

Theorem 9.6.5 If $\mathbf{x}^{\mathrm{T}}\mathbf{A}\mathbf{x}$ and $\mathbf{x}^{\mathrm{T}}\mathbf{B}\mathbf{x}$ are both n.n.d. or both n.p.d., they can be simultaneously reduced to diagonal forms by a non-singular transformation.

Proof It is enough to consider the case when **A** and **B** are n.n.d. Then by *Theorem* 9.4.6, $\mathbf{A}+\mathbf{B}$ is n.n.d. and $\mathcal{C}(\mathbf{B}) \subseteq \mathcal{C}(\mathbf{A}+\mathbf{B})$. So, by the preceding theorem, there exists a non-singular matrix **P** such that $\mathbf{P}^{\mathrm{T}}(\mathbf{A}+\mathbf{B})\mathbf{P}$ and $\mathbf{P}^{\mathrm{T}}\mathbf{B}\mathbf{P}$ and so $\mathbf{P}^{\mathrm{T}}\mathbf{A}\mathbf{P}$ are diagonal. ∎

Simultaneous diagonalization is a powerful technique and can be used to derive some inequalities involving n.n.d. matrices. We now give two applications. Some others can be found in Mirsky (1955) and Rao (1973).

Theorem 9.6.6 If $\mathbf{A}$ and $\mathbf{B}$ are n.n.d. matrices of order n,

$$|\mathbf{A}+\mathbf{B}| \geq |\mathbf{A}| + |\mathbf{B}| \tag{9.6.2}$$

Moreover, if $n \geq 2$ and $\mathbf{A}$ and $\mathbf{B}$ are p.d. then strict inequality holds in (9.6.2).

Proof By the preceding theorem there exists a non-singular $\mathbf{P}$ such that $\mathbf{P}^{\mathrm{T}}\mathbf{A}\mathbf{P} = \operatorname{diag}(\alpha_1, \alpha_2, \ldots, \alpha_n)$ and $\mathbf{P}^{\mathrm{T}}\mathbf{B}\mathbf{P} = \operatorname{diag}(\beta_1, \beta_2, \ldots, \beta_n)$ for some α's and β's. Then $\mathbf{P}^{\mathrm{T}}(\mathbf{A}+\mathbf{B})\mathbf{P} = \operatorname{diag}(\alpha_1+\beta_1, \ldots, \alpha_n+\beta_n)$. Since α's and β's are non-negative we get

$$|\mathbf{A}+\mathbf{B}| = \frac{(\alpha_1+\beta_1)\cdots(\alpha_n+\beta_n)}{|\mathbf{P}|^2} \geq \frac{\alpha_1\cdots\alpha_n}{|\mathbf{P}|^2} + \frac{\beta_1\cdots\beta_n}{|\mathbf{P}|^2} = |\mathbf{A}| + |\mathbf{B}|$$

If $\mathbf{A}$ and $\mathbf{B}$ are p.d. and $n \geq 2$, then α's and β's are strictly positive and strict inequality holds above. ∎

The preceding theorem can also be proved using Cauchy-Binet formula (*Theorem* 6.5.2). See *Exercise* 9.4.10.

Theorem 9.6.7 If $\mathbf{A}$ and $\mathbf{B}$ are p.d. matrices of the same order and if $\mathbf{A}-\mathbf{B}$ is n.n.d., then $\mathbf{B}^{-1}-\mathbf{A}^{-1}$ is n.n.d.

Proof By *Theorem* 9.6.1, there exists a non-singular matrix $\mathbf{P}$ such that $\mathbf{C} := \mathbf{P}^{\mathrm{T}}\mathbf{A}\mathbf{P}$ and $\mathbf{D} := \mathbf{P}^{\mathrm{T}}\mathbf{B}\mathbf{P}$ are diagonal. So by hypothesis $\mathbf{C}-\mathbf{D} = \mathbf{P}^{\mathrm{T}}(\mathbf{A}-\mathbf{B})\mathbf{P}$ is n.n.d. and $c_{ii} \geq d_{ii}$. Hence $1/d_{ii} \geq 1/c_{ii}$ and $\mathbf{D}^{-1}-\mathbf{C}^{-1}$ is n.n.d. So $\mathbf{B}^{-1}-\mathbf{A}^{-1} = \mathbf{P}(\mathbf{D}^{-1}-\mathbf{C}^{-1})\mathbf{P}^{\mathrm{T}}$ is n.n.d. ∎

In *Theorems* 9.6.1, 9.6.4 and 9.6.5, we gave only sufficient conditions for two quadratic forms to be simultaneously diagonalizable. (For a more extensive treatment, see Rao and Mitra (1971).) However, if we restrict ourselves to orthogonal transformations, we can give a *simple* necessary and sufficient condition. What is more, we can extend it to any finite number of quadratic forms. This result finds much use in linear statistical models.

Theorem 9.6.8 The n-ary quadratic forms $\mathbf{x}^{\mathrm{T}}\mathbf{A}_1\mathbf{x}, \mathbf{x}^{\mathrm{T}}\mathbf{A}_2\mathbf{x}, \ldots, \mathbf{x}^{\mathrm{T}}\mathbf{A}_p\mathbf{x}$ can be simultaneously reduced to diagonal forms by an orthogonal transformation iff $\mathbf{A}_1, \mathbf{A}_2, \ldots, \mathbf{A}_p$ commute pairwise.

Proof The *only if part* is trivial since if $\mathbf{P}$ is orthogonal and $\mathbf{P}^T\mathbf{A}_i\mathbf{P}$ and $\mathbf{P}^T\mathbf{A}_j\mathbf{P}$ are diagonal then $\mathbf{P}^T\mathbf{A}_i\mathbf{P}$ and $\mathbf{P}^T\mathbf{A}_j\mathbf{P}$ commute, so $\mathbf{A}_i$ and $\mathbf{A}_j$ commute.

We prove the *if part* by induction on n. We know the result for $n = 1$. So assume it for matrices of order $n-1$ and let $\mathbf{A}_1, \mathbf{A}_2, \ldots, \mathbf{A}_p$ be pairwise commuting real symmetric matrices of order n.

Let $\mathbf{x}_1$ be a real eigenvector of $\mathbf{A}_1$ (corresponding to λ_1, say). Then by *Theorem* 8.3.9 $\mathbf{A}_2$ has a real eigenvector $\mathbf{x}_2$ of the form $\sum_{i=0}^{k} \alpha_i \mathbf{A}_2^i \mathbf{x}_1$. Now

$$\mathbf{A}_1\mathbf{x}_2 = \sum_{i=0}^{k} \alpha_i \mathbf{A}_1\mathbf{A}_2^i\mathbf{x}_1 = \sum_{i=0}^{k} \alpha_i \mathbf{A}_2^i\mathbf{A}_1\mathbf{x}_1 = \sum_{i=0}^{k} \alpha_i \mathbf{A}_2^i\lambda_1\mathbf{x}_1 = \lambda_1\mathbf{x}_2$$

Thus $\mathbf{x}_2$ is a common eigenvector of $\mathbf{A}_1$ and $\mathbf{A}_2$, corresponding possibly to different eigenvalues. Next let $\mathbf{x}_3$ be a real eigenvector of $\mathbf{A}_3$ in the span of $\{\mathbf{x}_2, \mathbf{A}_3\mathbf{x}_2, \mathbf{A}_3^2\mathbf{x}_2, \ldots\}$. Then $\mathbf{x}_3$ is a common eigenvector of $\mathbf{A}_1$, $\mathbf{A}_2$ and $\mathbf{A}_3$. Proceeding thus we finally get a common real eigen vector $\mathbf{x}$ for $\mathbf{A}_1, \mathbf{A}_2, \ldots, \mathbf{A}_p$. Without any loss of generality we may assume that $\|\mathbf{x}\| = 1$. Let $\mathbf{Q}$ be an orthogonal matrix with $\mathbf{x}$ as the first column. Then

$$\mathbf{Q}^T\mathbf{A}_i\mathbf{Q} = \begin{bmatrix} \lambda_i & \mathbf{0} \\ \mathbf{0} & \mathbf{B}_i \end{bmatrix}, \quad i = 1, \ldots, p$$

for some real symmetric matrices $\mathbf{B}_1, \ldots, \mathbf{B}_p$, where λ_i is the eigenvalue of $\mathbf{A}_i$ corresponding to the eigenvector $\mathbf{x}$. Since $\mathbf{A}_i$'s commute pairwise it follows that $\mathbf{Q}^T\mathbf{A}_i\mathbf{Q}$'s and so $\mathbf{B}_i$'s commute pairwise. Hence by induction hypothesis, there exists an orthogonal matrix $\mathbf{S}$ such that $\mathbf{S}^T\mathbf{B}_i\mathbf{S}$ is diagonal for $i = 1, \ldots, p$. Now $\mathbf{P} := \mathbf{Q}\,\mathrm{diag}(1, \mathbf{S})$ is orthogonal and

$$\mathbf{P}^T\mathbf{A}_i\mathbf{P} = \begin{bmatrix} 1 & \mathbf{0} \\ \mathbf{0} & \mathbf{S}^T \end{bmatrix} \mathbf{Q}^T\mathbf{A}_i\mathbf{Q} \begin{bmatrix} 1 & \mathbf{0} \\ \mathbf{0} & \mathbf{S} \end{bmatrix} = \begin{bmatrix} \lambda_i & \mathbf{0} \\ \mathbf{0} & \mathbf{S}^T\mathbf{B}_i\mathbf{S} \end{bmatrix}$$

is diagonal for $i = 1, \ldots, p$. ∎

We now give an alternative proof of the *if part* of the preceding theorem which is more convenient for finding the matrix $\mathbf{P}$. This proof is by induction on p. We know the result for $p = 1$ (note that the hypothesis is satisfied vacuously). So assume it for $p-1$ and let $\mathbf{A}_1, \mathbf{A}_2, \ldots, \mathbf{A}_p$ commute pairwise, where $p \geq 2$. By induction hypothesis, there exists an orthogonal matrix $\mathbf{Q}$ such that $\mathbf{D}_k := \mathbf{Q}^T\mathbf{A}_k\mathbf{Q}$ is diagonal for $k = 1, \ldots, p-1$. Now partition the set $\{1, 2, \ldots, n\}$ by putting i and j in the same block iff $(\mathbf{D}_k)_{ii} = (\mathbf{D}_k)_{jj}$ for $k = 1, \ldots, p-1$. By replacing $\mathbf{Q}$ by $\mathbf{Q}$ times a permutation matrix if necessary, we may assume that the blocks of the above partition consist of the first n_1 elements, the next n_2 elements, ..., the last n_s elements. Denote $\mathbf{Q}^{\mathrm{T}}\mathbf{A}_p\mathbf{Q}$ by $\mathbf{C}$. Since $\mathbf{A}_k\mathbf{A}_p = \mathbf{A}_p\mathbf{A}_k$, we have $\mathbf{D}_k\mathbf{C} = \mathbf{C}\mathbf{D}_k$. Equating the (i,j)-th elements, we get $(\mathbf{D}_k)_{ii}c_{ij} = c_{ij}(\mathbf{D}_k)_{jj}$ for $k = 1, \ldots, p-1$. If i and j belong to different blocks of the partition then $(\mathbf{D}_k)_{ii} \neq (\mathbf{D}_k)_{jj}$ for some k, so $c_{ij} = 0$. Thus

$$\mathbf{C} = \operatorname{diag}(\mathbf{C}_1, \mathbf{C}_2, \ldots, \mathbf{C}_s)$$

where $\mathbf{C}_t$ is a matrix of order n_t for $t = 1, \ldots, s$. Clearly $\mathbf{C}$ and so each $\mathbf{C}_t$ is symmetric. Hence there exists an orthogonal matrix $\mathbf{U}_t$ of order n_t such that $\mathbf{U}_t^{\mathrm{T}}\mathbf{C}_t\mathbf{U}_t$ is diagonal. Taking $\mathbf{U} = \operatorname{diag}(\mathbf{U}_1, \ldots, \mathbf{U}_s)$ and $\mathbf{P} = \mathbf{QU}$, it is easy to see that $\mathbf{P}$ is orthogonal and

$$\mathbf{P}^{\mathrm{T}}\mathbf{A}_p\mathbf{P} = \mathbf{U}^{\mathrm{T}}\mathbf{Q}^{\mathrm{T}}\mathbf{A}_p\mathbf{QU} = \mathbf{U}^{\mathrm{T}}\mathbf{CU} = \operatorname{diag}(\mathbf{U}_1^{\mathrm{T}}\mathbf{C}_1\mathbf{U}_1, \ldots, \mathbf{U}_s^{\mathrm{T}}\mathbf{C}_s\mathbf{U}_s)$$

is diagonal. Next fix any k, $1 \le k \le p-1$, and let the first n_1 diagonal entries of $\mathbf{D}_k$ be δ_1, the next n_2 be $\delta_2, \ldots,$ the last n_s be δ_s. Then

$$\begin{aligned} \mathbf{P}^{\mathrm{T}}\mathbf{A}_k\mathbf{P} &= \mathbf{U}^{\mathrm{T}}\mathbf{Q}^{\mathrm{T}}\mathbf{A}_k\mathbf{QU} = \mathbf{U}^{\mathrm{T}}\mathbf{D}_k\mathbf{U} = \operatorname{diag}(\delta_1\mathbf{U}_1^{\mathrm{T}}\mathbf{U}_1, \ldots, \delta_s\mathbf{U}_s^{\mathrm{T}}\mathbf{U}_s) \\ &= \operatorname{diag}(\delta_1\mathbf{I}_{n_1}, \ldots, \delta_s\mathbf{I}_{n_s}) \end{aligned}$$

is also diagonal. Thus the orthogonal transformation $\mathbf{x} = \mathbf{Py}$ simultaneously diagonalizes $\mathbf{x}^{\mathrm{T}}\mathbf{A}_1\mathbf{x}, \ldots, \mathbf{x}^{\mathrm{T}}\mathbf{A}_p\mathbf{x}$. ∎

Exercises

1. Show that the quadratic forms $x_1^2 - x_2^2$ and $x_1^2 + x_2^2 + 2x_1x_2$ cannot be simultaneously diagonalized by a non-singular transformation.
2. If $\mathbf{B}$ is p.d. and $\mathbf{A}$ is symmetric, show that every root of $|\mathbf{A} - \lambda\mathbf{B}| = 0$ is real.
3. (a) Find a non-singular matrix $\mathbf{P}$ such that $\mathbf{P}^{\mathrm{T}}\mathbf{AP}$ is diagonal and $\mathbf{P}^{\mathrm{T}}\mathbf{BP} = \mathbf{I}$, where $\mathbf{A} = \begin{bmatrix} 2 & 10 & -6 \\ 10 & -14 & 10 \\ -6 & 10 & -7 \end{bmatrix}$ and $\mathbf{B} = \begin{bmatrix} 6 & 3 & -1 \\ 3 & 17 & -10 \\ -1 & -10 & 6 \end{bmatrix}$.

   (b) Do (a) with $\mathbf{A}$ replaced by $\begin{bmatrix} 1 & 2 & -1 \\ 2 & 4 & -2 \\ -1 & -2 & 1 \end{bmatrix}$.
4. Find an orthogonal matrix $\mathbf{P}$ such that $\mathbf{P}^{\mathrm{T}}\mathbf{A}_i\mathbf{P}$ is diagonal, $i = 1, 2, 3$, where $\mathbf{A}_1 = \begin{bmatrix} 9 & -3 & -6 \\ -3 & 9 & -6 \\ -6 & -6 & 0 \end{bmatrix}$, $\mathbf{A}_2 = \begin{bmatrix} 13 & 1 & 2 \\ 1 & 13 & 2 \\ 2 & 2 & 16 \end{bmatrix}$ and $\mathbf{A}_3 = \begin{bmatrix} 12 & 0 & 6 \\ 0 & 12 & 6 \\ 6 & 6 & 18 \end{bmatrix}$.

5. Let $\mathbf{A}_1, \mathbf{A}_2, \ldots, \mathbf{A}_k$ be symmetric matrices of order n and let $\mathbf{A}_1$ be p.d. Then show that there exists a non-singular matrix $\mathbf{P}$ such that $\mathbf{P}^T\mathbf{A}_i\mathbf{P}$ is diagonal for $i = 1, \ldots, k$ iff $\mathbf{A}_i\mathbf{A}_1^{-1}\mathbf{A}_j = \mathbf{A}_j\mathbf{A}_1^{-1}\mathbf{A}_i$ for all $i, j = 1, \ldots, k$.
6. Show that if two $n \times n$ (not necessarily symmetric) matrices $\mathbf{A}$ and $\mathbf{B}$ commute, then there exists a unitary matrix $\mathbf{P}$ such that $\mathbf{P}^*\mathbf{AP}$ and $\mathbf{P}^*\mathbf{BP}$ are upper triangular.
7. If $\mathbf{A}$ and $\mathbf{B}$ are p.d. matrices of the same order and $0 < \alpha < 1$, prove that $|\alpha\mathbf{A} + (1-\alpha)\mathbf{B}| \geq |\mathbf{A}|^\alpha |B|^{1-\alpha}$
8. Using the result that

$$(\alpha_1 \cdots \alpha_n)^{1/n} + (\beta_1 \cdots \beta_n)^{1/n} \leq \{(\alpha_1 + \beta_1) \cdots (\alpha_n + \beta_n)\}^{1/n}$$

for positive real numbers $\alpha_1, \ldots, \alpha_n, \beta_1, \ldots, \beta_n$, prove that if $\mathbf{A}$ and $\mathbf{B}$ are $n \times n$ p.d. matrices, then $|\mathbf{A} + \mathbf{B}|^{1/n} \geq |\mathbf{A}|^{1/n} + |\mathbf{B}|^{1/n}$.
9. If $\mathbf{A}$ and $\mathbf{B}$ are real symmetric matrices such that $\mathbf{A}$ is p.d. and each characteristic root of $\mathbf{A}^{-1}\mathbf{B}$ is 1, show that $\mathbf{A} = \mathbf{B}$.
10. Let $\mathbf{A}_1, \mathbf{A}_2, \ldots, \mathbf{A}_k$ be $n \times n$ complex matrices. Then show that there exists a unitary matrix $\mathbf{P}$ such that $\mathbf{P}^*\mathbf{A}_i\mathbf{P}$ is diagonal for $i = 1, \ldots, k$ iff $\mathbf{A}_1, \ldots, \mathbf{A}_k$ are normal and commute pairwise,
11. If $\mathbf{A}$ and $\mathbf{B}$ are n.n.d. matrices such that $\mathbf{A}^2 = \mathbf{B}^2$, prove that $\mathbf{A} = \mathbf{B}$. (See also *Exercise* 9.4.6.)
12. Let $\mathbf{A}$ and $\mathbf{C}$ be real symmetric, $\lambda_1 \geq \cdots \geq \lambda_n$ be the characteristic roots of $\mathbf{A}$ and $\mu_1 \geq \cdots \geq \mu_n$ be the characteristic roots of $\mathbf{C}$. Show that if $\mathbf{A} - \mathbf{C}$ is n.n.d., then $\lambda_i \geq \mu_i$ for all i and that the converse is false.
13. Let $\mathbf{A}$ be n.n.d., $\mathbf{B}$ be real symmetric and let $\mathcal{C}(\mathbf{B}) \subseteq \mathcal{C}(\mathbf{A})$. If $\mathbf{P}$ is a non-singular matrix such that $\mathbf{P}^T\mathbf{AP} = \text{diag}(\mathbf{I}_r, \mathbf{0})$ and $\mathbf{\Delta} := \mathbf{P}^T\mathbf{BP}$ is diagonal, show that the diagonal entries of $\mathbf{\Delta}$ are the characteristic roots of $\mathbf{A}^-\mathbf{B}$.
14. Let $\mathbf{A}$ and $\mathbf{B}$ be p.d. matrices of the same order. Show that $\mathbf{A}^{-1} + \mathbf{B}^{-1} - 4(\mathbf{A} + \mathbf{B})^{-1}$ is n.n.d. Do this in two ways: (i) by using *Theorem* 9.6.5 and (ii) by using *Exercise* 9.4.14.
15. Let $\mathbf{A}$ and $\mathbf{B}$ be n.n.d. matrices of the same order. Prove the following:
    (a) $\mathbf{A}(\mathbf{A} + \mathbf{B})^-\mathbf{B}$ is invariant under different choices of the g-inverse of $\mathbf{A} + \mathbf{B}$. (This is called the *parallel sum* of $\mathbf{A}$ and $\mathbf{B}$.)
    (b) $\mathbf{A}(\mathbf{A} + \mathbf{B})^-\mathbf{B}$ is n.n.d.
    (c) There exist n.n.d. g-inverses $\mathbf{A}^-$ and $\mathbf{B}^-$ of $\mathbf{A}$ and $\mathbf{B}$ respectively such that $\mathbf{A}^- + \mathbf{B}^-$ is a g-inverse of $\mathbf{A}(\mathbf{A} + \mathbf{B})^-\mathbf{B}$.
16. Let $\mathbf{A}$, $\mathbf{B}$ and $\mathbf{A} - \mathbf{B}$ be n.n.d. Show that there exist n.n.d. g-inverses $\mathbf{A}^-$ and $\mathbf{B}^-$ of $\mathbf{A}$ and $\mathbf{B}$ respectively such that $\mathbf{B}^- - \mathbf{A}^-$ is n.n.d.

17. Let $\mathbf{X} = \begin{bmatrix} \mathbf{X}_1 \\ \mathbf{X}_2 \end{bmatrix}$. Partition $\mathbf{P_X}$ accordingly as $\begin{bmatrix} \mathbf{R} & \mathbf{Q} \\ \mathbf{Q}^{\mathrm{T}} & \mathbf{S} \end{bmatrix}$. Show that $\mathcal{R}(\mathbf{X}_2) \subseteq \mathcal{R}(\mathbf{X}_1)$ iff $\mathbf{I} - \mathbf{S}$ is non-singular.
18. Prove *Theorem* 9.6.7 by considering $\begin{bmatrix} \mathbf{A} & \mathbf{I} \\ \mathbf{I} & \mathbf{B}^{-1} \end{bmatrix}$ and using the result of *Exercise* 9.4.13(b).

9.7 Square-root method

In *Section* 9.4 we saw that any n.n.d. matrix $\mathbf{A}$ can be written as $\mathbf{B}^{\mathrm{T}}\mathbf{B}$. In this section we prove that $\mathbf{B}$ can be chosen to be upper triangular, give a procedure for computing such a 'square-root' $\mathbf{B}$ and show how it can be used to solve $\mathbf{Ax} = \mathbf{b}$ when it is consistent, to find a g-inverse of $\mathbf{A}$, etc. The inner products considered in this section are the Euclidean inner products. As in *Theorem* 5.6.1, we start with a matrix $\widetilde{\mathbf{A}}$ instead of $\mathbf{A}$.

Theorem 9.7.1 Let $\widetilde{\mathbf{A}} = ((a_{ij}))$ be a real $n \times p$ matrix with $p \geq n$, such that the submatrix $\mathbf{A}$ formed by the first n columns is n.n.d. Then there exists a real $n \times p$ matrix $\widetilde{\mathbf{T}} = ((t_{ij}))$ such that

(i) the submatrix $\mathbf{T}$ of $\widetilde{\mathbf{T}}$ formed by the first n columns is upper triangular,

(ii) if any diagonal entry of $\mathbf{T}$ is 0, the corresponding row of $\mathbf{T}$ is null and

(iii) $\widetilde{\mathbf{A}} = \mathbf{L}\widetilde{\mathbf{T}}$. *Here and in what follows, given an upper triangular matrix* $\mathbf{T}$, *we will denote by* $\mathbf{L}$ *the non-singular lower triangular matrix obtained from* $\mathbf{T}^{\mathrm{T}}$ *by replacing the* 0*'s, if any, on the diagonal by* 1*'s.*

Proof Suppose for a moment that $\widetilde{\mathbf{T}}$ exists satisfying (i)–(iii). Then $\mathbf{LT} = \mathbf{T}^{\mathrm{T}}\mathbf{T}$ since the (i, j)-th element of both sides is $(\mathbf{T}_{*i})^{\mathrm{T}}\mathbf{T}_{*j}$ in view of (ii). Thus $\mathbf{A} = \mathbf{LT} = \mathbf{T}^{\mathrm{T}}\mathbf{T}$. Equating the (k, k)-th elements, we get

$$a_{kk} = \sum_{i=1}^{k-1} t_{ik}^2 + t_{kk}^2 \quad \text{for } k = 1, \ldots, n \tag{9.7.1}$$

Equating the (k, j)-th elements of $\widetilde{\mathbf{A}}$ and $\mathbf{L}\widetilde{\mathbf{T}}$ gives

$$a_{kj} = \sum_{i=1}^{k-1} t_{ik}t_{ij} + \ell_{kk}t_{kj} \quad \text{for } k = 1, \ldots, n \text{ and } j = k+1, \ldots, p \tag{9.7.2}$$

Conversely, suppose $\widetilde{\mathbf{T}}$ is a real $n \times p$ matrix such that (a) $t_{kj} = 0$ whenever $j < k$, (b) $t_{kk} = 0 \Rightarrow t_{kj} = 0$ for $j = k+1, \ldots, n$ and (9.7.1) and (9.7.2) hold, where ℓ_{kk} is 1 or t_{kk} according as t_{kk} is 0 or not. Let $\mathbf{L}$ be defined as in (iii) above. Then (9.7.1) and (9.7.2) give

$$(\widetilde{\mathbf{A}})_{kj} = (\mathbf{L}\widetilde{\mathbf{T}})_{kj} \tag{9.7.3}$$

whenever $k \leq j$. By (b), $\mathbf{LT} = \mathbf{T}^{\mathrm{T}}\mathbf{T}$. Thus $\mathbf{A}$ and $\mathbf{LT}$ are symmetric, so (9.7.3) holds even when $k > j$. Hence $\widetilde{\mathbf{A}} = \mathbf{L}\widetilde{\mathbf{T}}$ and $\widetilde{\mathbf{T}}$ is a matrix as required in the theorem. We also note that (9.7.1) and (9.7.2) involve only the first k rows of $\widetilde{\mathbf{A}}$ and $\widetilde{\mathbf{T}}$.

We now show how to construct $\widetilde{\mathbf{T}}$ row-wise. The first row is obtained by defining t_{11} to be $\sqrt{a_{11}}$ and t_{1j} to be a_{1j}/ℓ_{11} where ℓ_{11} is t_{11} or 1 according as $t_{11} \neq 0$ or $t_{11} = 0$. Then (9.7.1) and (9.7.2) hold for $k = 1$. Note also that $t_{11} = 0 \Rightarrow t_{1j} = 0$ since $a_{11} = 0 \Rightarrow a_{1j} = 0$ for $j = 1, \ldots, n$, $\mathbf{A}$ being an n.n.d. matrix. Finally, t_{11} and so t_{1j} are real since $a_{11} \geq 0$.

After the first m rows of $\widetilde{\mathbf{T}}$ are obtained with the required properties, we obtain the $(m+1)$-th row thus: define

$$t_{m+1,1} = \cdots = t_{m+1,m} = 0 \tag{9.7.4}$$

$$t_{m+1,m+1} = \sqrt{a_{m+1,m+1} - \textstyle\sum_{i=1}^{m} t_{i,m+1}^2} \tag{9.7.5}$$

and

$$t_{m+1,j} = \frac{1}{\ell_{m+1,m+1}}\left(a_{m+1,j} - \sum_{i=1}^{m} t_{i,m+1}t_{ij}\right) \quad \text{for } j = m+2, \ldots, p \tag{9.7.6}$$

Then, clearly, (9.7.1) and (9.7.2) hold for $k = m+1$.

We now prove (b) for $k = m+1$. Let $t_{m+1,m+1} = 0$ and $m+2 \leq j \leq n$. Let $\mathbf{B}$ and $\mathbf{W}$ be the leading principal $(m+1) \times (m+1)$ submatrices of $\mathbf{A}$ and $\mathbf{T}$ respectively. Also let $\mathbf{b}$ and $\mathbf{w}$ be the vectors formed by the first $m+1$ components of the j-th columns of $\mathbf{A}$ and $\mathbf{T}$ respectively. Finally, let $\mathbf{M}$ be the leading principal $(m+1) \times (m+1)$ submatrix of $\mathbf{L}$. (Note that $\mathbf{W}$, $\mathbf{w}$ and so $\mathbf{M}$ have been computed by now.) Then by (9.7.4), (9.7.5), (9.7.6) and the fact that $\mathbf{A}$ is n.n.d., we have $\mathbf{Mw} = \mathbf{b} \in \mathscr{C}(\mathbf{B}) = \mathscr{C}(\mathbf{MW})$. Since $\mathbf{M}$ is non-singular, we get $\mathbf{w} \in \mathscr{C}(\mathbf{W})$. Since the last row of $\mathbf{W}$ is null, it follows that the last component of $\mathbf{w}$, which is $t_{m+1,j}$, is 0. This proves (b) for $k = m+1$.

We finally prove that $t_{m+1,m+1}$ and so $t_{m+1,j}$ are real for all j. (Note that the proof in the preceding paragraph is valid since all the matrices can be considered to be over $\mathbb{C}$.) Clearly $t^2_{m+1,m+1}$ (i.e., the number under the square-root sign in (9.7.5)) is real. We have only to show that it is non-negative. So let $t^2_{m+1,m+1} \neq 0$. Let $N = \{k : 1 \leq k \leq m+1,\ t_{kk} \neq 0\}$, $\mathbf{X} = \mathbf{T}(N|N)$ and $\mathbf{C} = \mathbf{A}(N|N)$. Then $\mathbf{C} = \mathbf{X}^{\mathrm{T}}\mathbf{X}$. Since $\mathbf{X}$ is (upper triangular and) non-singular, it follows that $\mathbf{X}^{\mathrm{T}}$ and $\mathbf{C}$ are non-singular. Since $\mathbf{C}$ is n.n.d., it follows that $|\mathbf{C}| = \prod x^2_{ii} > 0$. Since $x^2_{ii} > 0$ except possibly for the last one, it follows that $x^2_{ii} > 0$ for all i. Thus $t^2_{m+1,m+1} > 0$. This proves that $t_{m+1,j}$ is real for all j.

Thus the $(m+1)$-th row of $\widetilde{\mathbf{T}}$ as computed has the required properties. Proceeding thus we finally get the $n \times p$ matrix $\widetilde{\mathbf{T}}$ as stated in the theorem. ∎

Remark The matrix $\widetilde{\mathbf{T}}$ in the preceding theorem is unique if we insist that $t_{ii} \geq 0$ for all i. This is because t_{kk} and t_{kj} are then uniquely determined from (9.7.1) and (9.7.2). In particular, if $\mathbf{A}$ is p.d., there exists a unique real upper triangular matrix $\mathbf{T}$ with positive diagonal entries such that $\mathbf{A} = \mathbf{T}^{\mathrm{T}}\mathbf{T}$.

Theorem 9.7.2 Let $\widetilde{\mathbf{A}}$, $\widetilde{\mathbf{T}}$ and $\mathbf{L}$ be as in the preceding theorem and let $\mathbf{P} = \mathbf{L}^{-1}$. Let the zero diagonal elements of $\mathbf{T}$ occur in the h_1-th, $\ldots, h_\nu$-th positions. Let $[\,\widetilde{\mathbf{A}}_{*j_1} : \widetilde{\mathbf{A}}_{*j_2} : \cdots : \widetilde{\mathbf{A}}_{*j_n}\,] = \mathbf{I}_n$. Then

(i) $\mathbf{P} = (\widetilde{\mathbf{T}}_{*j_1} : \widetilde{\mathbf{T}}_{*j_2} : \cdots : \widetilde{\mathbf{T}}_{*j_n})$,

(ii) $\mathbf{P}^{\mathrm{T}}\mathbf{P}$ is a p.d. g-inverse of $\mathbf{A}$,

(iii) the system $\mathbf{Ax} = \widetilde{\mathbf{A}}_{*j}$ is consistent iff $t_{h_1 j} = \cdots = t_{h_\nu j} = 0$,

(iv) If the system $\mathbf{Ax} = \widetilde{\mathbf{A}}_{*j}$ is consistent, a general solution is

$$\mathbf{P}^{\mathrm{T}}\widetilde{\mathbf{T}}_{*j} + \mathbf{Bz}$$

where $\mathbf{z}$ is arbitrary and $\mathbf{B} = [\mathbf{P}_{h_1 *}{}^{\mathrm{T}} : \cdots : \mathbf{P}_{h_\nu *}{}^{\mathrm{T}}]$. (If we do not have $\mathbf{P}$, a general solution can be obtained by solving the equivalent system $\mathbf{Tx} = \widetilde{\mathbf{T}}_{*j}$ by the method of back substitution. While doing so, $x_{h_1}, x_{h_2}, \ldots, x_{h_\nu}$ can be taken to be arbitrary.)

(v) If $\mathbf{b} = \widetilde{\mathbf{A}}_{*j}$ and $\mathbf{c} = \widetilde{\mathbf{A}}_{*s}$ belong to $\mathcal{C}(\mathbf{A})$ then $\mathbf{c}^{\mathrm{T}}\mathbf{A}^{-}\mathbf{b} = (\widetilde{\mathbf{T}}_{*s})^{\mathrm{T}}\widetilde{\mathbf{T}}_{*j}$

(vi) $|\mathbf{A}| = t^2_{11}t^2_{22}\cdots t^2_{nn}$ and

(vii) $\rho(\mathbf{A})$ is the number of non-zero t_{ii}'s.

Proof Statement (i) follows on observing that $\mathbf{P}\widetilde{\mathbf{A}} = \widetilde{\mathbf{T}}$. Since

$\mathbf{PA} = \mathbf{T}$, we have $\mathbf{A} = \mathbf{T}^{\mathrm{T}}\mathbf{T} = \mathbf{A}^{\mathrm{T}}\mathbf{P}^{\mathrm{T}}\mathbf{PA} = \mathbf{AP}^{\mathrm{T}}\mathbf{PA}$ and (ii) follows. To prove (iii), we first note that the systems $\mathbf{Ax} = \widetilde{\mathbf{A}}_{*j}$ and $\mathbf{Tx} = \widetilde{\mathbf{T}}_{*j}$ are equivalent. Now (iii) follows since $\mathbf{T}$ is upper triangular and the k-th row of $\mathbf{T}$ is null iff $t_{kk} = 0$. This also proves the statement in (iv) enclosed in parentheses. By (ii), a general solution of $\mathbf{Ax} = \widetilde{\mathbf{A}}_{*j}$ is $\mathbf{P}^{\mathrm{T}}\mathbf{P}\widetilde{\mathbf{A}}_{*j} + (\mathbf{I} - \mathbf{P}^{\mathrm{T}}\mathbf{PA})\mathbf{y} = \mathbf{P}^{\mathrm{T}}\widetilde{\mathbf{T}}_{*j} + (\mathbf{I} - \mathbf{P}^{\mathrm{T}}\mathbf{T})\mathbf{y}$ where $\mathbf{y}$ is arbitrary. Now $\mathcal{C}(\mathbf{I} - \mathbf{P}^{\mathrm{T}}\mathbf{T}) = \mathcal{C}(\mathbf{P}^{\mathrm{T}}(\mathbf{L}^{\mathrm{T}} - \mathbf{T})) = \mathcal{C}(\mathbf{B})$ and (iv) follows. To prove (v), let $\mathbf{b} = \mathbf{Au}$ and $\mathbf{c} = \mathbf{Av}$. Then $\mathbf{Tu} = \widetilde{\mathbf{T}}_{*j}$ and $\mathbf{Tv} = \widetilde{\mathbf{T}}_{*s}$, so

$$\mathbf{c}^{\mathrm{T}}\mathbf{A}^{-}\mathbf{b} = \mathbf{v}^{\mathrm{T}}\mathbf{AA}^{-}\mathbf{Au} = \mathbf{v}^{\mathrm{T}}\mathbf{Au} = \mathbf{v}^{\mathrm{T}}\mathbf{T}^{\mathrm{T}}\mathbf{Tu} = (\widetilde{\mathbf{T}}_{*s})^{\mathrm{T}}\widetilde{\mathbf{T}}_{*j}$$

and (v) follows. Statement (vi) follows from $\mathbf{A} = \mathbf{T}^{\mathrm{T}}\mathbf{T}$ since $\mathbf{T}$ is upper triangular. Statement (vii) follows since $\rho(\mathbf{A}) = \rho(\mathbf{LT}) = \rho(\mathbf{T})$ and the non-null rows of $\mathbf{T}$ form a row basis for $\mathbf{T}$. ∎

We now give the square-root method as

Algorithm 9.7.3 (*Square-root method*) Given: an $n \times n$ n.n.d. matrix $\mathbf{A}$ and $\mathbf{b}, \mathbf{c} \in \mathbb{R}^n$.

Step 1 Form $\widetilde{\mathbf{A}} = [\mathbf{A} : \mathbf{b} : \mathbf{c} : \mathbf{I}]$. Let p be the number of columns of $\widetilde{\mathbf{A}}$ and let a_{ij} denote the (i, j)-th element of $\widetilde{\mathbf{A}}$. Set $k = 1$ and $j = 1$.

Step 2 Compute t_{kk} using (9.7.1). Increase j by 1 and go to *Step* 3.

Step 3 Compute t_{kj} using (9.7.2). Note that if $t_{kk} = 0$ and $j \leq n$ then $t_{kj} = 0$. If $j < p$, increase j by 1 and go to *Step* 3. If $j = p$ go to *Step* 4.

Step 4 If $k < n$, increase k by 1, then put $j = k$ and go to *Step* 2. If $k = n$ go to *Step* 5.

Step 5 Complete the $n \times p$ matrix $\widetilde{\mathbf{T}} = ((t_{ij}))$ by putting $t_{ij} = 0$ whenever $i > j$.

Step 6 Let $\mathbf{T}$ be the submatrix of $\widetilde{\mathbf{T}}$ formed by the first n columns. Then $\mathbf{A} = \mathbf{T}^{\mathrm{T}}\mathbf{T}$.

Step 7 Compute $\mathbf{C} = \mathbf{P}^{\mathrm{T}}\mathbf{P}$ where $\mathbf{P}$ is the submatrix of $\widetilde{\mathbf{T}}$ formed by the last n columns. $\mathbf{C}$ is a g-inverse of $\mathbf{A}$.

Step 8 Check whether $t_{i,n+1} = 0$ for all i such that $t_{ii} = 0$. If no, declare the system $\mathbf{Ax} = \mathbf{b}$ to be inconsistent. If yes, compute the general solution as $\mathbf{P}^{\mathrm{T}}\widetilde{\mathbf{T}}_{*,n+1} + \mathbf{Bw}$ where $\mathbf{B}$ is obtained as follows. For each i such that $t_{ii} = 0$, find $(\mathbf{P}_{i*})^{\mathrm{T}}$. $\mathbf{B}$ is the matrix with these as the columns. It may be noted that $\mathbf{P}^{\mathrm{T}}\mathbf{T}$ is upper triangular. (Consistency of $\mathbf{Ax} = \mathbf{c}$ is similarly checked.)

Step 9 If $\mathbf{b}, \mathbf{c} \in \mathcal{C}(\mathbf{A})$ (this is checked as in *Step* 8), the value of $\mathbf{c}^{\mathrm{T}}\mathbf{A}^{-}\mathbf{b}$ is invariant under different choices of $\mathbf{A}^{-}$ and is computed as $\langle \widetilde{\mathbf{T}}_{*,n+1}, \widetilde{\mathbf{T}}_{*,n+2} \rangle$.

Step 10 Compute $|\mathbf{A}| = t_{11}^2 t_{22}^2 \cdots t_{nn}^2$.

Step 11 Find the rank of $\mathbf{A}$. This is the number of non-zero elements among $t_{11}, t_{22}, \ldots, t_{nn}$. Stop. ∎

Remark We can in fact use the above algorithm to check whether a real symmetric matrix $\mathbf{A}$ is n.n.d. or not. The matrix $\mathbf{A}$ is n.n.d. iff $\mathbf{T}$ can be computed by the algorithm, t_{kk} computed in *Step* 2 is real for each $k = 1, \ldots, n$ and $t_{k,k+1}, \ldots, t_{kn}$ computed in *Step* 3 using (9.7.2) are all 0 whenever $t_{kk} = 0$.

We now give a numerical example to illustrate the use of the square-root method.

Example 9.7.4 Consider the matrix

$$\widetilde{\mathbf{A}} = \begin{bmatrix} 4 & 2 & -2 & 0 & 6 & 4 & 1 & 0 & 0 & 0 \\ 2 & 2 & 1 & 3 & 5 & 2 & 0 & 1 & 0 & 0 \\ -2 & 1 & 5 & 6 & 1 & -2 & 0 & 0 & 1 & 0 \\ 0 & 3 & 6 & 18 & 15 & -1 & 0 & 0 & 0 & 1 \end{bmatrix}$$

Let $\mathbf{A}$ be the n.n.d. matrix formed by the first 4 columns and let $\mathbf{b}, \mathbf{c}$ be the fifth and sixth columns. We will find $\mathbf{A}^{-}$, solve $\mathbf{A}\mathbf{x} = \mathbf{b}$, find $\mathbf{c}^{\mathrm{T}}\mathbf{A}^{-}\mathbf{b}$, $|\mathbf{A}|$ and $\rho(\mathbf{A})$.

We start by computing $t_{11} = \sqrt{a_{11}} = 2$. We then compute $t_{1j} = a_{1j}/t_{11}$ for $j = 2, \ldots, 10$. We next compute

$$t_{22} = \sqrt{a_{22} - t_{12}^2} = \sqrt{2 - 1^2} = 1$$

and then $t_{2j} = (a_{2j} - t_{12}t_{1j})/t_{22}$ for $j = 3, \ldots, 10$. For example,

$$t_{25} = (5 - 1 \times 3)/1 = 2$$

We next compute

$$t_{33} = \sqrt{a_{33} - t_{13}^2 - t_{23}^2} = \sqrt{5 - (-1)^2 - 2^2} = 0$$

So we compute $t_{3j} = a_{3j} - t_{13}t_{1j} - t_{23}t_{2j}$ for $j = 4, \ldots, 10$. For example,

$$t_{37} = 0 - (-1)\left(\tfrac{1}{2}\right) - 2\left(-\tfrac{1}{2}\right) = \tfrac{3}{2}$$

Next; we compute

$$t_{44} = \sqrt{a_{44} - t_{14}^2 - t_{24}^2 - t_{34}^2} = \sqrt{18 - 0^2 - 3^2 - 0^2} = 3$$

and then $t_{4j} = (a_{4j} - t_{14}t_{1j} - t_{24}t_{2j} - t_{34}t_{3j})/t_{44}$ for $j = 5, \ldots, 10$. For example,

$$t_{48} = [0 - 0 \times 0 - 3 \times 1 - 0 \times (-2)]/3 = -1$$

We finally fill up the 0's below the diagonal to get

$$\widetilde{\mathbf{T}} = \begin{bmatrix} 2 & 1 & -1 & 0 & 3 & 2 & \frac{1}{2} & 0 & 0 & 0 \\ 0 & 1 & 2 & 3 & 2 & 0 & -\frac{1}{2} & 1 & 0 & 0 \\ 0 & 0 & 0 & 0 & 0 & 0 & \frac{3}{2} & -2 & 1 & 0 \\ 0 & 0 & 0 & 3 & 3 & -\frac{1}{3} & \frac{1}{2} & -1 & 0 & \frac{1}{3} \end{bmatrix}$$

We now compute a g-inverse of $\mathbf{A}$ as $\mathbf{P}^{\mathrm{T}}\mathbf{P}$ (note that the (i,j)-th element of this is the inner product of the i-th and the j-th columns of $\mathbf{P}$). It can be checked that

$$\mathbf{P}^{\mathrm{T}}\mathbf{P} = \begin{bmatrix} 3 & -4 & \frac{3}{2} & \frac{1}{6} \\ -4 & 6 & -2 & -\frac{1}{3} \\ \frac{3}{2} & -2 & 1 & 0 \\ \frac{1}{6} & -\frac{1}{3} & 0 & \frac{1}{9} \end{bmatrix}$$

We next solve the system $\mathbf{Ax} = \mathbf{b}$. We first note that t_{33} is the only zero diagonal element of $\mathbf{T}$ and $t_{35} = 0$. So the system $\mathbf{Ax} = \mathbf{b}$ is consistent. The vector $\mathbf{P}^{\mathrm{T}}\widetilde{\mathbf{T}}_{*5} = (2, -1, 0, 1)^{\mathrm{T}}$ is a particular solution. $\mathbf{B}$ is the transpose of $\mathbf{P}_{3*}$, viz. $(\frac{3}{2}, -2, 1, 0)^{\mathrm{T}}$ and a general solution of $\mathbf{Ax} = \mathbf{b}$ is $(2 + \frac{3}{2}\alpha, -1 - 2\alpha, \alpha, 1)^{\mathrm{T}}$ where α is arbitrary.

The vector $\mathbf{c}$ also belongs to $\mathcal{C}(\mathbf{A})$ since $t_{36} = 0$. So $\mathbf{c}^{\mathrm{T}}\mathbf{A}^{-}\mathbf{b}$ is the same for all g-inverses of $\mathbf{A}$ and its value is

$$(3 \times 2) + (2 \times 0) + (0 \times 0) + (3 \times (-\tfrac{1}{3})) = 5$$

Notice that $\mathbf{b}^{\mathrm{T}}\mathbf{A}^{-}\mathbf{b}$ can be obtained as the inner product of $\widetilde{\mathbf{T}}_{*5}$ with itself.

We next find that $|\mathbf{A}| = 2^2 \cdot 1^2 \cdot 0^2 \cdot 3^2 = 0$. Notice also that the m-th leading principal minor of $\mathbf{A}$ is $t_{11}^2 \cdots t_{mm}^2$ for all m. Finally since $\mathbf{T}$ has three non-zero diagonal entries, $\rho(\mathbf{A}) = 3$. ∎

Exercises

1. Let

$$\mathbf{A} = \begin{bmatrix} 9 & -3 & 12 & 3 & -6 \\ -3 & 1 & -4 & -1 & 2 \\ 12 & -4 & 20 & 2 & -2 \\ 3 & -1 & 2 & 2 & -5 \\ -6 & 2 & -2 & -5 & 14 \end{bmatrix}, \quad \mathbf{b} = \begin{bmatrix} 9 \\ -3 \\ 14 \\ 2 \\ -2 \end{bmatrix} \text{ and } \mathbf{c} = \begin{bmatrix} 1 \\ -2 \\ 5 \\ 3 \\ -2 \end{bmatrix}$$

Using *Algorithm* 9.7.3 find (i) $\mathbf{A}^-$, (ii) a general solution of $\mathbf{Ax} = \mathbf{b}$, (iii) whether $\mathbf{c} \in \mathscr{C}(\mathbf{A})$, (iv) $\mathbf{b}^T\mathbf{A}^-\mathbf{b}$ and (v) $\rho(\mathbf{A})$.

2. Let

$$\mathbf{A} = \begin{bmatrix} 4 & 2 & -2 & 6 \\ 2 & -8 & -7 & 6 \\ -2 & -7 & -4 & -3 \\ 6 & 6 & -3 & 8 \end{bmatrix}, \quad \mathbf{b} = \begin{bmatrix} -10 \\ 6 \\ 14 \\ -26 \end{bmatrix} \text{ and } \mathbf{c} = \begin{bmatrix} 2 \\ 1 \\ 3 \\ -1 \end{bmatrix}$$

Find using the square-root method: (i) $\mathbf{A}^{-1}$, (ii) the solution of $\mathbf{Ax} = \mathbf{b}$, (iii) the value of $\mathbf{c}^T\mathbf{A}^{-1}\mathbf{b}$ and (iv) $|\mathbf{A}|$. Also find the signature of the quadratic form $\mathbf{x}^T\mathbf{Ax}$.

3. Let $\widetilde{\mathbf{A}}, \widetilde{\mathbf{T}}$ and $\mathbf{P}$ be as in *Theorem* 9.7.2.
   *(a) If $\widetilde{\mathbf{A}}_{*j} \in \mathscr{C}(\mathbf{A})$, prove that $\mathbf{P}^T\mathbf{P}\widetilde{\mathbf{A}}_{*j}$ is the only solution of $\mathbf{Ax} = \widetilde{\mathbf{A}}_{*j}$ with the property that x_k is 0 whenever t_{kk}=0.
   (b) Show that $\mathbf{b} \in \mathscr{C}(\mathbf{A})$ iff $\mathbf{P}_{k*}\mathbf{b} = 0$ for all k such that $t_{kk} = 0$.

4. If $\widetilde{\mathbf{A}}$ is real and $\mathbf{A}$ is symmetric and if all the leading principal minors of $\mathbf{A}$ are non-zero, show that *Theorem* 9.7.1 and its proof hold except that the elements of $\widetilde{\mathbf{T}}$ may be complex numbers. Show also that in each row of $\widetilde{\mathbf{T}}$, either all elements are real or all elements are purely imaginary.

5. Let $\mathbf{A}$ be an $n \times n$ real symmetric matrix with all leading principal minors non-zero.
   (a) Show how the square-root method can be used to express $\mathbf{A}$ as $\mathbf{S}^T\mathbf{\Delta S}$ where $\mathbf{S}$ is an $n \times n$ real upper triangular non-singular matrix and $\mathbf{\Delta}$ is a diagonal matrix with diagonal entries 1's and −1's.
   (b) If m_k denotes the k-th leading principal minor of $\mathbf{A}$ and if there are d changes of sign in the sequence $1, m_1, m_2, \ldots, m_n$, show that the signature of $\mathbf{x}^T\mathbf{Ax}$ is $n - 2d$.

6. Prove that every n.n.d. matrix $\mathbf{A}$ has a lower triangular square-root (i.e., a lower triangular $\mathbf{S}$ such that $\mathbf{A} = \mathbf{S}^T\mathbf{S}$).

*7. If $\mathbf{A}$ is p.d. and is the dispersion matrix of a random vector X, show that the components of $Y = \mathbf{P}X$ are pairwise uncorrelated and have unit variance each, where $\mathbf{P}$ is as in *Theorem* 9.7.2.

9.8 Hermitian forms*

In this section we consider briefly the complex analogues of real quadratic forms, known as hermitian forms. All matrices and vectors in this section will be complex unless otherwise specified.

Definition 9.8.1 An *n-ary hermitian form* is an expression

$$\mathbf{x}^*\mathbf{A}\mathbf{x} = \sum_{i=1}^{n}\sum_{j=1}^{n} a_{ij}\overline{x}_i x_j = \sum_{i=1}^{n} a_{ii}|x_i|^2 + 2\sum_{i<j}\operatorname{Re}(a_{ij}\overline{x}_i x_j) \tag{9.8.1}$$

where $\mathbf{A}$ is a hermitian matrix of order n.

The values taken by $\mathbf{x}^*\mathbf{A}\mathbf{x}$ (as $\mathbf{x}$ runs over $\mathbb{C}^n$) are all real. This can be seen from (9.8.1) since a_{ii} is real for all i or from

$$\overline{(\mathbf{x}^*\mathbf{A}\mathbf{x})} = (\mathbf{x}^*\mathbf{A}\mathbf{x})^* = \mathbf{x}^*\mathbf{A}\mathbf{x}$$

Hence for hermitian forms (and so for hermitian matrices) the various categories of definiteness can be defined as in *Definition* 9.2.1 with $\mathbb{C}^n$ replacing $\mathbb{R}^n$. $\langle\mathbf{x},\mathbf{y}\rangle = \mathbf{y}^*\mathbf{A}\mathbf{x}$ is an inner product on $\mathbb{C}^n$ iff $\mathbf{A}$ is a p.d. hermitian matrix. The non-singular transformation $\mathbf{x} = \mathbf{P}\mathbf{y}$ takes the hermitian form $\mathbf{x}^*\mathbf{A}\mathbf{x}$ to the hermitian form $\mathbf{y}^*\mathbf{B}\mathbf{y}$ where $\mathbf{B} = \mathbf{P}^*\mathbf{A}\mathbf{P}$. We say that $\mathbf{P}^*\mathbf{A}\mathbf{P}$ is obtained from $\mathbf{A}$ by a *conjunctive transformation* if $\mathbf{P}$ is non-singular.

Using *Theorem* 8.7.3 instead of *Theorem* 8.7.2 we can see that any hermitian form $\mathbf{x}^*\mathbf{A}\mathbf{x}$ can be reduced to a diagonal form $\delta_1|y_1|^2 + \cdots + \delta_n|y_n|^2$ by a unitary transformation $\mathbf{x} = \mathbf{U}\mathbf{y}$. Note that $\delta_1, \ldots, \delta_n$ are the characteristic roots of $\mathbf{A}$ and are real. The δ's can be made 1's, -1's and 0's if any non-singular transformation can be used. Sylvester's law of inertia can be proved for hermitian forms by imitating the proof of *Theorem* 9.3.5. Hence rank and signature of a hermitian form can be defined as for a real quadratic form. *Theorems* 9.3.7, 9.3.8 and 9.3.9 remain valid for hermitian forms.

A hermitian matrix $\mathbf{A}$ is n.n.d. iff $\mathbf{A} = \mathbf{B}^*\mathbf{B}$ for some (complex) $\mathbf{B}$. $\mathbf{A}$ is p.d. iff $\mathbf{A} = \mathbf{B}^*\mathbf{B}$ for some non-singular $\mathbf{B}$. The proofs of these can be obtained from those of *Theorems* 9.4.1 and 9.4.2 by replacing transpose by adjoint throughout. In fact the statements and proofs of all results in *Sections* 9.4 and 9.5, with the exception of *Theorem* 9.4.7, go through for hermitian matrices with obvious changes. We leave it to the reader to work out the details. We only mention that $\det(\mathbf{B}^*\mathbf{B}) = |\det\mathbf{B}|^2$ for a complex square matrix $\mathbf{B}$.

The results of *Section* 9.6 also hold for hermitian matrices with obvious changes like replacing 'transpose' by 'adjoint' and 'orthogonal' by 'unitary'. However, if **A** and **B** are hermitian and **B** is p.d. then the roots of $|\mathbf{A} - \lambda\mathbf{B}| = 0$ are all real but the $\mathbf{x}_i$'s satisfying (9.6.1) may be complex.

Exercises

1. Extend the definitions and results of *Sections* 9.2 through 9.7 to hermitian matrices.
2. Let $\mathbf{H} = \mathbf{A} + i\mathbf{B}$ be a hermitian matrix where **A** and **B** are real.
   (a) Show that **A** is symmetric and **B** is skew-symmetric.
   (b) Show that **H** is p.d. iff **A** is p.d. and all the eigenvalues of $i\mathbf{A}^{-1}\mathbf{B}$ are real and less than 1.
3. (*Polar decomposition*) If **A** is a square matrix, show that there exists a unique n.n.d. matrix **H** and a unitary matrix **W** such that $\mathbf{A} = \mathbf{HW}$. (Hint: use singular value decomposition). Show also that **W** is uniquely determined by **A** if **A** is non-singular. What are **H** and **W** if $\mathbf{A} = (\alpha + i\beta)$ with α and β real?

References

Barnard, S. and J. M. Child, *Advanced Algebra*, Macmillan, London, 1939.

Barnett, Stephen, *Matrices — Methods and Applications*, Clarendon Press, Oxford, 1990.

Ben Israel, A. and T. N. E. Greville, *Generalized Inverses: Theory and Applications*, Wiley, New York, 1974.

Birkhoff, G. and S. MacLane, *A survey of Modern Algebra*, Fourth Edition, Macmillan, New York, 1977.

Gale, D., *Theory of Linear Economic Models*, McGraw-Hill, London, 1960.

Hadley, G., *Linear Programming*, Addison-Wesley, Reading, 1975.

Halmos, P. R., *Finite-dimensional Vector Spaces*, Second edition, Princeton University Press, Princeton, 1958.

Halmos, P. R., *Introduction to Hilbert Space and the Theory of Spectral Multiplicity*, Second edition, Chelsia Publ., New York, 1957.

Hoffman, K. and R. Kunze, *Linear Algebra*, Second edition, Prentice-Hall, New Delhi, 1984.

Jacobson, N., *Lectures in Abstract Algebra*, Vol. 2: Linear Algebra, D. Van Nostrand, London, 1953.

Mangasarian, O. L., *Non-linear Programming*, McGraw-Hill, London, 1969.

Marcus, M and H. Minc, *A Survey of Matrix Theory*, Prindle, Weber and Schmidt, Boston, 1964.

Mirsky, L., *Introduction to Linear Algebra*, Clarendon Press, Oxford, 1955.

Noble, Ben and J. W. Daniel, *Applied Linear Algebra*, Third Edition,

Prentice-Hall, Englewood Cliffs, 1988.

Prasada Rao, P. S. S. N. V., *Jordan canonical form of a matrix — an elementary derivation* (unpublished), 1984.

Rao, C. R., *Linear Statistical Inference and its Applications*, Second edition, Wiley, New York, 1973.

Rao, C. R. and S. K. Mitra, *Generalized Inverse of Matrices and its Applications*, Wiley, New York, 1971.

Rao, C. R. and M. B. Rao, *Matrix Algebra and Its Applications to Statistics and Econometrics*, World Scientific, 1998.

Searle, S. R., *Matrix Algebra useful for Statistics*, Wiley, New York, 1982.

Strang, G., *Linear Algebra and its Applications*, Third Edition, Harcourt, Brace, Jovanovich, San Diego, 1988.

Wilkinson, J. H., *The Algebraic Eigenvalue Problem*, Oxford University Press, London, 1965.

More Hints and Solutions

Section 1.2 (p. 21)

2(a). 1(c),1(d),4. (b). 4. (c). 3,4. (d). 3,4,5,6.
3. 1(d) and 1(e) will be violated. 4. See *Exercise* 1.8.10.
8. $P = B + C$, $Q = A + C$, $R = A + B$ and $S = A + B + C = P + Q - C$.
9. $(1, 0, 1)$, $(0, 2, 1)$ and $(2, 2, 1)$. 10. See page 60.

Section 1.3 (p. 28)

1(a). No. (b). No and No. (c). No, No and Yes. (d). Yes.
1(e). Yes. (f). No. (g). Yes. (h). No. (i). Yes.
1(j). Yes and No. 1(k). Yes.
2. If (b) holds and $\mathbf{x} \in S$, $\mathbf{0} = 0 \cdot \mathbf{x} \in S$. Note that $S = \emptyset$ satisfies (b) vacuously but is not a subspace.
4(d). Let $|X| \geq 2$ and $y \in X$. Then any $g \in F^X$ equals $f + h$ where $f(x)$ is 0 or $g(x)$ according as $x = y$ or not, and $h(x)$ is $g(y)$ or 0 according as $x = y$ or not. Next let $|F| \geq 3$. Let $\alpha \neq 0, 1$. Then any $g \in F^X$ equals $p + q$ where $p(x)$ is α or 1 according as $g(x) = 1$ or not, and $q(x)$ is $1 - \alpha$ or $g(x) - 1$ according as $g(x) = 1$ or not.
5. $\{\{1, 2, 3\}, \{2, 3, 4\}, \{1, 5\}, \{1, 4\}, \{2, 3, 5\}, \{1, 2, 3, 4, 5\}, \{4, 5\}, \emptyset\}$.
6(b). False. Take $A = \{(1, 0), (0, 1)\}$ and $B = \{(1, 1), (1, -1)\}$.
7. *Last part*: exactly two of $\mathbf{x}, \mathbf{y}, \mathbf{z}$ belonging to S is not possible.
9. Let $W = \cap\{S : S \text{ is a subspace and } S \supseteq A\}$. $\text{Sp}(A) \supseteq W$ since $\text{Sp}(A)$ is one S. Since every S contains $\text{Sp}(A)$, $W \supseteq \text{Sp}(A)$.
10. *Only if part.* $\mathbf{x} + \mathbf{y} \in S \cup T$. If $\mathbf{x} + \mathbf{y} \in S$ then $\mathbf{y} \in S$, a contradiction, etc.
11. The subspaces are $\{O\}$, $\{O, (1, 0), (2, 0)\}$, $\{O, (1, 1), (2, 2)\}$, $\{O, (0, 1), (0, 2)\}$, $\{O, (2, 1), (1, 2)\}$ and F^2.
12. The span of A is $\{f : f \text{ vanishes outside a finite subset of } X\}$.

Section 1.4 (p. 34)

1. $\{(1, 2), (4, 5)\}$, $\{(2, 4), (4, 5)\}$ and B.
2(a). Ind. (b). Dep. (c). Dep. (d). Ind.
4(a). Show $\{\mathbf{x}_1, \ldots, \mathbf{x}_{n-1}\}$ and $\{\mathbf{x}_2, \ldots, \mathbf{x}_n\}$ are independent and use symmetry. 4(b). $\sqrt{12} = 2\sqrt{3}$. (c). $\mathbf{x}_3 = 2\mathbf{x}_2 - 3\mathbf{x}_1$.
5(b). Let $\alpha(\mathbf{x} + \mathbf{y}) + \beta(\mathbf{y} + \mathbf{z}) + \gamma(\mathbf{z} + \mathbf{x}) = \mathbf{0}$. Then $\alpha + \gamma = 0$, $\alpha + \beta = 0$ and $\beta + \gamma = 0$, so $(1 + 1)(\alpha + \beta + \gamma) = 0$. If $1 + 1 \neq 0$, we get $\alpha + \beta + \gamma = 0$ and

$\alpha = \beta = \gamma = 0$. Over GF(2), $(\mathbf{x}+\mathbf{y})+(\mathbf{y}+\mathbf{z})+(\mathbf{z}+\mathbf{x}) = \mathbf{0}$.
6. Converse is false. Take for example, $k=1$, $n=2$ and $\mathbf{x}_1 = (0,1)$.
8. $\{\mathbf{x}_1,\mathbf{x}_2,\mathbf{x}_4\}$, $\{\mathbf{x}_1,\mathbf{x}_3,\mathbf{x}_4\}$, $\{\mathbf{x}_1,\mathbf{x}_4,\mathbf{x}_5\}$, $\{\mathbf{x}_2,\mathbf{x}_3,\mathbf{x}_4\}$, $\{\mathbf{x}_2,\mathbf{x}_4,\mathbf{x}_5\}$. (Note that $\mathbf{x}_3 = 2\mathbf{x}_1 - \mathbf{x}_2$, $\mathbf{x}_5 = 2\mathbf{x}_3$.)
9. *Second part*: No. Take $A = \{(1,0)\}$, $S = \mathrm{Sp}(A)$ and $B = \{(1,1),(1,2)\}$.
11. $O, \mathbf{x}_1, \mathbf{x}_2$ are not collinear and $O, \mathbf{x}_1, \mathbf{x}_2, \mathbf{x}_3$ are not coplanar.
12(a). $\alpha \neq 0, \pm\sqrt{2}$. (b). $\alpha = \beta$ or $\alpha + 3\beta = 0$.

Section 1.5 (p. 43)
1. Use *Theorems* 1.4.6 and 1.5.11. $3t^2 - 5t + 4 = 6f_1 + 7f_2 + 3f_3$.
3. Take $B = \{\mathbf{x}_1, \mathbf{x}_2 + \alpha\mathbf{x}_1, \ldots, \mathbf{x}_k + \alpha\mathbf{x}_1\}$, α large.
4. $\{\mathbf{x}\}$ is a basis if $\mathbf{x} \neq 1$. Dimension is 1.
5. Let $\mathbf{x} = \sum_{i=1}^n \alpha_i\mathbf{x}_i$ with $\mathbf{x}_i$'s in A. $\sum_{i=1}^n \beta_i\mathbf{x}_i = \mathbf{0} \Rightarrow \mathbf{x} = \sum_{i=1}^n (\alpha_i + \beta_i)\mathbf{x}_i$.
6. *Only if part*: If $\{\mathbf{x}_1, \ldots, \mathbf{x}_k\}$ is a basis and $k \geq 2$, $\{\mathbf{x}_1 + \mathbf{x}_2, \mathbf{x}_2, \ldots, \mathbf{x}_k\}$ is a different basis. If $d(V) = 1$, $|F| \geq 3$ and $\mathbf{x} \neq \mathbf{0}$, $\{\alpha\mathbf{x}\}$ is a basis for each $\alpha \neq 0$.
7. False. Take $A = \{(1,0)\}$, $B = \{(0,1)\}$ and $C = \{(1,1)\}$.
8. False. Take $B = \{(0,1),(1,1)\}$, $S = x$-axis.
9. *Algorithm* 1.5.14 chooses the first, second and fourth vectors.
10(a). $\{\{\omega\} : \omega \in \Omega\}$. (b). (i) $\{A\}$ (ii) $\{A, B\}$, $\{A, A \triangle B\}$, $\{B, A \triangle B\}$.
11. $\{\mathbf{x}_1, \mathbf{x}_2, (1,0,-1,0,0), (1,0,0,-1,0)\}$. 12. $\{(1,1,\ldots,1), \mathbf{e}_2, \ldots, \mathbf{e}_n\}$.
13. $\mathbf{x}_1, \mathbf{x}_2, \mathbf{x}_4, \mathbf{x}_3$ are independent by *Theorem* 1.4.6 and so form a basis.
14(a). $\ell_i(x_k) = 0$ iff $i \neq k$. If $f \in \mathcal{P}_n$ and $g(t) = \sum_{i=1}^n (f(x_i)/\ell_i(x_i))\ell_i(t)$, then x_i is a root of $f - g$ for $i = 1, \ldots, n$, so $f = g$. Thus $\ell_1, \ldots, \ell_n$ generate $\mathcal{P}_n$. 14(b). Linear independence is clear.
16(a). $\{(0,1,0,0),\ (2,0,1,0),\ (-1,0,0,1)\}$. (b). $\{(3,-2,1,0),\ (-1,1,0,1)\}$.
16(c). $\{(-\frac{5}{6}, \frac{1}{2}, -\frac{1}{3}, 1)\}$; $(a_1, \ldots, a_4) = (-5, 3, -2, 6)$. 17. $\alpha_i \neq 0$.
18. Functions $1, x, x^2, \ldots$ are lin. ind. since $x^k = \alpha_1 x^{k-1} + \alpha_2 x^{k-2} + \cdots + \alpha_k$ for all $x \in \mathbb{R}$ implies $t^n - \alpha_1 t^{n-1} - \cdots - \alpha_n$ has more than n roots.
19(a). Use *Theorem* 1.5.7(ii).
19(d). Find $|\{(S, X) : d(S) = k,\ X \text{ is an ordered basis of } S\}|$ in two ways.
19(e). Fix S with $d(S) = k$. Find $|\{(X, Y, T) : d(T) = \ell \text{ and } X \text{ and } (X, Y) \text{ are ordered bases of } S \text{ and } T\}|$, in two ways.
20. Use *Exercises* 1.2.11 and 1.5.19(a).
22. *Only if part*: Let $k = n$. Then $\mathbf{e}_i \in \mathrm{Sp}(A)$ in F^n and so in G^n for $i = 1, \ldots, n$, thus A is a basis of G^n. When $k < n$, extend A to a basis of F^n.

Section 1.6 (p. 50)
2. Converse is false. Take $S = W \neq \{\mathbf{0}\}$ and $T = \{\mathbf{0}\}$.
3. Iff $S \subseteq T$ or $T \subseteq S$ (see *Exercise* 1.3.10).
4. Use modular law and $d(S + T) \leq 3$. 5. min $= 3$ and max $= 5$.

6. Take three distinct lines through the origin in the plane.
8(a). $\{\mathbf{u}_1\}$, $\{\mathbf{u}_1, \mathbf{u}_2\}$, $\{\mathbf{u}_1, \mathbf{u}_3\}$ and $\{\mathbf{u}_1, \mathbf{u}_2, \mathbf{u}_3\}$ where $\mathbf{u}_1 = (0, -3, 2, 1)$, $\mathbf{u}_2 = (-2, 5, 0, 1)$ and $\mathbf{u}_3 = (\frac{2}{5}, -1, 0, 1)$. 8(c). $\{\mathbf{u}_1, \mathbf{u}_2, \mathbf{u}_3, \mathbf{e}_1\}$.
8(d). $S + T = \{(x_1, x_2, x_3, x_4) : 5x_1 + 2x_2 + 3x_3 = 0\}$.
9(a). $\sum_{i=1}^{k} \alpha_i \mathbf{x}_i = \beta \sum_{i=1}^{\ell} (\alpha_i/\beta)\mathbf{x}_i + \gamma \sum_{i=\ell+1}^{k} (\alpha_i/\gamma)\mathbf{x}_i$ if $\beta := \sum_{i=1}^{\ell} \alpha_i \neq 0$ and $\gamma := \sum_{i=\ell+1}^{k} \alpha_i \neq 0$. Show that the condition may be assumed without loss of generality.

Section 1.7 (p. 57)
1. *Only if part*: Take any bases. *If part*: It is clear that $d(S+T) = d(S) + d(T)$. *Last part*: Take $A = B$.
2. *Only if part*: Use $S \cap T = \{\mathbf{0}\}$. *If part*: If $\mathbf{x} \in S \cap T$ and $\mathbf{x} \neq \mathbf{0}$, $\{\mathbf{x}\}$ can be extended to bases of S and T.
3. $\{(x_1, \ldots, x_n) : x_{m+1} = \cdots = x_n = 0\}$. 4. $\{(\alpha\sqrt{-1}, 0, \ldots, 0) : \alpha \in \mathbb{R}\}$.
5. No. Take S, T and W to be three distinct lines through the origin in $\mathbb{R}^2$.
6. $\mathrm{Sp}(C - A)$ where A, B and C are bases of S, T and V with $C \supseteq A \cup B$.
7. *First part*: Extend a basis A of S to a maximal independent subset B of $S \cup T$ and take $W = \mathrm{Sp}(B - A)$. 8. $\mathrm{Sp}(\{\mathbf{e}_3\})$, $\mathrm{Sp}(\{\mathbf{e}_4\})$.
9. *If part*: Consider $\mathrm{Sp}(\{\mathbf{x}_{k+1}, \ldots, \mathbf{x}_n\})$ and $\mathrm{Sp}(\{\mathbf{x}_i + \mathbf{x}_{k+i} : 1 \leq i \leq n-k\})$ where $\{\mathbf{x}_1, \ldots, \mathbf{x}_k\}$ and $\{\mathbf{x}_1, \ldots, \mathbf{x}_n\}$ are bases of S and V. *Only if part*: Use modular law. 10. *Second part*: $3 + 2t^2 - 5t^4$.
11. $\{\mathbf{x}_1, \mathbf{x}_2\}$, $T = \mathrm{Sp}(\{\mathbf{u}, \mathbf{e}_1\})$, $W = \mathrm{Sp}(\{\mathbf{x}_1 + \mathbf{u}, \mathbf{e}_1\})$, $(-2, 0, -1, -3)$.
12. Use *Exercise* 1.7.1.
13. If $S \neq \{\mathbf{0}\}$ and $S \neq V$, $\mathrm{Sp}(\{\mathbf{x}_{k+1}, \ldots, \mathbf{x}_n\})$ and $\mathrm{Sp}(\{\mathbf{x}_1 + \mathbf{x}_{k+1}, \mathbf{x}_{k+2}, \ldots, \mathbf{x}_n\})$ are distinct complements of S with $\mathbf{x}$'s as in *Exercise* 1.7.9.
14. Use *Exercise* 1.7.6 with $T = \mathrm{Sp}(\{\mathbf{y} - \mathbf{x}\})$.
16. See the solution to *Exercise* 1.5.5.
17. Consider $\{\mathbf{x}_1, \ldots, \mathbf{x}_4\}$ where $\mathbf{x}_1 = 1, 0, 1, 0, -2)$, $\mathbf{x}_2 = (0, 1, 0, 1, -2)$, $\mathbf{x}_3 = (1, 1, 0, 0, -2)$ and $\mathbf{x}_4 = (0, 1, 0, 0, -1)$. Take $r_1 = 2$ and $r_2 = r_3 = 3$.

Section 1.8 (p. 61)
1. $\varphi(\mathbf{0}) = \varphi(\mathbf{0} + \mathbf{0}) = \varphi(\mathbf{0}) + \varphi(\mathbf{0})$, so $\varphi(\mathbf{0}) = \mathbf{0}$. For converse, use 1-1.
5. $d(f(0,0), f(1,0)) = \sqrt{10}$. $f(\mathbf{x} + S) = f(\mathbf{x}) + f(S)$ where S is a subspace.
7. No, since any vector space over $\mathbb{R}$ is either a singleton or infinite.
8. $(\alpha_1 + i\beta_1, \ldots, \alpha_n + i\beta_n) \mapsto (\alpha_1, \ldots, \alpha_n, \beta_1, \ldots, \beta_n)$ is an isomorphism.
9. $(x_1, x_2, x_3) \mapsto x_1$ is an isomorphism.
10. $x \mapsto \log x$ is an isomorphism. 11. $f(\alpha\mathbf{x}) \neq \alpha f(\mathbf{x})$ if $\alpha = x_1 = i$.
12. $(a_1, a_2, a_3, \ldots) \mapsto (a_1, a_2)$ is an isomorphism. 13. See Halmos (1958).

Section 1.9 (p. 66)

2. Use (a) of the lemma preceding *Theorem* 1.9.2.
3. Use *Theorem* 1.9.4 and *Exercise* 1.7.10.
4(a). Let $\mathcal{S} \subseteq V/S$ be a subspace of V/S. Take $W = \cup\mathcal{S}$.
4(b). If W is replaced by $W+S$, $\{\mathbf{w}+S : \mathbf{w} \in W\}$ does not change. Similarly W may be replaced by a T such that $(W \cap S) \oplus T = W$ (see *Exercise* 1.7.7).
5. The first statement follows from the proof of *Theorem* 1.9.3.
6. Show that $\mathbf{t} + (S \cap T) \mapsto \mathbf{t} + S$ is well-defined and is an isomorphism from $T/(S \cap T)$ onto $(S+T)/S$. Then use the corollary to *Theorem* 1.9.3.

Section 2.2 (p. 75)

1(a). Yes. (b). No. (c). Yes. (d). No. (e). Yes. (f). No.

1(g). Yes. 3. $f((\alpha,\beta)) = (\alpha - \beta, \frac{\beta}{2})$, $\begin{bmatrix} 1 & -1 \\ 0 & \frac{1}{2} \end{bmatrix}$.

4(a). $\begin{bmatrix} -1 & 0 \\ 0 & -1 \end{bmatrix}$. 4(b). $\begin{bmatrix} 1 & k \\ 0 & 1 \end{bmatrix}$. 5(a). $\begin{bmatrix} 2 & -3 \\ 1 & 0 \\ 5 & 1 \end{bmatrix}$. 5(b). $\begin{bmatrix} 1 & -4 \\ 4 & 12 \\ 0 & 8 \end{bmatrix}$.

6. $((\delta_{ij}))_{m\times n}$ and $((\delta_{ij}))_{n\times m}$, where δ_{ij} is 1 or 0 according as $i = j$ or $i \neq j$.
7(a). $((a_{ij}))_{3\times 4}$ where $a_{ij} = i$ if $j = i+1$ and $a_{ij} = 0$ if $j \neq i+1$.
7(b). $((b_{ij}))_{4\times 3}$ where $b_{ij} = 1/j$ if $i = j+1$ and $b_{ij} = 0$ if $i \neq j+1$.
7(c). $((\delta_{ij}))_{3\times 3}$ and $((c_{ij}))_{4\times 4}$ where c_{ij} is 1 if $i = j \geq 2$ and 0 otherwise.
8(a). The images of (x_1, x_2) are $(3x_1+3x_2, 3x_1-6x_2)$, $(8x_1-15x_2, 5x_2-x_1)$, $(2x_1+3x_2, 11x_2-x_1)$ and $(6x_1+9x_2, 3x_1-3x_2)$ respectively.

8(b). $\begin{bmatrix} 2 & 3 \\ 1 & -1 \end{bmatrix}, \begin{bmatrix} 1 & 0 \\ 2 & -5 \end{bmatrix}$.

8(c). $\begin{bmatrix} 3 & 3 \\ 3 & -6 \end{bmatrix}, \begin{bmatrix} 8 & -15 \\ -1 & 5 \end{bmatrix}, \begin{bmatrix} 2 & 3 \\ -1 & 11 \end{bmatrix}, \begin{bmatrix} 6 & 9 \\ 3 & -3 \end{bmatrix}$.

9. dim $= mn$. 10. $\mathbf{A} = \begin{bmatrix} 4/3 & 2/3 \\ -2/3 & -1/3 \end{bmatrix}$. Matrix of f^2 is also $\mathbf{A}$.

11. $f(x_1, x_2) = \frac{1}{2}(9x_1 + x_2, 5x_2 - 7x_1)$.
13(b). Since $f(\mathbf{0}) = \mathbf{0}$, $\sum \alpha_i \mathbf{x}_i = \mathbf{0} \Rightarrow \sum \alpha_i f(\mathbf{x}_i) = \mathbf{0} \Rightarrow \alpha_i = 0$ for all i.
14. $\sum \alpha_i \mathbf{x}_i \mapsto \alpha_i \mathbf{y}_i$ is the only linear operator with the required properties.
15. Show that $f(-\mathbf{x}) = -f(\mathbf{x})$, $f(m\mathbf{x}) = mf(\mathbf{x})$ and $nf(\frac{m}{n}\mathbf{x}) = f(m\mathbf{x})$ for all $m, n \in \mathbb{Z}$.

Section 2.3 (p. 80)

Note: In the following, $\mathbf{I}$ denotes a square matrix of the type $((\delta_{ij}))$ where δ_{ij} is 1 or 0 according as $i = j$ or $i \neq j$.

1. $\mathbf{A}_1 + \mathbf{A}_2$, $\mathbf{A}_1\mathbf{A}_4$ and $\mathbf{A}_2\mathbf{A}_4$ are

$$\begin{bmatrix} 3 & 1 \\ 3 & 6 \\ 3 & 3 \\ 1 & 1 \end{bmatrix}, \begin{bmatrix} 2 & 0 & -3 & 0 \\ 9 & 15 & -1 & 0 \\ 5 & -3 & -10 & 0 \\ 2 & 6 & 2 & 0 \end{bmatrix} \text{ and } \begin{bmatrix} 5 & 3 & -5 & 0 \\ 3 & 3 & -2 & 0 \\ 4 & 12 & 4 & 0 \\ 1 & -3 & -4 & 0 \end{bmatrix},$$

$$\mathbf{A}_4\mathbf{A}_1 = \begin{bmatrix} -7 & 3 \\ 10 & 14 \end{bmatrix}, \mathbf{A}_4\mathbf{A}_2 = \begin{bmatrix} 4 & -10 \\ 5 & 8 \end{bmatrix}, \mathbf{A}_4\mathbf{A}_3 = \begin{bmatrix} -13 \\ 1 \end{bmatrix},$$

$\mathbf{A}_3^{\mathrm{T}}\mathbf{A}_1 = \begin{bmatrix} 7 & 5 \end{bmatrix}$, $\mathbf{A}_3^{\mathrm{T}}\mathbf{A}_4^{\mathrm{T}} = \begin{bmatrix} -13 & 1 \end{bmatrix}$; $\mathbf{A}_2 + \mathbf{A}_1 = \mathbf{A}_1 + \mathbf{A}_2$.

2. We give $\mathbf{AB}$ and then $\mathbf{BA}$. 2(a). $\begin{bmatrix} 2 & -2 & 4 \\ 6 & 9 & -6 \\ 8 & 20 & 24 \end{bmatrix}, \begin{bmatrix} 2 & -3 & 8 \\ 4 & 9 & -8 \\ 4 & 15 & 24 \end{bmatrix}$.

2(b). $\mathbf{AB} = ((c_{ij}))_{4\times4}$ where $c_{ij} = \alpha_i\beta_{5-i}$ if $j = i$ and 0 otherwise; $\mathbf{BA}$ is the same as $\mathbf{AB}$ with α and β interchanged.

2(c). $\begin{bmatrix} 15 & 10 & 35 \\ 0 & 28 & 0 \\ 0 & 0 & 6 \end{bmatrix}, \begin{bmatrix} 15 & 9 & 38 \\ 0 & 28 & 11 \\ 0 & 0 & 6 \end{bmatrix}$. (d). $\begin{bmatrix} 4 & 5 & 6 \\ 7 & 8 & 9 \\ 1 & 2 & 3 \end{bmatrix}, \begin{bmatrix} 3 & 1 & 2 \\ 6 & 4 & 5 \\ 9 & 7 & 8 \end{bmatrix}$.

2(e). $\begin{bmatrix} 6 & 12 & 10 \\ -3 & -6 & -5 \\ 0 & 0 & 0 \end{bmatrix}, [0]$. (f). $\begin{bmatrix} 0 & 0 \\ 1 & 0 \end{bmatrix}, \begin{bmatrix} 0 & 0 \\ 0 & 0 \end{bmatrix}$.

3. $\mathbf{AA}^{\mathrm{T}} = \begin{bmatrix} 1 & -1 & 2 \\ -1 & 1 & -2 \\ 2 & -2 & 5 \end{bmatrix}, \mathbf{A}^{\mathrm{T}}\mathbf{A} = \begin{bmatrix} 6 & 2 \\ 2 & 1 \end{bmatrix}, \mathbf{B}^2 = \begin{bmatrix} 0 & 0 \\ 0 & 0 \end{bmatrix}$. 4. *mnp*.

5. $\mathbf{ABC} = \mathbf{A}(\mathbf{BC}) = \begin{bmatrix} 18 \\ -9 \\ 0 \end{bmatrix}$.

6(a) and (b). $\mathbf{A}^k = \mathbf{0}$ for $k \geq 3$. $\mathbf{A}^2$ has 6 in (1,3)-position in (a) and (3,1)-position in (b). All other elements are zero.

6(c). $\mathbf{A}^{2p} = \begin{bmatrix} 6^p & 0 & 0 \\ 0 & 6^p & 0 \\ 0 & 0 & 0 \end{bmatrix}$ and $\mathbf{A}^{2p+1} = \begin{bmatrix} 0 & 2^{p+1}3^p & 0 \\ 2^p3^{p+1} & 0 & 0 \\ 0 & 0 & 0 \end{bmatrix}$ for $p \geq 1$.

6(d). $\mathbf{A}, \mathbf{B}$ or $\mathbf{I}$ according as $k \equiv 1$, 2 or 0 (mod 3), where $b_{ij} = 1$ if $i + j = 5$ and 0 otherwise. 6(e). $\mathbf{I}$ or $\mathbf{A}$ according as k is even or odd.

6(f). $(\mathbf{A}^k)_{ij}$ is $\binom{k}{j-i}\alpha^{k-j+i}$ if $i \leq j$ and 0 otherwise.

6(g). $\mathbf{A}^{4q+r}$ is $3^{2q}\mathbf{I}$, $3^{2q}\mathbf{A}$, $3^{2q+1}\mathbf{B}$ or $3^{2q+1}\mathbf{C}$ according as $r = 0, 1, 2$ or 3, where $b_{11} = b_{23} = b_{31} = 1$ and all other b_{ij}'s are 0 and $\mathbf{C}$ is $\mathbf{A}$ with ω and ω^2 interchanged.

7(a). Either $a_{11} = a_{12} = a_{22} = 0$ or $a_{11} = a_{21} = a_{22} = 0$ or '$a_{11} = -a_{22} \neq 0$ and $a_{11}^2 + a_{12}a_{21} = 0$'.

7(b). Either $\mathbf{A} = \mathbf{I}$ or $\mathbf{A} = -\mathbf{I}$ or '$a_{11} = a_{22} = 0$ and $a_{12}a_{21} = 1$' or

'$a_{11} = -a_{22} \neq 0$ and $a_{11}^2 + a_{12}a_{21} = 1$'. 8. All entries 1.
10. $(f \circ f)(x_1, x_2, x_3) = (x_3, 0, 0)$ and $(f \circ f \circ f)(x_1, x_2, x_3) = (0, 0, 0)$. The three matrices are $\mathbf{A}, \mathbf{B}$ and $\mathbf{0}$ where $a_{ij} = 1$ if $i < j$ and 0 otherwise and $b_{ij} = 1$ if $(i, j) = (1, 3)$ and 0 otherwise.
12(e). The numbers of links from C_i's are the entries of $\mathbf{A}(1, 1, \ldots, 1)^{\mathrm{T}}$. The numbers of links to C_j's are the entries of $(1, 1, \ldots, 1)^{\mathrm{T}}\mathbf{A}$. The total number of links is $(1, 1, \ldots, 1)\mathbf{A}\mathbf{e}_j$. *Last part*: Direction of flow of traffic is reversed.

Section 2.4 (p. 89)

In what follows, we will use $\mathbf{U}_{ij}$ to denote a matrix of the appropriate order with 1 in the (i, j)-th place and 0's elsewhere.

3. $(\mathbf{Ax})(\mathbf{y}^{\mathrm{T}}\mathbf{B})$. 4. $\begin{bmatrix} -16 & -64 & -48 \\ -6 & -24 & -18 \end{bmatrix}$.

5(a). False. Take $\mathbf{A} = \mathbf{I}, \mathbf{B} = \mathbf{0}$ and $\mathbf{CD} \neq \mathbf{DC}$. 5(b). True.
5(c). True. (d). True. (e). True.
5(f). False (true if $n = 2$). Take $\mathbf{x} = (1, 1, 1)^{\mathrm{T}}, \mathbf{y} = (1, 0, -1)^{\mathrm{T}}$ and $\mathbf{z} = (0, 1, -1)^{\mathrm{T}}$. 8(c). Take $\mathbf{C} = \mathbf{A}^{k-1} + \mathbf{A}^{k-2}\mathbf{B} + \cdots + \mathbf{B}^{k-1}$.
10. $\{\mathbf{U}_{ij} : 1 \leq i \leq m, 1 \leq j \leq n\}$ is a basis, $\dim = mn$.
11(a). No. If $\mathbf{C} \in \mathrm{Sp}(B)$ then $c_{11} = c_{22}$. (b). $\{\mathbf{I}, \mathbf{A}\}$, $\{\mathbf{I}, \mathbf{A}, \mathbf{U}_{11}, \mathbf{U}_{21}\}$.
12. $\mathbf{I}, \mathbf{A}, \mathbf{A}^2, \ldots, \mathbf{A}^{n^2}$ are linearly dependent in $F^{n \times n}$ since $d(F^{n \times n}) = n^2$.

Section 2.5 (p. 94)

1. Let $i \neq j$ and $\mathbf{B} = \mathbf{U}_{ij}$. Then $(\mathbf{AB})_{ij} = a_{ii}$, $(\mathbf{BA})_{ij} = a_{jj}$, $(\mathbf{AB})_{jj} = a_{ji}$ and $(\mathbf{BA})_{jj} = 0$. 2. $a_{11} = a_{22}$ and $a_{21} = 0$. 3(a). Trace of $[\alpha]$ is α.
3(b). If $\mathbf{A} = \begin{bmatrix} 1 & 2 \\ 3 & 4 \end{bmatrix}, \mathbf{B} = \begin{bmatrix} 1 & 1 \\ 1 & 1 \end{bmatrix}$ and $\mathbf{C} = \begin{bmatrix} 1 & 1 \\ -1 & -1 \end{bmatrix}$, then $\mathrm{tr}(\mathbf{ABC}) = 0$ and $\mathrm{tr}(\mathbf{BAC}) = -4$. 3(d). Note that $\mathrm{tr}(\mathbf{AB} - \mathbf{BA}) = 0$.
4. $\mathbf{A}^k = \begin{bmatrix} a^k & \beta \\ 0 & 1 \end{bmatrix}$ where $\beta = b(1 + a + \cdots + a^{k-1})$.
7. $\mathbf{B} = \begin{bmatrix} 12 & 4 \\ 8 & 0 \end{bmatrix}, \mathbf{C} = \begin{bmatrix} -12 & -3 \\ -6 & -3 \end{bmatrix}$ and $\mathbf{BC} = \mathbf{CB} = \begin{bmatrix} -168 & -48 \\ -96 & -24 \end{bmatrix}$.
8. Projection is obtained by replacing the entry 3 by 0.
9. Let $\mathbf{B}_{ij} = \mathbf{U}_{ij} + \mathbf{U}_{ji}$. Then $\{\mathbf{U}_{11}, \mathbf{U}_{22}, \mathbf{U}_{33}, \mathbf{B}_{12}, \mathbf{B}_{13}, \mathbf{B}_{23}\}$ is a basis of S. The set of all matrices $\mathbf{A}$ with $a_{ij} = 0$ whenever $i \geq j$ is a complement.
11. No. Take $\mathbf{B}$ with $b_{11} = b_{22} = 1$, $b_{21} = b_{12} = 2$.
12. $(\mathbf{CAC}^{\mathrm{T}})^{\mathrm{T}} = (\mathbf{C}^{\mathrm{T}})^{\mathrm{T}}\mathbf{A}^{\mathrm{T}}\mathbf{C}^{\mathrm{T}} = \mathbf{CAC}^{\mathrm{T}}$. $\mathbf{U}_{12}\mathbf{U}_{12}^{\mathrm{T}} = \mathbf{U}_{11}$ and $\mathbf{U}_{12}^{\mathrm{T}}\mathbf{U}_{12} = \mathbf{U}_{22}$.
13. Let $\mathbf{C}_{m \times n} = \mathbf{U}_{1i} + \mathbf{U}_{2j}$. Then $(\mathbf{CAC}^{\mathrm{T}})_{12} = a_{ij}$ and $(\mathbf{CAC}^{\mathrm{T}})_{21} = a_{ji}$.
14. $\mathbf{x}^{\mathrm{T}}\mathbf{Ax}$ is a scalar.

Section 2.6 (p. 99)

1. Take '$\mathbf{u} = \mathbf{e}_2$ and $\mathbf{v} = \mathbf{e}_3$' or '$\mathbf{u} = \mathbf{e}_3$ and $\mathbf{v} = \mathbf{e}_1 + \mathbf{e}_2$'.
2. Let $u_k = \alpha \neq 0$. Take $\mathbf{A}_{*j}$ equal to $(1/\alpha)\mathbf{v}$ if $j = k$ and $\mathbf{0}$ otherwise.
3. If $\mathbf{P}$ is a permutation matrix, $(\mathbf{PA})_{i*} = \mathbf{A}_{k*}$ where $\mathbf{P}_{i*} = \mathbf{e}_k^{\mathrm{T}}$ and $(\mathbf{AP})_{*j} = \mathbf{A}_{*\ell}$ where $\mathbf{P}_{*j} = \mathbf{e}_\ell$. 4. $\operatorname{tr}(\mathbf{AU}_{ij}) = a_{ji}$.
5. Clearly $f(\mathbf{A}) = \sum_{i,j} c_{ij}a_{ij}$ for some c_{ij}'s $\in F$. Now $\mathbf{U}_{ij}\mathbf{U}_{k\ell} = \mathbf{e}_i(\mathbf{e}_j^{\mathrm{T}}\mathbf{e}_k)\mathbf{e}_\ell^{\mathrm{T}} = \mathbf{U}_{i\ell}$ if $j = k$ and $\mathbf{0}$ otherwise. So $f(\mathbf{U}_{ii}\mathbf{U}_{ij}) = f(\mathbf{U}_{ij}\mathbf{U}_{ii})$ gives $c_{ij} = 0$ and $f(\mathbf{U}_{ij}\mathbf{U}_{ji}) = f(\mathbf{U}_{ji}\mathbf{U}_{ij})$ gives $c_{ii} = c_{jj}$ if $i \neq j$. Thus $f(\mathbf{A}) = c \cdot \operatorname{tr}(\mathbf{A})$ for some constant c which is 1 if $f(\mathbf{I}) = n$.
6. Let $f(\mathbf{A}) = \sum_{i,j} c_{ij}a_{ij}$. Let $i \neq j$. Taking $\mathbf{A} = \mathbf{U}_{ij}$, we get $c_{ij} = 0$. Taking $\mathbf{A} = \mathbf{U}_{ii} + \mathbf{U}_{ij} - \mathbf{U}_{ji} - \mathbf{U}_{jj}$, we get $c_{ii} = c_{jj}$. 7. Use $[\,\alpha\,]^{\mathrm{T}} = [\,\alpha\,]$.
8(c). First take $\mathbf{x} = \mathbf{e}_i$ and then take $\mathbf{x} = \mathbf{e}_j + \mathbf{e}_k$.
9. The effects of premultiplication (postmultiplication) are addition of α times the first row (resp. second column) to the second row (resp. first column), addition of α times the second row (resp. first column) to the first row (resp. second column) and multiplying the first row (resp. first column) by α.
11(a) and (b). Show that $\mathbf{Ax} = \mathbf{0}$ and $\mathbf{Ax} = \mathbf{x}$ respectively for all $\mathbf{x} \in F^n$.
12(a). Let $x_j \neq 0$. If $\mathbf{A}_{*j} \doteq (1/x_j)\mathbf{e}_i$ and $\mathbf{A}_{*k} = \mathbf{0}$ for $k \neq j$ then $\mathbf{y}^{\mathrm{T}}\mathbf{Ax} = y_i$.
14(c). $(\mathbf{PAP})_{ij} = \mathbf{P}_{i*}\mathbf{AP}_{*j} = \mathbf{e}_{n-i+1}^{\mathrm{T}}\mathbf{Ae}_{n-j+1} = a_{n-i+1,n-j+1}$.

If $n = 3$, $\mathbf{PAP} = \begin{bmatrix} a_{33} & a_{32} & a_{31} \\ a_{23} & a_{22} & a_{21} \\ a_{13} & a_{12} & a_{11} \end{bmatrix}$. 14(d). Lower triangular.

Section 2.7 (p. 105)

1. $k \times (n - \ell)$, $(m - k) \times \ell$, $(m - k) \times (n - \ell)$.
2(a). $[\,\mathbf{P} : \mathbf{R}\,]\begin{bmatrix} \mathbf{Q} \\ \mathbf{S} \end{bmatrix}$. (b). Take $\mathbf{D} = \begin{bmatrix} \mathbf{I} \\ \mathbf{I} \end{bmatrix}$.
3. $\mathbf{A}_{\alpha\beta}^{\mathrm{T}} = \mathbf{A}_{\alpha,-\beta}$, $\mathbf{A}_{\alpha,\beta} + \mathbf{A}_{\gamma,\delta} = \mathbf{A}_{\alpha+\gamma,\beta+\delta}$ and $\mathbf{A}_{\alpha,\beta}\mathbf{A}_{\gamma,\delta} = \mathbf{A}_{\alpha\gamma-\beta\delta,\alpha\delta+\beta\gamma}$.

$\mathbf{B}_{\alpha,\beta,\gamma,\delta} = \begin{bmatrix} \mathbf{A}_{\alpha,\beta} & \mathbf{A}_{\gamma,\delta} \\ \mathbf{A}_{-\gamma,\delta} & \mathbf{A}_{\alpha,-\beta} \end{bmatrix}$. $\mathbf{BB}^{\mathrm{T}} = (\alpha^2 + \beta^2 + \gamma^2 + \delta^2)\mathbf{I}_4$.

4. With $\mathbf{A} = [\,\mathbf{A}_1 : \mathbf{A}_2\,]$ and $\mathbf{C} = \begin{bmatrix} \mathbf{C}_1 \\ \mathbf{C}_2 \end{bmatrix}$, $\mathbf{ABC} = \mathbf{A}_1\mathbf{B}_1\mathbf{C}_1 + \mathbf{A}_2\mathbf{B}_2\mathbf{C}_2$.

5(a) and (b). $\begin{bmatrix} \mathbf{AGP} & \mathbf{AGQ} \\ \mathbf{CGP} & \mathbf{CGQ} \end{bmatrix}$. 5(c). $\begin{bmatrix} \mathbf{B} & \mathbf{BC} \\ \mathbf{DB} & \mathbf{DBC} \end{bmatrix}$.

5(d). $\begin{bmatrix} \mathbf{AD} & \mathbf{0} \\ \mathbf{BD} + \mathbf{CE} & \mathbf{CG} \end{bmatrix}$. 5(e). $\begin{bmatrix} \mathbf{P}_1\mathbf{P}_2\mathbf{P}_3 & \mathbf{0} \\ \mathbf{0} & \mathbf{Q}_1\mathbf{Q}_2\mathbf{Q}_3 \end{bmatrix}$.

10. $(\mathbf{A}^3)_{*j} = \mathbf{A} \cdot (\mathbf{A}^2)_{*j}$.
12(a). $[\,1\;\;2\,] \otimes [\,3\;\;4\,] = [\,3\;\;4\;\;6\;\;8\,]$ whereas $[\,3\;\;4\,] \otimes [\,1\;\;2\,] = [\,3\;\;6\;\;4\;\;8\,]$.
12(h). The (i,j)-th elements are $\sum_j a_{ij}b_{jk}\mathbf{CD}$ and $\sum_j (a_{ij}\mathbf{C})(b_{jk}\mathbf{D})$.

Section 3.2 (p. 111)

1. Any two rows form a row basis and any two columns form a column basis.
3(a). The column space does not change. For row rank, use *Exercises* 1.8.2(c) and 1.8.3(b).
3(b). Use: row rank of $[\mathbf{B}\ \ \mathbf{0}]$ = row rank of $\mathbf{B} \leq$ number of columns of $\mathbf{B}$.
4. Let $\mathbf{B} = \mathbf{A}(i_1, \ldots, i_k \,|\, j_1, \ldots, j_\ell)$ and $\mathbf{C} = \mathbf{A}(i_1, \ldots, i_k \,|\, 1, \ldots, n)$. By *Exercise* 1.4.6, row rank of $\mathbf{B} \leq$ row rank of $\mathbf{C}$. But $\mathcal{R}(\mathbf{C}) \subseteq \mathcal{R}(\mathbf{A})$.
5. Take column bases of $\mathbf{A}$ and $\mathbf{C}$. The corresponding columns of the bigger matrix are linearly independent. For inequality, take $\mathbf{A} = \mathbf{C} = \mathbf{0}$ and $\mathbf{B} \neq \mathbf{0}$.
6(i). $\rho(\mathbf{A})$, $\rho(\mathbf{A}) - 1$ or $\rho(\mathbf{A}) + 1$.
6(ii). The rank may change by at most 2 (at most 1 if the two elements are in the same row or column).
7. $\rho(\mathbf{A})$, $\rho(\mathbf{A}) - 1$ or $\rho(\mathbf{A}) - 2$.
8. Take k linearly independent rows and then k linearly independent columns of the submatrix formed by them.
9. Use the proof of *Theorem* 3.2.3 and $\mathcal{C}(\mathbf{x}\mathbf{y}^{\mathrm{T}}) \subseteq \mathrm{Sp}(\{\mathbf{x}\})$.
10. Let $\mathbf{A}$ be the matrix with rows $\mathbf{x}_{ij}{}^{\mathrm{T}}$. Then $\sum_{j=2}^{p+1} \mathbf{A}_{*j} = \sum_{j=p+2}^{p+q+1} \mathbf{A}_{*j} = \mathbf{A}_{*1}$. The last $p+q-1$ columns of $\mathbf{A}$ form a column basis.

Section 3.3 (p. 115)

1(b). $(\frac{1}{2}, 0, 0, 0)^{\mathrm{T}}$, $(1, 1, -\frac{2}{3}, 0)^{\mathrm{T}}$, $(0, 1, 0, 0)^{\mathrm{T}}$ and $(1, 1, 1, -\frac{5}{4})^{\mathrm{T}}$. 2. $\begin{bmatrix} \mathbf{A}^{-1} \\ \mathbf{0} \end{bmatrix}$.
3(a). $\begin{bmatrix} 0 & 1 & 0 \\ 0 & 0 & 1 \end{bmatrix}$ and $\begin{bmatrix} 1 & -2 & -1 \\ 1 & -3 & 0 \end{bmatrix}$; $\begin{bmatrix} 0 & 1 \\ 1 & 0 \\ 0 & 0 \end{bmatrix}$ and $\begin{bmatrix} 1 & 1 \\ -1 & 0 \\ 1 & 0 \end{bmatrix}$.
4. $\mathbf{y}^{\mathrm{T}}\mathbf{A} = \mathbf{e}_1^{\mathrm{T}}$ (or $\mathbf{A}\mathbf{x} = \mathbf{e}_1$) has a unique solution. Now use *Exercise* 1.5.5.
6. *First part*: Take $\mathbf{A} = \begin{bmatrix} 1 & 1 \\ 3 & 1 \end{bmatrix}$ and $\mathbf{B} = \begin{bmatrix} 1 & -2 \\ -3 & 2 \end{bmatrix}$. *Second part*: Note that $\mathbf{AB} = \alpha\mathbf{I}$, $\alpha \neq 0 \Rightarrow \mathbf{B} = \alpha\mathbf{A}^{-1}$.
7. False: take $\mathbf{A} = [1\ \ 1]$ and $\mathbf{B} = [1\ \ -1]^{\mathrm{T}}$.
8. *First part*: Use *Theorem* 3.3.9. *Second*: Take $\mathbf{A} = \mathbf{e}_1\mathbf{e}_1^{\mathrm{T}}$.
9. Verify $-\frac{1}{5}(3\mathbf{A}^3 - 4\mathbf{A}^2 + 2\mathbf{I}) \cdot \mathbf{A} = \mathbf{I}$. Aliter: Verify $\mathbf{Ax} = \mathbf{0} \Rightarrow \mathbf{x} = \mathbf{0}$.
10(a). Let $\mathbf{A}$ have rank r, $\{\mathbf{A}_{i*} : i \in I\}$ be a row basis, $\{\mathbf{A}_{*j} : j \in J\}$ be a column basis and $\mathbf{B} = \mathbf{A}(I \,|\, 1, \ldots, n)$. Then $\rho(\mathbf{B}) = r$. Since $\{\mathbf{B}_{*j} : j \in J\}$ generates $\mathcal{C}(\mathbf{B})$, it forms a basis. (b). Take $\mathbf{A} = \mathbf{I}$, $I = \{1\}$ and $J = \{2\}$.
11(a). Use the preceding exercise. (b). False: Consider $\mathbf{e}_1\mathbf{e}_2^{\mathrm{T}} + \mathbf{e}_2\mathbf{e}_1^{\mathrm{T}}$.
12(i). If it exists and $\mathbf{A}$ is its matrix w.r.t. the canonical bases, $\mathcal{C}(\mathbf{A}) = F^m$, so $\rho(\mathbf{A}_{m\times n}) = m > n$.
12(ii). Here $\mathbf{Ax} = \mathbf{0} \Rightarrow \mathbf{x} = \mathbf{0}$, so $\rho(\mathbf{A}_{n\times m}) = m > n$.
13. *First part*: $\prod_{k=0}^{m-1}(q^n - q^k)$ by *Exercise* 1.5.19(b).

Section 3.4 (p. 119)
1(a). $\mathbf{A}^{-1} = \mathbf{I} - \frac{1}{3}\mathbf{J}$ where $\mathbf{J} = ((1))$. (b). Singular. (c). $\mathbf{A}^{-1} = \mathbf{A}$.
1(d). $\mathbf{A}$ is non-singular iff $\beta \neq 0$ and $\gamma \neq 0$. $\mathbf{A}^{-1} = \frac{1}{\beta\gamma}\begin{bmatrix} 0 & \beta \\ \gamma & -\alpha \end{bmatrix}$.
1(e). Singular. (f). $\mathbf{A}^{-1} = \mathrm{diag}(\frac{1}{3}, \frac{1}{2}, -1)$. (g). $\mathbf{A}^{-1} = \frac{1}{8}\begin{bmatrix} 4 & -2 & 3 \\ 0 & 4 & -6 \\ 0 & 0 & 4 \end{bmatrix}$.
1(h). Singular. 2. See *Exercise* 2.3.7(b); Yes.
3. Premultiply by $\mathbf{A}^{-1}$. 4. $\mathbf{I}$, $-\mathbf{I}$. 5(a). $\mathbf{B}^{-1}(\mathbf{AB})\mathbf{B}^{-1} = \mathbf{B}^{-1}(\mathbf{BA})\mathbf{B}^{-1}$.
6. Use $(\mathbf{A}^{-1})^{\mathrm{T}} = (\mathbf{A}^{\mathrm{T}})^{-1}$. 7. $\mathbf{U}_{11}$, $\mathbf{U}_{22}$, $\mathbf{U}_{12}$, $\mathbf{U}_{21}$ (recall $\mathbf{U}_{ij} = \mathbf{e}_i\mathbf{e}_j^{\mathrm{T}}$).
8. $c\mathbf{I} + d\mathbf{J}$ where $c = 1/\alpha$ and $d = -\beta/(\alpha(\alpha + n\beta))$.
9. $(\mathbf{A} \otimes \mathbf{B})(\mathbf{A}^{-1} \otimes \mathbf{B}^{-1}) = \mathbf{I} \otimes \mathbf{I} = \mathbf{I}$.
10. Let $\mathbf{B}$ and $\mathbf{C}$ be as in the solution to *Exercise* 3.2.4. $n = \rho(\mathbf{A}) \leq \rho(\mathbf{C}) + (n-k) \leq \rho(\mathbf{B}) + (n-\ell) + (n-k)$.
12(a). Identity element is $\mathrm{diag}(1,0)$ and the 'inverse' of $\mathrm{diag}(\alpha,0)$ is $\mathrm{diag}(\frac{1}{\alpha},0)$.
12(b). No identity element.

Section 3.5 (p. 126)
1. $\mathbf{A} = \mathbf{DB}$, so $\mathbf{AE} = \mathbf{DBE}$. If $\mathcal{C}(\mathbf{A}) \subseteq \mathcal{C}(\mathbf{B})$ then $\mathcal{C}(\mathbf{GA}) \subseteq \mathcal{C}(\mathbf{GB})$.
2. Use *Theorem* 3.5.7(i) with $\mathbf{B} = \mathbf{C} = \mathbf{A}$ and $\mathbf{D} = \mathbf{I}$. *Second part*: Take $\mathbf{A} = \mathbf{e1e}_2^{\mathrm{T}}$. 3. False. Consider $\mathbf{e}_1^{\mathrm{T}}(\mathbf{e}_1\mathbf{e}_2^{\mathrm{T}})\mathbf{e}_2$. 4. Take $\mathbf{B} = \mathbf{0}$ and $\mathbf{C} \neq \mathbf{D}$.
6(i). If $\mathbf{Ax} = \mathbf{0}$, $\mathbf{x} = (\mathbf{I} - \mathbf{A})\mathbf{x}$. (ii). $\mathbf{x} = \mathbf{Ax} + (\mathbf{I} - \mathbf{A})\mathbf{x}$.
7(a). $[\mathbf{A} : \mathbf{B}]\left[\begin{smallmatrix}\mathbf{u}\\ \mathbf{v}\end{smallmatrix}\right] = \mathbf{Au} + \mathbf{Bv}$. (b). $\left[\begin{smallmatrix}\mathbf{A}\\ \mathbf{C}\end{smallmatrix}\right]\mathbf{x} = \mathbf{0}$ iff $\mathbf{Ax} = \mathbf{0}$ and $\mathbf{Cx} = \mathbf{0}$.
8(a). Since $\mathcal{C}(\mathbf{A} : \mathbf{B}) \supseteq \mathcal{C}(\mathbf{A})$, $\rho(\mathbf{A} : \mathbf{B}) = \rho(\mathbf{A})$ iff $\mathcal{C}(\mathbf{B}) \subseteq \mathcal{C}(\mathbf{A})$.
9. $[\mathbf{I} \quad \mathbf{B}] = \mathbf{E}[\mathbf{A} \quad \mathbf{I}] \Leftrightarrow \mathbf{A}^{-1} = \mathbf{E} = \mathbf{B}$. Use *Exercise* 3.5.8(b).
10. *If part*: $\mathbf{B} = \mathbf{DA}$ for some $\mathbf{D}$, so $\mathcal{N}(\mathbf{A}) \subseteq \mathcal{N}(\mathbf{B})$. *Only if part*: By hypothesis, $\mathcal{N}(\mathbf{A}) = \mathcal{N}\left[\begin{smallmatrix}\mathbf{A}\\ \mathbf{B}\end{smallmatrix}\right]$, so $\rho(\mathbf{A}) = \rho\left[\begin{smallmatrix}\mathbf{A}\\ \mathbf{B}\end{smallmatrix}\right]$ and $\mathcal{R}(\mathbf{A}) \supseteq \mathcal{R}(\mathbf{B})$.
11(a). Use $\mathbf{MM}^{-1} = \mathbf{I}$.
11(b). $\mathbf{PD} = \mathbf{0}$, so $\mathcal{C}(\mathbf{D}) \subseteq \mathcal{N}(\mathbf{P})$. If $\mathbf{x} \in \mathcal{N}(\mathbf{P})$, then $\mathbf{x} = \mathbf{M}^{-1}\mathbf{Mx} = \mathbf{DQx} \in \mathcal{C}(\mathbf{D})$. Thus $\mathcal{C}(\mathbf{D}) = \mathcal{N}(\mathbf{P})$, so $\nu(\mathbf{P}) = n - k = n - \rho(\mathbf{P})$. Given $\mathbf{A}$, let the rows of $\mathbf{P}$ form a basis of $\mathcal{R}(\mathbf{A})$. Then by the preceding exercise, $\mathcal{R}(\mathbf{A}) = \mathcal{R}(\mathbf{P})$, so $\mathcal{N}(\mathbf{A}) = \mathcal{N}(\mathbf{P})$. Since $\rho(\mathbf{A}) = \rho(\mathbf{P})$ and $\mathbf{P}$ can be extended to a non-singular matrix, *Theorem* 3.5.9 follows.
12. *Only if part*: $\mathbf{AB} = \mathbf{0}$, so $\mathbf{B}^{\mathrm{T}}\mathbf{A}^{\mathrm{T}} = \mathbf{0}$ and $\mathcal{C}(\mathbf{A}^{\mathrm{T}}) \subseteq \mathcal{N}(\mathbf{B}^{\mathrm{T}})$. If $\mathbf{A}$ has n columns, $\nu(\mathbf{B}^{\mathrm{T}}) = n - \rho(\mathbf{B}) = n - \nu(\mathbf{A}) = \rho(\mathbf{A}) = \rho(\mathbf{A}^{\mathrm{T}})$. So $\mathcal{C}(\mathbf{A}^{\mathrm{T}}) = \mathcal{N}(\mathbf{B}^{\mathrm{T}})$. *If part* is similar. 13. $\mathcal{C}(\mathbf{B}) \subseteq \mathcal{N}(\mathbf{A})$, so $\rho(\mathbf{B}) \leq n - \rho(\mathbf{A})$. 14. $\begin{bmatrix} \mathbf{0} & \mathbf{I}_k \\ \mathbf{0} & \mathbf{0} \end{bmatrix}$.
15. Use *Theorems* 3.5.3 and 3.5.12. $|\rho(\mathbf{AB}) - \rho(\mathbf{BA})| \leq 2, 1$ or 0 according as $\min(\rho(\mathbf{A}), \rho(\mathbf{B}))$ belongs to $\{0,1\}$, $\{2,3\}$ or $\{4,5\}$. Consider $\mathbf{A} = \begin{bmatrix} \mathbf{I}_r & \mathbf{0} \\ \mathbf{0} & \mathbf{0} \end{bmatrix}$,

$\mathbf{B} = \begin{bmatrix} 0 & \mathbf{I}_s \\ 0 & 0 \end{bmatrix}$ for suitable r and s. 16. Use *Theorem* 3.5.12 $k-1$ times.
17. By *Theorem* 3.5.6, $\mathbf{A}^k = \mathbf{D}\mathbf{A}^{k+1}$ for some $\mathbf{D}$, so $\mathbf{A}^{k+1} = \mathbf{D}\mathbf{A}^{k+2}$.
18. Take $\mathbf{B} = \mathbf{A}^{k-1}$ and $\mathbf{C} = \mathbf{A}$ in *Theorem* 3.5.13.
19. *If part*: $\mathbf{A} = \mathbf{DPA}$ for some $\mathbf{D}$. Then $\rho(\mathbf{A}) = \rho(\mathbf{AQ}) = \rho(\mathbf{DPAQ}) \leq \rho(\mathbf{PAQ}) \leq \rho(\mathbf{A})$. 20(b). $\mathcal{K}(h) = W \cap \mathcal{K}(g)$.
21. $\mathbf{A} = \mathbf{BC}$ and $\mathbf{B} = \mathbf{AD}$ for some $\mathbf{C}$ and $\mathbf{D} \in \mathcal{G}$, etc. Also $\mathbf{A}^2 \in \mathcal{G}$.
22. *Only if part*: Let $\mathbf{B} = \mathbf{AC}$. Then $\mathbf{RBQx} = \mathbf{RASS}^{-1}\mathbf{CQx}$.
23. $\mathbf{AC} = -\mathbf{BD}$. So $\mathcal{C}(\mathbf{AC}) = \mathcal{C}(\mathbf{BD}) \subseteq \mathcal{C}(\mathbf{A}) \cap \mathcal{C}(\mathbf{B})$. Let $\mathbf{x} \in$ RHS and $\mathbf{x} = \mathbf{Au} = \mathbf{Bv}$. Then $\begin{bmatrix} \mathbf{u} \\ -\mathbf{v} \end{bmatrix} \in \mathcal{N}[\mathbf{A} : \mathbf{B}] = \mathcal{C}\begin{bmatrix} \mathbf{C} \\ \mathbf{D} \end{bmatrix}$. So $\mathbf{x} = \mathbf{Au} \in \mathcal{C}(\mathbf{AC})$.

Section 3.6 (p. 130)

We will use '$\mathbf{A} \rightsquigarrow (\mathbf{P}, \mathbf{Q})$' to denote '$(\mathbf{P}, \mathbf{Q})$ is a rank-factorization of $\mathbf{A}$' henceforth.

1. *Last part*: If $\mathbf{A} \rightsquigarrow (\mathbf{R}, \mathbf{S})$, then $\mathbf{R} = \mathbf{PT}$ for some $\mathbf{T}$. $\mathbf{T}$ is non-singular.
2. $\mathbf{P}$ can be cancelled on the left and $\mathbf{Q}$ on the right.
3. We give a general rank-factorization. For (a): $(\mathbf{T}, \mathbf{T}^{-1}\mathbf{A})$ where $\mathbf{T}$ is non-singular. For (b): $(\mathbf{PT}, \mathbf{T}^{-1}\mathbf{Q})$ where $\mathbf{P} = \begin{bmatrix} 5 & 3 \\ 0 & 2 \\ 10 & 4 \end{bmatrix}$, $\mathbf{Q} = \begin{bmatrix} 1 & \frac{1}{5} & 0 \\ 0 & 0 & 1 \end{bmatrix}$ and $\mathbf{T}$ is non-singular. 4. Use rank-factorization.
5. $\mathbf{A} = \mathbf{x}\mathbf{y}^{\mathrm{T}}$ and $\mathbf{y} \in \mathrm{Sp}(\{\mathbf{x}\})$. α cannot be dropped (consider $\mathbf{A} = (-1)$).
6(a). Let $\mathbf{A} \rightsquigarrow (\mathbf{P}, \mathbf{Q})$. Then $\mathbf{P} = \mathbf{BG}$ and $\mathbf{Q} = \mathbf{HD}$ for some $\mathbf{G}$ and $\mathbf{H}$.
6(b). Take $\frac{1}{2}(\mathbf{C} + \mathbf{C}^{\mathrm{T}})$.
7(a). Let $\mathbf{A} \rightsquigarrow (\mathbf{P}, \mathbf{Q})$. Consider $[\mathbf{P} : \mathbf{e}_i] \begin{bmatrix} \mathbf{Q} \\ \mathbf{e}_j^{\mathrm{T}} \end{bmatrix}$ with $\mathbf{e}_i \notin \mathcal{C}(\mathbf{P})$ and $\mathbf{e}_j^{\mathrm{T}} \notin \mathcal{R}(\mathbf{Q})$.
7(b). *First part*: W.l.g., let $\mathbf{A} = \begin{bmatrix} \mathbf{B} & \mathbf{u} \\ \mathbf{v}^{\mathrm{T}} & \alpha \end{bmatrix}$ where $\mathbf{B}$ is non-singular (see *Exercise* 3.2.8). Change α to $\mathbf{v}^{\mathrm{T}}\mathbf{B}^{-1}\mathbf{u}$. *Second part*: consider $((1))_{2\times 2}$.
8. Let $\mathbf{A} \rightsquigarrow (\mathbf{P}, \mathbf{Q})$ and $\mathbf{B} \rightsquigarrow (\mathbf{R}, \mathbf{S})$. Then $\rho(\mathbf{AB}) = \rho(\mathbf{QR})$. Let $\begin{bmatrix} \mathbf{Q} \\ \mathbf{K} \end{bmatrix}$ and $[\mathbf{R} : \mathbf{L}]$ be non-singular. $\mathbf{QR}$ is a submatrix of $\begin{bmatrix} \mathbf{Q} \\ \mathbf{K} \end{bmatrix}[\mathbf{R} : \mathbf{L}]$.
9. $\rho(\mathbf{A}) + \rho(\mathbf{B}) \leq \rho[\mathbf{A} : \mathbf{P}] + \rho\begin{bmatrix} \mathbf{B} \\ -\mathbf{Q} \end{bmatrix} \leq n + \rho(\mathbf{AB})$.
10. Let $\mathbf{B} \rightsquigarrow (\mathbf{P}, \mathbf{Q})$. Then $\rho(\mathbf{ABC}) = \rho(\mathbf{APQC}) \geq \rho(\mathbf{AP}) + \rho(\mathbf{QC}) - \rho(\mathbf{B}) = \rho(\mathbf{AB}) + \rho(\mathbf{BC}) - \rho(\mathbf{B})$.
11. *First part*: Given $\mathbf{PQPQ} = \mathbf{PIQ}$, $\mathbf{P}$ can be cancelled on the left and $\mathbf{Q}$ on the right. *Second part*: $\rho(\mathbf{A}^2) = \rho(\mathbf{PQPQ}) = \rho(\mathbf{QP})$.
13. Let $\mathbf{A} \rightsquigarrow (\mathbf{P}_1, \mathbf{Q}_1)$ and $\mathbf{P} = [\mathbf{P}_1 : \mathbf{P}_2]$ and $\mathbf{Q} = \begin{bmatrix} \mathbf{Q}_1 \\ \mathbf{Q}_2 \end{bmatrix}$ be non-singular.

Section 3.7 (p. 136)

1. Use: $\mathbf{A} = \sum \mathbf{P}_{*j}\mathbf{Q}_{j*}$ where $\mathbf{A} \rightsquigarrow (\mathbf{P}, \mathbf{Q})$. 2. Use *Theorem* 3.7.2.

3. $\rho(\mathbf{AB}-\mathbf{I}) = \rho(\mathbf{A}-\mathbf{I}+\mathbf{A}(\mathbf{B}-\mathbf{I})) \le \rho(\mathbf{A}-\mathbf{I}) + \rho(\mathbf{A}(\mathbf{B}-\mathbf{I}))$.
4. Let $\mathbf{A} \rightsquigarrow (\mathbf{P}_1, \mathbf{Q}_1)$, $[\mathbf{P}_1 \quad \mathbf{P}_2]$ and $[\mathbf{Q}_1^{\mathrm{T}} \quad \mathbf{Q}_2^{\mathrm{T}}]$ be non-singular. Take $\mathbf{B} = \mathbf{P}_2\mathbf{Q}_2$. 5(a). $\mathbf{A} \rightsquigarrow ((2,1,0)^{\mathrm{T}}, (1, \frac{1}{2}, 2))$ and $\mathbf{B} \rightsquigarrow ((1,0,1)^{\mathrm{T}}, (1,2,-1))$.
5(b). $\mathbf{e}_2\mathbf{e}_1^{\mathrm{T}}$ (use the preceding exercise). 5(c). $\frac{1}{18}\begin{bmatrix} -9 & 18 & 9 \\ 8 & -12 & 0 \\ 7 & -6 & -9 \end{bmatrix}$.
6. Let $\mathbf{A}_i \rightsquigarrow (\mathbf{P}_i, \mathbf{Q}_i)$ for each i and $\mathbf{P}, \mathbf{Q}$ be as in (3.7.4). $\mathbf{A}_i\mathbf{A}^{-1}\mathbf{A}_j = \mathbf{P}_i\mathbf{Q}_i\mathbf{Q}^{-1}\mathbf{P}^{-1}\mathbf{P}_j\mathbf{Q}_j = \delta_{ij}\mathbf{P}_i\mathbf{Q}_i$. Aliter: Since $\mathbf{A}_i\mathbf{A}^{-1}\mathbf{A} = \mathbf{A}_i$, we have $\mathbf{x}^{\mathrm{T}}(\mathbf{A}_i\mathbf{A}^{-1} - \mathbf{I})\mathbf{A}_i + \sum_{j \ne i} \mathbf{x}^{\mathrm{T}}\mathbf{A}_i\mathbf{A}^{-1}\mathbf{A}_j = \mathbf{0}$. Since $\mathcal{R}(\mathbf{A}_1) + \cdots + \mathcal{R}(\mathbf{A}_k)$ is direct, each term is $\mathbf{0}$ for all $\mathbf{x}$. 7(a). 3 since $\rho(\mathbf{A}) \le \rho(\mathbf{A}+\mathbf{B}) + \rho(-\mathbf{B})$.
7(b). $|\rho(\mathbf{A}) - \rho(\mathbf{B})| \le \rho(\mathbf{A}-\mathbf{B}) \le \rho(\mathbf{A}) + \rho(\mathbf{B})$.
8. *Only if part*: $\mathcal{C}(\mathbf{A}+\mathbf{B}) = \mathcal{C}(\mathbf{A}) + \mathcal{C}(\mathbf{B})$. *If part*: Let $\mathbf{A} = (\mathbf{A}+\mathbf{B})\mathbf{T}$. Then $\mathcal{C}(\mathbf{A}(\mathbf{I}-\mathbf{T})) = \mathcal{C}(\mathbf{BT}) \subseteq \mathcal{C}(\mathbf{A}) \cap \mathcal{C}(\mathbf{B})$, so $\mathbf{A} = \mathbf{AT}$ and $\mathbf{BT} = \mathbf{0}$. Hence $\mathbf{x}^{\mathrm{T}}\mathbf{A} = \mathbf{y}^{\mathrm{T}}\mathbf{B} \Rightarrow \mathbf{x}^{\mathrm{T}}\mathbf{A} = \mathbf{0}$. Thus $\mathcal{R}(\mathbf{A}) \cap \mathcal{R}(\mathbf{B}) = \{\mathbf{0}\}$.
9(a). *Only if part*: $\mathbf{B} = \mathbf{AC} \Rightarrow \mathbf{AB} = \mathbf{A}^2\mathbf{C} = \mathbf{AC} = \mathbf{B}$.
11. $\mathbf{H}^2 = \mathbf{H}$ iff $\mathbf{A}^2 = \mathbf{A}$, $\mathbf{C}^2 = \mathbf{C}$ and $\mathbf{AB} + \mathbf{BC} = \mathbf{B}$. Postmultiply the last equation by $\mathbf{C}$. Also note that $\mathbf{H}^2 = \mathbf{H} \Rightarrow (\mathbf{I}-\mathbf{H})^2 = \mathbf{I}-\mathbf{H}$.
12. By hypothesis, $\mathbf{BC} = 0$, $\mathbf{BD} = 0$, $\mathbf{DC} = \mathbf{0}$ and $\mathbf{D} = \mathbf{CB} + \mathbf{D}^2$.
13. $\mathbf{A} = \mathbf{I}$ or $\mathbf{A} = \mathbf{0}$ or '$a_{11}a_{22} = a_{12}a_{21}$ and $a_{11} + a_{22} = 1$'. Equivalently, $\mathbf{A} = \mathbf{I}$ or $\rho(\mathbf{A}) = \mathrm{tr}(\mathbf{A}) < 2$.
15. If $\mathbf{A}^2 = \mathbf{I}$, then $\mathbf{A} = 2\mathbf{B} - \mathbf{I}$ where $\mathbf{B} = \frac{1}{2}(\mathbf{A}+\mathbf{I})$ is a projector.
16. $\mathbf{x} = \mathbf{BGx} + \mathbf{CHx}$, $\mathbf{BGx} \in S$ and $\mathbf{CHx} \in T$ for all $\mathbf{x}$.
17. Take $k = 2$. *First part*: $\mathbf{A}_1 = \mathbf{U}_{21}$, $\mathbf{A}_2 = -\mathbf{A}_1$. *Second part*: $\mathbf{A}_1 = (2,1)^{\mathrm{T}}(1,-1)$, $\mathbf{A}_2 = \mathbf{U}_{11}$. *Third part*: $\mathbf{A}_1 = (1,1)^{\mathrm{T}}(2,1)$, $\mathbf{A}_2 = (1,-2)^{\mathrm{T}}(1,-1)$.
18. $\mathbf{A} = \mathbf{DA}^2$ for some $\mathbf{D}$. So $\mathbf{Ay} = \mathbf{Bz} \Rightarrow \mathbf{DA}^2\mathbf{y} = \mathbf{DABz} \Rightarrow \mathbf{Ay} = \mathbf{0}$, etc. *Second part*: take $\mathbf{A} = (1,1)^{\mathrm{T}}(1,1)$ and $\mathbf{B} = (1,1)^{\mathrm{T}}(1,-1)$ or $(1,-1)^{\mathrm{T}}(1,1)$.
19. Use the corollary to *Theorem* 3.7.6. Note $\mathcal{C}(\mathbf{A}) \subseteq \mathcal{N}(\mathbf{B})$ means $\mathbf{BA} = \mathbf{0}$.
20. (a) $\Rightarrow$ (c), (c) $\Rightarrow$ (d) and (d) $\Rightarrow$ (b) follow from *Theorem* 3.7.6 and *Exercises* 3.7.8 and 3.7.9 respectively. (b) $\Rightarrow$ (a) is trivial.
21. See Halmos (1958).

Section 3.8 (p. 141)

2(a). $\begin{bmatrix} \mathbf{A}^{-1} & 0 \\ -\mathbf{D}^{-1}\mathbf{C}\mathbf{A}^{-1} & \mathbf{D}^{-1} \end{bmatrix}$ (b). $\begin{bmatrix} -\mathbf{C}^{-1}\mathbf{D}\mathbf{B}^{-1} & \mathbf{C}^{-1} \\ \mathbf{B}^{-1} & \mathbf{0} \end{bmatrix}$.

3. $\frac{1}{6}\begin{bmatrix} 4 & -1 \\ -2 & 2 \end{bmatrix}$, $\frac{1}{22}\begin{bmatrix} 16 & 3 & -4 \\ -8 & 7 & 2 \\ -2 & -1 & 6 \end{bmatrix}$. 4. $\frac{1}{25}\left[\begin{array}{ccc|cc} 25 & 0 & 25 & -25 & -25 \\ -3 & 7 & 6 & 6 & -12 \\ 3 & -7 & 19 & -6 & -13 \\ \hline -12 & 3 & -26 & 24 & 27 \\ 1 & 6 & -2 & -2 & 4 \end{array}\right]$.

5. Suppose the first k leading principal submatrices are non-singular. Let $\left[\begin{smallmatrix} \mathbf{A} \\ \mathbf{C} \end{smallmatrix}\right]$ be the submatrix formed by the first $k+1$ columns, where $\mathbf{A}$ is of order

$k \times (k+1)$. Then $\rho(\mathbf{A}) = k$ and $\rho\left[\begin{smallmatrix}\mathbf{A}\\ \mathbf{C}\end{smallmatrix}\right] = k+1$. So there exists an i such that $\left[\begin{smallmatrix}\mathbf{A}\\ \mathbf{C}_{i*}\end{smallmatrix}\right]$ is non-singular. Interchange the $(k+1)$-th and $(k+i)$-th rows of $\mathbf{M}$.
6. Imitate the proof of *Theorem* 3.8.4.
7. $\mathbf{M}^{-1} = \begin{bmatrix} -\mathbf{K}^{-1}\mathbf{D}\mathbf{B}^{-1} & \mathbf{K}^{-1} \\ \mathbf{B}^{-1}+\mathbf{B}^{-1}\mathbf{A}\mathbf{K}^{-1}\mathbf{D}\mathbf{B}^{-1} & -\mathbf{B}^{-1}\mathbf{A}\mathbf{K}^{-1} \end{bmatrix}$. 8. $\frac{1}{7}\begin{bmatrix} 4 & -1 \\ -1 & 2 \end{bmatrix}$.
9. $\mathbf{A}+\mathbf{u}\mathbf{v}^{\mathrm{T}}$ is non-singular $\Leftrightarrow$ $\mathbf{M}$ is non-singular $\Leftrightarrow 1+\mathbf{v}^{\mathrm{T}}\mathbf{A}^{-1}\mathbf{u} \neq 0$.
10. Let $\mathbf{B}_{*j} = \mathbf{A}_{*j}+\mathbf{u}$, i.e., $\mathbf{B} = \mathbf{A}+\mathbf{u}\mathbf{e}_j^{\mathrm{T}}$. Then $\mathbf{B}^{-1} = \mathbf{A}^{-1}-\alpha\mathbf{A}^{-1}\mathbf{u}(\mathbf{A}^{-1})_{j*}$, where $\alpha = 1/(1+(\mathbf{A}^{-1})_{j*}\mathbf{u})$. 11(a). $\mathbf{A}_{\alpha,\beta} = (\alpha-\beta)\mathbf{I}+(\beta\mathbf{1})\mathbf{1}^{\mathrm{T}}$.
12. Use (3.8.2) and (3.8.8) for $\mathbf{M} = \begin{bmatrix} \mathbf{A} & -\mathbf{B} \\ \mathbf{C} & \mathbf{D}^{-1} \end{bmatrix}$.
13(a). $\Delta := 1-\theta^2(n-1) \neq 0$.
13(b). $\mathbf{A}_\theta^{-1} = \mathbf{I}+\frac{1}{\Delta}\mathbf{C}\mathbf{C}^{\mathrm{T}}$ where $\mathbf{C} = \begin{bmatrix} \theta\mathbf{1} & \mathbf{0} \\ -1 & \theta\sqrt{n-1} \end{bmatrix}$.
14(a). *Only if part*: $(\mathbf{I}-\mathbf{A}\mathbf{G})\mathbf{x} = \mathbf{0} \Rightarrow \mathbf{x} \in \mathcal{C}(\mathbf{A})$. *If part*: See (b).
14(b). $\mathbf{z}^{\mathrm{T}}\mathbf{x} = 1$ and $\mathbf{z}^{\mathrm{T}}\mathbf{A} = \mathbf{0}$ since $(\mathbf{I}-\mathbf{A}\mathbf{G})\mathbf{A} = \mathbf{0}$.

Section 3.9 (p. 145)
1. $\mathbf{A} = (1,i)^{\mathrm{T}}(1,1)$.
2(a). *If part*: Let $\mathbf{b} = \mathbf{A}\mathbf{y}$. Then $\mathbf{A}^{\mathrm{T}}\mathbf{A}\mathbf{u} = \mathbf{A}^{\mathrm{T}}\mathbf{b} \Rightarrow \mathbf{A}^{\mathrm{T}}\mathbf{A}\mathbf{u} = \mathbf{A}^{\mathrm{T}}\mathbf{A}\mathbf{y} \Rightarrow \mathbf{A}\mathbf{u} = \mathbf{A}\mathbf{y} = \mathbf{b}$. 2(b). $\mathbf{A}^{\mathrm{T}}\mathbf{b} \in \mathcal{C}(\mathbf{A}^{\mathrm{T}}) = \mathcal{C}(\mathbf{A}^{\mathrm{T}}\mathbf{A})$.
3. $\rho(\mathbf{A}\mathbf{B}) = \rho(\mathbf{B}^{\mathrm{T}}\mathbf{A}^{\mathrm{T}}\mathbf{A}\mathbf{B}) \leq \rho(\mathbf{A}^{\mathrm{T}}\mathbf{A}\mathbf{B}) \leq \rho(\mathbf{A}\mathbf{B})$.
4. If $\mathbf{C} = \mathrm{diag}(\sqrt{d_1},\ldots,\sqrt{d_m})$, $\rho(\mathbf{A}^{\mathrm{T}}\mathbf{D}\mathbf{A}) = \rho(\mathbf{A}^{\mathrm{T}}\mathbf{C}^{\mathrm{T}}\mathbf{C}\mathbf{A}) = \rho(\mathbf{C}\mathbf{A}) = \rho(\mathbf{A})$. The result remains true under (i) and is false under (ii) and (iii) (take $\mathbf{A} = \mathbf{I}$ and $\mathbf{D} = \mathbf{0}$). 5. Use *Exercise* 3.9.3. *Second part*: $\mathbf{A} = \mathbf{B} = \mathbf{e}_1(1,1)$.
6. $\mathrm{tr}(\mathbf{A}^{\mathrm{T}}\mathbf{A}) = \sum_i\sum_j a_{ji}^2$. 7. $\mathbf{y} = \overline{\mathbf{A}}\mathbf{x}$ iff $\overline{\mathbf{y}} = \mathbf{A}\overline{\mathbf{x}}$.
9. $\frac{1}{1167}\begin{bmatrix} -7 & -21 & -90 & 106 & 205 \\ 48 & 144 & 117 & -60 & -72 \end{bmatrix}^{\mathrm{T}}$ and its transpose.
10. $\mathbf{C}^{-1}$ exists since $[\,\mathbf{A} : \mathbf{B}\,]$ has full row rank. Also $\mathbf{A}\mathbf{A}^*\mathbf{C}^{-1}\mathbf{x} \in \mathcal{C}(\mathbf{A})$ and $(\mathbf{I}-\mathbf{A}\mathbf{A}^*\mathbf{C}^{-1})\mathbf{x} = \mathbf{B}\mathbf{B}^*\mathbf{C}^{-1}\mathbf{x} \in \mathcal{C}(\mathbf{B})$ for all $\mathbf{x}$.
11. $\mathbf{A} = \frac{1}{2}(\mathbf{A}+\mathbf{A}^*)+\frac{1}{2}(\mathbf{A}-\mathbf{A}^*)$. If $\mathbf{A} = \mathbf{B}_1+\mathbf{C}_1 = \mathbf{B}_2+\mathbf{C}_2$ then $\mathbf{B}_1-\mathbf{B}_2 = \mathbf{C}_2-\mathbf{C}_1$ is both hermitian and skew-hermitian and, so, is $\mathbf{0}$.

Section 3.10 (p. 151)
1. **PR**. 2(a). Use (3.10.3). 2(b). $\frac{1}{2}\begin{bmatrix} -2 & 2 \\ 1 & 0 \end{bmatrix}$. 3(a). $\frac{1}{2}\begin{bmatrix} 0 & 1 & 4 \\ 2 & 1 & -2 \\ -2 & 1 & -2 \end{bmatrix}$.
3(b). $\sigma(\alpha,\beta,\gamma) = \frac{1}{2}(\alpha+\beta+\gamma, -\alpha-\beta+3\gamma, 3\alpha-3\beta-\gamma)$ and $\sigma^{-1}(u,v,w) = \frac{1}{6}(5u-v+2w, 4u-2v-2w, 3u+3v)$. 3(c). The matrix of σ w.r.t. $\mathcal{Y}$ as well as w.r.t. $\mathcal{Y}'$ is $\mathbf{Q}$. The matrix of σ^{-1} is $\mathbf{Q}^{-1}$.

4. $\frac{1}{3}\begin{bmatrix} -3 & 6 \\ 6 & 10 \\ -6 & -1 \end{bmatrix}$. 5. $\{(11,-3),\ (3,-1)\}$ and $\begin{bmatrix} -45 & -13 \\ 181 & 52 \end{bmatrix}$.

7. $\mathrm{tr}(f)$ is well defined because $\mathrm{tr}(\mathbf{P}^{-1}\mathbf{AP}) = \mathrm{tr}(\mathbf{APP}^{-1}) = \mathrm{tr}(\mathbf{A})$.
8. $\mathbf{P}$ is the matrix of π w.r.t. $\mathcal{X}'$ since $\pi(\mathbf{x}'_j) = \sum_i p_{ij}\pi(\mathbf{x}_i) = \sum_i p_{ij}\mathbf{x}'_i$.

Section 3.11 (p. 155)

2(a). See *Theorem* 3.7.5 and the remark following it.
3(a). By *Exercise* 2.2.13, $\mathbf{x}_1, \ldots, \mathbf{x}_r$ are linearly independent, so $d(S) = r$. Now $f(S) \subseteq f(V_1)$ and $d(f(S)) \geq r = d(f(V_1))$, so $f(S) = f(V_1)$.
3(b). *First part*: Let $\mathbf{y}_1, \ldots, \mathbf{y}_r$ form a basis of $f(V_1) = f(S)$ and $\mathbf{x}_i \in S$ be such that $f(\mathbf{x}_i) = \mathbf{y}_i$ for $i = 1, \ldots, r$. Then $\mathbf{x}_1, \ldots, \mathbf{x}_r$ form a basis of S. *Second part*: If $\mathbf{x} \in V_1$ and $f(\mathbf{x}) = \sum \alpha_i\mathbf{y}_i$, then $\mathbf{x} - \sum \alpha_i\mathbf{x}_i \in K$. So $S + K = V_1$. If $\mathbf{u} = \sum \alpha_i\mathbf{x}_i \in K$, then $\sum \alpha_i\mathbf{y}_i = \mathbf{0}$, so $\alpha_i = 0$ for all i.

Section 4.2 (p. 161)

1(a). $\begin{bmatrix} 2 & 3 & -1 & 0 \\ 1 & 5 & 0 & 8 \\ 3 & 0 & 2 & -1 \end{bmatrix}$, $\begin{bmatrix} 1 & \frac{3}{2} & -\frac{1}{2} & 0 \\ 3 & 0 & 2 & -1 \\ 1 & 5 & 0 & 8 \end{bmatrix}$, $\begin{bmatrix} 2 & 3 & -1 & 0 \\ 0 & -\frac{9}{2} & \frac{7}{2} & -1 \\ 1 & 5 & 0 & 8 \end{bmatrix}$, $\begin{bmatrix} 2 & 3 & 1 & 0 \\ 3 & 0 & -2 & -1 \\ 1 & 5 & 0 & 8 \end{bmatrix}$,

$\begin{bmatrix} 2 & 3 & 0 & 0 \\ 3 & 0 & \frac{7}{2} & -1 \\ 1 & 5 & \frac{1}{2} & 8 \end{bmatrix}$. 1(b). $R_{31}(-\frac{5}{3})$, $\begin{bmatrix} 1 & 0 & 0 & 0 \\ 0 & 1 & 0 & 0 \\ -\frac{5}{3} & 0 & 1 & 0 \\ 0 & 0 & 0 & 1 \end{bmatrix}$ and $\begin{bmatrix} 2 & 3 & -1 & 0 \\ 3 & 0 & 2 & -1 \\ -\frac{7}{3} & 0 & \frac{5}{3} & 8 \end{bmatrix}$.

1(c). No, since rank is 3. 2. $\begin{bmatrix} 1 & 0 & 0 \\ -3/2 & 1 & 0 \\ -1/2 & 0 & 1 \end{bmatrix}$, $\begin{bmatrix} 2 & 3 & -1 & 0 \\ 0 & -9/2 & 7/2 & -1 \\ 0 & 7/2 & 1/2 & 8 \end{bmatrix}$.

3. R_{ij} is achieved by $R_{ij}(1)$, $R_{ji}(-1)$, $R_{ij}(1)$ and $R_j(-1)$ in that order.
4. C_{ij}, $C_i(\alpha)$ and $C_{ji}(\beta)$.
5. We give the condition for each type of pair to commute, assuming $\alpha \neq 1$ and $\beta \neq 0$. $\mathbf{E}_{ij}$ and $\mathbf{E}_{k\ell}$: $|\{i,j\} \cap \{k,\ell\}| \neq 1$. $\mathbf{E}_{ij}$ and $\mathbf{E}_k(\alpha)$: $k \notin \{i,j\}$. $\mathbf{E}_{ij}$ and $\mathbf{E}_{k\ell}(\beta)$: $\{i,j\} \cap \{k.\ell\} = \emptyset$. $\mathbf{E}_i(\alpha)$ and $\mathbf{E}_k(\gamma)$: always. $\mathbf{E}_i(\alpha)$ and $\mathbf{E}_{k\ell}(\beta)$: $i \notin \{k,\ell\}$. $\mathbf{E}_{ij}(\alpha)$ and $\mathbf{E}_{k\ell}(\beta)$: $i \neq \ell$ and $j \neq k$.
7. $\mathrm{Sp}(\{\alpha\mathbf{x}_i\}) = \mathrm{Sp}(\{\mathbf{x}_i\})$ and $\mathrm{Sp}(\{\mathbf{x}_i + \beta\mathbf{x}_j, \mathbf{x}_j\}) = \mathrm{Sp}(\{\mathbf{x}_i, \mathbf{x}_j\})$.
8. Row space and column space will change by column and row operations.
9. The transforming matrix of $\mathcal{S}$. 10. $\mathbf{E}_{23}\mathbf{E}_{52}(4)\mathbf{E}_{34}\mathbf{E}_4(-2)\mathbf{E}_{35}(-3)$.
11. *First part*: take $\mathbf{A} = \mathbf{B} = \mathbf{0}$. *Second*: $\mathbf{PA} = \mathbf{QA} \Rightarrow \mathbf{PAA}_R^{-1} = \mathbf{QAA}_R^{-1}$.
12. Converse: if $i \geq k+1$, $\left(\begin{bmatrix} \mathbf{I} & \mathbf{0} \\ \mathbf{Q} & \mathbf{I} \end{bmatrix}\mathbf{M}\right)_{i*} = q_{i-k,1}\mathbf{M}_{1*} + \cdots + q_{i-k,k}\mathbf{M}_{k*} + \mathbf{M}_{i*}$.
13(a). Take $\mathbf{Q} = -\mathbf{H}$. 13(b). *Second part*: $\mathbf{C} - \mathbf{HA} = \mathbf{0} \Leftrightarrow \mathbf{H} = \mathbf{CA}^{-1}$.
14(a). $\begin{bmatrix} \mathbf{A} & \mathbf{0} \\ \mathbf{0} & \mathbf{D} - \mathbf{CA}^{-1}\mathbf{B} \end{bmatrix} = \begin{bmatrix} \mathbf{I} & \mathbf{0} \\ -\mathbf{CA}^{-1} & \mathbf{I} \end{bmatrix}\begin{bmatrix} \mathbf{A} & \mathbf{B} \\ \mathbf{C} & \mathbf{D} \end{bmatrix}\begin{bmatrix} \mathbf{I} & -\mathbf{A}^{-1}\mathbf{B} \\ \mathbf{0} & \mathbf{I} \end{bmatrix}$.

14(b). *Second part*: Take $\mathbf{A} = \mathbf{B} = \mathbf{D} = \mathbf{0}$ and $\mathbf{C} = \mathbf{I}$.
15. We give the changes. (i): $\mathbf{y}_i' = \mathbf{y}_j$, $\mathbf{y}_j' = \mathbf{y}_i$, (ii): $y_i' = \frac{1}{\alpha}\mathbf{y}_i$, (iii): $\mathbf{y}_j' = \mathbf{y}_j - \beta\mathbf{y}_i$. For (iv) $\mathbf{C} = \mathbf{AE}_{ij}$, (v) $\mathbf{C} = \mathbf{AE}_i(\alpha)$ and (vi) $\mathbf{C} = \mathbf{AE}_{ji}(\beta)$, the changes are: (iv): $\mathbf{x}_i' = \mathbf{x}_j$, $\mathbf{x}_j' = \mathbf{x}_i$, (v): $\mathbf{x}_i' = \alpha\mathbf{x}_i$ and (vi): $\mathbf{x}_i' = \mathbf{x}_i + \beta\mathbf{x}_j$.

Section 4.3 (p. 164)

1(a). $\begin{bmatrix} 0 & -7 & -1 & -16 \\ 0 & -15 & 2 & -25 \\ 1 & 5 & 0 & 8 \end{bmatrix}$ and $\begin{bmatrix} 1 & 0 & -2 \\ 0 & 1 & -3 \\ 0 & 0 & 1 \end{bmatrix}$.

1(b). The third row is replaced by $(1,0,0,0)$. The transforming matrix is $\mathbf{I}_4$ with the $(1,2)$-th and $(1,4)$-th elements replaced by -5 and -8.
2. If $\mathbf{B}$ is the matrix obtained then b_{ij} is $a_{ij} - (a_{i\ell}a_{kj})/a_{k\ell}$ or 1 according as '$i \neq k$ and $j \neq \ell$' or '$i = k$ and $j = \ell$'. Other elements are 0.
3. $\mathbf{P}$ is of the type $\begin{bmatrix} 1 & \mathbf{0} \\ \mathbf{u} & \mathbf{I} \end{bmatrix}$ and $\mathbf{PAP}^{\mathrm{T}} = \mathbf{BP}^{\mathrm{T}}$ is symmetric.
5. The resulting matrix is of rank 1 and has $\mathbf{e}_2$ as the first column.
6. That $\mathbf{B} = \mathbf{UA}$ is clear. Let $\mathbf{W}$ be the matrix obtained from $\mathbf{I}_m$ by replacing the k-th column by $\mathbf{A}_{*\ell}$. Then $\mathbf{UW}$ is obtained by sweeping out the k-th column of $\mathbf{W}$ with the (k,k)-th element as the pivot and, so, is $\mathbf{I}$.

Section 4.4 (p. 172)

1(a) and (b). $\mathbf{B} = \begin{bmatrix} 1 & 0 & 1 & 0 & 2/3 \\ 0 & 1 & -2 & 0 & -1/3 \\ 0 & 0 & 0 & 1 & 2 \end{bmatrix}$ and $\mathbf{E} = \begin{bmatrix} 0 & 1 & 2 & 3/2 & 0 \\ 0 & 0 & 0 & 0 & 1 \\ 0 & 0 & 0 & 0 & 0 \\ 0 & 0 & 0 & 0 & 0 \end{bmatrix}$;
3 and 2; $\{\mathbf{A}_{1*}, \mathbf{A}_{2*}, \mathbf{A}_{3*}\}$ and $\{\mathbf{A}_{1*}, \mathbf{A}_{3*}\}$; $\{\mathbf{A}_{*1}, \mathbf{A}_{*2}, \mathbf{A}_{*4}\}$ and $\{\mathbf{A}_{*2}, \mathbf{A}_{*5}\}$; $([\mathbf{A}_{*1} : \mathbf{A}_{*2} : \mathbf{A}_{*4}], \mathbf{B})$ and $\left([\mathbf{A}_{*2} : \mathbf{A}_{*5}], \begin{bmatrix} \mathbf{E}_{1*} \\ \mathbf{E}_{2*} \end{bmatrix}\right)$.

2(a), (b) and (c). $\begin{bmatrix} 3 & 1 & 0 \\ 0 & \frac{8}{3} & 2 \\ 0 & 0 & \frac{1}{2} \end{bmatrix}$, $\begin{bmatrix} 1 & 0 & 2 \\ 0 & 1 & 2 \\ 0 & 0 & -8 \end{bmatrix}$ and $\begin{bmatrix} 3 & 1 & 2 & 4 \\ 0 & -\frac{2}{3} & -\frac{4}{3} & -\frac{2}{3} \\ 0 & 0 & 0 & 0 \\ 0 & 0 & 0 & 0 \end{bmatrix}$

are the upper triangular matrices obtained. The reduced echelon forms are $\mathbf{I}_3$, $\mathbf{I}_3$ and $\mathbf{E}$ where $\mathbf{E}_{1*} = (1,0,0,1), \mathbf{E}_{2*} = (0,1,2,1)$ and the other two rows of $\mathbf{E}$ are null. The matrix in (a) is $\mathbf{E}_1(3)\mathbf{E}_{21}(7)\mathbf{E}_{31}(2)\mathbf{E}_2(8/3)\mathbf{E}_{12}(1/3)\mathbf{E}_{32}(10/3)$ $\mathbf{E}_3(1/2)\mathbf{E}_{13}(-1/4)\mathbf{E}_{23}(3/4)$.
3. Non-singular part can be converted to $\mathbf{I}$ and the rest to $\mathbf{0}$.
4. $\begin{bmatrix} \mathbf{I} \\ \mathbf{0} \end{bmatrix}$. 5. Use *Algorithm* 4.4.6 and *Exercise* 4.2.3.

6(a). With $\mathbf{A}$ as in *Definition* 4.4.1, $\mathbf{x}^{(k)}$, $k \neq p_1, \ldots, p_r$, form a basis of $\mathcal{N}(\mathbf{A})$ where $\mathbf{x}_j^{(k)}$ is 1 if $j = k$, $-a_{ik}$ if $j = p_i$ for some i and 0 otherwise.
6(b). Use *Exercise* 3.5.10 and *Theorem* 4.2.3.

6(c). $\{(-1,2,1,0,0)^{\mathrm{T}}, (-\frac{2}{3}, \frac{1}{3}, 0, -2, 1)^{\mathrm{T}}\}$ and $\{(1,0,0,0,0)^{\mathrm{T}}, (0,-2,1,0,0)^{\mathrm{T}}, (0,-\frac{3}{2},0,1,0)^{\mathrm{T}}\}$.

7. The inverses are $\frac{1}{7}\begin{bmatrix} -8 & 6 & 5 & -2 \\ 6 & -8 & -2 & 5 \\ 5 & -2 & -4 & 3 \\ -2 & 5 & 3 & -4 \end{bmatrix}$ and $\frac{1}{4}\begin{bmatrix} 4 & -4 & -16 & 12 \\ -1 & -1 & 5 & 0 \\ -1 & 3 & 9 & -8 \\ -2 & 2 & 6 & -4 \end{bmatrix}$.

The solutions are $\frac{1}{7}(-18, 24, 20, -15)^{\mathrm{T}}$ and $\frac{1}{4}(52, -3, -31, -18)^{\mathrm{T}}$.

9. $\mathbf{E}_i(1/a_{ii})$ and $\mathbf{E}_{ij}(-a_{ij})$ are upper triangular if $i < j$.

10(a). Enough to show $\mathbf{B}$ is of full row rank iff $\mathbf{A}$ is of full row rank.

10(b). By *Exercise* 4.3.4, the row interchanges may be performed first and then the sweep-outs. Let $\mathbf{C}$ be the matrix obtained after the row interchanges. Since the first r rows of the final matrix are linearly independent, so are the first r rows of $\mathbf{C}$ by (a). But these are the s_1-th, s_2-th, $\ldots, s_r$-th rows of $\mathbf{A}$.

11(a). *Only if part*: W.l.g., let $\rho(\mathbf{C}) = m$. Let $\mathbf{C} = \mathbf{PD}$. Let the leading non-zero element in $\mathbf{C}_{i*}$ (resp. $\mathbf{D}_{i*}$) be in the p_i-th (resp. q_i-th) position. Show that $\rho(\mathbf{C}^{(p_i)}) = i$ and $\rho(\mathbf{C}^{(p_i-1)}) = i-1$ where $\mathbf{C}^{(k)}$ denotes the submatrix of $\mathbf{C}$ formed by the first k columns. Deduce that $p_i = q_i$ for all i and $\mathbf{P} = \mathbf{I}$.

12. $\begin{bmatrix}\mathbf{B}\\ \mathbf{0}\end{bmatrix}$ and $[\,\mathbf{B} : \mathbf{0}\,]$.

13. *If part*: use *Exercise* 4.2.13(b). *Only if part*: Let $\begin{bmatrix}\mathbf{A}\\ \mathbf{B}\end{bmatrix}$ be converted to $\begin{bmatrix}\mathbf{C}\\ \mathbf{D}\end{bmatrix}$. Then the first p_r columns of $\mathbf{D}$ are null with usual notation. So if $\mathbf{D}$ is non-null, $r\begin{bmatrix}\mathbf{C}\\ \mathbf{D}\end{bmatrix} > r$, a contradiction.

14(a). $p_i = i$ since the first i columns of $\mathbf{B}$ are linearly independent for $i = 1, \ldots, k$. *Second part*: Take $\mathbf{A}$ with $\mathscr{C}(\mathbf{A}) = S$ and the given vectors as the first few columns and consider the p_1-th, $\ldots, p_r$-th columns of $\mathbf{A}$ $(r = d(S))$.

15. Sweeping out a column requires at most m elementary row operations.

Section 4.5 (p. 179)

1(a) and (b). Rank-factorizations are

$$\left(\begin{bmatrix} 2 & 0 \\ 0 & 1 \\ 0 & 1 \\ 2 & 0 \end{bmatrix}, \begin{bmatrix} 1 & \frac{1}{2} & \frac{1}{2} & \frac{3}{2} & 2 \\ 0 & 1 & -1 & 1 & 2 \end{bmatrix}\right), \left(\begin{bmatrix} 0 & 0 & 2 & 0 \\ 0 & 0 & 0 & 1 \\ 2 & 0 & 0 & 0 \\ 0 & 2 & 0 & 0 \end{bmatrix}, \begin{bmatrix} 1 & 0 & 5/2 & 1/2 & 0 \\ 0 & 1 & 1 & 2 & 1/2 \\ 0 & 0 & 1 & 0 & 5/2 \\ 0 & 0 & 0 & 1 & 3 \end{bmatrix}\right).$$

4. Let $\mathbf{A} \rightsquigarrow (\mathbf{T}_1, \mathbf{W}_1)$ and $\mathbf{B} \rightsquigarrow (\mathbf{T}_2, \mathbf{W}_2)$. Let $\mathbf{R} = [\,\mathbf{T}_1 : \mathbf{T}_2 : \mathbf{T}_3\,]$ and $\mathbf{S}^{\mathrm{T}} = [\,\mathbf{W}_1^{\mathrm{T}} : \mathbf{W}_2^{\mathrm{T}} : \mathbf{W}_3^{\mathrm{T}}\,]$ be non-singular, $\mathbf{P} = \mathbf{R}^{-1}$ and $\mathbf{Q} = \mathbf{S}^{-1}$. Then $\mathbf{PAQ} = \operatorname{diag}(\mathbf{I}, \mathbf{0}, \mathbf{0})$ and $\mathbf{PBQ} = \operatorname{diag}(\mathbf{0}, \mathbf{I}, \mathbf{0})$.

5. By *Exercise* 3.6.6, $\mathbf{B} = \mathbf{AHA}$ for some $\mathbf{H}$. Let (4.5.2) hold. Then $\mathbf{PBQ} = (\mathbf{PAQ})(\mathbf{Q}^{-1}\mathbf{HP}^{-1})(\mathbf{PAQ})$. Last part: No ($\mathbf{A} = \mathbf{P} = \mathbf{Q} = \mathbf{I}$ and $\mathbf{B} = ((1))$).

6. First consider $\mathbb{Z}$. Get a non-zero entry of $\mathbf{A}$ with the smallest modulus to the $(1,1)$-position. If $a_{i1} \neq 0$ for some $i \neq 1$, make $|a_{i1}| < |a_{11}|$ by $R_{i1}(\beta)$. The case $a_{1j} \neq 0$ $(j \neq 1)$ is similar. Repeating, we make all a_{i1}'s and a_{1j}'s zero except a_{11}. Over $F[x]$, use degree instead of modulus.

Section 4.6 (p. 184)

1(a) and (b). $h_{11} = h_{33} = h_{44} = 1$, $h_{12} = 2$ and all other h_{ij}'s are 0.
2. The second statement is false: take $\mathbf{A} = \mathbf{B} = \mathbf{e}_1\mathbf{e}_2^{\mathrm{T}}$.
3. $\mathbf{H} - \mathbf{I}$ has exactly $n - \rho(H)$ non-null columns. Use *Theorem* 3.7.5(iii).
4. False: take $\mathbf{H}_1 = \mathbf{e}_1(1, 2)$ and $\mathbf{H}_2 = \mathbf{e}_2\mathbf{e}_2^{\mathrm{T}}$.
6. See page 171 and the discussion after *Theorem* 4.6.7.

Section 5.1 (p. 186)

1. $\beta\mathbf{u} + (1 - \beta)\mathbf{v}$ is a solution for all β.
2(a). False $(0x_1 + 0x_2 = 1)$. (b). False $(\mathbf{A} = \mathbf{b} = \mathbf{0}_{2\times 1})$.
3. Over GF(5), the solutions are $(0, 2)^{\mathrm{T}}$, $(1, 4)^{\mathrm{T}}$, $(2, 1)^{\mathrm{T}}$, $(3, 3)^{\mathrm{T}}$ and $(4, 0)^{\mathrm{T}}$.
4. The solution sets over $\mathbb{R}$ and over GF(5) are $\{(\frac{1}{2}, 0)^{\mathrm{T}}\}$ and $\{(0, 4)^{\mathrm{T}}, (1, 1)^{\mathrm{T}}, (2, 3)^{\mathrm{T}}, (3, 0)^{\mathrm{T}}, (4, 2)^{\mathrm{T}}\}$.
5. $\mathbf{y} = (1, 0, 4, -10)^{\mathrm{T}}$. Solving $\mathbf{Ux} = \mathbf{y}$ we get $\mathbf{x} = (23/2, -22/3, 11/3, -10/3)^{\mathrm{T}}$.

Section 5.2 (p. 188)

1(a) and (b). Use *Theorem* 5.2.2. 2. False (take $\mathbf{A} = \mathbf{0}_{2\times 1}$).
3. False (take $\mathbf{A}_{1*} = \mathbf{e}_1^{\mathrm{T}}$). 4. $\{(2, 3, 1)^{\mathrm{T}}\}$.
5. $(2\alpha + 2\beta, \beta, -\alpha, -\beta)$. Note that $\{\mathbf{A}_{*1}, \mathbf{A}_{*2}\}$ is a basis of $\mathcal{C}(\mathbf{A})$.
6. Both are equivalent to $\rho(\mathbf{A}) < n$ where $\mathbf{A}$ is of order n.
7. Use *Exercises* 3.5.7(b) and 3.5.10.
8. $d(\{\mathbf{x} : \mathbf{x}^{\mathrm{T}}\mathbf{A} = \mathbf{0}\}) = m - r$ and $\mathbf{P}_{i*}\mathbf{A} = \mathbf{0}$ for $r + 1 \le i \le m$.

Section 5.3 (p. 192)

1. $\mathcal{C}(\mathbf{A}) = F^m$ iff $\rho(\mathbf{A}) = m$.
2. $\mathbf{Ax} = \mathbf{b}$ has a unique solution for some $\mathbf{b} \Leftrightarrow \mathcal{N}(\mathbf{A}) = \{\mathbf{0}\} \Leftrightarrow \rho(\mathbf{A}) = n$.
3. $(18 + 2\alpha, 12 + 3\alpha, 1 + \alpha)^{\mathrm{T}}$ is a general solution.
4. $\mathbf{A} = \begin{bmatrix} 24 & -32 & -9 & 0 \\ 0 & -8 & -15 & 24 \end{bmatrix}$, $\mathbf{b} = \begin{bmatrix} -49 \\ -31 \end{bmatrix}$ $\left(\text{note } \mathcal{C}(\mathbf{A}^{\mathrm{T}}) = \mathcal{N}\left(\begin{bmatrix} 4 & 3 & 0 & 1 \\ 3 & 0 & 8 & 5 \end{bmatrix}\right)\right)$.
5(a). Use: (i) $\mathbf{A}_{*j} \in \mathrm{Sp}(\{\mathbf{A}_{*i} : i \neq j\})$ iff $\mathbf{Av} = \mathbf{0}$ for some $\mathbf{v}$ with $v_j \neq 0$ and (ii) $\mathbf{Aw} = \mathbf{b}$ iff $\mathbf{A}(\mathbf{w} - \mathbf{u}) = \mathbf{0}$, given that $\mathbf{Au} = \mathbf{b}$. *Last part*: No.
5(b). For $\mathbf{y} \in \mathcal{N}(\mathbf{A})$, let $J(\mathbf{y}) = \{j : y_j \neq 0\}$. If $k \notin J(\mathbf{y})$, $J(\mathbf{y} + \epsilon\mathbf{x}^{(k)})$ properly contains $J(\mathbf{y})$ for some $\epsilon \in \mathbb{R}$. Over GF(2): consider $\mathbf{A} = [\,1 : 1 : 1\,]$.
5(c). Let $\mathbf{A} = [\,\mathbf{B} : \mathbf{c}\,]$. Then $\rho(\mathbf{B}) = n - 1$ iff '$\mathbf{Ax} = \mathbf{0},\ \mathbf{x} \neq \mathbf{0} \Rightarrow x_n \neq 0$'.
6. See *Exercise* 3.9.2.
7. Note that each row of $\mathbf{A}$ is non-null. The lines have the same slope iff $\rho(\mathbf{A}) = 1$ and are identical iff $\rho(\mathbf{B}) = 1$.
8. *Only if part*: $\mathbf{A}(u, v, 1)^{\mathrm{T}} = \mathbf{0}$ for some u and v, so $\rho(\mathbf{A}) \le 2$. The other conclusion follows from the preceding exercise. *If part*: $\mathbf{A}(x, y, z)^{\mathrm{T}} = \mathbf{0}$ for some $(x, y, z) \neq \mathbf{0}$. Since $\rho[\,\mathbf{A}_{*1} : \mathbf{A}_{*2}\,] = 2$, z is non-zero and can be made 1.

9. Each equation has a non-zero coefficient. Use *Exercise* 5.3.7 to show distinctness. Concurrence follows since the last equation is the sum of the others.
10(a). (i)$\Leftrightarrow$(ii): $\mathbf{x}_1, \ldots, \mathbf{x}_k \in \mathbf{x}_1 + S$ iff $\mathbf{x}_2 - \mathbf{x}_1, \ldots, \mathbf{x}_k - \mathbf{x}_1 \in S$. (ii)$\Leftrightarrow$(iii): With the $(n+1,1)$-element as the pivot, sweep out the $(n+1)$-th row and the first column in (iii).
10(b). Replace RHS's of (ii) and (iii) by 2 and 3 respectively and the word 'collinear' by 'coplanar' in (i).
11(b). Solution exists since $\mathbf{b} \in \mathcal{C}(\mathbf{AA}^{\mathrm{T}})$ and is unique by rank-cancellation law. 11(c). $\rho(\mathbf{A}) = \rho(\mathbf{A}^2) \Leftrightarrow \mathcal{C}(\mathbf{A}) = \mathcal{C}(\mathbf{A}^2) \Leftrightarrow \mathbf{A} = \mathbf{A}^2\mathbf{U}$ for some $\mathbf{U}$.
12. *If part*: $\mathbf{Au} = \mathbf{0} \Rightarrow \mathbf{c}^{\mathrm{T}}\mathbf{u} = 0$, so $\mathbf{A}^{\mathrm{T}}\mathbf{x} = \mathbf{c}$ is consistent. Converse is trivial.
13(a). Let $\rho(\mathbf{A}) = \rho[\,\mathbf{A} : \mathbf{b}\,] = r$. Let $\mathbf{C}$ and $\mathbf{D}$ be the coefficient matrix and the augmented matrix of the new system. If $\mathbf{u}^{\mathrm{T}} \notin \mathcal{R}(\mathbf{A})$ then $\rho(\mathbf{C}) = r+1$, so $\rho(\mathbf{D}) = r+1$ for any β and $\nu(\mathbf{A}) \neq \nu(\mathbf{C})$. If $\mathbf{u}^{\mathrm{T}} \in \mathcal{R}(\mathbf{A})$ and the new system is consistent then $\rho(\mathbf{C}) = \rho(\mathbf{D}) = r$, so the new equation is redundant.
14. $\mathbf{w}^{\mathrm{T}}(\mathbf{A} + \mathbf{uv}^{\mathrm{T}}) = \mathbf{0} \Rightarrow \mathbf{w}^{\mathrm{T}}\mathbf{A} = -(\mathbf{w}^{\mathrm{T}}\mathbf{u})\mathbf{v}^{\mathrm{T}} \Rightarrow \mathbf{w}^{\mathrm{T}}\mathbf{A} = \mathbf{0} \Rightarrow \mathbf{w}^{\mathrm{T}}\mathbf{b} = 0$.

Section 5.4 (p. 198)

2. False. Take $\mathbf{A} = \mathbf{11}^{\mathrm{T}}$ and $\mathbf{G} = \mathbf{e}_1\mathbf{e}_2^{\mathrm{T}}$. 3. $\rho(\mathbf{A}) = \rho(\mathbf{GA}) = \mathbf{tr}(\mathbf{GA})$.
4(a). If $\rho(\mathbf{G}) = \rho(\mathbf{A})$ then $\rho(\mathbf{G}) = \rho(\mathbf{AG})$, so $\mathbf{AGAG} = \mathbf{AG}$ gives $\mathbf{GAG} = \mathbf{G}$. If $\mathbf{GAG} = \mathbf{G}$, $\rho(\mathbf{A}) \geq \rho(\mathbf{G})$.
4(b). *Only if part*: $\mathbf{AGA} = \mathbf{A}$ gives $\mathbf{QGP} = \mathbf{I}$. Now $\mathbf{G} = \mathbf{GAG} = \mathbf{GP} \cdot \mathbf{QG}$.
6. Since $n = \rho(\mathbf{I}_n) \leq \rho(\mathbf{I} - \mathbf{GA}) + \rho(\mathbf{GA}) \leq \rho(\mathbf{I} - \mathbf{GA}) + \rho(\mathbf{A}) = n$, we have $\rho(\mathbf{I} - \mathbf{GA}) + \rho(\mathbf{GA}) = n$ and $\rho(\mathbf{GA}) = \rho(\mathbf{A})$. Use *Theorems* 3.7.5 and 5.4.3.
8. *First part*: Take $\mathbf{A} = (1, 0)$, $\mathbf{A}^- = \left[\begin{smallmatrix} 1 \\ -1 \end{smallmatrix}\right]$, $\mathbf{B} = \left[\begin{smallmatrix} 1 \\ 0 \end{smallmatrix}\right]$, $\mathbf{B}^- = (1, 1)$. *Second part*: $\mathbf{AB}(\mathbf{AB})^-\mathbf{AB} = \mathbf{AB}$. Rank-cancel $\mathbf{B}$ on the right.
9(b). Use normal form and 9(a). Note $\mathbf{Q}^{-1}\mathbf{M}^-\mathbf{P}^{-1}$ is $(\mathbf{PMQ})^-$.
9(c). The given matrix may be taken to be $\mathbf{G}$ where the order of $\mathbf{A}$ is $\rho(\mathbf{G})$. Then $\mathbf{M}$ is a symmetric g-inverse of $\mathbf{G}$. 10. Take $\mathbf{V} = \mathbf{GAH}$.
11. $\mathbf{B} \neq \mathbf{0}$ and $\mathbf{BV}(\mathbf{I} - \mathbf{AA}^-)\mathbf{C} = \mathbf{0}$ for all $\mathbf{V}$. So $(\mathbf{I} - \mathbf{AA}^-)\mathbf{C} = \mathbf{0}$.
12. *Only if part*: $\mathbf{AA}^-\mathbf{C} = \mathbf{C} = \mathbf{CB}^-\mathbf{B}$. *Last part*: similar to *Exercise* 5.4.10.
13. x_j has the same value in all solutions iff $(\mathbf{I} - \mathbf{A}^-\mathbf{A})_{j*}\,\mathbf{z} = 0$ for all $\mathbf{z}$. *Second part*: take $\mathbf{A}^- = \mathbf{I}$.
14. $\mathbf{A} = \mathbf{D}(\mathbf{A} + \mathbf{B})$ for some $\mathbf{D}$, so $\mathbf{AG}(\mathbf{A} + \mathbf{B}) = \mathbf{A}$. Now imitate the second solution to *Exercise* 3.7.6.
15. Use the last expression in (3.8.2), $\mathbf{AA}^-\mathbf{B} = \mathbf{B}$ and $\mathbf{CA}^-\mathbf{A} = \mathbf{C}$.
16(a). If $\mathbf{C} - \mathbf{HA} = 0$, $\mathbf{D} - \mathbf{HB} = \mathbf{D} - \mathbf{HAA}^-\mathbf{B} = \mathbf{D} - \mathbf{CA}^-\mathbf{B}$. *Second part*: 'Sweep out' $\mathbf{C}$ and $\mathbf{B}$.
16(b). With $\mathbf{M}$ as in *Exercise* 3.8.9, $\rho(\mathbf{M}) = \rho(\mathbf{A}) + \rho(1 + \mathbf{v}^{\mathrm{T}}\mathbf{A}^-\mathbf{u}) = \rho(1) + \rho(\mathbf{A} + \mathbf{uv}^{\mathrm{T}})$. $(\mathbf{A} + \mathbf{uv}^{\mathrm{T}})^-$ is by direct verification.

17. $\mathbf{XA^-Au} = \mathbf{X(I - A^-A)v} \Rightarrow \mathbf{AX^-XA^-Au} = \mathbf{AX^-X(I - A^-A)v}$, i.e., $\mathbf{Au} = \mathbf{0}$. For row spaces, use postmultiplication by $\mathbf{A^-A}$. Finally, $\rho(\mathbf{A}) = \rho(\mathbf{AA^-A}) \leq \rho(\mathbf{XA^-A}) \leq \rho(\mathbf{A})$. 18. Note that $\mathbf{u}^\mathrm{T}(\mathbf{I} - \mathbf{A^-A}) \neq \mathbf{0}$.
19(a). Define $g(\sum_{i=1}^m \alpha_i y_i) = \sum_{i=1}^m \alpha_i x_i$.
19(c). Take $\mathbf{y}_1 = \mathbf{b}$ and $\mathbf{x}_1 = \mathbf{u}$ and use (a).
20. Let $\mathbf{Au} = \mathbf{b}$, $f : \mathbf{x} \mapsto \mathbf{Ax}$, $f \circ g \circ f = f$ and $g(\mathbf{b}) = \mathbf{u}$. Take $\mathbf{G}$ = the matrix of g w.r.t. the canonical basis. Aliter: $\mathbf{u} = \mathbf{A^-b} + (\mathbf{I} - \mathbf{A^-A})\mathbf{z}$ for some $\mathbf{z}$. Now $\mathbf{z} = \mathbf{Ub}$ for some $\mathbf{U}$ since $\mathbf{b} \neq \mathbf{0}$.

Section 5.5 (p. 208)
1(a). Use *Theorem* 5.5.3; $\mathbf{b} + (\mathbf{I} - \mathbf{H})\mathbf{z} = (2 + 4z_2 + 3z_4, z_2, 5 - 4z_4, z_4)^\mathrm{T}$.
1(b). Yes for both. Use (a) or *Exercises* 5.4.13 and 5.3.5(a).
1(c). Taking $\mathbf{H}^- = \mathbf{H}$ and $\mathbf{z} = (0, -17/11, 0, 0)^\mathrm{T}$, we get the solution $\mathbf{y} = (-46/11, -17/11, 5, 0)^\mathrm{T}$ in the notation of *Exercise* 5.4.18.
3. For the $\mathbf{A}$ in (a), (b), (d), (e) and (f), an $\mathbf{A}^-$ is:

$$\frac{1}{7}\begin{bmatrix} 1 & 2 & -1 \\ -7 & -7 & 14 \\ 4 & 1 & -4 \end{bmatrix}, \frac{1}{3}\begin{bmatrix} 2 & -1 & 0 & 0 \\ -1 & 2 & 0 & 0 \\ 1 & -2 & 3 & 0 \end{bmatrix}, \frac{1}{5}\begin{bmatrix} -2 & 3 & 1 \\ 3 & -2 & 1 \\ 0 & 0 & 0 \\ 1 & 1 & -3 \end{bmatrix}, \begin{bmatrix} -100 & -40 & 29 & -13 \\ -\frac{229}{2} & -46 & 33 & -15 \\ -31 & -12 & 9 & -4 \\ -7 & -3 & 2 & -1 \end{bmatrix}$$

and $\mathbf{I} - \frac{1}{5}\mathbf{J}$ respectively. Also $\rho(\mathbf{A})$ is: 3, 2, 3, 4 and 4 respectively. A general solution is: $(1, 2, -1)^\mathrm{T}$, $(1 - 2\gamma, -1 + 3\gamma, \gamma)^\mathrm{T}$, $(\frac{2}{5} - 3\alpha, \frac{2}{5} + 2\alpha, \alpha, -\frac{1}{5})^\mathrm{T}$, $(-1, -4, 2, -1)^\mathrm{T}$ and $(-1, 0, 1, 2)^\mathrm{T}$ respectively. (c) is not consistent.

4. $\frac{1}{8}\begin{bmatrix} 1 & 0 & 0 & 0 \\ -14 & 0 & -4 & 8 \\ -6 & -16 & 12 & 0 \\ 2 & 8 & -4 & 0 \end{bmatrix} = \mathbf{A}^-$. Consistent iff $8b_4 - 4b_3 - 7b_1 = 0$; then $\frac{1}{8}(b_1 - \gamma, 8\gamma, 12b_3 - 16b_2 - 6b_1, 2b_1 + 8b_2 - 4b_3)^\mathrm{T}$ is a general solution. $\rho(\mathbf{A}) = 3$.
5(a). Let $\mathbf{c}^\mathrm{T} = \mathbf{y}^\mathrm{T}\mathbf{A}$. If $\mathbf{Au} = \mathbf{b}$, then $\mathbf{c}^\mathrm{T}\mathbf{A^-b} = \mathbf{y}^\mathrm{T}\mathbf{AA^-Au} = \mathbf{y}^\mathrm{T}\mathbf{Au} = \mathbf{c}^\mathrm{T}\mathbf{u}$.
5(b). The transforming matrix is of the form $\begin{bmatrix} \mathbf{K} & \mathbf{0} \\ -\mathbf{h}^\mathrm{T} & 1 \end{bmatrix}$ where $\mathbf{h}^\mathrm{T}\mathbf{A} = \mathbf{c}^\mathrm{T}$. So $\mathbf{h}^\mathrm{T} = \mathbf{c}^\mathrm{T}\mathbf{A}^- + \mathbf{w}^\mathrm{T}(\mathbf{I} - \mathbf{AA}^-)$ and $\mathbf{h}^\mathrm{T}\mathbf{b} = \gamma$.
6. '$\alpha = 3$ and $\beta = 1$' or '$\alpha = -2$ and $\beta = 6$' or '$\alpha \neq 3, -2$'. Corresponding general solutions are $(-1 - 7x_3, 1 + 2x_3, x_3)^\mathrm{T}$, $(-1 + \frac{1}{2}x_3, 1 - \frac{1}{2}x_3, x_3)^\mathrm{T}$ and $\frac{1}{\Delta}((3\alpha+5)\beta - 4\alpha - 2, 2\alpha^2 - (\alpha+1)\beta - 14, 8 - 2\alpha - 2\beta)^\mathrm{T}$ where $\Delta = 3(\alpha-3)(\alpha+2)$.
7. We consider four cases. In each case we give the condition for consistency and then the general solution. *Case* (i). $\alpha = \beta = \gamma$: $\epsilon = \alpha$, $(1 - z_2 - z_3, z_2, z_3)^\mathrm{T}$. *Case* (ii). $\alpha = \gamma \neq \beta$: $\epsilon \in \{\alpha, \beta, -\alpha - \beta\}$, $(\frac{\beta-\epsilon}{\beta-\alpha} - z_3, \frac{\epsilon-\alpha}{\beta-\alpha}, z_3)^\mathrm{T}$. *Case* (iii). α, β and γ are distinct but $\alpha + \beta + \gamma = 0$: $\epsilon \in \{\alpha, \beta, \gamma\}$, $(\frac{\beta-\epsilon}{\beta-\alpha} - \frac{\beta-\gamma}{\beta-\alpha}z_3, \frac{\epsilon-\alpha}{\beta-\alpha} - \frac{\gamma-\alpha}{\beta-\alpha}z_3, z_3)^\mathrm{T}$. *Case* (iv). α, β and γ are distinct and $\alpha + \beta + \gamma \neq 0$: Then $x_1 = \frac{(\epsilon-\beta)(\epsilon-\gamma)(\epsilon+\beta+\gamma)}{(\alpha-\beta)(\alpha-\gamma)(\alpha+\beta+\gamma)}$. Other components can be written down by symmetry.

8. If $\alpha = \beta = \gamma$, the system is consistent iff $\delta = 0$ and then $(-z_2 - z_3, z_2, z_3)^{\mathrm{T}}$ is a general solution. Otherwise, the solution is $(\delta(\gamma - \beta)/\Delta, \delta(\alpha - \gamma)/\Delta, \delta(\beta - \alpha)/\Delta)^{\mathrm{T}}$ where $\Delta = -(\alpha^2 + \beta^2 + \gamma^2 - \alpha\beta - \beta\gamma - \gamma\alpha)$.
9(a). $\mathscr{C}(\mathbf{B}) = \{(x_1, \ldots, x_m)^{\mathrm{T}} : x_{r+1} = \cdots = x_m = 0\}$.
9(b). $\mathbf{Bu} = \sum_{i=1}^{r} d_i \mathbf{B}_{*p_i} = \sum_{i=1}^{r} d_i \mathbf{e}_i = \mathbf{d}$.
9(d). By (c), $\mathbf{KB}$ is in HCF and so idempotent and $\rho(\mathbf{KB}) = \rho(\mathbf{B})$. So $\mathbf{K} = \mathbf{B}^-$. If $P = \{p_1, \ldots, p_r\}$, a general solution is obtained by taking x_j arbitrary for $j \notin P$ and $x_j = d_i - \sum_{k \notin P,\, k > j} b_{ik} x_k$ if $j = p_i$.
9(e). Clearly $\mathbf{A} = \mathbf{P}^{-1}\mathbf{B}$. Hence $\mathbf{B}^-\mathbf{P}$ is an $\mathbf{A}^-$. Taking $\mathbf{B}^-$ to be $[\mathbf{e}_{p_1} : \cdots : \mathbf{e}_{p_r} : \mathbf{0} : \cdots : \mathbf{0}]$, we have $\mathbf{B}^-\mathbf{P} = \mathbf{G}$.

Section 5.6 (p. 216)

1. We give $\mathbf{L}$, $\widetilde{\mathbf{U}}$, $\mathbf{A}^{-1}$, $\mathbf{A}^{-1}\mathbf{b}$ and the last row of the extended $\mathbf{L}$.

1(a). $$\begin{bmatrix} 2 & 0 & 0 \\ 4 & -2 & 0 \\ 3 & -\frac{1}{2} & -\frac{7}{4} \end{bmatrix}, \begin{bmatrix} 1 & \frac{1}{2} & \frac{3}{2} & \frac{1}{2} & 0 & 0 & \frac{1}{2} \\ 0 & 1 & \frac{7}{2} & 1 & -\frac{1}{2} & 0 & -\frac{3}{2} \\ 0 & 0 & 1 & \frac{4}{7} & \frac{1}{7} & -\frac{4}{7} & -1 \end{bmatrix},$$

$$\frac{1}{7}\begin{bmatrix} 1 & 2 & -1 \\ -7 & -7 & 14 \\ 4 & 1 & -4 \end{bmatrix}, \begin{bmatrix} 1 \\ 2 \\ -1 \end{bmatrix} \text{ and } (1, \tfrac{1}{2}, -\tfrac{9}{4}, -2).$$

1(b). $$\begin{bmatrix} 2 & 0 & 0 & 0 \\ -1 & -1 & 0 & 0 \\ 6 & 0 & 1 & 0 \\ 1 & 3 & 2 & -1 \end{bmatrix}, \begin{bmatrix} 1 & -1 & 0 & 2 & \frac{1}{2} & 0 & 0 & 0 & 1 \\ 0 & 1 & -3 & -3 & -\frac{1}{2} & -1 & 0 & 0 & -7 \\ 0 & 0 & 1 & -4 & -3 & 0 & 1 & 0 & 6 \\ 0 & 0 & 0 & 1 & -7 & -3 & 2 & -1 & -1 \end{bmatrix},$$

$$\begin{bmatrix} -100 & -40 & 29 & -13 \\ -\frac{229}{2} & -46 & 33 & -15 \\ -31 & -12 & 9 & -4 \\ -7 & -3 & 2 & -1 \end{bmatrix}, \begin{bmatrix} -1 \\ -4 \\ 2 \\ -1 \end{bmatrix} \text{ and } (1, 2, 7, 33, 4).$$

1(c). $$\begin{bmatrix} 2 & 0 & 0 & 0 \\ 1 & \frac{3}{2} & 0 & 0 \\ 1 & \frac{1}{2} & \frac{4}{3} & 0 \\ 1 & \frac{1}{2} & \frac{1}{3} & \frac{5}{4} \end{bmatrix}, \begin{bmatrix} 1 & \frac{1}{2} & \frac{1}{2} & \frac{1}{2} & \frac{1}{2} & 0 & 0 & 0 & \frac{1}{2} \\ 0 & 1 & \frac{1}{3} & \frac{1}{3} & -\frac{1}{3} & \frac{2}{3} & 0 & 0 & 1 \\ 0 & 0 & 1 & \frac{1}{4} & -\frac{1}{4} & -\frac{1}{4} & \frac{3}{4} & 0 & \frac{3}{2} \\ 0 & 0 & 0 & 1 & -\frac{1}{5} & -\frac{1}{5} & -\frac{1}{5} & \frac{4}{5} & 2 \end{bmatrix},$$

$\mathbf{J} - \frac{1}{5}\mathbf{I}$, $(-1, 0, 1, 2)^{\mathrm{T}}$ and $(1, \frac{1}{2}, \frac{1}{3}, \frac{1}{4}, -2)$.

1(d). $$\widetilde{\mathbf{U}} = \begin{bmatrix} 1 & .1667 & .4167 & 8.3333 & 0 & 0 & -.0833 \\ 0 & 1 & .3571 & 1.4286 & 2.8571 & 0 & 4.2143 \\ 0 & 0 & 1 & -10.8336 & -.8339 & 6.2500 & 1.1991 \end{bmatrix}.$$

$\mathbf{L}$, $\mathbf{A}^{-1}$ and $\mathbf{A}^{-1}\mathbf{b}$ are

$$\begin{bmatrix} .12 & 0 & 0 \\ -.06 & .3500 & 0 \\ .20 & .0467 & .1600 \end{bmatrix}, \begin{bmatrix} 11.9646 & -.1784 & -2.2323 \\ 5.2973 & 3.1549 & -2.2319 \\ -10.8336 & -.8339 & 6.2500 \end{bmatrix}, \begin{bmatrix} -1.4998 \\ 3.5004 \\ 1.9991 \end{bmatrix}.$$

$(\ell_{41}, \ldots, \ell_{44}) = (1.0000, 0.8333, 0.2857, -3.9996)$.

2. $(\mathbf{A}^{-1})_{nn} = (\mathbf{U}^{-1}\mathbf{L}^{-1})_{nn}$ is $1/(a_{nn} - \mathbf{c}^{\mathrm{T}}\mathbf{B}^{-1}\mathbf{d})$ as well as $1/\ell_{nn}$.
3. If $\mathbf{A} = \mathbf{L}_1\mathbf{U}_1 = \mathbf{L}_2\mathbf{U}_2$ then $\mathbf{L}_2^{-1}\mathbf{L}_1 = \mathbf{U}_2\mathbf{U}_1^{-1}$ is diagonal and so $\mathbf{I}$.
4. Take $\mathbf{LD}$ and $\mathbf{D}^{-1}\widetilde{\mathbf{U}}$ as the new $\mathbf{L}$ and $\widetilde{\mathbf{U}}$ for a suitable diagonal matrix $\mathbf{D}$.
5. Take $\mathbf{LD}^{-1}$ as the new $\mathbf{L}$ where $\mathbf{D} = \operatorname{diag}(\ell_{11}, \ell_{22}, \ldots, \ell_{nn})$.
6. $\mathbf{A}_k = \mathbf{L}_k\mathbf{U}_k$ where $\mathbf{A}_k$ denotes $\mathbf{A}(1, 2, \ldots, k \,|\, 1, 2, \ldots, k)$.
7. With $\mathbf{P}$ as in *Exercise* 2.6.14, $\mathbf{A} = \mathbf{UL}$ iff $\mathbf{PAP} = \mathbf{PUP} \cdot \mathbf{PLP}$.

Section 6.1 (p. 218)

1. To get (the area of) $APSQ$, shift the origin to A. $APQ = \frac{1}{2}APSQ$.

Section 6.2 (p. 222)

1. $\begin{pmatrix} 1 & 2 & 3 & 4 & 5 & 7 \\ 2 & 3 & 7 & 1 & 4 & 5 \end{pmatrix}, \begin{pmatrix} 1 & 2 & 3 & 5 & 7 & 8 & 9 \\ 9 & 3 & 7 & 8 & 5 & 1 & 2 \end{pmatrix}, \begin{pmatrix} 1 & 2 & 3 & 4 & 5 & 6 & 7 & 8 & 9 \\ 9 & 3 & 7 & 1 & 8 & 6 & 5 & 2 & 4 \end{pmatrix},$
$\begin{pmatrix} 1 & 2 & 3 & 4 & 5 & 6 & 7 & 8 & 9 \\ 3 & 4 & 7 & 9 & 8 & 6 & 5 & 1 & 2 \end{pmatrix}, \begin{pmatrix} 1 & 2 & 3 & 4 \\ 4 & 1 & 3 & 2 \end{pmatrix}, \begin{pmatrix} 2 & 3 & 5 & 7 \\ 5 & 2 & 7 & 3 \end{pmatrix},$
$\begin{pmatrix} 1 & 2 & 3 & 4 & 5 & 7 \\ 4 & 5 & 1 & 2 & 7 & 3 \end{pmatrix}, \begin{pmatrix} 1 & 2 & 3 & 4 & 5 & 7 \\ 4 & 1 & 2 & 5 & 7 & 3 \end{pmatrix}$ and $\begin{pmatrix} 1 & 2 & 3 & 4 & 5 & 7 \\ 4 & 1 & 2 & 5 & 7 & 3 \end{pmatrix}$.

2(a). [1] [2]. (b). [1 2]. (c). [1 2 7] [3 8] [4] [5 9 6].

3. $\begin{pmatrix} s_1 & s_2 & s_3 & \cdots & s_{k-1} & s_k \\ s_k & s_1 & s_2 & \cdots & s_{k-2} & s_{k-1} \end{pmatrix}$.

4. $\sigma^p = \sigma^r$ where $p = kq + r$, $0 \le r < k$. Second part: Each n_i divides p.
5. If $\sigma_1 \cdots \sigma_p$ is the d.c.d. of π, the d.c.d. of π^{-1} is $\sigma_1^{-1} \cdots \sigma_p^{-1}$.
6. Let $\mathbf{P}_{*j} = \mathbf{e}_{k_j}$ for all j. Then $\varphi(\mathbf{P}) = (k_1, \ldots, k_n)$. If $\varphi(\mathbf{Q}) = (j_1, \ldots, j_n)$, then $\varphi(\mathbf{PQ}) = (k_{j_1}, \ldots, k_{j_n})$. 7(a). (2 4 6 3 5 1) and (2 4 1 6 5 3). 7(b). (2 4 3 5 1 6). 7(c). (6 1 3 2 5 4).
8. If θ is obtained from π by interchanging j_k and j_ℓ then $\theta = \pi[k\ \ell]$.
9. Let $j_k = n$ and $\theta = (j_1\ \cdots\ j_{k-1}\ j_{k+1}\ \cdots\ j_n\ j_k)$. Then $s(\theta) = s(\pi) - (n-k)$ Also $\theta = [j_k\ j_{k+1}\ \cdots\ j_n]\pi$, so $\epsilon(\theta) = (-1)^{n-k}\epsilon(\pi)$. View θ as a permutation of $\{1, 2, \ldots, n-1\}$ and use induction on n.

10(a). The given permutation is $\prod_{i=1}^{\ell}[i\quad n+1-i]$ where $\ell = \lfloor n/2 \rfloor$.
10(b). Even number of transpositions are performed since the 16-th square moves one square horizontally or vertically and returns to the original position.

Section 6.3 (p. 228)

1. $-$, $-$ and $+$. 2. $\alpha^n|\mathbf{A}|$. 3. $|\mathbf{A}| = |\mathbf{A}^{\mathrm{T}}| = |-\mathbf{A}| = (-1)^n|\mathbf{A}| = -|\mathbf{A}|$.

4. a_{12}^2 and $(a_{12}a_{34} - a_{13}a_{24} + a_{14}a_{23})^2$. 5. $|\mathbf{E}_{ij}| = -|\mathbf{I}|$, $|\mathbf{E}_i(\alpha)| = \alpha|\mathbf{I}|$, etc.
6. Perform $C_{1j}(1)$ for $j = 2, \ldots, n$ and then $R_{i1}(-1)$ for $i = 2, \ldots, n$.
8. Perform $R_{i,i+1}(-\rho)$ for $i = 1, \ldots, n-1$ in that order.
9. Write $\mathbf{P}$ in the form $\mathbf{E}_{i_1j_1} \cdots \mathbf{E}_{i_kj_k}\mathbf{I}$ and use $\varphi(\mathbf{E}_{ij}) = [i\ j]$.
10. $0 = |\mathbf{D}| = |\mathbf{A}| + |\mathbf{B}| + 0 + 0$.
11. $\frac{d}{dx}|\mathbf{A}| = \sum \epsilon(j_1 \cdots j_n)\{\frac{d}{dx}(a_{1j_1})a_{2j_2} \cdots a_{nj_n} + a_{1j_1}\frac{d}{dx}(a_{2j_2})a_{3j_3} \cdots a_{nj_n} + \cdots + a_{1j_1} \cdots a_{n-1,j_{n-1}}\frac{d}{dx}(a_{nj_n})\} = \sum |\mathbf{A}_k|$.
12(a). As in *Exercise* 6.3.10, $f(\mathbf{E}_{ij}\mathbf{A}) = -f(\mathbf{A})$. So $f(\mathbf{A}) = f(\sum_{j_1=1}^n a_{1j_1}\mathbf{e}_{j_1}, \ldots, \sum_{j_n=1}^n a_{nj_n}\mathbf{e}_{j_n}) = \sum_{j_1} \cdots \sum_{j_n} a_{1j_1} \cdots a_{nj_n} f(\mathbf{e}_{j_1}, \ldots, \mathbf{e}_{j_n}) = \sum_\pi a_{1\pi_1} \cdots a_{n\pi_n}\epsilon(\pi)f(\mathbf{I}) = f(\mathbf{I}) \cdot |\mathbf{A}|$. Aliter: Prove (iii) and (iv) and use (b).
12(b). By the solution to *Exercise* 4.2.3, $f(\mathbf{E}_{ij}\mathbf{A}) = -f(\mathbf{A})$. By (iii), $f(\mathbf{A}) = 0$ if a row of $\mathbf{A}$ is null. If $|\mathbf{A}| = 0$, some row can be made null by row operations, so $f(\mathbf{A}) = 0$. If $|\mathbf{A}| \neq 0$, reduce $\mathbf{A}$ to $\mathbf{I}$ by row operations and imitate the proof of *Theorem* 6.3.9 (for f).
13. Let $\mathbf{A}_{1*} = (\alpha_1, \alpha_2)$ and $\mathbf{A}_{2*} = (\beta_1, \beta_2)$. Area $OPRQ$ satisfies (iv) since $R_{21}(\delta)$ does not alter the base OP and the line QR.
14. Here $R_{31}(\delta)$ does not alter the base (parallelogram formed by OP and OQ) and the plane of the opposite face and so the volume.
15. $|\mathbf{A}|$ over F is the same as $|\mathbf{A}|$ over G. Use *Exercise* 3.3.10(a).

Section 6.4 (p. 235)

2. To evaluate the determinant, Subtract the first row from the others.
3(a). Expand by the first row to see that the equation represents a line. Observe that two rows become identical if (x, y) is replaced by (x_i, y_i).
3(b). The plane passing through the non-collinear points (x_1, y_1, z_1), (x_2, y_2, z_2) and (x_3, y_3, z_3) is $|\mathbf{A}| = 0$ where $\mathbf{A}_{1*} = (x, y, z, 1)$ and $\mathbf{A}_{i*} = (x_{i-1}, y_{i-1}, z_{i-1}, 1)$ for $i = 2, 3, 4$. Since this equation is consistent, it follows from *Exercise* 5.3.10(a) that at least one among the coefficients of x, y, z is non-zero.
3(c). By (b), the plane OP_1P_2 is $f(x, y, z) := |\mathbf{B}| = 0$ where $\mathbf{B}_{1*} = (x, y, z)$, $\mathbf{B}_{2*} = (u_1, v_1, w_1)$ and $\mathbf{B}_{3*} = (u_2, v_2, w_2)$. So the plane through P_3 parallel to OP_1P_2 is $f(x, y, z) = \alpha$ where $\alpha = f(u_3, v_3, w_3)$. Use *Theorem* 6.3.7.
4. $2\mathbf{A}_{2*} - \mathbf{A}_{1*} - \mathbf{A}_{3*} = \mathbf{0}$. 5(a). $x^3 + ax^2 + bx + c$.
5(b). Consider $\mathbf{A}_{n\times n}$ with $\mathbf{A}_{i*} = -\mathbf{e}_{i+1}^{\mathrm{T}}$ for $i = 1, \ldots, n-1$. Expand $|x\mathbf{I} + \mathbf{A}|$ by the first column and use induction.
6(i). -120 (Expand by third column twice and then by first column.)
6(ii). Expand by last two columns. (iii). Perform C_{13}, C_{24} and R_{24}.
8. To prove the second and third statements use induction.
9. $|\mathbf{A}| = a_{ij}A_{ij} + \sum_{k\neq j} a_{ik}A_{ik}$. Changing a_{ij} does not alter A_{ij}, a_{ik} and A_{ik}.
10. If $a_1, \ldots, a_{n+1}$ are roots of $c_0 + c_1t + \cdots + c_nt^n$, $\mathbf{A}_{n+1}(c_n, c_{n-1}, \ldots, c_0)^{\mathrm{T}} = \mathbf{0}$.

11. Similar to the preceding exercise.
13. Let $f(x,y,z) = \text{LHS} - \text{RHS}$. Then $f(x,y,z) = k(x-y)(y-z)(x-z)$ for some constant k since $f(y,y,z) = f(x,y,x) = f(x,y,y) = 0$. The coefficient of x^2y in $f(x,y,z)$ is 0, so $k = 0$.
14. Similar to the preceding exercise. Compare the coefficients of $abcd$.
16. $|\det \mathbf{A}| \le \sum_\pi |a_{1\pi_1}| \cdots |a_{n\pi_n}| \le (\sum_{\pi_1=1}^n |a_{1\pi_1}|) \cdots (\sum_{\pi_n=1}^n |a_{n\pi_n}|)$.
17(a). The new determinant is 0, so by *Theorem* 3.3.9, $|a_{11} - (\det \mathbf{A})/A_{11}| \le t_1$. Now $|a_{11}A_{11}| - |a_{11}A_{11} - \det \mathbf{A}| \le |\det \mathbf{A}|$, so $(|a_{11}| - t_1)|A_{11}| \le |\det \mathbf{A}|$. Use induction hypothesis for $|A_{11}|$. For the other inequality, use $|\det \mathbf{A}| \le |a_{11}A_{11}| + |\det \mathbf{A} - a_{11}A_{11}|$. *Last part*: Take $\mathbf{A} = (1,-1)^{\mathrm{T}}(2,-1)$.
17(b). $A_{11}(a_{11} - t_1) \le a_{11}A_{11} - |a_{11}A_{11} - \det \mathbf{A}| \le a_{11}A_{11} - (a_{11}A_{11} - \det \mathbf{A})$.

Section 6.5 (p. 239)
1(b). Use $|\mathbf{A}\mathbf{A}^{\mathrm{T}}| = |\mathbf{A}|^2$ and $|\mathbf{A}^2| = |\mathbf{A}|^2$.
2. LHS is a polynomial in a, b, c, d with coefficient of a^4 equal to 1 and, by *Exercise* 2.7.3, squares of LHS and RHS are equal. Note that $a^2+b^2+c^2+d^2$ is irreducible. 4. $|\mathbf{AB}| = 48$.
5. Use Laplace expansion by the last n columns. The initial determinant is $|\mathbf{A}| \cdot |\mathbf{B}|$. After reduction, it is $|\mathbf{AB}|$.
6. Take $n < p$. Let $N = \{1, \ldots, n\}$, $J = \{j_1, \ldots, j_n\} \subseteq P = \{1, \ldots, p\}$, $\overline{J} = P - J$, $K = \{1, \ldots, p-n\}$ and $L = \{p+1, \ldots, p+n\}$. Let $s(N)$ denote $\sum\{i : i \in N\}$. Then the cofactor of $\mathbf{A}(N\,|\,J)$ in $\mathbf{M}$ is, by Laplace expansion by the first $p-n$ columns, $(-1)^{s(N)+s(J)}|-\mathbf{I}_{p-n}|(-1)^{s(\overline{J})+s(K)}|\mathbf{B}(J\,|\,N)|$. After reduction, $|\mathbf{M}|$ becomes $|\mathbf{AB}|(-1)^{s(N)+s(L)}|-\mathbf{I}_p|$.
7. Let $\mathbf{C} = \mathbf{AB}$. If $\mathbf{A}_{i*} = \mathbf{A}_{k*}$ then $\mathbf{C}_{i*} = \mathbf{C}_{k*}$ and $|\mathbf{C}| = 0$. When $\mathbf{A}_{k*}$ is fixed for $k \ne i$, $|\mathbf{C}| = \sum_{j=1}^n c_{ij}C_{ij} = \mathbf{A}_{i*}\left(\sum_{j=1}^n \mathbf{B}_{*j}C_{ij}\right)$ is linear in $\mathbf{A}_{i*}$ (note C_{ij} does not depend on $\mathbf{A}_{i*}$). To get $c = |\mathbf{B}|$, put $\mathbf{A} = \mathbf{I}$ in $|\mathbf{AB}| = c|\mathbf{A}|$.
8. $|\mathbf{A}[\boldsymbol{\alpha}-\boldsymbol{\gamma} : \boldsymbol{\beta}-\boldsymbol{\gamma}]| = |\mathbf{A}| \cdot |[\boldsymbol{\alpha}-\boldsymbol{\gamma} : \boldsymbol{\beta}-\boldsymbol{\gamma}]|$. Use *Exercise* 6.1.1.

Section 6.6 (p. 245)
1. If $|\mathbf{A}| \ne 0$, $\mathbf{A}^{\circledast}/|\mathbf{A}|$ is $\mathbf{A}^{-1}$. If $|\mathbf{A}| = 0$ and $\mathbf{AB} = \mathbf{I}$, we get $0 \cdot |\mathbf{B}| = 1$.
2. $x_1 = \frac{1}{\Delta}(a_{22}b_1 - a_{12}b_2)$ and $x_2 = \frac{1}{\Delta}(a_{11}b_2 - a_{21}b_1)$ where $\Delta = a_{11}a_{22} - a_{12}a_{21}$.

4(a)–(c). $-\frac{1}{14}\begin{bmatrix} -3 & -1 \\ -8 & 2 \end{bmatrix}, \frac{1}{6}\begin{bmatrix} 3 & 3 & -4 \\ 0 & 6 & 0 \\ 0 & 0 & 2 \end{bmatrix}, -\frac{1}{6}\begin{bmatrix} -6 & 0 & 0 & 0 \\ 9 & -3 & 0 & 0 \\ 9 & -3 & 6 & 0 \\ -16 & 4 & -2 & -2 \end{bmatrix}$.

4(d). $A_{ii} = (a+(n-2)b)(a-b)^{n-2}$ and $A_{ij} = -b(a-b)^{n-2}$ if $i \ne j$. See also *Exercise* 3.4.8. 5. $(2,3,-4)^{\mathrm{T}}$. ($|\mathbf{A}| = -16$). 6. $x_4 = (-6)/(-6) = 1$.
7. Let the denominators be u, v and w respectively. If $(x,y,z)^{\mathrm{T}}$ is a solution, $b_iy + c_iz = -a_ix$ for $i = 1, 2$, so Cramer's rule gives $y = xv/u$ and $z = xw/u$.

8. $(\alpha\mathbf{A})^{\circledast} = \alpha^{n-1}\mathbf{A}^{\circledast}$. If $\mathbf{B} = \mathbf{A}^{\mathrm{T}}$, $(\mathbf{B}^{\circledast})_{ij} = B_{ji} = A_{ij} = (\mathbf{A}^{\circledast})_{ji} = ((\mathbf{A}^{\circledast})^{\mathrm{T}})_{ij}$. If $\mathbf{A}$ is non-singular, $\mathbf{A}^{\circledast}(\mathbf{A}^{-1})^{\circledast} = (\mathbf{A}^{-1}\mathbf{A})^{\circledast} = \mathbf{I}^{\circledast} = \mathbf{I}$.
9. Use *Exercise* 6.4.9. For (a), take $\mathbf{A}$ to be $(r+1) \times (r+1)$ with rank r, without loss of generality.
10. (a) $\Rightarrow$ (b): Take $\mathbf{b} = \mathbf{e}_j$. (b) $\Rightarrow$ (c): Both $1/|\mathbf{A}|$ and $|\mathbf{A}| \in \mathbb{Z}$. (c) $\Rightarrow$ (b): $\mathbf{A}^{-1} = \pm\mathbf{A}^{\circledast} \in \mathbb{Z}^{n\times n}$. (b) $\Rightarrow$ (a) is trivial.
11. (a) $\Rightarrow$ (b): Let $\mathbf{A} = \mathbf{PDQ}$ as in *Exercise* 4.5.6 and $j \le r = \rho(\mathbf{A})$. Then $\mathbf{e}_j = \mathbf{Du}$ for some $\mathbf{u}$, so $\mathbf{Pe}_j = \mathbf{AQ}^{-1}\mathbf{u} \in \mathcal{C}(\mathbf{A})$. By (a), $\mathbf{DQv} = \mathbf{e}_j$ for some integral $\mathbf{v}$. Thus Δ^{-1} is integral and $\mathbf{Q}^{-1}\mathrm{diag}(\Delta^{-1}, \mathbf{0})\mathbf{P}^{-1}$ is an integral g-inverse of $\mathbf{A}$. (b) $\Rightarrow$ (a) is trivial. (c) $\Rightarrow$ (a): Use *Theorem* 5.3.6 and the preceding exercise. (b) $\not\Rightarrow$ (c): If $\mathbf{A}_{2\times 2} = \mathbf{e}_1(1,2)$, $\mathbf{I}$ is $\mathbf{A}^{-}$.

Section 6.7 (p. 247)

1. Second part: $(-1)^{(n+1)k}|\mathbf{B}| \cdot |\mathbf{C} - \mathbf{DB}^{-1}\mathbf{A}|$ where $\mathbf{B}$ and $\mathbf{C}$ are of orders k and $n-k$.
2(a). Use 2(b). For *Exercise* 6.3.6, take $\mathbf{u} = \mathbf{1}$ and $\mathbf{v}^{\mathrm{T}} = (a_1, \ldots, a_n)$.
2(b). Use the preceding exercise and (6.7.1) for $\begin{bmatrix} \mathbf{A} & -\mathbf{B} \\ \mathbf{C} & \mathbf{I} \end{bmatrix}$.
4. Use the preceding exercise with $\alpha = -1$ and *Exercise* 6.7.1.
5. By the preceding exercise, the determinant is $abc + x(bc + ac + ab)$.
6. Use *Exercise* 6.7.2(a) with $\mathbf{A} = \mathrm{diag}(x_1, x_2, \ldots, x_n)$ and $\mathbf{b} = \mathbf{c} = \mathbf{1}$.
(i). $\prod_{i\neq k} x_i$ if $x_k = 0$. (ii). 0.
7(a). The determinant is $|\mathbf{A}|\cdot|\mathbf{D}-\mathbf{CA}^{-1}\mathbf{B}| = |\mathbf{A}|\cdot|\mathbf{D}-\mathbf{A}^{-1}\mathbf{CB}| = |\mathbf{AD}-\mathbf{CB}|$.
7(b). (i). $|\mathbf{DA} - \mathbf{BC}|$. (ii). $|\mathbf{DA} - \mathbf{CB}|$.
7(c). Take $\mathbf{A} = \mathbf{D} = \mathbf{J} - \mathbf{I}$ and $\mathbf{B} = \mathbf{C} = \mathbf{J} - \mathbf{e}_2\mathbf{e}_1^{\mathrm{T}}$.

Section 7.1 (p. 249)

2. $\langle \mathbf{x}, \mathbf{y} \rangle = 0$.
3. The line through O perpendicular to $O\mathbf{z}$; a line parallel to the earlier one.
4. Use $(x_1^2 + y_1^2)(x_2^2 + y_2^2) - (x_1x_2 + y_1y_2)^2 = (x_1y_2 - x_2y_1)^2$. Second part: Take $Q = \mathbf{0}$, $P = (x_1, y_1), R = (x_2, y_2)$. Then $(OP + OR)^2 \ge PR^2$ simplifies to the earlier inequality.

Section 7.2 (p. 252)

1. $\alpha_i > 0$ for all i. 2. If $\langle p, p \rangle = 0$, p has n distinct roots.
4(a). $\mathrm{tr}(\mathbf{B}^*\mathbf{A}) = \sum_{i,j} \bar{b}_{ij}a_{ij}$. 4(b). Second part of axiom 3.
5. $\mathbf{x}^{\mathrm{T}}\mathbf{x} = 0$ if $\mathbf{x} = (i, 1)^{\mathrm{T}}$; $(i\mathbf{x})^*\mathbf{y} \neq i(\mathbf{x}^*\mathbf{y})$.
6. *Only if part*: take $\mathbf{y} = \mathbf{x}$. 7. Real part of the canonical inner product.

Section 7.3 (p. 257)

1. (a). $\|\mathbf{0}\| = \|2 \cdot \mathbf{0}\| = 2\,\|\mathbf{0}\|$. (c). First inequality: $\|\mathbf{x}\| \le \|\mathbf{y}\| + \|\mathbf{x}-\mathbf{y}\|$.
2. $c|\mathbf{x}|$ where $c > 0$. ($c = \|\mathbf{1}\|$) 3. $\alpha_j > 0$ for all j.
4. No ($\|\mathbf{e}_j\|_p = 1$ for all p). 5. $\|\mathbf{e}_1 + \mathbf{e}_2\|_p > \|\mathbf{e}_1\|_p + \|\mathbf{e}_2\|_p$.
6. *Second part*: If $c = \max_i |x_i| = |x_k|$, $c = (|x_k|^p)^{1/p} \le (\sum_{i=1}^n |x_i|^p)^{1/p} = c(\sum_{i=1}^n (|x_i|/c)^p)^{1/p} \le cn^{1/p} \to c$ as $p \to \infty$. 8. See *Exercise* 7.2.4(a).
9(b). *Only if part*: take $\mathbf{x} = \mathbf{e}_1$ and $\mathbf{y} = \mathbf{e}_2$.
10(a). $\langle \mathbf{x}, \mathbf{y}\rangle = \frac{1}{4}\left(\|\mathbf{x}+\mathbf{y}\|^2 - \|\mathbf{x}-\mathbf{y}\|^2\right)$. 11. $d(6,0) \ne 2d(3,0)$.
12. Take $\mathbf{y} \ne \mathbf{0}$ and $\mathbf{x} = (\alpha/\|\mathbf{y}\|)\mathbf{y}$. 14(a). 10, $\sqrt{82}$ and 9.
14(b). To get $n^{-1/2}\|\mathbf{x}\|_1 \le \|\mathbf{x}\|_2$, take $y_i = x_i/|x_i|$ if $x_i \ne 0$ and 0 otherwise in (7.3.7), where $\langle \mathbf{x}, \mathbf{y}\rangle = \mathbf{y}^*\mathbf{x}$. The other inequalities are trivial.
15(a). Use $\|\mathbf{A}\| = 0 \Rightarrow \mathbf{A}\mathbf{x} = \mathbf{0}$ for all $\mathbf{x}$ and $N((\mathbf{A}+\mathbf{B})\mathbf{x}) \le N(\mathbf{A}\mathbf{x}) + N(\mathbf{B}\mathbf{x})$.
15(d). By (c), $N(\mathbf{A}\mathbf{B}\mathbf{x}) \le \|\mathbf{A}\|\,N(\mathbf{B}\mathbf{x}) \le \|\mathbf{A}\|\,\|\mathbf{B}\|\,N(\mathbf{x})$.
15(e). Let $c_j = \sum_i |a_{ij}|$ and $\max_j c_j = c_k$. Then $\|\mathbf{A}\mathbf{x}\|_1 = \sum_i \left|\sum_j a_{ij}x_j\right| \le \sum_j \sum_i |a_{ij}||x_j| \le c_k\|\mathbf{x}\|_1$ for all $\mathbf{x}$, so $\|\mathbf{A}\| \le c_k$. For equality, take $\mathbf{x} = \mathbf{e}_k$.
15(f). Let $r_i = \sum_j |a_{ij}|$ and $\max_i r_i = r_\ell$. Then $\|\mathbf{A}\mathbf{x}\|_\infty \le \max_i \sum_j |a_{ij}|\,|x_j| \le \|\mathbf{x}\|_\infty r_\ell$. For equality, take $x_j = 1$ or -1 according as $a_{\ell j} > 0$ or not.
15(g). If $(\mathbf{I}-\mathbf{A})\mathbf{u} = \mathbf{0}$ and $\mathbf{u} \ne \mathbf{0}$ then $\mathbf{A}\mathbf{u} = \mathbf{u}$ and $\|\mathbf{A}\| \ge 1$. Let $\mathbf{B} = (\mathbf{I}-\mathbf{A})^{-1}$. For the first inequality, use $\|\mathbf{B}\| \ge 1/\|\mathbf{I}-\mathbf{A}\|$. For the second, use $\|\mathbf{B}\| = \|\mathbf{I}+\mathbf{A}\mathbf{B}\| \le 1 + \|\mathbf{A}\|\,\|\mathbf{B}\|$. *Last part*: Same bounds.
15(h). $\mathbf{h} = (\mathbf{I}+\mathbf{A}^{-1}\mathbf{E})^{-1}\,\{\mathbf{A}^{-1}\mathbf{g} - \mathbf{A}^{-1}\mathbf{E}\mathbf{u}\}$. So by (c), (d) and (g),

$$r_{\mathbf{u}} = \frac{N(\mathbf{h})}{N(\mathbf{u})} \le \frac{1}{1-\|\mathbf{A}^{-1}\|\|\mathbf{E}\|}\left\{\|\mathbf{A}^{-1}\|\frac{N(\mathbf{g})}{N(\mathbf{u})} + \|\mathbf{A}^{-1}\|\|\mathbf{E}\|\right\}$$

Now use $(1/N(\mathbf{u})) \le (\|\mathbf{A}\|/N(\mathbf{b}))$ and $c(\mathbf{A})r_{\mathbf{A}} = \|\mathbf{A}^{-1}\|\|\mathbf{E}\|$.
16. Define $\langle \mathbf{x}, \mathbf{y}\rangle = \left(\|\mathbf{x}+\mathbf{y}\|^2 - \|\mathbf{x}\|^2 - \|\mathbf{y}\|^2\right)/2$. Prove that $\langle \mathbf{u}+\mathbf{v}, \mathbf{y}\rangle = \langle \mathbf{u}, \mathbf{y}\rangle + \langle \mathbf{v}, \mathbf{y}\rangle$. Use this to get $\langle \alpha\mathbf{u}, \mathbf{y}\rangle = \alpha\langle \mathbf{u}, \mathbf{y}\rangle$ for positive integral α, then rational α and finally real α by continuity.

Section 7.4 (p. 264)

2(a). Over $\mathbb{R}$, $\|\mathbf{x}+\mathbf{y}\|^2 = \|\mathbf{x}\|^2 + \|\mathbf{y}\|^2 + 2\langle \mathbf{x}, \mathbf{y}\rangle$. *Last part*: take $V = \mathbb{R}^1$, $x_1 = -1$, $x_2 = x_3 = 2$ and other x_i's 0.
2(b). Take $\mathbf{x} = (i, i)^{\mathrm{T}}$ and $\mathbf{y} = (1,1)^{\mathrm{T}}$ for the second part.
3. *Only if part*: Clearly $\mathcal{N}(\mathbf{1}^{\mathrm{T}}\mathbf{A}) = \mathcal{N}(\mathbf{1}^{\mathrm{T}}\mathbf{B})$, so $\mathbf{1}^{\mathrm{T}}\mathbf{B} = \alpha\,\mathbf{1}^{\mathrm{T}}\mathbf{A}$ and so $\mathbf{B} = a\mathbf{A}$ for some α. 4(a). $\|\sum \alpha_i\mathbf{x}_i\|^2 = \langle \sum \alpha_i x_i, \sum \alpha_j x_j\rangle = \sum_i \sum_j \alpha_i\bar{\alpha}_j\langle x_i, x_j\rangle$.
4(b). Note that $\mathbf{x} = \mathbf{z} + \sum\langle \mathbf{x}, \mathbf{x}_i\rangle\mathbf{x}_i$ and $\mathbf{z} \perp \sum\langle \mathbf{x}, \mathbf{x}_i\rangle\mathbf{x}_i$. 4(c). Use (b).
5. (a) $\Rightarrow$ (b): Otherwise, $B \cup \{\mathbf{x}/\|\mathbf{x}\|\}$ is orthonormal. (b) $\Rightarrow$ (c): Let $\mathbf{x} \in V$ and $\mathbf{z}$ be the residual of $\mathbf{x}$ w.r.t. B. By (b), $\mathbf{z} = \mathbf{0}$, so $\mathbf{x} \in \mathrm{Sp}(B)$. (c) $\Rightarrow$ (d): See *Theorem* 7.4.6. (d) $\Rightarrow$ (e): Similar to *Exercise* 7.4.4(a). (e) $\Rightarrow$ (f): Take

$\mathbf{x} = \mathbf{y}$. (f) $\Rightarrow$ (a): If $\{\mathbf{x}_1, \ldots, \mathbf{x}_k, \mathbf{x}\}$ is orthonormal, (f) gives $1 = 0$.
6. No. In $\mathbb{C}^1$, take $\mathbf{x}_1 = 1$ and $\mathbf{x} = \sqrt{-1}$.
8. (a). $\mathbf{y}_j \in S_j \subseteq S_k$ for $j = 1, \ldots, k-1$.
8(b). $\mathbf{z}_k \perp \mathbf{y}_j$ for $j = 1, \ldots, k-1$ by (7.4.4).
8(c). If $\mathbf{z}_k = \mathbf{0}$, then $S_k = S_{k-1}$ and $B_k = B_{k-1}$. So let $\mathbf{z}_k \neq \mathbf{0}$. By (a), $\mathrm{Sp}(B_k) \subseteq S_k$. By (7.4.4), $\mathbf{x}_k \in \mathrm{Sp}(B_k)$, so $S_k \subseteq \mathrm{Sp}(B_k)$.
9. $\{\frac{1}{\sqrt{6}}(2, -1, 0, 1), \frac{1}{\sqrt{18}}(2, 1, 2, -3)\}$.
10. $\{\frac{1}{2}(1, 1, 1, 1), \frac{1}{\sqrt{12}}(-3, 1, 1, 1), \frac{1}{\sqrt{6}}(0, -2, 1, 1), \frac{1}{\sqrt{2}}(0, 0, -1, 1)\}$; $\{\mathbf{e}_4, \ldots, \mathbf{e}_1\}$.
11. $\left\{\left(\frac{1}{\sqrt{2}}, 0\right), \left(\frac{1}{\sqrt{10}}, \frac{2}{\sqrt{10}}\right)\right\}$.
12. $\{(1, 0, -1), (1, -1, 0)\}$ is a basis of S. With $\mathbf{e}_1, \mathbf{e}_2, \mathbf{e}_3$ appended, *Algorithm* 7.4.8 gives the orthonormal basis $\{\frac{1}{\sqrt{7}}(1, 0, -1), \frac{1}{\sqrt{21}}(5, -7, 2), \frac{1}{\sqrt{6}}(1, 1, 1)\}$ of $\mathbb{R}^3$, the first two vectors forming B.
13. Use $\mathbf{A}_{*k} = \|\mathbf{z}_k\|\mathbf{y}_k + \sum_{i=1}^{k-1} \langle \mathbf{A}_{*k}, \mathbf{y}_i \rangle \mathbf{y}_i$.
14. Rank-factorizations are

$$\left(\frac{1}{\sqrt{2}}\begin{bmatrix} 1 & 0 \\ 0 & 1 \\ 0 & 1 \\ 1 & 0 \end{bmatrix}, \sqrt{2}\begin{bmatrix} 2 & 1 & 1 & 3 & 4 \\ 0 & 1 & -1 & 1 & 2 \end{bmatrix}\right); \left(\begin{bmatrix} 0 & 0 & 1 & 0 \\ 0 & 0 & 0 & 1 \\ 1 & 0 & 0 & 0 \\ 0 & 1 & 0 & 0 \end{bmatrix}, \begin{bmatrix} 2 & 0 & 5 & 1 & 0 \\ 0 & 2 & 2 & 4 & 1 \\ 0 & 0 & 2 & 0 & 5 \\ 0 & 0 & 0 & 1 & 3 \end{bmatrix}\right).$$

15. $\mathbf{Q}_1\mathbf{R}_1 = \mathbf{Q}_2\mathbf{R}_2 \Rightarrow \mathbf{Q}_1 = \mathbf{Q}_2\mathbf{T} \Rightarrow \mathbf{T}^*\mathbf{T} = \mathbf{I} \Rightarrow \mathbf{T} = \mathbf{I}$ where $\mathbf{T} = \mathbf{R}_2\mathbf{R}_1^{-1}$.

Section 7.5 (p. 271)
1. $(A^\perp)^\perp = \mathrm{Sp}(A)$. 2. $\langle \mathbf{x}+\mathbf{y}, \mathbf{x}-\mathbf{y} \rangle = \|\mathbf{x}\|^2 - \|\mathbf{y}\|^2$.
3. $\{(\xi_1, \ldots, \xi_4) : \xi_1 = \xi_2\}$, $\{(\xi_1, \ldots, \xi_4) : \xi_1 + \xi_2 + \xi_3 = 0, \xi_4 = 0\}$ and $\{(\xi_1, \ldots, \xi_4) : \xi_1 + \xi_2 = 0, \xi_3 = 0\}$. $(S+T)^\perp = \{(\xi_1, \ldots, \xi_4) : \xi_1 + \xi_2 = 0, \xi_3 = 0, \xi_4 = 0\}$. 4. $(\mathbf{A}\mathbf{A}^\circledast)_{i1} = 0$ for $i = 2, 3$ and $(A_{11}, A_{12}, A_{13}) \neq \mathbf{0}$.
5. Replace S and T by $S^\perp$ and $T^\perp$ in (7.5.1) and use *Theorem* 7.5.3.
6. Let $\mathbf{A} = [\mathbf{x}_1 : \cdots : \mathbf{x}_k]$. Then $\mathbf{A}\mathbf{A}^*\mathbf{x} = \sum_{i=1}^k \mathbf{x}_i\mathbf{x}_i^*\mathbf{x} = \sum_{i=1}^k \langle \mathbf{x}, \mathbf{x}_i \rangle \mathbf{x}_i$.
7. $\mathcal{C}(\mathbf{P}_\mathbf{A}) = \mathcal{C}(\mathbf{A})$ gives $\mathbf{P}_\mathbf{A} = \mathbf{AB}$. $\mathbf{P}_\mathbf{A}\mathbf{P}_\mathbf{A}^* = \mathbf{P}_\mathbf{A}$ gives $\mathbf{P}_\mathbf{A} = \mathbf{ABB}^*\mathbf{A}^*$. So $\mathbf{A}^*\mathbf{P}_\mathbf{A}\mathbf{A} = \mathbf{A}^*\mathbf{A}$ gives $\mathbf{BB}^* = (\mathbf{A}^*\mathbf{A})^-$.
8. Use $\mathbf{A}^*\mathbf{P}_\mathbf{A} = \mathbf{A}^*$ and rank-cancellation.

$$9.\ \mathbf{D} = \frac{1}{\sqrt{42}}\begin{bmatrix} 3\sqrt{3} & 1 \\ \sqrt{3} & 5 \\ -2\sqrt{3} & 4 \end{bmatrix} \text{ and } \mathbf{P}_\mathbf{A} = \mathbf{DD}^* = \frac{1}{3}\begin{bmatrix} 2 & 1 & -1 \\ 1 & 2 & 1 \\ -1 & 1 & 2 \end{bmatrix}.$$

10. $\{\mathbf{x}, \mathbf{y}\}$ is orthonormal, so $\mathbf{x}\mathbf{x}^\mathrm{T} + \mathbf{y}\mathbf{y}^\mathrm{T}$ is an orthogonal projector. Algorithm 7.4.8 applied to $\{\mathbf{x}, \mathbf{y}, \mathbf{e}_1, \mathbf{e}_2, \mathbf{e}_3, \mathbf{e}_4\}$ gives $\{\mathbf{x}, \mathbf{y}, \mathbf{u}, \mathbf{v}\}$ as an orthonormal basis of $\mathbb{R}^4$, so $\{\mathbf{u}, \mathbf{v}\}$ is an orthonormal basis of $\mathcal{C}(\mathbf{A})$ where $\mathbf{u} = \frac{1}{\sqrt{2}}(1, 0, -1, 0)^\mathrm{T}$ and $\mathbf{v} = \frac{1}{\sqrt{2}}(0, 1, 0, -1)^\mathrm{T}$. 11. $\mathbf{P} = \mathbf{P}^\mathrm{T}\mathbf{P}$ gives $p_{jj} = p_{jj}^2 + \sum_{i \neq j} p_{ij}^2$.
12(a). $\mathcal{C}(\mathbf{P}) \perp \mathcal{C}(\mathbf{Q}) \Leftrightarrow \mathbf{QP} = \mathbf{0}$. Use the corollary to *Theorem* 3.7.6 and $(\mathbf{P}+\mathbf{Q})^* = \mathbf{P}+\mathbf{Q}$. *Aliter for If part*: Let the columns of $\mathbf{C}$ and $\mathbf{D}$ form

orthonormal bases of $\mathcal{C}(\mathbf{P})$ and $\mathcal{C}(\mathbf{Q})$ respectively. Use $\mathbf{P} = \mathbf{CC}^*$, $\mathbf{Q} = \mathbf{DD}^*$ and $\mathbf{P} + \mathbf{Q} = [\mathbf{C} : \mathbf{D}][\mathbf{C} : \mathbf{D}]^*$.
12(b). $\mathcal{C}(\mathbf{X} : \mathbf{Y}) = \mathcal{C}(\mathbf{X} : \mathbf{Y} - \mathbf{P_X Y}) = \mathcal{C}(\mathbf{X} : \mathbf{Z}) = \mathcal{C}(\mathbf{X}) + \mathcal{C}(\mathbf{Z})$ and $\mathcal{C}(\mathbf{X}) \perp \mathcal{C}(\mathbf{Z})$, so $\mathbf{P}_{(\mathbf{X}:\mathbf{Y})} = \mathbf{P}_{(\mathbf{X}:\mathbf{Z})} = \mathbf{P_X} + \mathbf{P_Z}$.
13. $(1,1,1)$ and $(1,0,0)$.
14. Let $T = \mathrm{Sp}(S \cup \{\mathbf{u}\})$. Then $T \supseteq W$. First let $d(S) < d(V) - 1$. Take $\mathbf{x} \in T^{\perp}$, $\mathbf{x} \neq \mathbf{0}$. Then $\mathbf{s} = \mathbf{0}$. If $\mathbf{0}, \mathbf{w}$ and $\mathbf{x}$ are collinear, $\mathbf{x} \in T \cap T^{\perp}$. Next let $d(S) = d(V) - 1$. Then for any $\mathbf{x} \in V$, $\mathbf{x} - \mathbf{s}$ and $\mathbf{x} - \mathbf{w} \in S^{\perp}$, so $\mathbf{x} - \mathbf{s} = \alpha(\mathbf{x} - \mathbf{w})$ for some α and $\mathbf{s} = (1-\alpha)\mathbf{x} + \alpha\mathbf{w}$.
15. With $W = S$, *Theorem* 7.5.10 gives $\|\mathbf{x}\| \leq \|\mathbf{x} + \mathbf{z}\|$ for all $\mathbf{z} \in S$ iff the orthogonal projection of $\mathbf{x}$ into S is $\mathbf{0}$.
16(a)i. $\mathbf{B} = [\mathbf{u}_1 : \cdots : \mathbf{u}_n]$ is non-singular and $\sum \mathbf{u}_i \mathbf{u}_i^* = \mathbf{BB}^*$.
16(a)iii and ii. $\mathbf{A}^* = \mathbf{A}$ and $\mathbf{x}^*\mathbf{A}\mathbf{x} = (\mathbf{B}^{-1}\mathbf{x})^*\mathbf{B}^{-1}\mathbf{x}$. By *Exercise* 5.4.14, $\mathbf{u}_i\mathbf{u}_i^*\mathbf{A}\mathbf{u}_j\mathbf{u}_j^* = \mathbf{u}_i\delta_{ij}\mathbf{u}_j^*$. Now $\mathbf{u}_i$ has a left inverse and $\mathbf{u}_j^*$ a right inverse.
16(b). By (a), there exists a norm such that $\|2\mathbf{y}\| = \|\mathbf{z}\| = 1$.
17. The subspace of all odd polynomials. 18(a). 0 and $(-1,-1,2)/3$.
18(b). The line $x_2 = 2x_1$. 18(c). $\{\mathbf{A} \in \mathbb{C}^{n\times n} : a_{11} = \cdots = a_{nn} = 0\}$.
19(a). Imitate the proof of *Theorem* 7.5.12.
19(b). Use (a), *Theorem* 3.7.1 and *Exercise* 3.7.6.

Section 7.6 (p. 278)

1. $|a_{ij}| = \delta_{ij}$ ($\mathbf{A}^* = \mathbf{A}^{-1}$ is both lower and upper triangular.)
2. See *Exercise* 7.5.10.
4. $\frac{1}{\sqrt{2}}\begin{bmatrix} 1 & -1 \\ 1 & 1 \end{bmatrix}$, $\begin{bmatrix} 1 & 2 \\ 2 & 1 \end{bmatrix}$ and $\begin{bmatrix} 2 & -1 \\ 3 & -2 \end{bmatrix}$ satisfy only (i), only (ii) and only (iii) respectively. 5. Take $\mathbf{x} = (1, i)^{\mathrm{T}}$.
6. (a) $\Rightarrow$ (b): $\mathbf{e}_i^*\mathbf{A}^*\mathbf{A}\mathbf{e}_j = 0$ and $(\mathbf{e}_i + \mathbf{e}_j)^*\mathbf{A}^*\mathbf{A}(\mathbf{e}_i - \mathbf{e}_j) = 0$ whenever $i \neq j$. (b) $\Rightarrow$ (a) and (c) $\Leftrightarrow$ (b) $\Leftrightarrow$ (d) are easy.
8. $\mathrm{tr}(\mathbf{B}^{\mathrm{T}}\mathbf{B}) = \mathrm{tr}(\mathbf{Q}^{\mathrm{T}}\mathbf{A}^{\mathrm{T}}\mathbf{A}\mathbf{Q}) = \mathrm{tr}(\mathbf{A}^{\mathrm{T}}\mathbf{A}\mathbf{Q}\mathbf{Q}^{\mathrm{T}})$.
9. $(\mathbf{I} - \mathbf{A})\mathbf{x} = \mathbf{0} \Leftrightarrow \mathbf{A}(\mathbf{A}^* - \mathbf{I})\mathbf{x} = \mathbf{0} \Leftrightarrow (\mathbf{A}^* - \mathbf{I})\mathbf{x} = \mathbf{0} \Leftrightarrow \mathbf{x} \perp \mathcal{C}(\mathbf{A} - \mathbf{I})$.
10(a). Use $(\mathbf{A}^{\mathrm{T}})_{11} = (\mathbf{A}^{-1})_{11}$.
10(b). Use *Theorem* 6.6.8 and $\mathbf{A}^{\circledast} = \pm\mathbf{A}^{\mathrm{T}}$. *Aliter*: Let $\mathbf{A} = \begin{bmatrix} \mathbf{E} & \mathbf{B} \\ \mathbf{C} & \mathbf{D} \end{bmatrix}$. Then $|\mathbf{D}\mathbf{D}^{\mathrm{T}}| = |\mathbf{I} - \mathbf{C}\mathbf{C}^{\mathrm{T}}|$ and $|\mathbf{E}^{\mathrm{T}}\mathbf{E}| = |\mathbf{I} - \mathbf{C}^{\mathrm{T}}\mathbf{C}|$ are equal by *Exercise* 6.7.2(b).
12. Use $(\mathbf{B}+\mathbf{A})^* = \mathbf{B}-\mathbf{A}$, $((\mathbf{B}+\mathbf{A})^*)^{-1} = ((\mathbf{B}+\mathbf{A})^{-1})^*$ and $(\mathbf{B}-\mathbf{A})(\mathbf{B}+\mathbf{A}) = (\mathbf{B}+\mathbf{A})(\mathbf{B}-\mathbf{A})$.
13. $\mathbf{C}^{\mathrm{T}} = (\mathbf{I} - \mathbf{A}^{-1}\mathbf{B}^{-1})(\mathbf{I} + \mathbf{A}^{-1})^{-1}(\mathbf{I} + \mathbf{B}^{-1})^{-1}$ since $\mathbf{A}^{\mathrm{T}} = \mathbf{A}^{-1}$ and $\mathbf{B}^{\mathrm{T}} = \mathbf{B}^{-1}$. Multiply the three terms by $\mathbf{AB}, \mathbf{A}^{-1}$ and $\mathbf{B}^{-1}$ respectively.
14. $\mathbf{A}_\theta$ when $\cos\frac{\theta}{2} \neq 0$; $\widetilde{\mathbf{A}}_\theta = (\tan\frac{\theta}{2})\begin{bmatrix} 0 & 1 \\ -1 & 0 \end{bmatrix}$.
15. $\mathbf{A}_{\theta+\eta}$, $\mathbf{B}_{\theta+\eta}$, $\mathbf{B}_{\eta-\theta}$ and $\mathbf{A}_{\theta-\eta}$.

17. $\|\mathbf{Ax}\| = \|\mathbf{x}\|$ iff $\mathbf{x}^*(\mathbf{A}^*\mathbf{A} - \mathbf{I})\mathbf{x} = 0$. Now use *Exercise* 2.6.8(c).
18. Obtain orthogonal matrices $\mathbf{U}$ and $\mathbf{V}$ with $\mathbf{U}_{*1} = \mathbf{u}$ and $\mathbf{V}_{*1} = \mathbf{v}$. Take $\mathbf{C} = \mathbf{V}\mathbf{U}^{\mathrm{T}}$.
19. min $= 0$ (take $\mathbf{v} \perp \mathbf{x}$, $\|\mathbf{v}\| = \|\mathbf{u}\|$ and use the preceding exercise). max $= \|\mathbf{u}\|^2 \|\mathbf{x}\|^2$ (use (7.3.7); for equality take $\mathbf{v} = (\|\mathbf{u}\|/\|\mathbf{x}\|)\mathbf{x}$).

Section 8.2 (p. 283)
1(a). $\cos\theta \pm i\sin\theta$. 1(b). $(7 \pm \sqrt{21})/2$. 1(c). 0, $i\sqrt{14}$ and $-i\sqrt{14}$.
1(d). $1, 1, -5$. 1(e). $2, 3, -1$.
2. 2, 5 and 8. (Take them to be $\alpha - d, \alpha, \alpha + d$, and use *Theorem* 8.2.4.)
3(a). Expand $|\lambda\mathbf{I} - \mathbf{A}|$ by the first column and use induction on n.
3(b)i. Consider $\mathrm{diag}(\lambda_1, \ldots, \lambda_n)$.
4. By the preceding exercise, $\chi(\lambda) = \lambda^n - 1$. 5. $\mathbf{BA} = \mathbf{A}^{-1}(\mathbf{AB})\mathbf{A}$.
7. $\left([\mathbf{C}_{*1} : \mathbf{C}_{*2}], \begin{bmatrix} 1 & 0 & 1 & 1 & 0 \\ 0 & 1 & 0 & 1/2 & 1 \end{bmatrix}\right)$; The characteristic roots are $5, 2, 0, 0, 0$ by *Theorem* 8.2.7.
8. $\chi_{\alpha\mathbf{I}+\beta\mathbf{A}}(\lambda) = |(\lambda-\alpha)\mathbf{I} - \beta\mathbf{A}|$ is $(\lambda-\alpha)^n$ or $\beta^n\chi_{\mathbf{A}}(\frac{\lambda-\alpha}{\beta})$ according as $\beta = 0$ or not. In either case, $\chi_{\alpha\mathbf{I}+\beta\mathbf{A}}(\lambda) = \prod_i(\lambda - \alpha - \beta\lambda_i)$.
9. $|\beta\mathbf{I} - \mathbf{A}| = 0 \Leftrightarrow |\beta\mathbf{A}^{-1} - \mathbf{I}| = 0 \Leftrightarrow |(1/\beta)\mathbf{I} - \mathbf{A}^{-1}| = 0$ (note $\beta \neq 0$).
10. Consider $\mathbf{e}_1\mathbf{e}_2^{\mathrm{T}}$ and $\mathbf{0}$. If 0 occurs as a characteristic root of $\mathbf{A}$ exactly once, $\rho(\mathbf{A}) = n - 1$ by *Theorem* 8.5.6.
11. Iff $-1/\beta$ is a characteristic root of $\mathbf{A}^{-1}\mathbf{B}$.
12. $|\mathbf{I} + \mathbf{A}| = \chi_{\mathbf{A}}(-1) = 1 - \mathrm{tr}(\mathbf{A}) + |\mathbf{A}|$.
13. Evaluate $\frac{d^k}{d\lambda^k}|\lambda\mathbf{I} - \mathbf{A}|$ at $\lambda = 0$ using *Exercise* 6.3.11. (Try this out for $n = 3$.)

Section 8.3 (p. 289)
1. $\mathrm{ES}(\mathbf{A}, 2) = \mathrm{Sp}(\mathbf{1})$, $\mathrm{ES}(\mathbf{C}, 1) = \mathrm{Sp}(\mathbf{e}_1)$ and $\mathrm{ES}(\mathbf{C}, i) = \mathrm{Sp}((1/(1-i), -1)^{\mathrm{T}})$.
2. Let the rank-factorization given in the solution of *exercise* 8.2.7 be $(\mathbf{B}, \mathbf{A})$. Then $(1, 2)^{\mathrm{T}}$ and $(0, 1)^{\mathrm{T}}$ form bases of $\mathrm{ES}(\mathbf{AB}, 5)$ and $\mathrm{ES}(\mathbf{AB}, 2)$, so $\mathbf{B}(1, 2)^{\mathrm{T}} = (10, 5, 3, -8, 9)^{\mathrm{T}}$ and $\mathbf{B}(0, 1)^{\mathrm{T}} = (4, 2, 0, -4, 2)^{\mathrm{T}}$ form bases of $\mathrm{ES}(\mathbf{C}, 5)$ and $\mathrm{ES}(\mathbf{C}, 2)$. $\{(1, 0, -1, 0, 0)^{\mathrm{T}}, (1, -\frac{1}{2}, 0, -1, 0)^{\mathrm{T}}, (0, 1, 0, 0, -1)^{\mathrm{T}}\}$ is a basis of $\mathrm{ES}(\mathbf{C}, 0) = \mathcal{N}(\mathbf{C}) = \mathcal{N}(\mathbf{A})$.
3. $\mathbf{Ax} = \alpha\mathbf{x} \Rightarrow \mathbf{A}^2\mathbf{x} = \mathbf{A}(\alpha\mathbf{x}) = \alpha^2\mathbf{x}$, etc. *Second part*: $\mathbf{Ax} = \alpha\mathbf{x} \Leftrightarrow \frac{1}{\alpha}\mathbf{x} = \mathbf{A}^{-1}\mathbf{x}$. *Last part*: Take $\mathbf{A} = \mathbf{e}_1\mathbf{e}_2^{\mathrm{T}}$, $\alpha = 0$ and $k = 2$.
4. Let $\mathbf{B}$ be the $(\ell-k+1)\times(\ell-k+1)$ matrix with α's on the diagonal, 1's immediately above and 0's elsewhere. Take $\beta \neq \alpha$ and $\mathbf{A} = \mathrm{diag}(\alpha\mathbf{I}_{k-1}, \mathbf{B}, \beta\mathbf{I}_{n-\ell})$.
6(a). Use *Theorem* 8.3.11 for algebraic multiplicity. $\mathcal{N}((\alpha\mathbf{I}+\beta\mathbf{A}) - (\alpha+\beta\delta)\mathbf{I}) = \mathcal{N}(\mathbf{A} - \delta\mathbf{I})$. 6(b). Both are false: take $\delta = 1$, $\mathbf{A} = \mathrm{diag}(1, -1)$ and $f(t) = t^2$.
7. Take $\mathbf{P}$ as in *Exercise* 8.2.4 and use *Theorem* 8.3.11.

8. $\text{ES}(\mathbf{A},\alpha) \subseteq \mathcal{C}(\mathbf{A})$ if $\alpha \neq 0$, so by *Theorem* 8.3.7, $\rho(\mathbf{A}) \geq k-1$. (*Aliter*: use *Theorem* 8.3.5 for $\alpha = 0$.) Rank of $\text{diag}(0,0,\ldots,0,1,1,\ldots,1,2,3,\ldots,k-1)$ is ℓ if 1 occurs $\ell - k + 2$ times.
9. $[\beta_1\mathbf{x}_1 : \cdots : \beta_k\mathbf{x}_k]\mathbf{A}_k = \mathbf{0}$, so $\beta_i\mathbf{x}_i = \mathbf{0}$ and $\beta_i = 0$.
10(a) and (b). Use *Theorem* 8.2.7 (note that $\mathbf{u}^*\mathbf{u} \neq 0$).
10(c). $\text{ES}(\mathbf{A},\mathbf{u}^*\mathbf{u}) \subseteq \mathcal{C}(\mathbf{A}) = \mathcal{C}(\mathbf{u})$, so equality holds. Also, $\text{ES}(\mathbf{A},0) = \mathcal{N}(\mathbf{A}) = \mathcal{N}(\mathbf{u}^*) = \{\mathbf{u}\}^\perp$ has dimension $n-1$. So the algebraic multiplicities of 0 and $\mathbf{u}^*\mathbf{u}$ are $n-1$ and 1.
10(d). Let $\mathbf{P}$ be a unitary matrix with $\mathbf{u}/\|\mathbf{u}\|$ as the first column, Then $\mathbf{P}^{-1}\mathbf{A}\mathbf{P} = \text{diag}(\mathbf{u}^*\mathbf{u},0,\ldots,0)$.
11(i). Let $\beta \neq 0$. By the preceding exercise, the eigenvalues of $\mathbf{1}\mathbf{1}^{\mathrm{T}}$ are n and 0 and these are regular with multiplicities 1 and $n-1$. Use *Exercise* 7.3.6(a).
11(ii). Assume $\mathbf{u}$ is not a scalar multiple of $\mathbf{1}$. Then the two characteristic roots of $[\,\mathbf{1} : \mathbf{u}]^{\mathrm{T}}[\,\mathbf{u} : \mathbf{1}\,]$, viz., $\mathbf{1}^{\mathrm{T}}\mathbf{u} \pm \sqrt{n}\|\mathbf{u}\|$ are non-zero and simple. So by *Theorem* 8.2.7 the eigenvalues of $\mathbf{A} := \mathbf{u}\mathbf{1}^{\mathrm{T}} + \mathbf{1}\mathbf{u}^{\mathrm{T}}$ are $\mathbf{1}^{\mathrm{T}}\mathbf{u} \pm \sqrt{n}\|\mathbf{u}\|$ and 0 with algebraic multiplicities $1,1$ and $n-2$. All are regular since $\nu(\mathbf{A}) = \nu([\,\mathbf{1} : \mathbf{u}\,]^{\mathrm{T}}) = n-2$. Use *Exercise* 8.3.6(a).
12. $\lambda_1\mathbf{x}^{\mathrm{T}}\mathbf{y} = \mathbf{x}^{\mathrm{T}}\mathbf{A}\mathbf{y} = \lambda_2\mathbf{x}^{\mathrm{T}}\mathbf{y}$, so $\mathbf{x}^{\mathrm{T}}\mathbf{y} = 0$. 13. $\nu(\mathbf{A}-\lambda\mathbf{I}) = \nu(\mathbf{A}-\lambda\mathbf{I})^{\mathrm{T}}$.
14. Similar matrices have the same trace, the same determinant and the same characteristic roots. 15. See *Theorems* 7.5.4 and 7.5.5 of Mirsky (1955).
16. If $\mathbf{A}\mathbf{x} = \alpha\mathbf{x}$, $|\alpha| = \|\mathbf{A}\mathbf{x}\|/\|\mathbf{x}\| \leq \|\mathbf{A}\|$.
17(a). $\mathcal{C}(\mathbf{A}) = \{\mathbf{x} : \mathbf{A}\mathbf{x} = \mathbf{x}\}$, $\mathcal{N}(\mathbf{A}) = \mathcal{C}(\mathbf{I}-\mathbf{A})$ and $\mathcal{C}(\mathbf{A}) \oplus \mathcal{C}(\mathbf{I}-\mathbf{A}) = F^n$.
17(b). No: consider $\mathbf{e}_1\mathbf{e}_2^{\mathrm{T}}$.
18(a). Let $\mathbf{A}$ be the matrix of f w.r.t. $\mathcal{X} = \{\mathbf{x}_1,\ldots,\mathbf{x}_n\}$, $\mathbf{A}\mathbf{u} = \alpha\mathbf{u}$, $\mathbf{u} \neq \mathbf{0}$ and $\mathbf{x} = \sum_j u_j\mathbf{x}_j$. Then $f(\mathbf{x}) = \alpha\mathbf{x}$. Take $S = \text{Sp}(\{\mathbf{x}\})$.
18(b). Use induction on $d(V)$. Let S be as in (a). Define g on V/S by: $g(\mathbf{y}+S) = f(\mathbf{y})+S$. Let $\{\mathbf{z}_1+S,\ldots,\mathbf{z}_{n-1}+S\}$ be a basis of V/S w.r.t. which the matrix of g is upper triangular. Take $B = \{\mathbf{x},\mathbf{z}_1,\ldots,\mathbf{z}_{n-1}\}$.
19. $\mathbf{D} = \mathbf{P}^{\mathrm{T}}\mathbf{B}^{\mathrm{T}}\mathbf{B}\mathbf{P}$ where $\mathbf{B}$ is the Van der Monde matrix based on $\lambda_1,\ldots,\lambda_n$ and $\mathbf{P}$ is as in *Exercise* 2.6.14.

Section 8.4 (p. 295)

In the following, $m_{\mathbf{A}}(\lambda)$ will denote the minimal polynomial of $\mathbf{A}$.
2. $\chi_{\mathbf{A}}(\lambda) = \lambda^3 + 3\lambda^2 - 9\lambda + 5$, so $\mathbf{A}^{-1} = -(\mathbf{A}^2 + 3\mathbf{A} - 9\mathbf{I})/5$.
3. By C-H theorem, $\mathbf{A}^3 = -14\mathbf{A}$. So $\mathbf{A}^{10} = 38416\,\mathbf{A}^2$.
4. $\chi_{\mathbf{B}}(\lambda) = \lambda^3 - 1$, so $\mathbf{B}^{-1} = \mathbf{B}^2$. Now $h(\lambda) = \lambda + (\lambda+1)/(\lambda^2 - \lambda + 1) = \lambda + (\lambda+1)^2/(\lambda^3+1)$. So $h(\mathbf{B}) = \mathbf{B} + (\mathbf{B}+\mathbf{I})^2(2\mathbf{I})^{-1} = \begin{bmatrix} 3 & -1 & 5 \\ 0 & -3/2 & 3/2 \\ 0 & -3/2 & 0 \end{bmatrix}$.
5. If $h = f/g$, $h(\mathbf{A}) = f(\mathbf{A})[g(\mathbf{A})]^{-1}$ and $[g(\mathbf{A})]^{-1}$ is a polynomial in $g(\mathbf{A})$

and so in $\mathbf{A}$.
6. Let $\chi_{\mathbf{A}}(\lambda) = a_0 + a_1\lambda + \cdots + a_n\lambda^n$. Then $(\lambda\mathbf{I} - \mathbf{A})\mathbf{H} = \chi_{\mathbf{A}}(\lambda)\mathbf{I}$ over $\mathbb{C}[\lambda]$ gives $\mathbf{H}_{i-1} - \mathbf{A}\mathbf{H}_i = a_i\mathbf{I}$. Premultiplying this by $\mathbf{A}^i$ and summing over $i = 0, 1, \ldots, n$, we get $\mathbf{0} = \chi_{\mathbf{A}}(\mathbf{A})$.
7. For $\mathbf{A}$: $\lambda - 1$ or $(\lambda-1)^2$ according as $\alpha = 0$ or not. For $\mathbf{B}$: $(\lambda-1)(\lambda-2)$ or $(\lambda-1)^2(\lambda-2)$ according as $\alpha = 0$ or not. For $\mathbf{C}$: $\lambda - 1$, $(\lambda-1)^2$ or $(\lambda-1)^3$ according as '$\alpha = \beta = \gamma = 0$' or '$\alpha = \gamma = 0$ and $\beta \neq 0$' or 'β and exactly one of α and γ are 0'.
8. $\mathbf{A}^k = \mathbf{0}$ iff λ^k annihilates $\mathbf{A}$. Now use *Theorems* 8.4.2 and 8.4.3.
9(a). Use *Theorem* 8.2.6 and the preceding exercise. (b). λ^r for some $r \geq 1$.
10. Constant term is 0 iff 0 is a root.
11. Minimal polynomial has degree n since $(\mathbf{A}^k)_{1*} = \mathbf{e}_{k+1}^{\mathrm{T}}$ for $0 \leq k \leq n-1$.
12. $\lambda(\lambda-1)$ and $\lambda^2(\lambda-1)$.
13. By *Exercise* 8.4.8, trace of a nilpotent matrix is 0. Use *Theorem* 2.5.2.
14. λ or $\alpha^k m_{\mathbf{A}}(\lambda/\alpha)$ according as $\alpha = 0$ or not, where $k = \deg m_{\mathbf{A}}(\lambda)$.
15. $\lambda(\lambda - n)$. (Use *Theorems* 8.2.7 and 8.4.6. *Aliter*: Look at $\mathbf{J}^2$.)
16. $f(\mathrm{diag}(\mathbf{A}, \mathbf{B})) = \mathbf{0}$ iff $f(\mathbf{A}) = f(\mathbf{B}) = \mathbf{0}$. Now use *Theorem* 8.4.5.

Section 8.5 (p. 302)
1. By *Theorem* 8.3.10 and *Exercise* 8.4.8, $\mathbf{A}$ is similar to $\alpha\mathbf{e}_1\mathbf{e}_2^{\mathrm{T}}$ for some α. If $\alpha \neq 0$, $\mathbf{P}^{-1}(\alpha\mathbf{e}_1\mathbf{e}_2^{\mathrm{T}})\mathbf{P} = \mathbf{e}_1\mathbf{e}_2^{\mathrm{T}}$ where $\mathbf{P} = \mathrm{diag}(\alpha, 1)$.
2. Note that $(a-d)^2 + 4bc \neq 0$ iff the two characteristic roots are distinct.
3. $\lambda^k - 1$ annihilates $\mathbf{A}$ and has k distinct roots.
4. *If part*: $\mathbf{A}$ is idempotent iff $\mathbf{P}^{-1}\mathbf{A}\mathbf{P}$ is idempotent.
5. $\lambda^3 - \lambda^2$ annihilates $\mathbf{A}$, so each characteristic root is 0 or 1. *Second part*: Consider $\mathbf{e}_1\mathbf{e}_2^{\mathrm{T}}$.
6. *Only if part* follows from *Theorem* 8.5.2(ii). 7. $\mathbf{P}^{-1}f(\mathbf{A})\mathbf{P} = f(\mathbf{P}^{-1}\mathbf{A}\mathbf{P})$.
8. $\mathbf{P}\,\mathrm{diag}(3, 1)\mathbf{P}^{-1}$, where $\mathbf{P} = \mathbf{J} - 2\mathbf{e}_2\mathbf{e}_2^{\mathrm{T}}$, $\mathbf{P}\,\mathrm{diag}(3^{10}, 1)\mathbf{P}^{-1} = \frac{3^{10}-1}{2}\mathbf{J} + \mathbf{I}$, $\mathbf{P}\,\mathrm{diag}(\frac{1}{3}, 1)\mathbf{P}^{-1}$.
9(a). Let $\mathbf{P} = \begin{bmatrix} 1 & 1 - a_{11} \\ 1 & -a_{21} \end{bmatrix}$ and $\alpha = a_{11} - a_{21}$. Then $\mathbf{A} = \mathbf{P}\,\mathrm{diag}(1, \alpha)\mathbf{P}^{-1}$.
9(b). $\frac{1}{1-\alpha}\begin{bmatrix} a_{21} + (1-a_{11})\alpha^k & (1-a_{11})(1-\alpha^k) \\ a_{21}(1-\alpha^k) & 1 - a_{11} + a_{21}\alpha^k \end{bmatrix}$. 9(c). $\mathbf{e}_1\mathbf{e}_2^{\mathrm{T}} + \mathbf{e}_2\mathbf{e}_3^{\mathrm{T}} + \mathbf{e}_3\mathbf{e}_3^{\mathrm{T}}$.
10. $\mathbf{A} = \mathbf{P}\,\mathrm{diag}(3, -1, 0, 0)\mathbf{P}^{-1}$ where $\mathbf{P} = \begin{bmatrix} 1 & 3 & 1 & 1 \\ 1 & 1 & 1 & -1 \\ 2 & -2 & -1 & 0 \\ 0 & -2 & 0 & -1 \end{bmatrix}$.
11. $(\sum_i \alpha_i\mathbf{E}_i)(\sum_j (1/\alpha_j)\mathbf{E}_j)(\sum_\ell \alpha_\ell\mathbf{E}_\ell) = \sum_i \alpha_i\mathbf{E}_i$. $(i, j, \ell = 1, 2, \ldots, k-1.)$
12. If $\mathbf{P}^{-1}\mathbf{A}\mathbf{P} = \mathrm{diag}(\delta_1, \ldots, \delta_n)$, $(\mathbf{P}^{-1})_{i*}\mathbf{A} = \delta_i(\mathbf{P}^{-1})_{i*}$ for all i.
13. *Only if part*: use (8.5.9).

14. Let $\lambda_1, \ldots, \lambda_r$ be the non-zero characteristic roots of $\mathbf{A}$. By *Theorem* 8.5.6, $r = \rho(\mathbf{A})$. Use Cauchy-Schwarz inequality for $\mathbf{u} = (\lambda_1, \ldots, \lambda_r)^{\mathrm{T}}$ and 1.
15. Write (8.5.6) as $\mathbf{A} = \mathbf{P}_0\mathbf{\Delta}_0\mathbf{Q}_0$ where $\mathbf{\Delta}_0$ is non-singular. Then $\mathbf{E} = \mathbf{P}_0\mathbf{Q}_0$ is idempotent, $\mathcal{C}(\mathbf{A}) = \mathcal{C}(\mathbf{P}_0) = \mathcal{C}(\mathbf{E})$ and $\mathcal{N}(\mathbf{A}) = \mathcal{N}(\mathbf{Q}_0) = \mathcal{N}(\mathbf{E})$.
16. (a) $\Rightarrow$ (b): Let $\sum \alpha_i \mathbf{E}_i$ and $\sum \beta_j \mathbf{F}_j$ be the spectral forms of $\mathbf{A}$ and $\mathbf{B}$. By *Exercise* 8.5.13, $\mathbf{E}_i$ and $\mathbf{F}_j$ commute. So $(\mathbf{E}_i\mathbf{F}_j)^2 = \mathbf{E}_i\mathbf{F}_j$, $\mathbf{E}_i\mathbf{F}_j\mathbf{E}_k\mathbf{F}_\ell = \mathbf{0}$ if $(i,j) \neq (k,\ell)$ and $\sum_{i,j} \mathbf{E}_i\mathbf{F}_j = \mathbf{I}$. Also, $\mathbf{A} = \sum_{i,j} \alpha_i \mathbf{E}_i\mathbf{F}_j$ and $\mathbf{B} = \sum_{i,j} \beta_j \mathbf{E}_i\mathbf{F}_j$. So the proof of the *if part* of *Theorem* 8.5.5 gives (b).
(b) $\Rightarrow$ (c): Any $n \times n$ diagonal matrix is a polynomial of degree at most $n-1$ in diag$(1, 2, \ldots, n)$ since the Van der Monde matrix based on $1, 2, \ldots, n$ is non-singular.

Section 8.6 (p. 309)
1. The (i,j)-th element of $(\mathbf{J}_0(n))^\ell$ is 1 or 0 according as $j = i + \ell$ or not.
2. The vectors belong to $\mathcal{N}(\mathbf{B}^i)$ and their number $d(S_0) + \cdots + d(S_{i-1}) = n - \rho(B^i)$ by (8.7.1).
3 and 4. Let $\mathbf{T} = \mathbf{J}_0(n_1) \oplus \cdots \oplus \mathbf{J}_0(n_t)$. Extend $\mathbf{P}$ to a non-singular matrix $\mathbf{W} = [\,\mathbf{P} : \mathbf{Q}\,]$. Then $\mathbf{W}^{-1}\mathbf{B}\mathbf{W} = \begin{bmatrix} \mathbf{T} & \mathbf{R} \\ \mathbf{0} & \mathbf{S} \end{bmatrix}$ for some $\mathbf{R}$ and $\mathbf{S}$. Since 0 is a characteristic root of $\mathbf{T}$ a times, $\mathbf{S}$ is non-singular. So $g = \nu(\mathbf{B}) = \nu(\mathbf{T}) = t$. Also, $\mathbf{B}^n\mathbf{P} = \mathbf{P}\mathbf{T}^n = \mathbf{0}$. So $\mathcal{C}(\mathbf{P}) \subseteq \mathcal{N}(\mathbf{B}^n) = \mathcal{N}(\mathbf{B}^p)$. By *Theorem* 8.5.7, equality follows.
5. The Jordan forms are $\mathbf{J}_1(2) \oplus \mathbf{J}_0(1)$, $\mathbf{J}_1(2) \oplus \mathbf{J}_0(1)$ and $\mathbf{J}_1(1) \oplus \mathbf{J}_1(1) \oplus \mathbf{J}_0(1)$.
The transforming matrices are $\begin{bmatrix} 1 & 0 & 1 \\ 0 & \frac{1}{2} & -\frac{1}{2} \\ 0 & \frac{1}{2} & 0 \end{bmatrix}$, $\begin{bmatrix} 1 & 0 & 0 \\ 0 & 0 & 1 \\ 0 & \frac{1}{2} & 0 \end{bmatrix}$, $\begin{bmatrix} 1 & 0 & 0 \\ 0 & 2 & 1 \\ 0 & 1 & 0 \end{bmatrix}$.
6. *Theorem* 8.6.4: (i) $\Leftrightarrow$ (ii) is easy. (iii) $\Rightarrow$ (i) and (ii) $\Rightarrow$ (iii) follow from *Theorem* 8.6.3 and its proof. *Theorem* 8.6.5: Use *Theorems* 8.2.6 and 8.4.8 and *Exercise* 8.4.16. 7. Verify (ii) of *Theorem* 8.6.4.
8. Let $\mathbf{A} = \mathbf{P}\mathbf{J}\mathbf{P}^{-1}$ where $\mathbf{J} = \mathrm{diag}(\mathbf{J}_1, \ldots, \mathbf{J}_k)$, $\mathbf{J}_i = \oplus_{j=1}^{g_i} \mathbf{J}_{\alpha_i}(n_{ij})$ for all i and $\alpha_1, \ldots, \alpha_k$ are distinct. Let $n_i = \sum_j n_{ij}$, $n = \sum_i n_i$ and $\alpha = \sum_i \alpha_i$. Take $\mathbf{D} = \mathbf{P}\,\mathrm{diag}(\alpha_1\mathbf{I}_{n_1}, \ldots, \alpha_k\mathbf{I}_{n_k})\mathbf{P}^{-1}$. Let $\mathbf{K}_i = \mathrm{diag}(\mathbf{J}_1 - \alpha_i\mathbf{I}, \ldots, \mathbf{J}_{i-1} - \alpha_i\mathbf{I}, \mathbf{J}_{i+1} - \alpha_i\mathbf{I}, \ldots, \mathbf{J}_k - \alpha_i\mathbf{I})$. Now $(\mathbf{K}_i^n)^{-1} = f_i(\mathbf{K}_i)$ for some polynomial f_i. So $(\mathbf{J} - \alpha_i\mathbf{I})^n f_i(\mathbf{J} - \alpha_i\mathbf{I}) = \mathbf{I}_{(i)} := \mathrm{diag}(\mathbf{I}_{n_1}, \ldots, \mathbf{I}_{n_{i-1}}, \mathbf{0}, \mathbf{I}_{n_{i+1}}, \ldots, \mathbf{I}_{n_k})$. So $\mathrm{diag}(\alpha_1\mathbf{I}_{n_1}, \ldots, \alpha_k\mathbf{I}_{n_k}) = \sum_{i=1}^k \left(\frac{\alpha}{k-1} - \alpha_i\right)\mathbf{I}_{(i)}$ is a polynomial in $\mathbf{J}$ and $\mathbf{D}$ is a polynomial in $\mathbf{A}$. *Uniqueness*: Let $\mathbf{A} = \mathbf{D}_1 + \mathbf{N}_1 = \mathbf{D}_2 + \mathbf{N}_2$. By *Exercise* 8.5.16, $\mathbf{D}_1 - \mathbf{D}_2$ is diagonable. Now $(\mathbf{N}_2 - \mathbf{N}_1)^{2n} = \sum_j \binom{2n}{j} {\mathbf{N}_2}^j (-\mathbf{N}_1)^{2n-j} = \mathbf{0}$ since ${\mathbf{N}_1}^n = {\mathbf{N}_2}^n = \mathbf{0}$. Since $\mathbf{D}_1 - \mathbf{D}_2 = \mathbf{N}_2 - \mathbf{N}_1$, we get $\mathbf{D}_1 = \mathbf{D}_2$.
9. Let $\alpha_1 = 0$ in the preceding exercise and take $\mathbf{R} = \mathbf{P}\mathrm{diag}(\mathbf{0}, \mathbf{J}_2, \ldots, \mathbf{J}_k)\mathbf{P}^{-1}$.

Uniqueness: let $\mathbf{A} = \mathbf{R}_1 + \mathbf{M}_1 = \mathbf{R}_2 + \mathbf{M}_2$. Then $\mathbf{A}^n = \mathbf{R}_i{}^n + \mathbf{M}_i{}^n = \mathbf{R}_i{}^n$. So $\mathcal{C}(\mathbf{R}_1) = \mathcal{C}(\mathbf{R}_1{}^n) = \mathcal{C}(\mathbf{R}_2{}^n) = \mathcal{C}(\mathbf{R}_2)$. Similarly $\mathcal{R}(\mathbf{R}_1) = \mathcal{R}(\mathbf{R}_2)$. Now $\mathbf{R}_1{}^2 = \mathbf{R}_1\mathbf{A}$ and $\mathbf{A}\mathbf{R}_2 = \mathbf{R}_2{}^2$, so $\mathbf{R}_1\mathbf{R}_1\mathbf{R}_2 = \mathbf{R}_1\mathbf{R}_2\mathbf{R}_2$. Use rank cancellation.

Section 8.7 (p. 316)

1. $\frac{1+i}{\sqrt{2}}\begin{bmatrix} \frac{1}{2} & -\frac{1}{2i} \\ \frac{1}{2i} & \frac{1}{2} \end{bmatrix} + \frac{1-i}{\sqrt{2}}\begin{bmatrix} \frac{1}{2} & \frac{1}{2i} \\ -\frac{1}{2i} & \frac{1}{2} \end{bmatrix}$.

2. $P = \frac{1}{3}\begin{bmatrix} -1 & 2 & 2 \\ 2 & -1 & 2 \\ -2 & -2 & 1 \end{bmatrix}$ is orthogonal and $\mathbf{P}^{\mathrm{T}}\mathbf{A}\mathbf{P} = \begin{bmatrix} 9 & 9 & 9 \\ 0 & 9 & -9 \\ 0 & 0 & -9 \end{bmatrix}$.

3. $\mathbf{P} = \frac{1}{3\sqrt{2}}\begin{bmatrix} 2\sqrt{2} & -3 & -1 \\ 2\sqrt{2} & 3 & -1 \\ \sqrt{2} & 0 & 4 \end{bmatrix}$ is orthogonal and $\mathbf{P}^{\mathrm{T}}\mathbf{A}\mathbf{P} = \operatorname{diag}(0, 9, 9)$.

4. By the corollary to *Theorem* 8.7.1, the characteristic roots are $\pm\alpha_1 i, \ldots, \pm\alpha_p i$, $0, \ldots, 0$ for some $\alpha_1, \ldots, \alpha_p \in \mathbb{R} - \{0\}$. (a). $2p < n$. (b). $|\mathbf{A}| = 0$ or $\alpha_1^2 \cdots \alpha_p^2$ according as $p < n/2$ or not. (c). $|\mathbf{I} + \mathbf{A}| = (1 + \alpha_1^2) \cdots (1 + \alpha_p^2) \geq 1$.

5(a). $\mathbf{A} = \mathbf{P}\operatorname{diag}(22, 0, 0, 0)\mathbf{P}^{\mathrm{T}}$ where $P = \frac{1}{2\sqrt{11}}\begin{bmatrix} \sqrt{2} & \sqrt{11} & \sqrt{22} & 3 \\ 2\sqrt{2} & \sqrt{11} & 0 & -5 \\ \sqrt{2} & \sqrt{11} & -\sqrt{22} & 3 \\ 4\sqrt{2} & -\sqrt{11} & 0 & 1 \end{bmatrix}$.

5(b). $\sqrt{k}\,\frac{1}{2k}\begin{bmatrix} \mathbf{1}\mathbf{1}^{\mathrm{T}} & \sqrt{k}\mathbf{1} \\ \sqrt{k}\mathbf{1}^{\mathrm{T}} & k \end{bmatrix} - \sqrt{k}\,\frac{1}{2k}\begin{bmatrix} \mathbf{1}\mathbf{1}^{\mathrm{T}} & -\sqrt{k}\mathbf{1} \\ -\sqrt{k}\mathbf{1}^{\mathrm{T}} & k \end{bmatrix} + 0\begin{bmatrix} \mathbf{I} - \frac{1}{k}\mathbf{1}\mathbf{1}^{\mathrm{T}} & \mathbf{0} \\ \mathbf{0} & 0 \end{bmatrix}$.

(Imitate the solution of *Exercise* 8.5.10.)

6(a). $\mathbf{P}\operatorname{diag}(a\mathbf{I} + b\boldsymbol{\Delta}, a\mathbf{I})\mathbf{P}^{-1}$ in the form (8.7.1). $\sum_{i=1}^{k}(a + b\alpha_i)\mathbf{E}_i$ in the form (8.7.4) if $b \neq 0$.

6(b). $(a-b)(\mathbf{I} - \frac{1}{n}\mathbf{1}\mathbf{1}^{\mathrm{T}}) + \frac{a+(n-1)b}{n}\mathbf{1}\mathbf{1}^{\mathrm{T}}$ in the form (8.7.4). (See also *Exercise* 8.3.11(i).) 7(a). Check that $\mathbf{A}^{\mathrm{T}}\mathbf{A} = \mathbf{A}$.

7(b). $\sum_{i=1}^{3}\mathbf{y}_i\mathbf{y}_i^{\mathrm{T}}$ where $\mathbf{y}_1 = \frac{1}{\sqrt{165}}(11, -2, -2, -6)^{\mathrm{T}}$, $\mathbf{y}_2 = \frac{1}{\sqrt{110}}(0, 10, -1, -3)^{\mathrm{T}}$ and $\mathbf{y}_3 = \frac{1}{\sqrt{10}}(0, 0, 3, -1)^{\mathrm{T}}$. These $\mathbf{y}$'s form an orthonormal basis of $\mathcal{C}(\mathbf{A})$ and are obtained by *Algorithm* 7.4.8. 7(c). $\mathbf{y}_4\mathbf{y}_4^{\mathrm{T}}$ where $\mathbf{y}_4 = \frac{1}{\sqrt{15}}(2, 1, 1, 3)^{\mathrm{T}}$ forms an orthonormal basis of $\mathcal{C}(\mathbf{I} - \mathbf{A})$.

8. $\begin{bmatrix} 1 & 3+4i \\ 4+3i & 1 \end{bmatrix}$. However, every 2×2 real normal matrix is symmetric or skew-symmetric.

9. Let $\mathbf{A} = \mathbf{U}\boldsymbol{\Delta}\mathbf{U}^*$ where $\mathbf{U}$ is unitary and $\boldsymbol{\Delta}$ is diagonal. Then $\mathbf{A}\mathbf{A}^* = \mathbf{I} \Leftrightarrow \mathbf{U}\boldsymbol{\Delta}\boldsymbol{\Delta}^*\mathbf{U}^* = \mathbf{I} \Leftrightarrow \boldsymbol{\Delta}\boldsymbol{\Delta}^*\mathbf{U}^*\mathbf{U} = \mathbf{I} \Leftrightarrow |(\boldsymbol{\Delta})_{ii}| = 1$ for all i.

10. For hermitian (resp. real symmetric, skew-hermitian, unitary and orthogonal), add to (iii) the following: α_i's are real (resp. $\mathbf{E}_i$'s and α_i's are real, α_i's are purely imaginary, $|\alpha_i|$'s are 1 and $|\alpha_i|$'s are 1 and $\mathbf{A}$ is real).

11. False: consider $\mathbf{A}_{2\times 2}$ with $a_{11} = 1$, $a_{22} = -1$ and $a_{12} = a_{21} = i$.
12. The matrix is symmetric with characteristic roots $n, 0, \ldots, 0$.
13. Use *Theorem* 7.5.12 and *Exercise* 8.5.4.
14. $\mathbf{AA}^* = \mathbf{A}^*\mathbf{A} \Leftrightarrow \mathbf{x}^*\mathbf{AA}^*\mathbf{x} = \mathbf{x}^*\mathbf{A}^*\mathbf{Ax} \Leftrightarrow \|\mathbf{A}^*\mathbf{x}\| = \|\mathbf{Ax}\|$.
15(a). $\mathbf{A}$ is semi-simple with eigenvalues $\in \{1, -1\}$.
16(a). Let $\mathbf{A}^* = \mathbf{AW}$. Then $\rho(\mathbf{A}) = \rho(\mathbf{AA}^*) \le \rho(\mathbf{A}^2)$ and $\mathbf{A} = \mathbf{W}^*\mathbf{A}^*$, etc.
17(a). Reflection in the plane $x = c$ takes (x, y, z) to $(2c - x, y, z)$.
17(b). Rotation of space about the x-axis by θ is reflection in the plane $z = 0$ followed by reflection in the plane $z = (\tan\frac{\theta}{2})y$.
17(c). Resultant of reflections in x-y plane, y-z plane and x-z plane.
17(d). Rotation about the x-axis by θ followed by inversion is rotation about the x-axis by $\pi + \theta$ followed by reflection in the y-z plane since $-\mathbf{A}_\theta = \mathbf{A}_{\pi+\theta}$.

Section 8.8 (p. 322)

1. $$\begin{bmatrix} 0 & -\frac{2}{\sqrt{6}} \\ \frac{1}{\sqrt{2}} & \frac{1}{\sqrt{6}} \\ \frac{1}{\sqrt{2}} & -\frac{1}{\sqrt{6}} \end{bmatrix} \begin{bmatrix} \sqrt{11} & 0 \\ 0 & 3 \end{bmatrix} \begin{bmatrix} 0 & -\frac{2}{\sqrt{6}} \\ \frac{3}{\sqrt{22}} & -\frac{1}{\sqrt{6}} \\ \frac{2}{\sqrt{22}} & 0 \\ -\frac{3}{\sqrt{22}} & -\frac{1}{\sqrt{6}} \end{bmatrix}^*$$

3. False: Consider $\mathbf{e}_1\mathbf{e}_2^{\mathrm{T}}$. 4. $\rho(\mathbf{A}) = \rho(\mathbf{\Delta})$ with $\mathbf{\Delta}$ as in (8.8.1).
5. Use *Exercises* 5.4.4 and 5.4.9.
6(a). $\mathscr{C}(\mathbf{A}) = \mathscr{C}(\mathbf{R})$, so $\mathbf{P_A} = \mathbf{P_R} = \mathbf{RR}^* = \sum \mathbf{R}_i\mathbf{R}_i^* = \sum \mathbf{T}_i\mathbf{T}_i^*$.
6(b). By *Theorem* 7.5.12, $\mathbf{T}_i\mathbf{T}_i^*$ is an orthogonal projector. Also $\mathbf{T}_i\mathbf{T}_i^*\mathbf{T}_i\mathbf{T}_i^* = \mathbf{T}_i\mathbf{T}_i^* \Rightarrow \mathbf{T}_i^*\mathbf{T}_i\mathbf{T}_i^* = \mathbf{T}_i^*$. (ii): Take $\mathbf{T}_1 = (1, 0)$ and $\mathbf{T}_2 = (0, 1)$.
7. Let $\mathbf{A} = \mathbf{P\Delta P}^*$ where $\mathbf{P}$ is unitary and $\mathbf{\Delta} = \mathrm{diag}(\lambda_1, \ldots, \lambda_r, 0, \ldots, 0)$. Then the characteristic roots of $\mathbf{A}^*\mathbf{A}$ are the same as those of $\mathbf{\Delta}^*\mathbf{\Delta P}^*\mathbf{P}$.
8. $\sum \lambda_i^2 = \mathrm{tr}(\mathbf{A}^2)$ and $\sum d_i^2 = \mathrm{tr}(\mathbf{AA}^*)$. Use Cauchy-Schwarz inequality for $\langle \mathbf{A}, \mathbf{A}^*\rangle$, where $\langle \mathbf{A}, \mathbf{B}\rangle = \mathrm{tr}(\mathbf{B}^*\mathbf{A})$. Equality holds iff $\mathbf{A}^* = \alpha\mathbf{A}$ for some α.

Section 8.9 (p. 327)

1. $\mathbf{P}^{\mathrm{T}}\mathbf{AP} = \mathrm{diag}(7.03067, 4.24571, -3.53144, 0.25606)$ where

$$\mathbf{P} = \begin{bmatrix} 0.71300 & 0.28639 & -0.52073 & 0.37209 \\ -0.16868 & 0.77455 & 0.47135 & 0.38679 \\ 0.32037 & -0.53935 & 0.54176 & 0.55943 \\ -0.60045 & -0.16525 & -0.46169 & 0.63166 \end{bmatrix}$$

2. We get 7.03067 and the first column of $\mathbf{P}$.
3. $\mathbf{Cu} = \lambda_2\mathbf{u}$, so $\mathbf{AP}\begin{bmatrix} \beta \\ \mathbf{u} \end{bmatrix} = \mathbf{P}\begin{bmatrix} \lambda_1 & \mathbf{y}^{\mathrm{T}} \\ 0 & \mathbf{C} \end{bmatrix}\begin{bmatrix} \beta \\ \mathbf{u} \end{bmatrix} = \mathbf{P}\begin{bmatrix} \lambda_1\beta + \mathbf{y}^{\mathrm{T}}\mathbf{u} \\ \lambda_2\mathbf{u} \end{bmatrix} = \lambda_2\mathbf{P}\begin{bmatrix} \beta \\ \mathbf{u} \end{bmatrix}$

Section 9.1 (p. 329)

1(a)–(d): $\begin{bmatrix} 1 & 1 & -3 \\ 1 & 1 & 0 \\ -3 & 0 & -3 \end{bmatrix}$, $\frac{1}{2}\begin{bmatrix} 2 & -1 & 0 \\ -1 & 0 & 0 \\ 0 & 0 & 4 \end{bmatrix}$, $\frac{1}{2}\begin{bmatrix} 0 & 0 & 0 \\ 0 & 0 & 1 \\ 0 & 1 & 0 \end{bmatrix}$, $\begin{bmatrix} 4 & -2 & 6 \\ -2 & 1 & -3 \\ 6 & -3 & 9 \end{bmatrix}$.

1(e). $\mathbf{uu}^{\mathrm{T}}$ since $(\mathbf{u}^{\mathrm{T}}\mathbf{x})^2 = \mathbf{x}^{\mathrm{T}}\mathbf{uu}^{\mathrm{T}}\mathbf{x}$. 2. $\frac{1}{2}(\mathbf{A} + \mathbf{A}^{\mathrm{T}})$.
3. Use *Exercise* 2.6.8. 4. $\frac{1}{n}\mathbf{J}$ and $\mathbf{I} - \frac{1}{n}\mathbf{J}$.
5(a). Define $a_{ij} = \psi(\mathbf{e}_i, \mathbf{e}_j)$. 5(b). Use (a).
5(c). Define $\psi(\mathbf{x}, \mathbf{y}) = \mathbf{x}^{\mathrm{T}}\mathbf{Ay}$. Uniqueness: suppose $\varphi(\mathbf{x}, \mathbf{y}) = \mathbf{x}^{\mathrm{T}}\mathbf{By}$ is another. Then $\mathbf{B}$ is symmetric and $\mathbf{x}^{\mathrm{T}}\mathbf{Bx} = \mathbf{x}^{\mathrm{T}}\mathbf{Ax}$ gives $\mathbf{B} = \mathbf{A}$.

Section 9.2 (p. 331)

1. We give the type of definiteness of $\mathbf{A}, \mathbf{B}$ and then $\mathbf{A} + \mathbf{B}$. pd, pd: pd; pd, psd: pd; pd, nd: all; pd, nsd: pd, psd, ind; pd, ind: pd, psd, ind; psd, psd: pd, psd; psd, nd: nd, nsd, ind; psd, nsd: psd, nsd, ind; psd, ind: pd, psd, ind; nd, nd: nd; nd, nsd: nd; nd, ind: nd, nsd, ind; nsd, nsd: nd, nsd; nsd, ind: nd, nsd, ind; ind, ind: all.

2. $\mathbf{y}^{\mathrm{T}}\mathbf{By}$ where $\mathbf{B} = \begin{bmatrix} 1 & 0 & 0 \\ 0 & 1 & 0 \\ 1 & 1 & 1 \end{bmatrix}\begin{bmatrix} 1 & 1 & 0 \\ 1 & 2 & \frac{1}{2} \\ 0 & \frac{1}{2} & -1 \end{bmatrix}\begin{bmatrix} 1 & 0 & 1 \\ 0 & 1 & 1 \\ 0 & 0 & 1 \end{bmatrix} = \begin{bmatrix} 1 & 1 & 2 \\ 1 & 2 & \frac{7}{2} \\ 2 & \frac{7}{2} & 5 \end{bmatrix}$.
4. Consider $\mathbf{u}^{\mathrm{T}}\,\mathrm{diag}(\mathbf{A} : \mathbf{B})\mathbf{u}$ where $\mathbf{u}^{\mathrm{T}} = [\mathbf{x}^{\mathrm{T}} : \mathbf{y}^{\mathrm{T}}]$. *Second part*: indefinite.
5(a). $\mathbf{x}^{\mathrm{T}}\mathbf{PAP}^{\mathrm{T}}\mathbf{x} = (\mathbf{P}^{\mathrm{T}}\mathbf{x})^{\mathrm{T}}\mathbf{A}(\mathbf{P}^{\mathrm{T}}\mathbf{x}) \geq 0$.
5(b). Let $\mathbf{\Delta} := \mathbf{Q}^{-1}\mathbf{AQ}$ be diagonal where $\mathbf{Q}$ is orthogonal. Since $\mathbf{A}^3$ is n.n.d., $\mathbf{\Delta}^3$ and so $\mathbf{\Delta}$ and $\mathbf{A}$ are n.n.d. *Last part*: Take $\mathbf{P} = \mathbf{0}$.
6. *Only if part*: $\mathbf{x}^{\mathrm{T}}\mathbf{Ax} + \mathbf{x}^{\mathrm{T}}\mathbf{Bx} = 0$, so $\mathbf{x}^{\mathrm{T}}\mathbf{Ax} = 0$ for all $\mathbf{x}$, so $\mathbf{A} = \mathbf{0}$. *Last part*: Note $\mathbf{C}^2 = \mathbf{CC}^{\mathrm{T}}$ is n.n.d. Now use the first part or $\mathrm{tr}(\mathbf{CC}^{\mathrm{T}}) = \sum_{i,j} c_{ij}^2$.

Section 9.3 (p. 337)

1. $\mathbf{x}^{\mathrm{T}}\mathbf{Px} = \mathbf{x}^{\mathrm{T}}\mathbf{P}^{\mathrm{T}}\mathbf{Px} \geq 0$. 2(a). $\mathbf{B} = \mathbf{BAB}^{\mathrm{T}}$ since $\mathbf{B}$ is symmetric.
3. Extend $\mathbf{P}$ to a non-singular matrix $\mathbf{B} = (\mathbf{P} : \mathbf{Q})$. Then note that $\mathbf{PAP}^{\mathrm{T}} = \mathbf{B}\,\mathrm{diag}(\mathbf{A} : \mathbf{0})\mathbf{B}^{\mathrm{T}}$. If $\mathbf{A}$ is $n \times n$ and $\mathbf{P}$ is $k \times n$, $\nu(\mathbf{PAP}^{\mathrm{T}}) = k - n + \nu(\mathbf{A})$.
4. Use the identity: $(\mathbf{q}_1^{\mathrm{T}}\mathbf{x})^2 - (\mathbf{q}_2^{\mathrm{T}}\mathbf{x})^2 = ((\mathbf{q}_1 + \mathbf{q}_2)^{\mathrm{T}}\mathbf{x})((\mathbf{q}_1 - \mathbf{q}_2)^{\mathrm{T}}\mathbf{x})$.

6. $\mathbf{Q} = \begin{bmatrix} 1 & 0 & 1 \\ -\frac{1}{4} & 0 & -\frac{1}{4} \\ -1 & 2 & 0 \end{bmatrix}$ and $C = \begin{bmatrix} 2 & 0 & 0 \\ 0 & -\frac{1}{8} & 0 \\ 0 & 0 & 0 \end{bmatrix}$.

7. $\mathbf{Q}$ is $\begin{bmatrix} 1 & 0 & 0 \\ 3 & 0 & 1 \\ -\frac{1}{4} & 1 & \frac{1}{4} \end{bmatrix}$, $\begin{bmatrix} 1 & 0 & 0 \\ \frac{1}{2} & 1 & 0 \\ 0 & 0 & 1 \end{bmatrix}$, $\begin{bmatrix} 1 & 0 & 0 \\ 0 & 1 & 1 \\ 0 & -\frac{1}{2} & \frac{1}{2} \end{bmatrix}$, $\begin{bmatrix} 1 & 0 & 0 \\ \frac{1}{2} & 1 & 0 \\ -\frac{3}{2} & 0 & 1 \end{bmatrix}$ for (a), (b), (c) and (d) respectively. The corresponding $\mathbf{C}$'s are $\mathrm{diag}(1, -12, \frac{3}{4})$, $\mathrm{diag}(1, -\frac{1}{4}, 2)$, $\mathrm{diag}(0, 1, -\frac{1}{4})$ and $\mathrm{diag}(4, 0, 0)$. The first three are indefinite

and the last is p.s.d.
8(a). 3 and 1. (b). $2k$ and 0. (c). $n-1$ and $n-1$ ($\mathbf{A} = 2n(\mathbf{I} - \frac{1}{n}\mathbf{1}\mathbf{1}^{\mathrm{T}})$, so $\rho(\mathbf{A}) = n-1$. Signature is $n-1$ since $\mathbf{x}^{\mathrm{T}}\mathbf{A}\mathbf{x}$ is n.n.d.)
9(a)–(d). $Q = u_1^2 - u_2^2 - u_3^2$ where $u_1 = \frac{1}{2}x_1 + \frac{1}{2}x_2 + x_3, u_2 = \frac{1}{2}x_2 - \frac{1}{2}x_1$ and $u_3 = x_3$. Rank is 3, signature is -1. Indefinite.
10. *Theorem* 9.3.7 follows from *Theorem* 9.3.4.
11. Take α greater than the modulus of the minimum eigenvalue of $\mathbf{A}$.
12. Use the preceding exercise.
13. Let $\mathbf{P}$ be an orthogonal matrix such that $\mathbf{P}^{\mathrm{T}}\mathbf{A}\mathbf{P} = \operatorname{diag}(\lambda_1, \ldots, \lambda_n)$ and $\mathbf{x} = \mathbf{P}\mathbf{y}$. Then $\|\mathbf{x}\| = \|\mathbf{y}\|$, $\mathbf{x}^{\mathrm{T}}\mathbf{A}\mathbf{x} = \sum \lambda_i y_i^2$ and $(\min \lambda_j) \sum y_i^2 \leq \sum \lambda_i y_i^2$.
14. On suitably rotating the axes, $Q(x,y)$ becomes $a'x'^2 + b'y'^2$ and the rank and signature of Q are the same as those of $\operatorname{diag}(a', b')$.
15(a). Take $\mathbf{P} = \begin{bmatrix} \mathbf{I} & -\mathbf{A}^{-1}\mathbf{B} \\ \mathbf{0} & \mathbf{I} \end{bmatrix}$ and use $\mathbf{A}^{\mathrm{T}} = -\mathbf{A}$.
15(b). If the first column has a non-zero element, bring it to the (2,1)-position by symmetric row and column interchanges and use (a) with $\mathbf{A}$ of order 2×2.
15(c). Follows from (b).

Section 9.4 (p. 343)
1. $\mathbf{x}^{\mathrm{T}}\mathbf{B}^{\mathrm{T}}\mathbf{B}\mathbf{x} = 0 \Leftrightarrow \mathbf{B}\mathbf{x} = \mathbf{0} \Leftrightarrow \mathbf{B}^{\mathrm{T}}\mathbf{B}\mathbf{x} = \mathbf{0}$. *Last part*: take $\mathbf{x} = \mathbf{e}_i$ and $\mathbf{y} = \mathbf{e}_j$.
2. $\mathbf{P}^{\mathrm{T}}\mathbf{A}\mathbf{P}$ is n.n.d. If $\mathbf{A} = \mathbf{B}^{\mathrm{T}}\mathbf{B}$ with $\mathbf{B}$ non-singular, $\rho(\mathbf{P}^{\mathrm{T}}\mathbf{A}\mathbf{P}) = \rho(\mathbf{B}\mathbf{P}) = r$.
3. Let (8.7.3) hold with $\mathbf{R}$ real. Take $\mathbf{C} = [\,\mathbf{R}\boldsymbol{\Delta}^{1/2} : \mathbf{0}\,]$.
4(b). Let $\mathbf{v} = \mathbf{C}_L^{-1}\mathbf{u}$. Then $\mathbf{C}\mathbf{v} = \mathbf{u}$, $\mathbf{A} \pm \mathbf{u}\mathbf{u}^{\mathrm{T}} = \mathbf{C}(\mathbf{I} \pm \mathbf{v}\mathbf{v}^{\mathrm{T}})\mathbf{C}^{\mathrm{T}}$ and $\mathbf{v}^{\mathrm{T}}\mathbf{v} = \mathbf{u}^{\mathrm{T}}\mathbf{A}^{-}\mathbf{u}$. Also $\mathbf{T}\mathbf{T}^{\mathrm{T}} = \mathbf{T}^2 = \mathbf{I} + (2\alpha + \alpha^2\,\mathbf{v}^{\mathrm{T}}\mathbf{v})\mathbf{v}\mathbf{v}^{\mathrm{T}} = \mathbf{I} \pm \mathbf{v}\mathbf{v}^{\mathrm{T}}$.
5. Similar to *Exercise* 9.4.2.
6. Take $\mathbf{B} = \sum \alpha_i^{1/p}\mathbf{E}_i$. *Uniqueness*: Let $\mathbf{B}^p = \mathbf{A}$ and $\mathbf{B}$ be n.n.d. Then the distinct eigenvalues of $\mathbf{B}$ are $\alpha_1^{1/p}, \ldots, \alpha_k^{1/p}$. Let $\sum_{i=1}^k \alpha_i^{1/p}\mathbf{F}_i$ be the spectral form of $\mathbf{B}$. Then $\mathbf{A} = \mathbf{B}^p = \sum \alpha_i\mathbf{F}_i$. By *Theorem* 8.7.4(iv), $\mathbf{F}_i = \mathbf{E}_i$ for all i.
7. Any k-th order principal minor is $(1 + (k-1)\rho)(1-\rho)^{k-1}$. For the *only if part*, use $k = 2$ and $k = n$. *Second proof*: use *Exercise* 8.3.11.
8. *If part*: $\mathbf{A}\mathbf{G}\mathbf{A} = \mathbf{A}^{\mathrm{T}}\mathbf{G}\mathbf{A}$. *Converse*: $\mathbf{G} = (\mathbf{B}_L^{-1})^{\mathrm{T}}\mathbf{B}_L^{-1}$ where $\mathbf{A} \rightsquigarrow (\mathbf{B}, \mathbf{B}^{\mathrm{T}})$.
9. Use the solution to *Exercise* 3.7.4 with $\mathbf{Q}_i = \mathbf{P}_i^{\mathrm{T}}$ and *Exercise* 5.4.14.
10(a). By hypothesis, $\mathbf{x}^{\mathrm{T}}(\mathbf{A} - \mathbf{B})\mathbf{x} = 0$ for all $\mathbf{x}$. Use *Exercise* 9.1.3.
10(c) and (d). Let $\mathbf{B} = \mathbf{C}^{\mathrm{T}}\mathbf{C}$ and $\mathbf{A} - \mathbf{B} = \mathbf{D}^{\mathrm{T}}\mathbf{D}$ with $\mathbf{C}$ and $\mathbf{D}$ square. Then $\mathbf{A} = [\mathbf{C}^{\mathrm{T}} : \mathbf{D}^{\mathrm{T}}]\left[\begin{smallmatrix}\mathbf{C}\\ \mathbf{D}\end{smallmatrix}\right]$. For (d), use the fact that if $\mathbf{D} \neq \mathbf{0}$ then $[\,\mathbf{C}^{\mathrm{T}} : \mathbf{D}^{\mathrm{T}}\,]$ has a column basis other than the columns of $\mathbf{C}^{\mathrm{T}}$.
11. $|\mathbf{M}| = \alpha|\mathbf{A}|$ where $\alpha = d - \mathbf{b}^{\mathrm{T}}\mathbf{A}^{-1}\mathbf{b}$. If $\alpha > 0$, $\mathbf{M}$ is p.d. by *Theorem* 9.4.8. If $\alpha = 0$ and $\mathbf{A} = \mathbf{B}^{\mathrm{T}}\mathbf{B}$, then $\mathbf{M} = \begin{bmatrix} \mathbf{B}^{\mathrm{T}} \\ \mathbf{b}^{\mathrm{T}}\mathbf{B}^{-1} \end{bmatrix}[\mathbf{B} : (\mathbf{B}^{-1})^{\mathrm{T}}\mathbf{b}]$ is p.s.d. If $\alpha < 0$ and $\mathbf{y} = (-\mathbf{b}^{\mathrm{T}}\mathbf{A}^{-1} : 1)^{\mathrm{T}}$, then $\mathbf{y}^{\mathrm{T}}\mathbf{M}\mathbf{y} = \alpha < 0$ and $\mathbf{e}_1^{\mathrm{T}}\mathbf{M}\mathbf{e}_1 > 0$, so $\mathbf{M}$

is indefinite. 12. Use the preceding exercise.
13(a). *First part*: Note that $\mathbf{M} = \mathbf{P}^{\mathrm{T}} \operatorname{diag}(\mathbf{A}, \mathbf{D} - \mathbf{B}^{\mathrm{T}}\mathbf{A}^{-1}\mathbf{B})\mathbf{P}$ where $\mathbf{P} = \begin{bmatrix} \mathbf{I} & \mathbf{A}^{-1}\mathbf{B} \\ \mathbf{0} & \mathbf{I} \end{bmatrix}$. *Second part*: Note $\begin{bmatrix} \mathbf{0} & \mathbf{I} \\ \mathbf{I} & \mathbf{0} \end{bmatrix}^{\mathrm{T}} \begin{bmatrix} \mathbf{A} & \mathbf{B} \\ \mathbf{B}^{\mathrm{T}} & \mathbf{D} \end{bmatrix} \begin{bmatrix} \mathbf{0} & \mathbf{I} \\ \mathbf{I} & \mathbf{0} \end{bmatrix} = \begin{bmatrix} \mathbf{D} & \mathbf{B}^{\mathrm{T}} \\ \mathbf{B} & \mathbf{A} \end{bmatrix}$.
13(b). Imitate (a) noting that $\mathscr{C}(\mathbf{B}) \subseteq \mathscr{C}(\mathbf{A}) \Leftrightarrow \mathbf{A}\mathbf{A}^{-}\mathbf{B} = \mathbf{B}$.
13(c). By *Theorem* 3.8.1, $\mathbf{L} - \mathbf{A}^{-1} = \mathbf{A}^{-1}\mathbf{B}\mathbf{F}^{-1}\mathbf{B}^{\mathrm{T}}\mathbf{A}^{-1} = \mathbf{A}^{-1}\mathbf{B}\mathbf{F}^{-1}\mathbf{B}^{\mathrm{T}}(\mathbf{A}^{-1})^{\mathrm{T}}$.
13(d). If $|\mathbf{M}| = 0$, the inequality is trivial. So let $|\mathbf{M}| \neq 0$. Then $\mathbf{M}$ is p.d., so $|\mathbf{M}| = |\mathbf{A}| \cdot |\mathbf{D} - \mathbf{B}^{\mathrm{T}}\mathbf{A}^{-1}\mathbf{B}|$. Now use *Exercises* 9.4.13(a), 9.4.9(c) and (d).
14. Use *Exercise* 9.4.13(b). Note that $\mathbf{A}^{-1}[\,\mathbf{A} : \mathbf{I}\,] = [\,\mathbf{I} : \mathbf{A}^{-1}\,]$.
15. Use *Exercises* 9.4.13(a) and 9.4.13(b) with $\mathbf{B} = \mathbf{b}$ and $\mathbf{D} = [1]$.
16. Use the first part of *Exercise* 9.4.13(a).
17. If $\mathbf{B} = \mathbf{D}^{\mathrm{T}}\mathbf{D}$, $\mathbf{A} = \mathbf{D}^{\mathrm{T}}(\mathbf{I} + (\mathbf{D}^{\mathrm{T}})^{-1}\mathbf{C}\mathbf{D}^{-1})\mathbf{D}$ and $|\mathbf{I} + (\mathbf{D}^{\mathrm{T}})^{-1}\mathbf{C}\mathbf{D}^{-1}| \geq 1$.
18. *If part*: $\mathbf{P}\boldsymbol{\Delta}\mathbf{P}^{-1} = \mathbf{P}\mathbf{P}^{\mathrm{T}}((\mathbf{P}^{\mathrm{T}})^{-1}\boldsymbol{\Delta}\mathbf{P}^{-1})$. *Only if part*: Use the fact that $(\mathbf{C}\mathbf{C}^{\mathrm{T}})\mathbf{B} = \mathbf{C}(\mathbf{C}^{\mathrm{T}}\mathbf{B}\mathbf{C})\mathbf{C}^{-1}$ is similar to the real symmetric matrix $\mathbf{C}^{\mathrm{T}}\mathbf{B}\mathbf{C}$.
19. Let $\mathbf{A} = \mathbf{C}^{\mathrm{T}}\mathbf{C}$ and $\mathbf{B} = \mathbf{D}^{\mathrm{T}}\mathbf{D}$. Then $\mathbf{AB}$ and $\mathbf{E} := (\mathbf{C}\mathbf{D}^{\mathrm{T}})(\mathbf{C}\mathbf{D}^{\mathrm{T}})^{\mathrm{T}}$ have the same non-null eigenvalues and $\mathbf{E}$ is n.n.d.
20. Let $\mathbf{A} = \mathbf{C}\mathbf{C}^{\mathrm{T}}$. Then $\mathbf{A} + \mathbf{B}$ is p.d. $\Leftrightarrow \mathbf{I} + \mathbf{C}^{-1}\mathbf{B}(\mathbf{C}^{-1})^{\mathrm{T}}$ is p.d. $\Leftrightarrow$ each eigenvalue of $\mathbf{C}^{-1}\mathbf{B}(\mathbf{C}^{-1})^{\mathrm{T}}$ (i.e., of $(\mathbf{C}^{\mathrm{T}})^{-1}\mathbf{C}^{-1}\mathbf{B} = \mathbf{A}^{-1}\mathbf{B}$) is > -1.
21(a)i. Use Hadamard's inequality.
21(a)ii. Use *Theorem* 9.4.11. Note that if $|\det \mathbf{A}| = n^{n/2}$, $\mathbf{A}^{\mathrm{T}}\mathbf{A}$ is p.d.
21(b). Let $n > 2$. W.l.g., assume that $\mathbf{A}_{*1} = \mathbf{1}$. Let n_1, n_2, n_3 and n_4 be the number of times $(1,1)$, $(1,-1)$, $(-1,1)$ and $(-1,-1)$ occur in $[\mathbf{A}_{*2} : \mathbf{A}_{*3}]$. Using $\mathbf{A}_{*1}$, $\mathbf{A}_{*2}$ and $\mathbf{A}_{*3}$ are orthogonal, show that $n_1 = n_2 = n_3 = n_4 = n/4$.
22. Let $\mathbf{B} = \mathbf{D}\mathbf{D}^{\mathrm{T}}$ and $\mathbf{A} - \mathbf{B} = \mathbf{E}\boldsymbol{\Delta}\mathbf{E}^{\mathrm{T}}$ where $\mathbf{D}$ and $\mathbf{E}$ have full column rank and $\boldsymbol{\Delta}$ is diagonal and non-singular. Then $\mathscr{C}(\mathbf{A} - \mathbf{B}) = \mathscr{C}(\mathbf{E})$, so $\mathbf{G} = [\mathbf{D} : \mathbf{E}]$ has full column rank. Since $\mathbf{A} = \mathbf{G} \operatorname{diag}(\mathbf{I}, \boldsymbol{\Delta})\mathbf{G}^{\mathrm{T}}$ is n.n.d., so is $\boldsymbol{\Delta}$.
23. Let $\mathbf{A} = \left[\begin{smallmatrix} \mathbf{D} \\ \mathbf{E} \end{smallmatrix}\right]\left[\begin{smallmatrix} \mathbf{D} \\ \mathbf{E} \end{smallmatrix}\right]^{\mathrm{T}}$, $\mathbf{B} = \left[\begin{smallmatrix} \mathbf{F} \\ \mathbf{G} \end{smallmatrix}\right]\left[\begin{smallmatrix} \mathbf{F} \\ \mathbf{G} \end{smallmatrix}\right]^{\mathrm{T}}$ and $\mathbf{L} = \left[\begin{smallmatrix} \mathbf{H} \\ \mathbf{K} \end{smallmatrix}\right]$ where $\mathbf{H} = [\mathbf{D} : \mathbf{F}]$ and $\mathbf{K} = [\mathbf{E} : \mathbf{G}]$ so that $\mathbf{C} = \mathbf{L}\mathbf{L}^{\mathrm{T}}$ and $\mathbf{C}_{11} = \mathbf{H}\mathbf{H}^{\mathrm{T}}$. Then $\mathscr{R}(\mathbf{K}) \subseteq \mathscr{R}(\mathbf{H})$.
24. If $\begin{bmatrix} \mathbf{B} & \mathbf{C} \\ \mathbf{D} & \mathbf{E} \end{bmatrix}$ is orthogonal, then $\mathbf{B}\mathbf{B}^{\mathrm{T}} = \mathbf{I} - \mathbf{C}\mathbf{C}^{\mathrm{T}}$, $\mathbf{C}\mathbf{C}^{\mathrm{T}}$ is n.n.d. and $\nu(\mathbf{C}\mathbf{C}^{\mathrm{T}}) = k - \rho(\mathbf{C}) \geq k + \ell - n$. *Converse*: By hypothesis, $\mathbf{I} - \mathbf{B}\mathbf{B}^{\mathrm{T}}$ is n.n.d. with rank $k - \nu(\mathbf{I} - \mathbf{B}\mathbf{B}^{\mathrm{T}}) \leq n - \ell$. By *Exercise* 9.4.4, $\mathbf{I} - \mathbf{B}\mathbf{B}^{\mathrm{T}} = \mathbf{C}\mathbf{C}^{\mathrm{T}}$ for some $k \times (n - \ell)$ matrix $\mathbf{C}$. The rows of $[\mathbf{B} : \mathbf{C}]$ are orthonormal.
25(a). Let $\mathbf{V} = \mathbf{Y}\mathbf{Y}^{\mathrm{T}}$. Then $\mathscr{C}(\mathbf{V} + \mathbf{X}\mathbf{X}^{\mathrm{T}}) = \mathscr{C}[\mathbf{X} : \mathbf{Y}] = \mathscr{C}[\mathbf{X} : \mathbf{V}]$. For the second equality, note that $\mathbf{Vz} = \mathbf{P_X}\mathbf{Vz} + (\mathbf{I} - \mathbf{P_X})\mathbf{Vz}$ for all $\mathbf{z}$.
25(b). $\mathscr{C}(\mathbf{X}) \perp \mathscr{C}(\mathbf{I} - \mathbf{P_X}\mathbf{V})$, so $\rho(\mathbf{V} + \mathbf{X}\mathbf{X}^{\mathrm{T}}) = \rho(\mathbf{X}) + \rho((\mathbf{I} - \mathbf{P_X})\mathbf{V})$.
25(c). $\mathbf{y}^{\mathrm{T}}[\mathbf{V} : \mathbf{X}] = \mathbf{z}^{\mathrm{T}}[\mathbf{X}^{\mathrm{T}} : \mathbf{0}] \Rightarrow \mathbf{y}^{\mathrm{T}}\mathbf{V}\mathbf{y} = 0 \Rightarrow \mathbf{y}^{\mathrm{T}}\mathbf{V} = \mathbf{0}$ by *Exercise* 9.4.1.
25(d). Let $S =$ LHS and $T =$ RHS. If $\mathbf{y}^{\mathrm{T}}\mathbf{V} + \mathbf{z}^{\mathrm{T}}\mathbf{X}^{\mathrm{T}} = \mathbf{0}$ and $\mathbf{y}^{\mathrm{T}}\mathbf{X} = \mathbf{0}$ then $\mathbf{y}^{\mathrm{T}}\mathbf{V}\mathbf{y} = \mathbf{0}$, so $\mathbf{y}^{\mathrm{T}}\mathbf{V} = \mathbf{0}$ and $\mathbf{z}^{\mathrm{T}}\mathbf{X}^{\mathrm{T}} = \mathbf{0}$. Thus $T^{\perp} \subseteq S^{\perp}$, so $S \subseteq T$.

Section 9.5 (p. 350)

1. W.l.g., let $\mathbf{B} = \mathbf{A}(1,\ldots,k\,|\,1,\ldots,k)$. Then $\mathbf{z}^T\mathbf{B}\mathbf{z} = \mathbf{x}^T\mathbf{A}\mathbf{x}$ where $\mathbf{x} = \left[\begin{smallmatrix}\mathbf{z}\\0\end{smallmatrix}\right]$. Note that $\|\mathbf{x}\| = \|\mathbf{z}\|$. 2. Take $\mathbf{u} = \mathbf{1}$ and $\mathbf{B} = \mathbf{I}$ in *Theorem* 9.5.7.
3(a). Let $\delta = \max|a_{ij}|$ and $\mathbf{A}\mathbf{x} = \alpha\mathbf{x}$, $\|\mathbf{x}\| = 1$. Then $|\alpha| = |\mathbf{x}^T\mathbf{A}\mathbf{x}| \leq \delta\sum_{i,j}|x_i|\cdot|x_j| = \delta(\sum_i |x_i|)^2 \leq \delta n(\sum |x_i|^2) = \delta n$.
3(b). Let $\mathbf{T} = \mathbf{U}^*\mathbf{A}\mathbf{U}$ be upper triangular where $\mathbf{U}$ is unitary. Then $\operatorname{tr}(\mathbf{A}^*\mathbf{A}) = \operatorname{tr}(\mathbf{U}^*\mathbf{A}^*\mathbf{A}\mathbf{U}) = \operatorname{tr}(\mathbf{T}^*\mathbf{T}) \geq \sum |t_{ii}|^2 = \sum |\lambda_i|^2$.
4. *First part*: Use *Theorem* 9.5.1 and $\|\mathbf{A}\mathbf{x}\| = \sqrt{\mathbf{x}^T\mathbf{A}^T\mathbf{A}\mathbf{x}}$. *Second part*: The characteristic roots of $\mathbf{A}^T\mathbf{A} = \mathbf{A}^*\mathbf{A}$ are $|\lambda_i|^2$ by *Theorem* 8.7.8.
6(a). See *Exercises* 9.5.4 and 7.3.15. 6(b). See *Exercise* 8.3.16.
6(c). Note that $\|\mathbf{A}\|_2 = \sqrt{\lambda_1(\mathbf{A}^T\mathbf{A})}$ and $\|\mathbf{A}\|_E = \sqrt{\operatorname{tr}(\mathbf{A}^T\mathbf{A})}$.

Section 9.6 (p. 356)

1. Suppose $\mathbf{x} = \mathbf{P}\mathbf{y}$ diagonalises both. Then $p_{11}p_{12} - p_{21}p_{22} = 0$ and $(p_{11}+p_{21})(p_{12}+p_{22}) = 0$, so $\mathbf{P}$ is singular.
2. Let $\mathbf{B} = \mathbf{C}\mathbf{C}^T$. $|\mathbf{A} - \lambda\mathbf{C}^T\mathbf{C}| = 0 \Leftrightarrow |(\mathbf{C}^{-1})^T\mathbf{A}\mathbf{C}^{-1} - \lambda\mathbf{I}| = 0$.
3(a). $\mathbf{P}^T\mathbf{A}\mathbf{P} = \operatorname{diag}(0,1,-2)$ and $\mathbf{P}^T\mathbf{B}\mathbf{P} = \mathbf{I}$ if $\mathbf{P} = \begin{bmatrix} -1 & -1 & 0\\ 5 & 3 & 1\\ 8 & 5 & 2\end{bmatrix}$.
3(b). With the same $\mathbf{P}$, $\mathbf{P}^T\mathbf{A}\mathbf{P} = \operatorname{diag}(1,0,0)$. $\mathbf{P}$ is not unique here.
4. Taking $\mathbf{Q} = \begin{bmatrix} .89443 & -.18257 & .40825\\ 0 & .91287 & .40825\\ -.44721 & -.36515 & .81650\end{bmatrix}$ (obtained by Jacobi's method), we get $\mathbf{Q}^T\mathbf{A}_1\mathbf{Q} = \operatorname{diag}(12,12,-6)$. Now $\mathbf{Q}^T\mathbf{A}_2\mathbf{Q} = \operatorname{diag}(12,12,18)$. The final $\mathbf{P}$ is $\begin{bmatrix} .55735 & -.70711 & .40825\\ .55735 & .70711 & .40825\\ -.55735 & 0 & .81650\end{bmatrix}$, $\mathbf{P}^T\mathbf{A}_i\mathbf{P} = \mathbf{Q}^T\mathbf{A}_i\mathbf{Q}$ for $i = 1,2$ and $\mathbf{P}^T\mathbf{A}_3\mathbf{P} = \operatorname{diag}(6,12,24)$.
5. *Only if part* is easy. *If part*: imitate the proof of *Theorem* 9.6.1 (note that $\mathbf{M}^T\mathbf{A}_1\mathbf{M} = \mathbf{I}$ gives $\mathbf{M}\mathbf{M}^T = \mathbf{A}_1^{-1}$, so $\mathbf{M}^T\mathbf{A}_i\mathbf{M}$ and $\mathbf{M}^T\mathbf{A}_j\mathbf{M}$ commute).
6. Imitate the first proof of *Theorem* 9.6.8.
7. Let $\mathbf{P}$ be non-singular and $\mathbf{C} = \mathbf{P}^T\mathbf{A}\mathbf{P}$ and $\mathbf{D} = \mathbf{P}^T\mathbf{B}\mathbf{P}$ be diagonal. Then $|\mathbf{P}|^2|\alpha\mathbf{A} + (1-\alpha)\mathbf{B}| = |\alpha\mathbf{C} + (1-\alpha)\mathbf{D}| = \prod_i(\alpha c_{ii} + (1-\alpha)d_{ii}) \geq \prod_i(c_{ii}^{\alpha}d_{ii}^{1-\alpha}) = |\mathbf{C}|^{\alpha}|\mathbf{D}|^{1-\alpha} = |\mathbf{P}|^2|\mathbf{A}|^{\alpha}|\mathbf{B}|^{1-\alpha}$. (The inequality follows from *Lemma* 7.3.2.) 8. Use *Theorem* 9.6.1.
9. By *Exercise* 9.4.18, $\mathbf{A}^{-1}\mathbf{B}$ is semi-simple.
10. *If part*: Imitate the second proof of *Theorem* 9.6.8. *Only if part*: Use the fact that diagonal matrices are normal.
11. By *Theorem* 9.6.5, $\mathbf{A} = \mathbf{P}\boldsymbol{\Delta}\mathbf{P}^T$ and $\mathbf{B} = \mathbf{P}\boldsymbol{\Gamma}\mathbf{P}^T$ for some $\boldsymbol{\Delta}$ and $\boldsymbol{\Gamma}$ diagonal and $\mathbf{P}$ non-singular. $\mathbf{A}^2 = \mathbf{B}^2 \Rightarrow (\boldsymbol{\Delta}\mathbf{P}^T\mathbf{P}\boldsymbol{\Delta})_{ii} = (\boldsymbol{\Gamma}\mathbf{P}^T\mathbf{P}\boldsymbol{\Gamma})_{ii} \Rightarrow (\boldsymbol{\Delta})_{ii} = (\boldsymbol{\Gamma})_{ii}$.
12. Use (9.5.4). *Converse*: take $\mathbf{A} = \operatorname{diag}(2,1)$ and $\mathbf{B} = \operatorname{diag}(1,2)$.

13. By *Exercise* 3.4.22, $\mathbf{\Delta} = \text{diag}(\mathbf{D}, \mathbf{0})$ for some $r \times r$ matrix $\mathbf{D}$. By *Exercise* 5.4.9, $\mathbf{A}^{-}\mathbf{B} = \mathbf{P}\begin{bmatrix} \mathbf{I} & \mathbf{R} \\ \mathbf{S} & \mathbf{T} \end{bmatrix}\mathbf{P}^{\mathrm{T}}(\mathbf{P}^{\mathrm{T}})^{-1}\begin{bmatrix} \mathbf{D} & \mathbf{0} \\ \mathbf{0} & \mathbf{0} \end{bmatrix}\mathbf{P}^{-1} = \mathbf{P}\begin{bmatrix} \mathbf{D} & \mathbf{0} \\ \mathbf{SD} & \mathbf{0} \end{bmatrix}\mathbf{P}^{-1}$ for some $\mathbf{M}$.

14(i). Let $\mathbf{P}$, $\mathbf{\Delta}$ and $\mathbf{\Gamma}$ be as in the solution to *Exercise* 9.6.11. Note that $\mathbf{A}^{-1} = (\mathbf{P}^{\mathrm{T}})^{-1}\mathbf{\Delta}^{-1}\mathbf{P}^{-1}$ and $\mathbf{B}^{-1} = (\mathbf{P}^{\mathrm{T}})^{-1}\mathbf{\Gamma}^{-1}\mathbf{P}^{-1}$. Use $(a_i + b_i)^2 \geq 4a_ib_i$ where $a_i = (\mathbf{\Delta})_{ii}$ and $b_i = (\mathbf{\Gamma})_{ii}$.

14(ii). $\begin{bmatrix} \mathbf{A}+\mathbf{B} & 2\mathbf{I} \\ 2\mathbf{I} & \mathbf{A}^{-1}+\mathbf{B}^{-1} \end{bmatrix}$ is n.n.d. Use *Exercise* 9.4.13(b).

15(a). Use *Theorems* 9.4.6 and 5.4.9.

15(b) and (c). Let $\mathbf{P}$, $\mathbf{\Delta}$ and $\mathbf{\Gamma}$ be as in the solution to *Exercise* 9.6.11. Take $(\mathbf{A}+\mathbf{B})^{-} = (\mathbf{P}^{\mathrm{T}})^{-1}(\mathbf{\Delta}+\mathbf{\Gamma})^{-}\mathbf{P}^{-1}$, $\mathbf{A}^{-} = (\mathbf{P}^{\mathrm{T}})^{-1}\mathbf{\Delta}^{-}\mathbf{P}^{-1}$ and $\mathbf{B}^{-} = (\mathbf{P}^{\mathrm{T}})^{-1}\mathbf{\Gamma}^{-}\mathbf{P}^{-1}$ with $(\mathbf{\Delta}+\mathbf{\Gamma})^{-}$, $\mathbf{\Delta}^{-}$ and $\mathbf{\Gamma}^{-}$ diagonal.

16. Take $\mathbf{A}^{-}$ and $\mathbf{B}^{-}$ as in 15(c) with $\mathbf{\Delta}^{-} \leq \mathbf{\Gamma}^{-}$.

17. Let $\mathbf{X}_1^{\mathrm{T}}\mathbf{X}_1 = \mathbf{P\Delta P}^{\mathrm{T}}$ and $\mathbf{X}_2^{\mathrm{T}}\mathbf{X}_2 = \mathbf{P\Gamma P}^{\mathrm{T}}$ where $\mathbf{P}$ is non-singular and $\mathbf{\Delta}$ and $\mathbf{\Gamma}$ are diagonal. Then $\mathbf{I} - \mathbf{S}$ is non-singular iff 1 is not an eigenvalue of $\mathbf{S}$ $(= \mathbf{X}_2(\mathbf{X}^{\mathrm{T}}\mathbf{X})^{-}\mathbf{X}_2^{\mathrm{T}})$ iff 1 is not an eigenvalue of $(\mathbf{X}^{\mathrm{T}}\mathbf{X})^{-}\mathbf{X}_2^{\mathrm{T}}\mathbf{X}_2$ iff $(\mathbf{\Gamma})_{ii} \neq 0 \Rightarrow (\mathbf{\Delta})_{ii} \neq 0$ iff $\mathcal{R}(\mathbf{X}_1) \subseteq \mathcal{R}(\mathbf{X}_2)$.

Section 9.7 (p. 364)

1. $\widetilde{\mathbf{T}}$ and $\mathbf{A}^{-} = \mathbf{P}^{\mathrm{T}}\mathbf{P}$ are

$$\begin{bmatrix} 3 & -1 & 4 & 1 & -2 & 3 & \frac{1}{3} & \frac{1}{3} & 0 & 0 & 0 & 0 \\ 0 & 0 & 0 & 0 & 0 & 0 & -\frac{5}{3} & \frac{1}{3} & 1 & 0 & 0 & 0 \\ 0 & 0 & 2 & 1 & -3 & 1 & \frac{11}{6} & -\frac{2}{3} & 0 & \frac{1}{2} & 0 & 0 \\ 0 & 0 & 0 & 0 & 0 & 0 & \frac{9}{2} & -1 & 0 & \frac{1}{2} & 1 & 0 \\ 0 & 0 & 0 & 0 & 1 & 1 & -\frac{19}{6} & \frac{8}{3} & 0 & -\frac{3}{2} & 0 & 1 \end{bmatrix}, \begin{bmatrix} \frac{71}{9} & \frac{1}{3} & -\frac{29}{6} & -1 & \frac{8}{3} \\ \frac{1}{3} & 1 & 0 & 0 & 0 \\ -\frac{29}{6} & 0 & \frac{11}{4} & \frac{1}{2} & -\frac{3}{2} \\ -1 & 0 & \frac{1}{2} & 1 & 0 \\ \frac{8}{3} & 0 & -\frac{3}{2} & 0 & 1 \end{bmatrix}$$

$\mathbf{Ax} = \mathbf{b}$ is consistent and $(3 + \frac{1}{3}w_1 - w_2, w_1, -1 + \frac{1}{2}w_2, w_2, 1)$ is a general solution, $\mathbf{b}^{\mathrm{T}}\mathbf{A}^{-}\mathbf{b} = 11$, $\mathbf{Ax} = \mathbf{c}$ is not consistent and $\rho(\mathbf{A}) = 3$.

2. $\widetilde{\mathbf{T}}$ and $\mathbf{A}^{-1} = \mathbf{P}^{\mathrm{T}}\mathbf{P}$ are

$$\begin{bmatrix} 2 & 1 & -1 & 3 & -6 & 1 & \frac{1}{2} & 0 & 0 & 0 \\ 0 & 3i & 2i & -i & -4i & 0 & \frac{1}{6}i & -\frac{1}{3}i & 0 & 0 \\ 0 & 0 & i & 2i & 0 & -4i & -\frac{5}{6}i & \frac{2}{3}i & -i & 0 \\ 0 & 0 & 0 & 2 & -2 & -6 & -\frac{5}{3} & \frac{5}{6} & -1 & \frac{1}{2} \end{bmatrix}, \frac{1}{36}\begin{bmatrix} 83 & -28 & 30 & -36 \\ 28 & 5 & -6 & 15 \\ 30 & -6 & 0 & -18 \\ -30 & 15 & -18 & 9 \end{bmatrix}$$

The solution is $(1, -3, 2, -1)^{\mathrm{T}}$, $\mathbf{c}^{\mathrm{T}}\mathbf{A}^{-1}\mathbf{b} = 6$ and $|\mathbf{A}| = 144$. The signature is 0 (see *Exercise* 9.7.5(b) below).

3(a). Use *Theorem* 9.7.2(iv). $\mathbf{P}^{\mathrm{T}}\mathbf{P}\widetilde{\mathbf{A}}_{*j} = \mathbf{P}^{\mathrm{T}}\widetilde{\mathbf{T}}_{*j}$. Note that $\mathbf{P}_{*i} = \mathbf{e}_i$ for $i = h_1, \ldots, h_\nu$, so by the condition imposed on the solution, $\mathbf{z} = \mathbf{0}$.

3(b). If $\mathbf{b} = \widetilde{\mathbf{A}}_{*j}$, $\widetilde{\mathbf{T}}_{*j} = \mathbf{Pb}$. Now use *Theorem* 9.7.2(iii).

5(a). Let $\mathbf{T}$ be as in *Theorem* 9.7.1 (see the preceding exercise). Let $\mathbf{J} = \{j : t_{jj} \text{ is not real}\}$, $\mathbf{D} = \text{diag}(d_1, \ldots, d_n)$ and $\mathbf{\Delta} = \text{diag}(\delta_1, \ldots, \delta_n)$ where $d_j = 1/\sqrt{-1}$ and $\delta_j = -1$ for $j \in J$ and $d_j = \delta_j = 1$ for $j \notin J$. Take $\mathbf{S} = \mathbf{DT}$.
5(b). By (a), $\mathbf{A}$ and $\mathbf{\Delta}$ have the same signature $n - 2|J|$. Now $j \in J \Leftrightarrow t_{jj}$ is not real $\Leftrightarrow m_j$ and m_{j-1} are of opposite signs since $t_{jj} = \sqrt{m_j/m_{j-1}}$.
6. Imitate the proof of *Theorem* 9.7.1. Aliter: Imitate *Exercise* 5.6.7.
7. The dispersion matrix of $\mathbf{P}X$ is $\mathbf{PAP}^{\mathrm{T}} = \mathbf{I}$.

Section 9.8 (p. 366)
2(a). Use $\mathbf{H}^* = \mathbf{A}^{\mathrm{T}} - i\mathbf{B}^{\mathrm{T}}$.
2(b). Suppose $\mathbf{A}$ is p.d. Let $\mathbf{A} = \mathbf{CC}^{\mathrm{T}}$ where $\mathbf{C}$ is real non-singular. Then $\mathbf{S} := \mathbf{C}^{-1}\mathbf{B}(\mathbf{C}^{\mathrm{T}})^{-1}$ is skew-symmetric. Now $\mathbf{H}$ is p.d. $\Leftrightarrow \mathbf{I} + i\mathbf{S} = \mathbf{C}^{-1}\mathbf{H}(\mathbf{C}^{\mathrm{T}})^{-1}$ is p.d. $\Leftrightarrow \mathbf{I} - i\mathbf{S} = (\mathbf{I} + i\mathbf{S})^{\mathrm{T}}$ is p.d. $\Leftrightarrow$ all eigen values of $i\mathbf{S}$ are $< 1 \Leftrightarrow$ all eigenvalues of $i\mathbf{A}^{-1}\mathbf{B} = i(\mathbf{C}^{\mathrm{T}})^{-1}\mathbf{C}^{-1}\mathbf{B}$ are < 1. For *Only if part*: If $\mathbf{x}$ is real, $\mathbf{x}^{\mathrm{T}}\mathbf{B}\mathbf{x} = 0$ by (a), so $\mathbf{x}^{\mathrm{T}}\mathbf{A}\mathbf{x} = \mathbf{x}^*\mathbf{H}\mathbf{x}$ and $\mathbf{A}$ is p.d.
3. If $\mathbf{A} = \mathbf{UDV}^*$, take $\mathbf{H} = \mathbf{UDU}^*$ and $\mathbf{W} = \mathbf{UV}^*$. *Uniqueness*: Clearly $\mathbf{AA}^* = \mathbf{H}^2$. Thus $\mathbf{H}$ is an n.n.d. square-root of $\mathbf{AA}^*$ and is unique by the complex analogue of *Exercise* 9.4.6. If $\mathbf{A}$ is non-singular, so is $\mathbf{H}$ and $\mathbf{W}$ is unique. Note that we can also get an n.n.d. $\mathbf{K}$ such that $\mathbf{A} = \mathbf{WK}$. *Last part*: $\mathbf{H} = (r)$ and $\mathbf{W} = (e^{i\theta})$ where r and θ are the polar coordinates of (α, β).

List of symbols

F	Field or base field of a vector space
$\mathbb{R}$	Field of real numbers
$\mathbb{C}$	Field of complex numbers
$\mathbb{Q}$	Field of rational numbers
$\mathbb{Z}$	Ring of integers
$\mathcal{P}$	Space of polynomials
$\mathcal{P}_n$	Space of polynomials with degree $\leq n-1$
F^n	$F \times F \times \cdots \times F$ (n copies)
F^X	Space of all functions from X to F
$\mathrm{Sp}(A)$	Linear span of the set A of vectors
$d(S)$	Dimension of the subspace S
$\mathbf{e}_j$	The j-th vector in the canonical basis of F^n
$S+T$	The sum of S and T
$S \oplus T$	Direct sum of S and T
$V_1 \simeq V_2$	V_1 is isomorphic to V_2
V/S	The quotient of V modulo subspace S
$\mathbf{0}$	Null vector or null matrix
$\mathbf{I}$	Identity matrix
$(\mathbf{A})_{ij}$	The (i,j)-th element of matrix $\mathbf{A}$
$\mathbf{A}_{m \times n}$	Matrix $\mathbf{A}$ of order $m \times n$
$F^{m \times n}$	Vector space of all $m \times n$ matrices over F
$\mathbf{A}^{\mathrm{T}}$	Transpose of $\mathbf{A}$
$\mathbf{A}_{i*}$	The i-th row of $\mathbf{A}$
$\mathbf{A}_{*j}$	The j-th column of $\mathbf{A}$
$\mathrm{diag}(\alpha_1, \alpha_2, \ldots, \alpha_n)$	The diagonal matrix with diagonal entries $\alpha_1, \alpha_2, \ldots, \alpha_n$
$\mathbf{A}(i_1, \ldots, i_k \mid j_1, \ldots, j_\ell)$	The submatrix of $\mathbf{A}$ formed by i_1-th, $\ldots$, i_k-th rows and j_1-th, $\ldots$, j_ℓ-th columns
$\mathrm{diag}(\mathbf{A}_1, \ldots, \mathbf{A}_k)$	"Block-diagonal matrix" with $\mathbf{A}_1, \ldots, \mathbf{A}_k$ on the diagonal
$\mathcal{C}(\mathbf{A})$	Column space of $\mathbf{A}$
$\mathcal{R}(\mathbf{A})$	Row space of $\mathbf{A}$
$\rho(\mathbf{A})$	Rank of $\mathbf{A}$

$\mathbf{A}_L^{-1}$	A left inverse of $\mathbf{A}$
$\mathbf{A}_R^{-1}$	A right inverse of $\mathbf{A}$
$\mathbf{A}^{-1}$	The inverse of $\mathbf{A}$
$\mathcal{N}(\mathbf{A})$	Null space of $\mathbf{A}$
$\nu(\mathbf{A})$	Nullity of $\mathbf{A}$
$\mathrm{tr}(\mathbf{A})$	Trace of $\mathbf{A}$
$\overline{\mathbf{A}}$	Complex conjugate of matrix $\mathbf{A}$
$\mathbf{A}^*$	Adjoint (complex conjugate transpose) of $\mathbf{A}$
$\mathbf{1}$	$(1, 1, \ldots, 1)^{\mathrm{T}}$
$\mathbf{J}$	The matrix $((1))$
$\mathcal{K}(f)$	Kernel of the linear transformation f
R_{ij} (C_{ij})	Interchange of the i-th and j-th rows (columns)
$R_i(\alpha)$ $(C_i(\alpha))$	Multiplication of the i-th row (column) by α
$R_{ij}(\beta)$ $(C_{ij}(\beta))$	Addition of β times the j-th row (column) to the i-th row (column)
$\mathbf{E}_{ij}$, $\mathbf{E}_i(\alpha)$, $\mathbf{E}_{ij}(\beta)$	Elementary matrices obtained from $\mathbf{I}$ by the operations R_{ij}, $R_i(\alpha)$ and $R_{ij}(\beta)$ respectively
$\mathbf{A}^-$	A g-inverse of $\mathbf{A}$
$\begin{pmatrix} s_1 & \cdots & s_n \\ \pi_{s_1} & \cdots & \pi_{s_n} \end{pmatrix}$	Permutation
$[s_{i_1}\ s_{i_2}\ \cdots\ s_{i_k}]$	Cyclic permutation
$(i_1\ i_2\ \cdots\ i_n)$	A permutation of $\{1, \ldots, n\}$
$\epsilon(\pi)$	Sign of the permutation π
$\lvert\mathbf{A}\rvert$	Determinant of the matrix $\mathbf{A}$
A_{ij}	Cofactor of a_{ij} in $\mathbf{A}$
A_{IJ}	Cofactor of $\mathbf{A}(I \mid J)$ in $\mathbf{A}$
$\mathbf{A}^{\circledast}$	Classical adjoint of $\mathbf{A}$ (i.e., $((A_{ij}))^{\mathrm{T}}$)
$\langle \mathbf{x}, \mathbf{y} \rangle$	Inner product of $\mathbf{x}$ and $\mathbf{y}$
$\lVert\mathbf{x}\rVert$	Norm of $\mathbf{x}$
$\mathbf{x} \perp \mathbf{y}$	$\mathbf{x}$ is orthogonal to $\mathbf{y}$
$A \perp B$	$\mathbf{x} \perp \mathbf{y}$ for all $\mathbf{x} \in A$ and $\mathbf{y} \in B$
$S^{\perp}$	Orthogonal complement of S
$\mathbf{P}_{\mathbf{A}}$	Orthogonal projector into $\mathcal{C}(\mathbf{A})$
$\chi_{\mathbf{A}}(\lambda)$	Characteristic polynomial of $\mathbf{A}$
$\mathbf{G}(\lambda)$	A matric polynomial
$\mathbf{A}_1 \oplus \cdots \oplus \mathbf{A}_k$	$\mathrm{diag}(\mathbf{A}_1, \ldots, \mathbf{A}_k)$
$m_{\mathbf{A}}(\lambda)$	Minimal polynomial of $\mathbf{A}$
$\mathrm{ES}_{\alpha}(\mathbf{A})$	Eigenspace of $\mathbf{A}$ corresponding to α
$\mathbf{J}_{\alpha}(n)$	Jordan block of order n with each diagonal entry α
$P(\mathbf{A})$	Number of positive eigenvalues of (real symmetric) $\mathbf{A}$

$N(\mathbf{A})$	Number of negative eigenvalues of (real symmetric) $\mathbf{A}$
$\mathbf{A} \otimes \mathbf{B}$	Kronecker product of $\mathbf{A}$ and $\mathbf{B}$
$\mathbf{A} \odot \mathbf{B}$	Hadamard product of $\mathbf{A}$ and $\mathbf{B}$
$x \mapsto y$	The image of x is y
$\mathbf{A} \rightsquigarrow (\mathbf{P}, \mathbf{Q})$	$(\mathbf{P}, \mathbf{Q})$ is a rank-factorization of $\mathbf{A}$
$A \uplus B$	Disjoint union of sets A and B
$\emptyset$	Empty set
$:=$	Defined as
$\blacksquare$	End of proof, discussion, etc.
iff	If and only if
HCF	Hermite Canonical Form
n.n.d.	Non-negative definite
p.d.	Positive-definite
w.l.g.	Without loss of generality
w.r.t.	With respect to
LHS	Left hand side
RHS	Right hand side

Notes: In general, lower case Greek letters like α and β are used to denote scalars, lower case bold face Roman letters like $\mathbf{x}$ and $\mathbf{u}$ are used to denote vectors (column vectors from *Chapter* 2 onwards) and bold face capital letters like $\mathbf{A}$ and $\mathbf{\Sigma}$ are used to denote matrices. A null vector as well as a null matrix are denoted by $\mathbf{0}$.

However, in a few places (in *Sections* 2.2, 3.9) we have used bold face Greek letters like $\boldsymbol{\alpha}$ and $\boldsymbol{\beta}$ to denote vectors.

Index